Multifunctional Materials

Scrivener Publishing
100 Cummings Center, Suite 541J
Beverly, MA 01915-6106

Publishers at Scrivener
Martin Scrivener (martin@scrivenerpublishing.com)
Phillip Carmical (pcarmical@scrivenerpublishing.com)

Multifunctional Materials

Engineering and Biological Applications

Edited by

Divya Bajpai Tripathy

Dept. of Chemistry, School of Basic Sciences, Galgotias University, Greater Noida, India

Anjali Gupta

Dept. of Chemistry, School of Basic Sciences, Galgotias University, Greater Noida, India

and

Arvind Kumar Jain

Professor of Basic and Applied Sciences, Dean of Student Welfare IILM University, Greater Noida, India

Scrivener Publishing

WILEY

This edition first published 2025 by John Wiley & Sons, Inc., 111 River Street, Hoboken, NJ 07030, USA and Scrivener Publishing LLC, 100 Cummings Center, Suite 541J, Beverly, MA 01915, USA

For more information about Scrivener publications please visit www.scrivenerpublishing.com.

Wiley Global Headquarters
111 River Street, Hoboken, NJ 07030, USA

For details of our global editorial offices, customer services, and more information about Wiley products visit us at www.wiley.com.

Library of Congress Cataloging-in-Publication Data

ISBN 978-1-394-23412-7

Cover images courtesy of Adobe Firefly and Wikimedia Commons
Cover design by Russell Richardson

Set in size of 11pt and Minion Pro by Manila Typesetting Company, Makati, Philippines

Printed in the USA

10 9 8 7 6 5 4 3 2 1

Contents

Preface

Multifunctional materials have gained significant attention due to their ability to perform diverse applications and functionalities. Combining multiple functions into a single material offers numerous advantages, such as increased effectiveness, enhanced performance, and the potential for innovative applications. These materials can occur naturally or be engineered at the molecular or nanoscale level by modifying their mechanical, electrical, thermal, optical, and other properties to render them multifunctional. They may also be hybrids combining natural and synthetic components. The integration of various functionalities aims to create materials with unique features or improved performance not achievable with single-function materials. Additionally, this approach enhances system efficiency by reducing the volume and weight of individual components. Multifunctional materials should also be highly application-specific, leveraging their wide range of combinations and resulting characteristics. Multifunctional polymeric materials, with their adaptability and specific properties, find applications across industries, advancing materials science, healthcare, electronics, and drug delivery systems.

Chapters 1-3 discuss sustainable approaches to synthesizing multifunctional polymeric materials, such as biopolymers, using enzymatic methods and renewable feedstock, while summarizing their properties, characterization, and applications in biomedical, energy, structural, and environmental fields.

Chapters 4 and 7 explore graphene oxide-based and other nanocomposites in medical domains like drug delivery, tissue engineering, biosensing, and bioimaging, as well as their use in electronics, aerospace, defense, and energy sectors. Chapter 5 highlights multifunctional supramolecular polymers for energy storage, pollution sensors, diagnostics, medicine delivery, hygiene products, and self-repairing materials. Chapter 6 analyzes microbial-based biolubricants, including their industrial feasibility.

Chapters 8-9 provide an overview of polymeric emulsions, their classification, synthesis, and applications in energy and environmental science. Chapter 10 examines nanotechnology-based smart drug delivery systems using carbon nanotubes, quantum dots, silver nanoparticles, and more. Chapter 11 reviews multifunctional materials in engineering and processing applications. Chapter 12 details optical and electronic properties and advancements in multifunctional materials for optoelectronic devices like LEDs, solar cells, and photodetectors.

Chapter 13 describes analytical tools such as XRD, SEM, TEM, FTIR, XPS, DSC, TGA, and AFM for understanding material morphology and structure. Chapter 14 focuses on polysaccharides and their application in solar cells. Chapter 15 highlights the industrial applications of multifunctional biopolymers. Chapters 16-17 cover nanotechnology in agriculture and industrial sectors, including synthesis, properties, and potential applications.

Finally, Chapter 18 examines magnetization dynamics of ferromagnetic nanostructures for spintronics and biomedical applications.

The editors are grateful to all of the contributors to this book, and special thanks go to Martin Scrivener and Scrivener Publishing for their support and publication.

The Editors
January 2025

1

Multifunctional Polymer Chemistry: Sustainable Synthetic Procedures

Prem Shankar Mishra[1]*, Rakhi Mishra[2], Kabikant Chaurasiya[1] and Tanya Gupta[1]

[1]*Department of Pharmacy, School of Medical and Allied Sciences, Galgotias University, Greater Noida, Uttar Pradesh, India*
[2]*Noida Institute of Engineering and Technology (Pharmacy Institute), Greater Noida, Uttar Pradesh, India*

Abstract
Naturally occurring polymers such as DNA strands and polypropylene, which are widely used as plastic worldwide, are examples of polymers that surround us. Due to the required performance and poor recycling rates of polymers, there is a continuing demand for virgin polymers; nonetheless, this exacerbates serious challenges associated with the plastics sector, such as waste creation and greenhouse emissions. It is necessary to assess the sustainability effects of bio-based polymers to retain their biodegradation potential while maximizing their utilization in the functional use stage. The several green chemistry-based synthetic techniques used to produce multifunctional polymers are the main subject of this study. This chapter also includes information on the applications, challenges, and future possibilities of multifunctional polymers.

Keywords: Chemoenzymatic, polypropylene, controlled radical polymerization, microbial synthesis

1.1 Introduction

Sustainability is the responsible use of natural resources, preservation of the environment, waste minimization, and avoidance of dangerous materials so that future human generations can continue to live a decent standard of living on Earth (Figure 1.1) [1]. These days, sustainability and green chemistry are the primary strategies due to growing public knowledge of environmental deterioration, climate change, and the earth's diminishing resources. In order to be achieved by 2030, the United Nations adopted 17 sustainable development goals (SDGs) in 2015. Since polymers are ubiquitous, they are essential to achieving these objectives [2].

Another way to think of sustainability as a business strategy is to maximize the positive effects of an organization on the environment, communities, economy, and society [3].

**Corresponding author*: misra.prem@rediffmail.com

Divya Bajpai Tripathy, Anjali Gupta and Arvind Kumar Jain (eds.) Multifunctional Materials: Engineering and Biological Applications, (1–26) © 2025 Scrivener Publishing LLC

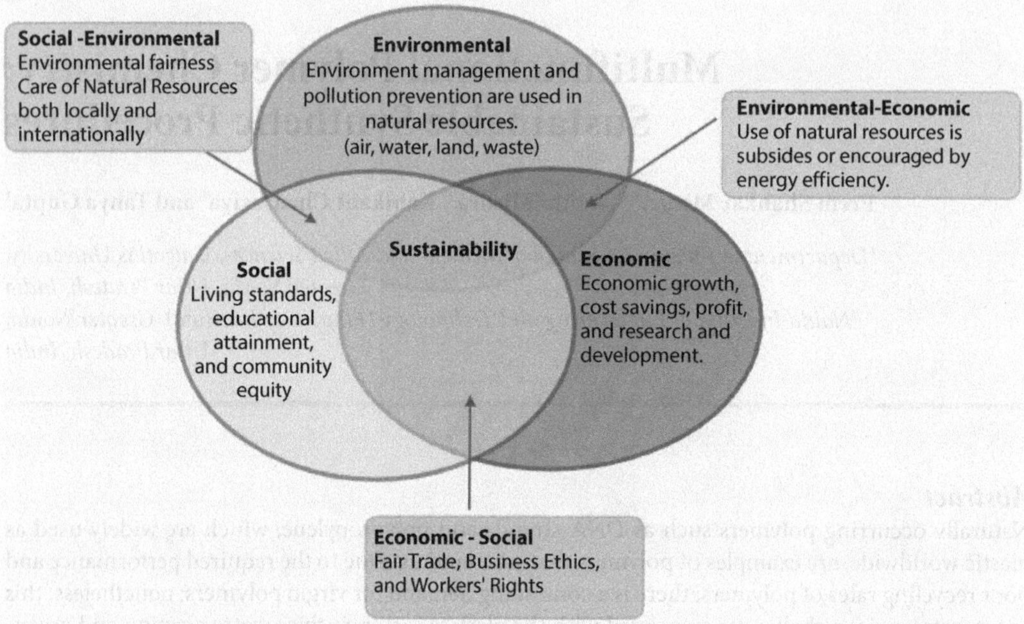

Figure 1.1 Impact of sustainability on different aspects.

Businesses are starting to see the benefits of sustainability, including increased exposure and lower manufacturing costs, energy use, risks, and dangers [4]. Sustainability can take on a variety of shapes depending on the enterprises involved, such as the following:

- The environmentally benign and non-hazardous production of commercial items using resources derived from agriculture.
- The use of "natural" components in commercial goods, particularly in applications related to food, cosmetics, and personal hygiene.
- The enhancement of existing industrial processes by reduction of non-recyclable plastic usage, reduction of needless byproducts, development of a greener supply chain, and/or reduction of carbon footprint.
- Enhancing the environmental, health, and safety aspects of existing product lines through prudent management [5, 6].

Today, disposing of plastic garbage is a major environmental issue since plastics are being produced in large quantities and are being used in more areas of our daily lives [7]. As a result, these problems contribute to the rising threat of the warming planet brought on by carbon dioxide emissions caused by burning conventional, non-biodegradable polymers, such as polyethylene, polypropylene, and polyvinyl chloride [8].

Polymers can be divided into two categories: natural polymers and synthetic polymers.

Biocatalysts, often enzymes, are invariably engaged in *in vivo* processes that produce natural polymers in the natural world. All living things, including humans, animals, and plants, include natural polymers. Among the natural polymers are lignocellulose, starch, protein, DNA, RNA, and polyhydroxyalkanoates (PHAs). Natural polymers often have clearly defined structures, while some, like lignocellulose, are an exception [9].

The most popular way to create synthetic polymers is to polymerize compounds with simple structures derived from petroleum. Synthetic polymer preparation often involves the use of chemical catalysts, particularly metal catalysts. The growth of the petrochemical industry, the simultaneous availability of cheap petroleum oils, the development of well-established and sophisticated polymerization techniques, and the availability of cheap petroleum oils have led to the development of many synthetic polymers, including phenol-formaldehyde resins, polyolefins, polyesters, polyvinyl chloride, polystyrene, and polyamides. Plastics, a broad category of synthetic polymers, gained popularity early in the 20th century and are today widely found in many different items, including textile fibers, films, bags, bottles, and cartons [10].

Biodegradable polymers are materials that, after a short time of use and under controlled circumstances, disintegrate into components that can be readily disposed of [11]. They can be made from a variety of wastes or bioresources, such as food, animal, and agro-waste waste, as well as cellulose and starch, Businesses are concentrating on generating bioplastics produced from renewable resources since they are often more cost-effective than those obtained from microbiological resources [12].

Using biodegradable polymers helps the environment by lowering the emissions of carbon dioxide, which contribute to global warming, promoting biodegradation, and regenerating raw resources [13]. Microorganisms such as bacteria and fungi can ingest biodegradable polymers, which subsequently undergo degradation into H_2O, CO_2, and methane. The substance's composition has an impact on the biodegradation process. Polymer shape, structure, molecular weight, and exposure to radiation and chemicals are some of the elements that affect how quickly a polymer degrades [14].

The market for biodegradable plastics is quite promising. However, as selective biowaste collection grows, they must be created simultaneously with a comprehensive analysis and worldwide integration with organic waste management and end-of-life treatment techniques. One advantage of biodegradable plastics is that they may be naturally decomposed at the end of their lives through processes like anaerobic digestion or composting. Biodegradable plastic composting is widely acknowledged and well-documented worldwide [14]. This chapter describes multifunctional polymers and the many environmentally friendly synthetic processes that are employed to create them using green chemistry. Details on the uses of multifunctional polymers, as well as their prospects and difficulties, are also included in this chapter.

1.1.1 Multifunctional Polymers

Materials known as polymer composites are created by adding fibers or other appropriate reinforcement to the polymer matrix [15, 16]. These are often described in various ways. The kind of matrix might be either natural or synthetic, depending on the intended use. Synthetic polymers are utilized to provide the matrix for most applications; biopolymers have only recently been employed in this capacity [17]. Fibers, either natural (like coir, bagasse, pine needles, hemp, flax, and sisal) or synthetic (such as carbon and glass), are the most common types of reinforcing materials. Both the polymer matrix and the reinforcement have been found to have a substantial impact on the overall physicochemical properties of the composites [18].

Due to the growing potential for polymer multifunctional application, researchers from many domains have recently focused a great deal of emphasis on two related essential topics: biopolymers and biomaterials (Figure 1.2) [19]. These days, biopolymers are a hot issue because of their prospective uses in the food, pharmaceutical, textile, medical, and other industries as well as in addressing the problems associated with rising environmental contamination [20].

The ecosystem has already suffered greatly from the careless use of easily accessible and reasonably priced synthetic plastic, and these materials are now posing a major danger to all life on Earth [21, 22]. Because synthetic plastic is used so extensively in everyday products, it is becoming a severe hazard to human health. In this regard, bio-based, sustainable, and biodegradable polymers hold great promise as a quick replacement for synthetic polymers generated from petrochemicals. To create bioplastics, a variety of biopolymers, including polysaccharides, proteins, and their mixtures, are frequently employed. The characteristics of biopolymer-based polymers are similar to those of synthetic ones [23, 24].

The addition of functional components such as nanomaterials, essential oils, phytochemicals, and bioactive components further enhances the physical and functional properties of the biopolymer-based materials. It has been demonstrated that one practical way to alter the mechanical characteristics of the polymer matrix is by adding fillers, as well as imparting thermal and electrical conductivity and increasing thermal resistance concerning glass transition and degradation temperatures. While the potential of biomaterials and biopolymers to create sustainable materials is great, there are still several issues that need to be resolved before further development can begin. These materials are more expensive than the ones that are typically used, which limits their application in many industries. To create future materials, a deeper understanding of biopolymers and biomaterials is needed [25–28].

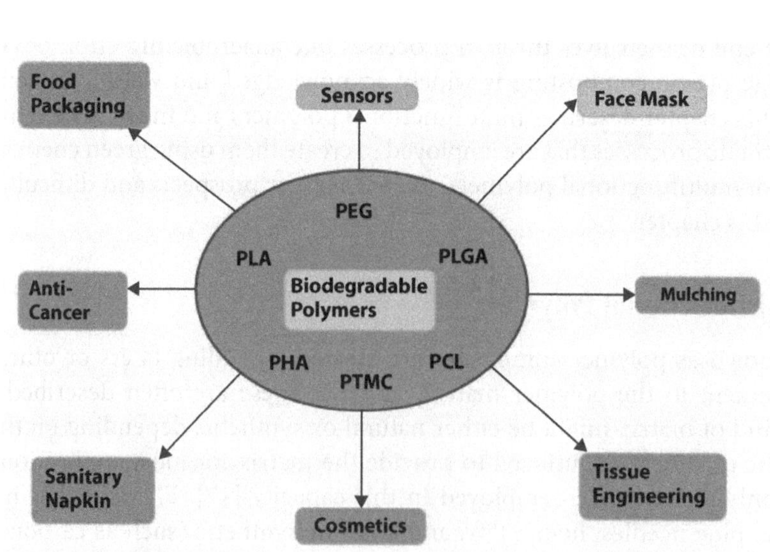

Figure 1.2 Examples of products formed by biodegradable polymers.

However, there has been a significant shift in the production of novel materials made from renewable resources in biorenewables in recent years as a result of increased environmental consciousness, health concerns, and the depletion of petroleum resources [19]. Well-known instances of ecologically friendly polymeric materials derived from biorenewable resources are among the biopolymers are polysaccharides, such as animal protein-based biopolymers, cellulose, alginate, starch, chitin/chitosan, and carbohydrate polymers, like collagen, silk, wool, and gelatin [29, 30]. Numerous industries, most notably the biomedical and automotive sectors, are currently using bio-based materials made from different natural resources. The product has several notable qualities, including biodegradability, low density, recyclability, easy separation, high toughness, superior thermal characteristics, reduced tool wear, non-skin irritation, and enhanced energy recovery are just a few of the many qualities that have made bio-renewable materials a material of choice [31–33].

There is a strong link between the rising number of interdisciplinary studies and the expanding utilization of innovative materials in current technologies [34]. Finding structurally well-defined polymers that surpass the state of the art is an ongoing objective in the case of polymers. To produce multifunctional polymers having two or more orthogonally accessible functions for use in several potential uses, such as gene therapy, tissue engineering, and medication delivery, application-driven research has been focused on these materials [35–37].

1.1.2 Importance of Sustainable Synthetic Procedures in Polymer Chemistry

Natural polymers like starch, cellulose, or chitin can be chemically modified to produce sustainable polymers from renewable resources. Comparing sustainable synthetic techniques to conventional synthetic procedures would result in lower pollution emissions, less water and non-renewable energy use, and a reduction in the synthesis of polymer chemistry. The following is a list of additional advantages provided by using sustainable synthetic processes using polymers:

- Environmentally friendly
- Lower energy and cost
- Toxicity reduction
- Efficiency of reactions and economy of atoms [38]
- Create and learn about new manufacturing techniques, chemicals, and product stewardship procedures
- Protection and enhancement of human health [39]

1.2 Sustainable Synthetic Procedures for Multifunctional Polymer Synthesis

High levels of environmental sustainability and energy efficiency are promoted by the design, preparation, and use of polymers made using sustainable resources for a variety of cutting-edge applications [40].

Because of this, the need for polymers will only increase along with the number of people on Earth, modernity, and advancements in technology. However, our understanding of the usage of polymers has changed because of the depletion of fossil fuels, the rise in plastic waste production, ocean pollution, and the corresponding increase in greenhouse gas emissions. A linear polymer economy is unsustainable, even if polymers were never intended to be recycled. The design for life-cycle assessments, recycling, and reuse will become more and more crucial components, and this may be accomplished by using a variety of techniques and approaches [41].

1.2.1 Green Chemistry Principles and Their Application to Polymer Synthesis

In polymer synthesis, the green chemistry method (Figure 1.3) aims for sustainability [42]. Herein, usually, (1) utilizing renewable resources as raw ingredients to produce polymers and (2) applying green methodologies for the synthesis of polymers are essential components of green polymer chemistry [43].

Three components are needed to produce polymers using green chemistry: (1) catalysts, solvents, and raw ingredients; (2) environmentally congenial synthesis procedures; and (3) low-carbon, sustainable polymers that can be recycled or disposed of with little harm to the environment include (bio) degradable polymers and others [44]. Below is a discussion of various techniques for creating sustainable biodegradable polymers using the green chemistry concept [45].

The application of ecologically friendly materials and sustainable procedures has been emphasized in the application of green chemistry concepts to polymer synthesis in a variety of ways. Utilizing renewable feedstock, atom economy, and waste reduction are a few of the fundamental tenets of green chemistry. These ideas have been incorporated into

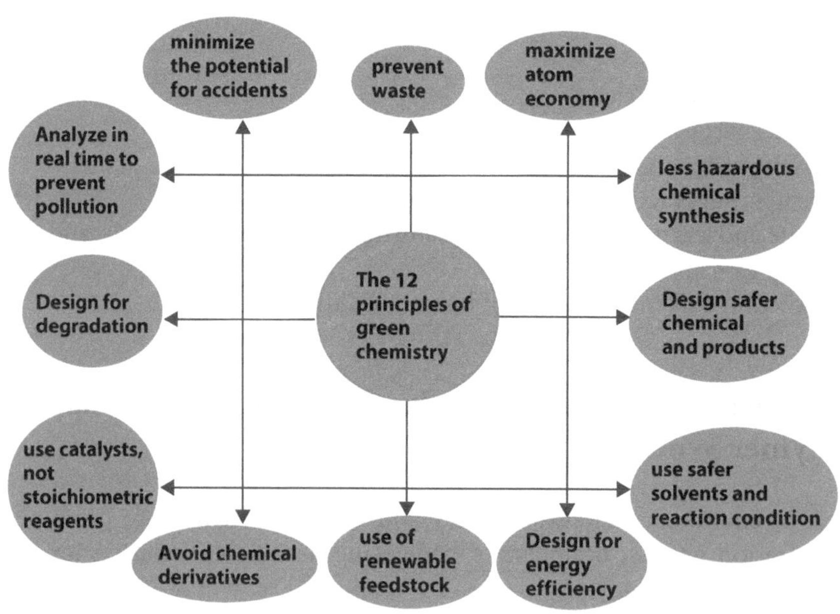

Figure 1.3 Green chemistry and its principles.

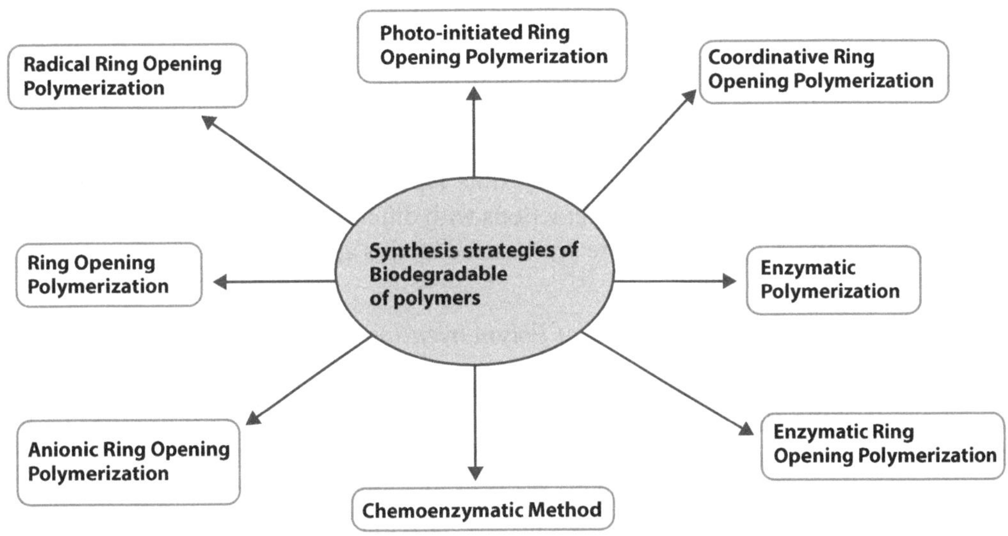

Figure 1.4 Different methods of synthesis under green chemistry.

the creation of multifunctional polymers, as demonstrated by the production of polymer Nanocomposite materials using environmentally benign thiol-ene chemistry [46].

Furthermore, the 12 green chemistry principles have been particularly examined and applied to the manufacture of polymers, emphasizing their applicability to present procedures and the possibility of sustainable progress in this area [47, 48].

Several of the techniques comprise (Figure 1.4)

- Ring-opening polymerization (ROP);
- Enzymatic ROP;
- Anionic ROP;
- Photo-initiated radical polymerization;
- Chemo enzymatic method;
- Enzymatic polymerization;
- ROP; and
- Coordinative ring opening polymerization.

1.2.1.1 Ring-Opening Polymerization (ROP)

The word "ROP" refers to the process by which cyclic monomers are polymerized into acyclic monomeric units (Figure 1.1). Here, the cyclic monomer ring system is opened to create a long polymer chain, with the final portion of the polymeric chain functioning as a center of reactivity. There are three recognized propagation centers: cationic, radical, and anionic. This procedure is thought to be among the most adaptable ways to synthesize large amounts of biopolymers [49].

1.2.1.2 Radical Ring-Opening Polymerization

Free radical ROP (FROP) is the progressive incorporation of free radicals as structural elements created by different methods that use distinct initiator chemicals to make polymers. Chain growth is caused by the newly created starting free radicals extending the polymer chain's monomer units. One of the most flexible types of polymerizations is free radical polymeric chain ends' (FROP) easy interactions with different substrates or chemicals [50]. Materials based on ε-Caprolactone are synthesized by this approach.

1.2.1.3 Chemo Enzymatic Method of Polymerization

It is widely acknowledged that chemoenzymatic synthesis yields active pharmacological ingredients (AAPIs) (Figure 1.5). Using this technique, researchers have produced a few prodrugs made of polymers that have optical activity. High molecular weight substrates of nonsteroidal anti-inflammatory drugs make up these prodrugs. This process combines traditional polymerization with a very effective enzymatic approach. Chemoenzymatic synthesis is believed to be highly beneficial in producing substantial molecular weight Biodegradable polymeric materials (BPMS). For instance, a chemoenzymatic procedure could be used to create poly lactic acid (PLA) monomer [51].

1.2.1.4 Photo-Initiated Radical Polymerization

The approach known as photo-initiated ROP (Figure 1.6) is intriguing due to its ability to achieve spatiotemporal control over fast polymerization rates under physiological settings. It gets rid of the need to add a hazardous cross-linker to the reaction medium. This more environmentally friendly method is frequently utilized to synthesize the majority of polymers that are frequently employed in biomedical applications [52].

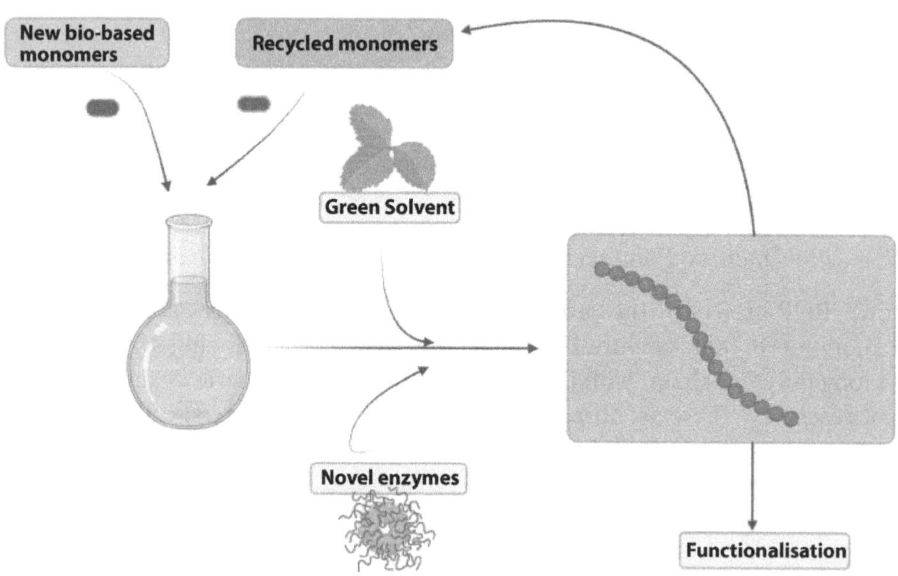

Figure 1.5 Chemo enzymatic polymerization.

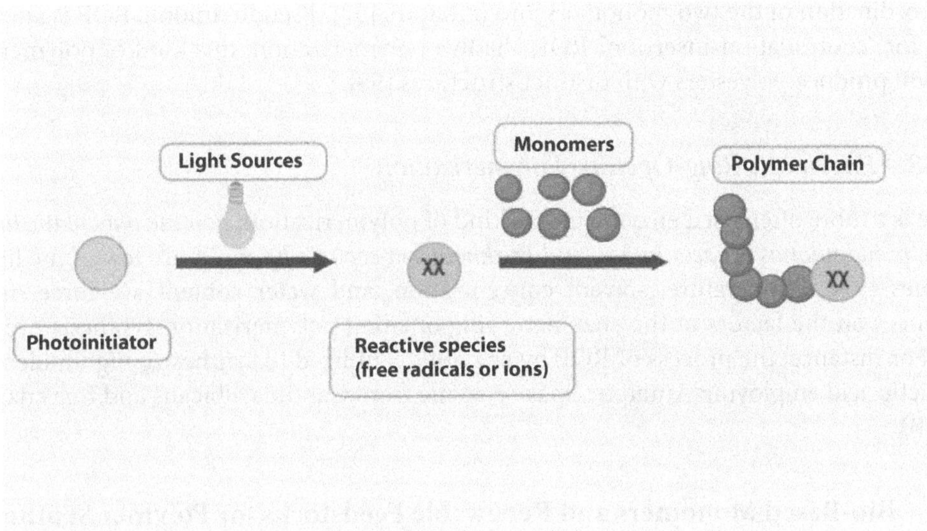

Figure 1.6 Photo-initiated ring-opening polymerization.

1.2.1.5 Enzymatic Polymerization

One innovative and environmentally friendly way of producing polymers is the enzymatic synthesis of biodegradable polymers. Biocatalysts with high catalytic activity that are renewable and non-toxic are called enzymes. Compared to traditional synthesis methods, less activation energy is needed when using enzymes to synthesize biodegradable polymers [53]. Even under milder circumstances, biodegradable polymers may be synthesized enzymatically without the usage of metal or hazardous organic pollutants. This broadens the range of domains in which it is applied, particularly in the biomedical sector [54]. For example, using methyl 12HS (12HS-Me), lipase is a catalyst used in the production of novel biodegradable polymers, such as polyester. It is a bio-based thermoplastic elastomer [55].

1.2.1.6 Anionic Ring-Opening Polymerization

Among the sophisticated techniques for creating telechelic polymers with a variety of topologies, including anionic ROP produces hyper branched, linear, star, and core-shell polymers. Because it offers an increased level of conversion of monomers, has no adverse effects, and has adjustable polymerization kinetics, anionic ROP is regarded as among the most beneficial techniques [56].

1.2.1.7 Coordinative Ring-Opening Polymerization

The ester cyclic molecule will be arranged by the metal atom in coordinative ROP. In this instance, the electrophilic catalytic center is the metal atom. For cyclic esters, this coordination insertion process activates the carbonyl group, while for cyclic phosphates, it activates the phosphorus atoms. The initiator, which might be amino, alkoxy, alkyl, etc., forms a direct link with the metal or a weak coordination with it. A reaction complex is formed by the coordination of the two monomers and initiators [57]. Pseudo-anionic ROP is another name for "coordination-insertion" ROP. Via live polymerization, this kind of polymerization will produce polyesters with distinct structures [58].

1.2.1.8 Enzymatic Ring-Opening Polymerization

Lipase is a more often used enzyme in this kind of polymerization. *Porcine pancreatic lipase (PPL), pseudomonas cepacia lipase, and Burkholderia cepacia lipase PS* are few of the lipase enzymes [59]. Temperature, solvent concentration, and water content are three major influences on the factors of the enzymatic ring opening polymerization synthesis process [60]. For instance, the process of ROP by enzymes is utilized to synthesize high molecular polylactic acid employing Antarctic lipase enzyme from Candida albicans and free enzyme (CALB).

1.2.2 Bio-Based Monomers and Renewable Feedstocks for Polymer Synthesis

Recent efforts have centered on the production of sustainable monomers and polymers, either to replace petroleum-based resources or to build multifunctional polymers, by using the structural variety of various biomass resources [61].

The bio-based monomers are renewable resources that are produced from biomass feedstock in a clean, energy-efficient method that leaves no harmful residues to contaminate the finished goods. As a result, the creation of green polymers and the prospects for a sustainable polymer sector are made possible by the polymerization of bio-based monomers into renewable polymers. These developments will ultimately be crucial to the realization and upkeep of a sustainable and bio-based society [62, 63].

Biotechnology and processes used in biorefineries are used to provide monomers that are replenishable and polymerized in very efficient traditional gas- or melt-phase polymerization techniques, rather than creating new pathways in biotechnology to produce biopolymers with somewhat laborious purification of polymers and challenging processing property tweaking. Monomers are significantly simpler to purify than polymers. Furthermore, the resultant bio-based polymers combine the benefits of gas-phase and solvent-free melt polymerization methods' high resource and energy efficiency with a minimal carbon footprint—a characteristic of feedstocks that are renewable. To prevent problems with food production, feedstocks, such as trash from forestry and agriculture, are employed preferentially. Numerous other bio-based monomers may be made through the fermentation of glucose, which is made from lignocelluloses and starch [64].

The use of renewable feedstocks and bio-based monomers in polymer synthesis is growing to lessen the environmental effect of polymer manufacturing and disposal. Among the important ideas and sources are as follows:

1.2.2.1 Renewable Energy Sources

Basic monomers that rely on coal, natural gas, and petroleum as starting materials are now included in the utilization of renewable energy sources in polymer research [65].

1.2.2.2 *Feedstocks from Agriculture and Forestry*

Feedstocks from agriculture and forestry, such as maize fiber, wheat, and other renewable resources, are the primary raw materials utilized to manufacture bio-based polymers [66, 67].

1.2.2.3 *Microbial Synthesis*

Using renewable resources, microbial synthesis has been employed to create bio-based monomers and polymers [68].

1.2.2.4 *Polymerization of Bio-Based Monomers*

Bio-based monomers, specifically 1,3-propanediol, have been used to polymerize bio-based polymers (PDO) [69, 70].

1.2.3 Catalysts and Reaction Conditions for Sustainable Polymerization Processes

Research and development on catalysts and reaction conditions for sustainable polymerization processes has been substantial. One technology that can help in the production of sustainable polymers is catalysis [71].

Many catalytic techniques have been included in green and sustainable chemistry during the past 20 years [72, 73]. The activity, content, and possible environmental effects of catalysts were assessed. The most useful catalysts frequently do away with the need for purification or synthesis stages and can withstand a broad variety of functional groups. In addition to being low in toxicity, desirable catalysts also reduce chemical waste and enable the replacement of dangerous chemicals with safe ones. Catalysts should ideally have long lives and only need small amounts of catalytic loadings [74].

Three major types of enzymes—oxidoreductases, transferases, and hydrolases—among the six primary classes have been used as catalysts in the production of multifunctional polymers. Similar to *in vivo* enzymatic processes, the proper design of a reaction involving an enzyme catalyst combined with a monomer produces macromolecules with precisely controlled structures. For the ability to select certain substances, including chemo-, regio-, stereo-, and choro-selectives, the reaction regulates the product structure [75].

Vinyl polymerizations and other oxidation polymerizations of aromatic compounds are catalyzed by oxidoreductases. Transferases are useful catalysts that may be used to create polyesters and polysaccharides with different architectures. The cleaving of macromolecules' bonds *in vivo* and the *in vitro* creation of new bonds, resulting in the formation of different polysaccharides and functionalized polyesters, is catalyzed by hydrolases [76]. The initial *in vitro* synthesis of complex natural polysaccharides such as chondroitin, cellulose, hyaluronan, amylose, xylan, and chitin, was made possible by enzymatic polymerizations. These polymerizations are considered "green" in several ways, including low byproduct production, excellent catalyst efficiency, somewhat selective reactions with renewable starting materials and environmentally friendly solutions, and nontoxicity of the enzymes.

To maintain a sustainable civilization, for environmental reasons and "green polymer chemistry," enzymatic polymerization is therefore beneficial [44].

1.3 Functionalization of Multifunctional Polymers

Many upcoming uses of polymeric materials need the design and production of multifunctionalized, structurally controlled polymers [77]. Multifunctional polymers may be made more unique by adding particular chemical groups to their structure, a process known as functionalization. Because of this modification, the polymers can acquire new characteristics or improve their current ones, which makes them beneficial for a variety of uses [78, 79]. As an illustration, atom transfer radical polymerization (ATRP) was applied to produce low molecular weight star, hyperbranched, and linear polymers (methyl acrylate). Radical addition processes were then used to end-functionalize the polymers. By adding allyl tri-n-butyl stannane at the end of the polymerization process, allyl groups were added to the polymer (methyl acrylate). Allyl alcohol or monomers of 1,2-epoxy-5-hexene that cannot be polymerized by ATRP, were introduced with high acrylate monomer conversion rates, and at the polymer chain terminus, the functionalities of epoxy and alcohol were present., respectively [80].

Different approaches for functionalization of multifunctional polymers are as follows:

Post-polymerization functionalization: This method entails modifying polymers to add particular functional groups after they are synthesized. In a group of uniform dispersity and molecular weight, it permits the adjustment of both physical and optoelectronic characteristics [81].
Polymerization-Induced Self-Assembly (PISA): Multifunctional polymersomes have been prepared using PISA, making it simple to create functionalized polymers with certain features [82].
Click Chemistry: A ubiquitous and simple method for creating multifunctional polymers that enable the controlled insertion of several functional groups is click chemistry [83].

Reactive handles may be easily introduced by using the Ugi reaction as a multifaceted click response to effectively create several multipurpose polymers by cross-synthesizing them or further functionalizing existing ones [84].

1.3.1 Sustainable Functionalization Reactions for Multifunctional Polymers

To create multifunctional polymers, several sustainable functionalization processes have been created. These processes emphasize the application of ecologically friendly materials as well as green chemistry concepts. For instance, a green approach method utilizing environmentally friendly thiol-ene chemistry has been developed for the production of multifunctional polymer nanocomposite material [85].

Several techniques, like the chemistry of clicks, sequential post-polymerization modification, and controlled radical polymerization (CRP) processes, can be used to create multifunctional polymer synthesis reactions. The following are some essential processes and reactions for the production of multifunctional polymers [47, 86–89].

1.3.1.1 *Controlled Radical Polymerization (CRP) Reactions*

For the production of multifunctional polymers, CRP is an effective tool. There is a great deal of promise for the creation of multifunctional polymers when click chemistry and CRP reactions are combined.

1.3.1.2 *Ugi Reaction*

Several multifunctional PEGylation polymers have been effectively synthesized using the Ugi reaction as a multicomponent click reaction [69].

1.3.1.3 *Sequential Post-Polymerization Modification*

Using this method, polymers having monomer units that have hidden functions are modified sequentially and selectively after polymerization to create multifunctional homopolymers.

1.3.1.4 *Polymerization-Induced Self-Assembly*

Multifunctional polymersomes, which are vesicles with special qualities and uses, have been created using this technique.

1.3.1.5 *Green Route Strategy*

Using environmentally friendly thiol-ene chemistry, a green approach method for the production of multifunctional polymer nanocomposite material has been developed.

Succinic acid (SuA), for instance, is a significant bio-based aliphatic acid monomer. It is an odorless, white substance that dissolves in acetone, ethanol, and water. The US Food and Drug Administration (FDA) has approved the use of glycerol and SuA for medical purposes and recognizes them as safe materials. When combined, these two naturally occurring substances provide a cheap and plentiful source of starting materials for the production of dendrimers [90].

PGSuc dendrimers were synthesized, according to Carnahan *et al.*, using the acetal of benzylidene as a protective group that, in some situations, can be removed selectively (Figure 1.7). It was discovered that these PGSuc oligomers were easily chemically degradable, quickly biodegradable, and non-ecotoxic [91].

PGSus dendritic formations have been used to repair corneal ulcers, as described by Luman *et al.* [92]. Other researchers have looked into their potential utility as transporters of tiny compounds, such as medications or dyes. An effective and high-yield divergent technique [93] was used to synthesize newly developed poly(glycerol succinate)-poly (ethylene glycol) composite dendritic-linear copolymers, which are made up of PEG dendritic blocks and polyethylene glycol (PEG) linear chains (Figure 1.8).

The possible methods for synthesizing bio-polypropylene from propylene were converted from methanol or ethanol with ZSM-5 catalysts. Engineered thermoplastics (ETP) and metaproterenol (MTP) conversions have the capacity to be sustainable bio-propylene sources given the abundance given the previously mentioned study on bio-EtOH generation and the studies currently available on bio-methanol (bio-MeOH) production [94–97].

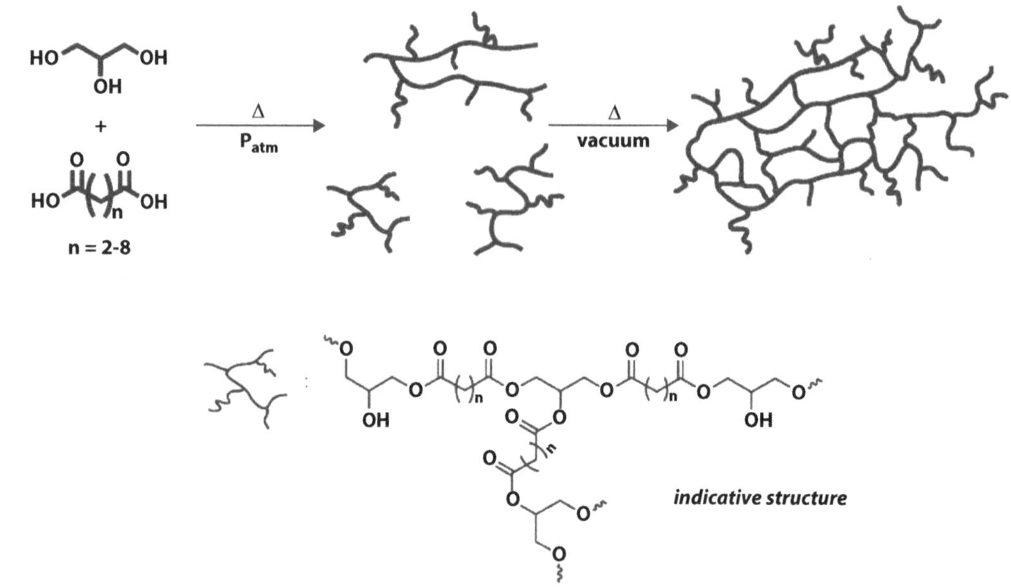

Figure 1.7 General synthesis of glycerol polyesters.

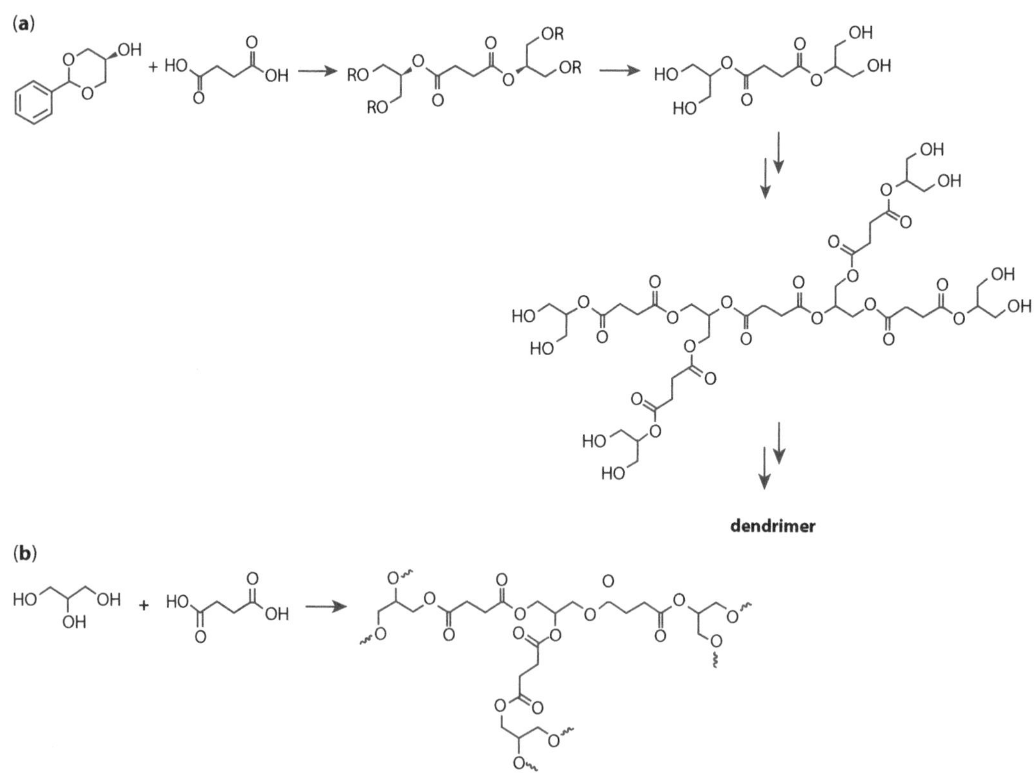

Figure 1.8 Synthesis of poly(glycerol succinate)-poly(ethylene glycol) hybrid dendritic-linear copolymers.

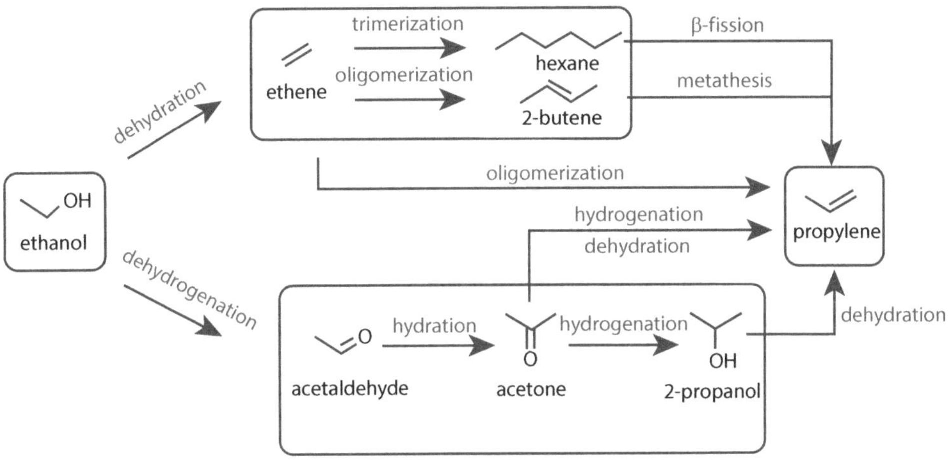

Figure 1.9 Pyrolysis of biomass.

As a result of biomass pyrolysis producing a complicated mixture that contains propylene, catalytic pyrolysis of biomass (Figure 1.9) offers another viable pathway to produce bio-propylene [64, 65]. Quick pyrolysis and conversion of pyrolysis vapors are the two stages of the two-step process of pyrolyzing biomass, certain catalysts, such as calcium oxide and ZSM-5 have been researched [98]. The most studied method uses dimerization, metathesis, and dehydration of bioethanol to produce bio-propylene. As of right now, it seems possible to synthesize polypropylene using bio-propylene as the monomer input [99].

1.4 Applications of Multifunctional Polymers

Multifunctional polymers have a variety of uses because of their special qualities and the ability to tailor their functionalities. Some of the most promising applications of multifunctional polymers include:

1.4.1 Biomedical Applications

Multifunctional polymeric nanocarriers have been studied for their potential in nanomedicine, where they may improve the pharmacokinetics, bioavailability, and effectiveness of certain drugs or contrast agents. It is possible to make these nanocarriers biocompatible and safe for patients to use while administering a range of medications, such as chemotherapy and immunotherapy [100].

1.4.2 Sensors and Actuators

The ability of multifunctional polymers to respond to different stimuli, such as light, pH, and temperature, makes them valuable for the creation of sensors and actuators. These substances can be engineered to identify certain analytes or to respond upon detection of a given stimulus [101].

1.4.3 Energy Applications

Multifunctional polymers can be applied to energy-related devices like batteries and supercapacitors and solar cells. These materials' design may improve the effectiveness of energy conversion and storage devices [102].

1.4.4 Environmental Applications

Environmental uses for multifunctional polymers include pollution removal and water treatment. These materials can be designed to boost the efficiency of wastewater treatment processes or to degrade or absorb contaminants [103].

1.4.5 Structural Applications

Multifunctional polymers can be used to create novel materials with enhanced electrical, thermal, and mechanical qualities. These materials can be applied to the construction, automotive, and aerospace sectors, among others, to increase the lifetime and performance of structures [104].

- There are a multitude of potential applications for multifunctional polymers across several industries and enterprises. The advancement of environmentally friendly and sustainable synthesis techniques has further expanded the potential of multifunctional polymers for wider use [105].

Other examples of multifunctional polymers with their applications are shown in Table 1.1.

1.5 Future Perspectives and Challenges

Although there are many possible uses for multifunctional polymers, there are a few issues that must be resolved. Developing environmentally friendly and sustainable processes for the synthesis of multifunctional polymers is one of the primary problems [117]. Considering the rise in emphasis on the circular economy and the need to move away from single-use disposable items, this is especially crucial.

Enhancing the biodegradability or recyclable nature of multifunctional polymers presents another difficulty. Although high polymers are typically thought of as being inherently non-toxic, the desire to create polymers that are readily recycled or biodegraded is developing [118].

Multifunctional polymers have several potential prospects despite these obstacles. The creation of multifunctional biodegradable polymers, for example, that may be used for a range of biological applications is gaining attention [119].

Table 1.1 Examples of multifunctional polymers with their applications [39, 106–116].

S. no.	Multifunctional polymer	Example	Application
1.	Sodium carboxymethyl cellulose (CMC)	Used in tissue engineering and medication delivery systems	Biomedical field. Paper, personal care/cosmetic, food, pharmaceutical, and other sectors.
2.	Nanocomposites	Nanocomposites for mechanical, tribological, and fire-protection applications	Applications in tribology, mechanics, fire safety, and the aerospace and automotive industries.
3.	Polymer composites	Various applications in several fields include bioengineering, automobile production, space exploration, and the development of organic solar cells	Bioengineering, automobile production, organic solar cell development, and space exploration.
4.	Lipase-catalyzed polyester	Bio-based thermoplastic elastomers	Biomedical field. Successfully used for the synthesis of polyamides derived from renewable plant oils.
5.	PCL-PEG co-polymers	Biomedical applications	Biomedical field. Preparation of hydrogels for drug encapsulation and drug delivery.
6.	Fluorescent polymer nanomaterials	Water-soluble and biocompatible fluorescent polymer nanomaterials	Biomedical field. Bioimaging, biodetection, cancer therapy, and medical imaging modalities.

1.5.1 Current Limitations and Challenges in Sustainable Multifunctional Polymer Chemistry

Current limitations and challenges in sustainable multifunctional polymer chemistry include the following.

1.5.1.1 Lack of Standardized Methods

Standardized procedures are required for the synthesis and characterization of multifunctional, sustainable polymers to promote uniformity and repeatability in the scientific community [120].

1.5.1.2 Limited Availability of Renewable Feedstocks

A limiting issue in the synthesis of sustainable multifunctional polymers may be the cost and availability of renewable feedstocks. Because of this, increasing the production of these materials may be challenging [120].

1.5.1.3 Environmental Impact

Multifunctional polymer manufacture and disposal can have a substantial negative influence on the environment. It is necessary to create environmentally friendly and sustainable processes for the production of multipurpose polymers in addition to plans for the appropriate utilization and elimination of these materials [121].

1.5.1.4 Performance Limitations

Since sustainable multifunctional polymers cannot always operate to the same standards as conventional polymers, this could be a restriction. This may restrict their uses and make it more challenging for them to compete with well-established materials [122].

1.5.1.5 Cost

The price of creating multifunctional, sustainable polymers may be a major barrier. This may restrict these materials' economic viability and make it challenging to scale up manufacturing [122].

1.6 Conclusion

Since their invention more than a century ago, polymers have significantly enhanced human well-being and are now essential to modern living and cutting-edge technology. They make advanced technology accessible to all people because of their intense energy-, economical and ecological, and resource-efficient manufacturing techniques, ease of processing, great adaptability regarding adjustable features, and an extensive range of uses. The world is currently moving away from an energy system reliant on fossil fuels and toward one that is powered by more sustainable and renewable resources.

An increasing number of people are interested in renewable polymer-based ecologically friendly products due to escalating environmental concerns and the exhaustion of petroleum-based materials. Throughout their life cycle, sustainable polymers also show a decreased influence on the environment. By decreasing the manufacture and using virgin plastics made from limited supplies, the creation of environmentally friendly polymers will hasten the age of eco-friendly polymers and bring about a fully circular plastics economy.

Many biopolymers made from transgenic plants or bacteria were not economically viable in the past due to their inadequate qualities and relatively low processability. Because biosynthesis requires such great accuracy, it might be challenging to modify their molecular structures to meet the requirements of polymer molding. Given the extremely quick cycle periods in injection molding, many biopolymers—like polyhydroxybutyrate, for example—crystallize too slowly. While commercial polymers require far more adaptable molar mass distributions with regulated branches on both short and long chains in addition to the ability for self-nucleation during the crystallization of polymers, all polymers in nature have equal chain lengths.

An additional issue with most synthetic polymers in biotechnology is the time-consuming task of extracting biopolymers that are produced as byproducts from cell proteins.

Solvents that contain chlorine are used to extract a variety of PHAs. All commercial polymers are manufactured in the gas or melt phase devoid of the use of wastes or solvents, fulfilling the requirements of polymer manufacturing as well as green chemistry, even if water is needed for biotechnological polymer biosynthesis and full purification of water.

Even with the amazing advances in biotechnology, fine-tuning biopolymers through enzyme modifications and genetic engineering to satisfy the diverse needs of particular uses for polymers would be time-consuming and costly. To manufacture commercially effective bio-based polymers, it is therefore far more practical to use biotechnology or biomass conversion processes to produce monomers, which may then be coupled with following gas- and melt-phase polymerization procedures. This group has produced several examples of how to produce environmentally friendly materials with appealing processing and material qualities. Conventional plastics such as PET and polyolefins may now be made greener and more renewable without sacrificing their appealing qualities or capacity for recycling thanks to the introduction of bio-based monomers. Plastics derived from petrochemistry and biotechnology will coexist for many years to come and continue to be crucial in the creation of affordable, ecologically favorable products that satisfy the demands of contemporary society.

References

1. Schleicher, J., Schaafsma, M., Vira, B., Will the Sustainable Development Goals address the links between poverty and the natural environment?, *Current Opinion in Lololo*, 47, 34, 43–47, 2018 Accessed 2020. https://www.un.org/sustainable development/development-agenda/.
2. Cheng, H.N., Gross, R.A., Chapter 1 Sustainability and Green Polymer Chemistry—An Overview, in: *ACS Symposium Series*, vol. 1372, pp. 1–11, 2020.
3. Cheng, H.N. and Gross, R.A., Sustainability and Green Polymer Chemistry—An Overview, in: *Sustainability & Green Polymer Chemistry Volume 1: Green Products and Processes*, pp. 1–1, 2020.
4. Whelan, T., and Fink, C., The comprehensive business case for Harvard Business Sustainability. *Review*, 21, 161–168, 2016.
5. Schmitz, E. A., Baum, M., Huett, P., Kabst, R., The Contextual Role of Regulatory Stakeholder Pressure in Proactive Environmental Strategies: An Empirical Test of Competing Theoretical Perspectives. *Organ. Environ.*, 32, 3, 281–308, 2019. Accessed 2020-09-01. https://saylordotorg.github.io/text_a-primer-on-sustainable-business/index.html.
6. Coons, R. and Westervelt, R., *ACI 2020: Sustainability in sharp focus*, January 31, 2020, Accessed 2020-09-01.https://chemweek.com/CW/Document/108823/ACI-2020-Sustainability-insharp-focus.
7. Pan, D., Su, F., Liu, C., Guo, Z., Research progress for plastic waste management and manufacture of value-added products. *Adv. Compos. Hybrid Mater.*, 3, 443–461, 2020.

8. Moharir, R.V. and Kumar, S., Challenges associated with plastic waste disposal and allied microbial routes for its effective degradation: A comprehensive review. *J. Clean. Prod.*, 208, 65–76, 2019.
9. Pushpamalar, J., Veeramachineni, A.K., Owh, C., Loh, X.J., Biodegradable polysaccharides for controlled drug delivery. *ChemPlusChem*, 81, 6, 504–514, 2016.
10. Ho, M.P. *et al.*, Critical factors on manufacturing processes of natural fibre composites. *Compos. Part B Eng.*, 43, 3549–3562, 2012.
11. Yang, J., Ching, Y.C., Chuah, C.H., Applications of Lignocellulosic Fibers and Lignin in Bioplastics: A Review. *Polymers (Basel)*, 11, 1–26, 2019.
12. Luckachan, G.E. and Pillai, C.K.S., Biodegradable polymers- a review on recent trends and emerging perspectives. *J. Polym. Environ.*, 19, 637–676, 2011.
13. Das, A., Ringu, T., Ghosh, S. *et al.*, A comprehensive review on recent advances in preparation, physicochemical characterization, and bioengineering applications of biopolymers. *Polym. Bull.*, 80, 7247–7312, 2023. https://doi.org/10.1007/s00289-022-04443-4.
14. Peng, X., Dong, K., Wu, Z., Wang, J., Wang, Z.L., A review on emerging biodegradable polymers for environmentally benign transient electronic skins. *J. Mater. Sci.*, 56, 16765–16789, 2021.
15. Taha, T.H., Abu-Saied, M.A., Elnouby, M., Hashem, M., Alamri, S., El Desouky, E.A., Morsy, K., Profitable exploitation of biodegradable polymer including chitosanblended potato peels' starch waste as an alternative source of petroleum plastics, *Biomass Convers. Biorefin.*, 14, 207–215, 2024.
16. Zhu, J. and Wang, C., Biodegradable plastics: Green hope or greenwashing? *Mar. Pollut. Bull.*, 161, 111774, 2020.
17. Ghanbarzadeh, B. and Hadi, A., Biodegradable polymers, in: *Biodegradation-life of science*, pp. 141–185, 2013.
18. Cazaudehore, G. *et al.*, Can anaerobic digestion be a suitable end-of-life scenario for biodegradable plastics? A critical review of the current situation, hurdles, and challenges. *Biotechnol. Adv.*, 56, 107916, 2022.
19. Baranwal, J., Barse, B., Fais, A., Delogu, G.L., Kumar, A., Biopolymer: A Sustainable Material for Food and Medical Applications. *Polymers*, 14, 983, 2022.
20. Prata, J.C., Airborne Microplastics: Consequences to Human Health? *Environ. Pollut.*, 234, 115–126, 2018.
21. Lau, W.W.Y., Shiran, Y., Bailey, R.M., Cook, E., Stuchtey, M.R., Koskella, J., Velis, C.A., Godfrey, L., Boucher, J., Murphy, M.B. *et al.*, Evaluating Scenarios toward Zero Plastic Pollution. *Science*, 369, eaba9475, 2020.
22. Bose, I., Roy, S., Pandey, V.K., Singh, R., A Comprehensive Review on Significance and Advancements of Antimicrobial Agents in Biodegradable Food Packaging. *Antibiotics*, 12, 968, 2023.
23. Zhang, W., Roy, S., Rhim, J.-W., Copper-Based Nanoparticles for Biopolymer-Based Functional Films in Food Packaging Applications. *Compr. Rev. Food Sci. Food Saf.*, 22, 1933–1952, 2023.
24. Webb, H.K., Arnott, J., Crawford, R.J., Ivanova, E.P., Plastic Degradation and Its Environmental Implications with Special Reference to Poly(ethylene terephthalate). *Polymers*, 5, 1–18, 2013.
25. Derraik, J.G.B., The pollution of the marine environment by plastic debris: A review. *Mar. Pollut. Bull.*, 44, 842–852, 2002.
26. Gregory, M.R., Environmental implications of plastic debris in marine settings—Entanglement, ingestion, smothering, hangers-on, hitch-hiking and alien invasions. *Philos. Trans. R. Soc. B*, 364, 2013–2025, 2009.

27. Abidi, N., Cabrales, L., Haigler, C.H., Changes in the cell wall and cellulose content of developing cotton fibers investigated by FTIR spectroscopy. *Carbohydr. Polym.*, 100, 9–16, 2014.
28. Christina, K., Subbiah, K., Arulraj, P., Krishnan, S.K., Sathishkumar, P., A sustainable and eco-friendly approach for environmental and energy management using biopolymers chitosan, lignin and cellulose—A review. *Int. J. Biol. Macromol.*, 257, 2, 128550, 2024.
29. Dziuba, R., Kucharska, M., Madej-Kiełbik, L., Sulak, K., Wiśniewska-Wrona, M., Biopolymers and Biomaterials for Special Applications within the Context of the Circular Economy. *Materials.*, 14, 7704, 2021.
30. Mazhandu, Z.S., Muzenda, E., Mamvura, T.A., Belaid, M., Nhubu, T., Integrated and Consolidated Review of Plastic Waste Management and Bio-Based Biodegradable Plastics: Challenges and Opportunities. *Sustainability*, 12, 8360, 2020.
31. Coppola, G., Gaudio, M.T., Lopresto, C.G., Calabro, V., Curcio, S., Chakraborty, S., Bioplastic from renewable biomass: a facile solution for a greener environment. *Earth Syst. Environ.*, 5, 231–51, Jun. 2021.
32. Thakur, V.K., Thakur, M.K., Raghavan, P., Kessler, M.R., Progress in Green Polymer Composites from Lignin for Multifunctional Applications: A Review. *ACS Sustain. Chem. Eng.*, 2, 1072–1092, 2014.
33. Faruk, O., Bledzki, A.K., Fink, H.-P., Sain, M., Biocomposites reinforced with natural fibers: 2000–2010. *Prog. Polym. Sci.*, 37, 1552–1596, 2012.
34. Fernandes, E.M., Pires, R.A., Mano, J.F., Reis, R.L., Bionanocomposites from lignocellulosic resources: Properties, applications and future trends for their use in the biomedical field. *Prog. Polym. Sci.*, 38, 1415–1441, 2013.
35. Pinto, M.A., Chalivendra, V.B., Kim, Y.K., Lewis, A.F., Effect of surface treatment and Z-axis reinforcement on the interlaminar fracture of jute/epoxy laminated composites. *Eng. Fract. Mech.*, 114, 104–114, 2013.
36. Ojijo, V. and Ray, S.S., Processing strategies in bionanocomposites. *Prog. Polym. Sci.*, 38, 1543–1589, 2013.
37. Cosutchi, A., II, Hulubei, C., Stoica, I., Ioan, S., Morphological and structural-rheological relationship in Epiclon-based polyimide/hydroxypropyl cellulose blend systems. *J. Polym. Res.*, 17, 541&ndash, 2010.
38. bk-2020-1372.ch001 (acs.org)
39. Mohanty, A.K., Wu, F., Mincheva, R., Hakkarainen, M., Raquez, J.M., Mielewski, D.F., Narayan, R., Netravali, A.N., Misra, M., Sustainable polymers. *Nat. Rev. Methods Primers*, 2, 1, 46, Jun. 16 2022.
40. Haque, F.M., Ishibashi, J.S., Lidston, C.A., Shao, H., Bates, F.S., Chang, A.B., Coates, G.W., Cramer, C.J., Dauenhauer, P.J., Dichtel, W.R., Ellison, C.J., Defining the macromolecules of tomorrow through synergistic sustainable polymer research. *Chem. Rev.*, 122, 6, 6322–73, Feb. 8, 2022.
41. Jasinska-Walc, L., Bouyahyi, M., Duchateau, R., Potential of functionalized polyolefins in a sustainable polymer economy: Synthetic strategies and applications. *Acc. Chem. Res.*, 55, 15, 1985–96, Jul. 18, 2022.
42. Anastas, P. and Eghbali, N., Green chemistry: principles and practice. *Chem. Soc. Rev.*, 39, 1, 301–12, 2010.
43. Kobayashi, S., Green polymer chemistry: new methods of polymer synthesis using renewable starting materials. *Struct. Chem.*, 28, 461–74, Apr. 2017.
44. Jiang, Y. and Loos, K., Enzymatic Synthesis of Biobased Polyesters and Polyamides. *Polymers (Basel)*, 8, 7, 243, Jun 25, 2016.
45. Anastas, P.T. and Warner, J.C., *Green Chemistry: Theory and Practice*, p. 30, Oxford University Press, New York, 1998.

46. Phan, H. *et al.*, Synthesis of Multifunctional Polymersomes Prepared by Polymerization-Induced Self-Assembly. *Polymers (Basel)*, 15, 14, 3070, Jul. 2023, doi: 10.3390/POLYM15143070/S1.
47. Mazumdar, P., Rattan, S., Singhal, P., Sharma, I., Gupta, B.K., A Green Route Strategy for the Synthesis of Multifunctional Polymer Nanocomposites for Environmental Sustainability. *ChemistrySelect*, 4, 4, 1491–1501, Jan. 2019. doi: 10.1002/SLCT.201803834.
48. Song, S., Ngo, K., Uhrich, K., Green chemistry principles in advancing hierarchical functionalization of polymer-based nanomedicines. *ACS Symp. Ser.*, 1372, 135–157, 2020.
49. Nuyken, O. and Pask, S.D., Ring-Opening Polymerization—An Introductory Review. *Polymers*, 5, 361–403, 2013.
50. Panchal, S.S. and Vasava, D.V., Biodegradable polymeric materials: Synthetic approach. *ACS Omega*, 5, 4370–4379, 2020.
51. Ferrari, R., Yu, Y., Morbidelli, M., Hutchinson, R.A., Moscatelli, D., E- Caprolactone-based macromonomers suitable for biodegradable Nanoparticles synthesis through free radical polymerization. *Macromolecules*, 44, 9205–9212, 2011.
52. Ahmed, S.T., Leferink, N.G.H., Scrutton, N.S., Chemo-enzymatic routes towards the synthesis of biobased monomers and polymers. *Mol. Catal.*, 467, 95–110, 2019.
53. Acik, G. and Karatavuk, A.O., Synthesis, properties and biodegradability of cross-linked amphiphilic poly(vinyl acrylate)- poly (tert-butyl acrylate)s by photo-initiated radical polymerization. *Eur. Polym. J.*, 127, 109602, 2020.
54. Yagihara, T. and Matsumura, S., Enzymatic synthesis and chemical recycling of novel polyester-type thermoplastic elastomers. *Polymers*, 4, 2, 1259–1277, 2012.
55. Chanfreau, S., Mena, M., Porras-Domínguez, J.R., Ramírez-Gilly, M., Gimeno, M., Roquero, P., Tecante, A., Bárzana, E., Enzymatic synthesis of poly-L-lactide and poly-L-lactide-co-glycolide in an ionic liquid. *Bioprocess. Biosyst. Eng.*, 33, 629–638, 2010.
56. Kobayashi, T. and Matsumura, S., Enzymatic synthesis and properties of novel biodegradable and biobased thermoplastic elastomers. *Polym. Degrad. Stab.*, 96, 2071–2079, 2011.
57. Mirmohammadi, S.A., Imani, M., Uyama, H., Atai, M., Teimouri, M.B., Bahri-Lale, N., Hybrid organic -inorganic nanocomposites based on poly(E caprolactone)/ polyhedral oligomeric silsesquioxane: synthesis and *In vitro* evaluations. *Polym. Int.*, 63, 479–485, 2014.
58. Nuyken, O. and Pask, S.D., Ring opening polymerization -an introductory review. *Polymers*, 5, 5, 361–403, 2013.
59. Zhao, H., Nathaniel, G.A., Merenini, P.C., Enzymatic ring-opening polymerization (ROP) of lactides and lactone in ionic liquids and organic solvents: digging the controlling factors. *RSC Adv.*, 7, 48639–48648, 2017.
60. Omay, D. and Guvenilir, Y., Synthesis and characterization of poly(d, l-lactic acid) via enzymatic ring-opening polymerization by using free and immomobilized lipase. *Biocatal. Biotransform.*, 31, 132–140, 2013.
61. Llevot, A., Dannecker, P.K., von Czapiewski, M., Over, L.C., Söyler, Z., Meier, M.A., Renewability is not enough: recent advances in the sustainable synthesis of biomass-derived monomers and polymers. *Chem. A Eur. J.*, 22, 33, 11510–21, Aug. 8, 2016.
62. Rosenboom, J.G., Langer, R., Traverso, G., Bioplastics for a circular economy. *Nat. Rev. Mater.*, 7, 117–137, 2022. https://doi.org/10.1038/s41578-021-00407-8.
63. Geyer, R., Jambeck, J.R., Law, K.L., Production, use, and fate of all plastics ever made. *Sci. Adv.*, 3, 25–29, 2017.
64. Tsang, Y.F., Kumar, V., Samadar, P., Yang, Y., Lee, J., Sik Ok, Y., Song, H., Kim, K.-H., Kwon, E.E., Jeon, Y.J., Production of bioplastic through food waste valorization. *Environ. Int.*, 127, 625–644, 2019.

65. Flaris, V. and Singh, G., Recent Developments in Biopolymers. *J. Vinyl Addit. Technol.*, 15, 1, 1–11, 2009.
66. Conrad, R., Bio-based polymers and plastics, 2024. Available: https://www.elastomer.kuraray.com/us/bio-based-polymers/.
67. Liwei, Y., Thompson, B.C., Improving the efficiency and sustainability of catalysts for direct arylation polymerization, *J. Polym. Sci.*, 1–36, 2021.
68. Zhang, X., Fevre, M., Jones, G.O., Waymouth, R.M., Catalysis as an Enabling Science for Sustainable Polymers. *Chem. Rev.*, 118, 2, 839–885, Oct. 2017.
69. Fiorentini, F., Diment, W.T., Deacy, A.C., Kerr, R.W.F., Faulkner, S., Williams, C.K., Understanding catalytic synergy in dinuclear polymerization catalysts for sustainable polymers. *Nat. Commun.*, 14, 1, 1–10, 2023.
70. Tarazona, N.A., Machatschek, R., Balcucho, J., Castro-Mayorga, J.L., Saldarriaga, J.F., Lendlein, A., Opportunities and challenges for integrating the development of sustainable polymer materials within an international circular (bio)economy concept. *MRS Energy Sustainability*, 9, 1, 28–34, Mar. 2022, doi: 10.1557/S43581-021-00015-7/FigureS/1.
71. Bultema, L.A., Huang, X., Brauer, D.D., Theato, P., Polymer Functionalization, in: Jafar Mazumder, M., Sheardown, H., Al-Ahmed, A. (eds) *Functional Polymers.* Polymers and Polymeric Composites: A Reference Series. Springer, Cham. 2019. https://doi.org/10.1007/978-3-319-95987-0_2
72. Maity, T., Paul, S., De, P., Designing functionalized polymers through post-polymerization modification of side chain leucine-based polymer. *Eur. Polym. J.*, 199, 112445, 2023.
73. Creamer, A. *et al.*, Post-polymerisation functionalisation of conjugated polymer backbones and its application in multi-functional emissive nanoparticles. *Nat. Commun.*, 9, 1, 1–11, 2018.
74. Schröder, K., Matyjaszewski, K., Noonan, K.J., Mathers, R.T., Towards sustainable polymer chemistry with homogeneous metal-based catalysts. *Green Chem.*, 16, 4, 1673–86, 2014.
75. Shoda, S., Uyama, H., Kadokawa, J., Kimura, S., Kobayashi, S., Enzymes as Green Catalysts for Precision Macromolecular Synthesis. *Chem. Rev.*, 116, 4, 2307–413, Feb. 24, 2016.
76. López-Iglesias, M. and Gotor-Fernández, V., Recent Advances in Biocatalytic Promiscuity: Hydrolase-Catalyzed Reactions for Nonconventional Transformations. *Chem. Rec.*, 15, 4, 743–59, Aug. 2015.
77. South, C.R., Burd, C., Weck, M., Modular and dynamic functionalization of polymeric scaffolds. *Acc. Chem. Res.*, 40, 1, 63–74, Jan. 16, 2007.
78. Lee, K.M., Kim, K.H., Yoon, H., Kim, H., Chemical Design of Functional Polymer Structures for Biosensors: From Nanoscale to Macroscale. *Polymers*, 10, 551, 2018.
79. Zarzar, L.D. and Aizenberg, J., Stimuli-Responsive Chemomechanical Actuation: A Hybrid Materials Approach. *Acc. Chem. Res.*, 47, 530–539, 2014.
80. Coessens, V., Pyun, J., Miller, P.J., Gaynor, S.G., Matyjaszewski, K., Functionalization of polymers prepared by ATRP using radical addition reactions. *Macromol. Rapid Commun.*, 21, 2, 103–9, Feb. 1, 2000.
81. Molino, P.J., Yue, Z., Zhang, B., Tibbens, A., Liu, X., Kapsa, R.M.I., Higgins, M.J., Wallace, G.G., Influence of Biodopants on PEDOT Biomaterial Polymers: Using QCM-D to Characterize Polymer Interactions with Proteins and Living Cells. *Adv. Mater. Interfaces*, 1, 1300122, 2014.
82. Ferguson, C.J., Hughes, R.J., Pham, B.T.T., Hawkett, B.S., Gilbert, R.G., Serelis, A.K., Such, C.H., Effective ab initio emulsion polymerization under RAFT control. *Macromolecules*, 35, 25, 9243-9245, 2002.
83. Geng, Z., Shin, J.J., Xi, Y., Hawker, C.J., Click chemistry strategies for the accelerated synthesis of functional macromolecules. *J. Polym Sci.*, 59, 963–1042, 2021.
84. Yang, B., Zhao, Y., Wei, Y., Fu, C., Tao, L., The Ugi reaction in polymer chemistry: syntheses{,} applications and perspectives. *Polym. Chem.*, 6, 48, 8233–8239, 2015.

85. Hu, C., Liu, C., Liu, Q., Zhang, H., Wu, S., Xiao, R., Effects of Steam to Enhance the Production of Light Olefins from *Ex-Situ* Catalytic Fast Pyrolysis of Biomass. *Fuel Proc. Technol.*, 210, 106562, 2020.
86. Wang, C., Yu, B., Li, W., Zou, W., Cong, H., Shen, Y., Effective strategy for polymer synthesis: multicomponent reactions and click polymerization. *Mater. Today Chem.*, 25, 100948, 2022.
87. Riedel, M. and Voit, B., Synthesis of multifunctional polymers by combination of controlled radical polymerization (CRP) and effective polymer analogous reactions. *Pure Appl. Chem.*, 85, 3, 557–571, 2013, doi: 10.1351/PAC-CON-12-04-09.
88. Yang, B. *et al.*, Synthesis of multifunctional polymers through the ugi reaction for protein conjugation. *Macromolecules*, 47, 16, 5607–5612, Aug. 2014, doi: 10.1021/MA501385M/SUPPL_FILE/MA501385M_SI_001.PDF.
89. Kubo, T., Easterling, C.P., Olson, R.A., Sumerlin, B.S., Synthesis of multifunctional homopolymers via sequential post-polymerization reactions. *Polym. Chem.*, 9, 37, 4605–4610, Sep. 2018.
90. Mangaraj, S., Yadav, A., Bal, L.M., Dash, S.K., Mahanti, N.K., Application of biodegradable polymers in food packaging industry: a comprehensive. *Rev. J. Packag. Technol. Res.*, 3, 77–96, 2019.
91. Tian K., and Bilal M., Chapter 15 - Research progress of biodegradable materials in reducing environmental pollution, Editor(s): Pardeep Singh, Ajay Kumar, Anwesha Borthakur, *Abatement of Environmental Pollutants*, 313–330, Elsevier, 2020. https://doi.org/10.1016/B978-0-12-818095-2.00015-1.
92. Carnahan, M.A. and Grinstaff, M.W., Synthesis and characterization of poly(glycerol-succinic acid) dendrimers. *Macromolecules*, 34, 7648–7655, 2001, doi: 10.1021/ma010848n.
93. Valerio, O., Pin, J.M., Misra, M., Mohanty, A.K., Synthesis of Glycerol-Based Biopolyesters as Toughness Enhancers for Polylactic Acid Bioplastic through Reactive Extrusion. *ACS Omega*, 1, 1284–1295, 2016.
94. Luman, N.R., Kim, T., Grinstaff, M.W., Dendritic polymers composed of glycerol and succinic acid: Synthetic methodologies and medical applications. *Pure Appl. Chem.*, 76, 1375–1385, 2007, doi: 10.1351/pac200476071375.
95. Immoos, C.E., Ribeiro, A.A., Morgan, M.T., Lee, S.J., Grinstaff, M.W., Carnahan, M.A., Finkelstein, S., Dendritic Molecular Capsules for Hydrophobic Compounds. *J. Am. Chem. Soc.*, 125, 15485–15489, 2003.
96. Phung, T.K., Pham, T.L.M., Vu, K.B., Busca, G., (Bio) Propylene Production Processes: A Critical Review. *J. Environ. Chem. Eng.*, 9, 105673, 2021, doi: 10.1016/j.jece.2021.105673.
97. Koempel, H. and Liebner, W., Lurgi's Methanol To Propylene (MTP®) Report on a successful commercialisation, Editor(s): Noronha, F.B., Schmal, M., Sousa-Aguiar, E.F., in: *Studies in Surface Science and Catalysis*, Volume 167, Pages 261–267, Elsevier, 2007.
98. Minteer, S.D., Biochemical Production of Other Bioalcohols: Biomethanol, Biopropanol, Bioglycerol, and Bioethylene Glycol, in: *Handbook of Biofuels Production*, pp. 258–265, 2011.
99. Zhang, H., Xiao, R., Jin, B., Xiao, G., Chen, R., Biomass Catalytic Pyrolysis to Produce Olefins and Aromatics with a Physically Mixed Catalyst. *Bioresour. Technol.*, 140, 256–62, 2013, doi: 10.1016/j.biortech.2013.04.094.
100. Karabasz, A., Bzowska, M., Szczepanowicz, K., Biomedical Applications of Multifunctional Polymeric Nanocarriers: A Review of Current Literature. *Int. J. Nanomed.*, 15, 8673–8696, 2020.
101. Ding, M., Jing, L., Yang, H., Machnicki, C.E., Fu, X., Li, K., Wong, I.Y., Chen, P.-Y., Multifunctional soft machines based on stimuli-responsive hydrogels: from freestanding hydrogels to smart integrated systems. *Mater. Today Adv.*, 8, 100088, 2020.
102. Zhang, J., Gu, M., Chen, X., Supercapacitors for renewable energy applications: A review. *Micro Nano Eng.*, 21, 100229, 2023.

103. Beni, A.A. and Jabbari, H., Nanomaterials for Environmental Applications. *Results Eng.*, 15, 100467, 2022.
104. Romero-Fierro, D., Bustamante-Torres, M., Bravo-Plascencia, F., Esquivel-Lozano, A., Ruiz, J.-C., Bucio, E., Recent Trends in Magnetic Polymer Nanocomposites for Aerospace Applications: A Review. *Polymers*, 14, 4084, 2022.
105. Kunduru, K.R., Basu, A., Domb, A.J., Biodegradable polymers: medical applications, in: *Encyclopedia of Polymer Science and Technology*, pp. 1–22, 2016.
106. Karolewicz, B., A review of polymers as multifunctional excipients in drug dosage form technology. *Saudi Pharm. J.*, 24, 5, 525–536, 2016, doi: 10.1016/j.jsps.2015.02.025.
107. Lopez, C.G., Rogers, S.E., Colby, R.H., Graham, P., Cabral, J.T., Structure of Sodium Carboxymethyl Cellulose Aqueous Solutions: A SANS and Rheology Study. *J. Polym. Sci. B. Polym. Phys.*, 53, 7, 492–501, 2015 Apr 1.
108. Bull, S.J., Multifunctional polymer nanocomposites for industrial applications, in: *Nanofibers and Nanotechnology in Textiles*, pp. 256–280, 2007.
109. Colorado, H.A., Gutierrez-Velasquez, E., Gil, L.D., de Camargo, I.L., Exploring the advantages and applications of nanocomposites produced via vat photopolymerization in additive manufacturing: A review. *Adv. Compos. Hybrid Mater.* 7, 1, 2023. Accessed: Feb. 04, 2024. [Online]. Available: https://link.springer.com/article/10.1007/s42114-023-00808-z.
110. Bu, Q., Li, P., Xia, Y., Hu, D., Li, W., Shi, D., Song, K., Design, Synthesis, and Biomedical Application of Multifunctional Fluorescent Polymer Nanomaterials. *Molecules*, 28, 9, 3819, 2023 Apr 29.
111. Dmitruk, A. and Kaczmar, J.W., Application of Polymer Based Composite Materials in Transportation, *Prog. Rubber Plast. Recycl. Technol.*, 32, 1, 1–24, 2016. Accessed: Feb. 04, 2024. [Online]. Available: https://nanografi.com/blog/applications-of-polymerbased-composites-/.
112. Finnveden, M., Hendil-Forssell, P., Claudino, M., Johansson, M., Martinelle, M., Lipase-Catalyzed Synthesis of Renewable Plant Oil-Based Polyamides. *Polymers*, 11, 1730, 2019. Accessed: Feb. 04, 2024. [Online]. Available: https://www.ncbi.nlm.nih.gov/pmc/articles/PMC6918247/.
113. Leng, J., Lau, A.K-T., *Multifunctional Polymer Nanocomposites*, 1–466, CRC Press, 1st Edition, 2017. Accessed: Feb. 04, 2024. [Online]. Available: https://www.routledge.com/Multifunctional-Polymer-Nanocomposites/Leng-Lau/p/book/9781138111806.
114. Dethe, M.R., A P, Ahmed, H., Agrawal, M., Roy, U., Alexander, A., PCL-PEG copolymer based injectable thermosensitive hydrogels. *J. Control. Release*, 343, 217–236, 2022. Web link: https://www.nanosoftpolymers.com/product/pcl-peg-pcl/
115. Friedrich, K. and Breuer, U., *Multifunctionality of Polymer Composites Challenges and New Solutions*, 945–964, William Andrew Publishing, 2015. Accessed: Feb. 04, 2024. [Online]. Available: https://www.sciencedirect.com/book/9780323264341/multifunctionality-of-polymer-composites.
116. Ganguly, S., Das, P., Parameswaranpillai J., Maity, P.P., Editorial: Fluorescent nanomaterials for biomedical applications. *Front. Mater.*, 10, 1328035, 1–3. 2023. [Online]. Available: https://www.frontiersin.org/research-topics/24585/fluorescent-nanomaterials-for-biomedical-applications.
117. Saalwächter, K., Grand challenges in polymers. *Front. Soft Matter*, 2, 1037349, 2022.
118. Shim, I.K., Jung, M.R., Kim, K.H., Seol, Y.J., Park, Y.J., Park, W.H., Lee, S.J., Novel Three-Dimensional Scaffolds of Poly(L-Lactic Acid) Microfibers Using Electrospinning and Mechanical Expansion: Fabrication and Bone Regeneration. *J. Biomed. Mater. Res. Part B Appl. Biomater.*, 95, 150–160, 2010.
119. Saleh, H.M. and Hassan, A.I., Synthesis and Characterization of Nanomaterials for Application in Cost-Effective Electrochemical Devices. *Sustainability*, 15, 10891, 2023.

120. Samir, A., Ashour, F.H., Hakim, A.A.A. *et al.*, Recent advances in biodegradable polymers for sustainable applications. *npj Mater. Degrad.*, 6, 68, 2022.
121. Rai, P., Mehrotra, S., Priya, S., Gnansounou, E., Sharma, S.K., Recent Advances in the Sustainable Design and Applications of Biodegradable Polymers. *Bioresour. Technol.*, 325, 124739, 2021.
122. Maraveas, C., Production of Sustainable and Biodegradable Polymers from Agricultural Waste. *Polymers*, 12, 1127, 2020.

2

Biopolymers: Green and Sustainable Approach in Polymer Science

Agrima Yadav and Shikha Yadav*

Department of Pharmacy, School of Medical and Allied Sciences, Galgotias University, Greater Noida, Uttar Pradesh, India

Abstract
In contemporary times, bio-based polymers derived from natural biomass have emerged as promising alternatives to conventional plastics derived from fossil fuels. The impetus for replacing conventional synthetic plastics primarily stems from numerous concerns regarding their adverse impacts on the environment and human well-being. One of the most cutting-edge approaches to bioplastic production entails applying basic components obtained from waste materials. Raw materials play a crucial role in promoting the well-being of both humans and animals, owing to their significant economic and environmental advantages. Different categories of biopolymers have significant importance. In this chapter, several applications of biopolymers are mentioned with the appropriate examples. This chapter aims to provide the significance of biopolymers, classification, and applications.

Keywords: Biopolymers, collagen, polyhydroxyalkanoates, wearing resistance

2.1 Introduction

In recent years, there has been a notable and swift increase in interest surrounding environmental preservation. This interest extends beyond the utilization of products derived from renewable resources, encompassing the preference for outcomes that break down into components that are beneficial to the environment. In nearly all developed nations, there has been a proliferation of initiatives, campaigns, green movements, different guidelines, etc, aimed at mitigating the quantity of solid polymer waste produced by consumers on an annual basis. Consumers have additionally conveyed their inclination for products that possess environmentally sustainable attributes, while concurrently delivering comparable outcomes to those derived from synthetic materials. Nevertheless, the inclination of consumers toward environmentally sustainable products may be impeded by the elevated price and substandard characteristics of these goods in comparison to their synthetically obtained counterparts. The term "biodegradable" pertains to the functional characteristic of a polymer known as "biodegradability." It is used to describe polymers that have the

**Corresponding author*: shikha.yadav@galgotiasuniversity.edu.in

Divya Bajpai Tripathy, Anjali Gupta and Arvind Kumar Jain (eds.) Multifunctional Materials: Engineering and Biological Applications, (27–46) © 2025 Scrivener Publishing LLC

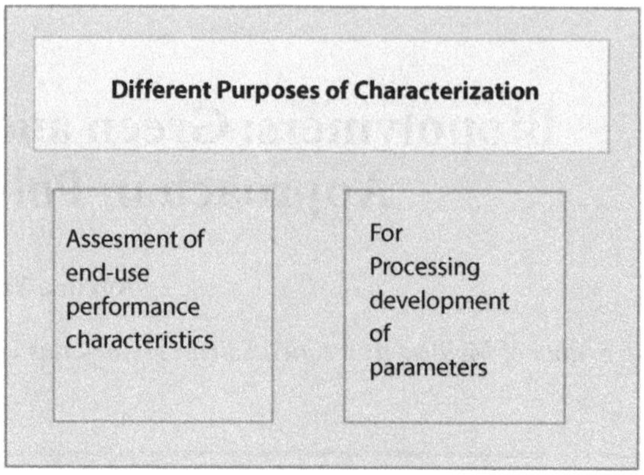

Figure 2.1 Different purposes of characterization of biopolymers.

ability to decompose when exposed to specific environmental conditions and microorganisms, including fungi, molds as well as microbes, within a designated time frame. There are several criteria by which the biopolymers can be defined such as by the polymer's biodegradability and other one the raw materials and their sources from where they originate or are derived. It has a large impact on the environment too, which is why the characterization of biopolymers requires a purpose (Figure 2.1). Biopolymers can be described as polymers that originate from renewable resources, encompassing both biodegradable polymers derived from biological sources and those derived from fossil-based materials [1].

2.1.1 Advantages of Biopolymers Over Traditional Polymers

Biopolymers have beneficial effects on the environment and a satisfactory outcome after having the product made by the biopolymers. Several advantages of biopolymers over traditional polymers are as follows:

1. These are environmentally friendly, carbon neutral as well as can be easily renewed.
2. These are less expensive because they can be easily obtained from renewable resources.
3. Different polysaccharides and their derivatives demonstrate the several compilations of polymers vastly utilized by the various pharmaceutical sections for the control and sustained drug delivery systems (DDSs).
4. Biopolymers are non-immunogenic and work efficiently in the agriculture sector.

Current research endeavors are primarily directed toward the exploration of novel technologies that harness the potential of biodegradable natural products, specifically natural polymers. The aim is to supplant conventional materials and thereby achieve a substantial reduction in the generation and accumulation of waste, particularly plastic waste. The efficient utilization of natural resources in a comprehensive manner for the purpose of preservation and recycling represents a commendable choice and a novel approach to the advancement of biodegradable products. Biopolymers possess a thermoplastic nature and exhibit properties akin to those of petroleum-based plastics. The significant advantage of natural biodegradation, facilitated by fungi, bacteria, and algae, is the production of CO2, water, and composted materials, as opposed to synthetic alternatives [2]. Polyhydroxyalkanoates (PHAs), a kind of biopolymer produced by microorganisms, are brittle, rigid, or malleable and have comparable physical qualities to plastics derived from petroleum. This has prompted greater study into PHA studies, as shown by Lemoigne (1926), who noted that over three hundred microbes can make PHA, with *Bacillus megaterium* producing the most of it [3]. Plants are an outstanding choice for making polymers since they can be cultivated on huge areas of land and produce a lot of biomasses because they rely on sunlight for energy.

2.1.2 Types of Biopolymers

There are several ways to classify biopolymers. On the basis of the characteristics of the repeating monomer units, biopolymers may be categorized into three major categories (Figure 2.2).

Polysaccharides, polynucleotides, and polypeptides are examples of these. When monosaccharide units are linked together by a glycosidic bond, a biopolymer called a polysaccharide is created. Well-known polysaccharides include cellulose, starch, glycogen, chitin/chitosan, pectin, and alginate (Table 2.1). Four more categories may be used to further break up polysaccharide biopolymers. These include lignin biopolymers, cellulose-based

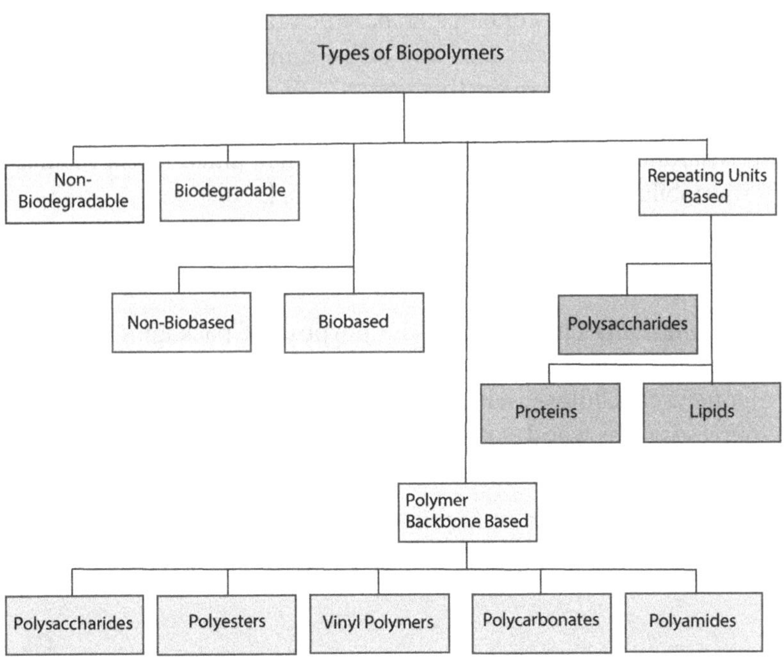

Figure 2.2 Classification of biopolymers.

Table 2.1 Biopolymers obtained from the seaweed and used as a renewable biopolymer [5].

Seaweed biopolymers	Their composites	Applications
Carrageenan	Carrageenan-based composites	Food industries, pharmaceuticals
Alginate	Alginate-based composites	Pharmaceuticals, food industries, dentistry
Agar	Agar-based composites	Biotechnology researches, agriculture, pharmaceuticals

biopolymers, starch-based biopolymers, and sugar-based biopolymers [4]. Seaweed biopolymers are renewable polymers and their applications can be seen in the vast and several fields [5].

2.1.2.1 *Biopolymer Derived from Sugar*

Either starch or sucrose can be utilized as a substrate in the production process of polyhydroxy butyrate. These objects can be manufactured through various processes such as injection, blowing, molding, vacuum forming, and extrusion. Lactic acid polymers, also known as polylactides, are derived from lactose, a sugar found in potatoes, sugar beet and wheat. Polylactides exhibit water resistance and can be fabricated through various techniques such as vacuum forming, blowing, and molding by injection [6].

2.1.2.2 *Biopolymer Derived from Starch*

Tapioca, wheat, and potatoes all contain starch, which functions as a natural polymer. The substance is kept in plant tissues as one-way carbs. It may be made by melting starch and glucose. Animal tissues don't contain this polymer. Vegetables including tapioca, maize, wheat, and potatoes contain it. Dextran is a kind of small-molecule carbohydrate formed by starch hydrolysis, and is created through enzymes when *Enterococcus faecalis* Esawy dextransucrase is immobilized on biopolymer transporters [7, 8].

2.1.2.3 *Biopolymer Derived from Cellulose*

These items are commonly employed for the purpose of packaging cigarettes, compact discs, and confectionery products. This polymer is composed of glucose and serves as the principal component of cellulose walls in plants. This substance is derived from naturally occurring resources such as wood, cotton, wheat, as well as corn [9].

2.1.2.4 *Biopolymer Derived from Lignin*

Lignin is one of the crucial biopolymers, which is not from the origin of polysaccharide.

2.1.2.5 *Polynucleotides*

DNA and RNA are the types of biopolymers, which are the units of nucleotides.

The bonds of peptides (amide linkage) hold amino acid monomer polymers together. Examples of biopolymers of these kinds of polypeptides include silk, collagen, and keratin.

2.1.2.6 *Biopolymers Derived from Synthetic Materials*

Making biodegradable polymers, which include aliphatic-aromatic copolyesters derived from petroleum, also uses biopolymers based on synthetic chemicals. Despite being made of synthetic materials, they are fully compostable and biodegradable.

There are considerably more bio-based biopolymers that are not biodegradable than those that are. A biopolymer, sometimes also referred to as "bioplastic," is defined by two distinct factors—the raw material supply and the polymer biodegradation. These two criteria differentiate the biopolymers easily [9].

2.1.2.7 *Biodegradable Biopolymers and Based on Renewable Basic Resources*

Polymers can be generated through two primary methods: biological systems, involving microorganisms, animals, and plants, or chemical synthesis utilizing biological feedstocks such as sugar, corn, and starch. One approach to rendering a polymer biodegradable entails the incorporation of a hydrolysable ester moiety within its structure. Biodegradable bio-based biopolymers encompass synthetic polymers derived from renewable sources, exemplified by polylactic acid (PLA). ii) Biopolymers that are synthesized through microorganisms, such as PHAs. iii) Naturally existing biological polymers, including as starch or proteins. Starch and PHAs are the most commonly employed bio-based biodegradable polymers [10].

2.1.2.8 *Non-Biodegradable Biopolymers and Based on Renewable Basic Resources*

Biopolymers can be created with the help of sources of renewable energy or biomass and possess non-biodegradable characteristics. Non-biodegradable bio-based biopolymers encompass two categories:

i) Synthetic polymers derived from renewable resources, including specific polyamides derived with castor oil, particular polyester fibers composed of biopropanediol, bio polypropylene (bio-pp), bio polyethylene (bio-LDPE, bio-HDPE) and bio poly (vinyl chloride) (bio-PVC) developed from ethanol sourced from sugarcane;
ii) Organic biopolymers such as amber and natural rubber [11].

2.1.2.9 *Biodegradable and Created from Fossil Fuels*

The biopolymers discussed in this context are derived from non-renewable sources, specifically fossil fuels such as natural gas or crude oil. These biopolymers are classified as artificial aliphatic polyesters and are acknowledged to possess biodegradable and compostable properties. PCL poly (butylene succinate) (PBS) and specific copolyesters known as "aliphatic-aromatic" copolyesters are polymers that are derived, to some extent, from fossil fuels. However, they possess the ability to undergo biodegradation through the action of microorganisms [11].

2.2 Biopolymer Synthesis

Biopolymer synthesis refers to the process of producing biopolymers, which are large molecules made up of repeating subunits derived from natural sources. These biopolymers are often used as alternatives to synthetic polymers because they are renewable and biodegradable and have lower environmental impact.

2.2.1 Microbial Synthesis

The process of biopolymer microbial synthesis involves the utilization of microorganisms as the primary means of producing biopolymers. Microorganisms, including yeasts, bacteria, and fungi, possess the ability to produce a diverse array of biopolymers via inherent metabolic pathways.

There are a few examples of biopolymers that are synthesized using different microbial synthesis processes: -

2.2.1.1 Polysaccharides

Microbial organisms possess the ability to synthesize a diverse range of polysaccharides, including chitosan, cellulose, and xanthan gum. Gluconacetobacter xylinus, a type of bacteria, possesses the capability to engage in cellulose synthesis, a process that finds extensive utilization in various domains including but not limited to paper production, textile manufacturing, and the development of healthcare products. Chitosan, a biopolymer obtained from chitin, can be synthesized through bacterial deacetylation of chitin utilizing microorganisms such as Aspergillus niger or Mucor rouxii. Xanthan gum, an exopolysaccharide derived from the bacterium Xanthomonas campestris, serves as a commonly employed thickening agent in various applications.

2.2.1.2 Biopolymers Based on Proteins

Microorganisms have the potential to undergo genetic modification in order to synthesize protein-based biopolymers, including silk and elastin-like polypeptides (ELPs). An example of genetic modification involves the utilization of bacteria such as Escherichia coli to express silk proteins that are sourced from spiders or silkworms. Subsequently, the recombinant proteins can be collected and subjected to further processing in order to produce fibers or films. Likewise, temperature-responsive ELPs, characterized by distinctive properties, can be synthesized through regenerated expression in microbes such as Escherichia coli [12].

2.2.2 Plant-Based Synthesis

The process of plant-based synthesis of biopolymers entails the utilization of nutrients or extracts derived from plants as the primary source for the manufacturing of biopolymers. Plants present a viable and environmentally friendly means of obtaining raw materials, rendering them an appealing choice for the synthesis of biopolymers. The synthesis of biopolymers derived from plants presents several notable benefits, including the utilization of plentiful and renewable feedstocks, the ability to biodegrade, and a reduced environmental footprint when compared to conventional synthetic polymers. Current research is centered on the optimization of extraction processes from plants, the modification of biopolymers to improve their properties, and the investigation of novel plant sources for the generation of biopolymers [13].

2.2.3 Animal-Based Synthesis

The process of animal-based synthesis of biopolymers entails the utilization of materials obtained from animal sources in order to manufacture biopolymers. Animals offer a wide array of biomaterials that possess the potential to be processed and converted into biopolymers, thereby presenting a multitude of applications. Biopolymers derived from animals exhibit distinct characteristics and functionalities that render them well-suited for a diverse range of applications within the biomedical, aesthetically pleasing, and industrial sectors. Nevertheless, it is crucial to take into account the ethical and sustainability implications associated with the usage of animal-derived materials. It is imperative to investigate different sources and develop synthetic analogs whenever feasible in order to address these concerns.

2.2.3.1 Collagen

Collagen, a biopolymer, is abundantly present in the connective tissue of various animal species, including but not limited to the integumentary system, skeletal structure, and tendinous regions. The protein in question possesses a robust structural composition and exhibits a notable degree of biocompatibility. It can be sourced from various animal origins, such as bovine or fish. Collagen exhibits various uses in the domains of dressings for wounds, tissue engineering, and cosmetic merchandize. The shape of these entities is cylindrical, and collagen possesses a molecular weight of 300 kilodaltons (kDa). The human body has the capacity to effectively assimilate it and exhibits remarkably minimal immunogenic response [14, 15].

2.2.3.2 Keratin

Keratin is a structural protein that is present in various animal-derived substances, such as hair, feathers, wool, as well as horns. The protein in question possesses a fibrous structure and exhibits a notable degree of tensile strength. Keratin-based biopolymers can be derived through the processing and modification of materials that are abundant in keratin. The biopolymers exhibit promising prospects for application in drug delivery, wound healing, and tissue engineering endeavors. The protein is classified into α and β keratin based on auxiliary order. Keratin exhibits an insoluble nature in water while can undergo partial assimilation via proteolysis, facilitated by enzymes such as papain, pepsin, or trypsin [16, 17].

2.2.3.3 Gelatin

Gelatin is a type of protein that is obtained through the process of partial hydrolysis of collagen. It is frequently derived from animal origins, specifically the skins and bones of pigs or cows. Gelatin is applicable in various industries, including the food industry, pharmaceutical industry, and photography industry, due to its wide range of applications in dietary supplements, healthcare capsules, and even photographic films. Gelatin is a heteropolysaccharide that is found in the cell walls of plants. Gelatin is composed of a complex arrangement of carbohydrates that are present in primary cell walls and particularly abundant in non-woody plants. In a commercial context, gelatins are derived from citrus peels or apple pomace, which are residual biomass generated from juice manufacturing facilities. Gelatin is believed to primarily consist of D-galacturonic acid (GalA) units, which are connected in chains through á-(1-4) glycosidic linkage [18].

2.3 Properties of Biopolymers

Biopolymers have garnered significant attention due to their potential to serve as substitutes for numerous consumer goods currently derived from petroleum-based materials. The variability in asset measurements of biopolymers can be attributed to multiple factors. The factors that influence the properties of polymers include the degrees of polymerization, categories, and the concentration of additives, and the inclusion of materials that are positive.

Limited information is available regarding the properties of biopolymers; however, extensive research has been conducted to explore their mechanical, physical, and thermal characteristics [19]. Certain biopolymers exhibit the ability to conduct electricity and ions, also referred to as electrically active biopolymers (EABPs). This development has endowed them with the capacity to supplant alternative synthetic materials. The electrical conductivity of various biopolymers, such as cellulose, starch, pectin, and chitosan, exhibits a broad spectrum [20].

2.3.1 Mechanical Properties

The mechanical characteristics of biopolymers pertain to their reaction and adaptation to mechanical stimuli, including tension, compression, bending, and stretching. The determination of the appropriateness of biopolymers in diverse applications, such as biomedical materials and packaging, relies significantly on the assessment of these properties. It is crucial to acknowledge that the mechanical characteristics of biopolymers can exhibit substantial variations, contingent upon factors such as the precise nature of the biopolymer, its molecular configuration, processing techniques, and prevailing environmental circumstances. Researchers and engineers persistently investigate strategies to augment or customize the mechanical characteristics of biopolymers via approaches such as mixing, cross-linking, or integrating reinforcing substances to enhance their efficacy in particular applications [21].

2.3.1.1 *Tensile Strength*

Tensile strength refers to the capacity of a biopolymer to withstand fracture when subjected to tension. The term "ultimate tensile strength" refers to the maximum stress that a biopolymer can endure prior to experiencing rupture. A higher tensile strength signifies an increased capacity to withstand and endure expanding and ripping forces.

2.3.1.2 *Flexibility and Ductility*

Flexibility pertains to the capacity of a biopolymer to undergo bending or flexion without experiencing fracture, whereas ductility denotes its capability to endure plastic deformation prior to fracturing. Biopolymers possessing notable flexibility and ductility exhibit the capacity to endure multiple instances of stretching or elongation without experiencing structural failure.

2.3.1.3 *Friction Phenomena and Wearing Resistance*

The wear resistance of a polymer is contingent upon several factors, including the value of the coefficient of abrasion, rigidity, resiliency, and level of brittleness. The surface friction coefficient of PLA fibers is high, while their wearing resistance is low. For example, the JIS L 0849 standard typically mandates a minimum grade 3 level of wearing resistance. However, conventional PLA fibers exhibit a significantly lower grade 1 level of wearing resistance. The insufficient durability of PLA poses a significant challenge when employed as a fabric material for both apparel and industrial applications. When conventional PLA fibers are utilized in multiple uses including clothing, their quality deteriorates due to issues such as fluffing, brightening, sparkle, and similar effects in areas prone to wear and tear. Additionally, color immigration to innerwear as well as related problems may also occur [21].

2.3.1.4 *Polyhydroxyalkanoates*

PHA polymers, specifically those that consist of monomers that are like short chains containing 3-5 carbon atoms, such as poly(3-hydroxybutyrate) (P3HB) and poly(3-hydroxybutyrate-co-3-hydroxyvalerate) (PHBHV), exhibit limited expansion, susceptibility to tearing when subjected to mechanical stress. PHBHV exhibits superior toughness with reduced brittleness in comparison to P3HB. As the length of the monomer chain increases, the material exhibits a greater degree of flexibility. Poly (3-hydroxy octanoate) (PHO) is an elastomer, exemplifying its elastic properties. Polymers of hydroxy alkanoates (PHAs) possessing elongated side chains exhibit properties akin to those of waxes. The physical and mechanical characteristics of P3HB exhibit a striking resemblance to (PP), with one notable difference of elongation at break. This is noteworthy considering the distinct chemical structures of these two polymers. P4HB demonstrates ductility and lacks brittle fracture properties, whereas the tensile strength is equivalent [22].

2.3.2 Thermal Properties

The thermal characteristics of biopolymers pertain to their mannerisms and reactions in the face of temperature fluctuations. The aforementioned properties are of utmost importance in assessing the appropriateness of biopolymers for diverse applications, encompassing manufacturing and thermal stability.

2.3.2.1 Thermal Stability

Thermal stability pertains to the capacity of a biopolymer to withstand thermal breakdown under conditions of elevated temperatures. The typical defining characteristic of this phenomenon is the temperature at which destruction or weight loss begins. Biopolymers exhibiting a notable degree of thermal stability are capable of enduring elevated temperatures without undergoing substantial chemical or structural alterations [22].

2.3.2.2 Thermal Conductivity

Thermal conductivity refers to the capacity of a biopolymer to conduct heat. The efficiency of heat transfer within a biopolymer's structure is determined by this factor. Biopolymers exhibiting low thermal conductivity function as thermal insulators, whereas those characterized by a high level of thermal conductivity demonstrate enhanced heat conduction capabilities [22].

2.3.3 Biodegradability

Biodegradability is one of the crucial properties a biopolymer should have. For the better application of the biopolymer in any of the sectors, it is essential that it will not cause any toxicity in the product of the healthcare system and so on. The property of biodegradability is of significant importance in the context of biopolymers, and it should be noted that this property is not directly related to the specific raw materials employed in their production. It is possible to derive a biodegradable polymer from fossil sources. Biodegradation is a naturally occurring natural process that takes place within the context of composting, leading to the emission of water, carbon dioxide, and non-organic particles, along with biomass. For a biopolymer to meet the criteria of being classified as biodegradable, it must undergo degradation at a comparable rate to that of certified compostable substances, while also ensuring the absence of any harmful residues [23]. There are a few methods of biodegradability or degradability that are carbon dioxide produced and loss of weight for the starch-based biopolymers and their environments are wheat (starch-oriented biopolymer), soil (wheat and starch-oriented biopolymer), and marine (neat starch) [24–27]. Biodegradation reactions are commonly facilitated by enzymes and typically take place in aqueous environments. Bio-based polymers that possess hydrolysable associations, such as aliphatic in polyester, amino acids, starch, and cellulose, are commonly prone to biodegradation through the action of hydrolytic enzymes produced by microorganisms. The biodegradability of polymers is significantly influenced by their hydrophilic/hydrophobic nature, whereby polymers with higher polarity tend to exhibit greater susceptibility to biodegradation in general. Crystallinity, along with chain versatility, is additional significant attributes of polymer compounds that exert an influence on their biodegradability. The process of biodegradation commences when microorganisms initiate growth on the biopolymer's surface and release enzymes that degrade the biopolymer into smaller units, including oligo- or monomeric hydroxy acids. In the context of aliphatic polyesters, microorganisms assimilate the hydroxy acids and utilize them as sources of carbon for promoting their growth [28, 29].

2.4 Applications of Biopolymers

In recent times, there has been a significant surge in the interest surrounding biopolymer materials due to their potential applications in the field of biomedicine. These applications include but are not limited to the engineering of tissues, pharmaceutical transporters, and medical equipment. Biopolymers possess inherent biocompatibility along with biodegradability, rendering them suitable for enhancing the efficacy of other compounds with biological activity within a given product [30]. Due to the significant significance of biomaterials, they possess numerous crucial applications that intersect across various domains in our daily existence (Figure 2.3).

2.4.1 Packaging

In recent years, there has been a significant advancement in the field of polymer modification through the application of nanomaterials. The utilization of natural or synthetic polymers in food packaging materials has led to significant enhancements in various properties such as durability, flexibility, resistance to heat, and barrier capabilities, along with others [31]. Starch nanocomposite materials, including montmorillonite/starch micron-sized–combined film, cellulose the nano-combined supplies, protein nano-combined substances, along with PLA nano-combined substances, have been utilized in food packaging. As a material utilized in food packaging, it has the potential to enhance various aspects such as the antimicrobial efficacy of the packaging, mechanical durability, mobility, thermal

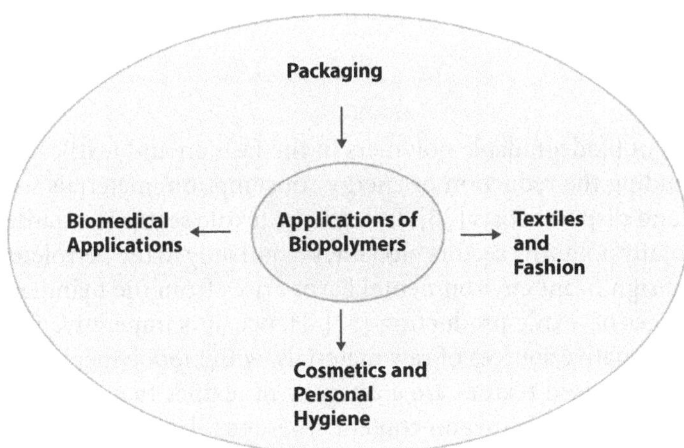

Figure 2.3 Different applications of biopolymers are used in several sectors on a daily basis.

resistance, and barrier characteristics, among others. Several nations in the United States and Europe conduct research using animal models, both *in vivo* and *in vitro*, to investigate the toxicity of nanoparticles on various organ systems, including the respiratory and cardiovascular systems. These experiments have provided evidence indicating that nanoparticles possess a certain level of toxic effects, leading to potential harm to animal organs [32, 33]. At present, the most economically feasible materials employed in the realm of food packaging are specific polyesters with biodegradability that can undergo processing using standard machinery. These materials have already been employed in various single-layer or multiple-layer functions within the food-packaging industry. Starch, PHA, and PLA are among the thermoplastics that have been extensively studied in the realm of biopolymers which are sustainable, and employed in monolayer packaging [34]. PLA and starch biopolymers are considered highly appealing as biodegradable materials. The aforementioned phenomenon can be attributed to the equilibrium of their characteristics and the advent of their availability in the commercial market. The primary objective for these particular biomaterials is to enhance their barrier along with thermal characteristics in order to achieve comparable performance to that of polyethylene terephthalate (PET). Additional materials derived with the help of biomass resources, including proteins (such as zein), lipids (such as waxes), and polysaccharides (such as chitosan), exhibit considerable promise in terms of their gas with aroma barrier properties. The primary limitations associated with these particular materials pertain to their inherent rigidity along with the challenges encountered when attempting to process them using traditional methods. In the context of bio-based packaging of food uses, the primary factor of utmost significance is the evaluation of the barrier's characteristics. Hydrophilic polymers typically exhibit suboptimal resistance to moisture, leading to the permeation of water vapor through packaging materials. Consequently, this phenomenon can have a detrimental impact on the general caliber of food products. Consequently, these outcomes manifest as reduced product longevity, heightened expenses, and ultimately augmented waste generation. One additional approach for enhancing the barrier capabilities of biopolymers involves the incorporation of diverse nanofillers such as metal oxide nanoparticles, nano-sized clays, and similar materials [35]. Polyglycolic acid (PGA) is considered to be a highly promising commercially available barrier polymer within the bioplastics category due to its exceptional barrier properties. The glyoxylate cycle has enabled the production of glycolic acid, its precursor, through a natural metabolic pathway [30].

2.4.2 Textiles

The incorporation of biodegradable polymers in the fashion and textile sectors offers several benefits, including the reduction of energy consumption, materials usage, and sourcing, production, and disposal costs [36]. Indeed, the textile sector is regarded as one of the most environmentally polluting sectors globally, second only to the petroleum industry. The primary source of significant environmental harm arises from the manufacturing, dyeing, and processing stages of textile production [37]. Hence, it is imperative for this particular industry to seek alternative sources of raw materials, with biopolymers emerging as a conscientious option. Bio-based textiles are a category of textiles that are required to possess a minimum of 20% sustainable carbon content. These textiles encompass a range of materials, including natural substances as well as filaments that are either natural, synthetic, or regenerated in nature. Polysaccharides such as cellulose and lignin, along with proteins and fatty acids derived from plants or animals, are utilized in the production of natural biopolymers [31]. Indeed, it is worth noting that fibers that are natural can also be derived from botanical origins, such as hemp, wool, and others. Among these options, cotton, along with silk and wool, is extensively used in the production of clothing due to its ability to satisfy aesthetically pleasing and durability criteria. Textiles commonly incorporate synthetic and regenerated fibers derived from bacterial processes, like PHAs, as well as through the synthesis of natural raw materials, including polyglycols, polylactides, polycaprolactones, and others [38]. In recent times, there has been a growing emphasis on sustainable production within the textile industry. This has resulted in a shift in focus toward materials like organic cotton, which is cultivated with no application of chemical pesticides, fertilizers as well as or other chemical substances. Additionally, there has been an increased interest in the development of synthetic recyclable textile fibers, commonly known as "biodegradable nonwovens." Among the various biopolymers employed in fiber spinning within the contemporary biodegradable textile sector, prominent examples include butyric acid (PHB), PLA, valeric acid (PHV), along caprolactone (PCL), among others [39]. In the textile sectors, some companies that employ 3D-printing methods and biopolymers like PLA, and melted PLA, as well as other materials such as BendLay, Ninjaflex, and TPE have successfully produced nonwoven textiles that exhibit enhanced structural and morphological features in comparison to conventional polymers. Loh *et al.* [40] conducted a study wherein they examined and analyzed the morphological along mechanical properties of three distinct polymer composites intended for application in the textile industry. The researchers employed a variety of combinations of nylon and PLA, alongside polyesters, and used direct extrusion techniques to fabricate these materials [41, 42].

2.4.3 Biomedical Applications

Biopolymers and their combined forms find application in the pharmaceutical and medical domains owing to their being biodegradable, affordability, abundant availability, processability, and notably, being compatible with biological tissues, cells, and organs [43–45]. These qualities find application in various contexts, such as serving as supplies for the transportation of medication molecules that are different, substances, including enzymes, anti-microbial agents, and anti-neoplastic pharmaceuticals, among others. They serve a purpose in systems pertaining to dental, ocular, nasal, and various domains [46]. Biopolymers derived from various sources, possessing pharmaceutically active constituents capable of modulating the drug delivery process, are employed in the fabrication of DDSs. These systems typically take the form of microspheres, microcapsules, hydrogels, liposomes, and nanogels. Polysaccharides in this particular domain have been applied for an extended period primarily due to their inherent characteristics that enable the formation of connections with lipids and proteins. Cellulose, along with starch, is widely used in the field of pharmaceuticals, making it one of the most commonly employed polysaccharides [47, 48]. The application of cellulose and its derivatives in the field of DDSs has witnessed significant growth owing to its remarkable properties, including its capacity for absorption and retention of water, its composability, and its structural attributes that enable the encapsulation of specific molecules. Furthermore, the advent of nanocellulose production from wood pulp, coupled with the integration of advanced technologies like 3D printing, has presented a promising avenue for recent and further years [47]. Furthermore, the potential for the synthesis of nanocellulose derived from wood pulp, in conjunction with the application of cutting-edge technologies like 3D printing, has presented a promising avenue for the advancement of novel materials intended for pharmacological purposes in recent times [48]. In contemporary times, the consumption of biopolymers has largely superseded conventional materials like ceramics and metals. This shift can be attributed to the propensity of the human body to mount an immune response against the latter materials, thereby leading to immunological rejection. Conversely, biopolymers demonstrate beneficial features such as effective degradability, biocompatibility, renewability, non-toxicity, and antibacterial properties throughout their lifespan [49–51].

2.5 Challenges and Future Perspectives

Despite the promising trends in applicability, biodegradable polymers derived from renewable sources exhibit certain drawbacks. These include inadequate mechanical characteristics, rapid breakdown rate, high hydrophobic capability, as well as in certain instances, inadequate mechanical performance, particularly in moist conditions. Consequently, these limitations render their application unfeasible [52, 53]. Within this particular context, there is a proliferation of divergent viewpoints regarding the level of acceptability associated with the usage of polymers that decompose within the industrial sector. While there are proponents who argue for the potential of alternative polymers to replace petroleum-based ones, there are also skeptical individuals who contend that various limitations, both technical as well as economic, may impede their widespread adoption, particularly in the foreseeable future [54]. The task at hand involves acquiring materials that possess properties comparable to those of products made from synthetic materials [55]. While using Casein and Biotin these substances few challenges arose in previous studies. Researchers mentioned that the present study focuses on the creation along characterization of innovative meltable polymers along with composite materials using soybean and casein as the primary constituents. This study aims to investigate the impact of inert (Al2O3) along with bioactive (tricalcium phosphate) ceramic additional support upon the tensile efficiency, absorption of water, and biological activity properties of the infusion-shaped thermoplastics, with a specific focus on their potential applications in the field of biomedicine. In conclusion, it can be inferred that thermoplastics derived from casein along with soy protein demonstrate an adequate spectrum of mechanical attributes along with deterioration features. Moreover, when strengthened with bone-like ceramics, these thermoplastics demonstrate a bioactive nature, potentially making them suitable for application to be biomaterials in the field of healthcare [56].

2.5.1 Economic Viability

Irrespective of the specific product being manufactured, minimizing operating costs is a crucial objective. The concept of enhancing the environmental sustainability of a product generally entails increased expenses for manufacturers, which in turn may result in higher costs for consumers. This assumption has traditionally prevailed. Nevertheless, this is no longer the prevailing circumstance. Eco Synthetix provides binder solutions that are environmentally sustainable and demonstrate comparable or superior performance in comparison to existing binder systems, while also offering a reduced total system cost [57]. In order to achieve a change in perspective, it is imperative for bioplastics to effectively compete with conventional petrochemical-based contaminants in various aspects such as cost-efficient production, large-scale manufacturing capabilities, integration with existing infrastructures, efficient waste management, successful end-of-life therapy, awareness among consumers, consumer-level waste sorting, and many other aspects [58, 59].

2.5.2 Large-Scale Production

Although bioplastics offer numerous advantages, they are also accompanied by several limitations and technical challenges. For instance, if intermediate substances produced during the production of bioplastics are not properly segregated at the origin, they have the potential to pollute the recycling process of conventional plastics [60]. Furthermore, a majority of the bioplastics currently manufactured prioritize the use of first-generation feedstock, including sugarcane, maize, wheat, potatoes, and cassava. It is important to acknowledge that the utilization of first-generation raw materials, particularly starch-based substances, in the production of biofuels, biological compounds, as well as biomaterials has faced significant criticism. This criticism primarily stems from concerns regarding the competition for limited food resources, arable land availability, agricultural energy requirements, and the expense of investments [61]. In order to address this issue, it is advisable to investigate the usage of unappealing or lignocellulosic feedstocks, including biomass from forests, agricultural crop residues, and intrusive crops, for the extraction of organic polymers, resins, and fibers [62–64].

The objective is to advance the production of designed biopolymers and novel bioplastics suitable for diverse sectors, including but not limited to biomedical, automobiles veterinary, medicinal products farming, along building sectors. Societal acceptance plays a crucial role in determining the lasting viability of bioplastics. Hence, it is imperative to provide education to urban and rural communities globally regarding fundamental aspects of bioplastics, including their application, structure, environmental consequences, as well as consumer-level practices such as composting as well as waste segregation. This has the potential to influence the trajectory of the future bioplastic industry and enhance the agricultural economy by facilitating the provision of raw materials. Furthermore, the amelioration of environmental issues can be achieved through the implementation of regulatory measures that prohibit the intentional use of disposable plastics or promote the adoption of biodegradable plastics, encompassing not only commercial wrapping and sealing materials but also extensive applications in the industry [65].

2.5.3 Innovations in Biopolymer Research in the Future

The development of recyclable polymers along with their combined forms has garnered significant attention, although it remains in its nascent stages. The demand for candidates with new qualifications from prospective producers of polymers that decompose is compelling. Specifically, there is a significant demand for a wide range of sustainable products. However, the production of these products is hindered by their poor performance characteristics as well as elevated development expenses, which are comparatively lower than those of traditional synthetic polymers. The adoption of new ecological standards, which prioritize the resolution of worries concentrated around the environment, has led to advancements in the field of contemporary polymeric substances and methods. These advancements are designed to align with the well-being of natural habitats. The development of biodegradable polymer products with high activity is imperative in order to harness the environmentally friendly, social, along industrial advantages they offer in recent years, there has been a notable increase in the attention given to bio-derived polyester-based materials due to their ability to degrade along with potential applications in the medical field. Several key factors include breakdown techniques, biological compatibility, and operating circumstances, alongside potential applications in healthcare and conservation of the environment, while agro-chemistry has garnered significant attention. The assessment of biological safety for established biological polymers and the evaluation of nano-safety for their composite materials remain limited in significance. This novel perspective highlights the importance of anticipating, evaluating, and illustrating potential issues associated with the application of advanced polymers. In future scenarios, the significant challenges related to biodegradable polymers will primarily revolve around the effective management of main materials, the efficacy evaluation of bio-derived resources, and the associated manufacturing expenses.

In addition, the production of bio-derived monomers or polymers coming from sources that are renewable poses a significant obstacle in terms of commercial manufacturing. The construction of industrial plants can present challenges due to limited familiarity with emerging technologies and the evaluation of stock and demand equilibrium. Despite the industrial-scale development of novel bio-derived polymers, there remain several unresolved concerns regarding the long-term viability of biodegradable polymers. There is an expectation that there will be going to be competition for feedstock due to the increasing global demand for both food and energy [66].

2.6 Conclusion

Herein, a short conclusion of the chapter consists of the importance of biopolymers with the whole aspects of its applications in the different sectors. Several properties are demonstrated with a wholesome view of the several types and importance of biopolymers in the existing world. A demonstration of the biodegradability and non-biodegradability concerns for the biopolymer. Recent research with more future innovations is also mentioned with suitable information used for learning purposes and acknowledging it for the future development of more biopolymers.

References

1. http://www.jbpaweb.net/english/
2. Bastioli, C., Global status of the production of biobased packaging materials. *Starch-Stärke*, 53, 8, 351–5, Aug. 2001.
3. Lemoigne, M., Products of dehydration and of polymerization of β-hydroxybutyric acid. *Bull. Soc. Chem. Biol.*, 8, 770–82, 1926.
4. https://www.jetir.org/papers/JETIR2201209.pdf

5. Jumaidin, R., Sapuan, S.M., Jawaid, M., Ishak, M.R., Sahari, J., Seaweeds as renewable sources for biopolymers and its composites: a review. *Curr. Anal. Chem.*, 14, 3, 249–67, June 1, 2018.
6. Fanun, M. (Ed.), *The role of colloidal systems in environmental protection*, Elsevier, Amsterdam, Netherland, Feb. 8, 2014.
7. Esawy, M.A., Gamal, A.A., Helal, M.M., Hassan, M.E., Hassanein, N.M., Hashem, A.M., Enzymatic synthesis using immobilised Enterococcus faecalis Esawy dextransucrase and some applied studies. *Int. J. Biol. Macromol.*, 92, 56–62, 2016.
8. Hashem, A.M., Gamal, A.A., Hassan, M.E., Hassanein, N.M., Esawy, M.A., Covalent immobilisation of Enterococcus faecalis Esawy dextransucrase and dextran synthesis. *Int. J. Biol. Macromol.*, 82, 905–12, Jan. 1, 2016.
9. Kaurav, M.S., *Engineering chemistry with laboratory experiments*, PHI Learning Pvt. Ltd, New Delhi, India, 2011.
10. Niaounakis, M., *Biopolymers: reuse, recycling, and disposal*, William Andrew, Norwich, NY, USA, Jun. 20, 2013.
11. Vroman, I. and Tighzert, L., Biodegradable polymers. *Materials*, 2, 2, 307–44, Apr. 1, 2009.
12. George, A., Sanjay, M.R., Srisuk, R., Parameswaran Pillai, J., Siengchin, S., A comprehensive review on chemical properties and applications of biopolymers and their composites. *Int. J. Biol. Macromol.*, 154, 329–38, Jul. 1, 2020.
13. van Beilen, J.B. and Poirier, Y., Prospects for biopolymer production in plants, in: *Green gene technology: research in an area of social conflict*, pp. 133–51, 2007.
14. León-López, A., Morales-Peñaloza, A., Martínez-Juárez, V.M., Vargas-Torres, A., Zeugolis, D.I., Aguirre-Álvarez, G., Hydrolyzed collagen—Sources and applications. *Molecules*, 24, 22, 4031, Nov. 7, 2019.
15. Lee, C.H., Singla, A., Lee, Y., Biomedical applications of collagen. *Int. J. Pharm.*, 221, 1-2, 1–22, Jun. 19, 2001.
16. Shah, A., Tyagi, S., Bhargava, R.N., Belhaj, D., Kumar, A., Saxena, G., Saratale, G.D., Mulla, S.I., Keratin production and its applications: current and future perspective, in: *Keratin as a Protein Biopolymer: Extraction from Waste Biomass and Applications*, pp. 19–34, 2019.
17. Sharma, S., Gupta, A., Chik, S.M., Kee, C.Y., Podder, P.K., Subramaniam, M., Thuraisingam, J., Study of different treatment methods on chicken feather biomass. *IIUM Eng. J.*, 18, 2, 47–55, Dec. 1, 2017.
18. Otoni, C.G., Moura, M.R., Aouada, F.A., Camilloto, G.P., Cruz, R.S., Lorevice, M.V. *et al.*, Antimicrobial and physical-mechanical properties of pectin/papaya puree/cinnamaldehyde nanoemulsion edible composite films. *Food Hydrocolloids*, 41, 188–94, 2014.
19. Van de Velde, K. and Kiekens, P., Biopolymers: overview of several properties and consequences on their applications. *Polym. Test.*, 21, 4, 433–42, Jan. 1, 2002.
20. Muthulakshmi, S., Chapter 2 Investigation on the Biopolymers Papain, Gum Acacia, Gum Tragacanth and Gum Guar; Physical and Antimicrobial Properties, in: *Structural and optical properties of biopolymers*, pp. 21–46, PhD Thesis Manonmaniam Sundaranar University, Tirunelveli, India, 2013.
21. Niaounakis, M., *Biopolymers: processing and products*, William Andrew, Norwich, NY, USA, Sep. 22, 2014.
22. Saito, Y. and Doi, Y., Microbial synthesis and properties of poly (3-hydroxybutyrate-co-4-hydroxybutyrate) in Comamonas acidovorans. *Int. J. Biol. Macromol.*, 16, 2, 99–104, Apr. 1, 1994.
23. Brizga, J., Hubacek, K., Feng, K., The unintended side effects of bioplastics: carbon, land, and water footprints. *One Earth*, 3, 1, 45–53, Jul. 24, 2020.
24. Ibrahim, N.I., Shahar, F.S., Sultan, M.T., Shah, A.U., Safri, S.N., Mat Yazik, M.H., Overview of bioplastic introduction and its applications in product packaging. *Coatings*, 11, 11, 1423, Nov. 22, 2021.

25. Zhao, X., Cornish, K., Vodovotz, Y., Narrowing the gap for bioplastic use in food packaging: an update. *Environ. Sci. Technol.*, 54, 8, 4712–32, Mar. 23, 2020.
26. Emadian, S.M., Onay, T.T., Demirel, B., Biodegradation of bioplastics in natural environments. *Waste Manage. (Oxford)*, 59, 526–36, Jan. 1, 2017.
27. Imam, S.H., Gordon, S.H., Shogren, R.L., Tosteson, T.R., Govind, N.S., Greene, R.V., Degradation of starch–poly (β-hydroxybutyrate-co-β-hydroxyvalerate) bioplastic in tropical coastal waters. *Appl. Environ. Microbiol.*, 65, 2, 431–7, Feb. 1, 1999.
28. Williams, S.F. and Peoples, O.P., Biodegradable plastics from plants. *Chemtech.*, 26, 9, 38–44, 1996.
29. De Wilde, B., International and national norms on biodegradability and certification procedures, in: *Handbook of biodegradable polymers*, p. 145, 2005.
30. Mohan, S., Oluwafemi, O.S., Kalarikkal, N., Thomas, S., Songca, S.P., Biopolymers–Application in Nanoscience and Nanotechnology [Internet], in: *Recent Advances in Biopolymers*, InTech, Rijeka, Croatia, 2016, Available from: http://dx.doi.org/10.5772/62225.
31. Hassan, M.E., Bai, J., Dou, D.Q., Biopolymers; definition, classification and applications. *Egypt. J. Chem.*, 62, 9, 1725–37, Sep. 1, 2019.
32. Bumbudsanpharoke, N., Choi, J., Ko, S., Applications of nanomaterials in food packaging. *J. Nanosci. Nanotechnol.*, 15, 9, 6357–72, 2015, doi: 10.1166/jnn.2015.10847.
33. Hashem, M., Refaie, R., Zaghloul, S., Ezzat, A., Ellaithy, A., A Shaaban, H., Bioactive jute fabric for packaging and storage of grains and legumes applications. *Egypt. J. Chem.*, 60, 4, 551–61, Aug. 30, 2017.
34. Siracusa, V., Blanco, I., Romani, S., Tylewicz, U., Rocculi, P., Rosa, M.D., Poly (lactic acid)-modified films for food packaging application: Physical, mechanical, and barrier behavior. *J. Appl. Polym. Sci.*, 125, S2, E390–401, Sep. 25, 2012.
35. Majeed, K., Jawaid, M., Hassan, A.A., Bakar, A.A., Khalil, H.A., Salema, A.A., Inuwa, I., Potential materials for food packaging from nano clay/natural fibres filled hybrid composites. *Mater. Des. (1980–2015)*, 46, 391–410, Apr. 1, 2013.
36. Younes, B., Classification, characterization, and the production processes of biopolymers used in the textiles industry. *J. Text. Inst.*, 108, 5, 674–82, May 4, 2017.
37. da Silva, C.J., de Medeiros, A.D., de Amorim, J.D., do Nascimento, H.A., Converti, A., Costa, A.F., Sarubbo, L.A., Bacterial cellulose textiles for the future of sustainable fashion: a review. *Environ. Chem. Lett.*, 19, 2967–80, Aug. 2021.
38. Patti, A. and Acierno, D., Towards the sustainability of the plastic industry through biopolymers: properties and potential applications to the textiles world. *Polymers*, 14, 4, 692, Feb. 11, 2022.
39. Fico, D., Rizzo, D., Casciaro, R., Corcione, C.E., A review of polymer-based materials for fused filament fabrication (FFF): focus on sustainability and recycled materials. *Polymers*, 14, 3, 465, Jan. 24, 2022.
40. Loh, G.H., Sotayo, A., Pei, E., Development and testing of material extrusion additive manufactured polymer–textile composites. *Fashion Text.*, 8, 1–21, Dec. 2021.
41. Cano-Vicent, A., Tambuwala, M.M., Hassan, S.S., Barh, D., Aljabali, A.A., Birkett, M., Arjunan, A., Serrano-Aroca, A., Fused deposition modelling: Current status, methodology, applications and future prospects. *Addit. Manuf.*, 47, 102378, Nov. 1, 2021.
42. Chakraborty, S. and Biswas, M.C., Fused deposition modeling 3D printing technology in textile and fashion industry: Materials and innovation, in: *Modern Concepts in Material Science*, Iris Publishers, Houston, TX, USA, Nov. 13, 2019.
43. Udayakumar, G.P., Muthusamy, S., Selvaganesh, B., Sivarajasekar, N., Rambabu, K., Banat, F., Sivamani, S., Sivakumar, N., Hosseini-Bandegharaei, A., Show, P.L., Biopolymers and

composites: Properties, characterization and their applications in food, medical and pharmaceutical industries. *J. Environ. Chem. Eng.*, 9, 4, 105322, 2021.
44. Abdelhak, M.J., A review: Application of biopolymers in the pharmaceutical formulation. *J. Adv. Bio Pharm. Pharmacovigil.*, 1, 1, 15–25, Mar. 31, 2019.
45. Gheorghita, R., Anchidin-Norocel, L., Filip, R., Dimian, M., Covasa, M., Applications of biopolymers for drugs and probiotics delivery. *Polymers*, 13, 16, 2729, Aug. 15, 2021.
46. Ferrari, F., Striani, R., Fico, D., Alam, M.M., Greco, A., Corcione, C.E., An Overview on Wood Waste Valorization as Biopolymers and Biocomposites: Definition, Classification, Production, Properties and Applications. *Polymers*, 14, 24, 5519, Dec. 16, 2022.
47. Bayón, B., Berti, I.R., Gagneten, A.M., Castro, G.R., Biopolymers from wastes to high-value products in biomedicine, in: *Waste to wealth*, pp. 1–44, 2018.
48. Xu, W., Wang, X., Sandler, N., Will for, S., Xu, C., Three-dimensional printing of wood-derived biopolymers: a review focused on biomedical applications. *ACS Sustainable Chem. Eng.*, 6, 5, 5663–80, Mar. 27, 2018.
49. Kumar, S.S., Biopolymers in Medical Applications, in: *Technical Textile*, pp. 1–9, CRC Press, Boca Raton, FL, USA, 2007.
50. Rebelo, R., Fernandes, M., Fangueiro, R., Biopolymers in medical implants: a brief review. *Procedia Eng.*, 200, 236–43, Jan. 1, 2017.
51. Islam, S.U. and Butola, B.S. (Eds.), *Advances in Functional and Protective Textiles*, Woodhead Publishing, Sawston, Cambridge, UK, Jun. 11, 2020.
52. Demirgöz, D., Elvira, C., Mano, J.F., Cunha, A.M., Piskin, E., Reis, R.L., Chemical modification of starch based biodegradable polymeric blends: effects on water uptake, degradation behaviour and mechanical properties. *Polym. Degrad. Stab.*, 70, 2, 161–70, Jan. 1, 2000.
53. Shankar, S. and Rhim, J.W., Effect of types of zinc oxide nanoparticles on structural, mechanical and antibacterial properties of poly (lactide)/poly (butylene adipate-co-terephthalate) composite films. *Food Packag. Shelf Life*, 21, 100327, Sep. 1, 2019.
54. Meraldo, A., Introduction to bio-based polymers, in: *Multilayer Flexible Packaging*, pp. 47–52, William Andrew Publishing, Norwich, NY, USA, Jan. 1, 2016.
55. El Knidri, H.B., Addaou, A., Laajeb, A., Lahsini, A., *Int., J., Biol. Macromol.*, 120, 1181–9, 2018.
56. Vaz, C.M., Fossen, M., Van Tuil, R.F., De Graaf, L.A., Reis, R.L., Cunha, A.M., Casein and soybean protein-based thermoplastics and composites as alternative biodegradable polymers for biomedical applications. *J. Biomed. Mater. Res. Part A Off. J. Soc. Biomater. Jpn. Soc. Biomater. Aust. Soc. Biomater. Korean Soc. Biomater.*, 65, 1, 60–70, Apr. 1, 2003.
57. Biopolymers improve economics and line efficiency [Internet], EcoSynthetix website. [cited 2023 Jul 7], 2016. Available from: https://ecosynthetix.com/improve-economics.
58. Patel, R., Nanda, S., Dalai, A.K., Conversion of municipal solid waste to biofuels, in: *Progressive thermochemical biorefining technologies*, pp. 43–61, CRC Press, Florida, USA, Sep. 1, 2021.
59. Patel, R., Nanda, S., Dalai, A.K., Conversion of plastic waste to fuels and chemicals, in: *Progressive Thermochemical Biorefining Technologies*, pp. 63–73, CRC Press, Florida, USA, Sep. 1, 2021.
60. Arikan, E.B. and Ozsoy, H.D., A review: investigation of bioplastics. *J. Civ. Eng. Arch.*, 9, 1, 188–92, 2015.
61. Magazù, S., Migliardo, F., Telling, M.T., Structural and dynamical properties of water in sugar mixtures. *Food Chem.*, 106, 4, 1460–6, Feb. 15, 2008.
62. Prasanth, S.M., Kumar, P.S., Harish, S., Rishikesh, M., Nanda, S., Vo, D.V., Application of biomass derived products in mid-size automotive industries: A review. *Chemosphere*, 280, 130723, Oct. 1, 2021.

63. Pattnaik, F., Nanda, S., Kumar, V., Naik, S., Dalai, A.K., Subcritical water hydrolysis of Phragmites for sugar extraction and catalytic conversion to platform chemicals. *Biomass Bioenergy*, 145, 105965, Feb. 1, 2021.
64. Okolie, J.A., Mukherjee, A., Nanda, S., Dalai, A.K., Kozinski, J.A., Next-generation biofuels and platform biochemicals from lignocellulosic biomass. *Int. J. Energy Res.*, 45, 10, 14145–69, Aug. 2021.
65. Nanda, S., Patra, B.R., Patel, R., Bakos, J., Dalai, A.K., Innovations in applications and prospects of bioplastics and biopolymers: A review. *Environ. Chem. Lett.*, 20, 1, 379–95, Feb. 2022.
66. Alaswad, S.O., Mahmoud, A.S., Arunachalam, P., Recent Advances in Biodegradable Polymers and Their Biological Applications: A Brief Review. *Polymers*, 14, 22, 4924, Nov. 15, 2022.

3

Multifunctional Polymeric Materials

Akshara Johari[1] and Pooja Agarwal[2]*

[1]*Division of Forensic Sciences, School of Biomedical Sciences, Galgotias University, Greater Noida, Uttar Pradesh, India*
[2]*Division of Chemistry, School of Basic Sciences, Galgotias University, Greater Noida, Uttar Pradesh, India*

Abstract
Multifunctional polymeric materials include a wide variety of compounds with specific properties and functionalities. Their adaptability enables applications in a variety of industries, contributing to advances in materials science, healthcare, electronics, electronics, and drug delivery systems. Multifunctional polymeric materials can be incorporated and altered according to the properties required for the applications. These materials can be synthesized using the polymerization method and detailed characteristics are analyzed using specific characterization techniques. Advances in material design, synthesis processes, and interdisciplinary collaborations are expected to accelerate the creation of multifunctional polymers with increasingly sophisticated and customized features, thereby contributing to scientific and technological progress.

Keywords: Multifunctional polymer, smart polymers, conducting polymer, polymerization, drug delivery

3.1 Introduction

Polymers featuring many types of functional groups or chemical moieties inside their molecular structure are referred to as multifunctional polymers. These functional groups can provide the polymer with a range of characteristics or functions, making it adaptable for a range of uses. The word "multifunctional" draws attention to the variety of functions that these polymers may provide. These materials are employed as electrolytes, dielectrics, and semiconductors and also have given significant improvements over traditional inorganic materials for many purposes. High specific energy, better energy density, leakage proof, flexibility, strong ionic conductivity, wide thermal and electrochemical stability, easy processing, low weight, and most significantly, cost efficiency are the qualities that make these materials ideal choices [1]. Polymer materials are still an interesting and popular topic for study due to their applications in a variety of industries, including healthcare, biomedical devices, agriculture, water purification, food, and textiles [2]. Polymers used

**Corresponding author*: poojaocm5@gmail.com

Divya Bajpai Tripathy, Anjali Gupta and Arvind Kumar Jain (eds.) Multifunctional Materials: Engineering and Biological Applications, (47–80) © 2025 Scrivener Publishing LLC

as structural materials can be categorized as organic or inorganic [3]. Organic-inorganic hybrid materials combine the benefits of organic and inorganic components, resulting in high-performance and highly useful materials. The goal of these hybrid materials is to generate substances that perform and operate better than those made entirely of organic or inorganic components [4]. Functional groups present in the backbone of the polymeric structure are responsible for a variety of qualities, including chemical stability, processability, mechanical strength, and flexibility. Chemical heterogeneity which appears when the bulk polymer is functionalized, offers several benefits, including increased compatibility, phase separation, reactivity, and association [5]. Multifunctional structural materials provide for both structural and non-structural functions without the use of external equipment. Structural function, such as tailored strength, stiffness, fracture toughness, and damping, can be combined with non-structural properties like noise and vibration control, electromagnetic interference (EMI) shielding, structural health monitoring, self-repair, thermochromism, self-cleaning, antibacterial, or energy harvesting/storage. Multifunctional lightweight materials are a concept that allows for considerable improvements in overall space system efficiency, performance, and environmental friendliness. Lightweight materials are often defined as polymers as host matrices or composites reinforced with fillers to satisfy the unique requirements of each application or device. With polymer composites as the material of choice, the inherent properties of the polymers (e.g., lightweight, flexible, and easy process) can be combined with unique properties, such as electrical conductivity, high dielectric properties, or high mechanical stiffness and strength [6]. Multifunctional structural batteries and supercapacitors can enhance performance and efficiency in sophisticated lightweight systems [7]. The use of multifunctional structures can result in significant weight savings and improved form factors in platforms ranging from electric vehicles to mobile phones and satellites. Composite materials are especially well-suited for multifunctional structures because the incorporation of many constituents into a single structure allows a large range of materials and stacking sequences that can be used for nonstructural functions [7]. Multifunctionality in materials refers to the purposeful combination of diverse material functionalities so that they coexist within the material yet are independent of one another. This notion is frequently investigated and applied in materials science and engineering to develop materials with a wide range of properties for specific purposes [8]. Various methods are employed to create multifunctional materials. Among these materials, one component materials system (OCPS) and multimaterial system (MS) are involved. OCPS involves the composition of a single type of polymers having multifunctional properties within a single type of polymer. On the other hand, MS is a composition of one or more types of materials in a system, and it can be a combination of polymers or a mixture of polymers and/or organic-inorganic materials [8]. Due to the feasibility of achieving various functionalities as needed for different purposes, there is a development of great interest in multifunctional polymeric material (MPM) for new research and scientific investigations.

3.2 Types of Multifunctional Polymeric Materials

Multifunctional polymers can be categorized based on variations in their processing, molecular structures, mechanical, physical, chemical, geological, and biological characteristics, preparation chemistry, and applications. Polymers may be divided into many other categories like plastics, elastomers, fibers, coatings, and adhesives. There are some overlapped groups of plastics, for instance, and polyamides are used for thermoplastic molding as well as for the purpose of creating synthetic fibers [3]. Additional categories are based on molecular chain structure and physical characteristics, like differences in the type of bonding between molecular chains in thermosets and thermoplastics. Since thermoplastic polymers may melt or dissolve, shaping procedures need heat or a solvent. They are flexible molecular architectures of linear, branching, or grafted structures with only secondary (physical) links between the main chains [3]. A novel class of materials known as multifunctional polymeric nanoparticles combines the benefits of nanotechnology with polymers.

3.2.1 Smart Polymers

In the recent several decades, natural science has become more interested in stimuli-responsive polymers, known as "smart materials." These materials are typically (macro)-molecules that alter the reaction to minute external changes in the surrounding environment [9]. There have been reports of numerous kinds of stimuli-responsive polymers. Proteins, nucleic acids, and polysaccharides make up a sizable class of stimuli-responsive biopolymers that are extensively widespread in the living environment. These materials do, in fact, fall within the oldest group of compounds in nature that exhibit stimuli-responsive qualities. However, for a number of years, the same attention has been given to the class of artificial polymers with stimuli-responsive qualities [9]. Synthetic polymers have been produced into a variety of functional forms to satisfy the needs of scientific and industrial applications. Classifying smart synthetic polymers in a distinctive way is challenging, although they may be divided into many groups based on their chemical or physical characteristics. Consequently, the scientific community has come up with a variety of terms to refer to smart polymers, including "stimuli-responsive polymers," "intelligent polymers," and "environmentally sensitive." The capacity to react to even minute alterations in their immediate surroundings is their defining feature [10]. The ability of smart polymers to alter macroscopically in their structure and have reversible transitions that restore the system to its starting state is a crucial feature to consider when dealing with them [10]. Their applications are reported in a wide range of fields, like biomedical, packaging industries, electronics industries, and textile industries, as shown in Figure 3.1.

These smart materials are more than just passive substances; they have intelligence. They can detect changes in their surroundings, act or respond to them, and have built-in control mechanisms to manage and regulate their behavior. Because of their sensing, actuation, and control capabilities, they are useful in a variety of sectors including robotics, aircraft, and healthcare [11]. The remarkable advancements in smart polymer nanocomposites (SPNs) have gained significant attention from both the scientific community and industry in the past few years. Shape memory, self-healing, self-sensing, self-healing, self-cleaning, and energy harvesting are among the possible uses of SPNs [12]. Smart materials can be used in a variety of environmental science and chemistry applications due to their unique features and capabilities. These uses include accelerating chemical reactions, sensing environmental elements, and taking part in decontamination procedures, all of which help to solve and mitigate environmental challenges [13]. With their distinctive structures, mixability, and adaptability, they are becoming increasingly significant and standard materials in a variety of industries for activities like as chiral separations, medication delivery, diagnostics, and

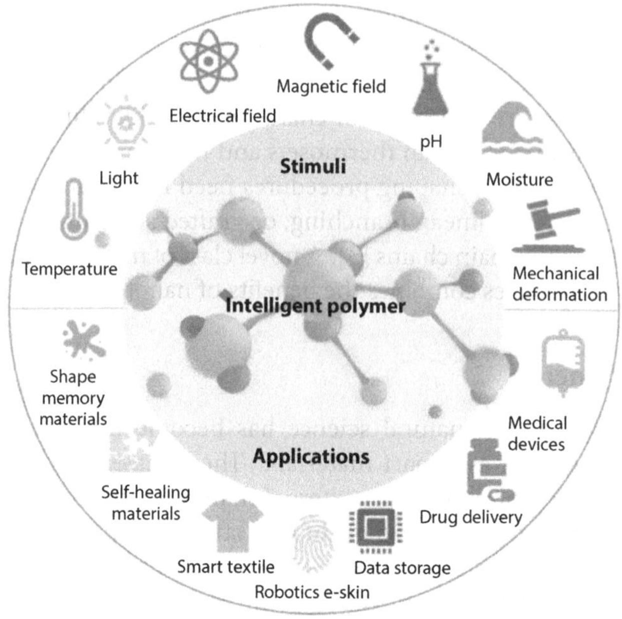

Figure 3.1 Showing smart functional polymers and their application [11].

others. These polymers play an important part in the biological revolution because they enable dynamic and programmable interactions at multiple levels within organisms and cells. These characteristics make them useful instruments for researchers and scientists seeking to understand and exploit biological processes for a variety of purposes [14]. A class of intelligent systems, specifically polymers, which can respond to electric fields, has potential uses in biology, where their capacity to respond precisely to electric stimuli can be used for a variety of functions, including modifying characteristics and sending electric signals. This characteristic could be used for a variety of applications, including signaling or inducing specific reactions in biological systems [15]. These polymers are biodegradable and offer versatility in creating a diverse selection of conductive materials with electrical conductivity and are flexible to be used in electronics and biological applications, like drug delivery, artificial muscles, and antibacterial materials. Further research is necessary to fully understand the potential applications of these intelligent polymeric materials for antimicrobial scaffolding and cell proliferation, even though the usage of electroactive smart polymers in conductive polymers for medication administration and artificial muscles has been widely established [15].

3.2.2 Self-Healing Polymers

A brand-new class of intelligent materials known as self-healing polymers is able to repair damage on its own without the need for any form of manual help. Expanding the range of uses for polymeric materials can be made possible by developing self-healing technologies that enable these materials to stop crack propagation at an early stage and prevent catastrophic failures. Materials with self-healing properties have been divided into two categories: autonomic healing and non-autonomic healing. Autonomic healing polymers are completely independent and devoid of any type of outside assistance [16]. Non-autonomic healing polymers are partially self-sufficient, the material has the ability to mend itself, but further external stimuli, like heat or ultraviolet light, are needed for the healing process to take place. These polymers set the degree of self-sufficiency in the healing mechanism apart from earlier methods like solvent welding, which needed the injury to be localized and solvent or heat to be manually applied [16]. Self-healing polymers have been created through physical and chemical methods such as shape-memory effects, diffusion and flow, heterogeneous self-healing systems, covalent-bond reformation and reorganization, supramolecular chemistry dynamics, or combinations of these. Materials that possess self-healing properties demonstrate the capacity to mend and regain functionality through their fundamental resources [17]. Materials with the potential to exhibit self-healing qualities include supramolecular polymers. These polymers are distinguished by their dynamic, non-covalent interactions, which enable the materials to respond to stimuli and be affected by contact pressure, light, or heat triggers. Therefore, the main benefit of supramolecular materials is that their inherent self-healing qualities eliminate the need for (reactive) additives like toxic monomers or catalysts or the engineering of particular chemical reactions to cause the healing process [18]. The polymers need to react with other groups in order to generate well-defined organic polymers with self-healing properties [19]. There are two types of self-healing polymers: extrinsic and intrinsic materials. When it comes to extrinsic systems, materials that have been damaged can be repaired by applying encapsulating materials, such as crosslinkers, organic monomers, or their liquid-like reaction mixture, to the damaged area [19].

Furthermore, intra- or intermolecular forces such as strong chemical covalent linkage connections, chemical/physical crosslinking networks, supramolecular interactions,

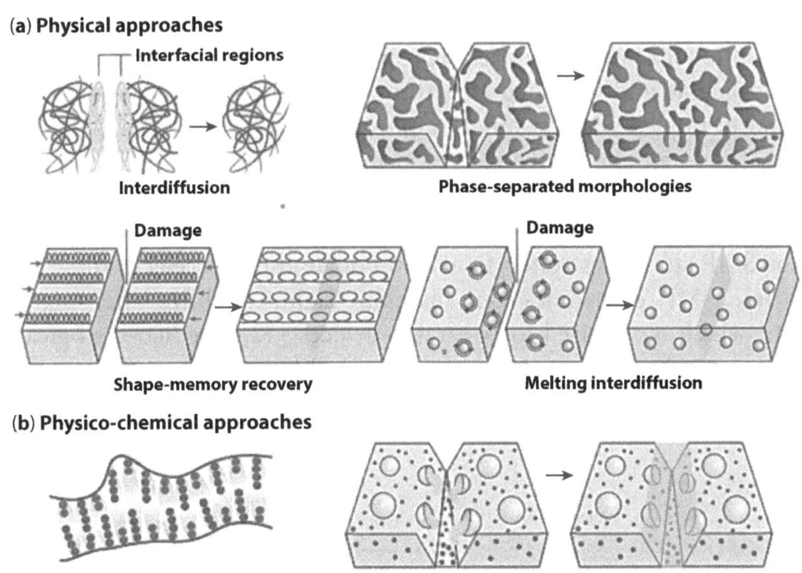

Figure 3.2 Self-healing mechanism: (a) the physical process of interdiffusion of the polymer chain and (b) the combined physical and chemical process for obtaining self-healing [20].

coordination, and hydrogen bonding cause an intrinsic self-healing mechanism in the injured material, as explained in the above Figure 3.2 [19]. Polymeric materials exhibit great potential for self-healing applications due to their complex and dynamic chemical structure. The ability to self-heal can be incorporated into polymers through the use of specific chemistries or by subjecting them to various physical, chemical, or thermal stimuli [19].

3.2.3 Shape Memory Polymers

Shape memory polymers are defined as polymers that may "memorize" a permanent shape and can be controlled in such a way that a certain temporary shape will be "fixed" in given appropriate conditions. Following that, a trigger (such as heat or light) will cause the temporary shape to convert into the memorized permanent shape [21]. In general, shape memory polymers have at least two distinct phases, a stable network and a second phase that can be changed by an external trigger. The former phase stabilizes the entire shape memory polymer (SMP) and is responsible for retaining the original shape, i.e., the deformation of this phase is the driving force for shape recovery, as given in Figure 3.3 [21]. The second phase fixes the temporary shape by crystallization (i.e., a melting transition leads to shape recovery), a glass transition, a transition between various liquid crystalline phases, or reversible covalent or non-covalent bonding [21]. The shape memory phenomenon is also well suited to use in biological applications. The polymer itself must be biocompatible and non-toxic for biomedical uses. Most of the structures of SMPs are biocompatible and interact with biological systems without any adverse reactions.

Shape memory polymers can be used in a variety of biological applications, including the repair of heart valves and endovascular stroke treatment (clot removal). Simple evacuation can be performed by introducing SMPs with the help of a catheter, photothermally activating it, and providing relief from ischemia [16]. Shape memory alloys are commercially accessible and have a wide variety of applications. Shape memory polymer composites (SMPCs) are lightweight, low-cost, and capable of producing active ability to change their shape intentionally, which are critical properties for usage in aircraft engineering, particularly in unfoldable and extended structures. SMPC hinges and reflector antennas are some common examples. SMPCs hinge reinforced with carbon fiber plain-wove fabrics have been created as an alternative to typical metallic hinges that take up a lot of space and weight [22]. Flexible electronics based on SMPs have become a popular field of study. SMPs are adaptable and can be used as the substrate for electrical devices. As the temperature rises, the SMP substrates soften to accommodate the rigid electronics. Furthermore, these metal-free SMP electronics can maintain stable qualities after many bending-recovery tests, which is required for wearable devices [22]. Medico-biological electronic devices require entirely different materials than SMP-based biological electronic devices.

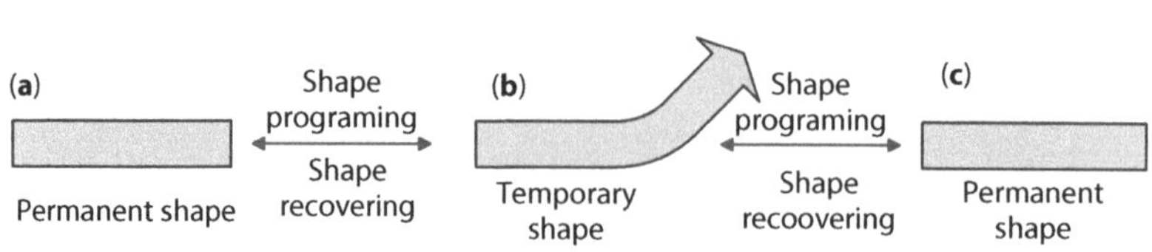

Figure 3.3 Mechanism of shape-recovering programming of polymers [21].

Because biological electronic devices are in contact with the human body, they require non-toxic and non-degradable materials [22]. A variety of unique materials have been found, encouraging the development of soft robotics and opening up new opportunities for solving problems that conventional rigid robots could not otherwise handle. These soft robots are built of complex elastomers and polymers and are lightweight, low in cost, flexible, and mechanically compliant. To date, a range of soft robots, including grippers, actuators, crawling robots, and other complex ones, have been constructed using SMPCs. These robots have the ability to do complex jobs like drug distribution, search and rescue, and serving as human assistants [22]. Silica molding, nanoindentation, or localized laser can be used to fabricate shape memory arrays made up of repetitive units like cylinders, cuboids, prisms, cones, and polyhedral. SMP topography is considerably affected by changing the layout and spacing of the repeated SMP units. The printed SMPs can change shape by responding to environmental inputs. In reality, 4D printing combines the benefits of 3D printing and SMPs. SMP complexes can be created and built in a relatively short period of time [22]. SMPs soften and get sticky when heated in the dry adhesion process, and they come into intimate contact with various types of objects under pressure. Because of the widespread scientific interest in SMPs and their impacts, material scientists have become more familiar with them [22, 23]. SMPs with improved functionality and portability are expected to emerge in the future. Controllable and deformable SMPs would cover the way for a new generation of soft stimuli-responsive materials to answer scientific problems [24, 25].

3.2.4 Conducting Polymers

Conducting polymers are those that conduct electricity as a result of electron delocalization. Such compounds may be metallic conductors or semiconductors. Organic materials include conducting polymers and insulating polymers [26]. They have great electrical conductivity but no mechanical characteristics comparable to other widely available polymers. Organic synthesis and advanced dispersion techniques can be used to modify the electrical characteristics. Currently, nanostructures of conductive polymers serve as the foundation for a wide range of innovative technical products, including electrochromic display devices, photovoltaic devices, and biosensors [26]. Because of the excellent electrical conductivity of these polymers, there has been a lot of interest in using organic chemicals in microelectronics. Many polymeric materials have been intensively investigated among the multiple conducting polymers because of their effortless synthesis. These polymers provide outstanding redox capabilities, stabilized oxidized form, high conductivity, water solubility, and commercially accessible and important electrical and optical qualities [26]. Poly (phenylene vinylene) (PPV) is a diamagnetic substance with electroluminescence and extremely low electrical conductivity. It is highly crystalline, mechanically strong, and environmentally stable. Electrical conductivity can be increased by doping with iodine, ferric chloride, alkali metals, or acid. Conducting polymers (CPs) have acquired a lot of attention in both fundamental and applied research because they have electrical and electrochemical properties that are similar to both traditional semiconductors and metals. CPs have outstanding properties such as low synthesis and processing temperatures, chemical and structural variety, adjustable conductivity, and structural flexibility. Nanotechnology advancements have enabled the manufacture of diverse CP nanomaterials with increased performance for a variety of applications such as electronics, optoelectronics, sensors, and energy devices [27]. Electrically conducting polymers with good biocompatibility, and their derivatives, find widespread use in biomedical fields such as bio actuators, biosensors, neural implants, drug delivery systems, and tissue engineering scaffolds. Conductive polymer demonstrates promising conductivity as bioactive scaffolds for tissue regeneration, and their conductive nature allows electrical signals to stimulate cells or tissue cultivated on them. Their mechanical brittleness and poor processability, however, limit their applicability [28]. Conducting polymers (e.g., polyaniline (PANI), polypropyrole (PPY), and polythiophene) promotes the adhesion, proliferation, and differentiation of a wide range of cell types *in vitro*, indicating that they are cytocompatibility [28]. Because CPs are very brittle and it is difficult to fabricate pure conducting polymer film from CPs, blending CPs with other degradable polymers such as PLA, Poly(lactic-co-glycolic acid) (PLGA), Polycarpolactone (PCL), chitosan, and silk fibroin is widely used as conductive biomaterials for tissue engineering [28]. The conducting composite film was formed by *in-situ* polymerization of pyrrole in an emulsion mixture, then pressed into a film and cured in a vacuum chamber. Surface properties and shape have a significant impact on cell behavior and have been widely employed as biomaterials in tissue engineering. Although CPs are not inherently biodegradable but can be modified into biodegradable polymers by the addition of functionalized copolymers like aniline/pyrrole [28].

3.2.5 Biodegradable Polymers

Biopolymers are polymers that are obtained and manufactured using natural resources such as animals and plants. Biocompatible polymers, on the other hand, travel with biological structures/systems and assess, treat, supplement, or replace any unit in the body. Biodegradable polymers are polymers that degrade into biologically acceptable compounds [29]. These polymers are often divided into two types: natural polymers and synthetic polymers. Natural biopolymers, such as polysaccharides and proteins, are derived from renewable or biological sources such as plant, animal, microbial, and marine sources, whereas synthetic polymers, such as polyesters and aliphatic polymers, are created chemically. Some of the biopolymers are given in Figure 3.4. Because of their outstanding biocompatibility and biodegradability, these polymers are gaining commercial importance. Enzymes are typically used to catalyze biodegradation, which can include both hydrolysis and oxidation [29]. Chemical processes are used to produce a wide range of biodegradable and biocompatible polymers. The framework of these polymers includes chemical components like esters, anhydrides, diacids, and amides. The weak hydrolysable linkages that compose the backbones of synthetic biopolymers are the primary source of biodegradability [29].

Biodegradable polymers are materials that can be degraded by various environmental microorganisms, such as bacteria and fungi, to form water and carbon dioxide. Bacteria can use some polyesters and polysaccharides for energy and carbon metabolism. Polyhydroxyalkanoates (PHAs) are biodegradable aliphatic polyesters derived exclusively from bacterial fermentation [30]. Gellan gum and curdlan are biodegradable polysaccharides made by bacteria. They have primarily been used as food additives, particularly gelling and thickening agents, due to their water absorbance and nontoxic behavior. Due to their high water vapor permeability, edible films made from gellan gum and curdlan have recently been created and used for food wrapping [30]. Chitin has been chemically changed by grafting with synthetic polymers to improve its compatibility with a variety of polymers. Chitin compounds with polyoxazoline side chains were synthesized, and synthetic polymers such

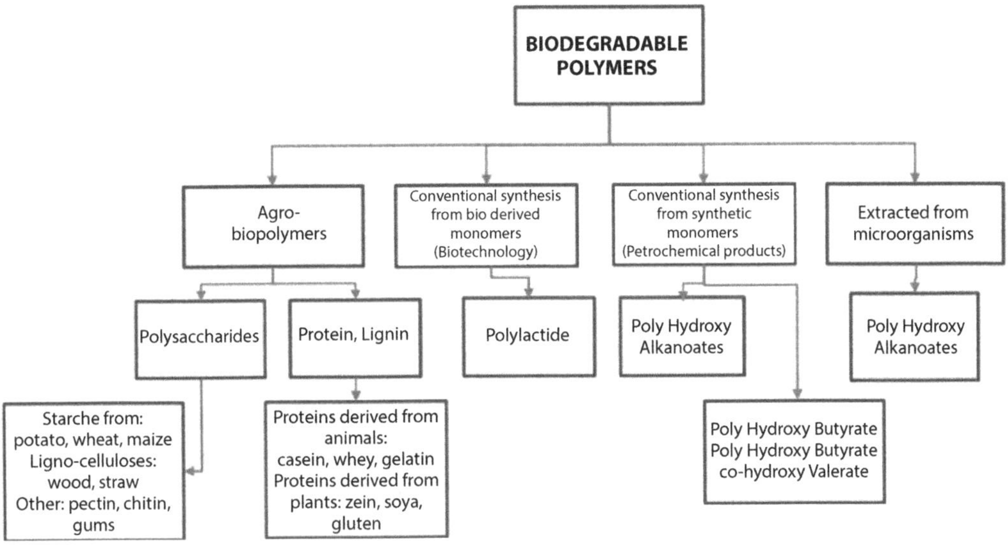

Figure 3.4 Types of biodegradable polymers [30].

as polyvinyl chloride and polyvinyl alcohol were used to create miscible blends. These blends are frequently used as new polymetric materials not only because they are biodegradable, but also because of their molding and mechanical qualities similar to conventional polymers [30]. Blending commercial synthetic polymers with natural materials like starch and chitin improves their biodegradability. Various types of biodegradable polymers are utilized in numerous fields ranging from agricultural to biomedical applications [30]. The manufacture and usage of bio-based and bio-degradable polymer materials are considerably increasing, which can help to reduce environmental problems associated with waste polymer materials [31]. One of the primary issues with biodegradable polymers derived from renewable sources is that they degrade quickly due to their water-attracting nature (hydrophilic), and in some situations, they may lack enough strength, particularly in damp conditions [31]. The usage of biodegradable products has the potential to reduce reliance on fossil fuels and contribute to a cleaner environment, and as a result, more research is being conducted on new biodegradable polymers and their improved qualities [31]. Several factors influence biodegradation, including polymer shape, structure, chemical treatment, and molecular weight. The usage of biodegradable polymers is quickly expanding, with a global market worth billions of dollars every year. Biodegradable polymers are utilized in a wide range of applications, including food packaging, computer keyboards, automobile interior parts, and medical applications such as implanted big devices, medicinal delivery, and tissue engineering [32].

3.3 Synthesis and Characterization of Multifunctional Polymeric Materials

A number of processes are involved in the synthesis and characterization of MPMs, and each is essential to obtaining the required functionalities and characteristics. Polymerization is one of the often-used methods used for the synthesis of polymeric materials.

3.3.1 Method of Polymerization

Polymerization is the process of combining small molecules, known as monomers, to form a larger, more complex structure termed a polymer. Multifunctional polymers contain more than one type of functional group inside their molecular structure. Multifunctional polymers are usually synthesized utilizing a variety of polymerization techniques. Co-polymerization is one of the polymerization methods [33].

3.3.2 Copolymerization

This method is carried out to obtain multifunctionality in the materials by combining different monomers. It allows the incorporation of a variety of properties in a single polymer formed after a combination of monomers. The polymer obtained during the copolymerization method is called copolymer. The polymer so formed combines the properties of each monomer within it. This is a versatile process frequently employed to give polymeric materials multifunctionality. Types of copolymerization involve random copolymerization, block copolymerization, alteration copolymerization, and graft co-polymerization [34]. Through a random copolymerization polymer chain, monomers are dispersed at random, and different monomers polymerize in a random sequence or order. Block copolymerization produces a series of monomer units within the polymer chain by polymerizing monomers in successive blocks. Different building blocks may have unique characteristics, resulting in materials with well-defined functional domains [35]. The process of alternating copolymerization involves the regular and alternating arrangement of monomers throughout the polymer chain [36]. Alteration can be difficult to achieve and frequently needs particular catalysts and reaction conditions. Graft copolymerization in which a certain kind of monomer serves as the primary polymer chain, and additional monomers are grafted onto it to generate side chains. The main chain's and side chains' features can be combined in this configuration [37]. The characteristics of the copolymer can be customized to fit specific requirements by varying the ratio of different monomers. This tunability makes a large variety of material properties possible. When distinct monomers are combined, collaborative effects can occur, resulting in a copolymer with superior qualities than that of the component homopolymers. This is particularly beneficial for reaching multifunctionality. The final material's compatibility and stability can be increased through copolymerization. This is especially crucial when mixing monomers that have dissimilar chemical or physical characteristics. Copolymerization allows the incorporation of monomers with specified functions, hence enabling the fine-tuning of material performances [34]. Copolymers, for instance, can be made to have a particular ratio of stiffness to flexibility. Gradient copolymers facilitate the all-in-one integration of a range of features by smoothly transitioning from one monomer to another throughout the polymer chain. Several functions can be included in a single polymeric substance through the process of copolymerization. A copolymer combines improved chemical resistance or biocompatibility with electrical conductivity. In order to produce polymer blends with better qualities like impact resistance, thermal stability, or chemical resistance, copolymerization is frequently utilized [38]. Copolymers are used to create smart materials that react to light, pH, temperature, and other environmental stimuli. When several functions are needed, copolymerization is employed in the design of polymers for medical implants, tissue engineering, and drug delivery systems.

Polymer electrolytes for batteries are made from copolymers, which combine mechanical strength and features like ion conductivity. To balance adhesion, flexibility, and chemical resistance in coatings and adhesives, copolymers can be customized for particular uses [39].

3.3.3 Incorporation of Functional Groups

The introduction of functional groups or reactive sites during polymerization enables further alterations or additions of specific functions. It is a crucial tactic in polymer chemistry for modifying the properties of the final polymer by the incorporation of functional groups during the polymerization process. A molecule's chemical reactivity and unique features are determined by the particular groupings of atoms that make up its functional groups [41, 40]. The resultant polymer can be used as a flexible platform for additional alterations or the inclusion of particular activities by purposefully adding these functional groups during polymerization. Selection of the monomer with appropriate versatility and functional group is the prior step of the polymerization process [43]. Monomers are polymer building blocks, and their structure influences the properties of the polymer that is to be obtained. Prior to polymerization, monomers can be functionalized with particular groups, which is commonly accomplished using chemical synthesis, which allows the alteration of functional groups on monomer molecules. Functional initiators can also be used to introduce specific functional groups. These initiators start polymerization by introducing their functional groups into the polymer chain [44]. Adding functional monomers to normal monomers during copolymerization is a typical method. This enables the introduction of specialized functionalities at chosen points along the polymer chain. Following the polymerization step, functional groups can be added via post-polymerization modifications [41]. This may need chemical processes that selectively link functional moieties to reactive spots on the polymer chain. Common functional groups added during polymerization are:

- Hydroxyl groups (OH): Allow reactions with isocyanates to generate polyurethanes and reactions with carboxylic acids to form esters.
- Amine groups (NH_2): Make it easier to produce polyurethane from isocyanates or amide from carboxylic acids.
- Carboxyl groups (COOH): Enable esterification processes and serve as sites for further modification via different events.

Other than the above-described method, introducing crosslinking agents also improves the material's stability and mechanical strength. The chemical and heat resistance of a material is also be affected by crosslinking. Also, depending on the polymerization procedure, selecting appropriate initiators or catalysts affects the reaction kinetics and the polymer's final structure [42].

The combination of different characterization techniques allows a detailed understanding of the MPMs including structure, composition, and specific performance. A thorough investigation is required to understand the physical, chemical, thermal, mechanical, structural, and behavioral functions of MPM.

Functional groups, polymeric chain structure, composition, and molecular mobility can be determined using different specific spectroscopic techniques like FTIR, NMR, and UV-visible spectroscopy. For the determination of molecular weight distribution gel permeation chromatography (GPC) is preferred. Based on the chemical properties, polymers are separated and analyzed using HPLC (high-performance liquid chromatography) [43]. Thermal analysis techniques like DSC (differential scanning calorimetry) and TGA (Thermogravimetric Analysis) are used to measure heat flow associated with phase transition, crystallization, and melting and analyze the thermal stability and decomposition ability of the polymers respectively [43]. High-resolution images of morphology and surface structure are provided by SEM (scanning electron microscopy) and TEM (transmission electron microscopy). AFM (atomic force microscopy) is used for the analysis of the topographic and mechanical properties of the materials at the nanoscale. Rheological techniques like rheometric and DMA (dynamic mechanical analysis) are used for the measurement of the flow and deformation behavior of the polymers, analyze the viscous-elastic properties, mechanical properties like stiffness and damping, and study the function of temperature and frequency of the polymers. Electrical and magnetic properties by Dielectric spectroscopy, which measures conductivity and dielectric constant, while the magnetic behavior analysis requires techniques like (SQUID) superconducting quantum interference device magnetometry [44]. XPS (X-ray photoelectron spectroscopy) and AES (Auger Electron Spectroscopy) are the characterization techniques that are used for the determination of surface composition and chemical state of the material [45]. To determine the size distribution and surface charges of the particles in a polymeric system, dynamic light scattering (DLS) and zeta potential analysis are preferred. Elemental analysis and mass spectrometry are used for the determination of the composition of polymers, identify and quantify fragments of polymers, and provide information on molecular weight and structure, respectively [46]. Table 3.1 explained below is the instance of the characterization techniques used for various types of polymers including natural and synthetic polymeric systems for their detailed information.

Table 3.1 Characterizations of the smart polymers.

Characterization	Polymeric materials	Analysis method	References
Morphology, thermal stability, functional group analysis	Chitosan, dextran, cellulose, alginate, derivatives of starch-coated polymers	TEM/SEM, TGA, FTIR, DLS	[43, 44, 46]
Functional group analysis, temperature response	Co-Polymers N-Nirtopropylacrylamide, (NIPAM)	FTIR, TGA	[44]
Structure, composition, morphology, functional group analysis	Graphene with PDMA (Polydimethylsiloxane)	XRD, FTIR, SEM/TEM	[45]
Thermal stability, composition, structure and morphology, functional group	PNIPAM (poly(N-isopropylacrylamide)) with azobenzene	TGA, XRD, EDS, XPS, FTIR<SET/ TEM	[45]

3.4 Properties and Applications of Multifunctional Polymeric Materials

3.4.1 Thermal and Mechanical Properties

Thermal analysis examines the chemical, physical, and structural responses to temperature changes [47]. The thermal stability of polymeric materials is often assessed by TGA. Inorganic nanofillers have been shown to improve the thermal stability of biodegradable polymers [48]. Studies on the thermal stability of biodegradable polymer nanocomposites are critical in determining their working temperature limit and operating conditions, which are connected to their thermal decomposition temperature and rate [48].

Thermal properties of multifunctional polymers include are following:

Glass Transition Temperature (Tg): The temperature at which an amorphous polymer changes from a hard and brittle to a more flexible or rubbery state. Multifunctional polymers may have variable Tg values to accommodate varying temperature needs.
Thermal conductivity: A material's capacity to conduct heat. Some multifunctional polymers can be engineered to have higher thermal conductivity for applications that require efficient heat transfer.
Thermal stability: Thermal stability refers to a polymer's ability to endure high temperatures without significant deterioration. This is critical for use in industries such as electronics, aircraft, and automotive.

Nanoparticle dispersibility and interfacial contact with the polymer matrix are crucial for achieving optimal thermal and mechanical properties. Modified carbon nanoparticles enhance the thermal and mechanical characteristics of polymers [49]. With the rapid advancement of miniaturization and integration of electronic equipment, the need for heat management does not need to be emphasized repeatedly. The most effective way to address this issue is to create novel thermal interface materials (TIMs) with high thermal conductivity (TC), insulative, and softness. Flexible polymeric composite-based heat pads are among the greatest options in contemporary electronic packaging technologies [50]. Thermal analysis examines the chemical, physical, and structural responses to temperature changes. Temperature affects chemical reactions, physical properties, and structural changes. Thermal analysis can be described as any scientific or technical assessment of a substance using temperature manipulation [51].

3.4.2 Electrical Properties

The electrical conductivity of multifunctional materials is critical, particularly when they are intended for use in electronic devices, sensors, actuators, or other electrical components. Several factors determine the electrical conductivity of such materials. The sort of elements and compounds included in the material has a considerable impact on its electrical conductivity. For example, materials containing metallic components or conductive polymers have increased electrical conductivity [52]. Doping is the purposeful introduction of impurities into a substance to modify its electrical characteristics. Doping can improve or reduce electrical conductivity depending on the type of dopant utilized. Well-organized structures frequently contribute to increased conductivity. Temperature affects the electrical conductivity of materials. Some materials may gain conductivity at higher temperatures (semiconductors), whereas others may decrease. Some multifunctional materials respond to external stimuli, such as light, pressure, or magnetic fields, altering their electrical conductivity. This characteristic is especially valuable in smart materials and electronics. The method utilized to manufacture or process multifunctional materials can influence their electrical characteristics. Thin-film deposition, sintering, and chemical vapor deposition are all techniques that can have an impact on conductivity. Incorporating certain chemicals or functional groups into the material can change its electrical conductivity. This is frequent in conductive polymers, where dopants or additives are used to increase conductivity [53, 54]. Materials with high electrical conductivity include graphene and carbon nanotubes. Incorporating these components into composites can improve the conductivity of multifunctional materials. When mechanical strain or deformation is applied to some materials, their electrical conductivity alters. This feature can be used in applications such as flexible electronics and sensors [53, 54].

3.4.3 Optical Properties

The capacity of a material to transmit or block light is critical for many purposes. Transparent materials are extensively utilized in optics, displays, and sensors, whereas opaque materials can be used to restrict light or absorb energy. Materials having controlled refraction are useful in lens systems, optical devices, and imaging applications. Multifunctional materials can be engineered to selectively absorb various wavelengths for use in applications such as solar cells and photodetectors. Mirrors, reflecting coatings, and optical communication devices all make use of reflective materials. The redirection of light in multiple directions and control over scattering qualities is critical in applications such as anti-glare coatings, light diffusers, and controlled illumination. The emission of light is caused by a substance absorbing photons and re-emitting them at a longer wavelength. This feature is used in applications such as LEDs, phosphors, and luminous materials. Due to light exposure, a change in electrical conductivity also takes place. This feature is essential for the development of photodetectors and solar cells. An optical property in which a material's refractive indices differ depending on the direction of light polarization, especially important in liquid crystal displays (LCDs) and other polarization-dependent devices. A nonlinear optical phenomenon, in which the refractive index changes in proportion to the intensity of light, this effect is used in products such as optical switches and modulators. Polarizers and LCDs depend heavily on optically anisotropic materials, which is a critical feature for the development of optical fibers and integrated photonic devices. Multifunctional material's optical properties can be adjusted using material design, engineering, and synthesis processes, making them useful in a variety of applications such as electronics, photonics, sensing, and energy harvesting [55]. Hybrid materials having optical properties are gaining popularity due to their potential applications in imaging, sensing, protection, therapy, and energy. Optical systems are commonly associated with human eyes, such as color changes, protection, and cameras, making them a major concern for humans. Researchers' great interest in the development of devices activated by light, creating light, or altering the optical signal is thus a logical outcome, and it is boosted by technological breakthroughs and the implementation of scientific knowledge [55]. Combining diverse dimensional materials creates unique and numerous optoelectronic capabilities that are impossible in typical pure low-dimensional materials [35]. Fluorescent-conductive-magnetic (FCM) multi-functional materials have found applications in medical diagnosis and treatment, microwave absorption, electromagnetic shielding, light-emitting devices, flexible electronics, sensors, and electronic packaging, among others. Fluorene polymer and fluorenone, being photoactive semiconductors, have a strong current response effect under light illumination and are hence good fluorescent materials [36, 37].

3.4.4 Biological Properties

The interaction of multifunctional materials with biological systems or their possible uses in biological and medical contexts is commonly brought up when talking about their biological qualities. Multifunctional materials must be biocompatible to avoid unpleasant reactions or toxicity. Biocompatibility means that the material does not trigger a strong immune response and can coexist with living tissues. Drug delivery applications frequently need the use of multifunctional materials. They can be designed to encapsulate and deliver medicinal substances in a regulated manner. Certain multipurpose polymers can stick to biological surfaces because of their bio-adhesive qualities. It is possible to engineer multifunctional polymers to have particular interactions with cells. Surface alterations that facilitate cell adhesion, proliferation, or differentiation could be implemented to render them valuable in the fields of tissue engineering and regenerative medicine. Diagnostic applications can benefit from the imaging capabilities of some multifunctional polymers that can be developed. Polymers could be engineered to contain contrast agents for imaging modalities such as fluorescence imaging or magnetic resonance imaging (MRI). Multifunctional polymers are sensitive to a variety of stimuli, including changes in pH, temperature, and enzyme activity. This responsiveness can be used to create materials that are intelligent enough to adjust to their surroundings or to release drugs under regulated circumstances. It is possible to alter some multifunctional polymers to give them antibacterial or antimicrobial qualities. Preventing infections linked to implants or medical devices is crucial in medical applications. The mechanical properties of multifunctional polymers, such as elasticity and strength, are critical for their biological performance. Polymers employed in tissue engineering, for example, must have mechanical properties similar to the target tissues. Researchers' attention has been focused more on multifunctional antibacterial materials. The development of sophisticated medical devices will be greatly aided by multifunctional antimicrobial materials since they have the potential to not only combat infections but also enhance the effectiveness of medical equipment. Because of the most recent developments in materials science, chemistry, and biomedical engineering, antimicrobial materials have come a long way in the last several decades. From bacteria to fungi and viruses, the antimicrobial range has expanded. Utilizing biomedical surfaces with combined antifouling and microbicidal capabilities is one way to increase biocompatibility. Another is the extensive research on self-adaptive antibacterial tactics [40]. Because of their special qualities, MPMs have found a wide range of uses in many industries. Multifunctional polymers are biocompatible, making them ideal for application in medical implants, medication delivery systems, and tissue engineering. These materials interact safely with biological systems, causing no harmful responses. This increases treatment efficiency while reducing negative effects. Polymers can be used in biosensor devices to detect biomolecules and track physiological data. These sensors have potential applications in diagnostics, illness monitoring, and personalized treatment. Some polymeric materials with their multifunctional properties and applications in various fields are shown in Table 3.2.

Table 3.2 Properties and applications of polymeric materials.

Polymeric materials	Properties and application	References
CNTs/Composites	• Mechanically strong and stiff, they provide toughness and can be used in impact-resistant materials. • Excellent conductors of electricity, can be used in flexible electronics, sensors, and conductive coatings • Possess remarkable thermal properties and can be used in thermal interface materials. • Due to their lightweight, they are employed in aerospace and automotive applications. • Due to their biocompatibility can be used in biomedical applications and drug delivery systems. • Gas absorbance property allows them in gas sensing tools. • Optical transparency and good for photoelectronic. • In alliance with self-healing composites, they can be used to repair damage autonomously. • Exhibit good magnetic properties used for sensing and actuation processes.	[56–58]
ZnO quantum dots	• Exhibit properties like photoluminescence and semiconductor behavior. • Antibacterial property, catalytic activity, piezoelectricity, gas sensing • UV -absorbance, optical sensing, biocompatibility, electrochemical properties. • Applications involve storage devices, flexible electronics, healthcare, drug delivery system UV-coatings, and sunscreens. • Sensors, solar cells, supercapacitors	[37, 39]
Graphene fillers/ graphene-based NCs	• Flexible materials have uses in a variety of industries, including electronics, aerospace, automotive, packaging, energy storage, and biomedicine. • Its incorporation greatly increases the mechanical strength of nanocomposites. • Graphene acts as a good gas barrier, making nanocomposites useful for applications that require improved barrier qualities, particularly relevant in packaging materials, where increased barrier performance is desired to lengthen product shelf life. • Graphene is thin, light, and flexible. When added to nanocomposites, it preserves or improves flexibility while reducing material weight. The aerospace and automobile industries use lightweight structural materials.	[37, 39]

(Continued)

Table 3.2 Properties and applications of polymeric materials. (*Continued*)

Polymeric materials	Properties and application	References
Carbon quantum dots	• CQDs have strong and tunable photoluminescence, which means they can emit light when triggered by external energy sources. • They are biocompatible and have low cytotoxicity, making them appropriate for usage in biological and medical fields. • CQDs can act as electron carriers and display good electron transport capabilities, making them useful in electronic devices, solar cells, and sensors. • Chemical sensing makes them appropriate to be used in chemical sensors for the detection of analytes. • The photocatalytic property makes them useful for environmental remediation processes. • They can be used as theranostics and energy storage devices. • They can be used for environmental sensing like monitoring water quality and detecting pollutants. • They are employed in antimicrobial applications.	[37]
PVA-PEG polymeric NCs	• PVA and PEG are hydrophilic polymers, and their combination in nanocomposites improves water absorption and retention. This hydrophilicity can be useful in applications like wound dressings and hydrogels. • PVA's adhesive qualities and PEG's flexibility make them ideal for coating, adhesive, and surface modification applications. • The combination of these polymers can improve processing qualities and produce a more homogeneous nanocomposite. • Other properties include electrical conductivity and environmental sensing.	[42, 57, 58]
Chitosan, alginate, starch, dextran, /NCs/hydrogels	• Ideal for wound dressings, medicine delivery, and food preservation. • They have electrical conductivity, which makes them helpful in electronic applications. • They are employed in tissue engineering because they are biocompatible and can simulate the extracellular matrix. • Material barrier qualities, such as those used in packaging to aid food preservation.	[59, 60]

(*Continued*)

Table 3.2 Properties and applications of polymeric materials. (*Continued*)

Polymeric materials	Properties and application	References
PLA, PLGA, PEG/ composites/NCs	• Transparency makes it suitable for packaging applications. • Controllable degradation rate. • Encapsulation of drugs and therapeutics. • Incorporation in nanoparticles can enhance the mechanical properties and barrier properties.	[61]
Melanin(poly-indolequinone)	• Melanin has been proven to scavenge a variety of free radicals, protecting cells and tissues from oxidative stress. This makes it a promising candidate for developing antioxidant medicines. • Thermoregulator properties, ability to absorb sunlight and convert it into heat aid in the maintenance of body temperature, particularly in cold situations.	[42]
Phosphorous-based nanomaterials	• Exhibit properties like photoluminescence, electrical properties, thermal, magnetic and photocatalytic properties, and antimicrobial properties. • Phosphorus-based nanoparticles are frequently integrated into polymers to provide flame-retardant characteristics, i.e., when heat is applied to phosphorus-containing compounds, flame-quenching gasses are released, lowering material flammability.	[62]

3.5 Application of Multifunctional Polymeric Materials

3.5.1 Applications in Electronics

Flexible and elastic electronic devices are made from multifunctional polymers. Wearable electronics, flexible screens, and conformal sensors can all be developed using these materials. They can tolerate mechanical deformation without losing any electrical properties. Polymers with several functions can be employed in printed electronics. To generate electronic circuits on a variety of substrates, they can be handled using printing processes like screen printing or inkjet printing. This makes manufacturing methods scalable and affordable. Energy storage systems such as supercapacitors and batteries employ MPMs. These substances can be utilized as parts of electrodes and electrolytes, which will increase cycle life, energy density, and performance. A revolutionary advancement in consumer goods is represented by multifunctional energy storage and conversion devices that integrate unique features and functions in intelligent and interactive modes. Examples of these products include wearable electronics, space satellites, electric vehicles, healthcare equipment, and artificial intelligence. Supercapacitors have a high-power density, a quick charge-discharge rate, a long cycling life, and other characteristics. They store charge via electrical double layers or redox processes. The device's self-healing mechanism includes both mechanical and electrical restoration; self-healing polymers repair the device's mechanical qualities, while electrode alignment restores its electrical function [63]. Multifunctional composites provide load-bearing qualities along with previously independent functionality like electrical energy storage and thermal, optical, chemical, and electromagnetic properties. A structural battery that combines load-bearing and electrochemical energy storage functions into a single construction is another illustration of a multifunctional composite. Because the structural batteries can support structural loads, a portion of the main structure can be removed, saving the system as a whole mass and volume. In the past, various methods of combining energy storage and load-bearing capabilities have been studied. These methods include structural supercapacitors, structural fuel cells, and structural capacitors [64].

3.5.2 Applications in Biomedical Field

The adaptability and programmable features of MPMs have led to a variety of biomedical applications, as shown in Figure 3.5. Because these materials may perform several tasks at once, they are useful for a variety of medical and healthcare-related applications. Nanomaterials are finding more and more applications in medicine, including medication delivery systems, vaccinations, prophylaxis, and diagnosis (such as the administration of contrast chemicals for imaging). By guaranteeing greater hydrophilicity, less interaction with plasma and cellular proteins, and better accumulation in target tissues, nanocarriers are used in medicine to improve the bioavailability, pharmacokinetics, and efficiency of certain medicines or contrast agents [65]. After application, a bio-adhesive needs to establish cohesive strength in addition to strong interfacial connections. An injectable bio-adhesive needs to be able to change under modest physiological conditions from a liquid to a solid state [66].

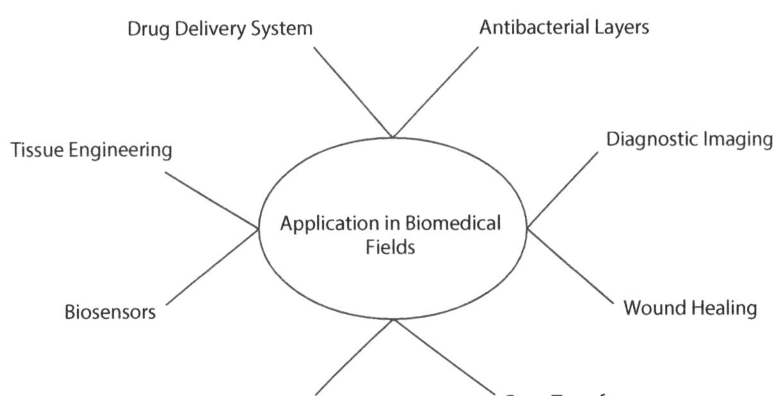

Figure 3.5 Various roles of multifunctional polymers in the biomedical field [17].

3.5.2.1 *Drug Delivery Systems*

Therapeutic drugs can be released in a targeted and controlled manner by using drug delivery systems designed with multifunctional polymers. Drugs can be delivered to particular tissues or cells, kept from degrading, and enclosed in these materials. Generally, drugs are dispersed throughout the material to produce nanoparticles or nanospheres, or they can be adsorbed in their pores or on their surface, or they can be conjugated into them [65].

3.5.2.2 *Tissue Engineering*

In tissue engineering, scaffolds that imitate the extracellular matrix (ECM) are made using multifunctional polymers. In addition to promoting cell adhesion and proliferation and offering structural support, these scaffolds have the ability to integrate growth factors for improved tissue regeneration. The bio-adhesive must have the bulk mechanical strength required to resist the forces exerted by the surrounding tissues after cure. Tissues that are frequently subjected to repeated stresses, such as cartilage, tendons, and ligaments, may be able to be repaired with a tissue adhesive whose adhesive strength and biomechanical characteristics correspond to those of native tissues [66].

3.5.2.3 *Diagnostic Imaging*

Imaging-capable polymers can be applied to medical conditions. Contrast agents for imaging modalities like computed tomography (CT), MRI, and fluorescence imaging can be carried by multifunctional polymeric nanoparticles. Targeted therapy development is becoming increasingly dependent on *in vivo* imaging, which is especially significant for medical diagnosis. Accurate biodistribution studies can be obtained by combining various imaging modalities. In terms of acquiring dynamic images at the organ or even cellular level, new generations of hybrid nanomaterials that integrate several imaging functions are very promising. Recently, significant advancements have been made through the combination of multiple techniques, including up-conversion luminescence (UCL), X-ray CT, two-photon excited fluorescence, MRI, and single-photon emission CT (SPECT). High-resolution cellular imaging, for example, has been accomplished with the use of two distinct contrast agents [67]. Since nanomaterials have become more widely available, imaging technologies have evolved from being only diagnostic to being potent "find, see, and treat" systems when used in conjunction with therapeutic compounds and target entities [67]. Nanomaterial-based therapy is not just for drug delivery. In fact, photothermal and photodynamic therapies complement chemotherapy [67]. Many polymers and functional polymeric materials have been created to increase medical diagnostic performance through a variety of methods, such as promoting molecular identification in advanced diagnostic evaluations and enhancing contrast in imaging technologies. Because polymers are dependable in providing quick and easy diagnoses, as well as in transporting and safeguarding medications that are immobilized inside their structures, they have gained significant interest from researchers and manufacturers in the field of diagnostic polymers Nanomedicine relies heavily on the development of nanoformulations capable of diagnosing and treating tumors. For theranostic applications, several nanomaterials are used, such as iron oxide nanoparticles, quantum dots, gold nanoparticles, and carbon nanotubes [68]. Polymer vesicles offer targeted distribution, controlled release, enhanced drug stability, and reduced cytotoxicity, making them a promising nano-system [69]. Crosslinked polymer nano-capsules have been extensively investigated as nanocarriers for the delivery of small molecule medicines. Aside from fundamental structural design requirements for CPNCs, structural elements that assist drug loading and regulated drug release should be carefully examined.

3.5.2.4 Biosensors

It is possible to develop polymeric materials to produce biosensors that can identify particular infections or proteins. Medical gadgets may incorporate these sensors for functions like disease diagnosis or glucose monitoring. A biosensor consists of a receptor layer and a transducer. The receptor layer is made of (bio)recognition material that recognizes the desired analyte, and the transducer translates the (bio)chemical signal produced by the analyte's contact with the receptor into a digital electronic signal. (Bio)sensors can be classified based on the type of transducer employed. Biosensor devices are typically made as micro-total analysis systems (μTAS) or lab-on-a-chip (LOC). A variety of materials, including glass, silicon, plastic, paper, hydrogels, polymers, and composites, are utilized to create miniaturized LOC devices based on microfluidic and electrochemical sensing techniques. Polymeric materials, on the other hand, stand out because of their low cost, flexibility, and biocompatibility. Polymer-based intelligent materials offer several benefits over low molecular weight materials, including biocompatibility, dispersion in water, structural stability, processing simplicity, and integration with detecting devices. The characteristics of conductive polymers, which include dynamic, chemical, electrical, and physical, can greatly enhance the sensors' and biosensors' performance [70]. For the development of functional biosensors, it is critical to choose an immobilization process that will ensure acceptable biosensor performance, such as strong operating and storage stability, high sensitivity, high selectivity, quick response time, and high repeatability [70]. The performance of the sensors is determined by the nanostructures with which the biomaterials interact. Using high aspect ratio 1D nanostructures as sensing components can dramatically boost sensor sensitivity. The huge surface area of these nanostructures allows for larger interaction areas with the biomolecules to be monitored, boosting the device's sensitivity. Because of their high sensitivity, biocompatibility, and ease of synthesis, polymer nanofibers, nanotubes, and nanorods are commonly utilized in biosensors [71]. Chitosan, a biodegradable and biocompatible polymer, is frequently employed in biosensing applications. Their capacity to immobilize enzymes is particularly useful, and chitosan and chitin nanofibers are frequently included in biosensors. Furthermore, the chitosan's amino and hydroxyl groups make it easier to crosslink with diverse substrates. PANI nanotube array had a considerable effect and was employed as a signal enhancement element to allow the DNA biosensor to identify a target oligonucleotide. In addition, the biosensor demonstrated a strong distinction between the target oligonucleotide and mismatched oligonucleotides. The proposed biosensor has the potential for use in single nucleotide polymorphism analysis and mutation detection. Using conducting polymers in chemical and biological sensors provides advantages as the interaction between analytes affects their redox and doping states, resulting in changes in current, resistance, and electrochemical potential. The high surface area-to-volume ratios of these functionalized nanostructures make them particularly useful for drug administration and biosensing applications, among other things [71]. Nanomaterials are highly effective sensors for several applications, including strain, gas, bio, and chemical detection. Advances in nanocomposite sensor technology have bridged the gap between the initial design and its practical use. In nanocomposite sensors, nanofiller type, nanofiller concentrations, polymer type, nanoparticle dispersion in matrix, and matrix-nanofiller interactions all play an important role in improving nanomaterial sensing performance. As a result, many sensors have been designed to detect strain-related impacts, electrical or optical changes, gasses, ions, hazardous gasses, chemical substances, proteins, DNA, RNA, and related biomolecules. As a result, nanocomposite sensors have found practical applications in wearable electronics, gas-sensing devices, and biological sensors [72].

3.5.2.5 Wound Healing

By utilizing multifunctional polymers in wound dressings, antibacterial agents, growth factors, and other therapeutic substances can be released under regulated conditions. These substances can also maintain a moist environment that promotes healing and act as a barrier against infections, as given in Table 3.3. Wound healing has been a hot topic of research because present goods, equipment, and techniques do not provide the need for fast recovery while maintaining skin function. The ideal healing system should be capable of speeding up wound closure, reducing infection, stimulating healing mechanisms, mimicking the ECM feature, moisturizing the wound, and decreasing scar formation [73]. Two main strategies have been developed for the application of nanomaterials of inorganic and organic nature, respectively: nanosized materials with antimicrobial properties are embedded in polymers, most commonly biopolymers abundantly available from natural sources, or nanocarriers are used to encapsulate active agents. In both circumstances, the goal is to effectively restrict the proliferation of bacteria near the wound site [73]. Hydrogel-based dressings with multifunctional features are excellent for wound healing, making them unique materials. Some contributing elements can affect the antibacterial, antioxidant, electrical, adhesiveness, mechanical, and tissue regeneration properties of the hydrogels. These aspects include the kind of polymer used to make the hydrogel, polymer modification, the incorporation of nanomaterials and nanostructures, and the addition of active components to the hydrogel network. Microbial infections have the potential to postpone the healing process in both acute and chronic wounds. Therefore, it's crucial to give the hydrogel dressings antimicrobial qualities. Various antibacterial agents, including conventional antibiotics, were included in hydrogels in earlier research to prevent wound infections [74]. When the epithelial continuity of the skin is disrupted, the underlying tissue's structure and function are disturbed, leading to the formation of wounds. Injuries may be the result of an illness, or they may be an unintentional or deliberate origin [75]. The kind of wound injury, the underlying illness, and the role of systemic and local mediators all influence how quickly a wound heals. There are two types of wounds: acute and chronic. Angiogenesis, epithelial migration, and cell proliferation are among the steps of the healing process that acute wounds, which are frequently brought on by trauma or surgery, certainly follow, as explained in Figure 3.6.

Hemostasis is the initial stage of wound healing, during which a blood clot is formed in an attempt to stop bleeding. Blood clots, or tiny pieces of cells, stick to the exposed collagen of the wound. Further boosting adhesion, adhered platelets produce substances that cause surrounding platelets to become sticky. The clinging platelets bind together to create a plug that closes off tiny holes in the damaged blood artery. This is a sequence of complicated

Table 3.3 Polymeric materials for wound healing.

Polymer	Polymeric system	Mechanism	Ref.
Chitosan	Chitosan/hydrogel Chitosan-nano-bioglass composite	In the form of powder, films, sponges, hydrogels, particles, fibers, and powders. Hydrogels: Facilitates barrier formation and prevents blood flow from the cavity.	[59, 76–78]
Dextran	Aldehyde dextran sponge	By rapid wound closure, cells' initiation, and aggregation, as well as coagulation factors' aggregation on the wound site.	[79, 80]
Collagen	(Collagen and thrombin) and (gelatin and thrombin)	Provides a site for platelet adherence, activation, and aggregation.	[66]
Gelatin	(Gelatin-thrombin granules). Graphene oxide-gelatin aerogels. Curcumin-alginate-gelatin sponge. m-TG-thrombin-gelatin.	Swell when they come in contact with the blood. These swollen granules stop hemorrhage by blocking the bleeding site.	[80, 81]
Alginate	PGA/alginate/AgNP alginate fibers, hydroxy apatite/alginate granules.	The released calcium ions promote prothrombin activation in the clotting cascade which leads to rapid hemostasis. Alginate granules swell enough to block the bleeding site.	[66, 67]
Cellulose	Cellulose-modified chitosan foam sponge. Oxidized cellulose patch	Facilitates hemostasis by quickly absorbing liquid, entrapping platelets and erythrocytes, and increasing, blood coagulating factors.	[67]
Hyaluronic Acid	Gel hydrogel HA/gelatin hydrogel, Serotonin hydrogel	HA-based hydrogels act as tissue sealants for hemorrhage control.	[67]
Starch	kCA-coated starch/cellulose nanofibers hydrogel TRAP-starch-PEG sponge	$Ca2^{+}$ CPSMs are applied to the bleeding sites and block the irregular bleeding. Starch-based sponge provides pressure to the wound and promotes hemostasis.	[68]

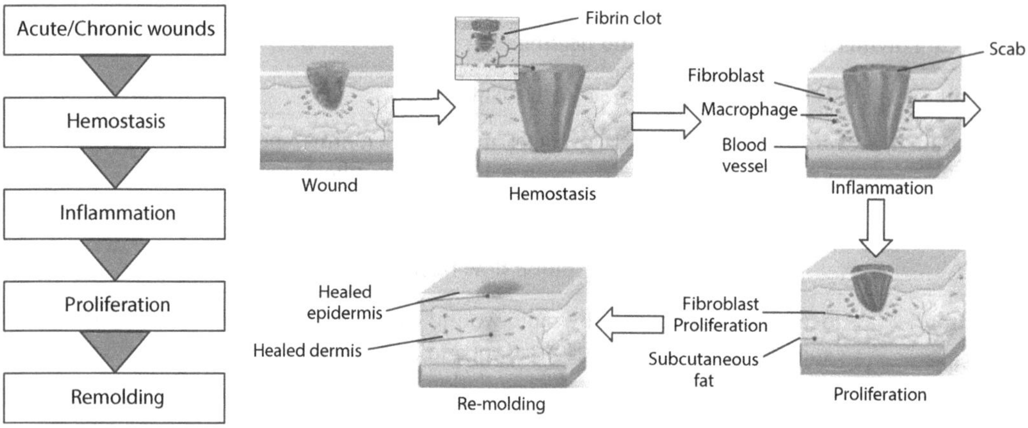

Figure 3.6 Mechanism of wound healing [61].

metabolic processes involving several blood clotting agents. Fibrin is a protein mesh that strengthens the platelet plug and forms a stable blood clot as a result [82]. A procedure known as fibrinolysis starts as soon as the incision has fully healed. Being a complex biological reaction to trauma, infection, or tissue damage, inflammation plays a vital role in the body's defense system. To start a healing process, a well-planned sequence of actions is taken in order to remove the source of the harm to the cells and remove damaged cells and tissues. The proliferative phase of the inflammatory response follows, during which new blood vessels grow to aid in healing and fibroblasts create collagen for tissue repair [83]. The proliferation phase is followed by the remodeling phase, which is essential to obtain repaired tissue and function. Even though the scar might never totally return to its original tissue, usefulness, and aesthetic appearance can be much enhanced by a well-planned and executed redesign. The size and type of the wound, the person's general health, and the standard of wound care given during the earlier stages of healing are some of the variables that affect how long and how successfully the remodeling phase lasts [84]. Chronic wounds typically take longer to heal and don't adhere to the standard healing protocol. Put another way, because the healing process is hampered and there is a high likelihood of infection and uncontrolled inflammation, chronic wounds do not heal and restore their functional and anatomic integrity [75].

Biomaterials are non-pharmaceutical materials that can be used to replace or enhance the function of bodily tissues and organs. Biomaterials have two different functions in the healing of acute and chronic wounds. First, they act as scaffolds to support the growth and wound closure of endogenous cells, and second, they can act as temporary dressings that cover the wound site and create an environment that is conducive to the healing process. When it comes to wound dressings, the perfect biomaterial should be able to cover the wound, retain its moisture content, let oxygen in, and stop the growth of harmful organisms while the wound heals. To facilitate their removal, potential scaffold materials should also be biocompatible, non-degradable, and resistant to cell adhesion and ingrowth [75]. Fibrin is a protein derived from fibrinogen. Polymerized fibrin is an important component in the coagulation process, which is important in the wound healing process. Other natural polymers such as fibrin, silk fibroin, dextran, elastin, apigenin, and nerolidol are considered ideal materials for enhancing wound healing. Some of the polymers with their mechanism of wound healing have been explained in Table 3.3. Since many bacteria found in soil, water, and the air can break down cellulose and its derivatives to produce enzymes known as cellulases, which are specific to cellulose, cellulose and its derivatives have drawn a lot of attention as natural polymers that are also environmentally friendly. Cellulose can be sensitive to changes in pH and ionic strength, as demonstrated by the esterification process that yields sodium carboxymethylcellulose (NaCMC). Derivatives of cellulose have a high biocompatibility, which has promoted their widespread application in clinical wound healing [85]. There are many products on the market for the treatment of wounds in their different phases; however, the use of any product in the treatment of burns or other types of wounds should follow strict protocol, seeking to identify the real contribution to the healing process [85].

3.5.2.6 *Implantable Devices*

Artificial joints, vascular stents, and controlled-release implants are among the devices made of polymeric materials. Multifunctional polymers have the potential to lower inflammation, increase biocompatibility, and deliver long-term medication release at the implant location. Implantable medical devices (IMDs) play a crucial role in modern healthcare, providing diagnosis, treatment, and rehabilitation, enhancing patients' quality of life. Cochlear implants restore hearing for the hearing impaired, deep brain stimulators reduce Parkinson's symptoms, and implantable spinal cord stimulators or drug delivery devices help those with neural system disorders. IMDs have three primary components: substrate, functional portion, and encapsulation. The substrate is the first component to consider as it serves as a supporting material for functional elements on all devices [86]. Functional portions in sensors and actuators are made up of electrical elements such as resistors, capacitors, and inductors constructed of conductive materials. Metals like Mg and Ca, transition metals like Mo, Zn, and Fe, and their alloys are examples of bioresorbable conductors [87]. Semiconducting materials are utilized to manufacture crucial electronic components like diodes and transistors. Semiconductors are essential for electronic devices like transistors and integrated circuits. These devices can be employed as medical implants or for other purposes before dissolving harmlessly within the body [87]. Implantable devices also include dielectric materials which serve as an insulating layer in sensors, protecting other functional components and preventing short circuits. They are also used to manufacture capacitors, which are commonly utilized as sensors in a variety of applications [72, 73]. The degradation mechanism, fabrication method, and mechanical and thermal properties are all important factors to consider when selecting a material for high-performance bioresorbable IMDs [86]. Biocompatible insulating materials (such as metal oxides, metal nitrides, and polymers) can serve as both a dielectric layer and an encapsulant in bioresorbable electronics [72]. Implantable drug delivery devices can be employed in various clinical applications, including women's health, oncology, ophthalmic illness, pain management, infectious disease, and central nervous system disorders [74].

3.5.2.7 *Gene Transfer*

In gene delivery systems, multifunctional polymers are employed to both protect genetic material against deterioration and enable its precise distribution into cells. Functional inorganic nanomaterials have recently emerged as reliable and adaptable nano scaffolds for gene delivery applications [88]. Plasmid DNA and siRNA are the two primary types of genetic material supplied to cells for gene therapy. Specific delivery and efficient release of genetic components from carriers are required for effective therapy. Specific gene delivery/targeting can be accomplished "actively" or "passively." "Active" targeting refers to the delivery of genetic materials that helped either internally by molecular-recognition-driven binding between a functional nanocarrier and receptors on a target cell's membrane or externally by a magnetic gradient, as in the case of magnetic nanoparticles.

3.5.2.8 Antibacterial Layers

Antimicrobial characteristics are inherent in some polymeric materials. This means that the material can prevent the growth of microbes like bacteria or fungi. Using these materials as coatings on medical devices can help prevent infections by limiting the presence of hazardous microbes on or around the implanted materials. Smart or responsive polymers can be programmed to respond to specific stimuli like pH, temperature, or enzyme activity [89]. Antimicrobial surfaces are meant to reduce bacterial attachment and growth on equipment. Materials-based techniques can be broadly divided into two categories, antibiofouling and bactericidal [76, 77]. Previous research has revealed that surface topography, hydrophobicity, surface charge, and surface chemistry all have an impact on antibiofouling performance. Antibiofouling surfaces typically feature hydrophilic, negatively charged, or low surface free energy properties, with the objective of repelling microorganisms, particularly bacteria, before they attach to the surface [90]. Although antibiotic-containing nanosystems improve antibacterial efficacy and slow the emergence of multidrug-resistant superbugs, there is still a need for new materials with intrinsic antibacterial properties. Polymers with intrinsic antimicrobial moieties, such as quaternary ammonium (QA) groups, phosphonium groups, 2-acrylamido-2-methylpropane sulfonic acid (AMPs), and their synthetic mimics, have received significant attention in recent years [91]. These antibacterial polymer nanoparticles provided an alternative way for managing multidrug-resistant bacteria or eradicating persistent biofilms. Multidrug-resistant bacteria are microorganisms that have developed resistance to numerous types of antibiotics, making conventional therapies ineffective. Antibacterial polymer nanoparticles offer an alternative strategy for controlling these resistant microorganisms. Antibacterial nano-systems with improved biocompatibility and increased blood circulation duration have the potential to be used in therapeutics [91].

3.5.3 Applications in Packaging Industry

The packaging industry's use of MPMs is always changing as a result of consumer demand for more cost-effective, sustainable, and functional products. The superior barrier qualities of multifunctional polymers can shield the contents inside from outside influences like light, moisture, oxygen, and pollutants. For fresh goods to have a longer shelf life, this property is essential. In order to reduce the need for extra preservatives, antimicrobials, antioxidants, and oxygen scavengers are a few examples of active ingredients of these materials. Smart packaging technology can help with product quality monitoring and assurance [77, 79].

There is growing concern over plastic pollution, which has encouraged the packaging sector to use more environmentally friendly polymers so there to great need to create multifunctional polymers that combine barrier qualities and biodegradability [66]. Nanoparticles can be added to polymeric matrices to improve their mechanical strength, thermal stability, and barrier qualities, and this can increase the durability and performance of the packaging materials, as shown in Table 3.4, for the PLA/composites. Multifunctional materials can be used for near field communication (NFC) and radio-frequency identification (RFID) technologies, this makes it easier to trace and monitor signals and gives logistics and inventory useful information. These materials can be used in labels and inks to improve adherence and printability. This makes packaging more visually pleasing and educational, supporting branding and marketing initiatives.

Utilizing polymers containing phase-change chemicals in packaging is a solution for managing temperature sensitivity in industries such as food and pharmaceuticals. This technology aids in maintaining the required temperature range, ensuring the quality and effectiveness of the products during storage and transport. It is possible to develop MPMs to have superior cushioning and shock absorption capabilities [78, 79]. This is necessary to safeguard fragile or sensitive goods while they are being transported. Considering recyclability while designing polymeric materials helps create packaging that is sustainable. Materials with several uses can continue to operate and still be acceptable for recycling [79].

Table 3.4 Composites of PLA in packaging industries [92].

PLA/Composites	Properties	Application
PLA/ZnO membranes	Mechanical properties, light visible performances	Food packaging
PLA/chitosan/basil olive oil films	Tensile, barrier, and antioxidant properties	Food packaging
PLA/lignin films	Toughness, antioxidant performance, water vapor, barrier properties, UV resistance behavior	Active food packaging
PLA/PVA/PEG blend/ thyme essential oils	Hydrophilicity, antibacterial properties	Antimicrobial packaging
PLA/chitosan/cellulose nanofibers	Mechanical, thermal, and wettability properties	Green packaging
PLA/lycopene nanocomposite film	Mechanical properties, oxidative and color stability of margarine	Margarine packaging
PLA/silica aerogel composites	Crystallinity, stretchability, elongation at break, high transparency	Food packaging

3.6 Future Prospects of Multifunctional Polymeric Materials and Conclusion

Due to the wide range of properties including morphological, topological, and chemical functions, multifunctional materials cover a vast variety of applications. Although self-healing polymers, shape-memory polymers, and conducting polymers can be classified under smart polymers because of their behavioral property, on separating them, they show property-specific functions. It is possible to construct sophisticated drug delivery systems, scaffolds for tissue engineering, and diagnostic tools with the help of MPMs. These materials can be made biocompatible and responsive to physiological [93]. Wearable electronics are the emerging field of multifunctional polymers. These materials provide a blend of flexibility, durability, and electrical conductivity, allowing for the development of cutting-edge electronic products with better functionality and user comfort. The development of energy storage and conversion technologies may be greatly aided by multifunctional polymers [94]. This covers the manufacture of solar cell materials, supercapacitors, and high-performance batteries. Improved conductivity, stability, and efficiency of polymer-based materials could make them important parts of energy storage systems. Water purification and pollution removal are two examples of environmental applications for which multifunctional polymers can be used [95]. Adsorption-related materials, catalytic functions, and self-healing materials have the potential to improve the effectiveness and sustainability of environmental remediation techniques. These polymeric materials can be effectively used in aerospace engineering, biomedical, and drug delivery systems. These materials have the potential to transform the production of customized goods across a range of sectors, such as healthcare, automotive, and aerospace. The development of new multifunctional polymers may be sped up by the application of AI and machine learning in materials design. Researchers can find the best combinations of monomers, additives, and processing conditions for particular applications with the use of predictive models and algorithms [96]. The development of MPMs is expected to significantly contribute to advancements across various industries, addressing complex challenges and creating new opportunities for technological innovation, as interdisciplinary research advances and collaborations between materials scientists, chemists, engineers, and other experts continue to grow. Although multifunctional polymers (MFPs) are proving to be effective in the detection and treatment of cancer in laboratories, there are still many obstacles ahead for these nanoparticles (NPs) in terms of translating from bench to bed. Importantly, the clinical effectiveness of smart mutifunctional nanopartilces (SMNPs) is determined by their capacity to overcome chemical and biological barriers such as toxicity, biodistribution, pharmacokinetics, and pharmacodynamics, as well as their industrial scale-up and reproducibility for dependable large-scale production. As a result, more substantial research is required in the future to solve the aforementioned problems in developing effective and feasible MFPs for cancer theranostics. The potential to create various therapeutic capabilities in a single drug delivery system is one of the most exciting opportunities in the development of polymer therapies. Because SMPs have generated so much interest in the study, material scientists have grown more familiar with them and the impacts they have. Better functioning and portable SMPs are anticipated to appear in the future. A new generation of soft stimuli-responsive materials to address scientific issues would result from the application of controlled and malleable SMPs.

References

1. Constantinescu, C., Rotaru, A., Nedelcea, A., Dinescu, M., Thermal behavior and matrix-assisted pulsed laser evaporation deposition of functional polymeric materials thin films with potential use in optoelectronics. *Mater. Sci. Semicond. Process.*, 30, 242–249, 2015.
2. Schulz, D.N., Patil, A.O., Company, E., Township, C., *Functional Polymers : An Overview*, pp. 1–14, 1998.
3. Akelah, A., *Functionalized Polymeric Materials in Agriculture and the Food Industry*, Springer New York, NY, 2013. https://doi.org/10.1007/978-1-4614-7061-8
4. Wang, F., Lu, X., He, C., Some recent developments of polyhedral oligomeric silsesquioxane (POSS)-based polymeric materials. *J. Mater. Chem.*, 9, 2775–2782, 2011, doi: 10.1039/c0jm02785e.
5. Hosseini, M., Amjadi, I., Mohajeri, M., Functional polymers: an introduction in the context of biomedical engineering, in: Advance Functional Polymers For Biomedical Applications, pp. 0–20, Elsevier, January 2019, doi: 10.1016/B978-0-12-816349-8.00001-1.
6. Costa, C.M., Costa, P., Lanceros-Mendez, S., Overview on lightweight multifunctional materials, in: , W *Advanced Lightweight Multifunctional Materials* oodhead Publishing in Materials, 2021.
7. Gienger, E.B., Nguyen, P.T., Chin, W., Behler, K.D., Snyder, J.F., Wetzel, E.D., Microstructure and multifunctional properties of liquid 1 polymer bicomponent structural electrolytes : Epoxy gels and porous monoliths. *J. Appl. Polym. Sci.*, 132, *42*, 42681, 1–9, 2015, doi: 10.1002/app.42681.
8. Polymers, M.S., Final Draft, *Adv. Mater.*, 22, *31, 3388–3410,* Willey online library 2010, doi: 10.1002/adma.).
9. Online, V.A., materials †. 7468–7483, 2013, doi: 10.1039/c2cs35191a.
10. Peponi, L., Arrieta, M.P., Lo, D., *6 Smart Polymers*, 2017, doi: 10.1016/B978-0-323-44353.
11. I., Intelligent Polymers, Fibers and Applications. *Polymers*, 13, *13(9)*, 1427, 2021.
12. Chow, W.S. and Ishak, Z.A.M., Smart polymer nanocomposites : A review. 14, 5, 416–435, 2020.
13. Gray, H.N. and Bergbreiter, D.E., Applications of Polymeric Smart Materials to Environmental Problems. 105, 55–63, February 1997.
14. Chang, B., Zhang, B., Sun, T., Smart Polymers with Special Wettability. 1–16, 2017, doi: 10.1002/smll.201503472.
15. Palza, H., Zapata, P.A., Angulo-Pineda, C., Electroactive Smart Polymers for Biomedical Applications, 2019.
16. Science, P., SELF-HEALING POLYMERS., 15, 2010.
17. Blaiszik, B.J., Kramer, S.L.B., Olugebefola, S.C., Moore, J.S., Sottos, N.R., White, S.R., Self-Healing Polymers and Composites, 2010, doi: 10.1146/annurev-matsci-070909-104532.
18. Van Gemert, G.M.L., Peeters, J.W., Söntjens, S.H.M., Janssen, H.M., Bosman, A.W., Self-Healing Supramolecular Polymers In Action. 234–242.
19. Raghava, K., El-zein, A., Airey, D.W., Alonso-marroquin, F., Schubel, P., Manalo, A., Self-healing polymers : Synthesis methods and applications. *Nano-Struct. Nano-Objects*, 23, 100500, 2020, doi: 10.1016/j.nanoso.2020.100500.
20. Kausar, A., Ahmad, I., Maaza, M., Bocchetta, P., Self-Healing Nanocomposites—Advancements and Aerospace Applications, 2023.
21. Hager, M.D., Bode, S., Weber, C., Schubert, U.S., Progress in Polymer Science Shape memory polymers : Past, present and future developments. *Prog. Polym. Sci.*, 49–50, 3–33, 2015, doi: 10.1016/j.progpolymsci.2015.04.002.
22. Xia, Y., He, Y., Zhang, F., Liu, Y., Leng, J., A Review of Shape Memory Polymers and Composites : Mechanisms, Materials, and Applications. 2000713, 1–33, 2020, doi: 10.1002/adma.202000713.

23. Mu, T., Liu, L., Lan, X., Liu, Y., Leng, J., Shape memory polymers for composites. *Compos. Sci. Technol.*, 160, 169–1988, 2018, doi: 10.1016/j.compscitech.2018.03.018.
24. Lendlein, A., Behl, M., Hiebl, B., Wischke, C., Shape-memory polymers as a technology platform for biomedical applications. 357–379, 2010.
25. Pilate, F., Toncheva, A., Dubois, P., Raquez, J., Shape-memory polymers for multiple applications in the materials world. *Eur. Polym. J.*, 2016, doi: 10.1016/j.eurpolymj.2016.05.004.
26. Awuzie, C., II, ScienceDirect Conducting Polymers. *Mater. Today Proc.*, 4, 4, 5721–5726, 2017, doi: 10.1016/j.matpr.2017.06.036.
27. Polymers, C., Electrical and Electrochemical Properties of Conducting Polymers, 2017, doi: 10.3390/polym9040150.
28. Guo, B., Sciences, M., Arbor, A., Arbor, A., Arbor, A., Arbor, A., HHS Public Access. 19, 6, 1764–1782, 2019, doi: 10.1021/acs.biomac.8b00276.Conducting.
29. Khalid, M., *Natural and synthetic biocompatible and biodegradable polymers*, January. 2018.
30. Baidurah, S., Methods of Analyses for Biodegradable Polymers : A Review, 2022.
31. Knez, Z., BIODEGRADABLE POLYMERS, CURRENT TRENDS OF RESEARCH AND THEIR APPLICATIONS, A REVIEW. 26, 4, 401–418, 2020.
32. Samir, A., Ashour, F.H., Hakim, A.A.A., Bassyouni, M., Recent advances in biodegradable polymers for sustainable applications, 2022, doi: 10.1038/s41529-022-00277-7.
33. John, G., Nagarajan, S., Vemula, P.K., Silverman, J.R., Pillai, C.K.S., Natural monomers: A mine for functional and sustainable materials–Occurrence, chemical modification and polymerization. *Prog. Polym. Sci.*, 92, 158–209, 2019.
34. Wang, Y., Ju, W., Chen, J., Liu, Z., Wang, J., One-Step Synthesis of Solid-State Photoluminescent Carbon Nanodots from Grains for Latent Fingerprint Detection. 8915–8923, 2020, doi: 10.1002/slct.202000712.
35. Keddie, D.J., A guide to the synthesis of block copolymers using reversible-addition fragmentation chain transfer (RAFT) polymerization. *Chem. Soc. Rev.*, 43, 2, 496–505, 2014.
36. Kubo, T., Easterling, C.P., Olson, R.A., Sumerlin, B.S., Synthesis of Multifunctional Homopolymers via Sequential Post-Polymerization Reactions, 2018, doi: 10.1039/x0xx00000x.
37. Zhai, L., Narkar, A., Ahn, K., Article in press, 2019.
38. Thakur, V.K. and Thakur, M.K., Recent advances in graft copolymerization and applications of chitosan: a review. *ACS Sustain. Chem. Eng.*, 2, 12, 2637–2652, 2014.
39. Su, X., Gao, Q., Wang, D., Han, T., Tang, B.Z., One-Step Multicomponent Polymerizations for the Synthesis of Multifunctional AIE Polymers. *Macromol. Rapid Commun.*, 42, 6, 2000471, 2021.
40. Chauhan, D., Afreen, S., Talreja, N., Ashfaq, M., Chapter 8-Multifunctional copper polymer-based nanocomposite for environmental and agricultural applications, in: *Multifunctional Hybrid Nanomaterials for sustainable Agri-food and Ecosystem*, Elsevier Inc, 2020.
41. Wang, S., Fu, C., Wei, Y., Tao, L., Facile One-Pot Synthesis of New Functional Polymers through Multicomponent Systems. *Macromol. Chem. Phys.*, 215, 6, 486–492, 2014.
42. Liu, J.F. *et al.*, 'Smart' Matrix Microneedle Patch Made of Self-Crosslinkable and Multifunctional Polymers for Delivering Insulin On-Demand. *Adv. Sci.*, 10, 30, 2303665, 2023.
43. Dümichen, E., Eisentraut, P., Celina, M., Braun, U., Automated thermal extraction-desorption gas chromatography mass spectrometry: A multifunctional tool for comprehensive characterization of polymers and their degradation products. *J. Chromatogr. A*, 1592, 133–142, 2019.
44. Lien, Y.-H. and Wu, T.-M., Preparation and characterization of thermosensitive polymers grafted onto silica-coated iron oxide nanoparticles. *J. Colloid Interface Sci.*, 326, 2, 517–521, 2008.

45. Shao, Z.-B., Zhang, M.-X., Li, Y., Han, Y., Ren, L., Deng, C., A novel multi-functional polymeric curing agent: synthesis, characterization, and its epoxy resin with simultaneous excellent flame retardance and transparency. *Chem. Eng. J.*, 345, 471–482, 2018.
46. Alfei, S. and Catena, S., Synthesis and Characterization of Versatile Amphiphilic Dendrimers Peripherally Decorated with Positively Charged Amino Acids. *Polym. Int.*, 67, 11, 1572–1584, 2018.
47. Wu, S., Peng, S., Wang, C.H., Multifunctional Polymer Nanocomposites Reinforced by Aligned Carbon Nanomaterials, 2020, doi: 10.3390/polym10050542.
48. Henrique, A., Rafael, C., De Oliveira, S., Mhike, W., Thermal, Morphological and Mechanical Properties of Multifunctional Composites Based on Biodegradable Polymers/Bentonite Clay : A Review, 2023.
49. Antonio, M., Benega, G., Marcos, W., Cristina, M., Jorge, R., Andrade, E., Improvements in thermal and mechanical properties of composites based on epoxy-carbon nanomaterials-A brief landscape. *Polym. Test.*, 98, March 2021, doi: 10.1016/j.polymertesting.2021.107180.
50. Fang, H., Zhao, Y., Zhang, Y., Ren, Y., Bai, S., 3D Graphene Foam Filled Elastomer Composites with High Thermal and Three-Dimensional Graphene Foam-Filled Elastomer Composites with High Thermal and Mechanical Properties, July 2017, doi: 10.1021/acsami.7b07650.
51. Ahn, C. *et al.*, Multifunctional Polymer Nanocomposites Reinforced by 3D Continuous Ceramic Nanofillers, 2018, doi: 10.1021/acsnano.8b03264.
52. Bao, W.S., Meguid, S.A., Zhu, Z.H., Pan, Y., Weng, G.J., A novel approach to predict the electrical conductivity of multifunctional nanocomposites. *Mech. Mater.*, 46, 129–138, 2012, doi: 10.1016/j.mechmat.2011.12.006.
53. Version, D., *Multifunctional Polymer Materials : From Synthesis to Disinfection*, Department of Chemistry, Lund University, 2021.
54. Pattarakunnan, K., Galos, J., Das, R., Mouritz, A.P., Tensile properties of multifunctional composites embedded with lithium-ion polymer batteries. *Compos. Part A*, 136, 105966, April 2020, doi: 10.1016/j.compositesa.2020.105966.
55. Parola, S., Julián-lópez, B., Carlos, L.D., Sanchez, C., Optical Properties of Hybrid Organic-Inorganic Materials and their Applications, 2016, doi: 10.1002/adfm.201602730.
56. Latif, Z., Ali, M., Lee, E., Zubair, Z., Lee, K.H., Thermal and Mechanical Properties of Nano-Carbon-Reinforced Polymeric Nanocomposites : A Review. 1–35, 2023.
57. Aram, E. *et al.*, Smart and Multi-Functional Magnetic Nanoparticles for Cancer Treatment Applications : Clinical Challenges and Future Prospects, 2022.
58. Vidakis, N. *et al.*, Carbon nanotube 3D printed nanocomposites with enhanced mechanical and electrical properties. *Adv. Compos. Mater.*, 31, 6, 630–654, 2022, doi: 10.1080/09243046.2022.2076019.
59. Singh, R., Shitiz, K., Singh, A., Chitin and chitosan : biopolymers for wound management, 2017, doi: 10.1111/iwj.12797.
60. Stability of Chitosan—A Challenge for Pharmaceutical and Biomedical Applications. *Mar. Drugs*, 13, 1819–1846, 2015, doi: 10.3390/md13041819.
61. Anisha, A.D. and Shegokar, R., Expert Opinion on Drug Delivery Polyethylene glycol (PEG): a versatile polymer for pharmaceutical applications. *Expert Opin. Drug Deliv.*, 13, 9, 1257–1275, 2016, doi: 10.1080/17425247.2016.1182485.
62. Tang, Z. *et al.*, Phosphorus Science-Oriented Design and Synthesis of Multifunctional Nanomaterials for Biomedical Applications. *Matter*, 2, 2, 297–322, 2020, doi: 10.1016/j.matt.2019.12.007.
63. Huang, Y. *et al.*, Multifunctional Energy Storage and Conversion Devices, 2016, doi: 10.1002/adma.201601928.
64. ce pte d M us pt, 2018.

65. Karabasz, A., Bzowska, M., Szczepanowicz, K., Biomedical Applications of Multifunctional Polymeric Nanocarriers : A Review of Current Literature, 2020, doi: 10.2147/IJN.S231477.
66. Pinnaratip, R. *et al.*, Multifunctional Biomedical Adhesives, *Adv. Healthc. Mater.*, 8, 2019, doi: 10.1002/adhm.201801568.
67. De Crozals, G., Bonnet, R., Farre, C., Chaix, C., Nanoparticles with multiple properties for biomedical applications : A strategic guide. *Nano Today*, 11, 4, 435–463, 2016, doi: 10.1016/j.nantod.2016.07.002.
68. Vijayan, V.M., Beeran, A.E., Shenoy, S.J., Muthu, J., Thomas, V., New Magneto-Fluorescent Hybrid Polymer Nanogel for Theranostic Applications. *ACS Appl. Bio Mater.*, 2, 757–768, 2019, doi: 10.1021/acsabm.8b00616.
69. Wang, F., Xiao, J., Chen, S., Sun, H., Yang, B., Jiang, J., Polymer Vesicles : Modular Platforms for Cancer Theranostics. 1705674, 1–8, 2018, doi: 10.1002/adma.201705674.
70. Spychalska, K., Zaj, D., Baluta, S., Halicka, K., Cabaj, J., Functional Polymers Structures for (Bio) Sensing Application—A Review, 2020.
71. Sayin, S., Ozdemir, E., Acar, E., Ince, G.O., Multifunctional one-dimensional polymeric nano-structures for drug delivery and biosensor applications, 2019.
72. Eisa, M.H., Multifunctional Polymeric Nanocomposites for Sensing Applications—Design, Features, and Technical Advancements, 2023.
73. Barroso, A. and Reis, C., Nanomaterials in wound healing : From material sciences to wound healing applications. 443–460, July 2020, doi: 10.1002/nano.202000055.
74. Asadi, N. *et al.*, Multifunctional hydrogels for wound healing : Special focus on biomacromolecular based hydrogels. *Int. J. Biol. Macromol.*, 170, 728–750, 2021.
75. Ijaola, A.O., Akamo, D.O., Damiri, F., John, C., Polymeric biomaterials for wound healing applications : a comprehensive review. *J. Biomater. Sci., Polym. Ed.*, 1–53, 2022, doi: 10.1080/09205063.2022.2088528.
76. Maria, O., Keshari, S., Dash, M., Lupascu, F., New antimicrobial chitosan derivatives for wound dressing applications. *Carbohydr. Polym.*, 141, 28–40, 2016, doi: 10.1016/j.carbpol.2015.12.078.
77. Moeini, A., Pedram, P., Makvandi, P., Malinconico, M., Gomez, G., Wound healing and antimicrobial effect of active secondary metabolites in chitosan-based wound dressings : A review. *Carbohydr. Polym.*, 233, 115839, January 2020. doi: 10.1016/j.carbpol.2020.115839.
78. Matica, M.A., Aachmann, F.L., Tøndervik, A., Sletta, H., Ostafe, V., Chitosan as a Wound Dressing Starting Material : Antimicrobial Properties and Mode of Action., 1, 1–33, 2019.
79. Turner, P.R., Murray, E., Mcadam, C.J., Mcconnell, M.A., Cabral, J.D., Peptide Chitosan/Dextran Core/Shell Vascularized 3D Constructs for Wound Healing, 2020, doi: 10.1021/acsami.0c07212.
80. Gaspar-pintiliescu, A., Journal Pr proof, 2019.
81. Ndlovu, S.P., Ngece, K., Alven, S., Aderibigbe, B.A., Gelatin-Based Hybrid Scaffolds : Promising Wound Dressings, 2021.
82. Sorg, H., Tilkorn, D.J., Hager, S., Hauser, J., Mirastschijski, U., Skin wound healing: an update on the current knowledge and concepts. *Eur. Surg. Res.*, 58, 1–2, 81–94, 2017.
83. Tottoli, E.M., Dorati, R., Genta, I., Chiesa, E., Pisani, S., Conti, B., Skin wound healing process and new emerging technologies for skin wound care and regeneration. *Pharmaceutics*, 12, 8, 735, 2020.
84. Singh, S., Young, A., McNaught, C.-E., The physiology of wound healing. *Surgery*, 35, 9, 473–477, 2017.
85. Ribeiro, A.M. and Flores-Sahagun, T.H.S., Application of stimulus-sensitive polymers in wound healing formulation. *Int. J. Polym. Mater. Polym. Biomater.*, 1–11, 2019, doi: 10.1080/00914037.2019.1655744.

86. Singh, R., Bathaei, M.J., Istif, E., Beker, L., A Review of Bioresorbable Implantable Medical Devices : Materials, Fabrication, and Implementation, 2021, doi: 10.1002/adhm.202000790.
87. Cha, G.D., Kang, D., Lee, J., Kim, D., Bioresorbable Electronic Implants : History, Materials, Fabrication, Devices, and Clinical Applications. 1801660, 1–20, 2019, doi: 10.1002/adhm.201801660.
88. Loh, X.J., Utilising inorganic nanocarriers for gene delivery. 1–38.
89. Manuscript, A., ce Ac pte d M us pt, 2020.
90. Liao, T., Easton, C.D., Thissen, H., Tsai, W., Aminomalononitrile-Assisted Multifunctional Antibacterial Coatings, 2020, doi: 10.1021/acsbiomaterials.0c00148.
91. Ding, X., Wang, A., Tong, W., Xu, F., Biodegradable Antibacterial Polymeric Nanosystems : A New Hope to Cope with Multidrug-Resistant Bacteria. 1900999, 1–29, 2019, doi: 10.1002/smll.201900999.
92. Balla, E. *et al.*, Poly (lactic Acid): A Versatile Biobased Polymer for the Future with Multifunctional Properties—From Monomer Synthesis, Polymerization Techniques and Molecular Weight Increase to PLA Applications, 2021.
93. Ekladious, I., Colson, Y.L., Grinstaff, M.W., Polymer–drug conjugate therapeutics : advances, insights and prospects. *Nat. Rev. Drug Discov.*, 18, 273–294, April 2019, doi: 10.1038/s41573-018-0005-0.
94. Goyal, S., An update on the future prospects of glycerol polymers, 2021, doi: 10.1002/pi.6209.
95. Saima, A., Rasool, M.S., Umar, F.K., Mohd, R., Alam, M., Phytochemical based nanomedicines against cancer : current status and future prospects. *J. Drug Targeting*, 1–22, 2017, doi: 10.1080/1061186X.2017.1408115.
96. Ortiz, T., Mart, G., Merinero, M., Argüelles-arias, F., Alcudia, A., Polymeric Nanoparticles for Drug Delivery : Recent Developments and Future Prospects.

4

Graphene-Based Polymer Composites for Aerospace, Electronic, Energy, and Biomedical Applications

Asha Panghal[1], Yogendra Kumar[2*], Prashant Kumar Mishra[2], Aakash Mathur[2] and Amit Kumar Srivastava[2]

[1]*Department of Applied Sciences, Delhi Technical Campus, Greater Noida, Uttar Pradesh, India*
[2]*Department of Physics, School of Basic Sciences, Galgotias University, Greater Noida, Uttar Pradesh, India*

Abstract
Multifunctional graphene-based nanocomposites are one class of nanocomposite materials that has spurred significant technological advancements across an extensive spectrum of applications. When compared to neat polymers and other nanomaterials, the multifunctional nanocomposites-derived applications sectors are aerospace, automotive industry, electronics, sensors, and biomedical exhibit improved electrical conductivity, sensitivity, selectivity, durability, and responsiveness. Owing to the significance of nanocomposites in several technical fields, the overview is being broadened to include an emphasis on the classification of multifunctional nanocomposites and cutting-edge applications of graphene/polymer-based nanocomposites. This chapter also emphasizes the fundamentals and current advancements in the field of multifunctional nanocomposites. Graphene oxide-based nanocomposites have uses in diverse medical sciences domains, including drug delivery systems, tissue engineering, biosensing, and bioimaging. In addition to being employed in biomedical systems, graphene and its composites are widely used as functional materials featuring applications in electronics, aerospace, defense, and energy sectors. This chapter will be valuable to researchers investigating the applications of multifunctional composites in various disciplines for the benefit of humanity, as well as others interested in this new area of materials science.

Keywords: Nanocomposites, fiber-reinforced polymers, polymer matrix nanocomposites, ceramic matrix nanocomposites

4.1 Introduction

Multifunctional nanocomposites are an emerging category of unique materials developed by mixing one or more dissimilar materials at the nanoscale using advanced micro/nano-processing processes to develop tunable structural functionality for enhancing performance [1, 2]. Nanocomposite refers to a broad class of materials, which includes zero-dimensional core shells that represent various forms of nano-mixed and layered materials,

**Corresponding author*: yogendra.iitp@gmail.com

Divya Bajpai Tripathy, Anjali Gupta and Arvind Kumar Jain (eds.) Multifunctional Materials: Engineering and Biological Applications, (81–104) © 2025 Scrivener Publishing LLC

two-dimensional lamellar composites, single-dimensional nanowires, and three-dimensional metal matrix composites (MMCs) [3–5]. The main goal for developing a nanocomposite is to combine one or more discontinuous nano-dimensional phases into a single continuous microphase and/or nanophase to produce "synergistic properties," or properties that are fundamentally different from, and preferably better than, those of the constituent material elements. Multifunctional materials and nanocomposites have become essential in current research and industry, indicating a paradigm shift in material science and engineering [4, 6–8] These advanced materials provide a synergistic blend of features, leveraging the benefits of nanotechnology to produce versatile and tailored alternatives for a wide variety of applications. Figure 4.1 depicts the various applications of nanocomposite material.

Blumstein published the first literature report on nanocomposites in 1961 [9]. His further research in 1965 also showed that polymethyl methacrylate (PMMA) layered silicate nanocomposites exhibited better thermal stability [10]. When comparing nanocomposites to commonly used polymers or pure polymer constituents, the nanoscale as well size dispersions result in superior properties. Lightweight, high modulus and strength, reduced gas permeability, enhanced solvent resistance, and enhanced thermal stability are the key benefits of multifunctional nanocomposites [11–13]. The two-dimensional reinforcement from the inorganic layers offers them improved durability than unidirectional fiber-reinforced polymers [14].

It is expected that the reducing nanomaterial costs, rising production volumes, and the resolution of technical challenges pertaining to their mass-scale synthesis will all contribute to the growth of multifunctional nanocomposites. Nanocomposites are expected to achieve a market value of US$9 billion with an amount of 5 million tons by 2025 [4]. The adaptable functionality of multifunctional materials enables focused applications in domains ranging

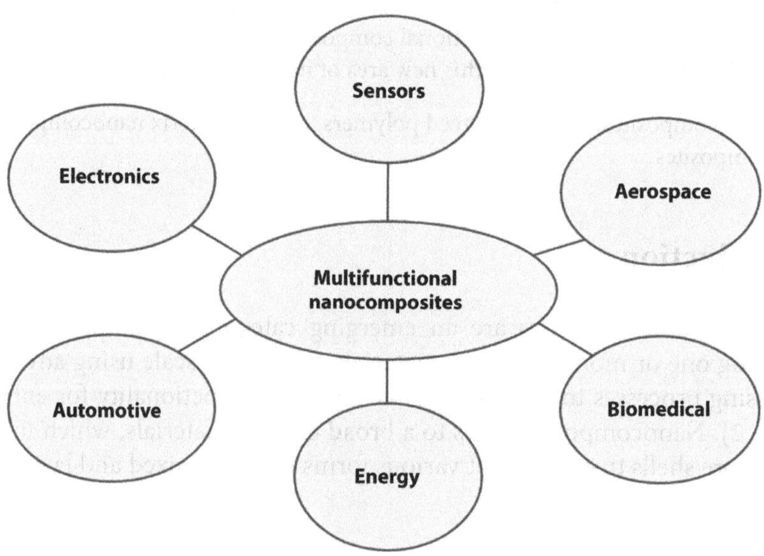

Figure 4.1 Schematic of the different applications of the multifunctional nanocomposites [5].

from electronics, energy, sensors, biomedical, aerospace, automotive, and environmental remediation [15–22].

4.2 Fundamentals of Multifunctional Composites/Nanocomposites

Nanocomposites are categorized into different categories based on the type of matrix material used. Figure 4.2 depicts the classifications of various types of nanocomposites.

4.2.1 Polymer Matrix Nanocomposites (PMNCs)

The polymer and nano-additives serve as matrix and reinforcing constituents in the polymer matrix nanocomposites (PMNCs). Polymer nanocomposites (PNCs) can have different characteristics depending on the type of reinforcing material used. PNCs possess multifunctional, high-performance polymer properties that go beyond those of conventionally filled polymeric materials as a result of the addition of either organic or inorganic nanofillers to polymer systems [23]. PNCs offer significant advantages such as enhanced stiffness, improved barrier effect, thermal conductivity, enhanced thermal and dimensional stability, and increased resistance to fire [24–26]. Moreover, PNC properties depend on features of individual components and various parameters such as the procedure employed in nanocomposite development, the extent of blending/interface of two phases, the dimension, characteristics of nanoparticles, and the nature of the architecture [23, 27]. Due to their exceptional qualities and potential uses, PNCs are used in a variety of cutting-edge industries, including the electrical, medicinal, catalytic, sensor, and environmental domains [16, 27, 28]. PMNCs include layered silicates, polymer/CNT, polyester/metal oxide, polymer/layered double hydroxides, and thermoplastic/thermoset polymers [23].

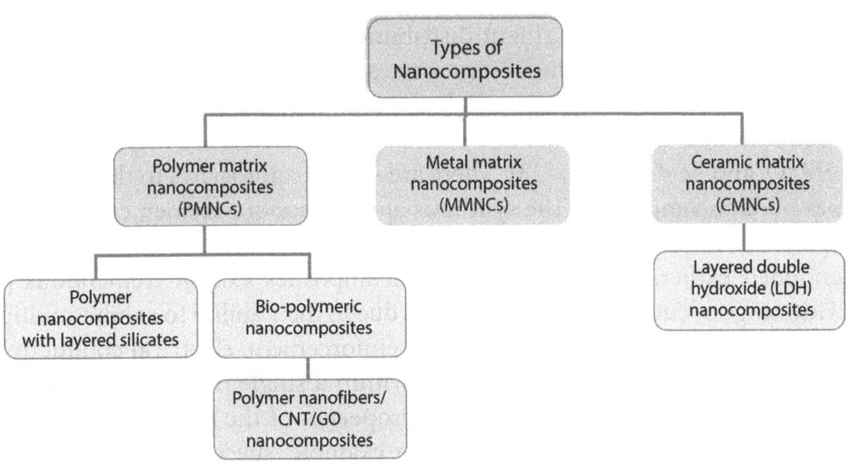

Figure 4.2 Illustrates the broad classification of nanocomposites [23, 30].

4.2.2 Ceramic Matrix Nanocomposites (CMNCs)

CMNCs have one or more phases in the nanometer range and are regarded as an emerging class of engineering materials with major possibilities for industrial applications. In comparison to conventional ceramic materials, CMNCs are capable of significant improvements in strength, toughness, and thermal stability through the integration of nanoscale reinforcements into ceramic matrices. The purpose of effectively reinforcing ceramics is to include components that can arrest the fractures, hence enhancing the resistance to crack development [29].

The continuous fibers are found to stop cracks by causing deflection at fiber/matrix contacts. SiC-based ceramic matrix composites are found in robust ceramics when fiber/matrix linking is effectively tuned by using a thin interphase [31]. SiC-matrix composites are extremely customizable materials in terms of a simple matrix, coatings, interphase ((BN-SiC), and fiber-type (Si-C-O). The SiC-matrix composites possess potential applications in an inner wall of the plasma chamber of the nuclear fusion reaction, gas turbines of electrical power/steam cogeneration, and advanced aerojet engines [31]. CMNCs include carbides, nitrides, and oxides with metallic or nonmetallic elements [32].

4.2.3 Metal Matrix Nanocomposites (MMNCs)

MMNCs have recently gained a lot of attention in different fields like automotive, aerospace, and biomedical engineering due to their unique functional and structural alteration [3]. Moreover, MMNCs possess high specific strength and stiffness, can withstand high temperatures, and are resistant to wear and corrosion. MMNCs have proven essential in the tribology sector owing to their exceptional mechanical characteristics and appropriate tribological performance [33]. MMNCs are made up of ductile metal/alloy matrix having nanosized reinforcing material integrated [23]. MMNCs have an obvious edge over metal matrix micro-composites (MMMCs) in numerous applications because of their superior combination of properties [3].

The processing techniques and the processing parameters play an imperative function in improving the characteristics of MMCs. Indeed, both processing techniques and parameters offer homogeneous reinforcement distribution and a durable matrix-reinforcement interface. When using mass production processes like casting, it can be difficult to incorporate reinforcement into the molten metal on an industrial scale without introducing it to a large amount of heat treatment. A few examples of MMNCs are Mg/CNT, Al/CNT, Ni/Al_2O_3, Fe/MgO, and Fe–Cr/Al_2O_3 [23]. There are several approaches to the synthesis of different types of nanocomposites. The synthesis approaches of polymer, ceramic, and metal nanocomposites are illustrated in Figure 4.3.

As mentioned earlier, multifunctional nanocomposites exhibit tremendous potential across a wide range of cutting-edge applications due to their ability to exhibit multiple functionalities simultaneously, such as mechanical reinforcement, electrical conductivity, thermal insulation, and even sensing capabilities, within a single material system [32, 33, 35, 36]. An important factor in defining the final properties of the nanocomposite is its form, size, aspect ratio, and surface chemistry. As an example, specific shapes such as nanowires or nanotubes can permit increased reinforcing and heat conductivity, whereas smaller nanoparticles often confer improved mechanical strength and electrical conductivity.

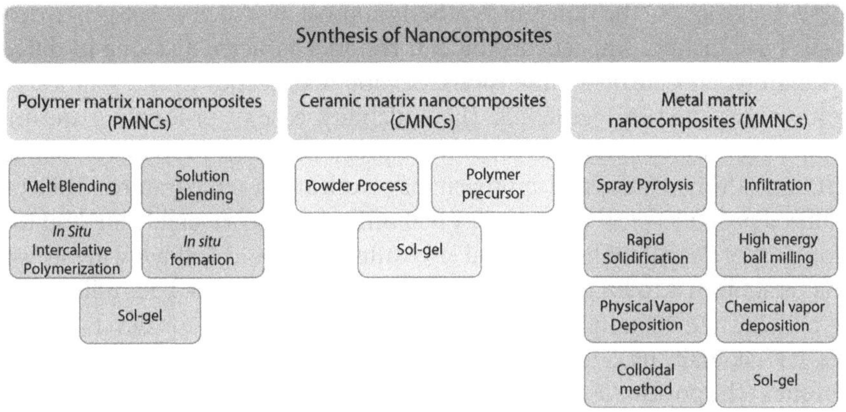

Figure 4.3 Fabrication approaches of nanocomposite [34].

Furthermore, the optimization of features like strength, toughness, and durability depends on the homogeneous dispersion and interfacial bonding of nanomaterials within the matrix. The dimension of the nanofiller/matrix interface, the aspect proportion of the nanofiller, and the percolation threshold are the primary drivers of the characteristics of PNCs [37].

Several other parameters are also crucial for the properties of multifunctional nanocomposites: (i) reaction temperature and time, (ii) annealing temperature, (iii) fabrication techniques (iv) pH values, and (v) volume fraction [37].

(i) Reaction Temperature and Time: The reaction temperature and time continue to be essential regulating factors for controlling the properties of nanocomposites. The enrichment in the reaction temperature accelerates breakdown, releases more free radicals, and promotes monomer diffusion, increasing the graft percentage [38]. The reaction time also affects the result, e.g., the grafting percentage decreases as the reaction time surpasses the optimum time.
(ii) Annealing Temperature: The annealing temperature is widely recognized as another significant factor that influences the characteristics of nanocomposites during the fabrication process. Samoei *et al.* explored the effect of thermal annealing on graphene/PVDF nanocomposites [38]. They recorded the impact of annealing on electrical conductivity at different temperatures within the range of 100–300°C. The electrical conductivity of the graphene/PVDF nanocomposites increased as the annealing temperature was raised to 150°C and thereafter approached saturation. Bignoli *et al.* fabricated the amorphous films of ZrCu/O nanocomposite and investigated the role of annealing on thermal stability and mechanical properties [39]. The ZrCu/O nanocomposite amorphous films exhibited noteworthy alterations of the mechanical properties and thermal stability, i.e., the nanocomposite films show superior thermal stability with crystallization phenomena upon annealing. Annealing enhances the interaction of the film architecture (nano-clustering/interfaces) which leads to unique mechanical properties. Moreover, the annealing can impact both the electrical and thermal conductivity of nanocomposites by modifying the configuration of nanostructures throughout the matrix. Increased annealing temperatures could enhance nanoparticle coordination, resulting in the alteration of the electrical conductivity of nanocomposites [40, 41].

(iii) Synthesis Techniques: The different synthesis approaches, such as sol-gel, hydrothermal, and aqueous, have distinct impacts on the end result's characteristics due to differences in shape, crystallinity, and particle dimensions. Several approaches to the synthesis of nanocomposites are presented in Figure 4.3. The scalability, ease of processing, and multifunctionality of sol-gel synthesis have led to its widespread adoption [42]. In comparison to the sol-gel technique, the hydrothermal approach may work at low temperatures while producing well-crystallized products [43]. High-energy ball milling is also a suitable method to fabricate the nanocomposites. A ball milling method also called mechanical alloying involves a powder mixture that is subjected to high-energy impacts with the balls. The ball milling/mechanical alloying is a facile, and economical top-down synthesis approach for developing the nanocomposites. It is generally used to synthesize the alloyed, metallic, multi-metallic, and ceramic nanocomposites. The positive aspect of mechanochemical alloying is the uniform distribution of particles in the matrix, which minimizes agglomeration [44, 45]. Zeng *et al.* fabricated a highly transparent nanocomposite film of ZnO/(PMMA-PS) with novel UV-shielding properties employing the solution mixing method [46]. In comparison to the film produced using the solution mixing method, the PMMA-PS film produced employing the slurry dispersion approach showed lower transparency and UV shielding properties. Figure 4.4 depicts the fabrication of the transparent hybrid film using a two-step solution mixing process.

Fathy *et al.* [45] explored the characteristics of ball-milled ZrO_2 nanostructures in the Cu matrix. They found that the microhardness and compressive strength enriched with the reduction in crystalline size. Similarly, Zawrah *et al.* [47] fabricated the embedded Cu

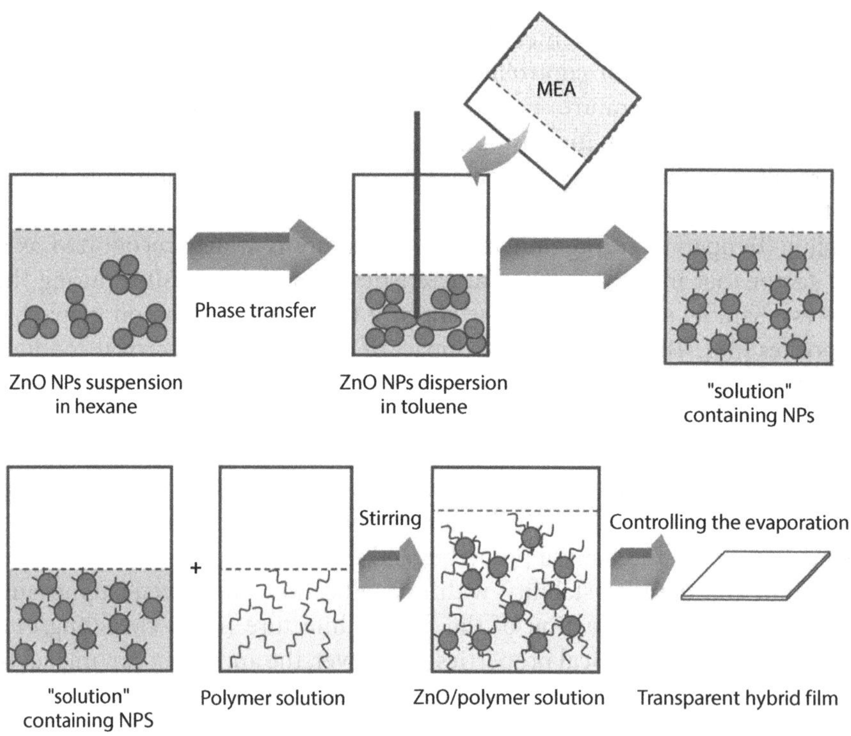

Figure 4.4 Synthesis of transparent hybrid film using the solution mixing method. Reproduced with permission [46]. Copyright 2011, ACS.

particles onto Al_2O_3 nanocomposite. The homogeneous distribution of Cu over Al_2O_3 exhibited a higher fracture strength. These investigations demonstrate that nanocomposites with varying chemical and physical properties can be effectively manufactured by employing the ball milling approach. Generally, melt blending is employed for preparing the polymer-based nanocomposites [34]. Melt blending is conducted in an inert atmosphere, i.e., in the presence of inert gases. Melt blending offers significant benefits over polymer solution intercalation or *in situ* intercalative polymerization. Since melt blending excludes organic solvents, it is considered safe for the environment. The procedure of melt blending has gained popularity because of its potential in industrial applications. Zeng *et al.* fabricated a highly transparent nanocomposite film of ZnO/(PMMA-PS) with novel UV-shielding properties employing the solution mixing method [46]. In comparison to the film produced using the solution mixing method, the PMMA-PS film produced employing the slurry dispersion approach showed lower transparency and UV shielding properties.

(iv) pH Value: It is noteworthy that the characteristics of fabricated films are altered with the function of pH range [48]. Therefore, the thickness, morphology, and optical characteristics of the produced thin films varied accordingly. Yang *et al.* [49] fabricated biodegradable, biocompatible, and non-toxic, pH-responsive cellulose-chitosan nanocomposite films. The nanocomposite films exhibited the high swelling as a function of pH. Yu and his co-workers [50] found that the Ag/TiO_2 composite thin film formation mechanisms might be accounted qualitatively by the pH. The influence of pH on the efficiency of thin film composites (TFCs) nanofiltration (NF) membranes was examined at various pH scales by Dalwani *et al.* [51]. This work demonstrated that the membrane porosity decreased under alkaline conditions. Azizun *et al.* prepared the bio-nanocomposite films using a casting technique [52]. The pH treatment has significant impacts on the antibacterial and gas-sensing characteristics of bio-nanocomposite films.

(v) Volume Fraction: The Al matrix composites reinforced with three distinct volume fractions of SiC nanoparticles have been produced by Iqbal *et al.* [53] employing the powder metallurgy process. It has been found that the properties of the low-volume fraction reinforced nanocomposite are not very desirable. Conversely, the high-volume fraction reinforced nanocomposite displayed remarkable hardness and superior microstructural characteristics [53]. As per Singh *et al.*, the volume fractions of Al_2O_3 nanoparticles in 6061-T6 aluminum alloy nanocomposites alter the microstructural, mechanical, and tribological properties [54]. The wear resistance and tensile strength of the nanocomposite matrix were significantly enhanced with the increase in the amount of Al_2O_3 nanoparticles. The incorporation of metallic particles into a polymer has the potential to improve electrical and thermal conductivity. The thermal conductivity performance of polymers modified through the incorporation of metallic particles depends on the metallic filler's thermal conductivity, the size, shape, and volume fraction of the particles [55].

4.3 Advancements and Current Research in Multifunctional Nanocomposites

The last few decades have seen a notable spike in multifunctional composite materials research activities. In terms of system efficiency, cost-effectiveness, and miniaturization, the use of multifunctional composite materials offers notable benefits, which makes them extremely important for a variety of industrial applications. In addition to load-bearing characteristics like toughness, strength, and stiffness, these materials are also intended for non-structural applications including actuation sensing, and self-healing [56].

Carbon fiber-reinforced polymer (CFRP) is highly demanded in automotive, aerospace, and various other applications in engineering because of its excellent unique stiffness and strength, lightweight, resistance to corrosion, and long-term reliability [57]. These composites are frequently incorporated with multiscale particles such as Al_2O_3, SiO_2, graphene, and carbon nanotubes (CNTs) to address issues with fiber-reinforced composites, such as their low off-axis behavior and inadequate electrical and thermal conductivity [1]. CFRP composites have gained popularity as being suitable for electromagnetic interference shielding (EMI) applications. Recently notable investigations demonstrate that polymer-based conductive composites are an attractive replacement to conventional metal-based shielding materials because of their distinct benefits, notably excellent unique strength, lightweight, stable chemical characteristics, and ease of development [58, 59]. Feni *et al.* hybrid biocomposites based on CFRP demonstrated exceptional EMI shielding performance. The shielding efficacy values of over 15 dB were achieved [1].

Functional nanocomposites contributed to the development of smaller and better-performing energy storage devices [17]. Si-based nanocomposite materials have shown potential in the fabrication of lithium-ion batteries [61]. The chemicals released from agriculture/industries and human activities contaminate water resources. Cantarella *et al.* [62] demonstrated that ZnO-MoS_2-PMMA polymeric nanocomposites synthesized employing the facile and economical solution casting approach are suitable for water purification. Cantarella and co-workers have used ZnO nanoparticles as photoactive, MoS_2 nanoflakes as co-catalyst, and poly (methyl methacrylate) (PMMA) as supporting matrix [62]. Recently, Carneiro *et al.* [60] examined the thermal properties of aluminum matrix nanocomposites

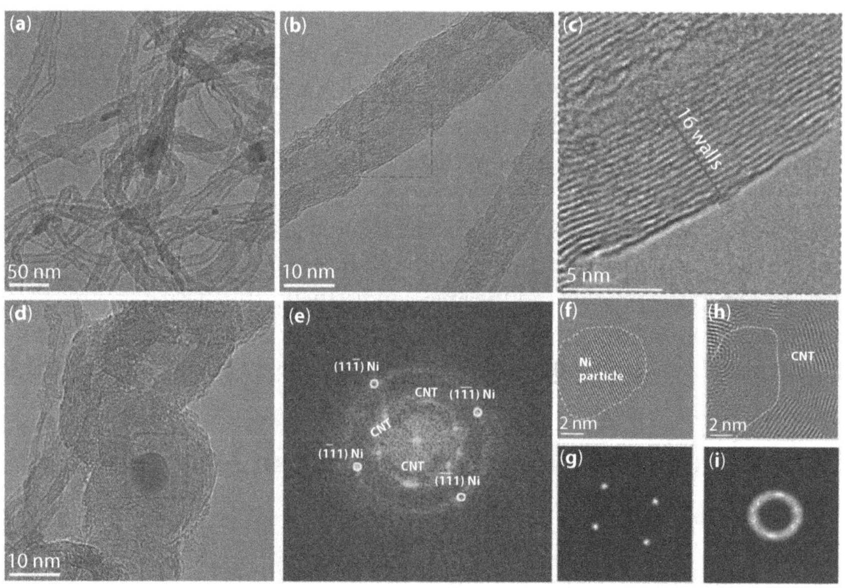

Figure 4.5 (a) TEM image, (b) HRTEM image of CNTs, (c) magnified view of (b), (d) HRTEM image of a CNT with a particle inside (d), (e) FFT, and (f to i) confirmation of the existence of the Ni particle and the CNT walls. Reproduced with permission [60]. Copyright 2023, Elsevier.

reinforced with multi-walled CNTs (MWCNTs). Figure 4.5(a) illustrates the image of transmission electron microscopy (TEM). The high-resolution TEM (HRTEM) image of functionalized CNTs is shown in Figure 4.5(b). Furthermore, the HRTEM image of CNT with a particle inside is shown in Figure 4.5(d). The outer diameter and number of walls of the prepared CNTs are evident. Functionalized nanocomposite nanotubes showed enhanced thermal stability and mechanical properties at high temperatures.

The optical characteristics of polymer composites, such as extinction coefficient (k) and refractive index (n) are essential for their real-world application including transistors, sensors, solar cells, and so on [63]. However, the optical properties of the polymer composites suffer from a high degree of variability. This insufficiency is ascribed to variations in the manufacturing or deposition factors, including substrate material, layer thickness, surface characteristics, temperature and pressure, and deposition rate [63]. Davoise and co-workers [63] prepared the PANI/HDI-GO nanocomposite film *via in situ* polymerization. It has been found that the optical characteristics of the PANI/HDI-GO nanocomposite film are highly intriguing for possible future applications in a wide range of fields, including photovoltaic cells and optoelectronic devices. One notable advancement of the integration of single-atomic-layered material, metallic nanoparticles, and polysulfone nanocomposites with exceptional antibacterial capability has shown promise for medical uses [64].

Dye-sensitized solar cells (DSSCs) are an aspiring renewable energy technology that harvests sunlight and generates power without emitting hazardous gasses. They are also noted for their decorative features, such as color tuning and transparency. However, a trade-off between power conversion efficiency (PCE) and transparency ought to be addressed securely to produce solar cells with transparency and high PCE. The nitrogen-doped reduced graphene oxide-$Sr_{0.7}Sm_{0.3}Fe_{0.4}Co_{0.6}O_{2.65}$ nanocomposites (N-RGO-SSFC) based DSSCs exhibited a PCE of 6.64%. These results suggest that the N-RGO-SSFC nanocomposites can be used as an applicable electrode material in DSSCs [65]. With a high content of nano-reinforcements achieving a uniform dispersion and high efficiency in MMNCs is still one of the main challenges [60]. To address the issue, flake dispersion is introduced as an assembly process that integrates flake powder metallurgy with significant-shear dispersion.

4.4 Applications of Multifunctional Composites/Nanocomposites

(i) Aerospace and Defense:

The extreme temperatures, vacuum, micrometeoroids, space debris, and significant changes brought on by sunspot activity are characteristics of the space environment. Materials with a lower erosion yield factor and the ability to tolerate temperature fluctuations of hundreds of degrees are deemed suitable for use in aerospace applications [19]. Polymer composites and nanocomposites have shown possibilities in the technological and industrial realms associated with aerospace [66]. Fiber-based composite structures were initially employed in the aerospace sector [18]. Furthermore, compared to heavy metal space structures, implementing lightweight nanocomposites in aerospace vehicles delivers the advantage of reducing fuel consumption and improving performance. Carbon-based nanomaterials such as graphene, CNTs, and nano-diamond nanofillers have been employed to develop nanocomposite materials for aerospace engineering and defense technology [8]. Polymer/graphene nanocomposites offer superior frivolity, resilience, processability, mechanical solidity, and thermal and electrical conductivity [67]. The unusual properties such as mechanical, thermal, and electrical conductivity of graphene and polymer/graphene nanocomposites have contributed to an exploration of possible aeronautical uses for these materials [68, 69]. To develop graphene-derived nanocomposites for aeronautical structures, a variety of thermosetting and thermoplastic polymers have been employed [70]. Epoxy resins classified as thermosetting polymers are widely employed in aerospace-related applications [71]. The fundamental conductive, thermal, and mechanical characteristics of epoxy matrices have been improved by reinforcing graphene and similar nanofillers [18]. Shadlou and coworkers [72] investigated epoxy/graphene nanocomposites for aeronautical applications that showed high tensile strength and Young's modulus. The nanofiller's superior dispersion and compatibility with the epoxy resin contributed to augmented mechanical performance. Bustero *et al.* [73] developed epoxy and graphene-nanoplatelet-derived nanocomposites for advanced aeronautical uses. Thermogravimetric analysis was employed to examine the epoxy and graphene nanoplatelet nanocomposite's thermal stability. The thermal stability of the epoxy resins improved through the incorporation of graphene nanoplatelets. The reinforcing effect of the graphene nanofiller was responsible for the aerospace matrix's enhanced thermal stability.

Yousefi *et al.* [75] prepared epoxy/reduced graphene oxide (rGO) nanocomposites for aeronautical applications. The prepared epoxy/rGo nanocomposites exhibited improved dielectric properties. With an exceptional 38 dB shielding efficiency, the highly dielectric rGO/epoxy nanocomposites with their tailored structure and characteristics offer high-performance EMI [75]. In another work, Yousefi *et al.* [74] fabricated the waterborne rGO/epoxy nanocomposite *via in situ* polymerization. Figure 4.6 illustrates the setup required to fabricate the rGo and uniformly dispersed nanocomposite. Effective interfacial interactions

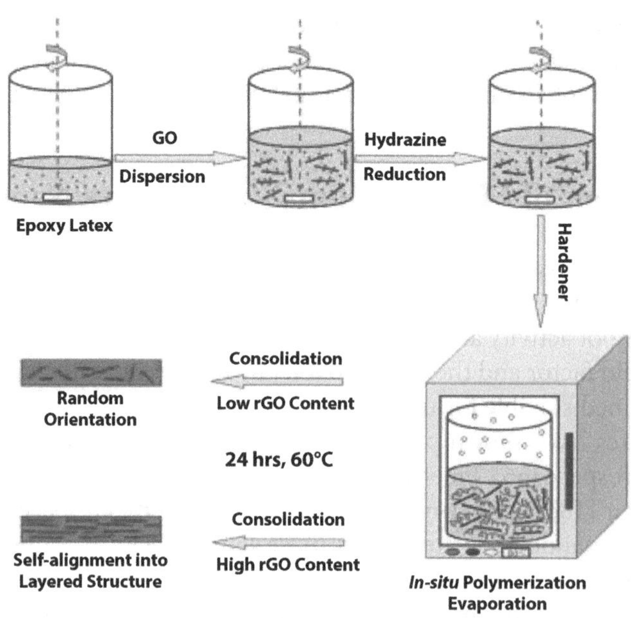

Figure 4.6 Schematic of nanocomposite preparation. Reproduced with permission [74]. Copyright 2013, Elsevier.

between the epoxy and rGO allowed for significant improvements in tensile strength and modulus of rGO/epoxy composites, at 1.5 weight percent rGO. CFRPs are used extensively in the aerospace industry [76]. The mechanical characteristics of carbon fibers have been regarded as crucial factors for the excellent performance of CFRP. Furthermore, carbon fiber's high specific surface area, chemical inertness, and interfacial strength are critical for improving CFRP properties [76].

Carbon fibers coated with graphene and composites based on TiAl alloys were developed by Cui *et al.* [77]. The aerospace-grade composites were produced *via* vacuum melting, melt spinning, and powder metallurgy. The incorporation of graphene-altered carbon fiber into the TiAl alloy matrix increased the strength from 1801 to 2312 MPa. Particularly in radome applications, where low dielectric abilities are essential to preserving radar transparency, moisture in composite systems can deteriorate mechanical and dielectric capabilities. Moisture absorption can be minimized by adding nanoclay, which could be an efficient method for enhancing structural and dielectric performance. The functioning of the epoxy matrix in aircraft radomes is enhanced upon the addendum of nanoclay particles [78]. It has been revealed that high-performance polymer/graphene nanocomposites are useful for both the creation of aerospace structural components and space defense technology.

(ii) Automotive Industry:

The demand for novel lightweight materials that are both affordable and highly effective is being driven by the current global expectations for low emissions and fuel efficiency in manufacturing as well as transportation. With their exceptional mechanical, thermal, and processing qualities, nanocomposites are a revolutionary class of materials that find application in the automotive industry [79]. It is believed that the use of nanocomposites in automotive systems as well as parts will boost manufacturing productivity, enhance environmental and thermal stability, stimulate recycling, and minimize weight [80]. Materials based on nanocomposite offer strength, stiffness, and dependability that are on par with or superior to those of metals. They offer higher modulus, thermal stability, dimensional stability, noise reduction, and corrosion resistance. For outdoor automotive applications, General Motors R&D and Montell USA have created thermoplastic olefin (TPO) clay nanocomposites. It exhibited the lightness and good dimensional stability [80]. For all of these factors, the upcoming generation of automobiles needs to be lighter, cleaner, and more fuel-efficient. Uses for automotive parts effective integration of graphene as polymer reinforcements in composite materials can be advantageous for energy-efficient and safe vehicles (EESVs). The aforementioned satisfies the growing requirement in the automotive industry for lightweight, safe, energy-efficient, and energy-efficient vehicles (EESVs) to lower CO_2 emissions and increase fuel efficiency [81]. It has been suggested that weight reduction of up to 50% is possible when graphene reinforcement is incorporated into composite materials employed for auto-structural applications [81].

Moreover, natural fiber composites that are strengthened have been commonly utilized in the automotive industry [1]. Scientists and engineers working today are finding new ways to purposefully produce materials at the nanoscale and take advantage of their enhanced properties [1]. Due to environmental concerns, it is thought to be crucial to use traditional fiber-reinforced plastics like glass, carbon, and aramid-reinforced thermoplastic material and thermoset resins. However, compared to synthetic fibers, naturally occurring fibers might also have positive environmental effects. There has been increased interest in examining the appropriateness of composites made of graphene as a filler material in automotive applications where robustness, reduced cost, and sustainable characteristics are required [82]. A manufacturer considers the following factors such as ultimate breaking force and elongation, dimensional stability, water absorption, impact resistance, acoustic absorption, temperature, and dwell time appropriateness for manufacturing [83]. The reduction in weight, or reducing the weight of the vehicle, is one technique to improve automotive performance. It has been proved that the remarkable property of graphene, its flexibility, is essential for designing technological equipment that can fit over curved surfaces [83]. Recently, Kiziltas *et al.* prepared the graphene nanoplatelets (GNP)/polyamide matrix nanocomposites and explored the thermal and mechanical properties of automotive applications [84]. GNP/polyamide matrix (PA610) nanocomposite demonstrated the enrichment in thermal stability and mechanical properties. According to this study, GNP fillers may be utilized in enhancing PA610's mechanical and thermal characteristics for use in automotive applications [84].

(iii) Electronics and Sensors:

There has been a notable surge in the quantity and variety of flexible electronics manufactured over the last ten years. It has led to a demand for flexible materials with exceptional performance. Graphene and its derivatives coated in polymers are used in the production of many different electronic products, including photovoltaic cells, photodetectors, sensors, energy storage and conversion, and biological applications [85, 86]. In accordance with a range of studies, the synergistic effects between the graphene-PNCs allow them to perform more effectively than the individual components [86, 87]. The poly (3,4-ethylenedioxythiophene): polystyrene sulfonate (PEDOT:PSS) is a highly conjugated polymer. It is regarded as an exceptional material for sensor devices, thermoelectric devices, flexible devices, DSSCs, OPV devices, and supercapacitors due to its many superior properties, including high work function, environmental stability, and solution processability [88]. However, there are still substantial obstacles preventing its employment in a variety of real-world applications due to its restricted structural and chemical properties [89]. Combining it with revolutionary carbon nanostructures is potentially an approach to addressing the limitations. The Au-nanoparticle incorporated PEDOT: PSS/rGO nanocomposite demonstrated a PCE of 9.34% more than that of PEDOT: PSS/rGO [6]. Recently, Kim and his coworkers developed a metal-free PEDOT: PSS/graphene nanocomposite which could potentially be employed as a counter electrode in DSSCs. They reported a photovoltaic performance of 8.33% for their nanocomposite, which equates to a better material than the platinum counter electrode, which exhibited a performance of 7.99% [90]. In contrast to the uses mentioned above, sensor devices have also made use of PEDOT: PSS/graphene nanocomposites. With the advent of printed electronics, inexpensive, flexible electronic devices made of organic materials can now be used in place of more conventional, inflexible ones based on silicon (Si). To construct the bendable and flexible temperature sensors on surfaces that can adapt to the curvature of the body; Soni *et al.* manufactured conductive and sensitive electrodes using PEDOT: PSS/GO nanocomposite [91].

The fabricated electrode demonstrated sensitivity for temperatures ranging from ambient temperature to 100°C. Seekaew *et al.* fabricated the low-cost and flexible graphene/PEDOT: PSS gas sensor using ink-jet printing [89]. Figure 4.7 exhibits the schematic diagram of the

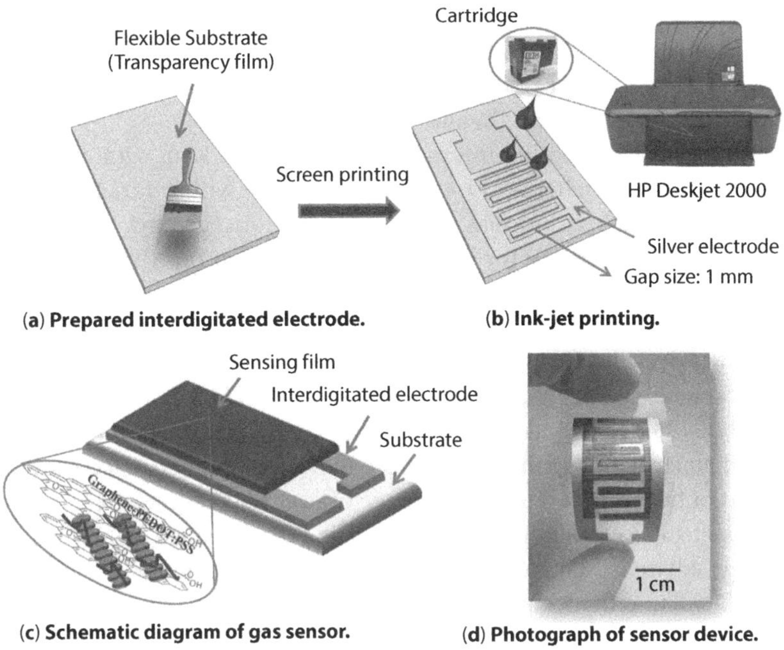

Figure 4.7 Schematic diagram of the structure process of the gas sensor device. Reproduced with permission [89]. Copyright 2014, Elsevier.

fabrication process of the gas sensor device. Interdigitated electrodes with a 1 mm interdigit spacing were screen printed with silver conductive paste over a flexible, transparent substrate to fabricate flexible gas sensors. The flexible graphene/PEDOT: PSS gas sensor has an excellent response and selectivity for NH_3 concentrations ranging from 25 to 1000 ppm at ambient temperature.

(iv) Energy Industry:

Since its first isolation from graphite in 2004 [37], graphene, a monolayer carbon sheet made of sp^2-hybridized carbon atoms has attracted a lot of attention. Its distinct electronic structure gives it extraordinary properties. Graphene, one of the strongest materials, demonstrates ultrahigh electrical conductivity and carrier mobility reaching up to 15,000 $cm^2/(Vs)$ at room temperature [17]. Graphene's high electrical conductivity makes it a great option for enhancing graphene-based composites conductivity in energy-storage devices. The energy-storage performance of numerous composites has significantly improved with the addition of graphene. Moreover, graphene has exceptional optical characteristics, including a transparency of approximately 97% across the visible range and an opacity of 2.3% over a broad wavelength range [92]. Graphene's optical properties render it appropriate for the creation of transparent energy-storage devices. Apart from its optical and electrical characteristics, graphene has remarkable thermal conductivity, reaching up to 5300 W m^{-1} K^{-1}. It has been demonstrated that the addendum of graphene to composite materials improves thermal conductivity, which helps to solve the problem of inconsistent heat dissipation under high current loads [17, 92]. Graphene also exhibits superior mechanical properties and its mechanical characteristics make it a promising material for wearable and flexible energy-storage devices. It has been claimed that the unique graphene layers are capable of functioning as stable interfacial layers to stabilize pure Li metal anodes due to graphene's unusual stability, mechanical strength, and flexibility [17].

Additionally, physical adsorption interactions like π-π stacking with other materials are made possible by graphene's chemical inertness. Graphene with a variety of properties can be created by adding defects, additional atoms, and functional groups. Moreover, graphene's substantial surface area is essential for energy-storage applications [92]. Graphene enriched with functional groups such as boron (B), nitrogen (N), sulfur (S), oxygen (O), and Si functioned more effectively as supercapacitor electrodes [17].

A graphene-based electrode developed from electrochemically reduced GO demonstrated a capacitance of approximately 165 F g^{-1} and a durability of over 1000 cycles in an aqueous electrolyte [93]. N-doped graphene displays a high specific capacitance of 151 F g^{-1} in aqueous KOH electrolyte. Furthermore, these N-doped graphene electrodes have a high rate capacity and outstanding cycle performance of over 5000 cycles [94].

Graphene-based supercapacitors frequently incorporate hybrid structures with electrically conducting polymers such as poly(thiophene), polypyrrole (PPy), (PTh)polyaniline (PANi), and PEDOT to enhance performance. The large energy storage capacity, superior electrical conductivity, affordability, and environmental stability of conductive polymers make them promising [95]. Although graphene-based functional composites show great promise for energy-storage devices, several issues must be resolved before they can be used in various fields. First and foremost, to increase the number of applications for graphene in energy-storage devices, it is imperative to investigate more effective methods for the large-scale synthesis of excellent graphene and its derivatives [17]. Second, the specific molecular interactions and working mechanisms of each component in graphene-based

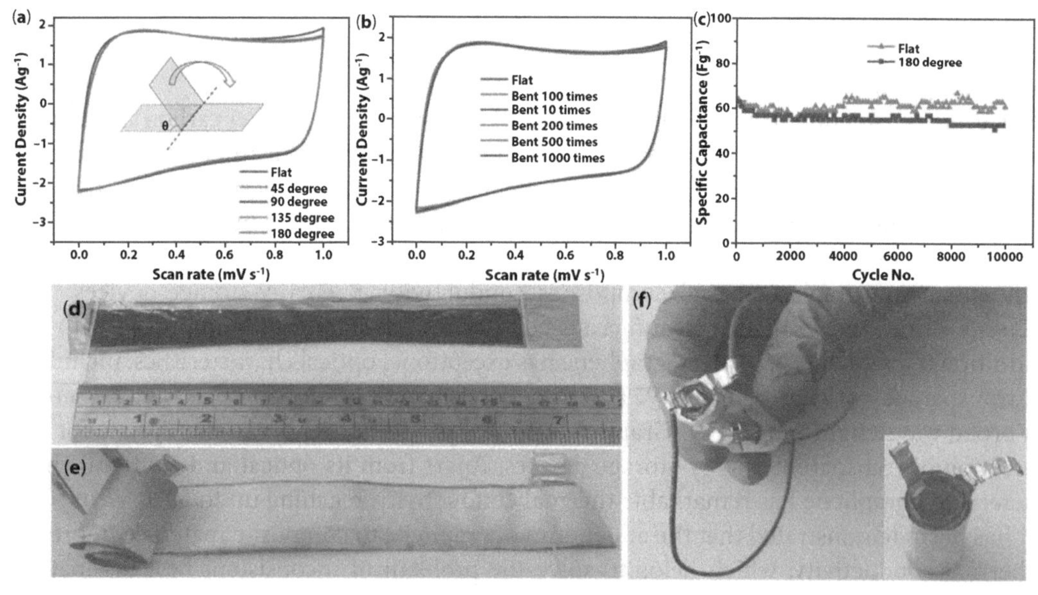

Figure 4.8 (a and b) CVs of rGO-PEDOT/PSS, (c) stability test and (d) electrolyte-coated flexible films, (e) rolled design, and (f) fabricated device. Reproduced with permission [99]. Copyright 2015, Springer Nature.

composites with enhanced electrochemical performances are still unknown despite a plethora of reports on the subject. Thirdly, to further improve the efficiency of energy-storage devices, multifunctional morphologies, and optimization of the ratios of graphene to other active materials are required [92]. The *in situ* oxidative polymerization of aniline in the presence of functionalized graphene yields PANi/graphene nanocomposites, which exhibit a large specific capacitance of 863.6 F g^{-1} and exceptional rates efficiency of 581.6 F g^{-1} [96]. In another study, the graphene/PANi composite that had been produced *via in situ* polymerization showed a remarkable specific capacitance of 1046 F g^{-1}, which was higher than the value of 115 F g^{-1} for pure PANi [97]. A specific capacitance and capacitance retention of 1126 F g^{-1} and 84% after 1000 cycles have been achieved by employing the flexible graphene/PANi nanocomposites [98]. Liu *et al.* employed an easy bar-coating approach to develop very flexible, bendable, and conductive rGO-PEDOT/PSS films [99].

The rGO-PEDOT/PSS electrode-based developed device may be bent and rolled up without compromising its electrochemical performance. The charge/discharge curves and cyclic voltammetry (CV) demonstrate that no significant shift occurred while bending at different angles [Figures 4.8(a and b)]. The long-term galvanostatic testing showed that, after 10,000 cycles, the capacitance evaluated at 180° only reduced by 11.6%, suggesting better functioning upon bending. The device consisted of a lengthy strip of electrodes (15 cm × 2 cm) that was wrapped up [Figures 4.8(d and e)] had sufficient power to run a light-emitting diode for 20 seconds when completely charged [Figure 4.8(f)]. The rGO-PEDOT/PSS composite electrode with a high mass loading (8.49 mg cm^{-2}), a comparatively high areal capacitance of 448 mF cm^{-2} was reached at a scan rate of 10 mV s^{-1}, suggested the potentiality for usage in real-world applications [99].

Alvi *et al.* obtained the specific capacitances of 800 to 1100 F g^{-1} in graphene-PHTh and graphene-PEDOT nanocomposites at a current density of 0.1 A g^{-1} [100]. The PEDOT and rGO composite films have greater capacitive properties than pure PEDOT and rGO, and even after 2000 cycles, they retain an excellent specific capacitance of up to 90% of their initial value [101]. In graphene/PNCs, the orientation of the polymer chains can improve charge/discharge rates and electron transport in supercapacitors. Using graphene/PNCs in supercapacitors also has the critical benefit of providing ions in the graphene composite structure with a tortuous path, which can result in irreversible specific capacitance losses. Conducting polymers limited cycle stability, which results from their structural breakdown during the intercalation/de-intercalation process, is the main drawback of using them in supercapacitors. According to this perspective, the synthesis and manufacturing of these nanocomposites are difficult, necessitating additional structural optimization and enhancement of their functionality.

(v) Biomedical and Healthcare:

The growing need and fascination with biological applications has resulted in several cutting-edge studies on a variety of materials, architectures, and production techniques. The numerous characteristics of graphene make it potentially appealing for bio applications. Its wide surface area, chemical purity, and ease of functionalization make it a great option for drug delivery. Its unusual mechanical qualities open up possibilities for tissue engineering applications [23]. Chemically functionalized graphene may also find applications in rapid and ultrasensitive measurement systems capable of detecting a variety of biological substances such as hemoglobin, cholesterol, glucose, and DNA [21]. Because graphene is lipophilic as well, it may be possible to overcome a further hurdle in drug delivery membrane barrier penetration. In recent times, biomedicine has made extensive use of graphene and its nanocomposites for tissue engineering, medication and gene delivery, cancer treatment, and biosensing [102]. The distribution of graphene layers within the polymer matrix and the interfacial connection between the graphene layers and polymer matrix determine the physicochemical characteristics of PMNC materials. Pristine graphene does not create homogenous composites since it is incompatible with organic polymers. On the other hand, organic polymers are compatible with graphene oxide (GO) sheets [103]. The sections cover some of the most current research in the disciplines of drug delivery, bioimaging, biosensing, and tissue engineering that have used graphene-based nanocomposites.

(a) Graphene Nanocomposite for Biosensing:
A wide range of biosensors that enable the precise, sensitive, and selective identification of biomarkers have demonstrated the vast potential applications of graphene-based composites. Consequently, these biosensors can detect a wide range of substances, including DNA, heavy metal ions, amino acids, glucose, nicotinamine, adenine dinucleotide, hemoglobin, adenosine triphosphate, cholesterol, catechol, and H_2O_2. [104–106] Feng and coworkers reported on another kind of biosensor, which is an electrochemical sensor based on GO that allows for the label-free detection of cancer cells [107]. Moreover, to improve cell-to-cell contacts in human neuroblastoma, Heo *et al.* developed graphene/PET electrodes using non-contact electric field stimulation [108].

(b) Graphene Nanocomposite for Tissue Engineering:
Graphene-based nanocomposites with favorable characteristics including high elasticity, flexibility, and adaptation to both flat and irregularly shaped surfaces, can be essential for prolonged proliferation, appropriate adhesion, and improved differentiation in this situation [2]. Hybrid hydrogels based on graphene may offer a practical means of promoting nerve regeneration. Huang and coworkers [109] synthesized graphene and polyurethane-based nanocomposite hydrogels. Neural stem cell proliferation and neuronal differentiation can be enhanced by these graphene-based polyurethane hydrogels. Through *in situ* polymerization, Qiao and coworkers [110] synthesized electro-responsive nanocomposite hydrogels by introducing GO into poly (acrylic acid) (PAA). Under the conditions of electrical stimulation, the surface shape of these nanocomposite hydrogels could have been altered, consequently promoting BMSCs neuronal differentiation.

(c) Graphene Nanocomposite for Bioimaging:
Bioactive polymer composites based on graphene quantum dots (GQDs) have a unique function in drug delivery [111]. Strong quantum confinement, broad absorption with narrow emission spectra, and relatively high quantum efficiencies with high molar extinction coefficients encourage the widespread use of bioactive GQD-based polymer composite in biological imaging. Utilizing GQD-based nanocomposites, bioassay, fluorescence imaging, and biological imaging has been carried out [111]. Due to their ability to emit near-infrared reflectance, bioactive GQD-based polymer composites are attractive options for imaging deeper tissue samples [112]. It has been suggested that imaging plays a particularly important role in the diagnosis of cancer since it can detect metastases and recurrences of the disease as well as early tumor detection through sensitive imaging. GQDs can be employed

as optical probes in fluorescence imaging without the need to further combine fluorescent dyes because of their intrinsic PLs. The development of bioactive GQD-based polymer composites in bioimaging has been prompted by their recent successes in MR imaging and NIR fluorescence imaging. Sheng *et al.* employed a hydrothermal technique to prepare the nitrogen-doped graphene quantum dots (N-GQDs). Employing raw ingredients which include glutamate, PVP K90, and citric acid, a quantum efficiency of 64.2% was achieved. These results demonstrated the potential of a novel fluorescent probe using nitrogen-doped GQDs for identifying MCF-7 cells and chromium (VI) detection [113].

(d) Graphene Nanocomposite for Drug Delivery:
Graphene-based composites are attractive materials for drug delivery because of the enormous sp^2 hybridized carbon area of graphene [21]. Liu and his coworkers [114] developed nanoscale GO (NGO) sheets functionalized with PEG and loaded with SN 38, an analog of camptothecin (CPT). The high potency and effectiveness of SN38 were retained by this combination (NGO–PEG–SN38), which showed good water solubility. The combination was shown to be roughly 1000 times more powerful than CPT and exhibited substantial cytotoxicity in HCT-116 cells. Pan and colleagues [115] have presented an effective method for functionalizing graphene sheets (GS) with well-defined poly(N-isopropyl acrylamide) (PNI PAM). They investigated the usage of PNIPAM-GS to load the water-insoluble anticancer drugs CPT and measured its release from the PNIPAM-GS-CPT complex in water and PBS at 37°C. Due to p–p stacking, hydrophobic interaction with graphene sheets, and a stronger anticancer activity than CPT, a superior CPT loading (18.5%) was achieved. Wang *et al.* [116] have created a GO-based drug delivery system that targets gliomas. With excellent effectiveness, doxorubicin was loaded onto chlorotoxin-conjugated GO (CTX-GO/DOX). The bacterial toxicity of GO and rGO against Gram-positive S. aureus and Gram-negative *E. coli* bacteria was studied by Akhavan *et al.* [117]. When it came to their antibacterial properties, GO and rGO were equally potent, however, rGO showed the greatest effectiveness.

4.5 Conclusion and Future Outlook

We have conducted a thorough analysis of the literature, concentrating especially on more recent studies on the development of multifunctional nanocomposites. This chapter provides an overview with emphasis on the classifications of composites, the current advancement in multifunctional nanocomposites, fabrication approaches, as well as factors that influenced nanocomposites characteristics. Graphene is widely recognized as one of the most promising materials for the future generation of technological devices due to its exceptional optical, mechanical, and various other characteristics. Graphene's remarkable mechanical and electrical properties provide enormous possibilities for structural and functional applications of graphene-based composites. Among the graphene-based composites, we discuss the developments in graphene-based PNCs in various fields such as aerospace and defense, automotive industry, energy, electronics, sensors, and biomedical. Recent advances in graphene/PNCs, as well as current research trends, are opening up new potential for the development of next-generation aerospace and defense structures. The graphene-based PNCs have also been widely explored as electrode materials for energy storage systems. Graphene/PNC composite electrodes have demonstrated promising electrochemical performance, good cycle life, and significant rate capability. Because of their exceptional optical, mechanical, and electrical properties, graphene/PNCs also have the potential to be one of the most promising materials for the upcoming generation of flexible electronic applications. Applications based on graphene/PNC have been envisioned as viable possibilities in biomedical applications because of the enormous diversity, benefits, and features of graphene. In conclusion, it has been noted that graphene/PNCs are an efficient approach for creating hybrid structures with exceptional mechanical resistance, cyclability, and chemical stability for many different kinds of applications, such as sensors, energy storage, electronics, and aerospace.

The primary obstacle confronting the real-world implementation of graphene could perhaps emerge from its exceptional characteristics. When utilized as an electrode material, graphene's high surface area and porosity can harm its performance. The main challenge is the large-scale manufacture of graphene and graphene-based composites, which will necessitate enormous effort. To overcome these issues, new techniques and engineering solutions must be developed to address these concerns and improve graphene-based composite characteristics and performance.

References

1. Feni, F., Jahan, M., Zhao, R., Li, G., Zhao, G.-L., Mensah, P.F., Multifunctional composite with hybrid carbon fiber and carbonaceous coconut particle reinforcement. *Front. Mater.*, 10, 1–11, December, 2023.
2. Silva, M., Alves, N.M., Paiva, M.C., Graphene-polymer nanocomposites for biomedical applications. *Polym. Adv. Technol.*, 29, 2, 687–700, 2018.
3. Malaki, M., Xu, W., Kasar, A., Menezes, P., Dieringa, H., Varma, R., Gupta, M., Advanced Metal Matrix Nanocomposites. *Metals*, 9, 3, 330, 2019.
4. Sahay, R., Reddy, V.J., Ramakrishna, S., Synthesis and applications of multifunctional composite nanomaterials. *Int. J. Mech. Mater. Eng.*, 9, 1, 25, 2014.
5. Sagadevan, S., Shahid, M.M., Yiqiang, Z., Oh, W.-C., Soga, T., Anita Lett, J., Alshahateet, S.F., Fatimah, I., Waqar, A., Paiman, S., Johan, M.R., Functionalized graphene-based nanocomposites for smart optoelectronic applications. *Nanotechnol. Rev.*, 10, 1, 605–635, 2021.
6. Adekoya, G.J., Sadiku, R.E., Ray, S.S., Nanocomposites of PEDOT:PSS with Graphene and its Derivatives for Flexible Electronic Applications: A Review. *Macromol. Mater. Eng.*, 306, 3, 1–21, 2021.
7. Lv, Y., Mei, L., Zhang, L., Yang, D., Yin, Z., Multifunctional graphene-based nanocomposites for simultaneous enhanced photocatalytic degradation and photothermal antibacterial activity by visible light. *Environ. Sci. Pollut. Res.*, 28, 36, 49880–49888, 2021.
8. Bilisik, K. and Akter, M., Graphene nanocomposites: A review on processes, properties, and applications. *J. Ind. Text.*, 51, 3_suppl, 3718S–3766S, 2022.
9. Kandola, B.K., Nanocomposites, in: *Fire Retardant Materials*, pp. 204–219, Elsevier, Sawston, Cambridge, 2001, ISBN 978-1-85573-419-7.
10. Blumstein, A., Polymerization of adsorbed monolayers. II. Thermal degradation of the inserted polymer. *J. Polym. Sci. Part A: Gen. Pap.*, 3, 7, 2665–2672, 1965.
11. Wang, Z., Zuo, J., Liu, C., Zhang, Z., Han, Y., Stress–Strain Properties and Gas Permeability Evolution of Hybrid Fiber Engineered Cementitious Composites in the Process of Compression. *Materials*, 12, 9, 1382, 2019.

12. Yin, H., Johnsen, S., Borup, K.A., Kato, K., Takata, M., Iversen, B.B., Highly enhanced thermal stability of Zn4Sb3 nanocomposites. *Chem. Commun.*, 49, 58, 6540, 2013.
13. Zabihi, O., Khayyam, H., Fox, B.L., Naebe, M., Enhanced thermal stability and lifetime of epoxy nanocomposites using covalently functionalized clay: experimental and modelling. *New J. Chem.*, 39, 3, 2269–2278, 2015.
14. Lee, J., Takekoshi, T., Giannelis, E.P., Fire Retardant Polyetherimide Nanocomposites. *MRS Proc.*, 457, c, 513, 1996.
15. Panghal, A., Kumar, Y., Singh, F., Singh, N.L., Role of structural ordering on the radiation response of Gd2Zr2O7 pyrochlore. *Ceram. Int.*, 49, 8, 12191–12200, 2023.
16. Díez-Pascual, A.M. and Rahdar, A., Graphene-Based Polymer Composites for Flexible Electronic Applications. *Micromachines*, 13, 7, 1123, 2022.
17. Ji, L., Meduri, P., Agubra, V., Xiao, X., Alcoutlabi, M., Graphene-Based Nanocomposites for Energy Storage. *Adv. Energy Mater.*, 6, 16, 7–16, 2016.
18. Kausar, A., Ahmad, I., Eisa, M.H., Maaza, M., Graphene Nanocomposites in Space Sector—Fundamentals and Advancements. *C*, 9, 1, 29, 2023.
19. Bhat, A., Budholiya, S., Aravind Raj, S., Sultan, M.T.H., Hui, D., Md Shah, A.U., Safri, S.N.A., Review on nanocomposites based on aerospace applications. *Nanotechnol. Rev.*, 10, 1, 237–253, 2021.
20. Chang, H. and Wu, H., Graphene-based nanocomposites: preparation, functionalization, and energy and environmental applications. *Energy Environ. Sci.*, 6, 12, 3483, 2013.
21. Pattnaik, S., Swain, K., Lin, Z., Graphene and graphene-based nanocomposites: biomedical applications and biosafety. *J. Mater. Chem. B*, 4, 48, 7813–7831, 2016.
22. Huang, Y. and Wan, C., Controllable fabrication and multifunctional applications of graphene/ceramic composites. *J. Adv. Ceram.*, 9, 3, 271–291, 2020.
23. Kendre, P.N., Gite, M., Jain, S.P., Pote, A., Nanocomposite polymeric materials: state of the art in the development of biomedical drug delivery systems and devices. *Polym. Bull.*, 79, 11, 9237–9265, 2022.
24. Peretz Damari, S., Cullari, L., Nadiv, R., Nir, Y., Laredo, D., Grunlan, J., Regev, O., Graphene-induced enhancement of water vapor barrier in polymer nanocomposites. *Composites, Part B*, 134, 218–224, 2018.
25. Huang, C., Qian, X., Yang, R., Thermal conductivity of polymers and polymer nanocomposites. *Mater. Sci. Eng.: R: Rep.*, 132, May 1–22, 2018.
26. Vassiliou, A., Bikiaris, D., Chrissafis, K., Paraskevopoulos, K.M., Stavrev, S.Y., Docoslis, A., Nanocomposites of isotactic polypropylene with carbon nanoparticles exhibiting enhanced stiffness, thermal stability and gas barrier properties. *Compos. Sci. Technol.*, 68, 3–4, 933–943, 2008.
27. Idumah, C.I. and Obele, C.M., Understanding interfacial influence on properties of polymer nanocomposites. *Surf. Interfaces*, 22, 100879, 2021.
28. Krishnan, S., Tadiboyina, R., Chavali, M., Nikolova, M.P., Wu, R.-J., Bian, D., Jeng, Y.-R., Rao, P.T.S.R.K.P., Palanisamy, P., Pamanji, S.R., Graphene-Based Polymer Nanocomposites for Sensor Applications, in: Hybrid Nanocomposites, pp. 1–62, Pan Stanford Publishing, 8 Temasek Boulevard, Singapore, Taylor & Francis, 2019, ISBN 978-0-429-00096-6.
29. Wing, B.L. and Halloran, J.W., Subsurface oxidation of boron nitride coatings on silicon carbide fibers in SiC / SiC ceramic matrix composites. *Ceram. Int.*, 44, 14, 17499–17505, 2018.
30. Rathod, V.T., Kumar, J.S., Jain, A., Polymer and ceramic nanocomposites for aerospace applications. *Appl. Nanosci.*, 7, 8, 519–548, 2017.
31. Naslain, R., Design, preparation and properties of non-oxide CMCs for application in engines and nuclear reactors : an overview §. *Compos. Sci. Technol.*, 64, March 2003, 155–170, 2004.

32. Muresan, L.M., Nanocomposite Coatings for Anti-Corrosion Properties of Metallic Substrates. *Materials*, 16, 14, 5092, 2023.
33. Pan, S., Jin, K., Wang, T., Zhang, Z., Zheng, L., Umehara, N., Metal matrix nanocomposites in tribology: Manufacturing, performance, and mechanisms. *Friction*, 10, 10, 1596–1634, 2022.
34. Rane, A.V., Kanny, K., Abitha, V.K., Thomas, S., Methods for Synthesis of Nanoparticles and Fabrication of Nanocomposites, in: *Synthesis of Inorganic Nanomaterials*, pp. 121–139, Elsevier, Sawston, Cambridge, 2018, ISBN 978-0-08-101975-7.
35. Kausar, A., Ahmad, I., Zhao, T., Aldaghri, O., Ibnaouf, K.H., Eisa, M.H., Multifunctional Polymeric Nanocomposites for Sensing Applications—Design, Features, and Technical Advancements. *Crystals*, 13, 7, 1144, 2023.
36. Ibrahim, A., Klopocinska, A., Horvat, K., Abdel Hamid, Z., Graphene-Based Nanocomposites: Synthesis, Mechanical Properties, and Characterizations. *Polymers*, 13, 17, 2869, 2021.
37. Yang, C., Wei, H., Guan, L., Guo, J., Wang, Y., Yan, X., Zhang, X., Wei, S., Guo, Z., Polymer nanocomposites for energy storage, energy saving, and anticorrosion. *J. Mater. Chem. A*, 3, 29, 14929–14941, 2015.
38. Thakur, V.K., Tan, E.J., Lin, M.-F., Lee, P.S., Polystyrene grafted polyvinylidenefluoride copolymers with high capacitive performance. *Polym. Chem.*, 2, 9, 2000, 2011.
39. Bignoli, F., Rashid, S., Rossi, E., Jaddi, S., Djemia, P., Terraneo, G., Li Bassi, A., Idrissi, H., Pardoen, T., Sebastiani, M., Ghidelli, M., Effect of annealing on mechanical properties and thermal stability of ZrCu/O nanocomposite amorphous films synthetized by pulsed laser deposition. *Mater. Des.*, 221, 1–10, 2022.
40. Samoei, V.K., Maharjan, S., Sano, K., Jayatissa, A.H., Effect of Annealing on Graphene/PVDF Nanocomposites. *ACS Omega*, 8, 15, 13876–13883, 2023.
41. Syurik, J., Ageev, O.A., Cherednichenko, D.I., Konoplev, B.G., Alexeev, A., Non-linear conductivity dependence on temperature in graphene-based polymer nanocomposite. *Carbon*, 63, 317–323, 2013.
42. Khan, A., Khan, A.A.P., Khan, I., Asiri, A.M., Rahman, M.M., Sol-Gel Synthesis and Characterization of Highly Selective Poly(N-methyl pyrrole) Stannous(II)Tungstate Nano Composite for Mercury (Hg(II)) Detection. *Crystals*, 12, 3, 371, 2022.
43. Chang, Q., Zhou, J.E., Wang, Y., Meng, G., Formation mechanism of zirconia nano-particles containing pores prepared via sol-gel-hydrothermal method. *Adv. Powder Technol.*, 21, 4, 425–430, 2010.
44. Joy, J., Krishnamoorthy, A., Tanna, A., Kamathe, V., Nagar, R., Srinivasan, S., Recent Developments on the Synthesis of Nanocomposite Materials via Ball Milling Approach for Energy Storage Applications. *Appl. Sci.*, 12, 18, 9312, 2022.
45. Ni, J., Li, J., Luo, W., Han, Q., Yin, Y., Jia, Z., Huang, B., Hu, C., Xu, Z., Microstructure and properties of *in-situ* TiC reinforced copper nanocomposites fabricated via long-term ball milling and hot pressing. *J. Alloys Compd.*, 755, 24–28, 2018.
46. Zeng, X.-F., Kong, X.-R., Ge, J.-L., Liu, H.-T., Gao, C., Shen, Z.-G., Chen, J.-F., Effective Solution Mixing Method to Fabricate Highly Transparent and Optical Functional Organic–Inorganic Nanocomposite Film. *Ind. Eng. Chem. Res.*, 50, 6, 3253–3258, 2011.
47. Zawrah, M.F., Essawy, R.A., Zayed, H.A., Abdel Fattah, A.H., Taha, M.A., Mechanical alloying, sintering and characterization of $Al2O_3$–20wt%–Cu nanocomposite. *Ceram. Int.*, 40, 1, 31–38, 2014.
48. Chen, S., Skordos, A., Thakur, V.K., Functional nanocomposites for energy storage: chemistry and new horizons. *Mater. Today Chem.*, 17, 100304, 2020.
49. Yang, J., Dahlström, C., Edlund, H., Lindman, B., Norgren, M., pH-responsive cellulose–chitosan nanocomposite films with slow release of chitosan. *Cellulose*, 26, 6, 3763–3776, 2019.

50. Yu, J., Xiong, J., Cheng, B., Liu, S., Fabrication and characterization of Ag–TiO2 multiphase nanocomposite thin films with enhanced photocatalytic activity. *Appl. Catal., B*, 60, 3–4, 211–221, 2005.
51. Dalwani, M., Benes, N.E., Bargeman, G., Stamatialis, D., Wessling, M., Effect of pH on the performance of polyamide/polyacrylonitrile based thin film composite membranes. *J. Membr. Sci.*, 372, 1–2, 228–238, 2011.
52. Azizun, N.N., Khairul, W.M., Daud, A.I., Sarbon, N.M., Effect of pH on functional, gas sensing and antimicrobial properties of bio-nanocomposite gelatin film for food packaging application. *J. Food Sci. Technol.*, 58, 9, 3338–3345, 2021.
53. Iqbal, A.A. and Amierah, N., Effect of reinforcement volume fraction on the mechanical properties of the Al-SiC nanocomposite materials. *IOP Conf. Ser.: Mater. Sci. Eng.*, 226, 1, 2017.
54. Singh, T., Tiwari, S.K., Shukla, D.K., Effects of Al_2O_3 nanoparticles volume fractions on microstructural and mechanical characteristics of friction stir welded nanocomposites. *Nanocomposites*, 6, 2, 76–84, 2020.
55. Han, Z. and Fina, A., Thermal conductivity of carbon nanotubes and their polymer nanocomposites: A review. *Prog. Polym. Sci. (Oxford)*, 36, 7, 914–944, 2011.
56. Gibson, R.F., A review of recent research on mechanics of multifunctional composite materials and structures. *Compos. Struct.*, 92, 12, 2793–2810, 2010.
57. Forintos, N. and Czigany, T., Multifunctional application of carbon fiber reinforced polymer composites: Electrical properties of the reinforcing carbon fibers – A short review. *Composites, Part B*, 162, June 2018, 331–343, 2019.
58. Pradhan, S.S., Unnikrishnan, L., Mohanty, S., Nayak, S.K., Thermally Conducting Polymer Composites with EMI Shielding: A review. *J. Electron. Mater.*, 49, 3, 1749–1764, 2020.
59. Song, P., Liu, B., Qiu, H., Shi, X., Cao, D., Gu, J., MXenes for polymer matrix electromagnetic interference shielding composites: A review. *Compos. Commun.*, 24, 100653, 2021.
60. Simões, S., Investigation of thermal stability of aluminum matrix nanocomposites using functionalized MWCNTs. *Mater. Today Commun.*, 36, July, 106845, 2023.
61. Kim, H., Lee, E.J., Sun, Y.K., Recent advances in the Si-based nanocomposite materials as high capacity anode materials for lithium ion batteries. *Mater. Today*, 17, 6, 285–297, 2014.
62. Cantarella, M., Spanò, V., Zimbone, M., Giuffrida, F., Lufrano, E., Strano, V., Franzò, G., Sfuncia, G., Nicotra, G., Alberti, A., Buccheri, M.A., Rappazzo, G., Scalisi, E.M., Pecoraro, R., Brundo, M.V., Impellizzeri, G., ZnO–MoS2-PMMA polymeric nanocomposites: A harmless material for water treatment. *Mater. Today Chem.*, 36, October 2023, 101912, 2024.
63. Velasco Davoise, L., Peña Capilla, R., Díez-Pascual, A.M., Optical properties of polyaniline/modified graphene oxide nanocomposites. *Mater. Today Commun.*, 36, 106733, 2023.
64. Bouchareb, S., Doufnoune, R., Riahi, F., Cherif-Silini, H., Belbahri, L., High performance of polysulfone/graphene oxide-silver nanocomposites with excellent antibacterial capability for medical applications. *Mater. Today Commun.*, 27, 102297, February, 2021.
65. Ndlovu, S., Muchuweni, E., Ollengo, M.A., Nyamori, V.O., Highly efficient nitrogen-doped reduced graphene oxide-Sr0.7Sm0.3Fe0.4Co0.6O2.65 nanocomposites utilized as a counter electrode in dye-sensitized solar cells. *Mater. Today Commun.*, 38, 107681, 2024.
66. Lebedev, O.V., Kurkin, T.S., Golubev, E.K., Vasiliev, A.L., Gatin, A.K., Goncharuk, G.P., Ozerin, A.N., Detonation Synthesis Nanodiamond Soot as a Promising Filler for Polymer Composites. *C*, 8, 4, 69, 2022.
67. Al Faruque, M.A., Syduzzaman, M., Sarkar, J., Bilisik, K., Naebe, M., A Review on the Production Methods and Applications of Graphene-Based Materials. *Nanomaterials*, 11, 9, 2414, 2021.
68. Garg, K.K., Pandey, S., Kumar, A., Rana, A., Sahoo, N.G., Singh, R.K., Graphene nanosheets derived from waste plastic for cost-effective thermoelectric applications. *Results Mater.*, 13, 100260, January, 2022.

69. Kausar, A., Shape memory polymer/graphene nanocomposites: State-of-the-art. *e-Polymers*, 22, 1, 165–181, 2022.
70. Sing, S.L., and Yeong, W.Y., Emerging Materials for Additive Manufacturing. *Materials*, 16, 1, 127, 2022.
71. Ahmadi, Z., Epoxy in nanotechnology: A short review. *Prog. Org. Coat.*, 132, 445–448, February, 2019.
72. Shadlou, S., Ahmadi-Moghadam, B., Taheri, F., The effect of strain-rate on the tensile and compressive behavior of graphene reinforced epoxy/nanocomposites. *Mater. Des.*, 59, 439–447, 2014.
73. Bustero, I., Gaztelumendi, I., Obieta, I., Mendizabal, M.A., Zurutuza, A., Ortega, A., Alonso, B., Free-standing graphene films embedded in epoxy resin with enhanced thermal properties. *Adv. Compos. Hybrid Mater.*, 3, 1, 31–40, 2020.
74. Yousefi, N., Lin, X., Zheng, Q., Shen, X., Pothnis, J.R., Jia, J., Zussman, E., Kim, J.-K., Simultaneous *in situ* reduction, self-alignment and covalent bonding in graphene oxide/epoxy composites. *Carbon*, 59, 406–417, 2013.
75. Yousefi, N., Sun, X., Lin, X., Shen, X., Jia, J., Zhang, B., Tang, B., Chan, M., Kim, J., Highly Aligned Graphene/Polymer Nanocomposites with Excellent Dielectric Properties for High-Performance Electromagnetic Interference Shielding. *Adv. Mater.*, 26, 31, 5480–5487, 2014.
76. Zhang, J., Chevali, V.S., Wang, H., Wang, C.-H., Current status of carbon fibre and carbon fibre composites recycling. *Composites, Part B*, 193, 108053, April, 2020.
77. Cui, S., Cui, C., Xie, J., Liu, S., Shi, J., Carbon fibers coated with graphene reinforced TiAl alloy composite with high strength and toughness. *Sci. Rep.*, 8, 1, 2364, 2018.
78. García, C., Fittipaldi, M., Grace, L.R., Epoxy/montmorillonite nanocomposites for improving aircraft radome longevity. *J. Appl. Polym. Sci.*, 132, 43, 1–9, 2015.
79. Komarneni, S., Feature article. Nanocomposites. *J. Mater. Chem.*, 2, 12, 1219, 1992.
80. Garcés, J.M., Moll, D.J., Bicerano, J., Fibiger, R., McLeod, D.G., Polymeric Nanocomposites for Automotive Applications. *Adv. Mater.*, 12, 23, 1835–1839, 2000.
81. Izzaty, N., Sastra, H.Y., Ilyas, The Implementation of Graphene Composites for Automotive: An Industrial Perspective. *IOP Conf. Ser.: Mater. Sci. Eng.*, 536, 1, 012133, 2019.
82. Müller, K., Bugnicourt, E., Latorre, M., Jorda, M., Echegoyen Sanz, Y., Lagaron, J., Miesbauer, O., Bianchin, A., Hankin, S., Bölz, U., Pérez, G., Jesdinszki, M., Lindner, M., Scheuerer, Z., Castelló, S., Schmid, M., Review on the Processing and Properties of Polymer Nanocomposites and Nanocoatings and Their Applications in the Packaging, Automotive and Solar Energy Fields. *Nanomaterials*, 7, 4, 74, 2017.
83. Elmarakbi, A. and Azoti, W., State of the Art on Graphene Lightweighting Nanocomposites for Automotive Applications, in: *Experimental Characterization, Predictive Mechanical and Thermal Modeling of Nanostructures and their Polymer Composites*, vol. 1, pp. 1–23, Elsevier, AE Amsterdam, Netherlands, 2018, ISBN 978-0-32-348061-1.
84. Kiziltas, A., Liu, W., Tamrakar, S., Mielewski, D., Graphene nanoplatelet reinforcement for thermal and mechanical properties enhancement of bio-based polyamide 6, 10 nanocomposites for automotive applications. *Composites, Part C: Open Access*, 6, 100177, 2021.
85. Jariwala, D., Sangwan, V.K., Lauhon, L.J., Marks, T.J., Hersam, M.C., Carbon nanomaterials for electronics, optoelectronics, photovoltaics, and sensing. *Chem. Soc. Rev.*, 42, 7, 2824–2860, 2013.
86. Lee, T., Min, S.H., Gu, M., Jung, Y.K., Lee, W., Lee, J.U., Seong, D.G., Kim, B.-S., Layer-by-Layer Assembly for Graphene-Based Multilayer Nanocomposites: Synthesis and Applications. *Chem. Mater.*, 27, 11, 3785–3796, 2015.

87. Koutsioukis, A., Georgakilas, V., Belessi, V., Zboril, R., Highly Conductive Water-Based Polymer/Graphene Nanocomposites for Printed Electronics. *Chem. A Eur. J.*, 23, 34, 8268–8274, 2017.
88. Liu, Q., Liu, Z., Zhang, X., Yang, L., Zhang, N., Pan, G., Yin, S., Chen, Y., Wei, J., Polymer Photovoltaic Cells Based on Solution-Processable Graphene and P3HT. *Adv. Funct. Mater.*, 19, 6, 894–904, 2009.
89. Seekaew, Y., Lokavee, S., Phokharatkul, D., Wisitsoraat, A., Kerdcharoen, T., Wongchoosuk, C., Low-cost and flexible printed graphene–PEDOT:PSS gas sensor for ammonia detection. *Org. Electron.*, 15, 11, 2971–2981, 2014.
90. Kim, J.C., Rahman, M.M., Ju, M.J., Lee, J.-J., Highly conductive and stable graphene/PEDOT:PSS composite as a metal free cathode for organic dye-sensitized solar cells. *RSC Adv.*, 8, 34, 19058–19066, 2018.
91. Soni, M., Bhattacharjee, M., Ntagios, M., Dahiya, R., Printed Temperature Sensor Based on PEDOT: PSS-Graphene Oxide Composite. *IEEE Sens. J.*, 20, 14, 7525–7531, 2020.
92. Georgakilas, V., Perman, J.A., Tucek, J., Zboril, R., Broad Family of Carbon Nanoallotropes: Classification, Chemistry, and Applications of Fullerenes, Carbon Dots, Nanotubes, Graphene, Nanodiamonds, and Combined Superstructures. *Chem. Rev.*, 115, 11, 4744–4822, 2015.
93. Wang, K., Li, L., Zhang, T., Liu, Z., Nitrogen-doped graphene for supercapacitor with long-term electrochemical stability. *Energy*, 70, 612–617, 2014.
94. Shao, Y., Wang, J., Engelhard, M., Wang, C., Lin, Y., Facile and controllable electrochemical reduction of graphene oxide and its applications. *J. Mater. Chem.*, 20, 4, 743–748, 2010.
95. Abdelhamid, M.E., O'Mullane, A.P., Snook, G.A., Storing energy in plastics: a review on conducting polymers & their role in electrochemical energy storage. *RSC Adv.*, 5, 15, 11611–11626, 2015.
96. Liu, X., Zheng, Y., Wang, X., Controllable Preparation of Polyaniline–Graphene Nanocomposites using Functionalized Graphene for Supercapacitor Electrodes. *Chem. A Eur. J.*, 21, 29, 10408–10415, 2015.
97. Yan, J., Wei, T., Shao, B., Fan, Z., Qian, W., Zhang, M., Wei, F., Preparation of a graphene nanosheet/polyaniline composite with high specific capacitance. *Carbon*, 48, 2, 487–493, 2010.
98. Wang, H., Hao, Q., Yang, X., Lu, L., Wang, X., A nanostructured graphene/polyaniline hybrid material for supercapacitors. *Nanoscale*, 2, 10, 2164, 2010.
99. Liu, Y., Weng, B., Razal, J.M., Xu, Q., Zhao, C., Hou, Y., Seyedin, S., Jalili, R., Wallace, G.G., Chen, J., High-Performance Flexible All-Solid-State Supercapacitor from Large Free-Standing Graphene-PEDOT/PSS Films. *Sci. Rep.*, 5, 1–11, October, 2015.
100. Alvi, F., Ram, M.K., Basnayaka, P.A., Stefanakos, E., Goswami, Y., Kumar, A., Graphene–polyethylenedioxythiophene conducting polymer nanocomposite based supercapacitor. *Electrochim. Acta*, 56, 25, 9406–9412, 2011.
101. Lehtimäki, S., Suominen, M., Damlin, P., Tuukkanen, S., Kvarnström, C., Lupo, D., Preparation of Supercapacitors on Flexible Substrates with Electrodeposited PEDOT/Graphene Composites. *ACS Appl. Mater. Interfaces*, 7, 40, 22137–22147, 2015.
102. Bellier, N., Baipaywad, P., Ryu, N., Lee, J.Y., Park, H., Recent biomedical advancements in graphene oxide- and reduced graphene oxide-based nanocomposite nanocarriers. *Biomater. Res.*, 26, 1, 65, 2022.
103. Vickery, J.L., Patil, A.J., Mann, S., Fabrication of Graphene–Polymer Nanocomposites With Higher-Order Three-Dimensional Architectures. *Adv. Mater.*, 21, 21, 2180–2184, 2009.
104. Hou, S., Kasner, M.L., Su, S., Patel, K., Cuellari, R., Highly Sensitive and Selective Dopamine Biosensor Fabricated with Silanized Graphene. *J. Phys. Chem. C*, 114, 35, 14915–14921, 2010.
105. Lian, H., Sun, Z., Sun, X., Liu, B., Graphene Doped Molecularly Imprinted Electrochemical Sensor for Uric Acid. *Anal. Lett.*, 45, 18, 2717–2727, 2012.

106. Kuilla, T., Bhadra, S., Yao, D., Kim, N.H., Bose, S., Lee, J.H., Recent advances in graphene based polymer composites. *Prog. Polym. Sci.*, 35, 11, 1350–1375, 2010.
107. Feng, L., Chen, Y., Ren, J., Qu, X., A graphene functionalized electrochemical aptasensor for selective label-free detection of cancer cells. *Biomaterials*, 32, 11, 2930–2937, 2011.
108. Heo, C., Yoo, J., Lee, S., Jo, A., Jung, S., Yoo, H., Lee, Y.H., Suh, M., The control of neural cell-to-cell interactions through non-contact electrical field stimulation using graphene electrodes. *Biomaterials*, 32, 1, 19–27, 2011.
109. Huang, C.-T., Kumar Shrestha, L., Ariga, K., Hsu, S., A graphene–polyurethane composite hydrogel as a potential bioink for 3D bioprinting and differentiation of neural stem cells. *J. Mater. Chem. B*, 5, 44, 8854–8864, 2017.
110. Qiao, K., Guo, S., Zheng, Y., Xu, X., Meng, H., Peng, J., Fang, Z., Xie, Y., Effects of graphene on the structure, properties, electro-response behaviors of GO/PAA composite hydrogels and influence of electro-mechanical coupling on BMSC differentiation. *Mater. Sci. Eng.: C*, 93, August, 853–863, 2018.
111. Mousavi, S.M., Hashemi, S.A., Kalashgrani, M.Y., Omidifar, N., Bahrani, S., Vijayakameswara Rao, N., Babapoor, A., Gholami, A., Chiang, W.-H., Bioactive Graphene Quantum Dots Based Polymer Composite for Biomedical Applications. *Polymers*, 14, 3, 617, 2022.
112. Miao, P., Han, K., Tang, Y., Wang, B., Lin, T., Cheng, W., Recent advances in carbon nanodots: Synthesis, properties and biomedical applications. *Nanoscale*, 7, 5, 1586–1595, 2015.
113. Sheng, L., Huangfu, B., Xu, Q., Tian, W., Li, Z., Meng, A., Tan, S., A highly selective and sensitive fluorescent probe for detecting Cr(VI) and cell imaging based on nitrogen-doped graphene quantum dots. *J. Alloys Compd.*, 820, 153191, 2020.
114. Liu, Z., Robinson, J.T., Sun, X., Dai, H., PEGylated Nanographene Oxide for Delivery of Water-Insoluble Cancer Drugs. *J. Am. Chem. Soc.*, 130, 33, 10876–10877, 2008.
115. Pan, Y., Bao, H., Sahoo, N.G., Wu, T., Li, L., Water-soluble poly(N-isopropylacrylamide)-graphene sheets synthesized via click chemistry for drug delivery. *Adv. Funct. Mater.*, 21, 14, 2754–2763, 2011.
116. Xu, Q., Wang, H., Gu, W., Xiao, N., Ye, L., Chlorotoxin-conjugated graphene oxide for targeted delivery of an anticancer drug. *Int. J. Nanomed.*, 9, 1, 1433, 2014.
117. Akhavan, O. and Ghaderi, E., Toxicity of graphene and graphene oxide nanowalls against bacteria. *ACS Nano*, 4, 10, 5731–5736, 2010.

5

Multifunctional Supramolecular Polymers

Ansar Ul Haq and Yasser Azim*

Department of Applied Chemistry, Zakir Hussain College of Engineering and Technology, Aligarh Muslim University, Aligarh, Uttar Pradesh, India

Abstract

This work discusses two related disciplines of research that deal with the fabrication and characteristics of complex structures from basic molecular building blocks: supramolecular chemistry and supramolecular polymers. The study of non-covalent interactions that result in high-order assemblies, and molecular recognition is known as supramolecular chemistry. Supramolecular polymers are made of monomeric units bound together by reversible and directed interactions such as electrostatic interactions, hydrogen bonding, van der Waals forces, and π-π stacking. The fundamental ideas and noteworthy traits of supramolecular polymers—such as their dynamic nature, flexibility, structural variety, hierarchical assembly, sustainability and recycling, and novel materials—are explained in the chapter. It also explains the ideas of supramolecular chemistry and molecular self-assembly that underpin the creation of supramolecular polymers. The article discusses the synthetic methods used to build supramolecular polymers, including metal-ligand coordination, hydrogen bonding, host-guest interactions, and π-π interactions. Analytical methods for characterizing supramolecular polymers are discussed, including mass spectrometry, electron microscopy, scanning probe microscopy, light scattering, size exclusion chromatography, viscometry, and vapor pressure osmometry. It emphasizes the uses of supramolecular polymers in a variety of industries, including energy storage, pollution sensors, diagnostic markers, targeted medicine delivery, personal hygiene products, and recyclable and self-repairing materials. It also provides an overview of current developments as well as directions for supramolecular polymer and supramolecular chemical research. The goal of the chapter is to provide readers with a thorough and understandable introduction to these intriguing cross-disciplinary subjects.

***Keywords*:** Supramolecular polymers, building blocks, self-assembly, host-guest interactions, supramolecular hydrogels

5.1 Introduction to Supramolecular Polymers

The term "supramolecular chemistry" was first coined by Jean-Marie Lehn in the 1970s. In 1987, Donald J. Cram, Jean-Marie Lehn, and Charles J. Pedersen shared the Nobel Prize in Chemistry for their development and use of molecules with structure-specific interactions

**Corresponding author*: yasser.azim@gmail.com; yasser.ac@amu.ac.in

Divya Bajpai Tripathy, Anjali Gupta and Arvind Kumar Jain (eds.) Multifunctional Materials: Engineering and Biological Applications, (105–132) © 2025 Scrivener Publishing LLC

of high selectivity. Supramolecular chemistry, also called "chemistry beyond the molecule," is the study of non-covalent interactions that generate high-order assemblies and molecular recognition [1]. Since supramolecular systems are made from building blocks linked together by weak non-covalent interactions, they respond tremendously to external stimuli [2]. Molecular machines [3], molecular sensors [4–6], gas absorption [7–9], nanoreactors [10, 11], chemical catalysis [12], and drug delivery [13, 14] are just a few of the fields in which supramolecular chemistry has been extensively studied.

Reversible and highly directed secondary interactions bind polymeric arrays of monomeric units together to form supramolecular polymers, which have polymeric characteristics in both bulk and diluted and concentrated solutions [15]. The essential characteristics of systems that fit the existing theories of polymer physics and may be classified as polymers are the directionality and strength of the supramolecular bonding [16]. Supramolecular polymers represent a fascinating class of materials that combine the principles of supramolecular chemistry and polymer science. Unlike traditional covalent polymers, which are held together by strong covalent bonds, supramolecular polymers are formed through reversible non-covalent interactions [17]. These interactions include hydrogen bonding, π-π stacking, van der Waals forces, and electrostatic interactions [18] as shown in the Figure 5.1. This unique approach to polymerization allows for dynamic and adaptable materials with properties that can be controlled and tuned by adjusting the supramolecular interactions. These supramolecular interactions exhibit different bond strengths as depicted in Table 5.1

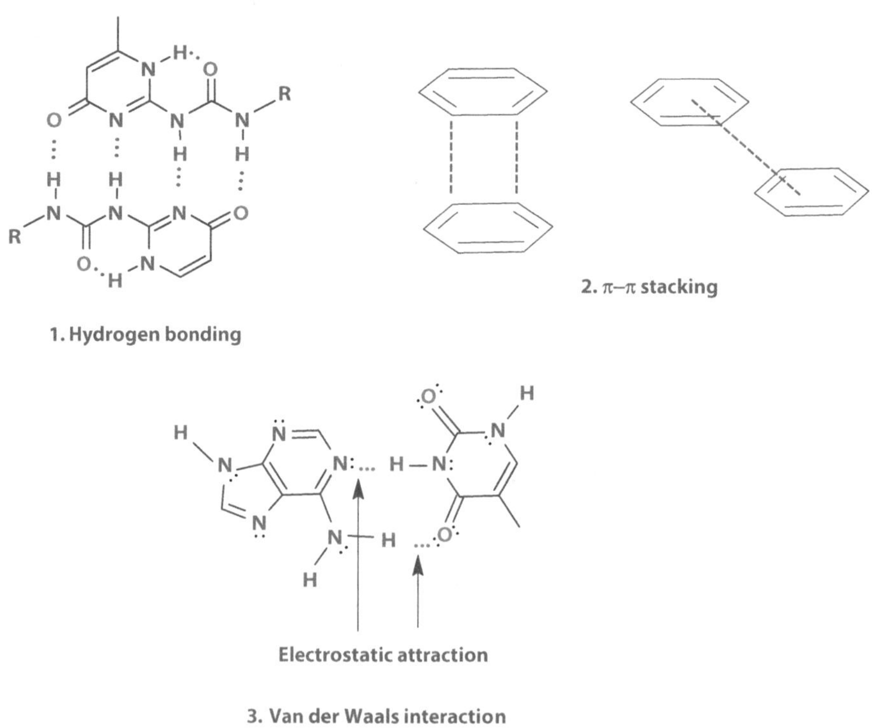

Figure 5.1 Main types of supramolecular interactions.

Table 5.1 Types of supramolecular interactions and their strength.

Interaction	Strength (kJ mol^{-1})
Covalent bond	180–410
Ion-ion	90–380
Ion-dipole	40–180
Dipole-dipole	4–40
Hydrogen bond	3–110
π-π	0–40
van der Waals	≤4 (surface area–dependent)

5.2 Supramolecular Chemistry Overview

The study of non-covalent interactions between molecules that develop more substantial, organized structures is known as supramolecular chemistry. It is the systematic study of the designs and activities of supermolecules that arise from the binding interactions between two or more molecules. Although these interactions are weaker than covalent connections, they are essential for many processes that occur in nature, including interactions between enzymes and their substrates, DNA base pairing, and protein folding [19]. Hydrogen (H) bonds are essential in supramolecular chemistry due to their directionality and adaptability [20], although they are not the strongest of non-covalent interactions. Cooperativity can solve this challenge, which is why several systems that integrate numerous H-bonds in a row have been devised. This improves the interaction's specificity as well as its strength. By using quadruple H-bonding units [21], very stable compounds may be produced. Supramolecular polymers were developed in the late 20th century due to the idea of utilizing these interactions to make new materials.

Generally speaking, "supramolecules" and "molecular assemblies" are the two domains that overlap most heavily in supramolecular chemistry. A supramolecule is a clearly defined distinct system that is created by interactions between molecules (either receptors or hosts) that have sites for hydrogen bond production, donor atoms, or large cavities as converging binding sites, and another molecule, either a guest or an analyte, with different binding sites, like hydrogen bond acceptor atoms. The analyte might be molecules or ions (cations and anions), while the receptor can be macromolecules or aggregates. In order to build a receptor molecule, a molecular-level understanding of several interactions is required, including structural and functional details, the energetics of the binding process, and the complementarity of molecular shapes. Multiple components spontaneously self-assemble under specific circumstances to form well-defined molecular assemblies. Molecular binding events serve as the guiding force for the self-assembly process [22]. To put it succinctly, supramolecular operation meets the molecular approach in creating organized molecular assemblies.

5.3 Basic Supramolecular Polymer Principles

Covalent bonds join monomers together to form conventional polymers, such as plastics. In contrast, reversible interactions between monomers or building blocks cause supramolecular polymers to form. It is unusual for covalent polymers to exhibit dynamic behavior, self-healing, or adaptation, yet its reversibility enables these properties [23].

The creation of polymers and polymeric materials devoid of macromolecular structure has been made possible by the remarkable advancements in supramolecular chemistry in recent times. Instead, highly directed secondary interactions combine the repeating units into a polymer array. Based on this idea, polymers have the potential to be a unique class of innovative materials, incorporating many of the desirable properties of traditional polymers with the reversibility that results from secondary interactions. Thus, distinct materials that react to outside stimuli may be produced by reversibly adjusting the architectural and dynamic factors that govern polymer characteristics, such as the degree of polymerization, longevity of the chain, and its conformation. These characteristics of supramolecular polymers have spurred interest in this intriguing family of molecules, encouraging us to combine supramolecular chemistry with materials research.

A well-designed functionality with strong and directed interactions is characterized by the existence of linear chains, which endure heat or dissolve in a substance. In a fundamental research context, strength and directionality are crucial since the aim is to comprehend the relationship between molecular structure and macroscopic attributes. Constructing extended chains through non-covalent contacts was first proposed in the 1990s. Researchers investigated the creation of linear and branching systems using hydrogen bonds [24–28] and π-π stacking [29–31]. These preliminary investigations set the groundwork for creating more advanced supramolecular polymer systems.

5.4 Significant Characteristics of Supramolecular Polymers

5.4.1 Dynamic Nature

Supramolecular polymers can react to external stimuli such as changes in temperature, pH, or the presence of particular molecules since the interactions are reversible. These polymers can create smart materials that can change their properties in response to their surroundings or carry out specific activities when needed [32]. This dynamic behavior is desirable for medication distribution, sensing, and responsive materials applications.

5.4.2 Adaptability

Supramolecular polymers can change their non-covalent connections to adapt to changing environmental conditions. The supramolecular polymers can go through self-assembly, exchange, and dissociation processes in a dynamic equilibrium because their non-covalent interactions are weaker than covalent bonds but stronger than random thermal fluctuations [33]. Materials with this feature can repair themselves after damage, resulting in longer lifetimes and less waste.

5.4.3 Structural Diversity

Supramolecular interactions are adaptable and can be adjusted to specific monomers or building blocks due to their structural diversity [34]. Due to its variability, various architectures, including 2D or 3D networks and linear and branched structures, can be made [33].

5.4.4 Hierarchical Assembly

Supramolecular polymers frequently develop through processes known as hierarchical assemblies. Initial interactions between small monomers result in discrete groups, which combine to form more significant structures. This hierarchical construction controls the characteristics and design of the finished material. Functional and complicated materials with adjustable qualities and uses can be created through hierarchical assembly [35].

5.4.5 Recycling and Sustainability

The reversible nature of non-covalent bonding in supramolecular polymers enables effective recycling and reprocessing of materials, making them more ecologically friendly than typical covalent polymers [33, 36].

5.4.6 Innovative Materials

Supramolecular polymers have the potential to produce innovative materials with unique combinations of features that regular covalent polymers cannot match [37].

5.5 Molecular Self-Assembly and Supramolecular Chemistry

The spontaneous synthesis of complex structures and materials from simpler molecular building blocks is the subject of two exciting fields of study: molecular self-assembly and supramolecular chemistry. Non-covalent interactions like hydrogen bonds, van der Waals forces, π-π interactions, and electrostatic interactions control these processes. Supramolecular chemistry includes studying these interactions and creating suitable molecule assemblies for various uses in nanotechnology, materials science, and medicine [38].

Different functional materials have been developed as a result of supramolecular chemistry. Host-guest systems created by inclusion complexes, in which a larger host molecule encloses a smaller guest molecule, have been the subject of substantial research. Cyclodextrins, cyclic oligosaccharides, are the archetypal example because they may form inclusion complexes with hydrophobic guest molecules through their hydrophobic cavities [39]. These host-guest interactions have benefited drug transport, sensing, and catalysis.

Designing molecular machines is a crucial development in supramolecular chemistry. Nobel laureates Jean-Pierre Sauvage, Fraser Stoddart, and Bernard L. Feringa pioneered the development of interlocking compounds known as rotaxanes and catenanes. Catenanes are rings that are linked together, whereas rotaxanes are rings that are threaded onto a molecular axle [40]. These molecular machines can perform regulated movements in response to the application of external stimuli, such as light or pH changes, opening the door to possible uses in molecular computing and nanotechnology.

Supramolecular chemistry has also been used to create functional materials with particular features. For instance, porous materials called metal-organic frameworks (MOFs) are made when coordination bonds between metal ions and organic ligands are established. Because of its adjustable porosity, MOFs may be used in gas storage, separation, and catalysis [41].

5.6 Synthetic Approaches for Supramolecular Polymer Formation

The following are four synthetic methods for producing supramolecular polymers.

5.6.1 Host-Guest Interactions

Host-guest interactions, such as inclusion complexes generated by cyclodextrins and guest molecules [42], can be exploited to synthesize supramolecular polymers. The reversible and non-covalent interactions between host and guest molecules form these polymers. The host molecules typically possess a cavity or binding site, while the guest molecules can fit into these cavities, creating supramolecular assemblies [43].

Supramolecular polymer production usually entails the following steps:

a. **Selection of Host Molecules:** The choice of host molecules involves considering certain factors, such as specific binding sites in synthetic host molecules or macrocycles like cyclodextrins or cucurbiturils [44].
b. **Selection of Guest Molecules:** Guest molecules are chosen based on their ability to interact with the host molecules. These can be small molecules, organic polymers, or nanoparticles [45].
c. **Supramolecular Polymer Assembly:** Host and guest molecules are combined in solution. The process of self-assembly leads to the development of supramolecular polymers, which are driven by the non-covalent interactions between the host and guest [46, 47].

5.6.2 Hydrogen Bonding

Hydrogen bonding is a typical non-covalent interaction in the creation of supramolecular polymers. One-dimensional polymer chains can be formed by self-assembling molecules

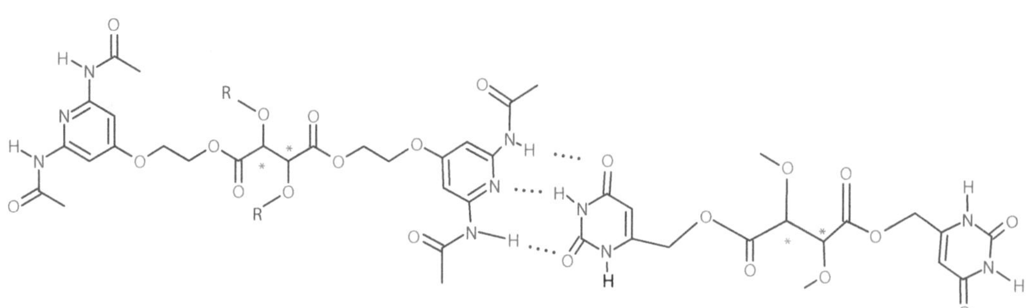

Figure 5.2 Lehn's diaminopyridine-uracil molecular assembly. Dotted lines and chiral centers indicate the hydrogen bonds between repeat units are marked *.

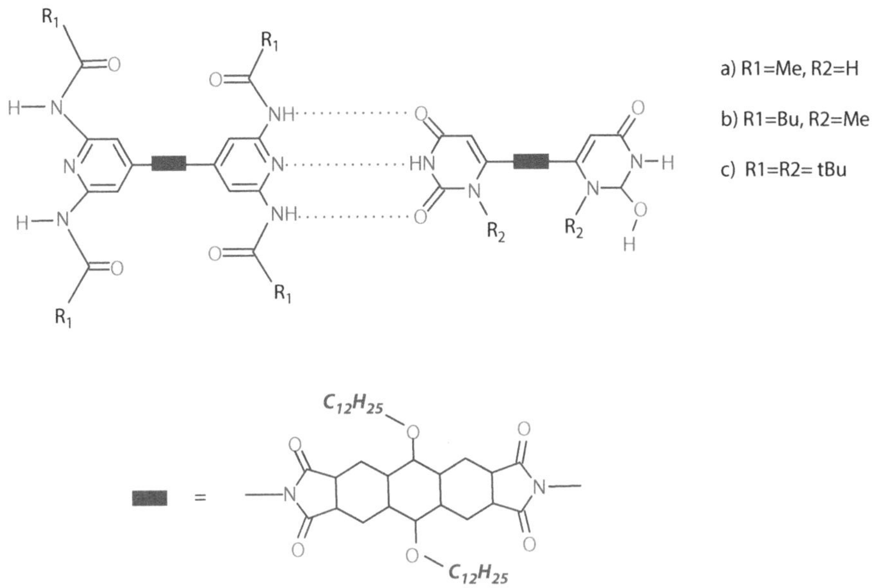

Figure 5.3 Supramolecular rigid rods. Dotted lines indicate the hydrogen bonds between repeat units.

with complementary hydrogen bonding sites [48]. Most agree that the original "supramolecular" polymer was an extended linear assembly made of 2,6-diaminopyridine and uracil in an equimolar ratio. This assembly was derivatized to create long aliphatic chains [49].

Unlike many of the liquid crystalline supramolecular polymers reported in the literature, Lehn adopted a well-defined structure (as shown in Figure 5.2), and no problems with its long-term stability have been observed. Supramolecular polymer liquid crystals are distinguished by their chirality, which is another feature of their ultimate structure. Afterward, supramolecular stiff rods (examples shown in Figure 5.3) exhibiting polymeric characteristics and being highly soluble in typical organic solvents were constructed using the same triple hydrogen bonding motif [50].

5.6.3 π-π Interactions

Supramolecular polymers are made via stacking interactions between aromatic molecules. Through π-π interactions, conjugated molecules with aromatic rings can self-assemble into one-dimensional or two-dimensional supramolecular polymers [51]. Creating a small area will have a greater impact on bonding strength than expanding the number of binding sites. The space restriction causes the bonding sites to come into closer touch with one another, significantly strengthening their power [52].

One of the host macrocycles that is frequently observed in host-guest interactions is Cucurbit[8]uril (CB[8]) [53]. Because confined spaces facilitate charge transfer between building blocks and stacking, their wide hydrophobic cavity (as shown in Figure 5.4) and ability to recognize a variety of guests make them excellent for generating a confined space to increase bonding strength. (CB[8]) shows tremendous property of radical stacking due to it's structure as shown in the Figure 5.5.

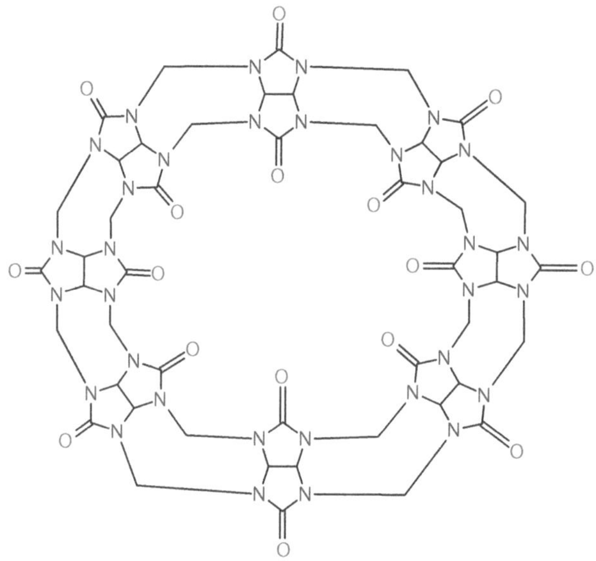

Figure 5.4 Structure of cucurbit[8]uril.

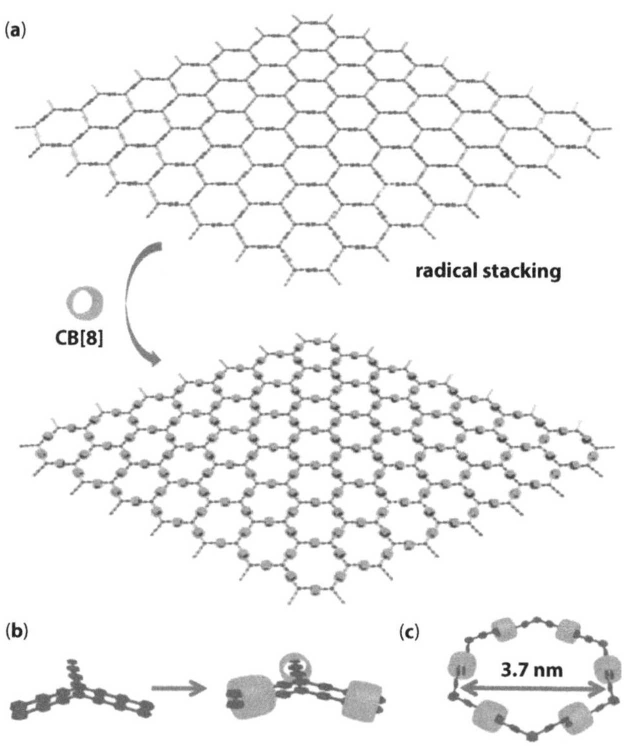

Figure 5.5 π-π stacking in **CB[8]** enhancement. Reprinted with permission from [53]. Copyright 2014 Royal Society of Chemistry.

5.6.4 Metal-Ligand Coordination

Supramolecular polymers can be made by coordination complexes containing metal ions and ligands with appropriate binding groups known as metallo-supramolecular polymers (MSPs). One-dimensional or two-dimensional polymer structures are formed via the coordination of metal ions with ligands [54].

A plethora of MSPs have been identified based on ligands with different donor atoms and denticities [55, 56]. However, nitrogen-based ligands with one or more pyridine or imidazole units arguably represent the most common binding motif [57, 58].

Among the transition metal ions that may form stable complexes with 2,2':6',2"-terpyridine (tpy) are ruthenium(II), iron(II), and zinc(II) [59]. MSPs with distinct optical, electrical, and magnetic characteristics may be created by bonding tpy to organic monomers, such as thiophene, styrene, or benzodipyrrolidone (BDP) [59]. Using ruthenium(II) ions, for example, BDP-tpy monomers can self-assemble to generate MSPs with weak donor-acceptor structures and absorption bands for intramolecular charge transfer or metal-to-ligand charge transfer [60]. Various methods have been proposed and utilized to synthesize MSP's as shown in the Figure 5.6 [61].

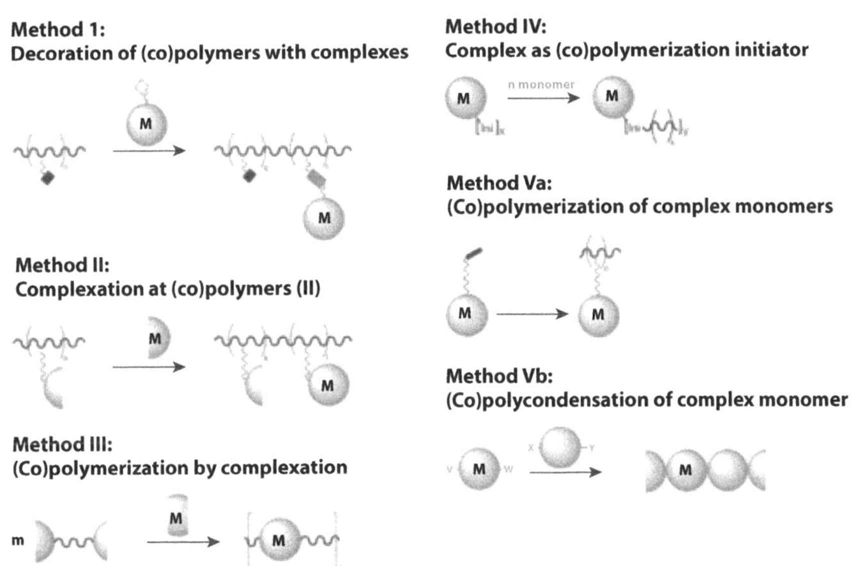

Figure 5.6 Schematic representation of the different general methods utilized to synthesize metal-containing (co)polymers [61]. Reproduced with permission from [61]. Reproduced with permission from [61]. Copyright © 2009 WILEY-VCH Verlag GmbH & Co. KGaA, Weinheim.

5.7 Analytical Techniques for Characterization of Supramolecular Polymers

Various analytical methods, such as theoretical estimation, SEC, viscometry, light scattering, VPO, MS, IM-MS, NMR, EM, SPM, and SAS, can characterize supramolecular polymers' structure, properties, and behavior.

5.7.1 Theoretical Estimation

Supramolecular polymerization is a thermodynamic equilibrium or self-assembly process. Based on the equilibrium constant, standard theoretical models may be utilized to determine the average molar mass of supramolecular polymers. R. B. Martin has described several indefinite self-association models applicable to supramolecular polymerization systems [62]. The most basic isodesmic model assumes that the relationship between the monomer end groups remains unchanged during the supramolecular polymerization process. This simplifying assumption allows a reasonably straightforward estimation of the degree of polymerization (DP) as DP ≈ $(K_a C)^{1/2}$, where C represents the total monomer concentration, and K_a is the equilibrium constant between the monomers [63]. This estimate approach states that a big binding constant or a high monomer concentration is needed to create long-chain supramolecular polymers.

Other elements must be considered in theoretical derivations for increasingly complex supramolecular polymerization systems. For instance, a supramolecular polymer's binding constant K_a may change as its chain lengthens. Expressing them in terms of generic fitting equations is challenging. Obtaining qualitative information requires combining the collected experimental data with a suitable theoretical estimation.

5.7.2 Size Exclusion Chromatography (SEC)

In traditional polymer characterization, size exclusion chromatography (SEC), particularly gel permeation chromatography (GPC), is extensively utilized and offers valuable insights into the molar mass distribution of a polymer. The retention time may be used to calculate the molecular size of the polymers in solution, which is often used to separate them. A molecular mass distribution, such as polystyrene, can be easily inferred using well-known standard samples. Nonetheless, it should be noted that the sample is diluted during the whole experimental procedure. The distribution acquired using SEC has a considerable tail as the supramolecular polymers are reversible and dynamic systems, which means that the degree of polymerization (DP) is always very concentration-dependent. As such, the SEC method is only somewhat appropriate on its own. Nevertheless, SEC has been utilized to get relevant data in a few instances, such as metal coordination and multiple hydrogen bonding arrays, where the non-covalent bonds have sufficiently sluggish association and disassociation kinetics [64, 65].

Schubert *et al.* described a supramolecular polymer in which the "monomers" had two terpyridine moieties at both the head and tail, and ruthenium(II) was used to coordinate them. GPC study demonstrated the production of high molecular weight supramolecular polymers [55].

The polymers created by mechanical bonding are another kind of supramolecular polymers [66]. The DP does not diminish under SEC processing conditions because the monomers are topologically interlocked and coupled by non-covalent interactions. Stoddart *et al.* synthesized a mechanically interlocked poly[c2]daisy chain as shown in the Figure 5.7 [67].

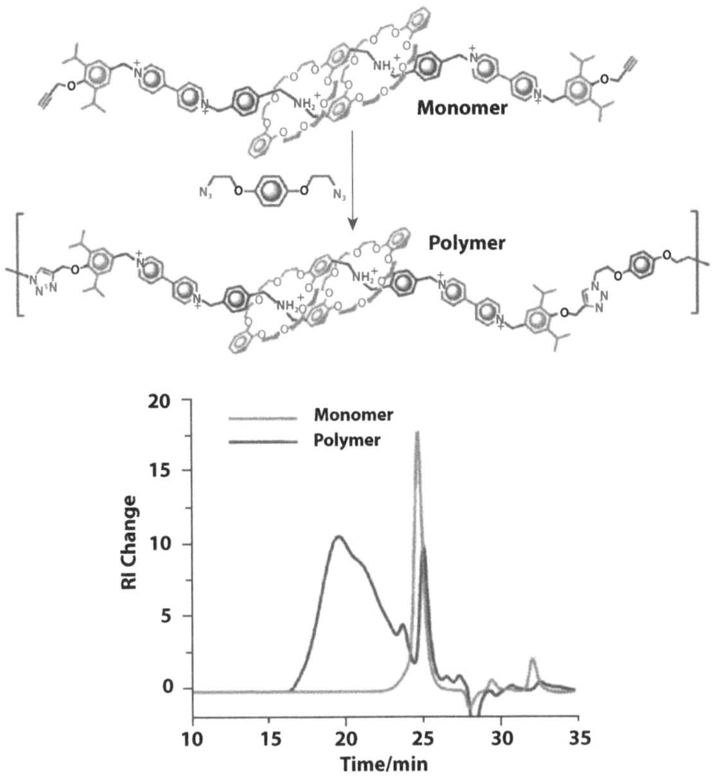

Figure 5.7 Synthesis of the poly[c2]daisy chain and its SEC chromatogram result [67]. Reprinted with permission from [67]. Copyright 2009 American Chemical Society.

SEC/multi-angle light scattering analysis was used to calculate the number average molecular weight (M_n).

Furthermore, Huang *et al.* described the production of a daisy chain at low temperatures by esterification of a well-planned crown ether–viologen complex [68]. Using polystyrene as the reference material for GPC analysis, M_n was estimated.

5.7.3 Viscometry

This technique gauges a supramolecular polymer solution's viscosity, which is correlated with the frictional resistance of the polymer to flow. A capillary viscometer, which calculates the time needed for a specific volume of solution to pass through a thin tube at a constant pressure, can be used to measure viscosity. The hydrodynamic volume of supramolecular polymers may be measured by computing the intrinsic viscosity. Using empirical formulae like the Mark-Houwink-Sakurada equation, i.e., $|\eta|=KM^{\alpha}$, in which K and α are both constants, the inherent viscosity of supramolecular polymers may be associated with their molar mass and shape [69].

Values of K and α for a particular polymer can be found by appropriate calibration tests using a sequence of sharp fractions. This formula may be used to calculate a polymer's molecular weight from its intrinsic viscosity. In principle, if the Mark-Houwink parameters are known, this approach may be used to calculate the molecular weight of supramolecular

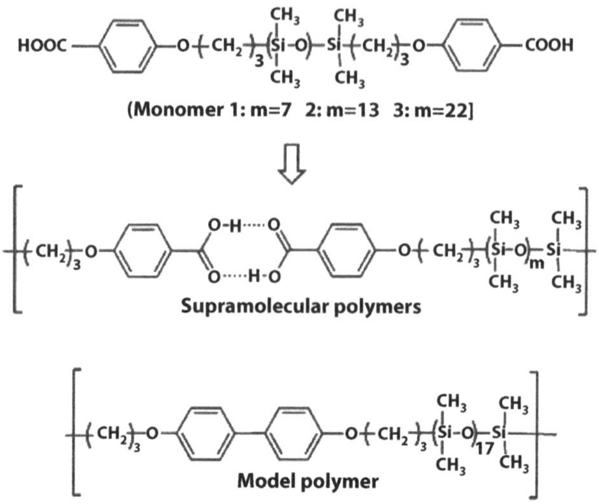

Figure 5.8 Structures of monomers, supramolecular polymers, and the model polymer [70]. Reprinted with permission from [70]. Copyright 2000 American Chemical Society.

polymers. To obtain these numbers, the problem here is to find a covalent model that matches this dynamic system.

A good example is described by Bouteiller *et al.* They created several supramolecular polymers by forming hydrogen bonds between the monomers of benzoic acid (1–3).

A covalent model of a model polymer with a comparable structure was created in the interim. The viscometric parameters can be calculated by examining the experimental data derived from this model. Because the model polymer was somewhat crude, the estimated curves fit the experimental data of the supramolecular polymer created by monomer 1 very well but not so well with the other two (as shown in Figure 5.8) [70].

5.7.4 Light Scattering

The intensity of light scattered by a supramolecular polymer solution is measured using this technique, which is correlated with variations in the supramolecular polymers' concentration and refractive index. A laser is pointed at different angles, and a detector can be used to measure the light scattering. The Rayleigh-Debye-Gans theory or the Zimm plot may be used to calculate the molar mass, radius of gyration, and form factor of supramolecular polymers from light scattering data [69, 71]. Dynamic light scattering (DLS) and static light scattering (SLS) are the two most prevalent techniques. In polymer science, SLS is often used to determine the molar mass.

DLS is a widely used method for figuring out how microscopic particles or aggregates are distributed in size. DLS is most frequently utilized to provide supplementary evidence and get size distribution information from aggregates made of supramolecular polymers [72, 73]. For instance, Weck and colleagues created a supramolecular alternating block copolymer using Pd and pyridine coordination. It was evident from the DLS data that a substantially increased size could be achieved by adding AgBF4, which promotes supramolecular polymerization [72].

5.7.5 Vapor Pressure Osmometry (VPO)

The idea behind vapor pressure osmometry (VPO), which measures M_n, is that at a given temperature and pressure, a solution's vapor pressure is lower than that of the pure solvent. Raoult's law allows for the relationship between the vapor pressure and M_n. Using this approach, it has also been possible to calculate the molar mass of supramolecular polymers.

Harada *et al.* frequently employed VPO to investigate the molar mass of supramolecular polymers [74, 75]. For instance, homoditopic adamantane derivatives (Ad dimers) and cyclodextrin dimers (CD dimers) were synthesized to construct supramolecular polymers. Measurements were made using a 1:1 combination of CD dimer and Ad dimer. It was discovered that the M_n rose with concentration, exceeding 6×10^4 at 5 mM and reaching 1×10^4 at 20 mM. As a control, the molecular weights of CD dimers were mainly unaffected by the concentration as is evident from Figure 5.9 [39].

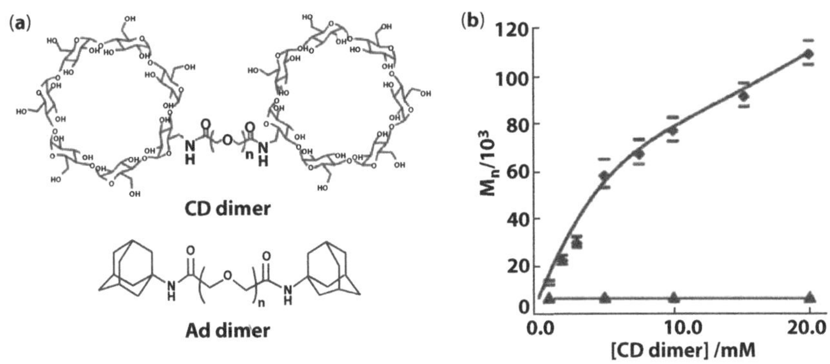

Figure 5.9 (a) Structure of the CD dimer and Ad dimer. (b) VPO results for 1:1 mixtures of the CD dimer, Ad dimer (rhombus), and CD dimer itself (triangle) [39]. Reprinted with permission from [39]. Copyright 2005 American Chemical Society.

5.7.6 Mass Spectrometry (MS)

Mass spectrometry (MS) has been used for at least 20 years for end-group identification, chemical composition analysis, and polymer analysis. This technique quantifies the mass-to-charge ratio of supramolecular polymers correlated with their molecular makeup and structure. Nonetheless, due to the poor volatility and thermal instability of biomolecules and polymers, the use of MS to determine their molar mass has been restricted. The development of "soft ionization" techniques, including matrix-assisted laser desorption/ionization time-of-flight mass spectrometry (MALDI-TOF-MS), has overcome these problems to a greater extent. An ion source, such as MALDI or electrospray ionization (ESI), is used to ionize the supramolecular polymers. A TOF or quadrupole mass analyzer separates the ionized supramolecular polymers. The mass spectrum may be utilized for supramolecular polymers to determine the molecular weight, fragmentation pattern, and formula [69].

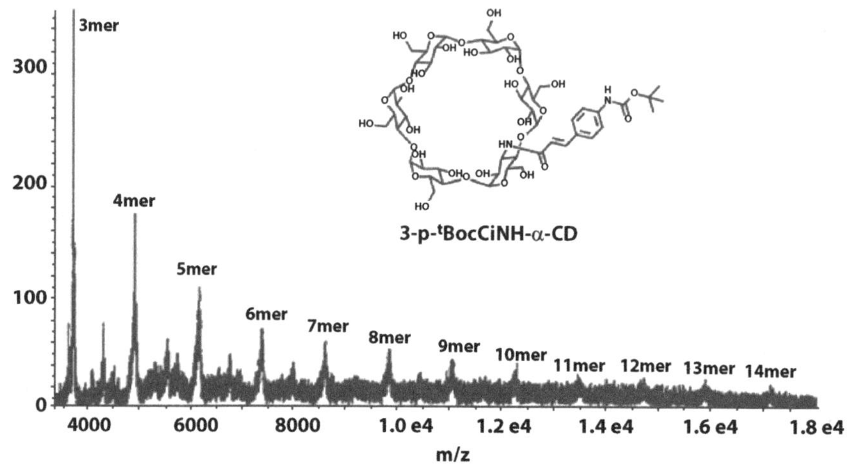

Figure 5.10 Positive turbo ion spray TOF mass spectra of 3-p-BocCiNH-a-CD in the range of 3,200 to 18,000. Reprinted with permission from [74]. Copyright 2005 American Chemical Society.

It is still challenging to determine the molecular weight of supramolecular polymers precisely. Usually, the only detectable species are short ones or oligomers [76, 77].

A study by Harada *et al.* provided an example of supramolecular polymers created by the host-guest interaction of the 3-*p*-*tert*-butoxyaminocinnamoylamino group and α-cyclodextrin that produced excellent mass spectra. Turbo ion spray TOF-MS was used to characterize these compounds, revealing the ability to identify up to 14-mer (as shown in Figure 5.10) [74].

5.7.7 Nuclear Magnetic Resonance (NMR) Spectroscopy

This technique detects the supramolecular polymer magnetic resonance, which is connected to the atom's nuclear spin and chemical surroundings. Radiofrequency pulses are applied to the supramolecular polymers under a magnetic field. Supramolecular polymer interactions, dynamics, conformation, and chemical structure may all be ascertained using the NMR spectra. Supramolecular polymer binding sites and affinities may be investigated using sophisticated NMR methods, such as saturation transfer difference (STD) NMR [78].

Furthermore, many supramolecular polymerization processes have broad distributions and resemble step-wise polymerization mechanisms. Concentration-dependent NMR may be used to investigate the supramolecular polymerization process. As the concentration of monomers increases, the NMR peaks broaden, signifying the creation of supramolecular polymers [79, 80]. For the characterization of supramolecular polymers, diffusion-ordered ^{1}H NMR spectroscopy (DOSY) is gaining popularity as it can quantify the diffusion coefficient directly [81–83]. The corresponding diffusion coefficients of supramolecular polymers can be used to compare their sizes qualitatively.

Haino and colleagues described the molecular recognition of trinitrofluorenone and bisporphyrin to form a supramolecular polymer [84] (as already depicted in Figure 5.11). Through a charge-transfer interaction, the electron-deficient guest moiety at the head, 4,5,7-trinitrofluorenone-2-carboxylate (TNF), can attach within the bisporphyrin cleft at the tail. This head-to-tail–type complexation results in supramolecular polymerization. The monomer and its analog,

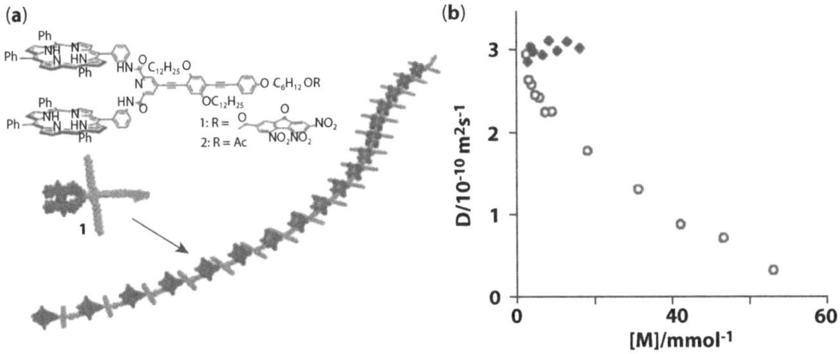

Figure 5.11 (a) Molecular structures of the heteroditopic monomer and its analog and a schematic illustration of the formation of supramolecular polymers. (b) Diffusion coefficients of the monomer (cycle) and its analog (rhombus) under different concentrations [84]. Reproduced with permission from [84]. Copyright © 2012 WILEY-VCH Verlag GmbH & Co. KGaA, Weinheim.

without the TNF moiety, were tested by DOSY. The diffusion coefficient of the analog $D_{avg} = 3.02 \times 10^{-10}\ m^2\ s^{-1}$ was not significantly affected by concentration. In other words, the independence of concentration on the diffusion coefficient infers that there is no supramolecular polymerization. Conversely, the monomer's diffusion coefficient was significantly dependent on concentration, showing a diffusion constant of $2.96 \times 10^{-10}\ m^2\ s^{-1}$ at 2.34 mM and $3.4 \times 10^{-11}\ m^2\ s^{-1}$ at 66 mM, respectively. Large polymeric aggregates are likely to have formed at greater concentrations, as indicated by the diffusion coefficient's considerable reduction with concentration.

For many systems, the correlation between supramolecular polymerization and diffusion that was previously described is valid. Wang *et al.* synthesized a linear supramolecular polymer with four hydrogen bonds, which they then cross-linked into networks using bisparaquat molecules [85]. According to the DOSY data, the diffusion coefficient steadily reduced as the monomer concentration rose, indicating the formation of longer supramolecular polymers. The diffusion coefficient of cross-linked polymers was much lower than that of linear supramolecular polymers.

5.7.8 Electron Microscopy (EM)

Using an electron beam, this technique visualizes the supramolecular polymers, which may be in a solution or a solid state. A variety of EM techniques, including cryogenic electron microscopy (cryo-EM), transmission electron microscopy (TEM), and scanning electron microscopy (SEM), can be used to examine supramolecular polymers. One may utilize the EM pictures to ascertain the supramolecular polymers' shape, size, and structure. High-resolution images may be obtained, and supramolecular polymers in solution can be preserved in their natural condition using cryo-EM [78].

Typically, aggregation made of supramolecular polymers may be seen using transmission electron microscopy (TEM). Schmuck *et al.* created switchable supramolecular polymers by using self-complementary Zwitterions, two orthogonal binding contacts, and metal coordination. Linear supramolecular polymer strands were discovered by cryo-TEM, which offers direct structural data from vitrified aqueous solution. The photos contrast nicely because of the metal ions in the aggregates [86].

5.7.9 Scanning Probe Microscopy (SPM)

Using a sharp tip, this technique scans the surface of supramolecular polymers, which can be in a solution or a solid state. Various modalities of scanning tunneling microscopy (STM), magnetic force microscopy (MFM), and atomic force microscopy (AFM) can all be used by the tip to interact with the supramolecular polymers. The topography, nanomechanics, and electrical characteristics of supramolecular polymers may all be ascertained using the SPM images. Supramolecular polymers' rupture forces and force-extension curves may be measured using AFM-based single-molecule force spectroscopy (SMFS) [87].

One of the most used techniques for characterizing two-dimensional structures is scanning tunneling microscopy (STM). There are various examples of supramolecular polymers visualized by STM [88, 89]. Supramolecular polymerization of a donor-acceptor bisterpyridine (BTP) furnished with an electron-rich carbazole unit was investigated using STM at the highly oriented pyrolytic graphite (HOPG)|solution interface [90]. It was shown that supramolecular (co)polymer two-dimensional crystals are formed via chain growth polymerization, which is further characterized by copolymerization statistics.

Another high-resolution SPM is AFM, which has a resolution of around one nanometer. The method, which has been extensively applied to scan the sample surfaces' morphology, might yield a three-dimensional surface profile [91]. AFM is an effective method for measuring growth height for surface-grafted supramolecular polymers. On a gold surface, Kim and colleagues created poly(pseudorotaxane). AFM research was able to determine that the poly(pseudorotaxane) was generated on the gold substrate with four repeating units because the height of the substrate grew by an average of 3.9 nm following the production of the poly(pseudorotaxane) [44].

5.8 Applications of Supramolecular Polymers

Some of the applications of supramolecular polymers in material science are the following:

5.8.1 Targeted Drug Delivery

Hydrogels made of supramolecular polymers can encapsulate medications and release them under regulated conditions in response to various stimuli, including pH, temperature, light, and enzyme [92]. For instance, it has been demonstrated that a supramolecular hydrogel derived from cucurbiturils and peptides has been used for targeted drug delivery [93].

5.8.2 Pollutant Sensors

Supramolecular polymers can function as fluorescence probes, changing their emission characteristics when bound to specific analytes such as pesticides, explosives, or heavy metals [94]. It has been demonstrated, for instance, that a supramolecular polymer based on a cyclodextrin and pyrene derivative may detect nitroaromatic explosives in water with selectivity [95].

5.8.3 Diagnostic Markers

Supramolecular polymers can be employed as contrast agents in bioimaging procedures, including positron emission tomography (PET), computed tomography (CT), and magnetic resonance imaging (MRI). A supramolecular polymer, for instance, based on a derivative of porphyrin and a gadolinium complex, has been demonstrated to improve the MRI signal of tumors in mice [96].

5.8.4 Energy Storage Devices

Batteries, solar cells, and supercapacitors may all employ supramolecular polymers as electrodes or electrolytes [97]. Conducting polymer hydrogels, or CPHs, have high electrical conductivity, electrochemical activity, and a micro-nanoscale three-dimensional network that makes them essential to energy storage [98].

5.8.5 Personal Care Products

To enhance the performance, stability, or functionality of cosmetics, shampoos, or lotions, supramolecular polymers can be added as ingredients or additives [99]. As an illustration, it has been demonstrated that a supramolecular polymer based on a urea derivative and a fatty acid may generate a thermoresponsive emulsion that can alter viscosity and appearance when heated or cooled [100].

5.8.6 Self-Repairing and Recycling Materials

Supramolecular polymers may be recycled by breaking and reforming their non-covalent bonds, or they can be employed as coatings, adhesives, or elastomers that can repair themselves after damage or degradation [101]. One supramolecular polymer demonstrated to generate a self-healing layer that may regain its mechanical and optical characteristics after scratching is based on a derivative of bipyridine and benzene-1,3,5-tricarboxamide [102].

5.9 Recent Advances in Supramolecular Chemistry

A central theme in recent supramolecular chemistry is the chemistry of fullerenes, which are spherical or ellipsoidal carbon allotropes. Due to their distinct electrical, optical, and magnetic characteristics, fullerenes are appealing for various uses, including drug delivery, spintronics, and solar cells [103–105]. However, fullerenes are challenging to synthesize and functionalize because of their excellent stability and limited reactivity. Supramolecular chemistry, which uses macrocycles and cages that can forcefully and selectively bind to fullerenes and control their addition patterns, can resolve these issues. Single-isomer addition products of fullerenes, such as fullerene bisadducts [59], fullerene monoadducts [68], and endohedral metallofullerenes, can be prepared using supramolecular templates [106].

Creating optical sensors for detecting and measuring diverse analytes, including ions, molecules, biomolecules, and environmental contaminants, is a significant field of study in supramolecular chemistry [107–110]. By using the selective and reversible binding of host-guest complexes, creating supramolecular aggregates, and controlling energy and electron transfer processes, supramolecular chemistry offers a flexible toolbox for designing and producing optical sensors [111].

Developing electrochemical energy storage devices, such as supercapacitors [112–114], which store and release energy by charging and discharging electrodes, is essential in supramolecular chemistry. Supercapacitors outperform traditional batteries in several ways, including high power density, quick charge/discharge rates, extended cycle lives, and environmental friendliness. Analysis has also been done on how supramolecular self-assembly affects the electrochemical performance of supercapacitors [115]. In addition to these projects, research is ongoing in several other areas, including sensing [116–119], catalysis [120–123], and medicine [124–126].

5.10 Future Aspects of Supramolecular Polymer Research

Supramolecular polymer research has a bright future, complete with challenges and opportunities. Future studies may focus on the following areas and objectives:

1. Establishing novel techniques and approaches, such as template-directed synthesis [127], self-sorting[128], and metal-containing monomers[129], for synthesizing supramolecular polymers.
2. Employing spectroscopic [130], microscopic [131], and computational methods [132] to investigate the structure-property relationship and the supramolecular polymerization mechanism.
3. Creating and manufacturing conductive [133, 134], magnetic [135, 136], luminous [133, 137], biocompatible [138, 139], and other functional supramolecular polymers [140–142].
4. Manufacturing and evaluating supramolecular polymer materials, including films [143–145], gels [146–149], crystals [150], and nanomaterials [151, 152].
5. Using supramolecular polymers in tissue engineering [153–155], medication delivery [156–158], energy storage [97, 159, 160], and molecular recognition [161–165], among other applications.

5.11 Conclusion

The numerous forces that drive supramolecular polymerization have been compiled in this chapter, along with information on how to prepare supramolecular polymers with various topologies, the most recent advancements in supramolecular polymerization under control, a synopsis of the techniques used to characterize supramolecular polymers, and a discussion of the possible uses for supramolecular polymers. Supramolecular polymers have made significant strides in the last 20 years, resulting in their emergence as a multidisciplinary topic of study that is gaining traction in both the academic and industrial domains.

The number of supramolecular polymers in the library has to be increased even if different non-covalent contacts have been employed to build supramolecular polymers. High binding continuous driving forces must be created. Other supramolecular polymer topologies, such as side chain, branched, and hyperbranched copolymers, should be prioritized more than linear main chain copolymers, which are more common. Following a few fundamental rules will help you produce supramolecular polymers with a high polymerization level. A high binding constant is the primary need for creating supramolecular polymers with a high degree of polymerization. A high concentration should be a valuable tactic to prevent oligomerization and cyclization and raise the degree of polymerization in systems with relatively low binding constants. The foundation of supramolecular polymer materials with highly customized functionalities is the research of the structure-function connection, which is greatly aided by supramolecular polymers with regulated structures. Though it is a formidable task, we are making progress. Scientists have worked hard and achieved a number of remarkable discoveries. We have reason to think that when the mechanism and kinetics of supramolecular polymerization [166, 167] are thoroughly studied, more and more methods can be created to accomplish regulated supramolecular polymerization. The applications of supramolecular polymers are definitely what has pushed this field of study forward. While metallosupramolecular polymers and supramolecular polymers based on numerous hydrogen bonds have significantly influenced material science, further research is still needed to explore the potential applications of supramolecular polymers based on other driving factors. Furthermore, to revitalize this relatively new field of study, we must expand into new application areas, including biomedical materials and smart responsive materials.

References

1. Huang, F. and Anslyn, E.V., Introduction: Supramolecular Chemistry. *Chem. Rev.*, 115, 15, 6999, 2015.
2. Ikeda, M., Stimuli-responsive supramolecular systems guided by chemical reactions. *Polym. J.*, 51, 4, 371–380, 2019.
3. Ariga, K., The evolution of molecular machines through interfacial nanoarchitectonics: From toys to tools. *Chem. Sci.*, 11, 39, 10594–10604, 2020.
4. Mako, T.L., Racicot, J., M., Levine, M., Supramolecular Luminescent Sensors. *Chem. Rev.*, 119, 1, 322–477, 2018.
5. Fabbrizzi, L. and Poggi, A., Sensors and Switches from Supramolecular Chemistry, 1995.
6. Minami, T., Design of supramolecular sensors and their applications to optical chips and organic devices. *Bull. Chem. Soc. Jpn.*, 94, 1, 24–33, 2020.
7. Rudkevich, D.M., Emerging Supramolecular Chemistry of Gases. *Angew. Chem. Int. Ed.*, 43, 5, 558–571, 2004.
8. Khariushin, I.V., Ovsyannikov, A.S., Baudron, S.A., Ward, J.S., Kiesilä, A., Rissanen, K., Kalenius, E., Kovalenko, K.A., Fedin, V.P., Solovieva, S.E., Antipin, I.S., Bulach, V., Ferlay, S., Selective gas adsorption by calixarene-based porous octahedral M32 coordination cages. *Chem. Commun.*, 58, 98, 13628–13631, 2022.
9. Lü, J., Perez-Krap, C., Trousselet, F., Yan, Y., H. Alsmail, N., Karadeniz, B., M. Jacques, N., Lewis, W., J. Blake, A., Coudert, F.-X., Cao, R., Schröder, M., Polycatenated 2D Hydrogen-Bonded Binary Supramolecular Organic Frameworks (SOFs) with Enhanced Gas Adsorption and Selectivity. *Cryst. Growth Des.*, 18, 4, 2555–2562, 2018.

10. Swaminathan, S., Fowley, C., Thapaliya, E.R., McCaughan, B., Tang, S., Fraix, A., Captain, B., Sortino, S., Callan, J.F., Raymo, F.M., Supramolecular nanoreactors for intracellular singlet-oxygen sensitization. *Nanoscale*, 7, 33, 14071–14079, 2015.
11. Capasso Palmiero, U., Sponchioni, M., Margani, F., Moscatelli, D., Galimberti, M., Barbera, V., A Graphene-Based Supramolecular Nanoreactor for the Fast Synthesis of Imines in Water. *Small*, 16, 40, 2001207, 2020.
12. Olivo, G., Capocasa, G., Del Giudice, D., Lanzalunga, O., Di Stefano, S., New horizons for catalysis disclosed by supramolecular chemistry. *Chem. Soc. Rev.*, 50, 13, 7681–7724, 2021.
13. Webber, M.J. and Langer, R., Drug delivery by supramolecular design. *Chem. Soc. Rev.*, 46, 21, 6600–6620, 2017.
14. Ailincai, D., Morariu, S., Rosca, I., Sandu, A.I., Marin, L., Drug delivery based on a supramolecular chemistry approach by using chitosan hydrogels. *Int. J. Biol. Macromol.*, 248, 125800, 2023.
15. Aida, T. and Meijer, E.W., Supramolecular Polymers – we've Come Full Circle. *Isr. J. Chem.*, 60, 1–2, 33–47, 2020.
16. Brunsveld, L., Folmer, B.J.B., Meijer, E.W., Sijbesma, R.P., Supramolecular polymers. *Chem. Rev.*, 101, 12, 4071–4097, 2001.
17. Wang, Y.J. and Tang, L.M., Supramolecular polymers formed through hydrogen-bonded non-covalent interactions. *Prog. Chem.*, 18, 308–316, 2006.
18. Aida, T., Meijer, E.W., Stupp, S.I., Functional Supramolecular Polymers.
19. Rowan, S.J., Cantrill, S.J., Cousins, G.R.L., Sanders, J.K.M., Stoddart, J.F., Dynamic covalent chemistry. *Angew. Chem. Int. Ed.*, 41, 6, 898–952, 2002.
20. Desiraju, G.R., Hydrogen Bonding, in: *Encyclopedia of Supramolecular Chemistry*, pp. 658–665, CRC Press, New York, 2004.
21. Sijbesma, R.P., Beijer, F.H., Brunsveld, L., Folmer, B.J.B., Hirschberg, J.H.K.K., Lange, R.F.M., Lowe, J.K.L., Meijer, E.W., Reversible Polymers Formed from Self-Complementary Monomers Using Quadruple Hydrogen Bonding. *Sci. (1979)*, 278, 5343, 1601–1604, 1997.
22. Whitesides, G.M. and Boncheva, M., Beyond molecules: Self-assembly of mesoscopic and macroscopic components. *PNAS April*, 16, (8), 2002.
23. Lehn, J.M., Towards complex matter: Supramolecular chemistry and self-organization. *Eur. Rev.*, 17, 2, 263–280, 2009.
24. Tan, S., Sha, Y., Zhu, T., Rahman, M.A., Tang, C., Photoresponsive supramolecular polymers based on quadruple hydrogen-bonding and a photochromic azobenzene motif. *Polym. Chem.*, 9, 44, 5395–5401, 2018.
25. Feng, Z., Zuo, H., Gao, W., Ning, N., Tian, M., Zhang, L., A Robust, Self-Healable, and Shape Memory Supramolecular Hydrogel by Multiple Hydrogen Bonding Interactions. *Macromol. Rapid Commun.*, 39, (20), 2018.
26. Neniškis, A., Račkauskaitė, D., Shi, Q., Robertson, A.J., Marsh, A., Ulčinas, A., Valiokas, R., Brown, S.P., Wärnmark, K., Orentas, E., A Tautoleptic Approach to Chiral Hydrogen-Bonded Supramolecular Tubular Polymers with Large Cavity. *Chem. Eur. J.*, 24, 53, 14028–14033, 2018.
27. Khazi, M.I., Balachandra, C., Shin, G., Jang, G.H., Govindaraju, T., Kim, J.M., Co-solvent polarity tuned thermochromic nanotubes of cyclic dipeptide-polydiacetylene supramolecular system. *RSC Adv.*, 10, 58, 35389–35396, 2020.
28. Lugger, J.D., S.Houben, S., J.A., Foelen, Y., Debije, M., G., Schenning, A., P.H.J., Mulder, D., J., Hydrogen-Bonded Supramolecular Liquid Crystal Polymers: Smart Materials with Stimuli-Responsive, Self-Healing, and Recyclable Properties. *Chem. Rev.*, 122, 5, 4946–4975, 2021.
29. Chen, T., Li, M., Liu, J., π–π Stacking Interaction: A Nondestructive and Facile Means in Material Engineering for Bioapplications. *Cryst. Growth Des.*, 18, 5, 2765–2783, 2018.

30. Burattini, S., Greenland, B., W., Hayes, W., Mackay, M., E., Rowan, S., J., Colquhoun, H., M., A Supramolecular Polymer Based on Tweezer-Type π–π Stacking Interactions: Molecular Design for Healability and Enhanced Toughness. *Chem. Mater.*, 23, 1, 6–8, 2010.
31. Hoeben, F., J.M., Jonkheijm, P., Meijer, E., W., Schenning, A., P.H.J., About Supramolecular Assemblies of π-Conjugated Systems. *Chem. Rev.*, 105, 4, 1491–1546, 2005.
32. Panja, S. and Adams, D.J., Stimuli responsive dynamic transformations in supramolecular gels. *Chem. Soc. Rev.*, 50, 8, 5165–5200, 2021.
33. Hema, K., Ravi, A., Raju, C., Sureshan, K.M., Polymers with advanced structural and supramolecular features synthesized through topochemical polymerization. *Chem. Sci.*, 12, 15, 5361–5380, 2021.
34. Sánchez-Fernández, J.A., Structural Strategies for Supramolecular Hydrogels and Their Applications. *Polym. (Basel)*, 15, (6), 2023.
35. Coste, M., Suárez-Picado, E., Ulrich, S., Hierarchical self-assembly of aromatic peptide conjugates into supramolecular polymers: It takes two to tango. *Chem. Sci.*, 13, 4, 909–933, 2022.
36. Niu, W., Zhu, Y., Wang, R., Lu, Z., Liu, X., Sun, J., Remalleable, Healable, and Highly Sustainable Supramolecular Polymeric Materials Combining Superhigh Strength and Ultrahigh Toughness. *ACS Appl. Mater. Interfaces*, 12, 27, 30805–30814, 2020.
37. Huo, Y., He, Z., Wang, C., Zhang, L., Xuan, Q., Wei, S., Wang, Y., Pan, D., Dong, B., Wei, R., Naik, N., Guo, Z., The recent progress of synergistic supramolecular polymers: preparation, properties and applications. *Chem. Commun.*, 57, 12, 1413–1429, 2021.
38. George M. Whitesides, B.G., Self assembly at all scales. *Sci. (1979)*, 295, 2418–2421, 2002.
39. Hasegawa, Y., Miyauchi, M., Takashima, Y., Yamaguchi, H., Harada, A., Supramolecular Polymers Formed from β-Cyclodextrins Dimer Linked by Poly(ethylene glycol) and Guest Dimers. *Macromolecules*, 38, 9, 3724–3730, 2005.
40. Stoddart, J.F., Mechanisch verzahnte Moleküle (MIMs) – molekulare Shuttle, Schalter und Maschinen (Nobel-Aufsatz). *Angew. Chem.*, 129, 37, 11244–11277, 2017.
41. Yaghi, O., M. and Li, H., Hydrothermal Synthesis of a Metal-Organic Framework Containing Large Rectangular Channels. *J. Am. Chem. Soc.*, 117, 41, 10401–10402, 2002.
42. Schmidt, B.V.K.J., Rudolph, T., Hetzer, M., Ritter, H., Schacher, F.H., Barner-Kowollik, C., Supramolecular three-armed star polymers via cyclodextrin host-guest self-assembly. *Polym. Chem.*, 3, 11, 3139–3145, 2012.
43. Yu, G. and Chen, X., Host–guest chemistry in supramolecular theranostics. *Theranostics*, 9, 11, 3041–3074, 2019.
44. Kim, K., Selvapalam, N., Ho Ko, Y., Min Park, K., Kim, D., Kim, J., Functionalized cucurbiturils and their applications. *Chem. Soc. Rev.*, 36, 2, 267–279, 2007.
45. Hu, J. and Liu, S., Engineering responsive polymer building blocks with host-guest molecular recognition for functional applications. *Acc. Chem. Res.*, 47, 7, 2084–2095, 2014.
46. Shi, B., Liu, Y., Zhu, H., T. Vanderlinden, R., Shangguan, L., Ni, R., Acharyya, K., Tang, J.-H., Zhou, Z., Li, X., Huang, F., Stang, P., J., Spontaneous Formation of a Cross-Linked Supramolecular Polymer Both in the Solid State and in Solution, Driven by Platinum(II) Metallacycle-Based Host–Guest Interactions. *J. Am. Chem. Soc.*, 141, 16, 6494–6498, 2019.
47. Harada, A., Supramolecular Polymers (Host-Guest Interactions), in: *Encyclopedia of Polymeric Nanomaterials*, pp. 1–5, Springer Berlin Heidelberg, 2013.
48. Armstrong, G., Supramolecular Polymers (Hydrogen Bonds), in: *Encyclopedia of Polymeric Nanomaterials*, pp. 1–10, Springer Berlin Heidelberg, 2013.
49. Lehn, J.-M., Supramolecular Chemistry-Molecular Information and the Design of Supramolecular Materials. *Makromol. Chem. Macromol. Symp.*, 69, 1–17, 1993.
50. Kotera, M., Lehn, J.-M., Vigneron, J.-P., Design and Synthesis of Complementary Components for the Formation of Self-Assembled Supramolecular Rigid Rods. *Tetrahedron*, 51, 7, 1995.

51. Deng, J.-H., Luo, J., Mao, Y.-L., Lai, S., Gong, Y.-N., Zhong, D.-C., Lu, T.-B., Stacking interactions: Non-negligible forces for stabilizing porous supramolecular frameworks, 2020.
52. Xu, J.F., Chen, L., Zhang, X., How to Make Weak Noncovalent Interactions Stronger. *Chem. Eur. J.*, 21, 34, 11938–11946, 2015.
53. Zhang, L., Zhou, T.Y., Tian, J., Wang, H., Zhang, D.W., Zhao, X., Liu, Y., Li, Z.T., A two-dimensional single-layer supramolecular organic framework that is driven by viologen radical cation dimerization and further promoted by cucurbit[8]uril. *Polym. Chem.*, 5, 16, 4715–4721, 2014.
54. R. Brocker, E., E. Anderson, S., H. Northrop, B., J. Stang, P., T. Bowers, M., Structures of Metallosupramolecular Coordination Assemblies Can Be Obtained by Ion Mobility Spectrometry–Mass Spectrometry. *J. Am. Chem. Soc.*, 132, 38, 13486–13494, 2010.
55. Winter, A. and Schubert, U.S., Synthesis and characterization of metallo-supramolecular polymers. *Chem. Soc. Rev.*, 45, 19, 5311–5357, 2016.
56. Schubert, U.S., Newkome, G.R., Winter, A., *Supramolecular Polymers and Assemblies, 1sted*, pp. 117–194, Wiley, VCH GmbH, Boschstr. Weinheim, Germany, 2021.
57. Dobrawa, R.,. and Würthner, F., Metallosupramolecular approach toward functional coordination polymers. *J. Polym. Sci. A Polym. Chem.*, 43, 21, 4981–4995, 2005.
58. Zhao, J., Yuan, J., Fang, Z., Huang, S., Chen, Z., Qiu, F., Lu, C., Zhu, J., Zhuang, X., One-dimensional coordination polymers based on metal–nitrogen linkages. *Coord. Chem. Rev.*, 4, 71, 214735 2022.
59. Chi, Z., Dong, H., Shi, G., Liu, P., Ma, C., Chen, X., Synthesis and Characterization of Metallo-Supramolecular Polymers Based on Benzodipyrrolidone. *Front. Chem.*, 9, 673834, 2021.
60. Winter, A., Friebe, C., Chiper, M., Hager, M.D., Schubert, U.S., Self-assembly of π-conjugated bis(terpyridine) ligands with zinc(II) ions: New metallosupramolecular materials for optoelectronic applications. *J. Polym. Sci. A Polym. Chem.*, 47, 16, 4083–4098, 2009.
61. Ulbricht, C., Beyer, B., Friebe, C., Winter, A., Schubert, U.S., Recent developments in the application of phosphorescent iridium(III) complex systems. *Adv. Mater.*, 21, 44, 4418–4441, 2009.
62. Bruce Martin, R., Comparisons of Indefinite Self-Association Models. *Chem. Rev.*, 96, 8, 3043–3064, 1996.
63. Brunsveld, L., Folmer, J.B., B., W., Sijbesma, P., R., Supramolecular Polymers. *Chem. Rev.*, 101, 12, 4071–4098, 2001.
64. Paulusse, J.M.J. and Sijbesma, R.P., A high molecular weight reversible coordination polymer of PdCl2 and 1,12-bis(diphenylphosphino)dodecane. *Chem. Commun.*, 3, 13, 1494–1495, 2003.
65. Meier, M.A.R., Hofmeier, H., Abeln, C.H., Tziatzios, C., Rasa, M., Schubert, D., Schubert, U.S., First GPC results of terpyridine based chain extended supramolecular polymers: comparison with viscosity and analytical ultracentrifugation., 016, 2006.
66. Raymo, F., M. and Fraser Stoddart, J., Interlocked Macromolecules. *Chem. Rev.*, 99, 7, 1643–1664, 1999.
67. Fang, L., Hmadeh, M., Wu, J., A. Olson, M., M. Spruell, J., Trabolsi, A., Yang, Y.-W., Elhabiri, M., Albrecht-Gary, A.-M., Fraser Stoddart, J., Acid–Base Actuation of [c2]Daisy Chains. *J. Am. Chem. Soc.*, 131, 20, 7126–7134, 2009.
68. Jie, K., Zhou, Y., Ji, X., A pH-responsive amphiphilic supramolecular graft copolymer constructed by crown ether based molecular recognition. *Polym. Chem.*, 6, 2, 218–222, 2015.
69. Liu, Y., Wang, Z., Zhang, X., Characterization of supramolecular polymers. *Chem. Soc. Rev.*, 41, 18, 5922–5932, 2012.
70. Abed, S., Boileau, S., Bouteiller, L., *Supramolecular association of acid terminated polydimethylsiloxanes. 3*, Viscosimetric study.
71. Characterization of Supramolecular Polymers, in: *Supramolecular Polymers and Assemblies*, pp. 453–502, John Wiley & Sons, Ltd, 2021.

72. Kyung Yang, S., Ambade, A., V., Weck, M., Supramolecular ABC Triblock Copolymers via One-Pot, Orthogonal Self-Assembly. *J. Am. Chem. Soc.*, 132, 5, 1637–1645, 2010.
73. Schmidt, R., Stolte, M., Grüne, M., Würthner, F., Hydrogen-Bond-Directed Formation of Supramolecular Polymers Incorporating Head-to-Tail Oriented Dipolar Merocyanine Dyes. *Macromolecules*, 44, 10, 3766–3776, 2011.
74. Miyauchi, M., Takashima, Y., Yamaguchi, H., Harada, A., Chiral Supramolecular Polymers Formed by Host–Guest Interactions. *J. Am. Chem. Soc.*, 127, 9, 2984–2989, 2005.
75. Ohga, K., Takashima, Y., Takahashi, H., Kawaguchi, Y., Yamaguchi, H., Harada, A., Preparation of Supramolecular Polymers from a Cyclodextrin Dimer and Ditopic Guest Molecules: Control of Structure by Linker Flexibility. *Macromolecules*, 38, 14, 5897–5904, 2005.
76. Huang, F. and Gibson, H., W., Formation of a Supramolecular Hyperbranched Polymer from Self-Organization of an AB2 Monomer Containing a Crown Ether and Two Paraquat Moieties. *J. Am. Chem. Soc.*, 126, 45, 14738–14739, 2004.
77. Huang, F. and Gibson, H.W., A supramolecular poly[3]pseudorotaxane by self-assembly of a homoditopic cylindrical bis(crown ether) host and a bisparaquat derivative. *Chem. Commun.*, 13), 1696–1698, 2005.
78. Denzer, B.R., Kulchar, R.J., Huang, R.B., Patterson, J., Advanced Methods for the Characterization of Supramolecular Hydrogels. *Gels*, 7, (4), 2021.
79. Yan, X., Xu, D., Chi, X., Chen, J., Dong, S., Ding, X., Yu, Y., Huang, F., A multiresponsive, shape-persistent, and elastic supramolecular polymer network gel constructed by orthogonal self-assembly. *Adv. Mater.*, 24, 3, 362–369, 2012.
80. Yan, X., Zhou, M., Chen, J., Chi, X., Dong, S., Zhang, M., Ding, X., Yu, Y., Shao, S., Huang, F., Supramolecular polymer nanofibers via electrospinning of a heteroditopic monomer. *Chem. Commun.*, 47, 25, 7086–7088, 2011.
81. Nally, R., and Isaacs, L., Toward supramolecular polymers incorporating double cavity cucurbituril hosts. *Tetrahedron*, 65, 35, 7249–7258, 2009.
82. Folmer, J.B., B., P., Sijbesma, R., Meijer, E., W., Unexpected Entropy-Driven Ring-Opening Polymerization in a Reversible Supramolecular System. *J. Am. Chem. Soc.*, 123, 9, 2093–2094, 2001.
83. Capici, C., Cohen, Y., D'Urso, A., Gattuso, G., Notti, A., Pappalardo, A., Pappalardo, S., Parisi, M.F., Purrello, R., Slovak, S., Villari, V., Anion-assisted supramolecular polymerization: From achiral AB-type monomers to chiral assemblies. *Angew. Chem. Int. Ed.*, 50, 50, 11956–11961, 2011.
84. Haino, T., Watanabe, A., Hirao, T., Ikeda, T., Supramolecular polymerization triggered by molecular recognition between bisporphyrin and trinitrofluorenone. *Angew. Chem. Int. Ed.*, 51, 6, 1473–1476, 2012.
85. Li, S.L., Xiao, T., Hu, B., Zhang, Y., Zhao, F., Ji, Y., Yu, Y., Lin, C., Wang, L., Formation of polypseudorotaxane networks by cross-linking the quadruple hydrogen bonded linear supramolecular polymers via bisparaquat molecules. *Chem. Commun.*, 47, 38, 10755–10757, 2011.
86. Gröger, G., Meyer-Zaika, W., Böttcher, C., Gröhn, F., Ruthard, C., Schmuck, C., Switchable supramolecular polymers from the self-assembly of a small monomer with two orthogonal binding interactions. *J. Am. Chem. Soc.*, 133, 23, 8961–8971, 2011.
87. Fukui, T., Uchihashi, T., Sasaki, N., Watanabe, H., Takeuchi, M., Sugiyasu, K., Direct Observation and Manipulation of Supramolecular Polymerization by High-Speed Atomic Force Microscopy. *Angew. Chem. Int. Ed.*, 57, 47, 15465–15470, 2018.
88. Ouchi, H., Kizaki, T., Yamato, M., Lin, X., Hoshi, N., Silly, F., Kajitani, T., Fukushima, T., Nakayama, K.I., Yagai, S., Impact of helical organization on the photovoltaic properties of oligothiophene supramolecular polymers. *Chem. Sci.*, 9, 15, 3638–3643, 2018.

89. Li, Z., Li, Y., Zhao, Y., Wang, H., Zhang, Y., Song, B., Li, X., Lu, S., Hao, X.Q., Hla, S.W., Tu, Y., Li, X., Synthesis of Metallopolymers and Direct Visualization of the Single Polymer Chain. *J. Am. Chem. Soc.*, 142, 13, 6196–6205, 2020.
90. Goll, F.D., Taubmann, G., Ziener, U., Static Scanning Tunneling Microscopy Images Reveal the Mechanism of Supramolecular Polymerization of an Oligopyridine on Graphite. *Angew. Chem. Int. Ed.*, 61, 16, 2022.
91. Zhu, L., Lu, M., Zhang, Q., Qu, D., Tian, H., Construction of Polypseudorotaxane from Low-Molecular Weight Monomers via Dual Noncovalent Interactions. *Macromolecules*, 44, 11, 4092–4097, 2011.
92. Torsten, R. and Seiffert, S., Supramolecular Polymer Networks: Preparation, Properties, and Potential, in: *Supramolecular Polymer Networks and Gels*, vol. pp, S. Seiffert, (Ed.), pp. 1–46, Springer International Publishing, Cham, 2015.
93. Gao, Y., Gao, Y., Ding, Y., Tan, H., Zou, A., Li, S., Polysaccharide-based supramolecular drug delivery systems mediated via host-guest interactions of cucurbiturils. *Chin. Chem. Lett.*, 32, 3, 949–953, 2021.
94. Dai, D., Li, Z., Yang, J., Wang, C., Wu, J.-R., Wang, Y., Zhang, D., Yang, Y.-W., Supramolecular Assembly-Induced Emission Enhancement for Efficient Mercury(II) Detection and Removal. *J. Am. Chem. Soc.*, 141, 11, 4756–4763, 2019.
95. Singh, G. and Singh, P.K., Stimulus-Responsive Supramolecular Host–Guest Assembly of a Cationic Pyrene Derivative with Sulfated β-Cyclodextrin. *Langmuir*, 35, 45, 14628–14638, 2019.
96. Shahbazi-Gahrouei, D., Novel MR imaging contrast agents for cancer detection, vol. 14.
97. Jena, S.R. and Choudhury, J., Solar cell-coupled metallo-supramolecular polymer-based electrochromic device in renewable energy storage and on-demand usage. *Sol. Energy Mater. Sol. Cells*, 239, 111660, 2022.
98. Alcaraz-Espinoza, J.J., Ramos-Sánchez, G., Sierra-Uribe, J.H., González, I., Supramolecular Assembly of Nanostructured Conducting Polymer Hydrogels by Hydrotropic Agents for Outstanding Supercapacitive Energy Storage. *ACS Appl. Energy Mater.*, 4, 9, 9099–9110, 2021.
99. Ahmadi, D., Ledder, R., Mahmoudi, N., Li, P., Tellam, J., Robinson, D., Heenan, R.K., Smith, P., Lorenz, C.D., Barlow, D.J., Lawrence, M.J., Supramolecular architecture of a multi-component biomimetic lipid barrier formulation. *J. Colloid Interface Sci.*, 587, 597–612, 2021.
100. Wang, Z., Liao, S., Wang, Y., Supramolecular Polymer Emulsifiers for One-step Complex Emulsions. *Chin. J. Polym. Sci.*, 36, 2017.
101. Hashim, P.K., Bergueiro, J., Meijer, E.W., Aida, T., Supramolecular Polymerization: A Conceptual Expansion for Innovative Materials. *Prog. Polym. Sci.*, 105, 101250, 2020.
102. Roosma, J., Mes, T., Leclère, P., Palmans, A.R.A., Meijer, E.W., Supramolecular Materials from Benzene-1,3,5-tricarboxamide-Based Nanorods. *J. Am. Chem. Soc.*, 130, 4, 1120–1121, 2008.
103. Yu, X., Wang, B., Kim, Y., Park, J., Ghosh, S., Dhara, B., Mukhopadhyay, R.D., Koo, J., Kim, I., Kim, S., Hwang, I.C., Seki, S., Guldi, D.M., Baik, M.H., Kim, K., Supramolecular Fullerene Tetramers Concocted with Porphyrin Boxes Enable Efficient Charge Separation and Delocalization. *J. Am. Chem. Soc.*, 142, 29, 12596–12601, 2020.
104. Diederich, F. and Gómez-López, M., Supramolecular fullerene chemistry. *Chem. Soc. Rev.*, 28, 5, 263–277, 1999.
105. Bao, L., Xu, T., Guo, K., Huang, W., Lu, X., Supramolecular Engineering of Crystalline Fullerene Micro-/Nano-Architectures. *Adv. Mater.*, 34, 52, 2200189, 2022.
106. Chang, X., Xu, Y., von Delius, M., Recent advances in supramolecular fullerene chemistry. *Chem. Soc. Rev.*, 2023.
107. Sasaki, Y., Lyu, X., Tang, W., Wu, H., Minami, T., Supramolecular optical sensor arrays for on-site analytical devices. *J. Photochem. Photobiol., C*, 51, 100475, 2022.

108. Martins, T.D., Ribeiro, A.C.C., Colmati, F., de Souza, G.A., de Camargo, H.S., Dias, D.L., da Costa Filho, P.A., de Sousa Cordeiro, D., Supramolecular Materials for Optical and Electrochemical Biosensors, in: *Biosensors*, T. Rinken, (Ed.), IntechOpen, Rijeka, 2015.
109. Charlesworth, J.M., Optical sensing of oxygen using phosphorescence quenching. *Sens. Actuators B Chem.*, 22, 1, 1–5, 1994.
110. Wolfbeis Otto, S., Huber, C., W.T., Solid State Supramolecular Optical Sensors, in: *Chemosensors of Ion and Molecule Recognition*, J.P. Desvergne and A.W. Czarnik, (Eds.), pp. 61–74, Springer, Netherlands, Dordrecht, 1997.
111. You, L., Zha, D., Anslyn, E.V., Recent Advances in Supramolecular Analytical Chemistry Using Optical Sensing. *Chem. Rev.*, 115, 15, 7840–7892, 2015.
112. Xu, S., Liang, X., Ge, K., Yuan, H., Liu, G., Supramolecular Gel Electrolyte-Based Supercapacitors with a Comparable Dependence of Electrochemical Performances on Electrode Thickness to Those Based on Bulk Electrolyte Solutions. *ACS Appl. Energy Mater.*, 5, 3, 2929–2936, 2022.
113. Peng, H., Montes-García, V., Raya, J., Wang, H., Guo, H., Richard, F., Samorì, P., Ciesielski, A., Supramolecular engineering of cathode materials for aqueous zinc-ion hybrid supercapacitors: novel thiophene-bridged donor–acceptor sp2 carbon-linked polymers. *J. Mater. Chem. A*, 11, 6, 2718–2725, 2023.
114. Kundu, S., Mogera, U., George, S.J., Kulkarni, G.U., A planar supercapacitor made of supramolecular nanofibre based solid electrolyte exhibiting 8 V window. *Nano Energy*, 61, 259–266, 2019.
115. Cheng, H., Liu, R., Zhang, R., Huang, L., Yuan, Q., Recent advances in supramolecular self-assembly derived materials for high-performance supercapacitors. *Nanoscale Adv.*, 5, 9, 2394–2412, 2023.
116. Hirao, T., Kishino, S., Haino, T., Supramolecular chiral sensing by supramolecular helical polymers. *Chem. Commun.*, 59, 17, 2421–2424, 2023.
117. Fukuhara, G., Polymer-based supramolecular sensing and application to chiral photochemistry. *Polym. J.*, 47, 10, 649–655, 2015.
118. Gole, B., Song, W., Lackinger, M., Mukherjee, P.S., Explosives Sensing by Using Electron-Rich Supramolecular Polymers: Role of Intermolecular Hydrogen Bonding in Significant Enhancement of Sensitivity. *Chem. A Eur. J.*, 20, 42, 13662–13680, 2014.
119. Lim, S., Kuang, Y., Ardoña, H.A.M., Evolution of Supramolecular Systems Towards Next-Generation Biosensors. *Front. Chem.*, 9, 2021.
120. P. Gerola, A., H. Wanderlind,.E., S. Gomes,.Y., A. Giusti,.L., García-Río, L., A. Nome,.R., J. Kirby, A., D. Fiedler, H., Nome, F., Supramolecular Polymer/Surfactant Complexes as Catalysts for Phosphate Transfer Reactions. *ACS Catal.*, 7, 3, 2230–2239, 2017.
121. Costa, P.F.A., de Abreu, R., Fontana, A.B., Fiedler, H.D., Kirby, A.J., Quina, F.H., Nome, F., Gerola, A.P., The role of hydrophobicity in supramolecular polymer/surfactant catalysts: An understandable model for enzymatic catalysis. *J. Colloid Interface Sci.*, 588, 456–468, 2021.
122. Miyamura, H., Bergman, G., R.N. Raymond, K., Dean Toste, F., Heterogeneous Supramolecular Catalysis through Immobilization of Anionic M4L6 Assemblies on Cationic Polymers. *J. Am. Chem. Soc.*, 142, 45, 19327–19338, 2020.
123. Neumann, L.N., Baker, M.B., Leenders, C.M.A., Voets, I.K., Lafleur, R.P.M., Palmans, A.R.A., Meijer, E.W., Supramolecular polymers for organocatalysis in water. *Org. Biomol. Chem.*, 13, 28, 7711–7719, 2015.
124. Ma, X. and Zhao, Y., Biomedical Applications of Supramolecular Systems Based on Host–Guest Interactions. *Chem. Rev.*, 115, 15, 7794–7839, 2015.
125. Ma,, Z. and Qi, J., Application of Supramolecular Polymer Nanoparticles in Controlled Release System of Anticancer Drugs. *J. Nanomater.*, 2022, 2219602, 2022.

126. Yu, G., Zhao, X., Zhou, J., Mao, Z., Huang, X., Wang, Z., Hua, B., Liu, Y., Zhang, F., He, Z., Jacobson, O., Gao, C., Wang, W., Yu, C., Zhu, X., Huang, F., Chen, X., Supramolecular Polymer-Based Nanomedicine: High Therapeutic Performance and Negligible Long-Term Immunotoxicity. *J. Am. Chem. Soc.*, 140, 25, 8005–8019, 2018.
127. Fasano, F., Bolgar, P., Iadevaia, G., Hunter, C.A., Supramolecular template-directed synthesis of triazole oligomers. *Chem. Sci.*, 13, 44, 13085–13093, 2022.
128. Sarkar, A., Sasmal, R., Empereur-mot, C., Bochicchio, D., Kompella, V.K., S. Sharma, K., Dhiman, S., Sundaram, B., Agasti, S., Pavan, G., S.M., George S., J., Self-Sorted, Random, and Block Supramolecular Copolymers via Sequence Controlled, Multicomponent Self-Assembly. *J. Am. Chem. Soc.*, 142, 16, 7606–7617, 2020.
129. Hofmeier, H., Hoogenboom, R., Wouters M., E.L., Schubert U., S., High Molecular Weight Supramolecular Polymers Containing Both Terpyridine Metal Complexes and Ureidopyrimidinone Quadruple Hydrogen-Bonding Units in the Main Chain. *J. Am. Chem. Soc.*, 127, 9, 2913–2921, 2005.
130. Frederix, P.W.J.M., Patmanidis, I., Marrink, S.J., Molecular simulations of self-assembling bio-inspired supramolecular systems and their connection to experiments. *Chem. Soc. Rev.*, 47, 10, 3470–3489, 2018.
131. Tan, J.C. and Cheetham, A.K., Mechanical properties of hybrid inorganic–organic framework materials: establishing fundamental structure–property relationships. *Chem. Soc. Rev.*, 40, 2, 1059–1080, 2011.
132. Maidur, R., Shankaragouda Patil, P., Krishnakanth Katturi, N., Rao Soma, V., Ai Wong,.Q., Kheng Quah, C., Ultrafast Nonlinear Optical and Structure–Property Relationship Studies of Pyridine-Based Anthracene Chalcones Using Z-Scan, Degenerate Four-Wave Mixing, and Computational Approaches. *J. Phys. Chem. B*, 125, 15, 3883–3898, 2021.
133. García, F., Buendía, J., Ghosh, S., Ajayaghosh, A., Sánchez, L., Luminescent and conductive supramolecular polymers obtained from an N-annulated perylenedicarboxamide. *Chem. Commun.*, 49, 81, 9278–9280, 2013.
134. Wang, Y., Shi, Y., Pan, L., Ding, Y., Zhao, Y., Li, Y., Shi, Y., Yu, G., Dopant-Enabled Supramolecular Approach for Controlled Synthesis of Nanostructured Conductive Polymer Hydrogels. *Nano Lett.*, 15, 11, 7736–7741, 2015.
135. Ali, A., Shah, T., Ullah, R., Zhou, P., Guo, M., Ovais, M., Tan, Z., Rui, Y.K., Review on Recent Progress in Magnetic Nanoparticles: Synthesis, Characterization, and Diverse Applications. *Front. Chem.*, 2021.
136. Armao IV, J., Maaloum, M., Ellis, T., Fuks, G., Rawiso, M., Moulin, E., Giuseppone, N., Healable Supramolecular Polymers as Organic Metals. *J. Am. Chem. Soc.*, 136, 32, 11382–11388, 2014.
137. Tao, L., Luo, Z.-W., Lan, K., Wang, P., Guan, Y., Shen, Z., Xie, H.-L., Stimuli-responsive luminescent supramolecular polymers based on hydrogen bonding: molecular fabrication, phase structure, and controllable-rewritable behavior. *Polym. Chem.*, 11, 39, 6288–6294, 2020.
138. Marić, I., Yang, L., Li, X., Santiago, G.M., Pappas, C.G., Qiu, X., Dijksman, J.A., Mikhailov, K., van Rijn, P., Otto, S., Tailorable and Biocompatible Supramolecular-Based Hydrogels Featuring two Dynamic Covalent Chemistries. *Angew. Chem. Int. Ed.*, 62, 14, e202216475, 2023.
139. Liu, S., Huang, Z., Li, F., Yan, T., Fu, S., Tian, R., Hou, C., Luo, Q., Xu, J., Liu, J., Supramolecular polymer nanocapsules by enzymatic covalent condensation: biocompatible and biodegradable drug-delivery systems for chemo-photothermal anticancer therapy. *Polym. Chem.*, 10, 26, 3566–3570, 2019.
140. Xia, D., Wang, P., Ji, X., M. Khashab, N., L. Sessler, J., Huang, F., Functional Supramolecular Polymeric Networks: The Marriage of Covalent Polymers and Macrocycle-Based Host–Guest Interactions. *Chem. Rev.*, 120, 13, 6070–6123, 2020.

141. Ghosh, S., Shankar, S., Philips, D.S., Ajayaghosh, A., Diketopyrrolopyrrole-based functional supramolecular polymers: next-generation materials for optoelectronic applications. *Mater. Today Chem.*, 16, 100242, 2020.
142. Wang, Y.-F., Zhang, D.-L., Zhou, T., Zhang, H.-S., Zhang, W.-Z., Luo, L., Zhang, A.-M., Li, B.-J., Zhang, S., A reversible functional supramolecular material formed by host-guest inclusion. *Polym. Chem.*, 5, 8, 2922–2927, 2014.
143. Fox, J., Wie, J., J.W. Greenland, B., Burattini, S., Hayes, W., Colquhoun, M., Mackay, M., H.E., Rowan, S., J., High-Strength, Healable, Supramolecular Polymer Nanocomposites. *J. Am. Chem. Soc.*, 134, 11, 5362–5368, 2012.
144. Yan, X., Wang, F., Zheng, B., Huang, F., Stimuli-responsive supramolecular polymeric materials. *Chem. Soc. Rev.*, 41, 18, 6042–6065, 2012.
145. Pernigoni, L. and Grande, A.M., Development of a supramolecular polymer based self-healing multilayer system for inflatable structures. *Acta Astronaut.*, 177, 697–706, 2020.
146. Chivers, P.R.A. and Smith, D.K., Shaping and structuring supramolecular gels. *Nat. Rev. Mater.*, 4, 7, 463–478, 2019.
147. Rossow, T. and Seiffert, S., Supramolecular polymer gels with potential model-network structure. *Polym. Chem.*, 5, 8, 3018–3029, 2014.
148. Rossow, T., Hackelbusch, S., van Assenbergh, P., Seiffert, S., A modular construction kit for supramolecular polymer gels. *Polym. Chem.*, 4, 8, 2515–2527, 2013.
149. Jones, C.D. and Steed, J.W., Gels with sense: supramolecular materials that respond to heat, light and sound. *Chem. Soc. Rev.*, 45, 23, 6546–6596, 2016.
150. Berger, G., Soubhye, J., Meyer, F., Halogen bonding in polymer science: from crystal engineering to functional supramolecular polymers and materials. *Polym. Chem.*, 6, 19, 3559–3580, 2015.
151. Datta, S., Takahashi, S., Yagai, S., Nanoengineering of Curved Supramolecular Polymers: Toward Single-Chain Mesoscale Materials. *Acc. Mater. Res.*, 3, 2, 259–271, 2022.
152. Kato, T., Gupta, M., Yamaguchi, D., Gan, K.P., Nakayama, M., Supramolecular association and nanostructure formation of liquid crystals and polymers for new functional materials. *Bull. Chem. Soc. Jpn.*, 94, 1, 357–376, 2021.
153. Pérez-Pedroza, R., Ávila-Ramírez, A., Khan, Z., Moretti, M., Hauser, C.A.E., Supramolecular Biopolymers for Tissue Engineering. *Adv. Polym. Technol.*, 2021, 8815006, 2021.
154. Saunders, L. and Ma, P.X., Self-Healing Supramolecular Hydrogels for Tissue Engineering Applications. *Macromol. Biosci.*, 19, 1, 1800313, 2019.
155. Goor, O.J.G.M., Hendrikse, S.I.S., Dankers, P.Y.W., Meijer, E.W., From supramolecular polymers to multi-component biomaterials. *Chem. Soc. Rev.*, 46, 21, 6621–6637, 2017.
156. Zhang, Z., Ai, S., Yang, Z., Li, X., Peptide-based supramolecular hydrogels for local drug delivery. *Adv. Drug Deliv. Rev.*, 174, 482–503, 2021.
157. Uchida, N., Design of supramolecular nanosheets for drug delivery applications. *Polym. J.*, 55, 8, 829–836, 2023.
158. Wang, Y., Chen, P., Luo, Q., Li, X., Zhu, W., Supramolecular polymeric prodrug micelles for efficient anticancer drug delivery. *Polym. Chem.*, 13, 20, 2964–2970, 2022.
159. Mondal, S., Yoshida, T., Maji, S., Ariga, K., Higuchi, M., Transparent Supercapacitor Display with Redox-Active Metallo-Supramolecular Polymer Films. *ACS Appl. Mater. Interfaces*, 12, 14, 16342–16349, 2020.
160. An, S.Y., Schon, T.B., McAllister, B.T., Seferos, D.S., Design strategies for organic carbonyl materials for energy storage: Small molecules, oligomers, polymers and supramolecular structures. *Eco.Mat.*, 2, 4, e12055, 2020.

161. Dong, S., Zheng, B., Wang, F., Huang, F., Supramolecular Polymers Constructed from Macrocycle-Based Host–Guest Molecular Recognition Motifs. *Acc. Chem. Res.*, 47, 7, 1982–1994, 2014.
162. Zheng, B., Wang, F., Dong, S., Huang, F., Supramolecular polymers constructed by crown ether-based molecular recognition. *Chem. Soc. Rev.*, 41, 5, 1621–1636, 2012.
163. Shi, B., Jie, K., Zhou, Y., Xia, D., Yao, Y., Formation of fluorescent supramolecular polymeric assemblies via orthogonal pillar[5]arene-based molecular recognition and metal ion coordination. *Chem. Commun.*, 51, 21, 4503–4506, 2015.
164. Sun, R., Xue, C., Ma, X., Gao, M., Tian, H., Li, Q., Light-Driven Linear Helical Supramolecular Polymer Formed by Molecular-Recognition-Directed Self-Assembly of Bis(p-sulfonatocalix[4]arene) and Pseudorotaxane. *J. Am. Chem. Soc.*, 135, 16, 5990–5993, 2013.
165. Haino, T., Designer supramolecular polymers with specific molecular recognitions. *Polym. J.*, 51, 3, 303–318, 2019.
166. Korevaar, P.A., George, S.J., Markvoort, A.J., Smulders, M.M.J., Hilbers, P.A.J., Schenning, A.P.H.J., De Greef, T.F.A., Meijer, E.W., Pathway complexity in supramolecular polymerization. *Nature*, 481, 7382, 492–496, 2012.
167. Ogi, S., Fukui, T., Jue, M.L., Takeuchi, M., Sugiyasu, K., Kinetic control over pathway complexity in supramolecular polymerization through modulating the energy landscape by rational molecular design. *Angew. Chem. Int. Ed.*, 53, 52, 14363–14367, 2014.

6

Microbial Based Biolubricants: In-Depth Analysis

Ninad Mhatre, Deepak Sonawane, Fatema Saiger, Prasad Sanap, Somesh Patil and Amit Pratap*

Department of Oils, Oleochemicals & Surfactants Technology, Institute of Chemical Technology, Mumbai, Maharashtra, India

Abstract
Biolubricants, a promising class of lubricants obtained from renewable material derivatization, have garnered significant attention as a sustainable replacement for conventional lubricants from petroleum. Environmental concerns and fossil fuel depletion have intensified the search for eco-friendly solutions, leading to the development of biolubricants. These biodegradable and non-toxic lubricants are derived from plant-based oils, fats from animal, or other biomass sources, making them more environmentally friendly and less harmful to human health. This chapter explores the key characteristics and advantages of biolubricants, highlighting their superior biodegradability, lower toxicity, and reduced greenhouse gas emissions compared to conventional lubricants. Furthermore, the impact of biolubricants on equipment performance and longevity is discussed, showcasing their ability to enhance machinery efficiency and reduce wear and tear. Various industrial applications, including automotive, marine, and aerospace sectors, are examined, revealing the potential for widespread adoption of biolubricants in diverse fields. While biolubricants present numerous benefits, certain challenges, such as their higher cost and limited temperature stability, are also addressed in this abstract. Researchers and industries are working together to overcome these obstacles through technological advancements and innovation. As governments and organizations worldwide intensify efforts to combat climate change and promote sustainable practices, biolubricants emerge as a viable solution for reducing the environmental impact of lubrication processes. This abstract concludes by emphasizing the need for continued research, development, and market adoption to explore the potential of biolubricants in creating a more sustainable and greener future for lubrication practices.

Keywords: Microbial enhanced oil recovery, interpenetrating polymeric network, life cycle assessment, biolubricants

List of Abbreviation

VO	vegetable oils
VI	Viscosity Index
WVO	Waste Vegetable Oils

**Corresponding author*: amitpratap2001@gmail.com

Divya Bajpai Tripathy, Anjali Gupta and Arvind Kumar Jain (eds.) Multifunctional Materials: Engineering and Biological Applications, (133–180) © 2025 Scrivener Publishing LLC

TMP	Trimethylolpropane
NADH	Nicotinamide adenine dinucleotide
HSA	Hydroxy Stearic Acid
EPS	Exopolysaccharide
HA	Hyaluronic Acid
PHA	Polyhydroxyalkanoates
PHB	Polyhydroxybutyarate
IPN	Interpenetrating polymeric network
Veg-oil	Vegetable Oil
CPS	Capsular Polysaccharides
MNP	Metallic Nanoparticles
MEOR	Microbial Enhanced Oil Recovery
EI	Emulsifying Index
MW	Molecular Weight
NASL	Natural Algae Based Synthetic Lubricant
MCC	Microcystallin Cellulose
HPMC	Hydroxypropyl methyl Cellulose
BOD	Biological Oxygen Demand
COD	Chemical Oxygen Demand
FAME	Fatty Acid Methyl Ester
FFA	Free Fatty Aid
WCO	Waste Cooking Oil
TEA	Techno-Economical Analysis
VPC	Vertical Pulsed Column
LCA	Life Cycle Assessment
MWF	Metal Working Fluid
CF	Cutting Fluid
MCF	Metal Cutting Fluid

6.1 Introduction

By creating a protective coating, lubricants can reduce wear and friction on mechanical parts as well as heating. In 2019, the market for lubricants reached US$126.5 billion, and by 2022, it is anticipated that production will have increased by 2.4% annually to reach 43.87 million tons [1]. Recently, traditional petroleum-based lubricants have lost ground to biolubricants in popularity [2]. This trend may be witnessed across all industries that use these products, regardless of the applications. There are a variety of elements that have an impact on this new trend, among these are the environmental issues brought on by traditional lubricants and the demand for renewable and biodegradable lubricants as a result of the depletion of oil supplies. In comparison to mineral oil lubricants, biolubricants have a number of advantages, such as they can be modified as per required lubricating performance, high biodegradability, and minimal toxicity, and neither the environment nor human health is affected. They are also particularly interesting in the context of green chemistry since they can be produced using a range of plant or animal-based renewable feedstocks and use of enzyme or microbial-catalyzed [3]. In contrast to finished lubricants with mineral bases, which have seen no change in demand over the past 10 years, the global demand for biolubricants has expanded to US$2.92 billion in 2023 from US$2.74 billion last year with a compounded average growth rate (CAGR) of 6.6% with the optimistic expectation to grow to US$3.63 billion by 2027 at a CAGR of 5.5%

Petroleum-based lubricants are currently extensively recognized and utilized as all-purpose lubricants in a variety of enterprises, trades, and private purposes. Depleting oil reserves and also growing transit costs and refining procedures have forced the development of more effective oil extraction methods, which has substantially altered the situation [4, 5]. For process industries as well as government organizations, environmental issues and accountability are vital. When it comes to lubricants, this worry is brought on by the serious issue of inappropriate disposal, which has a negative effect on the environment. An estimated 40% to 55% of all old lubricants are improperly disposed of, making their way into the land, water, and atmosphere [4, 6]. The fact that more than 95% of the lubricants that contribute to environmental degradation are made from petroleum is another worrying aspect of this issue. Due to their highly toxic nature and poor biodegradability, these conventional lubricants cause ecological harm when disposed of [1]. These lubricants, which are essential for the survival of aquatic ecosystems, when introduced into aquatic settings, produce a surface layer of oil that prevents sunlight and oxygen from penetrating, causing the ecosystems to suffer. In the same way, they impair soil fertility and reduce its capacity to filter and exchange ions when they are inappropriately disposed of in the soil. Additionally, when these lubricants are burned, they release harmful chemicals into the atmosphere [5, 6]. However, there are steps you may take to lessen the harmful effects lubricants have on the environment. Recent research suggests that biolubricants, which are synthetic biodegradable alternatives, can replace more than 90% of the petroleum-based lubricants that are now produced [1]. For lubricating oils made from fossil fuels, biolubricants offer an alluring alternative [1, 5, 7, 8]. These include reduced toxicity, improved lubrication efficiency, a higher flash point (the lowest temperature at which a substance produces enough flammable vapor to ignite), lower volatility, excellent pour point (cold flow properties, specifically a lower temperature), which indicates the temperature at which a liquid transitions into a semisolid state and loses its ability to flow and a higher viscosity index (VI; a measure of viscosity variation at 40°C and 100°C) [4, 5, 7, 9]. The large variety of renewable sources that can be used to produce biolubricants is another advantage of using them. These sources include waste oils as well as microbial oils produced by yeasts, filamentous fungi, microalgae, and vegetable and animal fats [1, 4, 5, 7]. Esterification, transesterification, epoxidation, and hydro-esterification are only a few examples of the chemical processes that can be used to create these compounds. Catalysts of many kinds, whether chemical or enzymatic, can speed up these reactions [7, 10, 11]. However, significant limitations have been linked to chemical pathways for the manufacturing of biolubricants, such as the need for high temperatures (resulting in higher energy expenditures), prolonged reaction durations, and the formation of chemical residues during downstream processing operations [12–14]. Utilizing enzymes, particularly lipases, has emerged as a possible alternative for the creation of these esters in order to overcome these difficulties. The growing demand for ecologically friendly goods and procedures is what motivates this strategy [15–18]. There are restrictions on the recovery and reusability of the biocatalyst from the reaction medium when using lipases in industrial processes, either in liquid or granulated/powder forms. Furthermore, lipases are extremely sensitive to temperature changes, which makes application more challenging [19]. By utilizing various immobilization procedures, these restrictions can be reduced or even removed [19–22].

Biolubricants can be derived from an array of veg-oils (VOs) like palm, soybean, sunflower, rapeseed, coconut, jatropha, and castor bean. The focus of this section is on modifying the chemical structure of these oils using different chemical derivatization methods. These modifications help address the limitations associated with directly using oil-rich raw materials as biolubricants. Furthermore, the chapter discusses the advancements in enzymatic catalysis and immobilization techniques for producing biocatalysts, which recent studies in the literature have identified as potentially valuable for generating various types of biolubricants. The chapter also delves into the future possibilities and challenges surrounding the commercial production of enzymatic biolubricants. Authorities worldwide are increasingly prioritizing pollution control and environmental preservation, with a specific focus on enhancing the reliability, durability, and energy efficiency of various industries. One potential solution lies in the use of alternate renewable resources, safer fuels, and controlled fuel combustion. Implementing these alternatives can help mitigate the environmental issues caused by vehicles and machinery. The detrimental effect on the environment due to the discharge of lubricants affecting water, soil, and air, poses significant concerns for the lubricant industry. Broad classification of lubricants based on raw material includes two categories - natural oils and synthetic oils as seen in Figure 6.1. Natural oils, also known as natural esters, encompass fats from animal sources, microbial oils, and VO. In contrast, synthetic oils include lubricants derived from synthetic chemical compounds such as perfluoroalkyl polyether, synthetic esters, and poly alpha-olefins. The discharge of these lubricants into the environment and their impact on water, soil, and air quality are particularly troubling for the lubricant industry. Vegetable and microbial oils are examples of natural oils commonly referred to as natural esters.

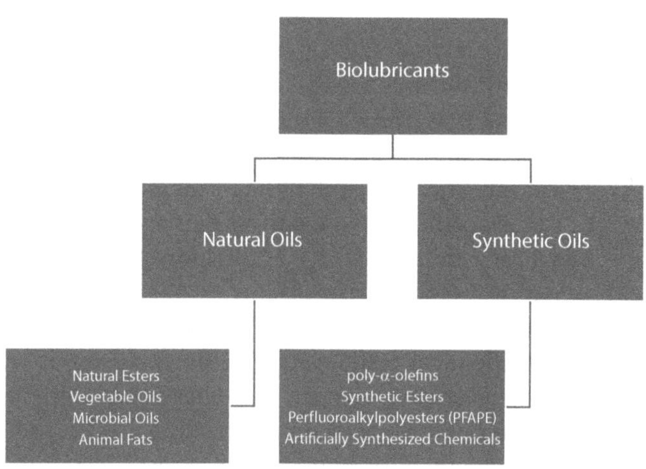

Figure 6.1 Classification of biolubricants.

6.2 Biolubricants: Substitutes for Conventional Lubricants

Biolubricants offer numerous advantageous physicochemical properties, providing several technical benefits. These include exceptional lubricity, a high VI, a high flash point, and minimal evaporative losses [23]. Numerous scientific studies have been carried out to enhance the functional characteristics of biolubricants. These studies have revealed that bio-based materials can be economically viable and effective substitutes. Researchers have observed that biolubricants exhibit superior lubricity in comparison to their conventional counterparts [24]. As a result, the favorable characteristics of biolubricants have led to their growing popularity. Key factors that have fueled their acceptance include their non-corrosive properties and the reactivity of unsaturated hydrocarbon chains [25–27]. Several researchers have explored the wear characteristics of biolubricants by utilizing a range of tribometer techniques.

Sulek *et al.* analyzed the tribological properties of biolubricants derived from rapeseed. The findings indicated a reduced coefficient of friction in comparison to petroleum-based oils [25]. The enhancement of lubricity across various low blend levels in biolubricants was observed [28]. Scientific study indicates there is a slight improvement in friction and wear as the temperature increases [29]. Kalam *et al.* focused on examining the wear and friction properties due to various lubricants including lubricants blended with WVO (waste VO). The findings indicated that lubricants containing WVO, along with the addition of amine phosphate as an additive for anti-wear, demonstrated a decrease in friction coefficients and wear properties, as well as an increase in viscosity [30]. Maleque *et al.* investigated the tribological behavior of a blended lubricant containing palm oil methyl ester (POME). The research specifically focused on examining a cast iron and steel pair and identified corrosive wear and the formation of pits as the primary wear observed on the damaged surface [37]. Another study highlighted that blending lubricants with POME higher than 5% by weight would result in corrosion and oxidation [31]. Moreover, this investigation emphasized that factors like moisture, oxidation, and temperature would affect the properties of biolubricants. Although mineral oils are frequently utilized in diverse applications, such as hydraulic oil, grease, insulating oil, engine oil, and metalworking fluid, biolubricants derived from VOs like neem, soybean, jatropha, palm, Karanja, coconut, sunflower castor, mahua, and olive have exhibited similar or even superior performance compared to petroleum or synthetic oil–based alternatives, all while being more economically advantageous [32]. In a study conducted by Asadauskas *et al.*, lubricating characteristics such as oxidative stability, lubricity, deposit formation, and viscosity of biodegradable lubricants were investigated. The results from the experiments revealed that VOs demonstrated higher VIs in comparison to mineral oils [33]. VOs are preferred over synthetic fluids because of their renewable nature, affordability, and environmentally friendly attributes. Furthermore, lubricants derived from VOs are biodegradable and non-toxic, unlike conventional mineral-based oils [34]. VOs display limited evaporative losses, mainly attributed to the triacylglycerol molecule's large molecular mass. Additionally, they exhibit a narrow range of viscosity variations in response to temperature fluctuations. These inherent technical characteristics, coupled with their biodegradability, make VOs a suitable substitute for conventional mineral oils in lubricant applications. Notably, VOs possess a high VI, impressive lubricating properties, high flash point, and minimal loss due to volatility in comparison to mineral oils [34]. Zulkifli *et al.*

investigated the effectiveness of a trimethylolpropane (TMP) ester of palm oil as a lubricant in mitigating wear in the engine using a four-ball tester to evaluate various lubrication conditions, including boundary lubricants and elastic. The lubricants tested were blends of conventional lubricants with varying concentrations (5% to 100%) of palm oil TMP esters. The findings revealed that samples containing TMP ester of palm oil notably enhanced wear prevention properties, as evidenced by reductions in wear-scar diameter and the coefficient of friction [35]. Arumugam *et al.* evaluated the tribological performance of biolubricants and biodiesel-contaminated lubricants in a cylinder liner piston ring combination engine. The research utilized a tribometer of pin-on-disc type, with the disc substance being alloy made from engine cylinder liner and white cast iron. Additionally, real top compression ring material was combined with white cast iron to create the pins. Samples of oil tested were a blend of 20W40 commercial lubricant with 10% biolubricants, 10% B20R, and 10% diesel blended. The performance of these three samples was compared with slandered 20W40 commercial lubricant [36]. The application of biolubricants led to a reduction in wear. Under comparable operating circumstances, the biolubricants demonstrated greater lubricating qualities in comparison to the other evaluated samples. This improved performance can be explained by the existence of long-chain fatty acids found in palm oil, which formed a protective layer of hydrocarbons, effectively minimizing wear and enhancing the wear resistance of the biolubricants [37]. Additionally, upon visual inspection, it was observed that the wear surfaces exhibited smoother characteristics when using the biodiesel and biolubricants contaminated lubricant oil, in contrast to the diesel-blended lube oil. According to the study's findings, VO-based motor lubricants are ecologically beneficial and have renewable properties [38]. In a similar experiment conducted by Bhale *et al.*, diesel and methyl esters of jatropha oil were bladed with motor lubricants to compare their lubrication properties. Biodiesel contamination performed better in some properties which could be due to the presence of unsaturation as well as oxygenated functionality in these molecules which contributed to an overall improvement in lubricating properties [39]. A study on a similar line takes advantage of the potential use of non-edible VO derivatives as alternative motor lubricants as it avoids the food versus fuel argument and offers all the advantages that edible oil offers on top of which, it is additionally economical than edible oil [40]. Biolubricants finds its application in various industrial and maintenance applications like two-stroke engine oil, metal working fluids, hydraulic fluids, grease etc. and are mainly used in machinery and equipment operating in environmentally sensitive areas. Biolubricants use may reduce costs and also enhance competitiveness. Moreover, adopting Biolubricants presents an opportunity to decrease dependence on petroleum-based lubricants in these sectors [41]. Observations made by Arumugam *et al.* on the coefficient of friction demonstrated that chemically modified biolubricants were approximately 23% lower when compared to commercial synthetic lubricants. However, petroleum-based commercial lubricants displayed approximately 12% lower wear compared to biolubricants. This outcome can be explained by the presence of anti-wear additives in commercial samples [42]. The findings of Ting *et al.* revealed that the viscosity of the epoxidized soy oil was considerably higher than that of both the engine lubricants and the unmodified soybean oil. Hence, the combination of epoxidized soy oil, hydrogenated soy oil, and unmodified soybean oil could achieve the desired viscosity for engine lubricants [43]. The research findings indicated that esters derived from synthetic oil and VO are highly suitable for formulating lubricants with eco-friendly properties which can be seen in Figure 6.2. These lubricants can be utilized in diverse applications either individually or in combination.

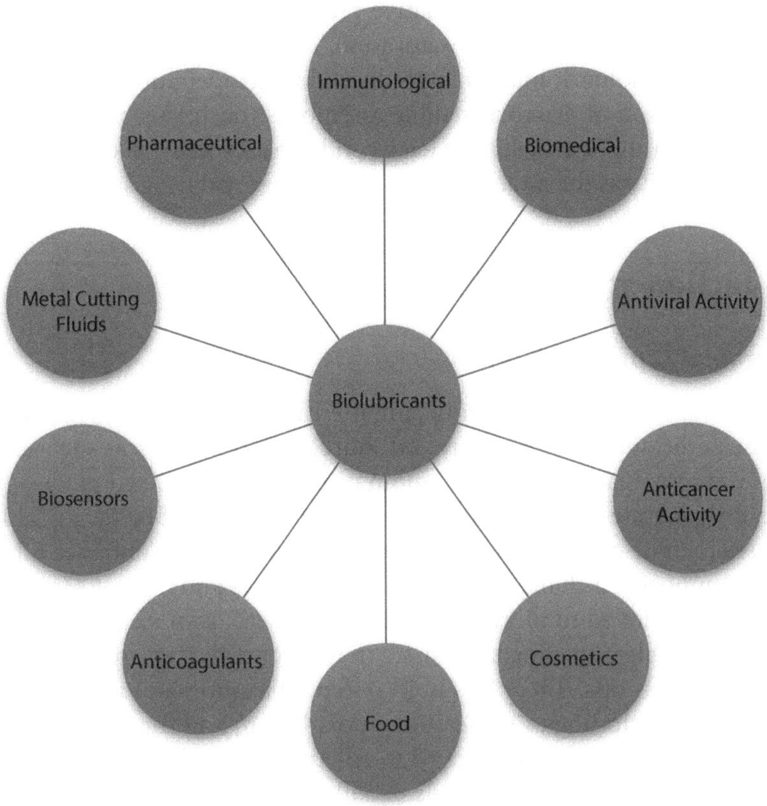

Figure 6.2 SEQ figure*ARABIC 2 applications of biolubricants.

There is a growing demand for biodegradable products, and it is projected that eco-friendly/biodegradable lubricants will capture a global volume share of approximately 15%, with even higher volume shares in specific regions, potentially reaching up to 30% within the next 10 to 15 years. The lubricant market is expected to undergo significant changes in the next 5 to 10 years, making it an intriguing and dynamic field worth exploring [44].

6.2.1 Advantages of Biolubricants

1. They are derived from renewable sources and hence offer eco-friendly alternatives.
2. The use of waste cooking could potentially be a good alternative.
3. Biolubricants derived from sources like coconut, palm kernel oil, and others possess natural antioxidants such as tocopherols, phenolics, and sterols, which make them less susceptible to oxidation, similar to conventional lubricants.

6.2.2 Disadvantages of Biolubricants

1. As previously mentioned, biolubricants have limited resistance to thermal and oxidative degradation due to the presence of acyl groups in their

molecules. The low thermo-oxidative property of VOs is attributed to their methylene intermittent polyunsaturation.

2. Additionally, biolubricants exhibit reduced low-temperature performance due to the formation of macrocrystalline assemblies (occurring at lower temperatures through aligned stacking of triglycerides' backbones), leading to restricted fluid flowability as the molecules lose kinetic energy during self-stacking.

6.3 Production of Biolubricants

6.3.1 Microbial Lipids and Oils

Animal oil and VO have garnered significant attention recently, while microbial oil presents itself as an additional and alternative oil source. Microorganisms such as fungi, yeast, cyanobacteria, microalgae, and their combinations can be utilized as oleaginous feedstocks to rapidly generate high oil yields. Few examples are listed below in Table 6.1. These microorganisms can be cultivated under favorable conditions to enhance oil production, and fatty acid composition changes with microbial species and habitat. Genetic modification can even be employed to increase oil yields further. From an economic standpoint, microbial oils are viable and affordable since they can be derived from low-cost or waste sources, such as excess oils. These oils hold potential as lubricant additives or base oils. Various techniques, including esterification, epoxidation, and the incorporation of complexes and nanoparticles, can be employed to enhance the tribological properties of these oils. Esterification processes for these oils usually occur in one or two steps and involve the use of polyols like polyvinyl alcohol and neopentyl glycol (NPG). Greener approaches utilizing enzymes like lipases have been developed for such processes. Several techniques, such as microwave treatment, have been described to facilitate improved, faster, and more efficient oil extraction. Additionally, advancements in bioprocesses have been reported to maximize oil production from these single-cell oleaginous sources. Hydroxy fatty acids derived from microbes offer renewable options for biodegradable oil–based lubricants. Due to the exclusive presence of hydratase and diol synthase in microbes, hydroxy fatty

Table 6.1 Biolubricant production from different microorganisms.

Sr. no.	Microorganism	Substrate	Yield	References
1	*Enterococcus faecalis*	Oleic acid (0.9 g/L)	0.8 g/L	[54]
2	*E. gallinarum*	Oleic acid (2.0 g/L)	1.9 g/L	[55]
3	*E. coli* with the gene of oleate hydratase from *S. maltophilia*	Oleic acid (50 g/L)	49 g/L	[46]
4	*Flavobacterium* sp. DS5	Oleic acid (5.4 g/L)	2.97 g/L	[56]
5	*Stenotrophmonas maltophilia*	Oleic acid (50 g/L)	40 g/L	[46]
6	*Sphingobacterium thalpophilum*	Soybean oil (18 g/L)	7.1 g/L	[57]
7	*Lactobacillus* sp.	Oleic acid (2.0 g/L)	1.5 g/L	[55]

acids from microbes exhibit more regional specialization and diversification compared to plant and animal hydroxy fatty acids [45]. Hydroxy fatty acids serve as essential building blocks in the production of lubricants. The tribological properties of hydroxy fatty acids derived from biosurfactants have also been examined. In contrast to plant-based oils, w-1 hydroxy branched fatty acids, specifically those produced by Candida Bombicola, offer various advantages. The researchers suggested that these fatty acids exhibit tribological characteristics similar to ricinoleic acid ester derivatives. The significant feature of this fatty acid is the occurrence of a hydroxy functional group at the terminal alkyl chain, which allows for several chemical modifications. Notably, 12-hydroxy stearic acid is commonly used as a thickening agent in lubricating greases [46]. Similar to 12-hydroxy stearic acid, 10-hydroxy stearic acid is employed in the production of lubricants and exhibits similar properties. Ricinoleic acid, on the other hand, finds application as a dispersant, lubricant, and emulsifier [47]. Additionally, it possesses laxative and anti-inflammatory properties [48]. Microorganisms can be utilized in fermentation, whole-cell conversion, and enzymatic processes to generate these fatty acids [49].

6.3.1.1 Production of Ricinoleic Acid

When compared to petroleum-based polymers, ricinoleic acid demonstrates enhanced flexibility, heat resistance, and chemical resistance. It provides a safe, non-toxic, and environmentally friendly alternative due to its biocompatibility and biodegradability [48]. 12-Hydroxylases (EC 1.14.13.26) found in fungal strains of *Claviceps purpurea*, a parasitic disease affecting cereal crops, are responsible for converting oleic acid into ricinoleic acid. These enzymes utilize oxygen and NADH to catalyze the hydroxylation process, specifically targeting the 12th position of oleic acid. Consequently, this conversion pathway can be employed to produce significant quantities of ricinoleic acid as an alternative approach [50]. Nevertheless, commercial biosynthesis of ricinoleic acid is unfeasible due to the synthesis of harmful ergot alkaloids by the fungus as byproducts. However, recombinant cells expressing the *C. purpurea* 12-hydroxylase gene avoid the production of ergot alkaloids, opening up possibilities for the industrial synthesis of ricinoleic acid [51]. A 52.6% total yield is achieved when oleic acid is converted to ricinoleic acid by recombinant *Schizosaccharomyces* pombe carrying *C. purpurea* oleate 12-hydroxylase gene [52]. This makes the recombinant cells possible novel biocatalysts for ricinoleic acid synthesis [45].

6.3.1.2 Production of Hydroxy Stearic Acid (HSA)

The properties of both 10-HSA and 12-HSA are similar. These substances are non-toxic, exist in a gel-like state at room temperature, and possess high melting points, excellent thermal stability, and impact resistance. They find applications in various industries, including cosmetics, greases, coatings for food, medicine, and automobiles, hydrophobic drug delivery systems, and coatings [53]. Numerous studies report the biosynthesis of 10-hydroxystearic acid by naturally occurring as well as genetically modified bacteria [47]. Genetically modified *Escherichia coli* cells containing the oleate hydratase gene from *Stenotrophomonas maltophilia* exhibit the highest levels of 10-hydroxystearic acid concentration and productivity when converting oleic acid. Several studies have investigated the generation of 10-HSA through cell conversion or fermentation processes [46]

Moreover, microbial oils have been extracted and their lubricating properties directly examined. Yeast strains like *Rhodotorula* spp. have demonstrated high oil production capabilities and potential use as base oils for lubricants. The production process was cost-effective as it utilized xylose derived from sugarcane bagasse as the raw material. Additionally, reports indicate that the esters derived from this microbial oil exhibit physicochemical characteristics comparable to ISO VG-46 grade, with an exceptionally low pour point of −45°C [58]. Additionally, it had superior tribological characteristics compared to petroleum-based lubricants of the same viscosity grade [59]. For the biosynthesis of these oils and their optimization, researchers have examined different microorganisms and

Table 6.2 Microbial studies for their lipid profile as the application in lubricants.

Sr. no.	Classification	Microbe	Modification	Reference
1	Fungi (yeast)	*Rhodotorula glutinis*	lipid converted to trimethylolpropane fatty acid triester	[59]
2	Fungi (yeast)	*Rhodotorula mucilaginosa* IIPL32	No modification	[58]
3	Yeast	*Mortierella isabellina*	Methyl ester, nano Schiff base complex	[62]
4	Marine microalgae	*Scenedesmus obliquus* and *C. vulgaris*	Transesterification using heteropolyacid supported on niobium	[60]
	Fungi-algae consortium	*Scenedesmus obliquus* and *Mucor circinelloides*		
5	Fungi	1) *Aspergillus awamori* 2B.361 U2/1 2) *Rhodosporidium toruloides* DSMZ 4444, 3) *Rhodosporium* spp. 4) *Rhodotorula glutinis* NRRL YB-252, 5) *Cryptococcus* spp. 6) *Cunninghamella echinulate* ATHUM 4411 7) *Mortierella ramanniana* MUCL 2935	Transesterification using polyol esters	[15]
6	Microalgae	1) *Auxenochlorella protothecoide* 2) *Chlorella sorokiniana*	No modifications	[63]
	Thraustochytrid	1) *Aurantiochytrium limacinum* SR2 2) *Aurantiochytrium* sp. T66 (ATCC PRA-276)		
	Fungi	1) *Cryptococcus curvatus* DSM-10103 2) *Rhodosporidium toruloides* NCYC 1576		

substrates [60]. Heterogeneous catalysts and several enzymes have esterified these oils [60]. In order to improve tribological properties, including reducing the friction coefficient, Wu *et al.* studied the impact of esterified and epoxidized microbial oils when interacting with copper complexes based on nano-Schiff technology [61].

Gavalás-Olea *et al.* extracted lipids and polysaccharides from *Dixoniella grisea*, a type of red algae, and examined their rheological properties using the biorefinery method. The findings portrayed these lipids could potentially be a valuable byproduct for the production of lubricant additives [62].

A few microbes studied for their lipid profile as an application in lubricants are given in Table 6.2.

6.3.2 Exopolysaccharides (EPS)

Polysaccharides have the ability to increase viscosity, maintain suspension, stabilize emulsions, and reduce friction. Originally obtained from plants, there is ongoing research to obtain polysaccharides from microbial sources. Some commercially significant polysaccharides include hyaluronic acid (HA), xanthan gum, dextran, pullulan, and more [64]. Lactic acid bacteria are utilized in the food and medicinal sectors, particularly in the production of dextran. Commercially, dextran is produced through fermentation of a sugar-rich medium. Dextran-producing strains include *Leuconostoc mesenteroides*, other lactic acid bacteria, and specific strains of *Gluconobacter oxydans*. Dextrans are glucose homopolysaccharides characterized by a substantial proportion of successive 1,6-linkages in their main chains, often exceeding 50% of total linkages. They also possess side chains, primarily consisting of (1,3)-branched connections but occasionally including (1,4)- or (1,2)-branched linkages. The structure of dextran varies depending on the microbial strain responsible for its production. It finds applications in ocular lubricant formulations.

HA is another biopolymer containing D-glucuronic acid (GlcUA) and *N*-acetylglucosamine (GlcNAc) disaccharide connected alternately by 1, 3, and 4 glycosidic linkages. HA is readily available on the market and is increasingly produced by streptococcus species utilizing glucose and starch as substrates. To address concerns, Novozyme is now using recombinant technology in *Bacillus subtilis* for large-scale commercial production. Streptococcus, although harmful, has found extensive use in the biomedical field as an adhesive and lubricant for joint replacement and degenerative joint conditions.

Exopolysaccharides (EPSs), including PHA, can serve as thickeners and binders in the production of rapidly degradable industrial lubricants. PHA exhibits lubricating and plasticizing effects and can be produced by various bacteria using different oil, sugar, and protein sources. Some examples include *Cupriavidus* sp. KKU38, P*seudomonas* sp., *Ralstonia eutropha*, *Haloferax mediterranei*, and *Streptomyces* (through sustainable manufacturing from waste). Biodegradable PHB has been added to the DMA-MPC-PHB coating for implantable medical devices to enhance lubrication and bacterial suppression [65]. Recombinant *E. coli*, *Burkholderia* spp., *Alcaligenes latus*, *Cupriavidus* sp., and other microbial species can be utilized for PHB production through microbiological fermentation [66].

Xanthomonas spp.'s gum, commonly known as xanthan gum, possesses several unique properties that distinguish it from other gums. These characteristics include a very high viscosity at lower doses, stability at high pH levels, and excellent compatibility with salts. These qualities make it highly desirable for a wide range of industrial applications. Xanthan gum finds use in various processes such as drilling fluids, polymer flooding for improved oil recovery, hydraulic fracturing in oil fields, flowable pesticides, metalworking fluids, cleaners, abrasives, polishing agents, liquid feed supplements, ceramics, foundry coatings, and slurry explosives. EPS (Exopolysaccharide) produced by *Alcaligenes* species exhibits high viscosity and stability under shear. Apart from being used as water-based thickeners in pesticide suspension formulations, they have several applications in the oil industry. Arad discovered that a sulfated polysaccharide derived from the red algae species *Porphyridium* exhibited improved characteristics as a biolubricant [67]. A comparison was made between the rheological characteristics of a sulfated polysaccharide from *Porphyridium* and HA, a commercially available biolubricant used in hydrogels. The study evaluated the rheological characteristics, friction, wear resistance, and elasticity of both substances. The EPS from *Porphyridium* demonstrated superior quality compared to most lubricants, as its viscosity remained stable while maintaining lubricity during simulated joint movements associated with walking and running. This can be attributed to its rheological properties, which exhibited stability even at higher temperatures.

Limited work is done on the drag-reducing properties of polysaccharides. Gasljevic *et al.* conducted research suggesting that certain marine microalgal polysaccharides have the potential to act as drag-reducers. Among these polysaccharides, those derived from *R. maculata* and *P. cruentum* exhibited the highest drag-reducing capabilities at lower dosages. Interestingly, the polysaccharide from *R. maculata* required 25% more quantity compared to *P. cruentum* and nearly triple the concentration of *S. capsulate* to achieve similar levels of drag-reduction efficiency, as observed by the researchers.

In a study conducted by Xiaqing Sun in 2020 [245], a novel microbial EPS lubricant produced by *Paenibacillus* sp. ZX1905 was investigated, demonstrating remarkable lubricating capabilities. The researchers compared its rheological properties and moisture absorption characteristics to those of glycerol and HA. The results revealed that the microbial EPS exhibited comparable properties to HA. However, the production of the microbial EPS was deemed more economically viable due to its higher yield, lower production cost, and simplified downstream processes.

6.3.3 Microbial Polysaccharides

Microbial polysaccharides are complex carbohydrate polymers with large molecular weight (MW) that are biosynthesized by various microorganisms, such as fungi, yeast, microalgae, and bacteria [68–70]. Microbial polysaccharides encompass various types, including cell-wall polysaccharides, intracellular polysaccharides, and extracellular polysaccharides. Cell-wall polysaccharides contribute to the structural stability of cells, as observed in the case of chitin. Intracellular polysaccharides act as energy reserves stored within the cytoplasm of cells, similar to glycogen. On the other hand, extracellular polysaccharides are secreted by cells as a loose slime layer on the cell surface (known as EPSs) or as a capsule closely associated with the cell surface (known as capsular polysaccharides [CPS]) [71]. Among the various types mentioned earlier, capsular polysaccharides (CPS) are primarily linked to the pathogenicity of bacteria and play a crucial part in promoting virulence [72]. On the other hand, EPSs have been suggested to serve multiple functions, including offering defense against habitat stresses, facilitating cell attachment to surfaces, and as well serving as reserves for carbon and water storage [73]. Polysaccharides can be used in two main areas of application:

(i) Structuring agents possess the capacity to create polymeric structures, including nanoparticles, films, microparticles, gels, and emulsions.
(ii) Biologically active substances have the potential to be utilized in the creation of novel pharmaceutical drugs or as substitutes for certain existing products [74, 75]. Other potential applications encompass utilizing them as valuable sources of monomers

with high worth, such as rare sugars (e.g., glucuronic acid, rhamnose, ribose, and fucose), to synthesize oligosaccharides (e.g., fucooligosaccharides and galactooligosaccharides). These oligosaccharides find utility in the field of nutraceuticals [76].

6.3.3.1 Functional Properties and Applications of Microbial Polysaccharide

Microbial polysaccharides are primarily carbohydrates, with mannose, galactose, and glucose being the most prevalent monomers. However, these biopolymers also often contain other neutral sugars like arabinose, fucose, and rhamnose, as well as amino sugars and uronic acids. Additionally, microbial polysaccharides can integrate diverse acyl substituents within their molecular chains, such as pyruvate ketals and esters [71, 77]. Microbial polysaccharides exhibit an extensive range of components, including diverse sugar monomers and noncarbohydrate groups, leading to a range of potential molecular configurations for these macromolecules. As a result, polysaccharides obtained from microbes display diverse chemical and physical properties. The monomeric units combined with specific glycosidic bonds (α or β anomers) create intricate molecular structures, varying from linear homopolysaccharides to complex heteropolysaccharides. Key factors, such as the presence of ionizable groups, chemical composition, and molecular mass distribution that impart polyelectrolyte behavior, significantly influence the properties of microbial polysaccharides. Additionally, intra-/intermolecular interactions are influenced by the nature and number of these components. By combining microbial polysaccharides with other substances, such as polymers, salts, or cross-linking agents, their properties can be tailored. At an industrial level, certain microbial polysaccharides have been successfully developed and are commercially available. One notable example is xanthan, a heteropolysaccharide produced *by Xanthomonas* sp., comprising pyruvate, mannose, acetate, glucuronic acid, and glucose. Xanthan finds its use in the personal care, food, and pharmaceutical, industries, as well as in oil recovery processes [78]. A hydrophilic glucan, called dextran is synthesized by lactic acid bacteria, such as *Lactobacillus Streptococcus*, *Weissella*, *Pediococcus*, and *Leuconostoc*. This versatile polysaccharide has diverse applications across several industries, including food, cosmetics, and medicine [79]. HA is a linear polymer composed of *N*-acetylglucosamine and glucuronic acid and units. It is biosynthesized by *Streptococcus zooepidemicus* and finds applications in the fields of cosmetics, medicine, and pharmaceuticals [80]. Pullulan is a linear α-glucan that is produced by microorganism *Aureobasidium* pullulans which is a black yeast. It serves as a thickening mediator and is also utilized as an edible coating in various applications [81]. Scleroglucan is a linear hydrophilic β-glucan that is produced by the *Sclerotium* sp. which is a plant pathogen. It finds its use in numerous fields including enhanced oil recovery, the food industry, cosmetics, and pharmaceuticals [82]. Moreover, there is an increasing body of research focusing on newly identified polysaccharides with unique molecular structures derived from various taxonomic groups. These polysaccharides exhibit diverse biological activities and hold potential for novel applications. Recently, microbial biopolymers have garnered attention because of their favorable characteristics and ease of biosynthesis in contrast to other naturally produced biopolymers [68]. Microbes generally exhibit faster growth rates compared to plants or algae, and their biosynthesis procedures can be effortlessly modified to enhance productivity [83]. Furthermore, the biosynthesis procedure is not influenced by climate variations. Also, inexpensive waste or byproducts could be used as starting material [84]. Microbial Polysaccharides obtained by microbial fermentation possess exclusive characteristics. Notably, these biopolymers have found extensive use in the pharma, medical, food, and personal care industries mainly because of their excellent performance as binding, thicker, and stabilizing agents. Specific biopolymers like microbial alginate, FucoPol, and gellan can engage in intermolecular interactions, facilitating the formation of polymeric matrices. This property enables the physical manipulation of polysaccharides, leading to the development of structured formation for instance gels (e.g., hydrogels) or film, which hold potential applications in the biomedical field [71, 85]. Apart from these features, polysaccharides' ability to interact with diverse inorganic substances is a crucial aspect for applications like encapsulating bioactive compounds (e.g., drug encapsulation) to achieve sustainable or targeted release and integrating nanostructures to create advanced biomaterials. This interaction enables the improvement of bio-based material properties, like synergistic conductive or magnetic traits, depending on the incorporated nanostructures [86].

6.3.3.2 Commercially Relevant Microbial Polysaccharides

Microbial polysaccharides offer a diverse array of compositions and functional characteristics, making them applicable in several industries, like food products, pharmaceuticals, and biomedicine. Although only a limited number have been commercially developed, prominent examples include HA, scleroglucan, dextran, pullulan, gellan gum, xanthan gum, and levan.

6.3.3.2.1 Curdlan

Curdlan, a glucan synthesized by *Alcaligenes*, *Bacillus*, and *Rhizobium* species, is characterized by its ability to form alkaline-soluble and water-insoluble gels. However, this property restricts its potential industrial applications [87–89]. As a result, curdlan has traditionally found utility in the food industries where it serves as a stabilizing agent, texturizer, and thickener [90, 91]. Moreover, there have been studies indicating interest in synthesizing oligosaccharides from curdlan with increased solubility, which have demonstrated significant biomedical functions like immunological and antitumor activities [92]. Also, curdlan-derived oligosaccharides have demonstrated promise in promoting sustainable agriculture by efficiently stimulating the innate immune defense system in plants. Ongoing research is exploring curdlan and its derivatives as potential eco-friendly substitutes for petroleum-based plastics [91].

6.3.3.2.2 Gellan Gum

The gellan gum molecule is industrially synthesized using the *Sphingomonas paucimobilis* ATCC 31461 [95]. Gellan gum is a heteropolysaccharide made up of D-glucose monomers and L-rhamnose, D-glucose, D-glucuronic acid accompanied by acetyl as well as glyceryl side chains. It possesses a distinct ability to create gels, and the properties of these gels are the effect of acyl group content. The low acyl form of gellan gum results in thermally stable, brittle non-elastic, and rigid gels; in contrast, the high content of acyl form yields thermo-reversible elastic, non-brittle, and soft gels [94, 95]. Gellan gum is widely employed in the food industries, primarily serving as a binding, thicker, and stabilizing agent. It finds frequent use in numerous food items, including desserts, drinking jellies, fruit preparations, low-calorie jams, sauces, nonfat salad dressings, and yogurt, and in the production of edible films [96]. Gellan is recognized as a versatile additive with multiple functions in various pharmaceutical products, particularly in controlled-release formulations. It is commonly employed in oral, ophthalmic, nasal, and other types of formulations to enhance their performance and achieve controlled-release properties [97]. Gellan-based materials, such as those found in ophthalmic formulations like Timoptol, have already been utilized in the field. Recent studies indicate that these gellan-based materials have potential applications in areas like gene transfer technology, regenerative engineering, and tissue engineering [98].

6.3.3.2.3 Levan

Levan is a type of homo-fructan, containing fructose units and is produced using various microorganisms (such as *Lactobacillus* sp., *Acetobacter* sp., *Zymomonas sp.*, and *Halomonas* sp.) or synthesized by plants [99, 100]. It exhibits unique properties such as low intrinsic viscosity. It is both the solubility of levan in oil and water and its compatibility with surfactants and salts. Also, it has properties like adhesive ability, emulsifying capacity, and biological activity. As a result of its functional characteristics, Levan has diverse applications in multiple industries, including food, cosmetics, and pharmaceuticals. In the food sector, it is used as a prebiotic agent, while in cosmetics, it serves as a dermal filler. Its low viscosity makes Levan suitable for capsule production and coatings in the pharmaceutical field, enabling various therapeutic uses. Levan nanoparticles show potential for delivering peptides and protein drugs, and they also contribute to eco-friendly gold and silver and gold nanoparticle synthesis [101].

6.3.3.2.4 Hyaluronic Acid

HA is a linear polymer containing repetitive disaccharide units of *N*-acetylglucosamine and glucuronic acid [102]. HA has historically been biosynthesized by *S. zooepidemicus*. But, due to apprehensions about the pathogenic nature of Streptococcus species, the attention to biosynthesis of HA has shifted to using nonpathogenic organisms that fall under the generally recognized as safe (GRAS) category, for example, *Lactococcus lactis* and *Bacillus subtilis* [102]. HA exhibits remarkable attributes such as excellent non-immunogenicity, biodegradability, swelling capacity, viscoelasticity, and biocompatibility. These exceptional physical and biological properties make HA highly suitable for a range of applications in different domains like the cosmetic industry, pharmaceuticals, and medicine. HA has demonstrated its potential in the treatment of conditions such as osteoarthritis and tissue engineering applications [103–107].

6.3.4 Hydrogels Derived from Microbial Polysaccharides

Microbial polysaccharides have the unique capability to form hydrogels, which are versatile materials that exhibit a semi-liquid and semi-solid behavior. Hydrogels are cross-linked polymers that have three-dimensional networks that have a large water absorption and retention capacity. They find extensive use in diverse fields because of their remarkable properties and applications [108, 109]. In the past few decades, various polysaccharides such as dextran, alginate, cellulose, HA, chitin/chitosan, xanthan, and gellan gum, obtained microbial biosynthesis, have been extensively employed for the development and production of the hydrogel. Depending on the desired properties and fabrication process, hydrogels are synthesized by either a chemical or physical cross-linking approach [110]. Cross-linking of hydrogels by physical method exhibits reversible behavior under certain conditions, where the polymeric chains are held together by weak secondary forces like hydrophobic interaction, hydrogen bonding, or ionic interaction. Although these hydrogels may have instability in shape due to bond reversibility, they typically possess higher rigidity compared to chemically cross-linked hydrogels [110]. The use of divalent ions like Ca^{2+} to cross-link alginate could be a good example to demonstrate the use of ionic interaction to form hydrogel [111]. Contrarily, chemical cross-linking of hydrogels is distinguished by their permanence and robustness, attained through the establishment of strong covalent bonds between a cross-linking mediator and polymer backbone. Chemically hydrogel can be synthesized using diverse methods, including graft copolymerization, incorporation of a cross-linking mediator, or radiation. In the context of polysaccharides, a frequently employed approach involves using a cross-linking agent that interacts with the reaction active sites, commonly the -OH (hydroxyl) group, found on polysaccharides [108]. Hydrogels can be categorized into various categories according to the composition of their monomers. Homopolymer hydrogels consist of only one monomer unit, while two or more monomeric units are found in copolymer hydrogel, with at least one being hydrophilic. The use of a cross-linker to interlock two distinct cross-linked networks gives us an interpenetrating polymeric network (IPN). Furthermore, if one of the components is a non–cross-linked polymer, then hydrogels are categorized as semi-IPN [109].

Hydrogels by polysaccharides could be grouped into various classes depending on their ionic charge. They are classified as ionic hydrogel (either cationic or anionic), nonionic (neutral) hydrogel, or ampholytic hydrogel, based on the overall charge they carry. Neutral hydrogels do not possess any charge group, while cationic hydrogels have a positively charged group, and anionic hydrogels have negatively charged groups. Ampholytic hydrogel, on the contrary, contains both positive as well as negative charge groups within its structure [108, 111, 112]. Polyelectrolytes, specifically ionic and ampholytic hydrogels, are responsive to stimuli and are commonly referred to as smart hydrogels. Depending on their intended use, hydrogels can be engineered to exhibit unique responses to conditions like ionic strength, temperature, pH, and magnetic or electric fields [109, 113]. The capacity of hydrogels to exhibit response to external stimuli has earned them the labels of "smart" or "stimuli-sensitive" hydrogels. This unique characteristic makes them highly valuable in numerous biomedical uses, including controlled delivery of drugs. Additionally, their stimuli-sensitive properties also render them beneficial for applications in agriculture [114–116]. An instance of a naturally occurring microbial polysaccharide that exhibits thermo-responsive behavior is gellan gum. An anionic polysaccharide is biosynthesized by microbial fermentation, and it possesses the ability to form gels that undergo reversible changes with temperature. The mechanical characteristics of these gels differ depending on the composition of the gellan gum. Specifically, the acetylated form of gellan gum leads to the formation of elastic and soft gel, while the gellan gum containing deacetylated material leads to brittle or hard gels [117]. Generally, gellan gum displays characteristics of an upper critical solution temperature (UCST) behavior. This implies that, upon increasing the temperature, a polymer solution is formed, and it transforms into a gel as it cools. The temperature range at which gelation occurs for gellan gum usually lies between 35°C and 42°C. However, this range can be subject to variations depending on factors like the MW of gellan gum, processing conditions, and the presence of cations [118]. The food industry commonly utilizes gellan gum gels as an additive for food, primarily for its thickening and gelling properties [119]. Researchers have explored the potential of gellan gum gels for numerous biomedical uses, like tissue engineering and drug delivery [120]. Gellan hydrogels, with their unique properties, are well suited for utilization as injectable systems aimed at facilitating long-term cartilage regeneration [121]. The biosynthesis of hydrogels using polysaccharides of microbial origin is gaining prominence primarily due to their reduced toxicity, biocompatibility, biodegradability, and satisfactory mechanical strength [108]. Chitosan, a prominent polysaccharide, has been extensively investigated for its potential in hydrogel development. Microbial chitosan, derived from the deacetylation of chitin from fungi and yeast cell walls, exhibits a cationic nature and semi-crystalline structure. Hydrogels based on chitosan have been successfully created through physical or chemical cross-linking techniques, aiming to produce materials appropriate for biomedical uses, like drug delivery systems and wound healing dressings [122]. The limited water solubility of chitosan is widely recognized, and it can remain in solution under acidic conditions. Therefore, when neutralizing a solution of chitosan above 6.2 pH (the amine pKa), a gelation process occurs [123]. A binary hydrogel was developed to include curcumin, aiming to utilize it for cutaneous and topical wound treatments. The research outcomes indicated that prolonged skin exposure to this hydrogel did not cause any irritation, making them a viable option for addressing skin conditions such as psoriasis. Additionally, another study involved chemically cross-linking xanthan with starch to create hydrogels intended for drug delivery purposes [124]. Hydrogels produced from the combination of xanthan gum and starch revealed selective permeability, depending on the charge of the released drugs. This feature renders them a favorable option for drug release control of diverse medications. A variety of polysaccharides has been employed as appealing materials in the fabrication and design of hydrogel. In particular, microbial polysaccharides possess exceptional properties that make them appealing materials for various biomedical uses.

6.3.5 Bio-Nanocomposites Derived from Microbial Polysaccharides

Research and development are currently focused on bio-nanocomposites that incorporate microbial polysaccharides. These materials combine the unique properties of microbial polysaccharides with the advantages of nanomaterials, resulting in innovative bio-nanocomposite materials. Figure 6.3 provides details about such bio-nanocomposites finding their applications in diverse fields, including biomedicine, packaging, electronics, and environmental remediation. Among the valuable nanoscale structures are magnetic nanoparticles (MNPs), which have proven their usefulness in various fields, particularly in molecular

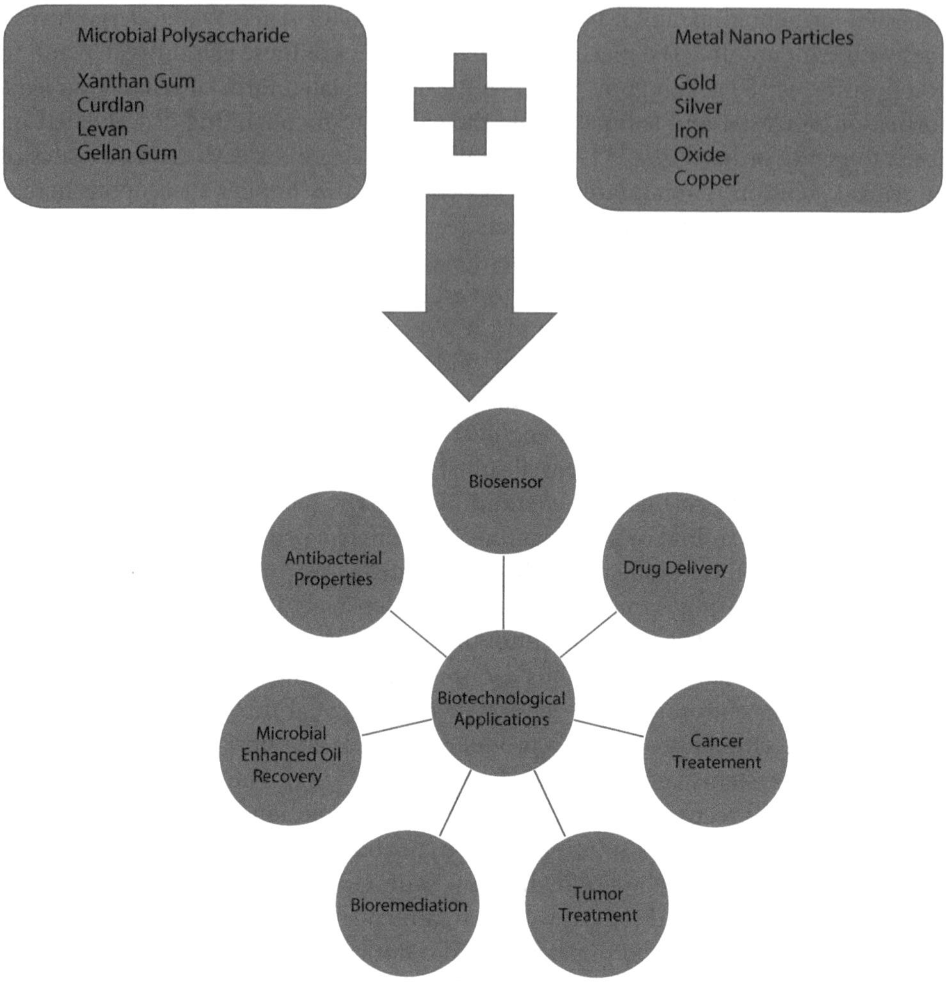

Figure 6.3 Microbial polysaccharide–based nanocomposites with metal nanoparticles for biotechnological applications.

diagnostics and biomedicine. MNPs possess distinctive physical characteristics that can be tailored by adjusting their size and composition. This composition may involve noble metals like silver and gold, magnetic elements such as iron and cobalt, or semiconductors like carbon nanotubes. The tunable properties of MNPs allow for a variety of uses and make them versatile tools in scientific and medical research [125]. The combination of MNPs with organic (e.g., biopolymeric) or inorganic (e.g., ceramic) matrices leads to the creation of nanocomposites, which are materials comprising multiple phases. The synergistic interactions between the components in these nanocomposites give rise to unique properties and functionalities [126]. Nanocomposite can be created using different matrix materials, such as ceramics, metals, or polymers. By integrating various nanostructures like metal nanoparticles, ceramics, carbon nanotubes, or even biologically active substances, these matrices can be further enhanced in terms of their properties and functionalities. This blending of diverse materials at the nanoscale level facilitates the improvement and customization of the overall characteristics of the nanocomposite [126]. Nanocomposites have opened up diverse possibilities for application across various fields. A notable example is the development of biocompatible hydrogels with conductive attributes for biomedical use. This is accomplished by integrating conductive inorganic materials, like carbon nanotubes, into a polymer-based matrix. The resulting nanocomposite shows potential in the creation of electrochemical biosensors, contributing to advancements in research and medical applications [127]. Another noteworthy example entails the utilization of magnetic nanoparticles to fabricate nanocomposites. Magnetic nanoparticles possess well-documented medical potential, and their applications often exploit three distinct inherent properties. These properties encompass field-induced mobility, exploited in drug delivery system development [128, 129].

6.4 Bioactive Polysaccharides from Microalgae

Microalgae and cyanobacteria possess a remarkable capacity to utilize carbon dioxide (CO_2) through photosynthesis, converting it into complex organic compounds. This process serves a crucial role in capturing CO_2 from the atmosphere, with microalgae and cyanobacteria responsible for approximately half of the annual CO_2 uptake by photosynthetic organisms. These microorganisms have relatively straightforward growth requirements, mainly requiring a seawater growing medium to which few micronutrients are supplemented. These nutrients are vital for their metabolic processes and the synthesis of essential biomolecules. The capability of cyanobacteria and microalgae to thrive in diverse aquatic environments, including oceans, freshwater bodies, wastewater treatment plants, and photobioreactors, highlights their adaptability. Their efficient CO_2-captured capability and adaptability make them attractive candidates for sustainable applications, such as biofuel production, wastewater treatment, and carbon capture and utilization. Overall, these microorganisms significantly contribute to CO_2 mitigation by utilizing this greenhouse gas as a carbon source and converting it into valuable organic compounds. Their simple growth requirements and CO_2-capturing abilities make them promising contenders for various environmental and industrial applications [130]. Because of their photosynthetic nature, microalgae and cyanobacteria do not require external carbon sources in their cultivation medium. This attribute offers several benefits, such as the reduction of wastewater produced during the process. Unlike other cultivation methods that rely on carbon-rich inputs, microalgae cultivation results in minimal waste generation.

The flexibility of microalgae cultivation extends to its water requirements, allowing for the use of non-potable water sources like seawater or wastewater. This characteristic lessens the strain on freshwater resources, making it especially advantageous in regions with limited access to freshwater or where using potable water for cultivation is impractical. Moreover, the reliance on natural sunlight as the primary energy source contributes to an ecologically and economically sustainable process. Natural sunlight eliminates the need for artificial lighting and energy-intensive systems, leading to reduced energy consumption and costs. Combining these factors, including the absence of an external carbon source, the use of non-potable water, and reliance on sunlight, results in a more environmentally friendly and cost-effective cultivation process. Microalgae cultivation provides a promising solution for sustainable production systems with minimal impact on water resources and energy consumption, making it an appealing choice for various applications in biotechnology, bioenergy, and environmental management [131]. Approximately 30,000 species of microalgae have been reported [132] of different evolutionary lines and with enormous metabolic diversity. In fact, microalgae production is an emergent market with an expected 10% yearly growth in this emergent market is expected for microalgae production [73]. Microalgae serve as a lucrative source of various high-value products that are in demand in industries such as nutraceuticals and skin care. These products include natural polyunsaturated fatty acids, astaxanthin β-carotene, proteins, and antioxidants. The commercialization of these compounds derived from microalgae has gained significant traction because of their beneficial characteristics and potential uses in the nutraceutical and skin care sectors [133–135]. In addition, there is growing interest in exploring microalgae as an eco-friendly and sustainable substitute for fossil fuels. This is because microalgae possess the ability to accumulate significant quantities of carbohydrates and lipids which can serve as a valuable feedstock for the synthesis of biofuels. Researchers are increasingly studying microalgae as a promising source of renewable energy, considering their potential to mitigate environmental impacts and reduce dependence on traditional fossil fuel resources [136, 137]. Nevertheless, the focus of commercial microalgae products has primarily been on biomass extraction, with the recognition of their potential as producers of valuable polysaccharides emerging only in recent times.

Microalgae and cyanobacteria are types of photosynthetic microorganisms characterized by their thalloid structure and ability to produce oxygen through photosynthesis. While microalgae are eukaryotic, cyanobacteria are prokaryotic and known for their remarkable adaptability to extreme environments such as saltpans. These organisms can also form extensive blooms in both freshwater and marine environments, covering vast distances [138–140]. Microalgae and cyanobacteria play significant roles in various biotechnological applications, including food production, animal feed, phycoremediation, pharmaceuticals, biofertilizers, and biofuel production [141, 142]. Exopolysaccharides (EPS) can be obtained as byproducts in industries that utilize cyanobacteria and microalgae for biodiesel production and other applications. These EPSs, which contain sulfate groups, exhibit bioactivity and can be extracted and utilized as biolubricants to reduce drag and dampen turbulence during fluid movement [143, 144]. The versatile properties of polysaccharides derived from microalgae and cyanobacteria have the potential to serve as alternatives to synthetic lubricants [145]. The research on aqueous polysaccharide lubrication focuses on studying the adsorption, tribology, and rheological behavior of these substances [146]. The oil industry has been implementing microbial enhanced oil recovery (MEOR) techniques, primarily utilizing microbes such as *Bacillus subtilis* [147]. Biolubricants have the potential to function as drag reducers in various types of pipe systems, including both closed and open configurations [148].

6.4.1 Microalgae

Starch from microalgae is generated as a side product in an oil refinery, holding potential for utilization in the production of biolubricants. It has been found that *Chlorella pyrenoidosa*, for instance, can yield a minimum of 44.8 g of polysaccharides per kilogram of dry weight through extraction [149]. The polysaccharides derived from *Schizochlamydella capsulata*, a microalga, have demonstrated effective drag-reduction properties (type B) within a narrow range of Reynolds numbers, specifically between 9,000 and 10,000, with minimal variation [150]. The distribution of glucose polymers, such as polyglucans (α/β), varies among algae of various taxonomic positions. Paramylon bodies, found in numerous *Euglena* and *Heterokontophyta* species, store β-1,3 glucan, which serves as an effective lubricant for the eyes. Cellulose glucan, which is valuable for biolubricant production, is present in the algal cell walls in a β-1,4-linked and β-1,4/β-1,3-linked configuration [151]. Cyanobacteria and algae are known to produce extracellular polysaccharides (EPS). In the case of the unicellular and rapidly growing alga *Porphyridium* sp. from the Rhodophyceae family, an acidic and sulfated heteropolysaccharide was successfully extracted as EPS [152]. The extracted EPS, with a concentration ranging from 0.5% to 2% (w/w), demonstrated its effectiveness as a lubricant for viscosupplementation. This property is beneficial in reducing friction and alleviating pain when the EPS is injected into human joints such as the knee. This alternative treatment option can be particularly advantageous as it provides a non-invasive solution, potentially avoiding the need for costly and temporary total knee replacements, which typically last around 10 to 15 years [153]. Sulfated polysaccharides are identified as a potential lubricant for arthritis, offering an alternative to HA [153]. The adhesive polysaccharide hydrogels, known for their low friction coefficient, have multiple applications as drug delivery systems, biolubricants, and biocompatible material in tissue engineering [154]. These biolubricants exhibit comparable properties to lubricin and are effective in mitigating the allergic reactions associated with HA.

6.4.2 Cyanobacteria

Cyanobacteria are known to secrete slime, which is attributed to the presence of sheaths and capsules that encase them. The extracellular polysaccharides (EPSs) produced by cyanobacteria have shown potential applications in MEOR and soft tribology and could serve as a potential substitute for synthetic surfactants [155]. By optimizing various factors such as the growth medium composition and physical parameters including light duration, light intensity, and temperature, it is possible to maximize the biosynthesis of extracellular polysaccharides (EPS) from cyanobacteria and microalgae (as shown in Table 6.2). Closed photobioreactors prove to be valuable in maintaining the quality and consistency of EPS during large-scale production processes [156, 157]. The production of carbohydrate-based biolubricants for oil drilling can be cost-effective by utilizing biomass obtained from open outdoor ponds. The use of raceway ponds proves to be an economical and scalable method for industrial production, allowing for production at any desired scale [158]. The composition of EPS derived from these organisms and their potential applications can be diverse. In contrast to bacteria, the EPSs produced by cyanobacteria are anionic and highly hydrophobic heteropolymers consisting of 10 to 12 monomers. Novel polysaccharides such as nostofan, cyanofan, and sacran have been isolated from *Nostoc flagelliforme*, *Cyanothece* sp., and *Aphanothece sacrum*, respectively [161–163]. Approximately 75% of the synthesized carbohydrates can potentially be excreted into the surrounding medium as polysaccharides. By modifying simple factors, the biosynthesis of EPS can be maximized, as shown in Table 6.2. An example of such EPS is emulcyan, an emulsifier produced by the *Cyanobacterium Phormidium* J-1, which reaches its peak production during the stationary phase [164]. Cyanofan, a polymer derived from *Cyanothece* sp., exhibits droplet formation with sizes of up to 200 μm, resembling the behavior of xanthan gum. It possesses an emulsification index (EI) of at least 50% and maintains stability for over a month. Additionally, an emulsifier biosurfactant obtained from *Nostoc* sp. shows varying EIs ranging from 21.4% to 37.5% when tested with different hydrocarbons and oils [140]. The extracted EPSs from cyanobacteria such as *Cyanothece* sp. CCY 0110, *Limnothrix redekei*, and *N. flagelliforme* exhibit a high viscosity level and can be compared to commercially available xanthan gum [163–165]. The EPS obtained from *Cyanothece epiphytica* exhibits lubricating properties comparable to grease, with significant emulsifying activity. Additionally, the EPS derived from *N. muscorum*, which contains a substantial amount of uronic acid, shows potential applications in paint emulsions and oil drilling [69]. The EI of the sulfated EPS obtained from *C. epiphytica* was found to range between 58.24% and 72.37% at its lowest concentration (0.2%) [143]. The EPS obtained from *N. flagelliforme* at a concentration of 0.45% exhibited a higher EI ranging from 27.3% to 73.8% compared to xanthan gum in hydrocarbons and oil. Furthermore, the emulsion formed by the EPS from *N. muscorum* remained stable for a period of 3 months, with an EI greater than 75% and an E24 value exceeding 80% [166]. The emulsifying properties of these substances have practical applications in the field of oil drilling.

6.4.3 Chemical and Physical Properties of Polysaccharides

The examination of the chemical and physical characteristics of polysaccharides can uncover their potential as biolubricants and inspire entrepreneurs to recognize algal productions as lucrative opportunities for producing various products. Several key properties should be considered, including as follows:

6.4.3.1 Molecular Weight

The MW of a substance plays a vital role in determining its viscosity and, by extension, its lubricating properties. Polysaccharides obtained from algal and cyanobacterial are characterized by their high MWs. One example is fucoidan, a polysaccharide extracted from *Turbinaria turbinata*, which possesses a high MW exceeding 10,000 kDa and can be extracted using gentle enzymatic methods [167] and the high MW of the cell-bound polysaccharide (CPS) allows for its efficient extraction, similar to the released polysaccharide (RPS) [166, 168]. The natural algae–based synthetic lubricant (NASL), obtained from the slime of *Macrocystis pyrifera*, has a MW exceeding 300,000. This high MW surpasses that of regular lubricants and contributes to its exceptional drag-reduction properties, making it a superior choice in this regard [169]. The stability of the emulsion is achieved by utilizing a low concentration (1:50 to 1:1000) of bio-emulsion, which has a high MW. This high MW allows for the formation of a stable emulsion [170]. The use of ultra-high MW polyethylene in hip and knee replacements can sometimes lead to inflammation. However, the application of algal biolubricant has shown the potential to reduce this inflammation.

6.4.3.2 Carbohydrate Composition

Carbohydrate-based biolubricants exhibit hydrophobic as well as hydrophilic properties, which depend on the composition of monomers. Polysaccharides obtained from algal and cyanobacterial typically consist of various monomeric units. For instance, xanthan gum is composed of glucuronic acid mannose and glucose. EPS derived from *cyanobacterium*

Cyanothece epiphytica, contains sulphated polysaccharides [143]. Sulfated polysaccharides have proven to be valuable biolubricants in biomedical applications. The hydrophobicity of sugars can be determined based on their partition coefficient (Kav), which reflects the partitioning of sugars between polystyrene gel and aqueous solvents. In terms of hydrophobicity, the order of the sugar sequence is as follows: D-ribose > D-arabinose > D-xylose > D-mannose > D-glucose > D-galactose [171]. In the development of oral capsules and tablets, hydrophobicity is not desirable, whereas it is beneficial for coating oil drilling pipes. For the production of hydrophilic tablets, a formulation was prepared by co-processing microcrystalline cellulose (MCC), starch, and chitin polymers, along with the addition of magnesium silicate as an additive [172]. Hydroxypropyl methylcellulose (HPMC) is a hydrophilic derivative of cellulose that is utilized in the formulation of tolcapone tablets. This is attributed to its notable characteristics, such as high viscosity, gelling and swelling capabilities, inertness, nonionic nature, absence of odor, and its role as an effective filler. Moreover, HPMC significantly influences the release kinetics of the active compound within the system [173]. Nanomaterials are commonly employed to develop superhydrophobic coatings for oil and gas pipelines. These coatings protect against corrosion, fouling agents, and other potentially damaging substances [174]. A green superhydrophobic film was created by applying a mixture of myristic acid (0.1 M) and cerium chloride (0.038 M) to the surface of carbon steel [175]. The potential of hydrophobic lubricants derived from carbohydrates is not widely acknowledged. The presence of ester-linked acetyl groups and peptide moieties, along with deoxy-hexoses such as fucose and rhamnose, contributes to the hydrophobic properties of EPS from cyanobacteria [176].

6.4.3.3 *Thermal Stability*

Thermal stability is an essential characteristic for effective lubrication. Algal/cyanobacterial polysaccharides exhibit a remarkable level of stability across a wide temperature range due to the intricate nature of their polymer structures. In the case of excreted polysaccharides obtained from *Cyanothece* sp. CCY 0110, a 65% reduction in stability was observed within the temperature range of 248°C to 300°C [164]. RPS derived from *N. carneum* exhibited a significantly lower loss of only 39% within the temperature range of 237°C to 378°C. The differential scanning thermogram analysis indicated a latent energy of crystallization of 108.67 mJ and a crystallization temperature of 107.4°C. In the case of EPS obtained from *C. epiphytica*, a degradation of 50% occurred at 288°C, while the EPS from *Arthrospira maxima* experienced a loss in weight of 66.6% at temperatures exceeding 500°C. Remarkably, even at 700°C, a residual mass of 34% was still present [177]. EPS derived from *Scenedesmus* sp. SB1 underwent a significant mass loss of 80% at a temperature of 167°C [178].

6.4.3.4 *Crystallinity*

Carbohydrate biolubricants generally exhibit an amorphous structure. In contrast, sodium alginate displays a crystalline nature with prominent peaks at 2θ angles of 13.5° and 21.9°. Calcium alginates, on the other hand, exhibit distinct semi-crystalline peaks at varying angles [83]. The polysaccharides derived from cyanobacteria, such as *C. epiphytica*, *N. flagelliforme Cyanothece* sp. CCY 0110, and *N. muscorum*, have an amorphous structure. However, the analysis by X-ray diffraction of EPS obtained from microalga *Scenedesmus* sp.

SB1 revealed that they exhibit both amorphous (84.8%) and crystalline (15.2%) characteristics. It is worth noting that amorphous materials have commonly been employed for lubrication purposes in numerous applications, including tablet formulations and earthquake lubrication [179, 180].

6.4.3.5 *Rheological Property*

In the field of tribology, which studies friction, lubrication, and wear, the role of aqueous polysaccharide lubrication is significant, particularly in hydrodynamic, boundary, and mixed regimes. The effectiveness of this lubrication method relies on various factors, including viscosity, adsorption, and their combined effects [146]. Water alone is commonly used as a lubricant in drilling muds, but the efficiency of this approach can be improved by incorporating carbohydrate biolubricants. These biolubricants, derived from microalgae and cyanobacteria, exhibit thixotropic, pseudoplastic, and non-Newtonian behavior. They are similar to commercially available polysaccharides such as microbially synthesized xanthan gum and seaweeds. Newtonian lubricants, on the other hand, are suitable for hydrodynamic, elastohydrodynamic, and mixed regimes. The rheological properties of these biolubricants are influenced by factors such as concentration, temperature, extraction processes, pH, MW, and chemical structure, which are carefully evaluated to ensure suitability for different applications. Notably, the flow index of CPSs and RPSs from *Anabaena* sp. CCC 745, as well as the high MW cyanofan from *Cyanothece* sp. *CCY* 0110, displayed superior rheological properties compared to xanthan gum [161]. An example illustrating the difference in viscosity between cyanofan and xanthan gum is as follows: A 1% solution of cyanofan has a viscosity of 1594.0 mPa.s, whereas xanthan gum (1%) has a lower viscosity of 1113.0 mPa.s. In terms of rheological behavior, xanthan gum exhibited more plasticity than viscosity compared to cyanofan. Additionally, the EPS from *Limnothrix redekei* PUPCCC 116 (0.2%) and xanthan gum demonstrated similar shear-thinning characteristics [163]. It is worth noting that the viscosity of xanthan gum decreased as the temperature increased within the range of 15°C to 55°C. However, there was no significant change observed in the viscosity of the EPS under the same temperature conditions.

6.4.4 Advantages of Microalgae and Cyanobacteria

In general, carbohydrate-based biolubricants derived from seaweed materials such as carrageenans, alginic acid, and agar are commonly used. However, the cultivation of macroalgae presents a critical factor, along with associated environmental concerns. The cultivation process is affected by seasonal variations in temperature and light, particularly during autumn and spring. Overharvesting from the natural habitat also poses a threat as it reduces the genetic diversity of macroalgae. Cultivation of macroalgae occurs on the surface of seawater, and this can have an impact on the productivity of other marine resources, including phytoplankton and benthic microalgae, due to shading effects [181]. Phytoplankton play a vital role in the primary productivity of marine ecosystems [139]. The availability of NO_3-N in tropical seawater is limited, which can have implications for seaweed cultivation. Cultivating seaweeds in natural seawater may also deplete the nutrient levels and impact marine organisms. Since nitrogen is crucial for algal growth, urea is commonly used as a nitrogen source in offshore algae cultivation [182]. The production of microalgal/cyanobacterial biomass can be integrated with the treatment of industrial effluents. Microalgae species such as *C. protothecoides* and *Chlorella vulgaris* exhibited a small lag phase in the production of biomass and effectively removed NH_{4+} (40 mg/L) from anaerobic digestion effluent within a period of 10 days [183]. Cultivation of *C. minutissima* in saline aquaculture wastewater resulted in a significant five times increase in cell density [184]. The total nitrogen (N) and phosphorus (P) content decreased significantly, with reductions of up to 88% and over 99%, respectively. Moreover, there were substantial reductions in NO_3-N (88.6%), NO_2-N (74.3%), and dissolved orthophosphates (99%). When cultivated in "dairy wastewater," *Scenedesmus abundans*, *C. pyrenoidosa*, and *Anabaena ambigua* showed increased biomass productivity and effectively reduced various pollutants, including biological oxygen demand (BOD) by 56%, chemical oxygen demand (COD) by 77%, NO3-N by 88%, and phosphate by 85% within a span of 25 days [185]. *Neochloris* sp. SK57, *Chlorella* sp. SL7A, and *Chlorococcum* sp. SL7B were grown in river water contaminated with pharmaceutical effluent. *Neochloris* sp. SK57 demonstrated a remarkable ability to reduce COD by 90% within a period of 10 days [186]. In the case of *Chlorococcum humicola* cultivated in undiluted textile mill effluent, the levels of NO_3-N and NO_2-N were effectively reduced to below detectable limits within a span of 3 days [187]. The highest biomass yield was achieved using a combination of TE (90%) and $NaNO_3$ (0.15%), resulting in a growing rate of 0.37 per day. Cyanobacteria and Microalgae are utilized in bioremediation, known as phycoremediation to generate biomass that contains EPS [187–189]. Interestingly, microalgae and cyanobacteria have the ability to thrive in environments with low nutrient levels as well as some industrial effluents. This not only helps in reducing excess nutrients but also decreases the water requirement for cultivation. Cultivating these organisms in wastewater proves to be the most cost-effective method for biomass production [187]. It has been reported that the cyanobacterium *Oscillatoria boryana* (BDU 92181) has the ability to degrade "melanoidin," pigment found in the effluent of distilleries. Moreover, it can utilize this pigment as a nitrogen and carbon source [190]. The conversion into biomass is facilitated by inorganic carbon and light. Additionally, microalgae and cyanobacteria can utilize flue gasses from industries, leading to the conversion of approximately 513 tons of CO_2 into 100 tons of dry algal biomass [191]. The cultivation of algal/cyanobacterial biomass offers several advantages, making it an attractive option for various biotech industries, including biolubricant production. One significant benefit is that it does not require arable land, and it can grow in wastewater, enabling the use of unconventional cultivation areas. Additionally, microalgae/cyanobacteria exhibit faster growth rates compared to higher plants, ensuring a continuous supply of raw materials throughout the year. Moreover, marine cyanobacteria/microalgae can be harvested from seawater, which reduces the demand for freshwater resources. For instance, successful cultivation of *Porphyridium* sp. has been achieved using a 3% CO_2 supplement and continuous illumination of 15 $\mu E/m^2/s$ in an artificial seawater medium (ASW). Photobioreactors can be employed to enhance biomass production, and raceway ponds offer a cost-effective option for large-scale biomass cultivation. Furthermore, harvesting harmful algal/cyanobacterial blooms presents an opportunity to utilize different components of the biomass for various purposes. These advantages make algal/cyanobacterial biomass production an environmentally friendly and economically viable choice for multiple biotech applications [143]. Microalgae and cyanobacteria have faster growth rates in comparison to seaweeds, and their carbohydrate properties closely resemble those found in seaweeds. Cultivating selected marine microalgal strains can be achieved indoors or outdoors in seawater, depending on the intended products. Thanks to their photosynthetic nature, unlike fungi and bacteria, cyanobacteria and algae can be harvested from outdoor natural daylight conditions. When considering the production of energy by algal biorefineries, carbohydrates and pigments can serve as valuable byproducts. Extracting and utilizing the carbohydrate components for biolubricants production can significantly reduce the overall cultivation costs for both the primary product as well as its byproducts. Additionally, mild stress conditions, such as nutrient deprivation (e.g., nitrogen or phosphorus starvation), increased salinity, or mild ozonation, can be employed to maximize the biosynthesis of polysaccharides by cyanobacteria and microalgae [192–194]. Cultivating macroalgae in natural sea surface water does not offer the same level of control and manipulation as seen with microalgae and cyanobacteria. Unlike seaweeds, microalgae, and cyanobacteria are more resistant to diseases. When cultivated in open ponds or raceways, algal and cyanobacterial strains can effectively compete with environmental microbes by releasing metabolites that have an allelopathic influence [185].

6.4.5 Challenges in the Production and Application of Biolubricants from Microalgae and Cyanobacteria

Overall, the advantages of producing lubricants from biomass of microalgae lie in its sustainable and environmentally friendly nature, and its ability to control the composition of favorable fatty acids through the manipulation of microalgae growth parameters. However, a major challenge in this field is the scalability of production. Once a promising strain of microalgae has been identified for a certain desired application, the next phase is to achieve cultivation of these strains at a larger scale, if possible by the use of waste sources. This process involves various challenges and complexities. The composition of microalgae biomass typically includes carbohydrates (5% to 64%), proteins (6% to 71%), and lipids (7% to 23%) with percent composition changing depending on the condition of growth and specific species. Few strains of microalgae can accumulate over 50% lipids by dry weight, but achieving high concentrations of target fatty acids within these lipids requires specific stress conditions. This necessitates extensive research in both laboratory and field settings. In the cultivation of biomass of microalgae, certain requirements must be addressed, such as the use of sterile raw materials for production, the guarantee of uncontaminated and unialgal cultures in photobioreactors, and the production of uncontaminated lipids/oils acquired from the biomass for biolubricant production. Efficient techniques for downstream processing and extracting oil must be applied based on the lipid content, growth stage, and species of the microalgae biomass. However, these processes currently pose challenges to scalability, as they tend to be expensive and unsuitable for large-scale production. The lack of exploration of microalgae biomass producing high-value lipids specifically for biolubricants further hinders the commercialization of microalgae-based lubricants on a large scale. Addressing these issues will be crucial for the future development of microalgae biolubricant production [195]:

1. The design and construction of photobioreactors, both open and closed systems, play a crucial role in microalgae biomass production. However, the lack of experience among engineering companies and biologists in building

these novel facilities often leads to significant delays in initiating microalgae biomass production projects. These delays occur due to deviations from the specifications of the design of the photobioreactors.

2. The process of harvesting, seed culture preparation, and inoculation becomes more challenging as the scale of microalgae production increases, especially when dealing with volumes up to 300,000 liters. The lack of expertise in handling large-scale microalgal cultivation, and its downstream processing operations poses a significant obstacle to achieving efficient operations in this regard.
3. Climate changes and environmental regulations, as well as land use regulations, pose challenges to microalgae production. Site-specific issues such as power and water outages, seepages, contamination, water evaporation, and unpredictable weather variations can disrupt production and even lead to the loss of the culture broth. Additionally, special land use permissions may be required, further complicating the process.
4. Ensuring product quality and consistency through the implementation of environmentally friendly processes is crucial. Regular sampling and analysis of the culture are necessary to maintain the desired level of quality and consistency. However, achieving an acceptable level of product quality and consistency remains a research challenge, particularly in terms of incorporating green methods for oil extraction and harvesting.
5. Similarly, to other VOs being considered for biolubricant production, there are certain challenges that need to be addressed in order to confidently and extensively utilize microalgae biolubricants in industrial and automobile applications. One of the prime challenges is ensuring product consistency and continuous availability. This aspect is influenced by factors such as the supplier, feedstocks, and production methods, which align with the overall challenges faced in microalgae-based oil production.
6. Guarantee of proper compatibility with machine materials [198–200]. Achieving satisfactory thermal oxidation stability and ensuring optimal performance in cold weather conditions are essential requirements for microalgae biolubricants. These properties need to be thoroughly assessed and enhanced for each oil/lubricant to meet stringent specifications prior to market release. Another crucial aspect is guaranteeing the acceptable low toxicity and biodegradability of biolubricants obtained from microalgae, particularly when chemical modifications are applied to base stock oils. The biodegradability of biolubricants can potentially be compromised by such chemical modifications and the addition of certain additives. Therefore, it is necessary to conduct further evaluations and implement appropriate treatments to ensure that the resulting biolubricants possess eco-friendly characteristics.
7. Gaining acceptance from machine manufacturers is a critical factor. It is essential to demonstrate the feasibility and widespread applicability of microalgae biolubricants in various machines and sizes, showcasing their impact on machine performance, emissions, durability, and resistance to oxidation. This demonstration is crucial in order to instill confidence in both

consumers and manufacturers regarding the effectiveness and reliability of biolubricants.

Furthermore, it is important to raise awareness about the environmental advantages of microalgae oil compared to petroleum lubricating oil and other VOs. By addressing the aforementioned challenges, microalgae oils have the potential to become the most sustainable and environmentally friendly sources for producing advanced biolubricants for a wide range of industrial applications.

6.4.6 Direct Use of Cell Cultures as a Potential Lubricating Fluid

To make the process easier by lowering the frictional force between tool and object, providing heat dissipation, and facilitating flushing, several lubricating chemicals are employed in cutting, drilling, milling, grinding, and forming activities [196]. They are referred to as cutting fluids or metalworking fluids. Up to 10% of water-based fluids may be used in many machining operations, or the lubricants may be formed of mineral oil. Water-based lubricants are very susceptible to microbiological contamination, which might be harmful to people and the environment as well as a deteriorating factor. Therefore, microbial contamination and propagation are a frequent and inevitable issue, especially in water-based cutting fluids, causing unrestrained variation of the fluid chemistry and raising serious concerns due to the risk of health risks for the machine operators, such as hypersensitivity pneumonitis, eye irritation skin dermatitis, and allergic reactions. The mineral oil–based lubricants are likewise risky and bad for the environment throughout their life cycle. Since they do not degrade and contaminate the environment, even how to dispose of them becomes a problem. The Metal Working Fluids sector is actively searching for other raw materials in anticipation of the declining oil sources. Algal oils and renewable vegetable raw materials have either previously been developed or are in use [197]. Using microbial suspensions rather than customary cutting fluids is an alternative strategy. It has less of an adverse effect on the environment and resources than other sustainable greener alternatives. To start experiments, preliminary research was offered. Future options that might improve performance and lessen the impact on the environment include bio-integration manufacturing. A definition of bio-integration manufacturing is the fusion of components from the biosphere and the technosphere in the production environment. *Pseudomonas oleovorans*, *Meyerozyma guilliermondii*, and *Saccharomyces cerevisiae* demonstrated strong friction reduction ability and performed better than standard fluids according to Daniel Meyer's 2017 milling experiment and tribological examination of several cell suspensions [198]. The milling studies and tribological tests both showed that at constant cell counts, the lubricating ability was influenced by cell size [197]. Teti *et al.*, on the other hand, conducted a cylindrical turning machining technique using pre-harvested cell matter of the algae spirulina arthrospores maximum, commonly known as *Spirulina platensis* [196]. Results were contrasted with those obtained using water and industrial cutting fluids. It was discovered that the surface quality and geometrical correctness for microbial-based cutting fluids showed superior values compared to dry cutting and equivalent values for conventional cutting fluids. Such attempts to combine technical processes with live systems have been made on several occasions. This strategy includes various benefits of the circular bioeconomic, including a smaller carbon footprint, applying the biorefinery idea and obtaining new products with added value, converting sustainable energy, reusing waste materials, and recycling carbon.

6.5 Biolubricants Synthesis Using Esterification and Transesterification Process

The process of creating an ester by reacting a carboxylic acid with an alcohol is known as esterification. In the case of bio-based oils like VOs, their chemical structure consists of natural esters called triglycerides, which require transesterification to enhance their lubricity properties. For biolubricant production, the trans-esterification procedure involves two steps. In the first step, short-chain primary alcohol reacts with triglyceride molecules, like methanol or ethanol, resulting in the formation of glycerol and mixtures of ethyl or methyl esters of fatty acid (FAME/FAEE). Methanol also known as "wood alcohol" is commonly preferred due to its favorable physicochemical properties, such as ease of solubility with sodium hydroxide (NaOH) solution and its rapid reaction with triglycerides [199, 200]. In the second step, the methyl esters derived from VO react with polyols to generate di- or polyester depending on the choice of alcohol in the second step, which would serve as biolubricants. The use of polyols offers a significant advantage as they lack α-hydrogens, which enhances the thermal stability of elevated temperatures of biolubricants by inhibiting self-polymerization and the formation of free fatty acids (FFAs). Figure 6.4 illustrates the transesterification process (i.e., the two-stage reaction process) using mono-alcohol.

6.5.1 Factors that Affect the Esterification and Transesterification Process

The transesterification process is influenced by various factors including the amount and type of catalyst used, the moisture content of alcohol, reaction temperature, molar ratio of reactants (primary alcohol to oil and FAME to polyol), reaction time, stirring rate, and the composition of fatty acids in the biomass oil. These factors play a critical role in determining the efficiency and outcome of the transesterification reaction [201]. However, this discussion will focus specifically on the important parameters, namely, the molar ratios of

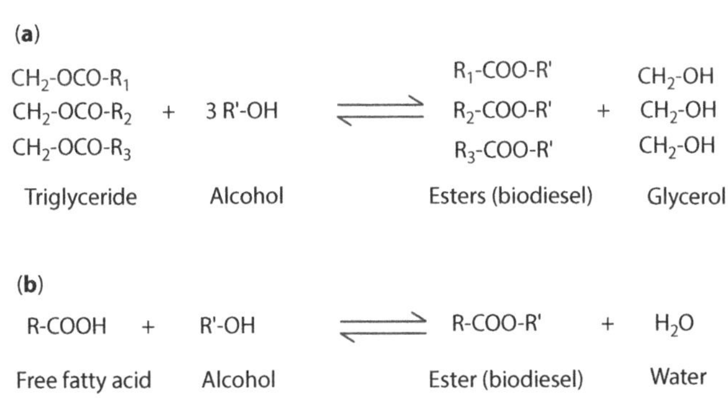

Figure 6.4 Schematic representation of transesterification reaction.

oil/primary alcohol and FAME/polyol, the temperature of the reaction, the reaction time, and the type of catalyst and catalyst ratio. The molar ratio between the primary alcohol and oil is a crucial factor in both esterification and transesterification processes. It significantly impacts the conversion efficiency and yield of biodiesel (FAME), as well as the overall production yield and cost [199]. In the initial stage of the process, close attention is paid to monitoring the molar ratio. To ensure effective interaction between the alcohol and triglyceride molecules, as well as improve their miscibility, it is important to use molar ratios higher than the stoichiometric ratio of 3:1 of alcohol to oil. This approach promotes better contact and enhances the overall reaction efficiency [202]. In fact, for a successful reaction, molar ratios of 4–6:1 or higher are required. These higher molar ratios ensure optimal conditions for the reaction to take place effectively [202]. The presence of a large quantity of methanol has been linked to the disruption of the bonds that hold the fatty acid and glycerin together [203]. Numerous research studies have demonstrated that increasing the molar ratio of the primary alcohol to oil can improve both the purity and yield of the resulting biodiesel. In an optimization study conducted by Eevera *et al.*, it was observed that the highest yield of methyl ester was achieved at methanol concentrations of 180 mL for edible oils and 210 mL for non-edible oils, within a range of 120 to 240 mL [204]. In a separate investigation conducted by Patil *et al.*, an esterification reaction catalyzed by acid demonstrated the highest conversions of esters. For jatropha oil, the conversion reached a maximum of 90% to 95%, while for Karanja oil, it reached 80%. These optimal conversions were achieved using an oil-to-methanol molar ratio of 1:6 [205]. The molar ratio between the polyol and the produced FAME plays an important role in the second stage of biolubricant synthesis. The esterification and transesterification reactions reach an equilibrium state without achieving complete conversion, as these reactions are reversible. According to Le Chatelier's principle, using a molar ratio higher than the stoichiometric ratio can improve the overall yield by driving the reaction toward completion. However, there is a discrepancy in the literature regarding whether increasing the alcohol relative to the FAME leads to an increase in biolubricant yield. Some researchers choose to use excess FAME due to its lower cost compared to alcohol [205]. Furthermore, in certain cases, a substantial yield of triester was attained when the molar ratio of FAME to alcohol was increased [206]. In the study conducted by Menkiti *et al.*, the percentages of triester yield were recorded for various combinations of Jatropha methyl ester and TMP molar ratios at different temperatures. The results showed that at the highest temperature of 160°C and a molar ratio of 7:1, an impressive 84% of triesters (biolubricant) was obtained. This observation suggests a positive correlation between the molar ratio of FAME to alcohol and the triester yield [201].

6.5.1.1 *Reaction Temperature and Time*

The two-step process of esterification and transesterification for the production of biolubricants is significantly impacted by the reaction temperature and duration. From a kinetic perspective, the reaction temperature directly influences the reaction rate and the yield of the resulting biolubricants. To prevent the evaporation of the primary alcohol employed in FAME synthesis, it is crucial to ensure that the temperature employed is below its boiling point [207]. In the initial stage of the overall process, raising the temperature has been observed to result in higher production of FAME, as demonstrated by Istiningrum *et al.*

Specifically, in the enzymatically catalyzed transesterification of waste cooking oil (WCO), the content of methyl ester increased significantly (from 60% to 80%) when the temperature was increased from 45°C to 55°C [208]. This evaluation aligns with numerous studies documented in the literature [208]. Similar trends can be observed in the second step of the process. Increasing the temperature leads to higher yields and improved purity of polyester (biolubricants). Studies conducted by Delgado *et al.* demonstrated that a temperature increase from 100°C to 140°C resulted in a biolubricant yield exceeding 90%. These findings were obtained using a reaction setup consisting of a 1:1 molar ratio of 2,2-dimethyl-1,3-propanediol and FAME derived from cardoon oil, along with a 1.5% concentration of Ti (IV) isopropoxide catalyst [209]. Similarly, Resul *et al.* observed a similar pattern in their study, where the 47.38% yield of jatropha triester was achieved at 200°C reaction temperature. The reaction conditions included the use of a 1.0% (wt/wt) concentration of sodium methoxide catalyst and a reaction time of 3 hours [210]. Increased temperatures can accelerate the rate of esterification and transesterification reactions; however, if not properly controlled, they can also lead to a higher likelihood of reverse reactions, resulting in a decrease in triester yield. Like many other chemical processes, the reaction time also plays a significant role in esterification and transesterification reactions. Generally, longer reaction times promote higher conversions of triglycerides into FAME and polyol esters in the two-stage esterification and transesterification process. However, there are conflicting findings, especially when enzymatic catalysis is involved. Some studies have reported higher FAME yields at shorter reaction times, while others have observed the opposite. A study by Amini *et al.* demonstrated that a longer reaction time of 68 hours in a lipase-catalyzed transesterification of Ocimum basilica L. resulted in a higher FAME yield (>92%) [211], (sweet basil) seed oil at conditions of 40°C, 12:1 M ratio of methanol to oil. The previous findings align with the studies conducted by Afifah *et al.*, who investigated the enzymatic catalysis of palm stearin transesterification [212]. Enzymatic transesterification reactions are influenced by factors such as the enzyme's specificity, the composition of the oil, and the selection of alcohol. To mitigate the longer reaction times associated with enzymatic catalysis, continuous reactor systems such as packed bed reactors (PBRs) and fluidized bed reactors (FBRs) have been utilized. These reactor designs help to improve the overall efficiency of the enzymatic reactions [211]. Moreover, for most acid- and base-catalyzed reactions, longer reaction times have been found to result in higher yields of fatty acid methyl esters (FAME). In a kinetic study conducted by Narayan *et al.*, investigating the esterification reaction of ricinoleic acids and 10-undecenoic acids with supercritical methanol, conversion rates exceeding 80% were reported at various molar ratios (such as 1:1, 2:1, 5:1, and 40:1) [213]. In both reactions involving 10-undecenoic acid and methanol, as well as ricinoleic acid and methanol, the reaction conversion was observed to increase with time across different temperature ranges (523 K, 573 K, 623 K, and 673 K). Similar to the first-stage process, the relationship between conversion and reaction time also holds true for the second-stage process. Idrus *et al.* conducted a study where they reported a higher conversion of POME using pentaerythritol (PE) alcohol after a reaction time of 2 hours. The reaction conditions for this study included a temperature of 160C, a molar ratio of POME to PE of 4.5:1, a pressure of 10 mbar, and a NaOCH3 catalyst concentration of 1.25% w/w [214]. At the conclusion of the reaction, a higher percentage (36%) of PE tetraester was produced, and there was no detectable presence of PE alcohol. The optimal reaction yield, exceeding 79%, was achieved after a reaction time of 4.61 hours, with a selectivity of tetraester reaching 91%. This highlights the significant role and relevance of this particular reaction parameter.

6.5.1.2 Catalyst Type and Catalyst Loading

Enzymatic, heterogeneous, and homogeneous forms of catalysts can all be used to enhance transesterification (acid/base). By removing a proton from the alcohol and increasing its reactivity, homogeneous base catalysts help to increase the rate of the process. On the other hand, acid catalysts speed up the reaction by giving the carbonyl group a proton, resulting in increasing its reactivity [215]. To utilize a base catalyst for biodiesel production, it is necessary for the fats and oil source to be free of fatty acids or have negligible levels of FFAs [216]. Soap formation can occur when a base catalyst is employed in FFA presence [216]. Non-edible VOs, animal fats, WCOs, and grease are examples of low-quality feedstocks. Compared to base catalysts, acid catalysts like H_2SO_4, HCl, and H_3PO_4 decrease the rate of reaction. FFAs initially transformed into esters using acid catalysis for the synthesis of biodiesel from low-quality feedstocks. This process results in the production of triglycerides and fatty acid alkyl esters [216]. A base catalyst is then used to transesterify the esters produced in the preceding step in the presence of methanol or ethanol as the alcohol reactant. Nevertheless, regardless of the type of feedstock, it is important to keep in mind that the acid-catalyzed transesterification process is comparatively slow and requires higher temperatures. Recent research on acid-catalyzed esterification and transesterification, including that of Moreira *et al.*, has shown noteworthy conversions and yields [217]. Furthermore, it is important to acknowledge that homogeneous catalysts have certain drawbacks. They consume a significant amount of energy, are corrosive, and pose challenges in terms of separation and recovery from the reaction solution. Because of these drawbacks, the use of enzymatic and heterogeneous solid catalysts is more often used in the transesterification process. These catalysts overcome the drawbacks of homogeneous catalysts and provide improved conversion rates for the synthesis of biolubricants. Heterogeneous acid catalysts, including ion-exchange resins, sulfated inorganic oxides, and inorganic super-acids such as WO_3/ZrO_2, as well as basic catalysts like alkaline earth oxides (CaO and MgO), alkali-supported catalysts (KF/K_2CO_3- or KNO_3-supported Al_2O_3), zeolites, and guanidine-supported catalysts, are actively being employed in various research endeavors [218]. In addition, Oh and associates carried out a study in which they used sulfated zirconia catalysts ($Zr(OCH_2CH_2CH_3)_4$) to create biolubricants from soybean oil. The reaction was conducted under the following conditions: catalyst weight of 100 mg, reaction period of 4 hours, and temperature of 140°C [219]. The acquired outcomes showed an oleic acid conversion rate of over 90% and an 84% biolubricant yield as output. In a separate study by Cavalcanti *et al.*, soybean oil biolubricants were synthesized using three commercially available lipases (*Candida rugosa*, *Rhizomucor miehei*, and *Candida antarctica*) that were immobilized onto microporous polypropylene beads. The enzymatic reactions were carried out in the presence of various polyhydric alcohols, including NPG, TMP, and PE [220]. As stated in their most recent study, Afifah *et al.*'s enzymatic transesterification of palm oil biolubricant ultimately produced great results. In addition, *Candida rugosa* demonstrated the highest conversions for all three polyols in the enzymatic reactions involving polyhydric alcohols, with 97% for NPG, 100% for TMP, and 55% for PE. Following *Candida rugosa*, the enzymatic reactions catalyzed by *Candida antarctica* and *Rhizomucor miehei* yielded lower conversion rates [212]. *Candida antarctica* lipase was used in the work by Afifah *et al.* as the enzymatic transesterification catalyst. They were able to reach a maximum yield of 95% under the following conditions: 60°C temperature, 8 hours of reaction period, and a lipase concentration of 6.0 wt%. Furthermore, Ma *et al.* conducted an enzymatic transesterification process using microbial oil (*R. glutinis* lipid). They utilized the *Candida* sp. 99 to 125 enzyme as the catalyst and achieved an 89.5% yield of TMP triester. The reaction conditions included a reaction time of 26 hours, temperature of 50°C, FAME:TMP ratio of 3.4:1, and enzyme loading of 15 wt% (5250 U/g substrates). The resulting biolubricant exhibited tribological characteristics comparable to petroleum- and mineral-based alternatives with the same viscosity grade (ISO VG46). Notably, catalyst loading or concentration has shown positive effects on the two-stage esterification and transesterification reactions in several studies [59]. Ghafar *et al.* conducted a study on the production of biolubricants using WCO as the feedstock. They employed a solid heterogeneous catalyst made from waste cockle shell to catalyze the reaction [14]. They observed that the conversion of triesters increased from 94% to 97% as the concentration of the catalyst (3% to 5% w/w) was increased in the two-stage esterification and transesterification reaction. This was observed at a constant molar ratio of FAME to TMP (3:1). In another study by Hussein *et al.*, it was found that increasing the CaO catalyst loadings from 0.8% to 1.5% (w/w) resulted in an increment in conversion from 62% to 98%. These results were obtained under reaction conditions of 130°C, 90 minutes, and a molar ratio of FAMEs to ethylene glycol of 3.5:1 [221]. Consistent findings were observed in the studies conducted by Afifah *et al.* and several other researchers. In the two-stage esterification and transesterification reaction process used to synthesize biolubricants, these investigations emphasize the importance of catalyst loading [212].

6.5.2 Epoxidation of Oils

VOs and their derivatives undergo epoxidation, which is followed by the opening of the oxirane ring, creating a complex ester with elements of both saturated and branched fatty acids. The oxidative stability and cold flow characteristics of the resultant ester are known to be improved by this chemical rearrangement. Epoxidation of carbon-carbon double bonds results in the formation of the oxirane ring, which is made up of three atoms grouped as carbon, oxygen, and carbon. VOs are frequently epoxidized using the Prileshajev procedure, which uses peracids and enables the production of the oxirane ring from the double bonds contained in the fatty acid components [222, 223]. Conventionally, By progressively introducing a cold carboxylic acid, such as formic acid, with the reactivity order of H_3PO_4 > HNO_3 > HCl, peracids are generated *in situ* [224]. Heterogeneous catalysts (e.g., SnO_2, Amberlite IR-120) [225–227] and lipase enzymes (e.g., Novozymes 435, an immobilized *Candida antarctica* lipase) [228] have also been used to produce peracids. It normally takes 5 to 16 hours for the reaction to complete completion in the case of peracid-based epoxidation, which is slow in rate and exothermic reaction [229]. According to Petrovic *et al.*, the type of peracid utilized affects the rate of epoxidation of soybean oil, with performic acid showing a higher rate than peracetic acid. This technique produces epoxidized VO, which has better oxidative stability and increased kinematic viscosity and may be used as a biolubricant [227]. The reaction of the epoxidized oil or its derivatives with a carboxylic acid, which takes place in an inert gas environment at a temperature of 100°C, opens the oxirane ring [229] or under ambient air with a catalyst present [230, 231]. Butanol, hexanol, or hexadecanol are examples of alcohols that can be used to open the oxirane ring in some circumstances. Acetic anhydride can also be used in some cases [226].

6.5.3 Fatty Acid Condensation: Estolide Synthesis

In order to create oligomers containing estolide connections, fatty acids must be connected at both the carboxylic acid group and the fatty acid chain. These unique compounds can be naturally found in various sources, including trees like kamala (*Mallotus philippinensis*) and Chinese tallow (*Sapium sebiferum*), flowering plants primarily belonging to the *Lesquerella* species, as well as the sclerotia of the ergot fungus (*Claviceps purpurea*) [232]. Natural estolides are usually present in combination with oils that have a hydroxyl group in the fatty acid chain, with the exception of oils generated from the ergot fungus. One such example is ricinoleic acid (12-hydroxyoctadec-cis-9-enoic acid) [232]. The first documented synthetic estolide was synthesized by condensing fatty acids obtained from the castor oil plant (*Ricinus communis*). This procedure required the interaction of two fatty acids, one of which contained hydroxyls, at a temperature of 220°C in a vacuum [233]. Due to its high ricinoleic acid content, which makes up around 90% of the oil's makeup, castor oil is often chosen as the lipid feedstock for these applications [234]. Fatty acids, such as oleic acid, or fatty acid alkyl esters are frequently used to end synthetic estolides [235]. The transesterification of ricinoleic and oleic acids with branched alcohols, such as 1-ethylhexanol, produces alkyl esters as capping groups. The extent of polymerization is generally controlled at a low level, with sulfuric acid, p-toluenesulfonic acid, BF_3•Et2O, montmorillonite K10, H_3PO_4, and HNO_3 being ranked in order of decreasing effectiveness as catalysts [236]. The standard procedure in the literature calls for carrying out the synthesis of estolides under ideal circumstances. This usually entails reacting unsaturated fatty acids at a temperature of 60°C for a period of 24 hours while using $HClO_4$ as the catalyst. Currently, Biosynthetic Technologies, a business that Calumet Specialty Products bought in 2018, produces estolides produced from VOs for commercial use [237]. Their production procedure was inspired by the US Department of Agriculture's processing technologies. They produce estolide-containing oils for 5W-30 grade motor oil as part of their line of goods. These oils are approved to include 35% bio-based components and comply with all applicable rules and regulations.

6.6 Biolubricants Physical and Chemical Properties

6.6.1 Viscosity

The viscosity of lubricants is a critical factor in determining their quality because it refers to their capacity to flow. The viscosity of the VO is an important factor to take into account when deciding on the final use. Viscosity is substantially influenced by temperature, pressure, and lubricant layer thickness. Generally speaking, thicker lubricant sheets come from higher viscosity. The VI is an essential component because it demonstrates how lubricants respond to temperature variations. Because they display minimal viscosity change when exposed to various temperatures, high VI lubricants are favored because they can maintain a constant thickness of the lubrication layer on the sliding surface of the engine while it is running.

6.6.2 Foam Resistance

Foaming is one of the undesirable physical properties of lubricants. Air is added to the lubricant by mechanical stirrers such as motors, compressors, and gearboxes. These air bubbles eventually collect to produce foam. This foam creates barriers between the liquid and the metal surface, reducing lubricity, obstructing heat transfer, accelerating oxidation, and hastening the wear and tear of machinery. Antifoaming chemicals are widely used to minimize lubricants' foaming capabilities.

6.6.3 Lubricity (Friction and Wear)

According to Appeldorn and Dukek, "lubricity" is defined as "the property of one liquid to create reduced friction, wear, or scuffing when two liquids of equal viscosity are combined." When selecting lubricants for end-use applications, the lubricity, friction, and wear requirements must be satisfied. Generally speaking, lubricity is connected to the development of the lubrication layer: the more the lubricity, the lower the friction and energy loss, since the moving surface's direct contacts are minimized. The boundary, hydrodynamic, and mixed lubrication regimes are the three types of lubrication regimes that may be seen on systems without an external pumping unit, according to the Stribeck curve.

6.6.4 Pour Point

The pour point is essential in applications that need lower operating temperatures. One of the fundamental properties of lubricants is their ability to remain liquid throughout a wide temperature range. The pour point denotes the lowest temperature limit, while the flash point denotes the highest. In lower-temperature applications like engines, oils should have a low pour point; otherwise, the wax crystals they produce may clog filters and jam engines. The pour point of a bio-oil product, like that of biodiesel, is frequently higher than that of petroleum-based diesel and varies greatly depending on the composition of fatty acids.

6.7 Expansion and Practical Viability on an Industrial Scale

This section provides a summary of the few articles that have concentrated on process scale-up, techno-economic analysis (TEA), and life cycle assessment (LCA) to evaluate the commercial viability and environmental effects of bio-based lubricants. Despite the abundance of experimental studies in recent years, assessing the potential for large-scale biolubricant production requires the development of kinetic and process models for reactor design and conceptual process flow sheeting [238]. Several authors conducted experiments to investigate the reaction kinetics of different biolubricant compounds obtained through various reaction pathways. Many of these studies focused on a two-step transesterification route, starting from VO as the feedstock [201, 239–241]. In particular, the first step involves the production of FAMEs from triglycerides and methanol, which has been extensively studied for biodiesel production.

Determining kinetic parameters is crucial for the development of reactor design and scale-up processes. In addition to kinetics, an efficient reactor design should also consider aspects of mass transfer and energy efficiency to improve reaction performance and reduce production costs. For instance, Souf *et al.* suggest the utilization of a vertical pulsed column (VPC) reactor to enhance mass transfer by increasing the interfacial area between immiscible reactants during the transesterification of WCO methyl ester with TMP [242]. From a different standpoint, Diaz and colleagues introduced a pervaporation-assisted reactor concept that focuses on the *in situ* removal of methanol during the transesterification process of castor oil methyl ester with TMP. The purpose of this approach is to shift the equilibrium toward the desired reaction direction and enhance the yield of biolubricants [243]. Ultimately, in order to evaluate the industrial viability of a particular biolubricant production technology, it is essential to perform conceptual process design and conduct mass and energy balances for large-scale processes. Hussein and colleagues addressed this by developing a comprehensive process simulation using Aspen HYSYS for the production of dioleoyl ethylene glycol ester biolubricant starting from WCO as the feedstock [221]. The authors conducted an experimental study to determine the optimal operating conditions for biolubricant production and subsequently performed process simulation to establish material balances for a large-scale process. The production capacity considered was 88,700 tons per year, with an achieved molar purity of 87%. However, the economic feasibility of the process was not assessed by the authors. In a separate study, Riazi and colleagues conducted a comprehensive TEA and LCA using the SuperPro Designer process simulator. Their analysis focused on the production of isostearic acid (IA) lubricant from soybean oil and tall oil, considering a production capacity of 4,500 tons per year [244].

6.8 Future Aspects

Because of the continued growth of manufacturing and industrial operations, fueled by rapid industrialization and rising vehicle ownership rates, the need for lubricants is anticipated to increase globally. This growth trend is particularly favorable for regions like Africa, the Middle East, and Latin America. Notably, lubricant consumption is increasing most quickly in the industrial industry and other areas. Due to its comparatively low labor costs and political stability, which draw businesses from throughout the world, the Asia-Pacific region, led by China, will continue to be a major driver of this expansion. The specifications for base oils have changed greatly over time. Correct viscosity and the lack of acidic components were essential in the 1950s. Base oils changed into additive carriers in the ensuing decades, and synthetic fluids with better performance started to appear in the 1970s. The 1980s brought quasi-synthetic hydro-cracked oils that closely resembled synthetic hydrocarbons. The 1990s emphasized lubricant performance, environmental concerns, and safety, leading to chemically pure oils and oleochemical derivatives with rapid biodegradability. Performance and compatibility improvement continued into the new millennium. Lubricant quality trends show a significant shift toward viscosity grades and product specifications, with biolubricants exceeding mineral lubricants in viscosity, low carbon-forming tendency, stability, oxidation stability, volatility requirement, and response to additives. The lubricant industry faces the challenge of developing better-performing lubricants for specific applications. The demand for lubricated automotive equipment that reduces environmental impact through lower emissions and increased biodegradability and non-toxicity has led to the development of new-generation heavy-duty lubricants. Biolubricants are gaining widespread acceptance for modern automotive engine oils due to their inherent performance advantages over conventional petroleum-based oils. These advancements in lubricant technology aim to address environmental concerns while providing enhanced performance for various applications.

In the past three decades, there has been a renewed interest in natural and bio-based oils as lubricants; however, their commercial application remains limited due to performance shortcomings and production limitations compared to mineral oils, which are preferred for cost reasons. Consequently, extensive research has been conducted to improve the properties and synthesis processes of biolubricants, as well as their renewable content and environmental impact. Finding renewable oil sources that do not compete with the food chain in terms of feedstock or land requirements is essential for developing a sustainable market. Additionally, the development of more environmentally friendly and efficient catalysts and additives can further promote biolubricant applications. The commitment of more countries to achieving zero carbon emissions by 2050 may lead to increased investment in biolubricant research, as it reduces the lubricant industry's dependence on fossil sources. In regions like North America and Europe, strict emission standards influence the advancement of internal combustion engines, particularly in sectors like industrial diesel engines and marine engines, which cannot be efficiently decarbonized through electrification. This will drive the development and adoption of biolubricants. The hydraulic fluids market is also expected to witness growth due to the growing environmental concerns associated with petroleum-based hydraulic fluids, leading to the potential entry of bio-based hydraulic fluids. Despite significant efforts to enhance biolubricant synthesis, certain areas lack comprehensive data, such as investigations into different chemical compounds and reaction pathways, and assessments of the physico-chemical properties of the product in relation to its applications. Industrial-scale production feasibility and overall life cycle impacts also require further examination. Therefore, the development of more comprehensive process simulations, conceptual process design, TEA, and LCA will be crucial in guiding future research directions.

References

1. Reeves, C.J., Siddaiah, A., Menezes, P.L., A Review on the Science and Technology of Natural and Synthetic Biolubricants. *J. Bio- Tribo-Corros.*, 3, 1, 11, 2017.
2. Salimon, J. *et al.*, Synthesis, reactivity and application studies for different biolubricants. *Chem. Cent. J.*, 8, 1, 16, 2014.
3. Bolina, I.C.A., Gomes, R.A.B., Mendes, A.A., Biolubricant Production from Several Oleaginous Feedstocks Using Lipases as Catalysts: Current Scenario and Future Perspectives. *Bioenergy Res.*, 14, 4, 1039–1057, 2021.
4. Zainal, N.A. *et al.*, A review on the chemistry, production, and technological potential of bio-based lubricants. *Renew. Sustain. Energy Rev.*, 82, 80–102, 2018.
5. Mobarak, H.M. *et al.*, The prospects of biolubricants as alternatives in automotive applications. *Renew. Sustain. Energy Rev.*, 33, 34–43, 2014.
6. Syahir, A.Z. *et al.*, A review on bio-based lubricants and their applications. *J. Clean. Prod.*, 168, 997–1016, 2017.
7. Ho, C.K., McAuley, K.B., Peppley, B.A., Biolubricants through renewable hydrocarbons: A perspective for new opportunities. *Renew. Sustain. Energy Rev.*, 113, 109261, 2019.

8. Chan, C.-H. *et al.*, Tribological behavior of biolubricant base stocks and additives. *Renew. Sustain. Energy Rev.*, 93, 145–157, 2018.
9. Cecilia, J.A., Ballesteros Plata, D., Alves Saboya, R.M., de Luna, F.M.T., Cavalcante, Jr. C.L., and Rodríguez-Castellón, E., An Overview of the Biolubricant Production Process: Challenges and Future Perspectives. *Processes,* 8, 3, 257, 2020, https://doi.org/10.3390/pr8030257.
10. Panchal, T.M. *et al.*, A methodological review on bio-lubricants from vegetable oil based resources. *Renew. Sustain. Energy Rev.*, 70, 65–70, 2017.
11. McNutt, J. and He, Q., Development of biolubricants from vegetable oils via chemical modification. *J. Ind. Eng. Chem.*, 36, 1–12, 2016.
12. Saboya, R.M.A. *et al.*, Synthesis of biolubricants by the esterification of free fatty acids from castor oil with branched alcohols using cationic exchange resins as catalysts. *Ind. Crops Prod.*, 104, 52–61, 2017.
13. Ivan-Tan, C.T. *et al.*, Screening of solid base catalysts on palm oil based biolubricant synthesis. *J. Clean. Prod.*, 148, 441–451, 2017.
14. Ghafar, F. *et al.*, Study on The Potential of Waste Cockle Shell Derived Calcium Oxide for Biolubricant Production. *Mater. Today: Proc.*, 19, 1346–1353, 2019.
15. Papadaki, A. *et al.*, Development of Microbial Oil Wax-Based Oleogel with Potential Application in Food Formulations. *Food Bioprocess Technol.*, 12, 6, 899–909, 2019.
16. Lage, F.A.P. *et al.*, Preparation of a biocatalyst via physical adsorption of lipase from Thermomyces lanuginosus on hydrophobic support to catalyze biolubricant synthesis by esterification reaction in a solvent-free system. *Enzyme Microb. Technol.*, 84, 56–67, 2016.
17. da Silva, NL. *et al.*, *Optimization of Biodiesel Production From Castor Oil, in Twenty-Seventh Symposium on Biotechnology for Fuels and Chemicals*, J.D. McMillan (Ed.), pp. 405–414, Humana Press, Totowa, NJ, 2006.
18. Bassi, J.J. *et al.*, Interfacial activation of lipases on hydrophobic support and application in the synthesis of a lubricant ester. *Int. J. Biol. Macromol.*, 92, 900–909, 2016.
19. Adlercreutz, P., Immobilisation and application of lipases in organic media. *Chem. Soc. Rev.*, 42, 15, 6406–6436, 2013.
20. Zhong, L. *et al.*, Production and use of immobilized lipases in/on nanomaterials: A review from the waste to biodiesel production. *Int. J. Biol. Macromol.*, 152, 207–222, 2020.
21. Zhao, Z., Zhou, M.-C., Liu, R.-L., Recent Developments in Carriers and Non-Aqueous Solvents for Enzyme Immobilization. *Catalysts*, 9, 647, 2019. DOI: 10.3390/catal9080647.
22. Mulinari, J., Oliveira, J.V., Hotza, D., Lipase immobilization on ceramic supports: An overview on techniques and materials. *Biotechnol. Adv.*, 42, 107581, 2020.
23. Usta, N. *et al.*, Properties and quality verification of biodiesel produced from tobacco seed oil. *Energy Convers. Manage.*, 52, 5, 2031–2039, 2011.
24. Erhan, S.Z. and Asadauskas, S., Lubricant basestocks from vegetable oils. *Ind. Crops Prod.*, 11, 2, 277–282, 2000.
25. Sulek, M.W., Kulczycki, A., Malysa, A., Assessment of lubricity of compositions of fuel oil with biocomponents derived from rape-seed. *Wear*, 268, 1, 104–108, 2010.
26. Lebedevas, S. *et al.*, Oxidation stability of biofuel containing Camelina sativa oil methyl esters and its impact on energy and environmental indicators of diesel engine. *Energy Convers. Manage.*, 65, 33–40, 2013.
27. Haseeb, A.S.M.A. *et al.*, Effect of temperature on tribological properties of palm biodiesel. *Energy*, 35, 3, 1460–1464, 2010.
28. Anastopoulos, G. *et al.*, Lubrication properties of low-sulfur diesel fuels in the presence of specific types of fatty acid derivatives. *Energy Fuels*, 15, 1, 106–112, 2001.
29. Jayadas, N.H., Prabhakaran Nair, K., Ajithkumar, G., Tribological evaluation of coconut oil as an environment-friendly lubricant, in: *Tribology International*, vol. 40, pp. 350–354, 2007.

30. Kalam, M.A. *et al.*, Friction and wear characteristics of waste vegetable oil contaminated lubricants. *Int. J. Mech. Mater. Eng.*, 6, 3, 431–436, 2011.
31. Masjuki, H.H. and Maleque, M.A., Investigation of the anti-wear characteristics of palm oil methyl ester using a four-ball tribometer test. *Wear*, 206, 1, 179–186, 1997.
32. Singh, Y., Garg, R., Kumar, S., Aspects of Non-edible Vegetable Oil-Based Bio-lubricants in the Automotive Sector, vol. 5, pp. 59–72, 2015.
33. Asadauskas, S., Perez, J.M., Duda, J.L., Lubrication properties of castor oil - Potential basestock for biodegradable lubricants©. *Lubr. Eng.*, 53, 12, 35–40, 1997.
34. Adhvaryu, A., Liu, Z., Erhan, S.Z., Synthesis of novel alkoxylated triacylglycerols and their lubricant base oil properties. *Ind. Crops Prod.*, 21, 1, 113–119, 2005.
35. Zulkifli, N.W.M., Kalam, M.A., Masjuki, H.H., Shahabuddin, M., Yunus, R., Wear prevention characteristics of a palm oil-based TMP (trimethylolpropane) ester as an engine lubricant, *Energy*, 54, 167–173, 2013 , ISSN 0360-5442, https://doi.org/10.1016/j.energy.2013.01.038
36. Arumugam, S. and Sriram, G., Effect of Bio-Lubricant and Biodiesel-Contaminated Lubricant on Tribological Behavior of Cylinder Liner–Piston Ring Combination. *Tribol. Trans.*, 55, 4, 438–445, 2012.
37. Maleque, M.A., Masjuki, H.H., Ishak, M., Bio-Fuel-Contaminated Lubricant and Hardening Effects on the Friction and Wear of AISI 1045 Steel. *Tribol. Trans.*, 41, 1, 155–159, 1998.
38. Bekal, S. and Bhat, N.R., Bio-lubricant as an Alternative to Mineral Oil for a CI Engine—An Experimental Investigation with Pongamia Oil as a Lubricant. *Energy Sources Part A*, 34, 11, 1016–1026, 2012.
39. Bhale, P.V., *Simulation of wear characteristics of cylinder liner ring combination with diesel and biodiesel*, 2008.
40. Suhane, A. and Khaira, H., Potential of non edible vegetable oils as an alternative lubricants in automotive applications. *Int. J. Eng. Res. Appl.*, 2, 5, 1330–5, 2012.
41. Jain, A. and Suhane, A., Capability of Biolubricants as Alternative Lubricant in Industrial and Maintenance Applications. 2013.
42. Arumugam, S. and Sriram, G., Synthesis and characterisation of rapeseed oil bio-lubricant – its effect on wear and frictional behaviour of piston ring–cylinder liner combination. *Proc. Inst. Mech. Eng., Part J: J. Eng. Tribol.*, 227, 1, 3–15, 2012.
43. Ting, C.-C. and Chen, C.-C., Viscosity and working efficiency analysis of soybean oil based bio-lubricants. *Measurement*, 44, 8, 1337–1341, 2011.
44. Nagendramma, P. and Kaul, S., Development of ecofriendly/biodegradable lubricants: An overview. *Renew. Sustain. Energy Rev.*, 16, 1, 764–774, 2012.
45. Kim, K.R. and Oh, D.K., Production of hydroxy fatty acids by microbial fatty acid-hydroxylation enzymes. *Biotechnol. Adv.*, 31, 8, 1473–85, 2013.
46. Joo, Y.-C. *et al.*, Production of 10-hydroxystearic acid from oleic acid by whole cells of recombinant Escherichia coli containing oleate hydratase from Stenotrophomonas maltophilia. *J. Biotechnol.*, 158, 1-2, 17–23, 2012.
47. Kim, H., Kuo, T., Hou, C., Production of 10, 12-dihydroxy-8 (E)-octadecenoic acid, an intermediate in the conversion of ricinoleic acid to 7, 10, 12-trihydroxy-8 (E)-octadecenoic acid by Pseudomonas aeruginosa PR3. *J. Ind. Microbiol. Biotechnol.*, 24, 3, 167–172, 2000.
48. Tunaru, S. *et al.*, Castor oil induces laxation and uterus contraction via ricinoleic acid activating prostaglandin EP3 receptors. *Proc. Natl. Acad. Sci.*, 109, 23, 9179–9184, 2012.
49. Redetzky, M. *et al.*, The Lubrication Properties of Microbial Cells and their Biopolymers. *Appl. Mech. Mater.*, 794, 285–291, 2015.
50. Morris, L., S. Hall, James, A., The biosynthesis of ricinoleic acid by Claviceps purpurea. *Biochem. J.*, 100, 3, 29C, 1966.

51. Mantle, P. and Nisbet, L., Differentiation of Claviceps purpurea in axenic culture. *Microbiology*, 93, 2, 321–334, 1976.
52. Holic, R. *et al.*, Engineered high content of ricinoleic acid in fission yeast Schizosaccharomyces pombe. *Appl. Microbiol. Biotechnol.*, 95, 179–187, 2012.
53. Gow, G., Lubricating grease, in: *Chemistry and technology of lubricants*, pp. 411–432, 2010.
54. Hudson, J.A., MacKenzie, C.A.M., Joblin, K.N., Conversion of oleic acid to 10-hydroxystearic acid by two species of ruminal bacteria. *Appl. Microbiol. Biotechnol.*, 44, 1, 1–6, 1995.
55. Morvan, B. and Joblin, K.N., *Hydration of oleic acid by enterococcus gallinarum, Pediococcus acidilactici andLactobacillus sp. Isolated from the Rumen. Anaerobe*, 5, 6, 605–611, 1999.
56. Heo, S.-H., Hou, C.T., Kim, B.S., Production of oxygenated fatty acids from vegetable oils by Flavobacterium sp. strain DS5. *New Biotechnol.*, 26, 1, 105–108, 2009.
57. Kuo, T.M. and Levinson, W.E., Biocatalytic production of 10-hydroxystearic acid, 10-ketostearic acid, and their primary fatty amides. *J. Am. Oil Chem. Soc.*, 83, 8, 671–675, 2006.
58. Bandhu, S. *et al.*, Influence of cytosolic malic enzyme in oleaginous yeast Rhodotorula mucilaginosa IIPL32 for lipid biosynthesis. *J. Bioanal. Biomed.*, 9, 100–106, 2017.
59. Ma, X. *et al.*, Enzyme-catalyzed synthesis and properties of polyol ester biolubricant produced from Rhodotorula glutinis lipid. *Biochem. Eng. J.*, 173, 101–108, 2021.
60. Zorn, S.M.F.E. *et al.*, *In Situ* Transesterification of Microbial Biomass for Biolubricant Production Catalyzed by Heteropolyacid Supported on Niobium. *Energies*, 15, 4, 1591, 2022.
61. Gao, X. *et al.*, Rapid screening and cultivation of oleaginous microorganisms, 2012.
62. Gavalas-Olea, A. *et al.*, Potential of the Red Alga Dixoniella grisea for the Production of Additives for Lubricants. *Plants (Basel)*, 10, 9, 1836, 2021.
63. Patel, A. *et al.*, Assessment of fatty acids profile and omega-3 polyunsaturated fatty acid production by the oleaginous marine thraustochytrid Aurantiochytrium sp. T66 cultivated on volatile fatty acids. *Biomolecules*, 10, 5, 694, 2020.
64. Baird, J., Sandford, P., Cottrell, I., Industrial applications of some new microbial polysaccharides. *Bio/Technology*, 1, 9, 778–783, 1983.
65. Suo, D. *et al.*, A universal biocompatible coating for enhanced lubrication and bacterial inhibition. *Biomater. Sci.*, 10, 13, 3493–3502, 2022.
66. Sharmila, G. *et al.*, Food waste valorization for biopolymer production, in: *Current Developments in Biotechnology and Bioengineering*, pp. 233–249, 2020.
67. Arad, M.R.A.S.M., Ginzberg, A., Huleihel, M., *Antiviral activity of sulfated polysaccharides of marine red algae*, in: Biomaterials from Aquatic and Terrestrial Organisms, 49–74, 2006.
68. Rühmann, B., Schmid, J., Sieber, V., Methods to identify the unexplored diversity of microbial exopolysaccharides. *Front. Microbiol.*, 6, 565, 2015.
69. Moscovici, M., Present and future medical applications of microbial exopolysaccharides. *Front. Microbiol.*, 6, 1012, 2015.
70. Becker, A. *et al.*, Microbial production of biopolymers and polymer precursors, in: *Xanthan Biosynthesis by Xanthomonas Bacteria: an Overview of the Current Biochemical and Genomic Data*, pp. 1–13, Caister Academic Press, UK, 2009.
71. Freitas, F., Alves, V.D., Reis, M.A.M., Advances in bacterial exopolysaccharides: from production to biotechnological applications. *Trends Biotechnol.*, 29, 8, 388–398, 2011.
72. Cress, B.F. *et al.*, Masquerading microbial pathogens: capsular polysaccharides mimic host-tissue molecules. *FEMS Microbiol. Rev.*, 38, 4, 660–697, 2014.
73. Freitas, F., Torres, C.A.V., Reis, M.A.M., Engineering aspects of microbial exopolysaccharide production. *Bioresour. Technol.*, 245, 1674–1683, 2017.
74. Yu, Y. *et al.*, Biological activities and pharmaceutical applications of polysaccharide from natural resources: A review. *Carbohydr. Polym.*, 183, 91–101, 2018.

75. Yildiz, H. and Karatas, N., Microbial exopolysaccharides: Resources and bioactive properties. *Process Biochem.*, 72, 41–46, 2018.
76. Roca, C. *et al.*, Exopolysaccharides enriched in rare sugars: bacterial sources, production, and applications. *Front. Microbiol.*, 6, 288, 2015.
77. Rehm, B.H.A., Bacterial polymers: biosynthesis, modifications and applications. *Nat. Rev. Microbiol.*, 8, 8, 578–592, 2010.
78. Kumar, A. and Han, S.S., Application of xanthan gum as polysaccharide in tissue engineering: A review. *Carbohydr. Polym.*, 180, 128–144, 2018.
79. Patel, A. and Prajapati, J., Food and Health Applications of Exopolysaccharides produced by Lactic acid Bacteria. *Advances in Dairy Research.*, 1, 1–7, 2013, 10.4172/2329-888X.1000107
80. Amado, I.R. *et al.*, Cheese whey: A cost-effective alternative for hyaluronic acid production by Streptococcus zooepidemicus. *Food Chem.*, 198, 54–61, 2016.
81. Prajapati, V.D., Jani, G.K., Khanda, S.M., Pullulan: An exopolysaccharide and its various applications. *Carbohydr. Polym.*, 95, 1, 540–549, 2013.
82. Schmid, J., Meyer, V., Sieber, V., Scleroglucan: biosynthesis, production and application of a versatile hydrocolloid. *Appl. Microbiol. Biotechnol.*, 91, 4, 937–947, 2011.
83. Alves, V.D. *et al.*, Rheological and morphological characterization of the culture broth during exopolysaccharide production by Enterobacter sp. *Carbohydr. Polym.*, 81, 4, 758–764, 2010.
84. Suresh Kumar, A., Mody, K., Jha, B., Bacterial exopolysaccharides – a perception. *J. Basic Microbiol.*, 47, 2, 103–117, 2007.
85. Manivasagan, P. and Oh, J., Marine polysaccharide-based nanomaterials as a novel source of nanobiotechnological applications. *Int. J. Biol. Macromol.*, 82, 315–327, 2016.
86. Aimé, C. and Coradin, T., Nanocomposites from biopolymer hydrogels: Blueprints for white biotechnology and green materials chemistry. *J. Polym. Sci., Part B: Polym. Phys.*, 50, 10, 669–680, 2012.
87. Yang, M. *et al.*, Production and optimization of curdlan produced by Pseudomonas sp. QL212. *Int. J. Biol. Macromol.*, 89, 25–34, 2016.
88. Liu, Y. *et al.*, Isolation and characterization of curdlan produced by Agrobacterium HX1126 using α-lactose as substrate. *Int. J. Biol. Macromol.*, 81, 498–503, 2015.
89. Cao, L. *et al.*, Determination of Curdlan Oligosaccharides with High-Performance Anion Exchange Chromatography with Pulsed Amperometric Detection. *J. Anal. Methods Chem.*, 2018, 3980814, 2018.
90. Miwa, M. and Nara, K., Food applications of curdlan, in: *Food hydrocolloids, structures, properties, and functions*, pp. 119–124, 1994.
91. Chien, C.-Y. *et al.*, Synthesis and characterization of regioselectively substituted curdlan hetero esters via an unexpected acyl migration. *Carbohydr. Polym.*, 155, 440–447, 2017.
92. Miyanishi, N. *et al.*, Induction of TNF-α production from human peripheral blood monocytes with β-1,3-glucan oligomer prepared from laminarin with β-1,3-glucanase from Bacillus clausii NM-1. *J. Biosci. Bioeng.*, 95, 2, 192–195, 2003.
93. Bajaj, I.B. *et al.*, Gellan gum: fermentative production, downstream processing and applications. *Food Technol. Biotechnol.*, 45, 4, 341–354, 2007.
94. Zia, K.M. *et al.*, Recent trends on gellan gum blends with natural and synthetic polymers: A review. *Int. J. Biol. Macromol.*, 109, 1068–1087, 2018.
95. Freitas, F. *et al.*, Production and food applications of microbial biopolymers, in: *Engineering Aspects of Food Biotechnology. Part I: Use of Biotechnology in the Development of Food Processes and Products*, pp. 61–88, 2013.
96. Saha, D. and Bhattacharya, S., Hydrocolloids as thickening and gelling agents in food: a critical review. *J. Food Sci. Technol.*, 47, 587–597, 2010.

97. Paolicelli, P. *et al.*, Effect of glycerol on the physical and mechanical properties of thin gellan gum films for oral drug delivery. *Int. J. Pharm.*, 547, 1, 226–234, 2018.
98. Osmałek, T., Froelich, A., Tasarek, S., Application of gellan gum in pharmacy and medicine. *Int. J. Pharm.*, 466, 1, 328–340, 2014.
99. Srikanth, R. *et al.*, Review on production, characterization and applications of microbial levan. *Carbohydr. Polym.*, 120, 102–114, 2015.
100. Silbir, S. *et al.*, Levan production by Zymomonas mobilis in batch and continuous fermentation systems. *Carbohydr. Polym.*, 99, 454–461, 2014.
101. González-Garcinuño, A. *et al.*, Levan and levansucrases: Polymer, enzyme, micro-organisms and biomedical applications. *Biocatal. Biotransform.*, 36, 3, 233–244, 2018.
102. Jia, Y. *et al.*, Metabolic engineering of Bacillus subtilis for the efficient biosynthesis of uniform hyaluronic acid with controlled molecular weights. *Bioresour. Technol.*, 132, 427–431, 2013.
103. Berkó, S. *et al.*, Advantages of cross-linked versus linear hyaluronic acid for semisolid skin delivery systems. *Eur. Polym. J.*, 49, 9, 2511–2517, 2013.
104. Benedini, L.J. and Santana, M.H.A., Effects of soy peptone on the inoculum preparation of Streptococcus zooepidemicus for production of hyaluronic acid. *Bioresour. Technol.*, 130, 798–800, 2013.
105. Allemann, I.B. and Baumann, L., Hyaluronic acid gel (Juvéderm™) preparations in the treatment of facial wrinkles and folds. *Clin. Interventions Aging*, 3, 4, 629–634, 2008.
106. Tiwari, S., Patil, R., Bahadur, P., Polysaccharide Based Scaffolds for Soft Tissue Engineering Applications. *Polymers*, 11, 1, 2019. DOI: 10.3390/polym11010001.
107. Rezaeeyazdi, M. *et al.*, Injectable Hyaluronic Acid-co-Gelatin Cryogels for Tissue-Engineering Applications. *Materials*, 11, 1, 2018. DOI: 10.3390/ma11081374.
108. Bhatia, J.K., Kaith, B.S., Kalia, S., *Polysaccharide Hydrogels: Synthesis, Characterization, and Applications, in Polysaccharide Based Graft Copolymers*, S. Kalia and M.W. Sabaa (Eds.), pp. 271–290, Springer Berlin Heidelberg, Berlin, Heidelberg, 2013.
109. Ahmed, E.M., Hydrogel: Preparation, characterization, and applications: A review. *J. Adv. Res.*, 6, 2, 105–121, 2015.
110. Barbucci, R., Spera, V., Armenia, E., Quagliariello, V., Microarchitecture of Water Confined in: *Hydrogels*, 2018, 10.1201/9781315152226-1.
111. Gacesa, P., Alginates. *Carbohydr. Polym.*, 8, 3, 161–182, 1988.
112. Oliveira, J.T. and Reis, Rui L., Hydrogels from polysaccharide-based materials: Fundamentals and applications in regenerative medicine, in: *Natural-Based Polymers for Biomedical Applications.*, pp. 485–514, 2008, 10.1533/9781845694814.4.485.
113. Usta, U. and Asmatulu, R., Hydrogels in various biomedical applications, in: *Polymer Science: Research Advances, Practical Applications and Educational Aspects*, pp. 248–257, 2015.
114. Soppimath, K.S. *et al.*, Stimulus-Responsive "Smart" Hydrogels as Novel Drug Delivery Systems. *Drug Dev. Ind. Pharm.*, 28, 8, 957–974, 2002.
115. Rudzinski, W.E. *et al.*, Hydrogels as controlled release devices in agriculture. *Des. Monomers Polym.*, 5, 1, 39–65, 2002.
116. Deen, G.R. and Loh, X.J., Stimuli-Responsive Cationic Hydrogels in Drug Delivery Applications. *Gels*, 4, 1, 13, 2018.
117. Ferris, C.J., Gilmore, K.J., Wallace, G.G., Modified gellan gum hydrogels for tissue engineering applications. *Soft Matter*, 9, 14, 3705–3711, 2013.
118. Graham, S., Marina, P.F., Blencowe, A., Thermoresponsive polysaccharides and their thermoreversible physical hydrogel networks. *Carbohydr. Polym.*, 207, 143–159, 2019.
119. Giavasis, I., Harvey, L.M., McNeil, B., Gellan gum. *Crit. Rev. Biotechnol.*, 20, 3, 177–211, 2000.
120. Matricardi, P. *et al.*, Preparation and characterization of novel gellan gum hydrogels suitable for modified drug release. *Molecules*, 14, 9, 3376–3391, 2009.

121. Fan, J. *et al.*, *In vitro* engineered cartilage using synovium-derived mesenchymal stem cells with injectable gellan hydrogels. *Acta Biomater.*, 6, 3, 1178–1185, 2010.
122. Jayakumar, R. *et al.*, Biomaterials based on chitin and chitosan in wound dressing applications. *Biotechnol. Adv.*, 29, 3, 322–337, 2011.
123. G Tahrir,Ganji, F.F. and Ahooyi, T.M., Injectable thermosensitive chitosan/glycerophosphate-based hydrogels for tissue engineering and drug delivery applications: a review. *Recent Pat. Drug Deliv. Formul.*, 9, 2, 107–120, 2015.
124. Shalviri, A. *et al.*, Novel modified starch–xanthan gum hydrogels for controlled drug delivery: Synthesis and characterization. *Carbohydr. Polym.*, 79, 4, 898–907, 2010.
125. Larguinho, M. *et al.*, Gold nanoprobe-based non-crosslinking hybridization for molecular diagnostics. *Expert Rev. Mol. Diagn.*, 15, 10, 1355–1368, 2015.
126. Camargo, P.H.C., Satyanarayana, K.G., Wypych, F., Nanocomposites: synthesis, structure, properties and new application opportunities. *Mater. Res.*, 12, 1–39, 2009.
127. Wang, S.-F. *et al.*, Preparation and Mechanical Properties of Chitosan/Carbon Nanotubes Composites. *Biomacromolecules*, 6, 6, 3067–3072, 2005.
128. Swierczewska, M. *et al.*, Polysaccharide-based nanoparticles for theranostic nanomedicine. *Adv. Drug Deliv. Rev.*, 99, 70–84, 2016.
129. Dias, A.M.G.C. *et al.*, A biotechnological perspective on the application of iron oxide magnetic colloids modified with polysaccharides. *Biotechnol. Adv.*, 29, 1, 142–155, 2011.
130. Rossi, F. and De Philippis, R., Exocellular polysaccharides in microalgae and cyanobacteria: chemical features, role and enzymes and genes involved in their biosynthesis, in: *The physiology of microalgae*, pp. 565–590, 2016.
131. Bafana, A., Characterization and optimization of production of exopolysaccharide from Chlamydomonas reinhardtii. *Carbohydr. Polym.*, 95, 2, 746–752, 2013.
132. Abd El-Hack, M.E. *et al.*, Microalgae in modern cancer therapy: Current knowledge. *Biomed. Pharmacother.*, 111, 42–50, 2019.
133. Roy, S.S. and Pal, R., Microalgae in Aquaculture: A Review with Special References to Nutritional Value and Fish Dietetics. *Proc. Zool. Soc.*, 68, 1, 1–8, 2015.
134. Nicoletti, M., Microalgae Nutraceuticals. *Foods*, 5, 3, 54, 2016.
135. Stolz, P. and Obermayer, B., Manufacturing microalgae for skin care. *Cosmet. Toiletries*, 120, 3, 99–106, 2005.
136. Chen, C.-Y. *et al.*, Microalgae-based carbohydrates for biofuel production. *Biochem. Eng. J.*, 78, 1–10, 2013.
137. Markou, G. and Nerantzis, E., Microalgae for high-value compounds and biofuels production: A review with focus on cultivation under stress conditions. *Biotechnol. Adv.*, 31, 8, 1532–1542, 2013.
138. Thajuddin, N., Marine cyanobacteria of the southern East coast of India_survey and ecobiological studies, 1991.
139. Sahoo, D. and Pandey, P.C., *Advances in Marine and Antarctic Science*, APH Publishing, 2002.
140. Borah, D., Gopalakrishnan, S., Nooruddin, T., Carbohydrate Biolubricants from Algae and Cyanobacteria. *J. Polym. Environ.*, 29, 11, 3444–3458, 2021.
141. Thajuddin, N., and Subramanian, G., Cyanobacterial biodiversity and potential applications in biotechnology. *Curr. Sci.*, 89, 47–57, 2005.
142. Baldev, E. *et al.*, Unveiling algal cultivation using raceway ponds for biodiesel production and its quality assessment. *Renew. Energy*, 123, 486–498, 2018.
143. Borah, D. *et al.*, Biolubricant potential of exopolysaccharides from the cyanobacterium Cyanothece epiphytica. *Appl. Microbiol. Biotechnol.*, 102, 8, 3635–3647, 2018.
144. Raposo, MFJ., De Morais, R.M.S.C., Bernardo de Morais, A.M.M., Bioactivity and Applications of Sulphated Polysaccharides from Marine Microalgae. *Mar. Drugs*, 11, 1, 233–252, 2013.

145. Paniagua-Michel, J.,Olmos-Soto, J., Morales-Guerrero, E., Algal and microbial exopolysaccharides: New insights as biosurfactants and bioemulsifiers. *A Adv. Food Nutr. Res.*, 73, 221–57, 2014, 10.1016/B978-0-12-800268-1.00011-1.
146. Stokes, J.R. *et al.*, Lubrication, Adsorption, and Rheology of Aqueous Polysaccharide Solutions. *Langmuir*, 27, 7, 3474–3484, 2011.
147. Park, T. *et al.*, Modification of Interfacial Tension and Wettability in Oil–Brine–Quartz System by *in Situ* Bacterial Biosurfactant Production at Reservoir Conditions: Implications for Microbial Enhanced Oil Recovery. *Energy Fuels*, 33, 6, 4909–4920, 2019.
148. Yanuar *et al.*, Effect of Agar Jelly Coating in Rectangular Pipe to Flow Drag Reduction. *J. Appl. Fluid Mech.*, 10, 4, 1161–1166, 2017.
149. Shi, Y. *et al.*, Purification and identification of polysaccharide derived from Chlorella pyrenoidosa. *Food Chem.*, 103, 1, 101–105, 2007.
150. Gasljevic, K. *et al.*, Drag-reducing polysaccharides from marine microalgae: species productivity and drag reduction effectiveness. *J. Appl. Phycol.*, 20, 3, 299–310, 2008.
151. Aikawa, S. *et al.*, Improving polyglucan production in cyanobacteria and microalgae via cultivation design and metabolic engineering. *Biotechnol. J.*, 10, 6, 886–898, 2015.
152. Arad, S., Adda, M., Cohen, E., The potential of production of sulfated polysaccharides from Porphyridium. *Plant Soil*, 89, 1, 117–127, 1985.
153. Arad, S. and Atar, D., Viscosupplementation with algal polysaccharides in the treatment of arthritis, 2015, Google Patents.
154. Yang, Q. *et al.*, Friction of sodium alginate hydrogel scaffold fabricated by 3-D printing. *J. Biomater. Sci., Polym. Ed.*, 28, 5, 459–469, 2017.
155. Smyth, T.J. *et al.*, Directed microbial biosynthesis of deuterated biosurfactants and potential future application to other bioactive molecules. *Appl. Microbiol. Biotechnol.*, 87, 4, 1347–1354, 2010.
156. Fischer, D., Schlösser, U.G., Pohl, P., Exopolysaccharide production by cyanobacteria grown in closed photobioreactors and immobilized using white cotton towelling. *J. Appl. Phycol.*, 9, 3, 205–213, 1997.
157. Soanen, N. *et al.*, Improvement of exopolysaccharide production by Porphyridium marinum. *Bioresour. Technol.*, 213, 231–238, 2016.
158. Liu, Q. *et al.*, Production and structural characterization of a new type of polysaccharide from nitrogen-limited Arthrospira platensis cultivated in outdoor industrial-scale open raceway ponds. *Biotechnol. Biofuels*, 12, 1, 131, 2019.
159. Kanekiyo, K. *et al.*, Isolation of an Antiviral Polysaccharide, Nostoflan, from a Terrestrial Cyanobacterium, Nostoc flagelliforme. *J. Nat. Prod.*, 68, 7, 1037–1041, 2005.
160. Okajima, M.K., Miyazato, S., Kaneko, T., Cyanobacterial Megamolecule Sacran Efficiently Forms LC Gels with Very Heavy Metal Ions. *Langmuir*, 25, 15, 8526–8531, 2009.
161. Mota, R. *et al.*, Cyanoflan: A cyanobacterial sulfated carbohydrate polymer with emulsifying properties. *Carbohydr. Polym.*, 229, 115525, 2020.
162. Fattom, A. and Shilo, M., Production of emulcyan by Phormidium J-1: its activity and function. *FEMS Microbiol. Lett.*, 31, 1, 3–9, 1985.
163. Khattar, J.I.S. *et al.*, Isolation and Characterization of Exopolysaccharides Produced by the Cyanobacterium Limnothrix redekei PUPCCC 116. *Appl. Biochem. Biotechnol.*, 162, 5, 1327–1338, 2010.
164. Mota, R. *et al.*, Production and characterization of extracellular carbohydrate polymer from Cyanothece sp. CCY 0110. *Carbohydr. Polym.*, 92, 2, 1408–1415, 2013.
165. Han, P.p. *et al.*, Emulsifying, Flocculating, and Physicochemical Properties of Exopolysaccharide Produced by Cyanobacterium Nostoc flagelliforme. *Appl. Biochem. Biotechnol.*, 172, 1, 36–49, 2014.

166. Borah, D. *et al.*, Ozone enhanced production of potentially useful exopolymers from the cyanobacterium Nostoc muscorum. *Polym. Test.*, 84, 106385, 2020.
167. Monsur, H.A. *et al.*, Chemical structure of sulfated polysaccharides from brown seaweed (Turbinaria turbinata). *Int. J. Food Prop.*, 20, 7, 1457–1469, 2017.
168. Tiwari, O.N. *et al.*, Purification, characterization and biotechnological potential of new exopolysaccharide polymers produced by cyanobacterium Anabaena sp. CCC 745. *Polymer*, 178, 121695, 2019.
169. Copp, E.A. and Glantz, D., Environmentally-friendly kelp-based energy saving lubricants, biofuels, and other industrial products, 2012, Google Patents.
170. Rosenberg, E. and Ron, E.Z., Bioemulsans: microbial polymeric emulsifiers. *Curr. Opin. Biotechnol.*, 8, 3, 313–316, 1997.
171. Janado, M. and Yano, Y., Hydrophobic nature of sugars as evidenced by their differential affinity for polystyrene gel in aqueous media. *J. Solution Chem.*, 14, 12, 891–902, 1985.
172. Assaf, S.M. *et al.*, Preparation and characterization of co-processed starch/MCC/chitin hydrophilic polymers onto magnesium silicate. *Pharm. Dev. Technol.*, 24, 6, 761–774, 2019.
173. Nardi-Ricart, A. *et al.*, Formulation of Sustained Release Hydrophilic Matrix Tablets of Tolcapone with the Application of Sedem Diagram: Influence of Tolcapone's Particle Size on Sustained Release. *Pharmaceutics*, 12, 7, 674, 2020.
174. Ijaola, A.O., Farayibi, P.K., Asmatulu, E., Superhydrophobic coatings for steel pipeline protection in oil and gas industries: A comprehensive review. *J. Nat. Gas Sci. Eng.*, 83, 103544, 2020.
175. Hassan, H. *et al.*, Super-hydrophobic green corrosion inhibitor on carbon steel, in: *IOP Conference Series: Materials Science and Engineering*, IOP Publishing, 2017.
176. Shepherd, R. *et al.*, Novel bioemulsifiers from microorganisms for use in foods. *J. Biotechnol.*, 40, 3, 207–217, 1995.
177. Kaplan Can, H., Gurbuz, F., Odabaşı, M., Partial characterization of cyanobacterial extracellular polymeric substances for aquatic ecosystems. *Aquat. Ecol.*, 53, 3, 431–440, 2019.
178. Angelaalincy, M. *et al.*, Enhanced Extracellular Polysaccharide Production and Self-Sustainable Electricity Generation for PAMFCs by Scenedesmus sp. SB1. *ACS Omega*, 2, 7, 3754–3765, 2017.
179. Démuth, B. *et al.*, Lubricant-Induced Crystallization of Itraconazole From Tablets Made of Electrospun Amorphous Solid Dispersion. *J. Pharm. Sci.*, 105, 9, 2982–2988, 2016.
180. Rowe, C.D. *et al.*, Earthquake lubrication and healing explained by amorphous nanosilica. *Nat. Commun.*, 10, 1, 320, 2019.
181. Campbell, I. *et al.*, The Environmental Risks Associated With the Development of Seaweed Farming in Europe - Prioritizing Key Knowledge Gaps. *Front. Mar. Sci.*, 6, 107, 2019.
182. Roleda, M.Y. and Hurd, C.L., Seaweed nutrient physiology: application of concepts to aquaculture and bioremediation. *Phycologia*, 58, 5, 552–562, 2019.
183. Yu, H., Kim, J., Lee, C., Nutrient removal and microalgal biomass production from different anaerobic digestion effluents with Chlorella species. *Sci. Rep.*, 9, 1, 6123, 2019.
184. Hawrot-Paw, M. *et al.*, Production of Microalgal Biomass Using Aquaculture Wastewater as Growth Medium. *Water*, 12, 1, 106, 2020.
185. Brar, A., Kumar, M., Pareek, N., Comparative Appraisal of Biomass Production, Remediation, and Bioenergy Generation Potential of Microalgae in Dairy Wastewater. *Front. Microbiol.*, 10, 678, 2019.
186. Singh, A., Ummalyma, S.B., Sahoo, D., Bioremediation and biomass production of microalgae cultivation in river water contaminated with pharmaceutical effluent. *Bioresour. Technol.*, 307, 123233, 2020.
187. Borah, D. *et al.*, Bioremediation and Biomass Production with the Green Microalga Chlorococcum humicola and Textile Mill Effluent (TE). *Proc. Natl. Acad. Sci., India, Sect. B Biol. Sci.*, 90, 2, 415–423, 2020.

188. Shashirekha, S., Uma, L., Subramanian, G., Phenol degradation by the marine cyanobacterium Phormidium valderianum BDU 30501. *J. Ind. Microbiol. Biotechnol.*, 19, 2, 130–133, 1997.
189. Baldev, E. *et al.*, Degradation of synthetic dye, Rhodamine B to environmentally non-toxic products using microalgae. *Colloids Surf., B*, 105, 207–214, 2013.
190. Francisca Kalavathi, D., Uma, L., Subramanian, G., Degradation and metabolization of the pigment—melanoidin in distillery effluent by the marine cyanobacterium Oscillatoria boryana BDU 92181. *Enzyme Microb. Technol.*, 29, 4, 246–251, 2001.
191. Bilanovic, D. and *et al.*, Freshwater and marine microalgae sequestering of CO2 at and N concentrations - Response surface methodology analysis. *Energy Convers. Manage.*, 50, 2, 262–267, 2009.
192. Jindal, N., Pal Singh, D., Singh Khattar, J., Optimization, characterization, and flow properties of exopolysaccharides produced by the cyanobacterium Lyngbya stagnina. *J. Basic Microbiol.*, 53, 11, 902–912, 2013.
193. Tiwari, O.N. *et al.*, Characterization and Optimization of Bioflocculant Exopolysaccharide Production by Cyanobacteria Nostoc sp. BTA97 and Anabaena sp. BTA990 in Culture Conditions. *Appl. Biochem. Biotechnol.*, 176, 7, 1950–1963, 2015.
194. Han, P.-. *et al.*, Influence of culture conditions on extracellular polysaccharide production and the activities of enzymes involved in the polysaccharide synthesis of Nostoc flagelliforme. *RSC Adv.*, 7, 71, 45075–45084, 2017.
195. da Silva, T.L. and Reis, A., Scale-up problems for the large scale production of algae, in: *Algal biorefinery: An integrated approach*, pp. 125–149, 2015.
196. Teti, R., D'Addona, D.M., Segreto, T., Microbial-based cutting fluids as bio-integration manufacturing solution for green and sustainable machining. *CIRP J. Manuf. Sci. Technol.*, 32, 16–25, 2021.
197. Padmanaban, V. *et al.*, Development of Bio based Semi-Synthetic Metal Working Fluid from Industrial Waste Water. *Procedia Eng.*, 64, 1436–1444, 2013.
198. Meyer, D., Redetzky, M., Brinksmeier, E., Microbial-based metalworking fluids in milling operations. *CIRP Ann.*, 66, 1, 129–132, 2017.
199. Ma, F. and Hanna, M.A., Biodiesel production: a review1Journal Series #12109, Agricultural Research Division, Institute of Agriculture and Natural Resources, University of Nebraska–Lincoln.1. *Bioresour. Technol.*, 70, 1, 1–15, 1999.
200. Demirbas, A., Studies on cottonseed oil biodiesel prepared in non-catalytic SCF conditions. *Bioresour. Technol.*, 99, 5, 1125–1130, 2008.
201. Menkiti, M.C., Ocheje, O., Agu, C.M., Production of environmentally adapted lubricant basestock from jatropha curcas specie seed oil. *Int. J. Ind. Chem.*, 8, 2, 133–144, 2017.
202. Jin-Suk, L. and Shiro, S., Biodiesel production by heterogeneous catalysts and supercritical technologies. *Bioresour. Technol.*, 101, 19, 7191–7200, 2010.
203. Miao, X. and Wu, Q., High quality biodiesel production from heterotrophic microalgal oil. *Bioresour. Technol.*, 97, 841–846, 2006.
204. Eevera, T., Rajendran, K., Saradha, S., Biodiesel production process optimization and characterization to assess the suitability of the product for varied environmental conditions. *Renew. Energy*, 34, 3, 762–765, 2009.
205. Patil, P.D. and Deng, S., Optimization of biodiesel production from edible and non-edible vegetable oils. *Fuel*, 88, 7, 1302–1306, 2009.
206. Resul, M.F.M.G., Ghazi, T.I.M., Idris, A., Temperature Dependence on The Synthesis of Jatropha Biolubricant. *IOP Conf. Ser.: Mater. Sci. Eng.*, 17, 1, 012032, 2011.
207. Abbah, E. *et al.*, Effect of reaction temperature on the yield of biodiesel from neem seed oil. *Am. J. Energy Sci.*, 3, 3, 16–20, 2016.

208. Istiningrum, R.B., Aprianto, T., Pamungkas, F.L.U., Effect of reaction temperature on biodiesel production from waste cooking oil using lipase as biocatalyst. *AIP Conf. Proc.*, 1911, 1, 020031, 2017.
209. Nogales-Delgado, S., Sánchez, N., Encinar, J.M., *Valorization of Cynara Cardunculus L. Oil as the Basis of a Biorefinery for Biodiesel and Biolubricant Production. Energies*, 13, 19, 5085, 2020.
210. Gunam Resul, M., Mohd. Ghazi, T.I., Idris, A., Kinetic Study of Jatropha Biolubricant from Transesterification of Jatropha curcas Oil with Trimethylolpropane: Effects of Temperature. *Ind. Crops Prod.*, 38, 1, 87–92, 2012.
211. Amini, Z. and *et al.*, Biodiesel production by lipase-catalyzed transesterification of (sweet basil) seed oil. *Energy Convers. Manage.*, 132, 82–90, 2017.
212. Afifah, A.N. *et al.*, Biolubricant production from palm stearin through enzymatic transesterification method. *Biochem. Eng. J.*, 148, 178–184, 2019.
213. Narayan, R.C. *et al.*, Kinetics of esterification of 10-undecenoic and ricinoleic acids with near-critical and supercritical methanol. *Sustain. Energ. Fuels*, 1, 6, 1425–1436, 2017.
214. Rashid, U. and Rahman, N.A., High Oleic Pentaerythritol Tetraester Formation via Transesterification: Effect of Reaction Conditions, Nor Faeqah Idrus1, 2, Robiah Yunus1, 3, Zurina Zainal Abidin1.
215. Hotha, K.K., Roychowdhury, S., Subramanian, V., Drug-excipient interactions: case studies and overview of drug degradation pathways. *Am. J. Anal. Chem.*, 7, 1, 107–140, 2016.
216. Biermann, U. and Metzger, J.O., Synthesis of alkyl-branched fatty acids. *Eur. J. Lipid Sci. Technol.*, 110, 9, 805–811, 2008.
217. Moreira, D.R. *et al.*, Moringa polyesters as eco-friendly lubricants and its blends with naphthalenic lubricant. *Ind. Crops Prod.*, 158, 112937, 2020.
218. Yunus, R. *et al.*, Development of optimum synthesis method for transesterification of palm oil methylolpropane to environmentally acceptable palm oil-based lubricant. *J. Oil Palm Res.*, 15, 2, 35–41, 2003.
219. Oh, J. *et al.*, Synthesis of biolubricants using sulfated zirconia catalysts. *Appl. Catal., A*, 455, 164–171, 2013.
220. Cavalcanti, EAC. *et al.*, Improved production of biolubricants from soybean oil and different polyols via esterification reaction catalyzed by immobilized lipase from Candida rugosa. *Fuel*, 215, 705–713, 2018.
221. Hussein, R.Z.K. *et al.*, Experimental investigation and process simulation of biolubricant production from waste cooking oil. *Biomass Bioenergy*, 144, 105850, 2021.
222. Tan, S.G. and Chow, W.S., Biobased Epoxidized Vegetable Oils and Its Greener Epoxy Blends: A Review. *Polym. Plast. Technol. Eng.*, 49, 15, 1581–1590, 2010.
223. Hilker, I. *et al.*, Chemo-enzymatic epoxidation of unsaturated plant oils. *Chem. Eng. Sci.*, 56, 2, 427–432, 2001.
224. Dinda, S. *et al.*, Epoxidation of cottonseed oil by aqueous hydrogen peroxide catalysed by liquid inorganic acids. *Bioresour. Technol.*, 99, 9, 3737–3744, 2008.
225. Somidi, A.K.R., Sharma, R.V., Dalai, A.K., Synthesis of Epoxidized Canola Oil Using a Sulfated-SnO2 Catalyst. *Ind. Eng. Chem. Res.*, 53, 49, 18668–18677, 2014.
226. Sharma, R.V. and Dalai, A.K., Synthesis of bio-lubricant from epoxy canola oil using sulfated Ti-SBA-15 catalyst. *Appl. Catal., B*, 142-143, 604–614, 2013.
227. Petrović, Z.S. *et al.*, Epoxidation of soybean oil in toluene with peroxoacetic and peroxoformic acids — kinetics and side reactions. *Eur. J. Lipid Sci. Technol.*, 104, 5, 293–299, 2002.
228. Sun, S. *et al.*, Enzymatic Epoxidation of Corn Oil by Perstearic Acid. *J. Am. Oil Chem. Soc.*, 88, 10, 1567–1571, 2011.
229. Moser, B.R. *et al.*, Diesters from Oleic Acid: Synthesis, Low Temperature Properties, and Oxidation Stability. *J. Am. Oil Chem. Soc.*, 84, 7, 675–680, 2007.

230. Salih, N., Salimon, J., Yousif, E., The physicochemical and tribological properties of oleic acid based triester biolubricants. *Ind. Crops Prod.*, 34, 1, 1089–1096, 2011.
231. Erhan, S.Z. *et al.*, Lubricant Base Stock Potential of Chemically Modified Vegetable Oils. *J. Agric. Food. Chem.*, 56, 19, 8919–8925, 2008.
232. Zhang, H. *et al.*, Rapid identification of triacylglycerol-estolides in plant and fungal oils. *Ind. Crops Prod.*, 37, 1, 186–194, 2012.
233. Modak, S.N. and Kane, J.G., *Studies in estolides. I. Kinetics of estolide formation and decomposition. J. Am. Oil Chem. Soc.*, 42, 5, 428–432, 1965.
234. Cermak, S.C., Brandon, K.B., Isbell, T.A., Synthesis and physical properties of estolides from lesquerella and castor fatty acid esters. *Ind. Crops Prod.*, 23, 1, 54–64, 2006.
235. García-Zapateiro, L.A. *et al.*, Viscous, thermal and tribological characterization of oleic and ricinoleic acids-derived estolides and their blends with vegetable oils. *J. Ind. Eng. Chem.*, 19, 4, 1289–1298, 2013.
236. Isbell, T.A., Kleiman, R., Plattner, B.A., Acid-catalyzed condensation of oleic acid into estolides and polyestolides. *J. Am. Oil Chem. Soc.*, 71, 2, 169–174, 1994.
237. Vijayendran, B., Biobased Motor Oils Are Ready for Primetime. *Ind. Biotechnol.*, 10, 2, 64–68, 2014.
238. Saha, B. and Vlachos, D.G., Synthesis of (hemi)cellulosic lubricant base oils via catalytic coupling and deoxygenation pathways. *Green Chem.*, 23, 14, 4916–4930, 2021.
239. Zheng, T. *et al.*, Biolubricant Production of 2-Ethylhexyl Palmitate by Transesterification Over Unsupported Potassium Carbonate. *J. Am. Oil Chem. Soc.*, 95, 1, 79–88, 2018.
240. Elmelawy, M.S. *et al.*, Synthesis and kinetics study of trimethylolpropane fatty acid triester from oleic acid methyl ester as potential biolubricant. *Biomass Convers. Biorefin.*, 13, 3, 1645–1657, 2023.
241. Encinar, J.M., Nogales-Delgado, S., Pinilla, A., Biolubricant Production through Double Transesterification: Reactor Design for the Implementation of a Biorefinery Based on Rapeseed. *Processes*, 9, 7, 1224, 2021.
242. Dehghani Soufi, M. *et al.*, Valorization of waste cooking oil based biodiesel for biolubricant production in a vertical pulsed column: Energy efficient process approach. *Energy*, 189, 116266, 2019.
243. Diaz, P.A.B., Kronemberger, FA., Habert, A.C., A pervaporation-assisted bioreactor to enhance efficiency in the synthesis of a novel biolubricant based on the enzymatic transesterification of a castor oil based biodiesel. *Fuel*, 204, 98–105, 2017.
244. Riazi, B. *et al.*, Life Cycle Environmental and Cost Implications of Isostearic Acid Production for Pharmaceutical and Personal Care Products. *ACS Sustain. Chem. Eng.*, 7, 18, 15247–15258, 2019.
245. Sun, X., Wang, L., Fu, R., Yang, Y., Cheng, R., Li, J., Wang, S., Zhang, J., The chemical properties and hygroscopic activity of the exopolysaccharide lubcan from Paenibacillus sp. ZX1905, *Int. J. Biol. Macromol.*,164, 2641–2650, 2020, ISSN 0141-8130, https://doi.org/10.1016/j.ijbiomac.2020.08.129.

7

Multifunctional Materials for Nanotechnology

Aakash Mathur[1]*, Ankita Mathur[2], Prashant Kumar Mishra[1], Amit Kumar Srivastava[1] and Yogendra Kumar[1]

[1]*School of Basic Sciences, Galgotias University, Greater Noida, Uttar Pradesh, India*
[2]*Schulich Faculty of Chemistry, Technion - Israel Institute of Technology, Haifa, Israel*

Abstract
The emergence of nanotechnology has found a phenomenal advancement in the research field pertaining to assorted fields. As a cornerstone of nanotechnology, multifunctional materials provide previously unheard-of efficiency and adaptability in a wide range of applications. This chapter explains the various functions that multifunctional materials have across several disciplines and examines their critical significance in the advancement of nanotechnology. Many industries, such as electronics, healthcare, energy, catalysis, and environmental remediation, use multifunctional materials. Several multifunctional nanomaterials such as smart materials, composites, and self-healing materials are discussed. The difficulties and possibilities involved in the creation, synthesis, and characterization of multifunctional materials are also highlighted in this chapter, with a focus on the significance of multidisciplinary methods and cutting-edge manufacturing processes for achieving the full potential of these materials. It also explores potential future developments, such as the use of AI and machine learning to find and optimize multifunctional materials, which might lead to new directions in nanotechnology innovation and technological advancement.

Keywords: Nanotechnology, multifunctional materials, composites, smart materials, self-healing materials

7.1 Introduction

7.1.1 Overview of Multifunctional Materials and Nanotechnology

Multifunctional materials have gained widespread attention because of their ability to participate in diverse applications and functionalities [1]. Materials that display several unique functions within a single material or system are referred to as multifunctional materials. Because these materials can accomplish numerous tasks at once, they have drawn a lot of interest from scientists and engineers in a variety of fields. Combining various functions into one material has many benefits, including increased effectiveness, better performance, and the possibility of creative application [2]. They can exist naturally or can be engineered

**Corresponding author*: dr.aakash1288@gmail.com

Divya Bajpai Tripathy, Anjali Gupta and Arvind Kumar Jain (eds.) Multifunctional Materials: Engineering and Biological Applications, (181–206) © 2025 Scrivener Publishing LLC

at the molecular or nanoscale level by altering their mechanical, electrical, thermal, optical, etc. properties to render them multifunctional, or can be a hybrid composed by combining some parts of natural and some parts of synthetic materials. In the context of multifunctional materials, "functionalities" refers to particular attributes or capacities that are built into the material. Among other things, these could be biological, mechanical, electrical, magnetic, thermal, or optical qualities. The goal of integrating various functionalities is to produce materials with unique combinations of features or improved performance that are not possible with single-function materials. Also, such an integrating approach should also focus on increasing the efficiency of the system through savings on the volume and weight of individual components (Figure 7.1). They should also be significantly more application-specific than the current unifunctional materials due to the wide range of material combinations and the ensuing characteristics and functions.

Nanotechnology is an emerging technology with applications in a wider sector of science and technology. It can vary from electronics to biotechnology, defense to energy, etc. It involves the manipulation of materials at the nanoscale level, typically resulting in structures and devices in the size range of 1–100 nm. When the materials reach the order of wavelength of an electron, it causes quantum confinement effects. This causes the quantization of energy levels and restricts the movement of electrons in quantized energy levels [3]. It leads to alteration of the properties, lending them much different from their bulk-counterparts. Breaking down bulk materials into nanoscale endows a higher surface area-to-volume ratio, leading to improved properties. Nanotechnology endows materials with

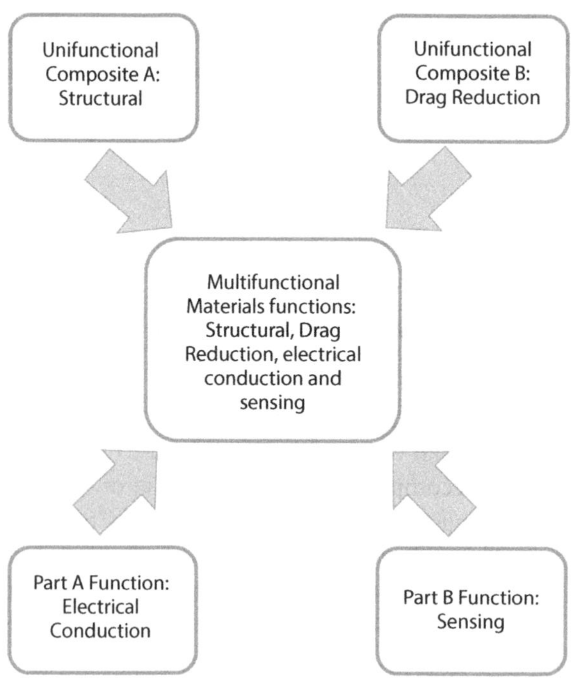

Figure 7.1 Figure illustrating that multifunctional materials should integrate the specific functions of individual material, composites, components in order to increase the total efficiency of the system.

excellent physical and chemical properties. Hence, tailoring multifunctional materials into their nano form leads to tremendous potential toward technological performance due to reduction in weight, size, cost, and power consumption while improving efficiency, safety, and versatility. This in turn leads to improvement in economic growth, a healthier environment, and open avenues for sustainable development.

7.1.2 Importance of These Materials in Modern Science and Technology

High functionality areas may be attained by multifunctional materials and composites, which can enhance operations and output, open up new pathways for increasing sustainability, and directly and favorably affect the environment, economic growth, and standard of living. These materials have been used extensively in the building, manufacturing, and development of interdisciplinary devices related to biology, the environment, and energy. The efficiency and miniaturization of nanostructures and nanodevices were enhanced by the use of multifunctional materials. For instance, because nanomaterials are more soluble in water, they can be used as carriers for medications that are insoluble in water to cure illnesses [4]. Carbon nanotubes (CNTs) can purify water in a much-improved manner as compared to bulk carbon particles. Multifunctional biomaterials that release, alter, or deploy therapeutic action are now being investigated as viable implant materials for the human body. Biomaterials that can change their form are being researched for applications like implants that might be used in less invasive surgery and medical equipment that could help elderly people with housework and other everyday chores [5]. Accelerated industrialization and the creation of a vast array of goods for consumption with the constant rise in the consumption of power are the results of readily available and reasonably priced energy. Alternatives to fossil fuels must now be used to supply this expanding need. Future energy paths that are sustainable and reliant on renewable resources depend heavily on multifunctional materials. *In vitro* diagnostics, robotics, energy harvesting, displays, and improved medications are just a few of the fields in which flexible electro-mechanical sensing devices have shown tremendous promise [6]. The use of multifunctional materials has shown advantageous for improved sensors with the potential to have inbuilt healing and powering capacity, electronic skins, and wearable and implantable devices. The use of very flexible, highly sensitive, and transparent multifunctional materials, such as graphene, hybrid composites, nano architectures in the form of tubes or wires, and organic-inorganic matrix arrays, has enabled these astounding advancements. The development of multifunctional materials has provided opportunities for the development of lighter aircraft, the convenience of medical diagnostic equipment, and the retention of an enormous quantity of data in a single device [6]. Smaller, lighter batteries that have a longer energy/power lifetime can be carried into space by satellites. Nanosensors are another tool that helps with early illness diagnosis since they can find even minute amounts of biomarkers in the body. Medication administered at an early stage of the disease is significantly more successful than medication administered at a later time. The entire landscape of research and innovation has changed as a result of multifunctional materials, which have also given the research sectors a new focus and strategy for solving contemporary issues in a way that is sustainable, effective, and cutting-edge.

7.2 Multifunctional Nanomaterials

7.2.1 Definition and Types of Multifunctional Nanomaterials

Materials between 1 nm and 100 nm in size that have many activities for carrying out various tasks are known as multifunctional nanomaterials, or MFNs. These qualities are what these materials are designed to have. Numerous uses are dependent on factors such as size, dimensions, structure, and chemical composition. Usually, nanomaterials are classified based on dimensions as: 0-D/1-D/2-D and 3-D, based on degree of freedom or degree of confinement in the nanostructure. Several MFNs can be grouped according to the functionalities they possess such as:

A. Smart Materials:

Smart materials are materials of the future that outperform traditional structural and functional materials. These materials are intelligent and have the ability to adjust to external inputs like loads and surroundings. Materials that may alter their physical characteristics in a predetermined way in response to a predetermined stimuli input are referred to as smart materials. Nuclear radiation, chemicals, hydrostatic pressure, electric and magnetic fields, pressure, and temperature are some examples of the stimuli. Potentially varying physical attributes in this context might include damping, shape, rigidity, and viscosity. Humanity will be greatly impacted by this new era of smart materials. Some of them have the ability to change their properties in response to their surroundings, while others have remarkable sensory capabilities, auto-repairing abilities, or self-degradation potential. These remarkable properties of smart materials will affect every facet of civilization. Various smart materials already exist like piezoelectric materials, shape memory alloys, magneto-rheological materials, electro-rheostat materials, and many more to be discovered.

Piezoelectric materials are those that, when mechanical stress is applied, produce an electric potential, which may be utilized for energy harvesting or sensing (Figure 7.2). In another form, an electric field is supplied to a material and causes it to either change form or produce mechanical changes. This property may be employed for both shape control and actuation. They have comparatively linear behavior when actuating or sensing at frequencies ranging from around 1Hz to MHz [1]. Additionally, the high stiffness of piezoelectric materials gives them powerful voltage-dependent actuation. Polycrystalline ceramics (PZT, $PbTiO_3$, $BaTiO_3$), single crystals (SiO_2, $LiNbO_3$, LiTaO), and polymerics (PVDF, co-polymer) are the most commonly used forms of piezoelectric materials [7–9]. These

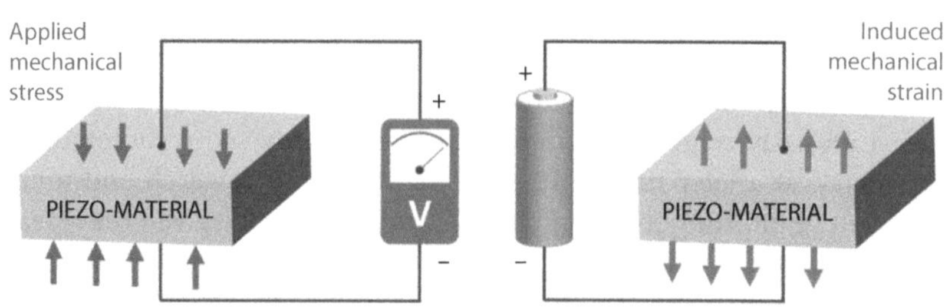

Figure 7.2 The piezo-electric effect: generation of charge due to mechanical stress (direct effect, left), and strain caused by the application of an electric field (converse effect, right).

materials can be (nano)fibers/wires, monolithic, wafers, or thin films, and they can be hollow and arranged as macro and active fiber composites.

B. Nanocomposite Materials:

Composites consist of two or more components, each with distinct physical or chemical properties. One type of composite that has piqued the interest of researchers are nanocomposites (NCs), which are composites made with at least one nanoscale component. NCs possess exceptional qualities that make them versatile structural and functional materials with applications in catalysis, sorption, separation, fuel cells, and more [10]. High-end applications including space travel, heavy machinery manufacture, and aviation technology employ composite materials. Composite materials have been used, as demonstrated by the latest advances in the field of multifunctional composite materials. Applications for composite materials in the design, manufacturing, and development of interdisciplinary equipment are diverse. Numerous uses, requisite characteristics, and significant role of multifunctional materials in use presently are depicted in Figure 7.3 [11]. Conventional composite materials can have specific qualities added to them or their composition changed by adding different stiffeners and strengtheners. Scientists and researchers are still faced with the difficulty of developing a multifunctional composite material integrated with various structural functionalities. While the NCs have many practical uses, the precursors and/or hazardous chemicals involved in their synthesis raise concerns about toxicity and environmental sustainability. Therefore, exploring more environmentally friendly methods for synthesizing NCs is essential. One solution is to use biogenic sources and green solvents.

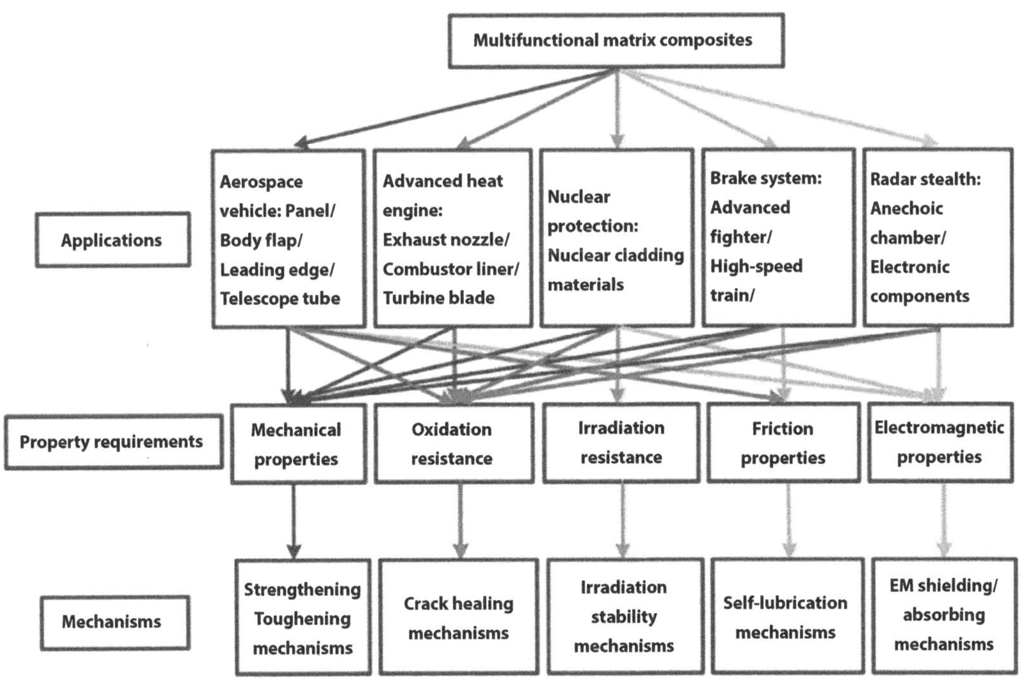

Figure 7.3 Block diagram illustrating the main characteristics, methods, and multifunctional matrix composite applications.

C. Multiferroic (MF) Materials:

MF materials are highly stimulating multifunctional materials that are used in a wide variety of systems. They integrate two or more distinct ferroic orders—ferromagnetic, ferroelectric, and ferroelastic—in a single phase [12]. The idea of manipulating one ferroic property with the conjugate field of another is particularly exciting since it can result from the coupling of ferroic orders in MFs. This is because an electric field can be used to change the orientation of magnetization [13]. The coexistence of magnetic and electric order is especially intriguing because it combines characteristics that may be used for the processing, transmission, and storage of information. It permits the interaction of magnetic and electric order with magnetic and electric fields. Nonetheless, most materials—MF materials in particular—face formidable technological obstacles when it comes to integration and shrinking, which drives researchers to look for low-dimensional MF materials of atomic thickness. Compared to three-dimensional (3D) MF materials, low-dimensional materials often have flat surfaces and high dielectric constants, which enable the creation of nanoscale devices [14].

D. Conductive Polymers:

Because of their exceptional qualities—such as their tunable electrical characteristics, excellent optical and mechanical properties, ease of synthesis and manufacturing, and superior environmental stability compared to traditional inorganic materials—conducting polymers are the subject of much research. Hybridization with other materials overcomes the many constraints that conducting polymers have in their unaltered state. Because of their synergistic effects, conducting polymer composites have extensive applications in the electrical, electronics, and optoelectronic sectors. Conducting polymers are primarily soluble and processable due to their connected side chains, and they possess mechanical, electrical, and optical characteristics due to the attached dopant ions. Conducting polymers are crystalline and partially amorphous. Both localized and delocalized states are present in conducting polymers. P-bond delocalization is strongly dependent on the disorder, and this delocalization is crucial for the production of charge carriers such as polarons, bipolarons, solitons, and so on, which are in charge of converting an insulator into a metal. In their pure state, conjugate polymers' conductivity functions as an insulator against semiconductors, and it rises with dopant concentration.

E. Self-Healing Materials:

Self-healing materials possess a remarkable ability to autonomously repair damage, restoring structural integrity without external intervention [15]. Inspired by biological systems, these materials exhibit a unique property where cracks, fractures, or other forms of damage trigger a self-repair mechanism. Microcapsules containing healing agents or reversible chemical reactions within the material contribute to this remarkable capability [15]. Self-healing materials find applications in diverse fields, from aerospace and automotive industries to everyday consumer products, reducing maintenance costs and extending the lifespan of materials. Their innovative nature continues to drive research toward more resilient and sustainable solutions in material science [6].

7.2.2 Properties and Applications of Multifunctional Materials

A. Synergistic Integration of Functionalities:
Multifunctional materials are defined by their capacity to integrate and synergize various functionalities. This integration is not merely additive; instead, it often results in emergent properties that enhance overall material performance. For instance, a material may combine mechanical strength with electrical conductivity, creating a composite that surpasses the individual properties of each component.

B. Tailored and Tuneable Properties:
The design of multifunctional materials is a meticulous process that involves tailoring the material at the atomic or molecular level to achieve specific properties. These materials are tunable, allowing researchers to fine-tune their characteristics based on the desired application. This tailored design ensures that multifunctional materials meet the precise requirements of diverse industries.

C. Interdisciplinary Nature:
The development of multifunctional materials often requires interdisciplinary collaboration, drawing expertise from fields such as materials science, chemistry, physics, and engineering. The combination of insights from different disciplines facilitates a holistic approach to material design and optimization.

D. Dynamic Response to External Stimuli:
Many multifunctional materials exhibit dynamic responses to external stimuli, such as temperature, pressure, light, or magnetic fields. This dynamic behavior enables reversible changes in the material's properties, adding a layer of adaptability that is advantageous in various applications.

E. Versatility Across Industries:
Multifunctional materials find applications across a broad spectrum of industries. From electronics and healthcare to energy and aerospace, their versatility makes them suitable for a wide range of innovative solutions. For example, these materials may be utilized in electronic devices, medical implants, structural components, and environmental sensors.

F. Nanotechnology Integration:
Multifunctional materials often leverage advancements in nanotechnology. Nanomaterials, with dimensions at the nanoscale, exhibit unique properties, and their integration into multifunctional materials contributes to enhanced mechanical, electrical, or optical characteristics.

Graphene is a perfect example to explain the correlation between the properties of MFNs with their applications. Graphene, which belongs to the 2D family of nanomaterials, is considered as world's thinnest and strongest nanomaterial with extraordinary physicochemical properties [16]. Pristine graphene exhibits high optical transparency, specific surface area, thermal conductivity, flexibility as a thin film, electron mobility, etc. [16]. It also possesses high mechanical strength and Young's modulus. Based on these properties, graphene can be employed for different applications. As such graphene is used for fluorescence imaging and therapy, magnetic resonance imaging, etc., owing to its excellent optical properties.

Owing to its high surface area, it can be employed for the detection of trace targets as electrochemical sensors for chemicals, electrodes for energy storage devices, such as batteries, supercapacitors, fuel cells, and solar cells. Other applications of graphene include drug or gene delivery owing to its excellent biocompatibility, high surface area, and high cargo capability, where it acts as a "carrier" for medicines and therapeutics [17]. Owing to its outstanding electrical properties, it is also used either as filler in polymer composites or in the form of aerogels as electromagnetic interference (EMI) shielding material to block the interference of electromagnetic signals emitted by electronic devices [18].

It can be wrapped into a 0-D spherical structure called fullerene or rolled into one-dimensional CNT (CNT, 1-D) structures and can also be stacked to get 3-D layered graphite structures. So, the planar sp_2 structure can be rearranged into several structures to give rise to multiple applications. The properties of graphene can be tuned by introducing alterations in the surface of the graphene structure, e.g., by introducing defects, converting graphene to reduced graphene oxide (rGO) or graphene oxide (GO). The functional group on the surface provides more active sites for chemical reactions and converts graphene hydrophobicity to hydrophilic carbon. The functionalization gives rise to numerous oxygen-containing hydroxyl groups, carboxylic acid, and epoxy groups in the basal plane.

This illustrates the versatile and multifunctional nature of graphene, which can be further by functionalizing the structure or introducing some defects. The same can be done with other MFNs and similarly, their properties correlate with their applications.

7.2.3 Examples of Multifunctional Nanomaterials in Different Industries

MFNs have been employed in broad spectrum applications, so it has sparged wide attention in industries also. These have also grabbed the focus of many start-ups, small-scale industries, and academia-industrial collaboration projects. Some notable industries include:

A. Chemical Industry:

The chemical industry involves the extraction, purification, and synthesis of new chemicals. These processes mainly involve chemical reactions catalyzed by nanomaterials. Due to their large aspect ratio, better chemical selectivity, and higher catalytic efficiency compared to their bulk counterparts, nanomaterials widely participate in various chemical reactions varying from petroleum refining to fuel cell reactions. BASF, Reliance Industries, LG Chemicals, DuPont, etc. are some of the chemical industries that manufacture chemicals.

Inorganic nanomaterials are popular for their oxidative and hydrogenation properties, such as iron (Fe) and nickel (Ni) have binding sites leading to the formation of carbon-carbon (C-C) bonds, whereas copper (Cu) is utilized where rearrangement of C-C bonds is needed [19, 20]. Based on this property, Cu nanoparticle mediates the synthesis of various organic molecules, such as the Ullman reaction and Diels Alder reaction [21]. Fe and Ni nanoparticles mediate the formation of CNTs from ethylene. CNTs are industrially very important compounds because of the exceptional properties that they possess. In addition, low cost and ability for mass production have also played a major role in its usage in industries. However, recent efforts by large companies like LG Chem as well as start-ups like Nanocomp, Tortech,

OCSiAl, DexMat, and Jeio have made every effort to mass-produce CNTs or CNTFs, so it is anticipated that CNTs will soon be adapted into some application fields like batteries and composites [22]. Oxides of iron (e.g., Fe_3O_4) catalyze the formation of NH3 from N_2 and hydrogen, called the Haber Bosch method. NH_3 is a very important compound for the chemical industry because of its wide applications in fertilizers as a source of nitrogen for yeast and microorganisms, as a refrigerant in industrial refrigeration systems and in air conditioning equipment, cleaning agent in window and floor cleaners, etc.

Organic nanomaterials such as metal-organic frameworks (MOFs) are also excellent catalysts for the synthesis of various industrially important compounds, such as aminated carbon used as active pharmaceutical ingredients (APIs) [23]. MOFs possess 3D architecture, high surface area, and pore structure; therefore, they are used for gas storage, purification and separation, other than catalysis, and energy storage. They offer diverse chemical functionality and hence, their properties can be tuned by modifying either the metal particle or the organic functional group.

B. Pharmaceutical and Healthcare Industry:

MFNs have been employed in various sectors of the pharmaceutical and healthcare industry, ranging from drug or gene delivery to developing organ-on-a-chip technology. Nanomaterials act as carriers for encapsulating and delivering drugs to a specific target organ in the human body. This reduces side effects and enhances the effectiveness and bioavailability of the drug. Nanomaterials such as gold, iron, and silica have been widely used as carriers. These inorganic nanomaterials have free electrons on their surface that oscillate at different frequencies depending on the shape and size of the nanomaterial and give photothermal properties. Iron oxide, in the forms of magnetite (Fe_3O_4) or maghemite (Fe_2O_3), possesses additional superparamagnetic properties and has also got approval from the FDA to carry various medicines, such as INFeD, DexFerrum, and Feraheme [24]. These nanomaterials are also used to enhance contrast in imaging using MRI, CT, and ultrasound. These are used primarily in *in vitro* imaging but show promise for *in vivo* diagnostics [25]. Certain polymers are also used for these applications, such as polyethylene glycol (PEG) and polydimethylsiloxane (PDMS). Lipid-based nanoparticles, such as liposomes or lipid nanoparticles are widely used for the delivery of nucleic acids [26]. In a mouse model, fluorescent carbon nanodots (CDs) smaller than 10 nm have demonstrated trackable administration of medications for targeted cancer therapy [27]. A non-covalent bonding (electrostatic interactions or hydrogen bonds) is formed between the amine groups of drug molecules and carboxyl groups present on the surface of CDs. This bond is sufficient to stabilize the complex biological system [28]. The fluorescence helps in enabling highly sensitive imaging of the nanomaterial and for monitoring the disease progression.

Certain nanomaterials have been employed as biosensors for the early detection of diseases and monitoring treatment response. These can also be used as implantable sensors for monitoring physiological parameters, thus allowing a personalized healthcare infrastructure.

Polymers such as PDMS are also useful for building microfluidic-based organs on a chip device for providing more accurate and physiologically relevant models for drug testing and disease research [29].

These applications highlight the versatility of MFNs in addressing various challenges in pharmaceuticals and healthcare, ranging from drug delivery to diagnostics and therapeutics.

As research in nanotechnology progresses, we can expect further innovations and advancements in these areas.

C. Energy Sector:
For advancing energy generation or conversion and storage, MFNs play a crucial role. Nanomaterials are a great approach to shifting focus from fossil fuels to renewable energy for energy sources. Their unique properties enable high energy and power density output. Some nanomaterials with a bandgap in the range of the visible spectrum can absorb light and catalyze light-mediated chemical reactions (also called photocatalysis) such as titanium dioxide (TiO_2). This is a method of directly converting sunlight to electricity and using that energy for mediating chemical reactions. Solar cells also rely on using crystalline silicon (Si), or GaAs wafer-based solar cells with a maximum possible light-to-electricity conversion of 32% [30].

Fuel cells are also a choice for the generation of clean energy, as an alternative to fossil fuels. Many companies rely on fuel cells for the generation of power in cars such as Toyota Mirai and Hyundai NEXO. It consists of two major reactions: hydrogen oxidation reaction (HOR) at the anode and oxygen reduction reaction (ORR) at the cathode. So far, platinum has been shown as the best catalyst for ORR. Ion-selective membranes employed in fuel cells are also made of polymer nanomaterials.

Energy generated can be stored within the layers of nanomaterials by fabricating lithium-ion battery (LIB) devices for which the work of Goodenough, etc. was awarded a Noble Prize in 2019. LIB works on the insertion and de-insertion of Li ions into various host materials, leading to the charging and discharging respectively of the battery [31]. Goodenough and Mizushima *et al.* used $LixCoO_2$ or $LixNiO_2$ as cathode material for LIB [32]. Since then, rapid advancements in LIB are going on. The transport industry relies on LIB for electric vehicles such as BMW, Tesla, Audi, etc.

Energy can also be stored electrostatically by forming a double layer of ions like a parallel plate capacitor, called a supercapacitor. Supercapacitors are a choice of energy storage when there is a need for a quick spurge of power [33]. Carbon-based nanomaterials like graphene and activated carbon are mostly used as electrodes for supercapacitors and present high specific capacitance and coulombic efficiency [34].

Other than these devices for energy, nanomaterials have been used for harnessing wind energy, nuclear energy, smart grids, thermal energy in thermoelectric materials, and producing hydrogen which is considered as fuel for the future, through HER, photocatalysis, water splitting, etc. reactions. MFNs are used in many more industries including the food industry, electronics, and transport.

7.3 Synthesis and Characterization Techniques

7.3.1 Techniques for Synthesizing and Characterizing Multifunctional Materials and Nanomaterials

The synthesis of nanomaterials requires the use of sophisticated equipment, and a lot of parameters to optimize. However, typically, there are two major approaches to the synthesis of nanomaterials:

A. Top-down - It involves breaking a micro-sized particle into nano-sized particles or devices. For example, ball milling, focussed ion beam, photolithography, electron-beam lithography, etc.

B. Bottom-up - It involves the synthesis of nanomaterials by combining ions or atoms into nano-sized particles. For example, chemical synthesis, chemical vapor deposition, sol-gel process, etc.

The most popular methods for creating multifunctional materials and nanomaterials are depicted in Figure 7.4 and summarized as follows: Low-dimensional structures along with coating can be done by solvothermal processes and hydrothermal processes. Nanofibers of polymers, oxides, and carbon material can be done by electrospinning techniques. Likewise, vapor deposition techniques such as atomic layer deposition (ALD) and chemical vapor deposition (CVD), including physical vapor deposition techniques are also used for assorted nanostructures. Mostly for 3D and porous nanostructures, among others, template- or pore-forming aided synthesis and self-assembly (including constructing complicated structures from diverse building components) are performed.

Lithography: Lithography is a technique that may be used to produce nanoarchitectures using a focused light or electron beam. Lithography can be done either using a mask or a without mask [34]. In masked nanolithography, specific templates are used for the fabrication of nanopatterns. Masked lithography includes techniques including photolithography [35], soft lithography [36], and nanoimprint lithography. Maskless lithography includes techniques including scanning probe lithography [37], focussed ion beam lithography [38], and electron beam lithography. A method called maskless lithography allows for arbitrary nanopattern writing without the need for a mask. 3D freeform micro-nanofabrication may

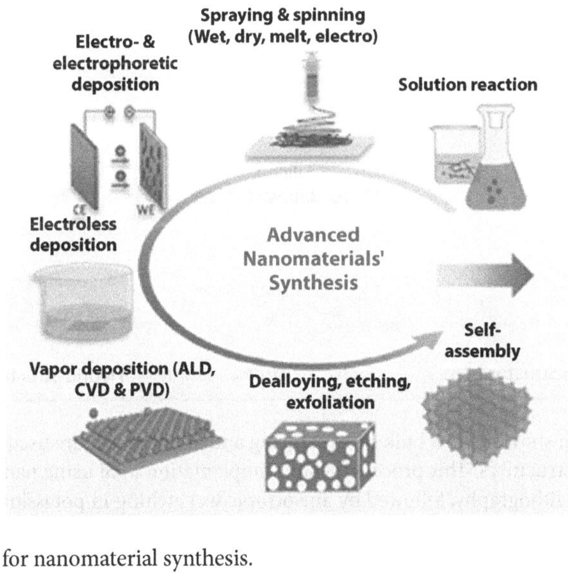

Figure 7.4 Approaches for nanomaterial synthesis.

be achieved by combining focused ion beam implantation with chemical etching, as shown in Figure 7.5 [39].

The technique of sputtering involves subjecting surfaces of solid material to high-energy ions like gas or plasma in order to make nanomaterials. It is believed that sputtering is a practical technique for nanomaterial fabrication. During the sputtering deposition process, which includes bombarding the surface with powerful gaseous ions, small atom clusters are physically expelled from the target surface depending on the incident gaseous-ion energy (Figure 7.6). Sputtering can be done in a number of ways, such as radio-frequency diode, magnetron, and direct current (DC) diode sputtering [40]. Sputtering is usually carried out in a chamber that has been emptied and filled with sputtering gas. When free electrons collide with gas at a high voltage applied to the cathode target, gas ions are produced. The high energy speedy ions perform repeated striking on the cathode material, causing it to be ruptured and ejecting atoms from its surface. The sputtering method is attractive because it yields nanomaterials with compositions equal to the target material but with fewer impurities, and it is less costly than electron-beam lithography [40].

For the production of carbon-based nanomaterials, chemical vapor deposition methods are crucial. In CVD, vapor-phase precursors react chemically to generate a thin coating on the topmost layer of the substrate. The following characteristics of a precursor make it appropriate for CVD: high chemical purity, good stability during evaporation, cheap cost, non-hazardous nature, and long shelf life [41]. Moreover, once it decomposes, there ought to be no residual pollutants [41]. In CVD, vapor-phase precursors react chemically to generate a thin coating on the substrate surface [42]. The following characteristics of a precursor make it appropriate for CVD: high chemical purity, good stability during evaporation, cheap cost, non-hazardous nature, and long shelf life. Additionally, once it breaks down,

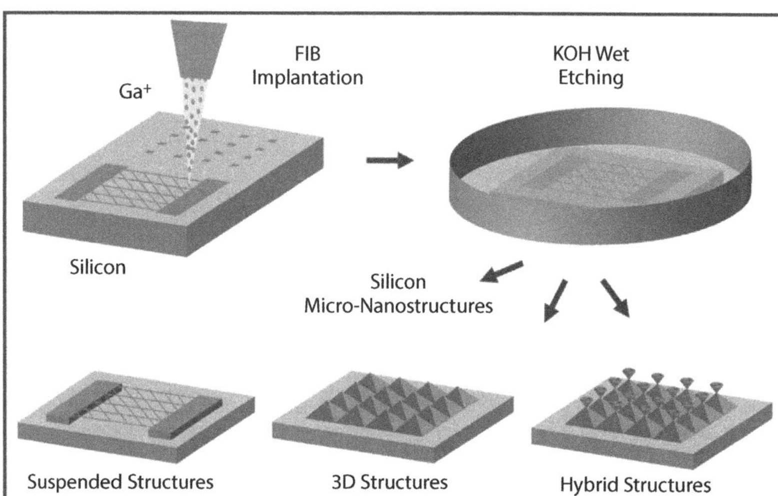

Figure 7.5 An illustration showing how bulk Si structuring and an ion beam are used to create three-dimensional micro-nanostructures. This process includes implantation in Si using nanometer-resolution mask-writing and Ga FIB lithography, followed by anisotropic wet etching in potassium hydroxide (KOH) solution, and the creation of Si structures of micro-nano level by selectively removing the unimposed area. Reprinted with permission [39].

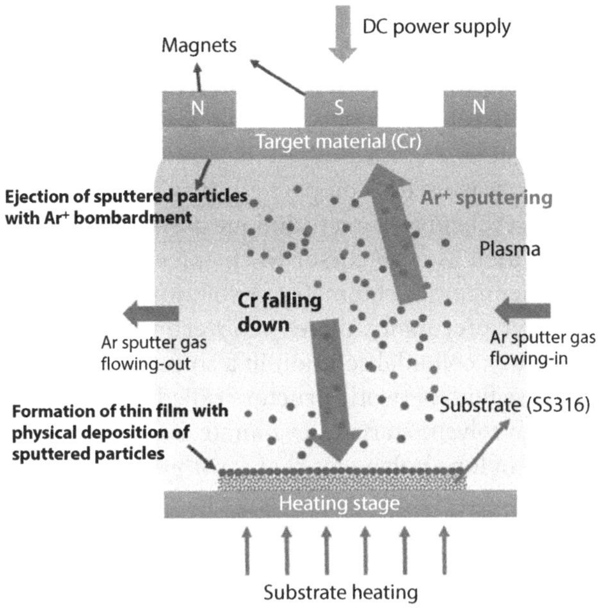

Figure 7.6 A schematic diagram of the DC magnetron sputtering process. Reprinted with permission [40].

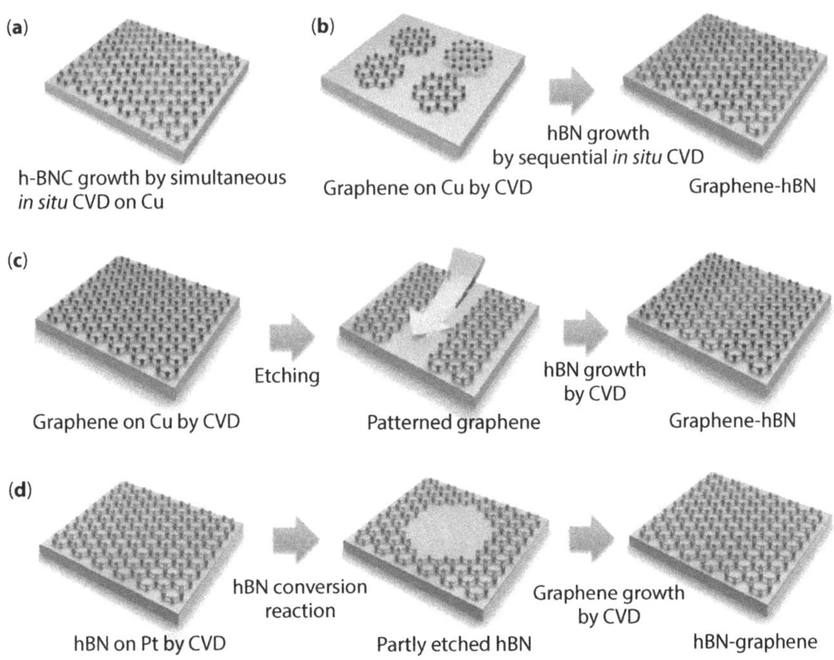

Figure 7.7 An illustration showing how in-plane graphene and hBN heterostructures have grown using several methods, including conversion growth, lithography-assisted growth, sequential *in situ* CVD growth, and simultaneous *in situ* growth. Reprinted with permission [43].

there shouldn't be any leftover contaminants. For example, during the CVD process, a substrate is heated to a high temperature in an oven to produce CNTs. This process is widely recognized for producing two-dimensional nanomaterials, as shown in Figure 7.7 [43].

Sol-Gel Method:

A sol-gel synthesis process was used to prepare all the Ga-doped and Ga-co-doped ZnO samples in this study. Sol-gel chemistry is a technique used for making inorganic polymers. This technique is further used as a precursor to chemical homogeneity for the synthesis of ceramics of complex chemical stoichiometry. A solution of component elements is prepared from different soluble precursors, thereby converting the liquid mixed precursor to sol. Solid ions suspended in a colloidal solution in a solvent are called a sol. Subsequently, the sol is converted to a less fluid network structure called a "gel." A gel is the formation of a semirigid mass when the solvent starts to evaporate from the sol and the particles/ions left behind. The formation of long polymeric chains with the required elements attached to the polymers in a less hydrated state is the actual reality of these gels. Oxides are formed by joining the elements with oxo or hydroxy chains. This results in the formation of -oxo or -hydroxy polymers in the solution. Next, the gel is dried to remove the remnant liquid phase. This sometimes forms a sticky consolidated mass or a porous powder. After this, calcination and annealing may be carried out to attain the desired ceramic composition in the maximum possible chemical homogeneity. Generally, water-soluble precursors such as metal nitrates and metal acetates are used for material synthesis. Figure 7.8 below represents the same.

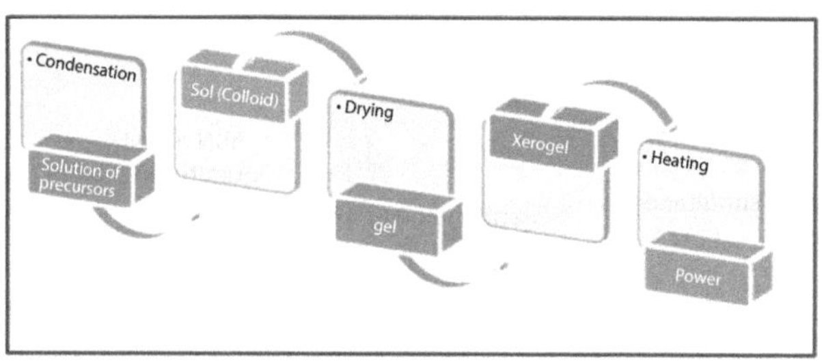

(a)

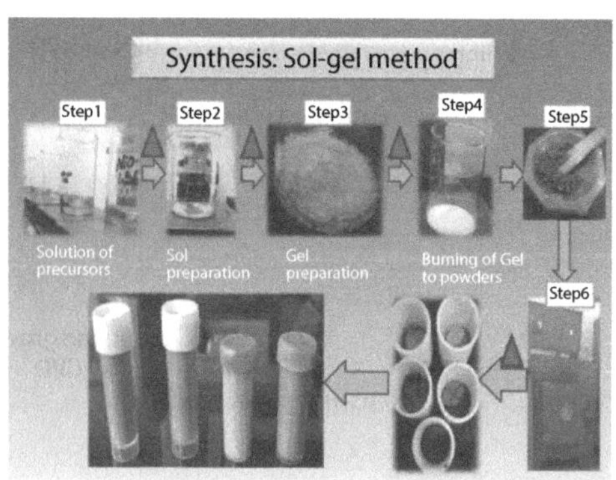

(b)

Figure 7.8 (a) Sol-gel process steps. (b) Experimental steps of sol-gel process.

Atomic Layer Deposition (ALD)

A process called ALD may deposit several kinds of thin-film materials from the vapor phase [44]. ALD has demonstrated significant promise in recently developed energy conversion and semiconductor technologies. Due to its superior conformality and control over the material thickness and composition, ALD has shown potential advantages over alternative deposition processes as the device needs to push toward smaller and more spatially demanding structures. The cyclic, self-saturating nature of the ALD process is the source of these special qualities [45]. Figure 7.9 depicts a general ALD process schematic. Sequential alternating pulses of gaseous chemical precursors reacting with the substrate surface make up a typical ALD process. "Half-reactions" are the name given to these gas-surface reactions. In order to enable the precursor to fully react with the substrate surface through a self-limiting process, precursor A is first pulsed into the reactor chamber under vacuum for a predetermined amount of time. Subsequently, an inert carrier gas (often N_2 or Ar) is used to purge the chamber in order to eliminate any unreacted precursor and undesirable reaction byproducts. To create up to one layer of the required material, the counter-reactant precursor (precursor B) pulse and inert gas purge come next. Until the required film thickness is reached, this procedure is repeated.

Both soft and hard template methods are heavily used in the production of nanoporous materials. The soft template approach is a simple conventional method for producing nanostructured materials. The soft template technique is thought to offer advantages because of its ease of application, wide range of morphologies of developed materials, and mild testing conditions [46]. The soft templating technique uses a variety of soft templates to build nanoporous materials, such as cationic, anionic, and non-ionic surfactants, block copolymers, and flexible organic molecules [47]. The main modes of interaction between

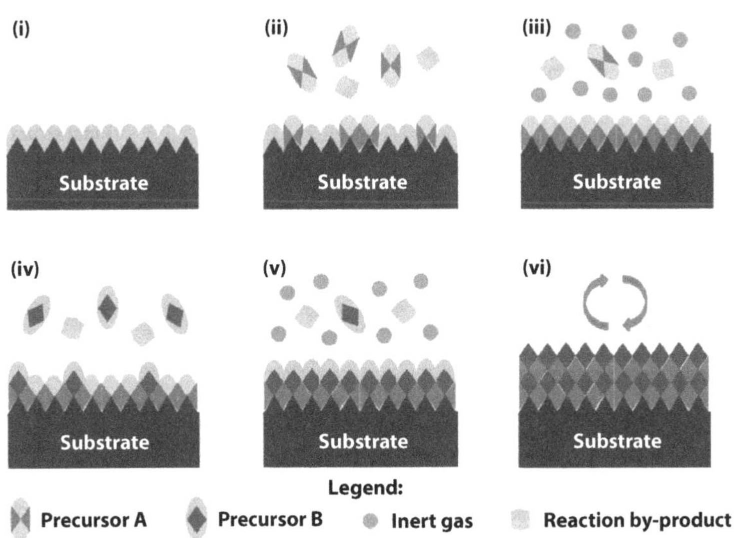

Figure 7.9 Diagram illustrating the ALD process in steps. The process involves functionalizing the substrate, pulsing precursor A into the ALD chamber so that it reacts with the substrate surface, purging excess precursor and reaction by-products with an inert carrier gas, pulsing precursor B into the ALD chamber, purging excess precursor and reaction by-products with an inert carrier gas, and repeating steps ii–v until the desired film thickness is reached.

the soft templates and the precursors are electrostatic, van der Waals, and hydrogen bonding. Organized mesoporous structures in three dimensions are built using soft templates of highly structured liquid crystalline micelles. Using alkyltrimethylammonium surfactant to produce mesoporous solids, such as cubic, lamellar, and hexagonal-structured mesoporous silicas, is a well-known example [48].

Nano-casting is another term for the hard template process. Carefully created solid materials are used as templates. Precursor molecules are then inserted into the pores of the templates to form the nanostructures required for the desired applications [49]. There are three main steps in the synthetic process of employing templating methods to create nanostructures. In the first phase, the appropriate original template is made or selected. To convert the template mesopores into an inorganic solid, a certain precursor is then introduced into them. The mesoporous replica is created in the last step by removing the original template. Mesoporous templates may be applied to create a variety of unique nanomaterials, including nano-dimensional wires, rods, metal, and several other nanoparticles (Figure 7.10).

After having summarized some important methodologies for the synthesis of numerous multifunctional materials, appropriate characterization techniques that are used to determine the 2D morphology will be briefly commented on. To determine the physical characteristics, such as morphology, chemical composition, functional groups, band gap, etc. of the nanomaterials, they are characterized using microscopic and spectroscopic techniques. Spectroscopic techniques involve studying the interaction of electromagnetic radiations such as X-ray and infrared with the material and studying the spectral lines or the energy of

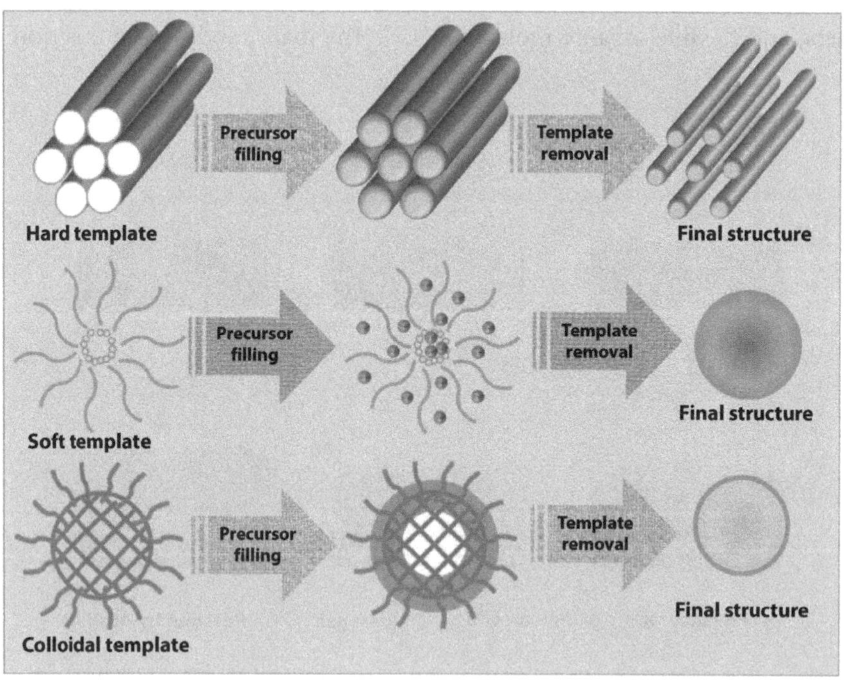

Figure 7.10 An illustration of the synthesis of materials utilizing several template types. Reprinted with permission [49].

the material. Microscopic studies involve visualizing materials by magnifying the surface of the material using microscopes that are too small to be seen through the naked eye.

Some of the spectroscopic techniques to characterize nanomaterials are:

1. UV-Vis Spectroscopy - It is used to identify the band gap of the nanomaterials. It involves irradiating ultraviolet and visible light on the nanomaterials and studying the electronic transitions within the material to identify its optical properties.
2. Fourier Transform Infrared Spectroscopy (FTIR) - It is used to identify the presence of functional groups in the nanomaterials. When infrared radiations pass through the materials, changes in the dipole moment of the molecules occur corresponding to specific vibrational energy. The wavenumber and radiations absorbed are correlated using the mathematical method of Fourier transformation and plotted in a 2-D plot.
3. Raman spectroscopy - It involves the identification of molecules and study of chemical bonds and intramolecular bonds, based on the study of vibrational frequencies when the sample is irradiated with light. It is used to calculate the interlayer distance in 2D materials and is an important tool to characterize graphene.
4. X-Ray Diffraction (XRD) - XRD is based on the diffraction of X-rays through the different planes of the materials. Based on the diffraction angle, a plot is obtained between 2 theta (where theta is the angle of diffraction) and the intensity of diffraction (in arbitrary unit), various planes are recognized, and the material is identified. A database has also been built to help discover future nanomaterials.

Other than these, other equipments are also important for the characterization of nanomaterials, such as

a. BET Measurements - to study the surface area, pore size, and pore volume of the samples,
b. ICP-MS Analysis - to calculate the compositional elemental percentage,
c. X-ray Photoelectron Spectroscopy (XPS) - to find out the composition and oxidation states of the constituent elements,
d. Nuclear Magnetic Resonance (NMR) Measurement - to find out the structure of organic molecules,
e. Vibrating Sample Magnetometry (VSM) - to study the magnetic properties of the sample, etc. These techniques are mostly used in combination with each other for a comprehensive understanding of the chemical nature of the sample and the results obtained are correlated with results from other equipment to verify the true nature of analysis/interpretation.

Some of the microscopic techniques to characterize nanomaterials are:

1. Scanning Electron Microscopy (SEM)
It is used to record the morphology of the sample by visualizing its surface. The electrons scattered from the surface of the sample are detected and interpreted to construct the image of the sample surface. It is often attached with other tools, such as Elemental dispersive spectroscopy (EDS) for elemental identification and mapping, focussed ion beam (FIB) for etching and designing structures on the sample surface, electron beam lithography (EBL) for writing or etching from the surface of the sample for designing structures.

2. Transmission Electron Microscopy (TEM)
It is also used for visualizing the surface of the sample by bombarding high-speed electrons on the surface of the sample, but the sample should be ultrathin because it is based on detecting electrons transmitted through the sample. As such, TEM provides information on the inner structure of the sample, such as crystal structure, morphology, defects, and dislocations. The resolution and magnification of TEM are much better than SEM; hence, it can be used with nanomaterials with even smaller sizes. EDS attached with TEM is used for identifying elements and for elemental mapping. It can also be used to study the behavior of materials at elevated and cryogenic temperatures.

3. Atomic Force Microscopy (AFM)
It is based on scanning the surface of the sample using a physical probe and creating images. It is one of the methods of scanning probe microscopy (SPM) and is based on the measurement of the force between the probe and sample surface as a function of the distance between them. These measurements can be used to study the mechanical and electrical properties of the sample, design a 3D topography of the sample surface, measure the thickness of deposited thin film, and change the properties of the sample surface in a controlled manner.

Numerous analytical techniques are used to characterize the chemical composition of nanomaterials. For the examination of C, N, O, S, and X (halide, F, Cl, Br, and I), elemental analysis (EA) is frequently employed. Another name for it is "CHNX" analysis. Typically, it provides the elemental ratio of organic-based nanomaterials with a 0.3% allowable standard deviation (SD) [50]. Both organic and inorganic species with a high ionization affinity can be studied using mass spectrometry [51]. It is difficult to ionize inorganic species. As a result, strong methods are frequently employed to ionize inorganic species by turning the material's element into an atom [52]. The components in the solution are determined using the majority of these techniques. Solid-state samples, however, may be assessed using techniques like electrothermal vaporization. Techniques including energy dispersive X-ray (EDX), X-ray fluorescence (XRF), and X-ray photoelectron spectroscopy (XPS) can be used to assess the surface's chemical composition. The depth of penetration of these methods is a crucial component of these investigations. Analyses using XPS and EDX are appropriate for thicknesses between 10–100 Å and 2 μm, respectively. Light elements like C, N, and O are not suited for surface analysis methods like EDX. The majority of the time, they offer semi-quantitative analysis or the ratio between the metals.

When characterizing materials, particle size is a crucial factor. TEM, SEM, AFM, and other microscopy methods can be used to determine it. The solid-state form employs these methods. Laser diffraction spectroscopy (LDS) and dynamic light scattering (DLS) are two methods used to analyze the size of a particle distributed in a liquid. Other approaches such as nanoparticle tracking analysis (NTA) provide a way to see and examine particles in liquids following Brownian motion. The particle size distribution for a diameter ranging from 10–1000 nm is determined via NTA [53]. Brownian motion serves as the foundation for both DLS and NTA. The analytical methodologies are where the biggest differences lie. While DLS uses a digital correlator to depict the total particles, NTA is based on changes in the location of each particle [53].

7.3.2 Advancements in Synthesis and Characterization Techniques

Due to advancing technology and demand for tailored nanomaterials, there has been a huge spurge in the advancement of synthesis and characterization techniques for various applications.

Some of the advanced synthesis techniques are the following:

A. Plasma-Enhanced Chemical Vapor Deposition (PECVD):
Plasma is the fourth state of matter and recently it has been employed for the synthesis of nanomaterials. Using the energy of plasma and under low-temperature vacuum conditions, it is used to prepare thin film for the semiconductor industry. This technique has been used to deposit materials that are sensitive to temperature, such as aluminum (Al), polymers, metal alloys, etc. Other advantages of PECVD are low cost, high mass yield, and toxin-free synthesis of nanomaterials, which makes it popular in biological applications and medicines.

B. Green Synthesis:
Chemical synthesis involves the usage of harsh chemicals, acids, and bases. Therefore, green synthesis has got the attention which involves using environmentally friendly synthesis of nanomaterials. It employs sustainable reactants and production of non-hazardous by-products. Bacteria, algae, and other microorganisms have been used as catalysts in some of the reactions.

C. Continuous Flow Synthesis:
This technique is particularly useful for batch synthesis of nanomaterials on an industrial scale, such as petrochemicals and pharmaceuticals. It is achieved by continuously flowing the reactants through the reactor and the product also emerges as continuous flow. The catalyst is usually situated inside the reactor. It offers a solution for safely handling hazardous species, escaping the need to remove impurities and by-products during multi-step reactions, and turning complex reactions into a continuous flow. Synthesis of chemicals such as ibuprofen, lidocaine hydrochloride, and controlled polymer growth has been achieved using a continuous flow synthesis technique.

These protocols for the advanced synthesis of materials are a result of the demand for advanced nanomaterials. For example, high entropy alloys (HEA) need unconventional methods for their synthesis. HEA is a crystalline solid solution containing five or more metals uniformly mixed and stabilized due to high mixing entropy. They need high temperatures and sudden cooling, which is difficult in conventional furnaces. So, microwave heating could be used to synthesize PtPdFeCoNi HEA-nanoparticles. In another strategy, a fast-moving bed pyrolysis strategy was used to prepare FeCoPdIrPt HEA nanoparticles.

Additionally, certain thin films need accurate synthesis, necessitating the use of methods like ALD or molecular beam epitaxy (MBE) to precisely regulate thickness and composition at the atomic level. Some of the advanced characterization techniques are:

D. *In Situ* and Operando Characterization:
It is important to study the catalyst under reaction conditions to gain insights into the complex reaction kinetics, active sites, and reaction mechanisms. It can be further used to deduce molecular structure-activity relationship and underlying catalytic mechanism to design further catalysts with desirable activity, stability, and selectivity. The various *in-situ* characterization tools that have been developed are- *in-situ* TEM and STM- for studying the growth mechanism of graphene using iron single-atom catalyst. *In situ* XPS probe the oxidation state of Pt in Pt1Com during catalytic reduction of NO with H2 at high temperature. Similarly, *in situ* MS, XRD, FTIR, etc. have also been employed to study the behavior of materials during a reaction.

E. Advanced Microscopy Techniques:
A combination of both SEM and TEM, called STEM (scanning tunneling electron microscopy), scans the sample surface (just like in SEM) as well as detects and creates images through the transmitted electrons (just like in TEM). Electron energy loss spectroscopy (EELS) attached to STEM measures the energy lost by electrons due to inelastic scattering to identify the atomic and chemical composition and the electronic properties of the sample. STEM could be used to observe 2-3 nm sized Pt nanoparticles loaded on 20-25 nm CeO2 nanoparticles. In addition, it is a good idea to combine TEM with AFM in order to get structural as well as mechanical properties information.

F. Machine Learning–Assisted Characterization:
Combining microscopy characterization techniques with machine learning (ML) algorithms can offer insights into the structure and properties of 2D materials, with the advantage of automation, accuracy, and high throughput. There are reports wherein an ML-based random forest algorithm has been combined with data obtained experimentally through XPS to create a database of the Co(III)/Co(II) ratio in Co-rGO materials for supercapacitor applications.

These advancements collectively contribute to the development of novel nanomaterials with tailored properties and improved performance, while also providing more sophisticated tools for their characterization and analysis.

7.4 Challenges and Opportunities

7.4.1 Challenges in Developing and Commercializing Multifunctional Materials and Nanomaterials

While multifunctional materials and nanomaterials hold great promise for various industries, their development and commercialization face several challenges. Addressing these challenges is crucial for realizing the full potential of these advanced materials. Here are some key challenges:

A. The synthesis, upscaling, and re-production of the nanomaterials pose challenges. It is difficult to up-scale a lab protocol to large-scale production of the nanomaterial. It is important to precisely control the quality, size, shape surface functionalization, etc. properties because it directly affects the performance. In addition, the characterization equipment involved is all bulky and costly, and it is crucial to perform the measurements with utmost accuracy, to achieve reproducibility in the results. Some nanomaterials degrade fast under ambient conditions, so it is important to preserve them to ensure their durable usage.
B. Another significant issue is the production of nanomaterials in an economical manner. Large-scale manufacturing of high-quality nanomaterials is usually limited since they must be created under severe circumstances and with complex equipment. The majority of low-cost processes typically result in low-quality products with faults. The production of nanomaterials under control remains a difficult task. More concentrated efforts are needed to create novel synthesis techniques that get beyond the drawbacks of traditional techniques.
C. Defects in nanomaterials have the potential to impair both their intrinsic properties and performance. As an example, one of the known strongest materials is CNTs. Nonetheless, flaws, random orientations, discontinuous tube lengths, and contaminants can significantly reduce the tensile strength of CNTs.
D. Precise Control of Properties: Among the main difficulties in creating 2D ultrathin materials are their production and stability. Except in the instance of graphene, virtually little experimental examination of 2D ultrathin materials—an exceptional class of nanomaterials with attractive theoretical properties—has been conducted. It is expected that in the future, their synthesis and practical application will receive greater attention.
E. Durability and Stability: Particle aggregation at the low dimensional level is a fundamental problem that significantly impairs performance in related domains. Most nanomaterials cluster together as soon as they come into touch with one another. Agglomeration can occur as a result of electrostatic interactions, high surface energy, or physical entanglement. These difficulties make it difficult to use composite materials or multifunctional materials in real-world applications.
F. Integration with Existing Technologies: A multidisciplinary yet methodical approach to the entire materials system is needed in the field of multifunctional materials; theoretical, computational, and experimental methodologies must be integrated into a unified methodology. Another requirement is the evolution of new manufacturing techniques, moving away from traditional processes that 'merge' functionalities at the macroscale and toward processing platforms that, upon demand, integrate various local building block functionalities into the global architecture. The fact that multifunctional materials frequently demand in-depth expertise from a variety of domains must have some bearing on the material's ultimate cost. As the

material itself will fulfill the criteria of numerous functions of different applications, engineers from different fields must be involved.

7.4.2 Opportunities for Future Research and Development in These Fields

MFNs have gained attention in all fields of science and technology. With their unique properties and versatile applications, there are immense opportunities for nanomaterials in future research and development. Future research in these areas has the potential to revolutionize various industries, addressing current challenges and paving the way for the development of innovative technologies and applications based on MFNs.

1. Artificial Intelligence (AI): Using AI and ML algorithms to fabricate nanomaterials and optimize their composition. These tools can accelerate the discovery and enhance the understanding of structure-property relationships.
2. Bio/Nanomedicine: Nucleic acid nanostructures, multiple functionalities, magnetic nanoparticles, soft nanoparticles, etc. can be employed as carriers for drug delivery, numerous medical diagnostics and therapeutic applications, cell targeting and sorting, etc. Miniature dielectrophoresis device (DEP) can be developed for the capture of virus and bacterial cells.
3. Electronics: The development of nanoscale devices for photovoltaics, energy storage, photocatalysis, ferroelectricity, and optics, can greatly accelerate research and development in electronic devices.
4. 2D Materials: These materials, including graphene and beyond, have the properties to be incorporated into advanced devices for sensing, electronics, and energy storage.
5. Integrated Functionalities: By inserting the desired functionalities in the nanomaterials they can be tailored according to the need and smart and responsive toward the environment. Develop materials that adapt dynamically according to their environment based on pH, temperature, light, etc. external stimuli. These materials are used in making sensors, for drug delivery, etc.

7.5 Conclusion

7.5.1 Future Outlook for Multifunctional Materials and Nanotechnology

While the future of multifunctional materials and nanotechnology holds great promise, addressing challenges such as ethical concerns, environmental impact, and regulatory uncertainties will be crucial to ensuring the responsible and sustainable growth of these fields. Ongoing research, collaboration, and innovation will be key drivers in shaping this future outlook. We will be able to increase our control and expand the spectrum of physical characteristics of materials even as we shrink the integration scale as our understanding of materials at the nanoscale grows. Even if we are getting close to comprehending and using nanophysics, there is still a great deal of effort to be made in order to discover how to create large-scale materials from nanoscale components. While biological processing and self-assembly techniques appear promising, they are not yet developed sufficiently for the manufacturing of multicomponent systems.

True materials integration has potential in two-phase multifunctional systems. But like biological systems, genuinely smart materials systems will need to combine at least three different functions: structure, actuation, logic, sensing, and energy storage. On a modest scale, biological systems have mastered multifunction. These ideas will be expanded into large-scale constructions by designing a priori various functionalities into a materials system. Because of the complexity of these higher-order systems, developing new, maybe less singly optimum methods of reaching function and multivariable optimization techniques will require a deep understanding of how fundamental physical mechanisms might be controlled. For instance, research into the ionic conduction mechanism for new logic capabilities or the electrical conductivity processes for potential effects on mechanical strength might both lead to an enhancement in the mechanical strength of energy storage systems.

With the advancements of these multifunctional materials, extensive research is dedicated to achieving specific application goals. A multidisciplinary yet methodical approach to the entire materials system is needed in the field of multifunctional materials; theoretical, computational, and experimental methodologies must be integrated into a unified methodology. Another requirement is the development of new manufacturing techniques that incorporate functionalities at a large level and toward processing platforms that, upon request, integrate several individual functionalities into the integrated architecture. Fabrication schemes, which have more in common with programming than with establishing a final configuration, will go well beyond the shape of materials, as process engineering is essential for material functionalization.

References

1. Ferreira, A.D.B.L., Nóvoa, P.R.O., Marques, A.T., Multifunctional Material Systems: A state-of-the-art review. *Compos. Struct.*, 151, 3–35, 2016.
2. Saritha, G.N.G., Anju, T., Kumar, A., Nanotechnology - Big impact: How nanotechnology is changing the future of agriculture? *J. Agric. Food Res.*, 10, 100457, 2022.
3. Song, J.-H., Min, S.-H., Kim, S.-G., Cho, Y., Ahn, S.-H., Multi-functionalization Strategies Using Nanomaterials: A Review and Case Study in Sensing Applications. *Int. J. Precis. Eng. Manuf. Green Technol.*, 9, 323–347, 2022.
4. Srinivasan, M., Rajabi, M., Mousa, S.A., Multifunctional Nanomaterials and Their Applications in Drug Delivery and Cancer Therapy. *Nanomaterials*, 5, 1690–1703, 2015.
5. Jung, F., Wischke, C., Lendlein, A., Degradable, Multifunctional Cardiovascular Implants: Challenges and Hurdles, in: *MRS Bulletin*, vol. 35, pp. 607–613, 2010.
6. Lendlein, A. and Trask, R.S., Multifunctional materials: concepts, function-structure relationships, knowledge-based design, translational materials research. *Multifunct. Mater.*, 1, 010201, 2018.
7. Muralt, P., Polcawich, R.G., Trolier-McKinstry, SJM., Piezoelectric thin films for sensors, actuators, and energy harvesting, in: *MRS Bulletin*, vol. 34, pp. 658–664, 2009.
8. Cook, A.C. and Vel, S.S., Multiscale thermopiezoelectric analysis of laminated plates with integrated piezoelectric fiber composites. *J. Eur. J. Mechanics-A/Solids*, 40, 11–33, 2013.

9. Probst, T., Kochersberger, K., Stiltner, B., Hickling, C., Ohanian, O., Karni, E., Olien, C., Blain, A., Smart material actuators as a means of UAV flight control. *50th AIAA Aerospace sciences meeting including the new horizons forum and aerospace exposition*, p. 486, 2012.
10. Liu, F., Diercks, D., Hussain, A.M., Dale, N., Furuya, Y., Miura, Y., Fukuyama, Y., Duan, C., Nanocomposite catalyst for high-performance and durable intermediate-temperature methane-fueled metal-supported solid oxide fuel cells. *J. ACS Appl. Mater. Interfaces*, 14, 53840–53849, 2022.
11. Ali, A. and Andriyana, A., Properties of multifunctional composite materials based on nanomaterials: a review. *RSC Adv.*, 10, 16390–16403, 2020.
12. Lone, I.H., Aslam, J., Radwan, N.R.E., Bashal, A.H., Ajlouni, A.F.A., Akhter, A., Multiferroic ABO3 Transition Metal Oxides: a Rare Interaction of Ferroelectricity and Magnetism. *Nanoscale Res. Lett.*, 14, 142, 2019.
13. Schmid, H.J.F., The dice-Stone der Würfelstein: Some personal souvenirs around the discovery of the first ferromagnetic ferroelectric. *Ferroelectrics*, 427, 1–33, 2012.
14. Kumar, M., Kumar, A., Anshul, A., Sharma, S., Advances and future challenges in multifunctional nanostructures for their role in fast, energy efficient memory devices. *Mater. Lett.*, 277, 128369, 2020.
15. White, S.R., Sottos, N.R., Geubelle, P.H., Moore, J.S., Kessler, M.R., Sriram, S.R., Brown, E.N., Viswanathan, S., Autonomic healing of polymer composites. *Nature*, 409, 794–797, 2001.
16. Zhang, Y., Wu, M., Wu, M., Zhu, J., Zhang, X., Multifunctional Carbon-Based Nanomaterials: Applications in Biomolecular Imaging and Therapy. *ACS Omega*, 3, 9126–9145, 2018.
17. Liu, J., Cui, L., Losic, D., Graphene and graphene oxide as new nanocarriers for drug delivery applications. *Acta Biomater.*, 9, 9243–9257, 2013.
18. Das, P., Deoghare, A.B., Ranjan Maity, S., Exploring the Potential of Graphene as an EMI Shielding Material – An Overview. *Mater. Today: Proc.*, 22, 1737–1744, 2020.
19. Sun, M., Dong, Y., Zhang, G., Qu, J., Li, J., α-Fe_2O_3 spherical nanocrystals supported on CNTs as efficient non-noble electrocatalysts for the oxygen reduction reaction. *J. Mater. Chem. A*, 2, 13635–13640, 2014.
20. Yahyazadeh, A. and Khoshandam, B., Carbon nanotube synthesis via the catalytic chemical vapor deposition of methane in the presence of iron, molybdenum, and iron–molybdenum alloy thin layer catalysts. *Results Phys.*, 7, 3826–3837, 2017.
21. Tang, X., Wu, W., Zeng, W., Jiang, H., Copper-Catalyzed Oxidative Carbon–Carbon and/or Carbon–Heteroatom Bond Formation with O2 or Internal Oxidants. *Acc. Chem. Res.*, 51, 1092–1105, 2018.
22. Lee, S.-H., Park, J.H., Kim, S.M., Synthesis, property, and application of carbon nanotube fiber. *J. Korean Ceram. Soc.*, 58, 148–159, 2021.
23. Pascanu, V., González Miera, G., Inge, A.K., Martín-Matute, B., Metal–Organic Frameworks as Catalysts for Organic Synthesis: A Critical Perspective. *J. Am. Chem. Soc.*, 141, 7223–7234, 2019.
24. Abo-zeid, Y., Ismail, N.S.M., McLean, G.R., Hamdy, N.M., A molecular docking study repurposes FDA approved iron oxide nanoparticles to treat and control COVID-19 infection. *Eur. J. Pharm. Sci.*, 153, 105465, 2020.
25. Chen, Y. and Hou, S., Recent progress in the effect of magnetic iron oxide nanoparticles on cells and extracellular vesicles. *Cell Death Discov.*, 9, 195, 2023.
26. Mitchell, M.J., Billingsley, M.M., Haley, R.M., Wechsler, M.E., Peppas, N.A., Langer, R., Engineering precision nanoparticles for drug delivery. *Nat. Rev. Drug Discov.*, 20, 101–124, 2021.
27. Li, S., Su, W., Wu, H., Yuan, T., Yuan, C., Liu, J., Deng, G., Gao, X., Chen, Z., Bao, Y., Yuan, F., Zhou, S., Tan, H., Li, Y., Li, X., Fan, L., Zhu, J., Chen, A.T., Liu, F., Zhou, Y., Li, M., Zhai, X.,

Zhou, J., Targeted tumour theranostics in mice via carbon quantum dots structurally mimicking large amino acids. *Nat. Biomed. Eng.*, 4, 704–716, 2020.
28. Zeng, Q., Shao, D., He, X., Ren, Z., Ji, W., Shan, C., Qu, S., Li, J., Chen, L., Li, Q., Carbon dots as a trackable drug delivery carrier for localized cancer therapy *in vivo*. *J. Mater. Chem. B*, 4, 5119–5126, 2016.
29. Stavrou, M., Phung, N., Grimm, J., Andreou, C., Organ-on-chip systems as a model for nanomedicine. *Nanoscale*, 15, 9927–9940, 2023.
30. Kalyanasundaram, K. and Grätzel, M., Themed issue: nanomaterials for energy conversion and storage. *J. Mater. Chem.*, 22, 24190–24194, 2012.
31. Whittingham, M.S., Electrical energy storage and intercalation chemistry. *Science*, 192, 1126–1127, 1976.
32. Fujita, T. and Toda, K., Microdisplacement Measurement Using a Liquid-Delay-Line Oscillator. *Jpn. J. Appl. Phys.*, 42, 6131, 2003.
33. Julien, C.M. and Mauger, A., Nanostructured MnO_2 as Electrode Materials for Energy Storage. *Nanomaterials*, 7, 396, 2017.
34. Singh, G., Sharma, M., Mathur, A., Halder, A., Vaish, R., Diesel Soot as a Supercapacitor Electrode Material. *J. Electrochem. Soc.*, 168, 050551, 2021.
35. Szabó, Z., Volk, J., Fülöp, E., Deák, A., Bársony, I.J.P., Nanostructures-Fundamentals, Applications, Regular ZnO nanopillar arrays by nanosphere photolithography. *Photonics Nanostruct. Fundam. Appl.*, 11, 1–7, 2013.
36. Kuo, C.W., Shiu, J.Y., Cho, Y.H., Chen, P., Fabrication of large-area periodic nanopillar arrays for nanoimprint lithography using polymer colloid masks. *Adv. Mater.*, 15, 1065–1068, 2003.
37. Xu, K. and Chen, J.J.A.N., High-resolution scanning probe lithography technology: a review. *Appl. Nanosci.*, 10, 1013–1022, 2020.
38. Matsumoto, R., Adachi, S., Sadki, E.H.S., Yamamoto, S., Tanaka, H., Takeya, H., Takano, Y.J.A.A.E.M., Maskless Patterning of Gallium-Irradiated Superconducting Silicon Using Focused Ion Beam. *ACS Appl. Electron. Mater.*, 2, 677–682, 2020.
39. Garg, V., Mote, R.G., Fu, J.J.A.S.S., Facile fabrication of functional 3D micro-nano architectures with focused ion beam implantation and selective chemical etching. *Appl. Surf. Sci.*, 526, 146644, 2020.
40. Son, H.H., Seo, G.H., Jeong, U., Shin, D.Y., Kim, S.J., Capillary wicking effect of a Cr-sputtered superhydrophilic surface on enhancement of pool boiling critical heat flux. *Int. J. Heat Mass Transfer*, 113, 115–128, 2017.
41. Jones, A.C. and Hitchman, M.L.J.C.V.D.P., Processes, Applications, Overview of chemical vapour deposition. *Chem. Vapour Deposition: Precursors, Processes Appl.*, 1, 1–36, 2009.
42. Shah, K.A. and Tali, B.A., Synthesis of carbon nanotubes by catalytic chemical vapour deposition: A review on carbon sources, catalysts and substrates. *Materials Science in Semiconductor Processing*, 41, 67–82, 2016.
43. Wu, Q., Wongwiriyapan, W., Park, J.-H., Park, S., Jung, S.J., Jeong, T., Lee, S., Lee, Y.H., Song, Y.J., *In situ* chemical vapor deposition of graphene and hexagonal boron nitride heterostructures. *Curr. Appl Phys.*, 16, 1175–1191, 2016.
44. Johnson, R.W., Hultqvist, A., Bent, SFJM., A brief review of atomic layer deposition: from fundamentals to applications. *Mater. Today*, 17, 236–246, 2014.
45. George, SMJC., Atomic layer deposition: an overview. *Chem. Rev.*, 110, 111–131, 2010.
46. Liu, Y., Goebl, J., Yin, Y.J.C.S.R., Themed issue: Chemistry of functional nanomaterials. *Chem. Soc. Rev.*, 42, 2610–2653, 2013.
47. Li, W. and Zhao, D.J.C.C., An overview of the synthesis of ordered mesoporous materials. *Chem. Commun.*, 49, 943–946, 2013.

48. Kresge, C., Leonowicz, M.E., Roth, W.J., Vartuli, J., Beck, JJ., Ordered mesoporous molecular sieves synthesized by a liquid-crystal template mechanism. *Nature*, 359, 710–712, 1992.
49. Poolakkandy, R.R. and Menamparambath, M.M.J.N.A., Soft-template-assisted synthesis: A promising approach for the fabrication of transition metal oxides. *Nanoscale Adv.*, 2, 5015–5045, 2020.
50. Kandioller, W., Theiner, J., Keppler, B.K., Kowol, C.R., Elemental analysis: an important purity control but prone to manipulations. *Inorg. Chem. Front.*, 9, 412–416, 2022.
51. Cotter, D.R., Sabherwal, S., Schubert, K.O., Chapter 25-Proteomics for diagnostic and therapeutic blood biomarker discovery in schizophrenia and other psychotic disorders, in: *Personalized Psychiatry*, B.T. Baune (Ed.), pp. 307–317, Academic Press, San Diego, 2020.
52. Ortiz Ortega, E., Hosseinian, H., Rosales López, M.J., Rodríguez Vera, A., Hosseini, S., Characterization Techniques for Chemical and Structural Analyses, in: *Material Characterization Techniques and Applications*, E. Ortiz Ortega, H. Hosseinian, I.B. Aguilar Meza, M.J. Rosales López, A. Rodríguez Vera, S. Hosseini (Eds.), pp. 93–152, Springer Singapore, Singapore, 2022.
53. Ribeiro, LNM., Couto, V.M., Fraceto, L.F., de Paula, E., Use of nanoparticle concentration as a tool to understand the structural properties of colloids. *Sci. Rep.*, 8, 982, 2018.

8

Multifunctional Materials Surface Science

Mansi Sharma

School of Engineering and Technology, Manav Rachna International Institute of Research and Studies, Faridabad, Haryana, India

Abstract

With the progressive scientific interest approaching advanced materials, the scope of their applicability in various fields of science has globally evolved and the materials depicting such advancement in terms of their properties are known as multifunctional materials. Multifunctionality is considered as the part of the material phase where the properties of the same are observed to have variations depending on the specific requirements. The tailored material synthesis led to alter the properties. The chapter presents a compiled overview of various multifunctional materials, their properties, methods of synthesis, and application in various fields of science including energy and environment from the reported literature. An attempt has been made to provide insights into the challenges in the fabrication and designing of multifunctional materials, their industrialization based on analytical as well as numerical views of optimizations, and the scope for future applications.

***Keywords*:** Multifunctional materials, nanoparticles, multifunctional surfaces, bioactive materials, sustainable energy material, self healing, plasmonic, energy storage

8.1 Introduction

Presently, the research and development of materials with specific thermal, magnetic, and optoelectrical properties are of great interest in view of technological advancement. The production of advanced materials target to meet the rising demands for better optoelectronic properties, thermal stability, structural compatibility, and ability to cope with the surrounding electromagnetic variations as required to have potential applications for energy storage devices, sensing, robotics, automobiles, and healthcare systems [1]. The materials with tailored properties to meet the desirable device applications are considered as multifunctional materials. Multifunctionality is the property of a material to perform a variety of functions [2]. Further classification of these materials is in the form of multifunctional composites [3] and multifunctional nanostructure is based on the requirement of the system.

Email: mansisharma663@gmail.com

Divya Bajpai Tripathy, Anjali Gupta and Arvind Kumar Jain (eds.) Multifunctional Materials: Engineering and Biological Applications, (207–224) © 2025 Scrivener Publishing LLC

One of the significant factors responsible for achieving the multifunctionality of a material is the tailored size, i.e., to design materials from micro to macroscales or from macro to microscales. The present chapter aims to provide an insight to understand the multifunctional materials and their applications as advanced materials in various fields of research and innovation. It briefly outlines the significant properties of the materials with the multiphasic structure and the scope of their growing global applicability in research and industries.

8.1.1 Background and Importance of Multifunctional Materials

In general, all the composites work as multifunctional in view of the combination of their distinct physical and chemical nature. In nature or in biological systems, there exist a number of organic/inorganic composites that have specific functionality. For decades, in view of rising technological dependence, science has evolved with various materials ranging from bulk conductors to semiconductors that could meet the specified requirements. With the growing interest in the need for new materials, research and development led to the evolution of advanced materials at the nanoscale to fulfill the need for miniature device structures. In this, nanoparticles have gained much attention due to their distinct properties as a result of the quantum size effect. The surface of nanoparticles is considered to be of high reactivity in view of the high surface-to-volume ratio. As the size of the particle decreases, the surface atoms are known to become significant and dominate the overall structural as well as optoelectronic properties of the materials. Therefore, the surface modifications are considered to be much more approachable with the tailored size.

Multifunctional materials are known to have a complex structure having the combination of two or more material nanoparticles, which can be organic or inorganic in nature [4]. In the application of drug delivery and DNA separation, the use of functionalized multifunctional nanoparticles is found to exhibit magnetic as well as plasmonic properties in a biocompatible matrix [5]. The combination of magnetic and plasmonic nanomaterials in such applications results in a complex structure of the material.

8.2 Surface Science Principles and Techniques

The surface science of multiphasic structures is mainly governed by the principles of quantum confinement effects at the nanoscale. With the surface modifications and the reduction of particle size, the properties of the material are highly influenced due to the dominance of the quantum confinement effect at low dimensions. The materials are assumed to have a confined geometry to exhibit significant variations at the nanoscale. Thus, the classification of such multifunctionality at the nanoscale extends with the processing of 1D and 2D nanostructures. To have the required multiphasic nature of the materials at the nanoscale, there are a variety of approaches in material processing. These approaches are known as the bottom-up approach and top-down approach, respectively, and a detailed discussion of these methods of designing has been given in the subsequent sections.

8.3 Multifunctional Surfaces

In device designing, the external surface as well as the surface of the materials in contact at the interface both have a significant role in the applicability of the device depending on the surface characteristics. The surface could have a rough, smooth, or textured surface, which drastically affects the atomic level interactions at the interfaces and contacts. Thus, the surface properties and the texture are one of the important parameters in structural designing used in nano-engineered devices. Multifunctional surfaces are surfaces that could depict a variety of functions depending on the special nanoscale design. These modified surfaces are mainly considered to be having a hydrophobic or hydrophilic nature. This behavior of the surface is so classified on the basis of surface response to the water droplet. Such surface-wetting characteristics have gained so much technological interest. Other than the conditions of surface wettability, the multifunctionality of the surface incorporates specific strength, healing, and stimuli-responsive surface properties in the device structure.

8.3.1 Superhydrophobic and Superhydrophilic Surfaces

In material science and engineering, the study of surperhydrophobic and super hydrophilic materials is one of the underlined topics of research in the past few decades in view of their significant application based on the material functionality in drag reduction, self-cleaning, self-healing, etc. [6]. These material properties were analyzed with the condition of artificial rough surfaces with low surface energy. The design and fabrication of these artificial surfaces are based on the concept of the multifunctionality of the material [7].

A fluorine-free ultra-stretchable superhydrophobic coating was reported by Wang, Fang *et al.* (2020) [6]. The surface so formed exhibits strong resistance and stability in terms of hydrophobic nature to different mechanical and chemical treatments including temperature variations and photon irradiation. Such coatings have also shown exceptional antibacterial activity against gram-positive and gram-negative bacteria. In one of the reported literature by Deng *et al.* (2011), the porous silica capsule was used to fabricate a transparent thermally stable super hydrophobic coating [8]. The combination of transparency and mechanical stability showed the advantage of the material to be used in organic photovoltaics in order to meet the demand for cost-effective device maintenance.

8.3.2 Self-Healing and Anti-Corrosion Surfaces

During the material processing, the strategies of fabrication as well as the intrinsic properties of the material play a crucial role in defining the possible structural healing and the repair capacity of the material for the smooth functioning of the device. The specific internal damages are barely identified and are being repaired manually, which highly impacts the lifetime of the devices based on such materials.

The multifunctionality of a material is significantly reflected via the internal repair system of the material, which is considered to induce the property of self-healing. Thus, the class of materials with a self-healing property is of great importance during the classification of the significant multiphasic properties. These materials are known to have the ability to self-repair unexpected damages. In living beings, the networking of various tissues and cells is the best example of having a structural complexity in terms of not holding a single phasic nature and evident the self-healing of the bones and skin. Therefore, nature is one of the inspirations of the self-healing and self-cleaning process [9].

In artificial materials/nano-engineered materials, the advanced multiphasic structure exhibits the properties of self-healing based on their design. The healing mechanism can be based on external triggering and without triggering. This is known as the non-autonomic and autonomic response [10].

Hexamethylene di-isocyanate is one of the multifunctional materials with good anti-corrosion and self-healing properties. In one of the reports by Wu *et al.* (2014), the material microcapsule is well dispersed in a polymeric matrix and the self-healing and anti-corrosion functions of the encapsulated material activate on its own when the microcapsule breaks and hexamethylene diisocyanate comes into contact with the surrounding moist content [11]. Here, the trigger has been provided by the internal damage, not due to any external factors such as thermal or photonic excitations. This mechanism is considered one of the examples of the autonomic system. Several reports are available concerning the self-healing ability of metals, ceramics, polymers, and various composite materials [7, 12–14]. The characteristic so observed is considered to be the result of the multiphasic nature of the materials and its design shows a wide scope in aviation and spaceship applications. Continuous carbon fiber composite material incorporated into polyurethanes has been reported to depict the self-healing properties where polyurethanes are one of the healable materials in response to thermal variations [15]. The testing of healing efficiency for the same has been done by short beam shear testing, which results in the observed stability of the composite as a result of the repetitive healing process. The average healing efficiencies of ~85% and ~73 % were concluded and maintained the intrinsic properties of the composite. Figure 8.1 presents the self-healing of the multiphasic matrix with an encapsulated healing agent.

The development of soy protein microfibrillated cellulose composite materials with self-healing properties [16]. The composite contains green solvent-based microcapsules, which depicts the self-healing mechanism. Thus, the green composite with an extended lifetime due to the self-healing mechanism is suggested as a better replacement for petroleum-based conventional composite materials [17]. The application of advanced green composites in automotive parts and aerospace, medicine, etc has also been explored to replace conventional industrial composites [18].

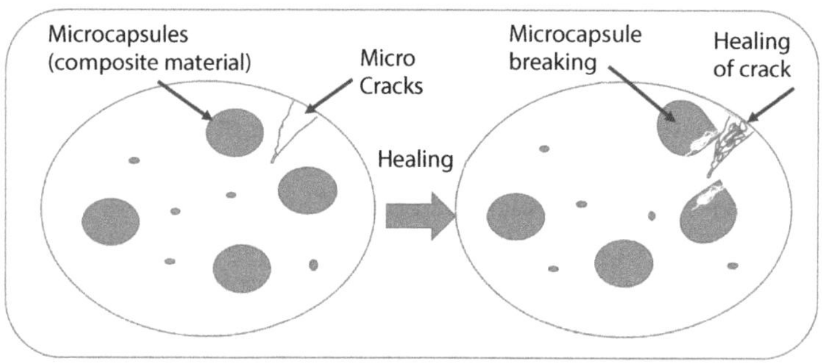

Figure 8.1 Illustration for self-healing within a multiphasic structure containing composite microcapsule.

8.3.3 Stimuli-Responsive Surfaces

One of the significant properties of the multiphasic materials discussed above is the self-healing nature of the materials. This self-healing as stated above can be autonomic or non-autonomic. The non-autonomic system of materials is further identified as stimuli-responsive materials. Stimuli-responsive materials belong to a class of materials that are made to adapt the external changes easily. In the era of advanced materials, there are a number of researched materials available for smart devices; however, most of them are found to have a limited response toward external stimuli. These external responsive materials are well illustrated in terms of their nature and their classified materials. They specifically display their adaptable characteristics in response to the variable parameters toward the environmental changes. The study of such materials is found to extend to the field of biometrics. These materials are one of the naturally evolved materials that have the ability to sense and respond to environmental changes, which is why they are also known as smart materials.

The multifunctionality of these materials guides them to represent the variable characteristics, physical appearance, or response in terms of electrical and optical variations. Such materials are found to have vast applications in industry and infrastructure and have the potential to expand their requirement in the repair and maintenance field [19].

8.3.4 Biocompatible and Bioactive Surfaces

The multifunctionality of materials is well showcased and illustrated by nature. Natural materials are known to present significant illustrations of multifunctionality [20]. The technology nowadays is being inspired by most of these bio as well as economical materials. These are being used in several diagnosis techniques, therapies, and bio imaging. Moreover, the use of multiphasic nanoparticles is found to be effective in allowing the tagging and separation of biomolecules during drug delivery and the tracking of the same during imaging [21].

In view of biomedical analysis, the multifunctional surfaces undergo specific treatment to produce an active desirable functional surface, which renders the compatibility, and conjugate with the DNA of viruses and antibodies in the body. The functionalized nanoparticles with surface modification using an active function group are found to exhibit compatibility for conjugating with various biological objects [22]. Such surfaces display potential applications in the detection and treatment of various cancer cells. Multifunctional nanoparticles have gained lots of research attention with their application in the diagnosis and optical tagging and separation of biomolecules.

8.3.5 Conductive and Electroactive Surfaces

In the era of smart materials, the need for material designing also desires to depict the unique properties of charge transport as well as the surface reactivity in terms of conduction. The research of multifunctional 2D materials has proven as a breakthrough in materials that are different from the bulk and exhibit unexceptional application in the field of electronics and healthcare. Several thin layers of 2D multifunctional structure including metal oxides, layered double hydroxides, and metal oxide frameworks have been identified as depicting the unique electrical and chemical properties [23].

The 2D conductive metal oxide frameworks have also depicted their potential application as electroactive materials with high conductivity. These materials result in the enhanced capacity of electrical capacitors [24]. Sheberla *et al.* (2017) have reported the MOF-based supercapacitor using Ni_2HITP_2. The use of such material with a porous framework showed a high degree of diffusion of electrolytic ions, avoiding the need for conductive additives [25].

Recently, the application of conductive polymer gel-based multifunctional structures has also been reported [26]. Their wide applications have been identified in view of tuneable physical and chemical properties, high flexibility, and ionic conductivity. Such electroactive surface offers diverse applications in energy storage, sensors energy conversions, etc.

8.3.6 Optical and Photonic Surfaces

Advances in material research from bulk to nanoscale led to the discovery of the extraordinary properties of 2-D and 1-D materials in optics. For decades, the research for exploring the compatible materials for dealing with IR-based photonic for spectroscopic sensing, thermal imaging, optical communication, etc. is considered to be one of the challenges in terms of achieving the desired optical properties based on the material bandgap. For example, in photonic devices, the optimized values (mid-IR) of bandgap and carrier concentration are considered to have significant effects on the grating modulations and consequently the device performance. The use of multifunctional surfaces is found to be one of the solutions for such limitations where the introduction of the combination of energy bandgap due to the multifunctionality of material at the nanoscale depicts the improved and versatile properties of the same device. Skylar *et al.* (2019) have carried out investigations for tellurene-based solutions for mid-IR-integrated optoelectronic devices [27]. The materials showed great potential for optoelectronics with the property of thickness-dependent bandgap and the ability to reduce dark current output effectively in photodetectors.

The advantage of modified properties based on the tailored size of the particles in the given material with the high surface area-to-volume ratio also led to depicting surface-enhanced plasmon resonance [28].

8.4 Synthesis and Fabrication of Multifunctional Surfaces

In order to preserve the significant properties in terms of the multifunctionality of the material, their fabrication processing needs special approaches. These approaches are much different from the synthesis methods used for the fabrication in the macroscopic range (bulk). In this section, a discussion has been made to highlight different approaches (Figure 8.2) and understand the processing methods for the fabrication of multifunctional surfaces within nanoscales.

8.4.1 Physical and Chemical Methods

The synthesis and fabrication of multifunctional materials and based devices is one of the significant concerns in view of the design and the quantum confinement without impacting the original properties and functionality of the material. The nano-engineering of materials with the inclusion of multiphasic structures requires an advanced approach and a

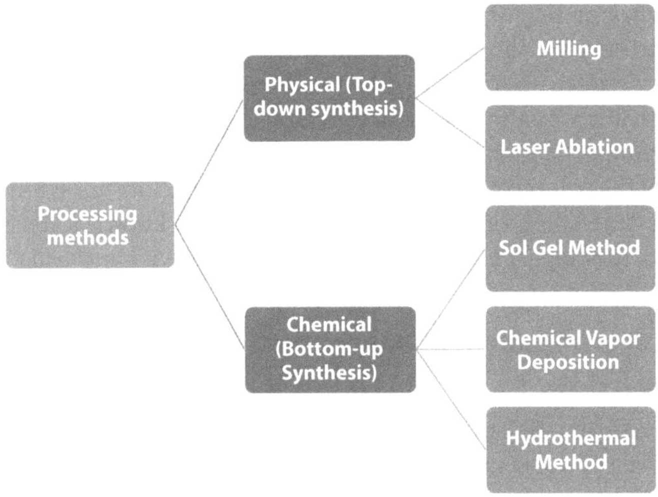

Figure 8.2 Multifunctional processing techniques.

sophisticated processing environment. Such processing can be done via (i) physical methods where the fabrication is merely dependent on the atomic interactions based on the chemical reactivity of materials and the surface. In these techniques, the ball milling method is one of the traditional ways of producing nano-powders of the desirable range. The powders so obtained will bind together for the final device. The technique is highly suitable for material application in ceramics and based devices. (ii) Chemical methods are where the processing is based on the chemical reactivity of the surface and the target material. The methods are based on the growth of nanoscale particles followed by self-assembly or positional assembly approaches. The overview of both approaches has been extended in the subsequent section.

8.4.2 Top-Down and Bottom-Up Approaches

The synthesis approaches based on the physical and chemical methods of the fabrication of multifunctional nanoparticles could be precisely classified as top-down and bottom-up approaches.

The top-down approach in a simpler way includes the methods of the shaping and framing of the structure at the nanoscale [29]. The method employs a transformational approach where the progressive scaling down of the materials takes place to transform the structure larger scale of the structure to miniature. The approaches utilize the cutting and milling process to shape the desired structure at the nanoscale. The techniques involved in such processing are mainly mechanical-processing techniques. Other than these, the advanced techniques include lithography, etching, and laser beams for the processing of bulk to have a nanoscale structure. These methods are based on a destructive approach and are known to be more expensive than the other techniques.

The bottom-up synthesis is based on the chemical processing of nanoscale structures. It is also known as the synthetic approach where the formulation of the structure takes place from the molecular or atomic level. The approach utilizes the chemical properties of the material for the growth of the required nanostructure in a positional assembly or self-assembly environment. In self-assembly growth, the molecules self-assemble themselves based on their chemical properties to form a nanostructure. On the other hand, positional assembly is a guided approach to growth where specific growth patterns are provided to the molecules to form a nanoscale structure. The chemical-processing techniques such as sol-gel, contact printing, colloidal aggregation, spin coating, vapor deposition, and chemical epitaxy are some of the common deposition techniques for bottom-up synthesis. These methods of growth are based on a constructive approach that is known to be cost effective and can produce smaller nanoparticles compared to the top-down approach.

8.4.3 Nanostructuring and Nanofabrication Techniques

Under the above-mentioned approaches, there are various available techniques for multiphasic nano-structuring. At the nanoscale, the range of particle size is defined as 1–100 nm. The fabrication of such nanoscale materials for device applications has been well explored with the use of various available material-processing techniques under the different approaches discussed above. In consideration of the complex structure of multiphasic materials, the fabrication in specific conditions also demands the combination of top-down and bottom-up syntheses. Some of the significant techniques are:

Chemical Vapor Deposition: The method is so termed deposition techniques for the materials at the nanoscale. The process involves a chemically assisted vapor phase reaction where a solid material gets deposited on the given heated substrate. Thus, with precise control over the deposition parameters like the temperature of the substrate, the pressure of the chamber, and the flow precursors used, the required coating of the multiphase compound, metal, semiconductor, etc. can be obtained [30]. The process offers the advantage of high deposition rates and has attracted a lot of scientific interest in view of its mass production capacity for nano-coatings. The technique also provides films with high precision in terms of their thickness; however, there is a scope to understand the growth kinematics in order to identify and control the residual/powder formation under various circumstances during deposition.

Laser Ablation: The method is also known as the method of surface reduction. In this, the target is being struck through a high beam laser in a high vacuum chamber with background gasses, and the result of which leads to the formation of nanoparticles [31]. The need for background gasses is to provide the reactive plasma environment nearer to the target, which is advantageous to control the size of the nanoparticles and their purity. The process is highly recommended for minimal waste production and particles with fine size control. Liquid laser ablation is also employed for the growth of nanoparticles in the liquid phase with the help of confined plasma plum in liquid [32, 33]. However, it is still challenging for efficient long-term usage, the requirement of high energy input, and the limitation of exposer area [30].

Sol-Gel Technique: Sol-gel is one of the conventional methods for the production of metal oxide nanoparticles and is also known as wet chemical synthesis. The metal-based precursor is dissolved and stirred in specific reactive conditions, which results in the formation of a gel-based product. Depending on the required nanoparticle, the drying mechanism can be chosen for the transformation of the gel. The methods allow precise control over the process parameters, which results in the formation of a variety of nanoparticles/composite structure-based metal, or semiconductor oxides [34–36]. Compared to other conventional methods of production, the technique allows the processing at lower temperatures with high purity [37–40].

Hydrothermal Method: It is considered one of the simplest methods of obtaining desired nanoparticles directly from the solution. In this, the nanoparticle formation is well controlled under various temperatures and pressure conditions to process the desired composite in a chemical reaction [41]. The method is highly preferred for the synthesis of single-phase nanoparticles as well as two phasic nanoparticles. In multifunctional material synthesis, the required formation of material composites with multiple phases is possible with this method [42]. Moreover, the technique is highly suitable for the growth of single crystals, and the so obtained nanoparticle is found to be explored in various device applications [43]. Studies have explored the advantages of the hydrothermal method in the processing of metal oxide/metal/semiconductor nanoparticles [44, 45].

8.4.4 Surface Modification and Functionalization Methods

The surface modifications of multiphasic materials or nanoparticles are the inclusion of any functional group on the surface or the exclusion of the impurities elements from the surface to obtain the desirable functional materials [21]. Such modified functional nanoparticle surfaces are found to depict potential applications in device designing and to have required compatibility with biological enzymes and antibodies in biomedicine applications. Some of the well-known functional groups for such functional surfaces are carboxyl and biotin.

Other than the active functional surfaces, the modified surfaces are required to have a protective layer against corrosion, oxidation, etc. The purpose of such a protective coating is to attain better stability. The surface modification methods include surface treatment/enhancement techniques like plasma treatment, surface passivation processing physically or chemically, and chemical grafting. Lu *et al.* [22] have reported the use of silica coating on magnetic nanoparticles in order to avoid unnecessary interactions of these particles with the surroundings. The protective layer of silica provides an impenetrable layer to the nanoparticles, which results in perfect shielding from oxidation and helps in maintaining the stability of particles [21].

8.5 Applications of Multifunctional Surfaces

Multifunctional materials and surfaces offer a much wider scope for their application in various sectors of material science. Figure 8.3 highlights some of the significant applications of multifunctional surfaces. The present section provides a brief overview of the applications.

8.5.1 Biomedical and Healthcare Applications

In advanced healthcare, there are numerous techniques available for diagnostics in view of specific diseases. However, in recent years, light-based therapeutic instruments have gained significant interest in view of their functionality as wearable devices [46]. These devices are found to work on the basic concept of optogenetics, where the optical interaction of neurons can be mapped. The manipulation of the functions of the nervous system at various

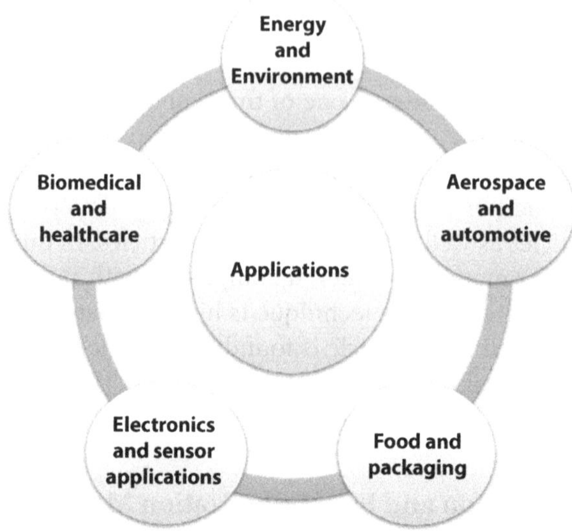

Figure 8.3 Applications of multifunctional materials.

levels can be effectively made with this approach. The basic principle involved is molecular imaging to visualize the absorption of light (photons) by the cells and its scattering at different depths depending on the penetration offered by the incident radiation and the cells. This photo-physical interaction carries the necessary information on the spatial distribution of tissues within the target area. Therefore, multifunctional materials also evolved as one of the highly suitable functional materials in wearable photonic for light-based healthcare devices [46].

8.5.2 Energy and Environment Applications

The advanced multifunctional materials were also found to play a significant role in facilitating and promoting renewable and sustainable energy development. In the present era, the rising demand for energy has oriented the research and industrial ground to have accessible green energy alternatives. The progress toward sustainable energy resources is specifically based on the energy output in terms of efficiency, energy storage, transportation, distribution, and energy management in view of electronic waste and its reuse [1]. Moreover, such materials are found to exhibit much scope in managing energy storage issues for transportation systems with electric vehicle technology. The use of multifunctional composites is found to be more reliable compared to conventional Li-ion batteries. Ladpali *et al.* (2016) have demonstrated the design, development, and characterizations of multifunctional composites for energy storage. The study presents a systematic comparison of conventional Li-ion batteries with multifunctional composites for energy storage [47].

In the transportation system, the obstruction for electric vehicles mainly relies on the charging and discharging time cycle. The use of hybrid multifunctional composite materials results in improved structural and electrical properties of supercapacitors for energy storage devices. The devices are found to evolve with the increased current density per unit area. Natasha *et al.* (2013) demonstrated the superior properties of composites with carbon fiber reinforcement for energy storage applications [48]. Furthermore, the effect of crystal structural variations on the capacitance of supercapacitors has also been demonstrated by Staaf *et al.* [49]. The work presents an improved charge/discharge cycle with high capacitance and dielectric constant with the use of a CNT-based composite. The variations were presented as a result of microstructure formation within the composite.

Furthermore, multifunctional nanoscale particle-based detections are also considered a highly sensitive tool for observing the specific environmental contaminations or the presence of any pollutant in view of their enhanced surface reactivity due to their small size.

8.5.3 Electronics and Sensor Applications

In soft matter engineering, multifunctional materials have demonstrated potential applications in robotics and wearable electronic devices. The advances in electronics and the age of artificial intelligence have come up with innovative solutions for interfacing such devices with the human body. The multifunctionality of the materials (natural/artificial) offers a device that can work in variable environmental conditions with their tuneable optoelectronic properties [50]. The recent technological development in soft robotics requires the soft actuators as a primary component which provides the system with desirable actuation patterns. The multifunctionality of the materials used for soft actuation at the nanoscale opens a wide scope for various responsive actions like electrical response, thermal response, and magnetic response and with applications for soft grippers, medical devices, artificial muscles, etc.

8.5.4 Food and Packaging Applications

In view of environmental concerns, advanced research led to bioplastics as an alternative eco-friendly sustainable solution to balance the rising demand and supply for plastic-based packaging. Moreover, globally, the food industries are also seeking and adopting such solutions to meet the requirements of food safety in terms of transportation, quality, and storage along with economic feasibility.

The idea of bio-based packaging and suitable materials has been researched to balance the demands of sustainability in combination with health and economic benefits [51].

Till 2020, nearly 99% share of plastic packaging was carried by petroleum-based plastic with a lifetime of over 100 years [52]. Bio-plastic includes biodegradable or plastic based on plant extracts. Trees have shown advanced applications based on their complex structure in terms of multifunctionality and are one of the abundant bio-resources. Chao *et al.* (2021) reported the lignocellulose-based food packaging films [53]. The method of extraction of cellulose is completely dependent on the plant-based physical and chemical treatment. These bioproducts are further classified as cellulose, lignin, and hemicellulose, which exhibit altered properties based on structural variability. As a product of waste biomass, cellulose nanocrystals are found to exhibit numerous applications in the food industry as food thickeners, emulsion stabilizers, packaging, and so on. Cellulose nanofiber is one of the potential multifunctional materials with attractive mechanical, regeneration, strength, and degradability, etc, which acts as a reinforcing agent to enhance the mechanical properties of hemicellulose films to overcome certain barriers for packaging purposes [54].

8.5.5 Aerospace and Automotive Applications

In view of accessing the applications of a material in the dynamical system, it is essential to investigate its static properties thoroughly. The multifunctional composite materials are found to depict a suitable design for high-speed structural applications in the aerospace and automobile industry [55].

In vehicle designing, the use of multifunctional materials helps in managing the produced with advanced mechanical properties along with the lightweight design [56]. In 1999, the use of a mixed-material approach including steel grade, plastic, aluminum, and carbon fiber designing of the BMW M6, which offered reduced vehicle weight and enhanced strength to provide the required vehicle balance, was discussed [56, 57]. The advantage of using nanocomposites in the automobile industry provides dimensional stability, better corrosion resistance, stiffness, and noise damping [58]. Moreover, for a few decades, the potential applications of nano clay compounds have been explored for exhibiting improved rheological properties and reduced weight [58].

In aerospace designing, the structures are required to be tested in view of damages and strength to tolerate high-velocity impacts of unexpected weather conditions during flight. The underwater tolerance of shock waves during sea mining and the strain rate in variable conditions are found to be suitably balanced with the use of polymer matrix composite structures. This strain rate balancing includes variable loading rates, which significantly affect the mechanical properties at the interface. Therefore, hybrid composites with variable strength and mechanical properties in repose to the loading rate were observed to be better in place of conventional composites [59]. Several theoretical as well as experimental studies present the comparative assessment of the loading parameters and the rheological properties of the composites used at the interface and coating. Navier's approach was used for the numerical solution of the frequency equation to establish the viscoelastic behavior of the nanoplate. The solutions for non-dimensional geometric parameters and non-local parameters were examined and used analytical methods based on Hamilton's principle [60].

8.6 Challenges and Future Prospects

8.6.1 Materials Design and Selection

One of the setbacks for the rising potential of multifunctional materials is the complex methodology and fabrication processing where the growth and evolution of the material are required to be precisely controlled in view of preserving desirable structure properties [1]. The designing of multifunctional materials/surfaces with the typical merging of the basic materials and the functional materials at their transition phases is one of the researched challenges at the nanoscale. The inclusion of functional material in the existing material for specific surface activity could result in the loss of basic structural symmetry [61]. The encapsulation of functional nanoparticles in the composites has significant effects on their structural integrity when integrated with device application. The theoretical modeling of stress analysis and crack propagation based on the finite element model was presented in one of the reports by Xiao *et al.* [62]. In this study, a 2-D model was developed to understand the tensile strength profile of epoxy laminates and address the crack propagation at different loading stages.

Other than the difficult encapsulated design, the joining methods play a significant role in the process of manufacturing and designing multifunction materials or composite structures to provide multifunctionality. These methods typically include adhesive bonding, mechanical bonding, solid-state welding, and friction stirring. Although with the availability of both conventional and non-conventional joining approaches, the precise control over the properties at the joining interface still has a lot of scope for future research. Konstantinos *et al.* (2010) have addressed some of the common joining issues in the designing of composite materials to exhibit multifunctional properties. The report concludes some of the suitable joining methods and techniques for the processing of various materials and surface modifications [19].

8.6.2 Surface Stability and Durability

Another challenge in designing a multifunctional surface is maintaining surface stability in terms of avoiding the formation of precipitation and the clustering of nanoscale structures, which could result in agglomeration. Such changes are more significant during the functionalization of nanoparticles or the processing of surface modifications in order to alter the given properties of the material. These reactive surfaces are found to have high surface reactivity; thus, the processing itself is challenging to provide precise control for the required self-organization of particles nearer to the transition phases during growth.

8.6.3 Scale-Up and Commercialization

<u>Limited Mass Production:</u> Considering the rising demand for multifunctional materials, it is expected that their mass production of primary materials and, hence, the rapid commercialization of the composite-based devices practically is still a challenge for the research in view of limited production rates from the current processing.

<u>Surface Reproduction with Accuracy:</u> Furthermore, with rapid production, it is very difficult to preserve and maintain the original configuration of the material because of the limited accuracy levels of attaining the repeatability of processing.

8.6.4 Multifunctional Integration and Optimization

In view of the discussed challenges, it can further be concluded that the overall designing demands the significant support of computational modeling and simulation techniques to establish such multiscale material designs [63]. The way of theoretical modeling for exploring the material joining and removal issues and optimizations for mass production at acceptable accuracy levels could provide some relevant solutions for these production and accuracy challenges to experiment with materials in cost-effective processing.

8.7 Conclusion and Outlook

8.7.1 Implications for Future Research

The alternative to plastic food packaging material, polysaccharide from the class of multiphasic material-based packaging has been found to exhibit interesting properties for the same. In this, the hemicellulose-based films are yet to be explored for it is attractive to replace conventional food packaging plastics.

However, the packaging based on hemicellulose films as discussed in Section 8.5.4 requires further attention for improved compatibility and mechanical strength, which requires having surface modifications through various processing. Thus, the structural improvements with modified surface compatibility of hemicellulose films without causing degradation of the material are one of the future concerns.

8.7.2 Final Thoughts and Recommendations

The chapter attempted to present the variety of applications of multifunctional materials and the involved processing. The design and the performance of materials have been well explored to provide a primary understanding of the experimentation and the techniques. However, the research has yet to be explored in terms of establishing a theoretical view in interpreting the experimental results. Therefore, it is required have more analytical investigational optimizations based on numerical simulations in multifunctional surfaces for exploring their properties experimentally. The future demands smart multifunctional materials with a potential scope of integrating functionalities to meet the requirements of various applications.

References

1. Lendlein, A. and Trask, R.S., Multifunctional materials: concepts, function-structure relationships, knowledge-based design, translational materials research. *Multifunct. Mater.*, 1, 1, 010201, 2018.
2. Christodoulou, L. and Venables, J.D., Multifunctional material systems: The first generation. *Jom*, 55, 39–45, 2003.
3. Amini, E. and Atack, P.A., A survey of composite materials. *Sheet Met. Ind.*, 7–15, 1974, January.
4. Ghosh Chaudhuri, R. and Paria, S., Core/shell nanoparticles: classes, properties, synthesis mechanisms, characterization, and applications. *Chem. Rev.*, 112, 4, 2373–2433, 2012.
5. Chu, T.D., Phi, T.H., Dinh, T.T.D., Tran, T.H., Luu, M.Q., Nguyen, H.N., Combination of 4-ATP coated silver nanoparticles and magnetic Fe3O4 nanoparticles by inverse emulsion method. *VNU J. Sci.: Mathematics-Physics*, 30, 1–9, 2014.
6. Wang, F., Pi, J., Song, F., Feng, R., Xu, C., Wang, X.L., Wang, Y.Z., A superhydrophobic coating to create multi-functional materials with mechanical/chemical/physical robustness. *Chem. Eng. J.*, 381, 122539, 2020.
7. Ferreira, A.D.B., Nóvoa, P.R., Marques, A.T., Multifunctional material systems: a state-of-the-art review. *Compos. Struct.*, 151, 3–35, 2016.

8. Deng, X., Mammen, L., Zhao, Y., Lellig, P., Müllen, K., Li, C., Butt, H.J., Vollmer, D., Transparent, thermally stable and mechanically robust superhydrophobic surfaces made from porous silica capsules. *Adv. Mater.*, 23, 26, 2962–2965, 2011.
9. Patrick, J.F., Hart, K.R., Krull, B.P., Diesendruck, C.E., Moore, J.S., White, S.R., Sottos, N.R., Continuous self-healing life cycle in vascularized structural composites. *Adv. Mater.*, 26, 25, 4302–4308, 2014.
10. White, S.R., Sottos, N.R., Geubelle, P.H., Moore, J.S., Kessler, M.R., Sriram, S.R., Brown, E.N., Viswanathan, S., Autonomic healing of polymer composites. *Nature*, 409, 6822, 794–797, 2001.
11. Wu, G., An, J., Tang, X.Z., Xiang, Y., Yang, J., A versatile approach towards multifunctional robust microcapsules with tunable, restorable, and solvent-proof superhydrophobicity for self-healing and self-cleaning coatings. *Adv. Funct. Mater.*, *24*, 43, 6751–6761, 2014.
12. Ghosh, S.K. (Ed.), *Self-healing materials: fundamentals, design strategies, and applications*, vol. 18, Wiley-vch, Weinheim, 2009.
13. Yuan, Y.C., Yin, T., Rong, M.Z., Zhang, M.Q., Self healing in polymers and polymer composites. Concepts, realization and outlook: A review. *eXPRESS Polym. Lett.*, 2, 4, 238–250, 2008.
14. Tavangarian, F. and Li, G., Bio-inspired crack self-healing of SiC/spinel nanocomposite. *Ceram. Int.*, 41, 2, 2828–2835, 2015.
15. Heo, Y. and Sodano, H.A., Thermally responsive self-healing composites with continuous carbon fiber reinforcement. *Compos. Sci. Technol.*, 118, 244–250, 2015.
16. Kim, J.R. and Netravali, A.N., Self-healing green composites based on soy protein and microfibrillated cellulose. *Compos. Sci. Technol.*, *143*, 22–30, 2017.
17. Kilanko, O., Babaremu, K.O., Fayomi, O.S.I., Akande, I.G., February. Overview of Composite Materials from Green Product in Mechanical Application, in: *IOP Conference Series: Earth and Environmental Science*, vol. 655, p. 012043, IOP Publishing, 2021.
18. Netravali, A.N., Advanced green composites: New directions. *Mater. Today: Proc.*, 8, 832–838, 2019.
19. Salonitis, K., Pandremenos, J., Paralikas, J., Chryssolouris, G., Multifunctional materials: engineering applications and processing challenges. *Int. J. Adv. Manuf. Technol.*, 49, 803–826, 2010.
20. Subbiah, R., Veerapandian, M., S Yun, K., Nanoparticles: functionalization and multifunctional applications in biomedical sciences. *Curr. Med. Chem.*, 17, 36, 4559–4577, 2010.
21. Nam, N.H. and Luong, N.H., Nanoparticles: Synthesis and applications, in: *Materials for biomedical engineering*, Elsevier, Amsterdam, 211-240, 2019.
22. Lu, A.H., Salabas, E.E., Schüth, F., Magnetic nanoparticles: synthesis, protection, functionalization, and application. *Angew. Chem. Int. Ed.*, 46, 8, 1222–1244, 2007.
23. Ko, M., Mendecki, L., Mirica, K.A., Conductive two-dimensional metal–organic frameworks as multifunctional materials. *Chem. Commun.*, 54, 57, 7873–7891, 2018.
24. Sheberla, D., Sun, L., Blood-Forsythe, M.A., Er, S., Wade, C.R., Brozek, C.K., Aspuru-Guzik, A., Dincă, M., High electrical conductivity in Ni3 (2, 3, 6, 7, 10, 11-hexaiminotriphenylene) 2, a semiconducting metal–organic graphene analogue. *J. Am. Chem. Soc.*, 136, 25, 8859–8862, 2014.
25. Sheberla, D., Bachman, J.C., Elias, J.S., Sun, C.J., Shao-Horn, Y., Dincă, M., Conductive MOF electrodes for stable supercapacitors with high areal capacitance. *Nat. Mater.*, 16, 2, 220–224, 2017.
26. Zhao, F., Shi, Y., Pan, L., Yu, G., Multifunctional nanostructured conductive polymer gels: synthesis, properties, and applications. *Acc. Chem. Res.*, 50, 7, 1734–1743, 2017.
27. Deckoff-Jones, S., Wang, Y., Lin, H., Wu, W., Hu, J., Tellurene: a multifunctional material for midinfrared optoelectronics. *ACS Photonics*, 6, 7, 1632–1638, 2019.
28. Melaine, F., Roupioz, Y., Buhot, A., Gold nanoparticles surface plasmon resonance enhanced signal for the detection of small molecules on split-aptamer microarrays (small molecules detection from split-aptamers). *Microarrays*, 4, 1, 41–52, 2015.

29. Taniguchi, N., On the basic concept of 'nano-technology'. *Proc. Intl. Conf. Prod. Eng. Tokyo, Part II, 1974*, Japan Society of Precision Engineering, 1974.
30. Iguchi, T., Takenaka, S., Nakagawa, K., Orita, Y., Matsune, H., Kishida, M., Production of carbon nanotube by ethylene decomposition over silica-coated metal catalysts. *Top. Catal.*, 52, 563–570, 2009.
31. Naser, H., Alghoul, M.A., Hossain, M.K., Asim, N., Abdullah, M.F., Ali, M.S., Alzubi, F.G., Amin, N., The role of laser ablation technique parameters in synthesis of nanoparticles from different target types. *J. Nanopart. Res.*, 21, 1–28, 2019.
32. Amendola, V. and Meneghetti, M., Laser ablation synthesis in solution and size manipulation of noble metal nanoparticles. *PCCP*, 11, 20, 3805–3821, 2009.
33. Semaltianos, N.G., Nanoparticles by laser ablation. *Crit. Rev. Solid State Mater. Sci.*, 35, 2, 105–124, 2010.
34. Tian, N., Zhou, Z.Y., Sun, S.G., Ding, Y., Wang, Z.L., Synthesis of tetrahexahedral platinum nanocrystals with high-index facets and high electro-oxidation activity. *Science*, 316, 5825, 732–735, 2007.
35. Xu, R., Wang, D., Zhang, J., Li, Y., Shape-dependent catalytic activity of silver nanoparticles for the oxidation of styrene. *Chemistry–An Asian J.*, 1, 6, 888–893, 2006.
36. Rahman, I.A. and Padavettan, V., Synthesis of silica nanoparticles by sol-gel: size-dependent properties, surface modification, and applications in silica-polymer nanocomposites—a review. *J. Nanomater.*, 1, 132424, 2012.
37. Niederberger, M. and Pinna, N., *Metal oxide nanoparticles in organic solvents: synthesis, formation, assembly and application*, Springer Science & Business Media, Heidelberg, 2009.
38. Collinson, M.M., Wang, H., Makote, R., Khramov, A., The effects of drying time and relative humidity on the stability of sol–gel derived silicate films in solution. *J. Electroanal. Chem.*, 519, 1-2, 65–71, 2002.
39. Vijayalakshmi, R. and Rajendran, V., Synthesis and characterization of nano-TiO_2 via different methods. *Arch. Appl. Sci. Res.*, 4, 2, 1183–1190, 2012.
40. Verma, R., Mantri, B., Kumar Srivastava, A., Shape control synthesis, characterizations, mechanisms and optical properties of large scaled metal oxide nanostructures of ZnO and TiO_2. *Adv. Mater. Lett.*, 6, 4, 324–333, 2015.
41. Gan, Y.X., Jayatissa, A.H., Yu, Z., Chen, X., Li, M., Hydrothermal synthesis of nanomaterials. *J. Nanomater.*, 2020, 1–3 , 2020.
42. Ye, J., Zhang, H., Yang, R., Li, X., Qi, L., Morphology-controlled synthesis of SnO_2 nanotubes by using 1D silica mesostructures as sacrificial templates and their applications in lithium-ion batteries. *Small*, 6, 2, 296–306, 2010.
43. Zhou, M., Gao, Y.A., Wang, B., Rozynek, Z., Fossum, J.O., Carbonate-assisted hydrothermal synthesis of nanoporous CuO microstructures and their application in catalysis, 2010.
44. Van Bui, H., Nguyen, H.N., Hoang, N.N., Truong, T.T., Pham, V.B., Optical and magnetic properties of Mn-doped ZnS nanoparticles synthesized by a hydrothermal method. *IEEE Trans. Magn.*, 50, 1–4, 2014.
45. Kim, M., Son, W.S., Ahn, K.H., Kim, D.S., Lee, H.S., Lee, Y.W., Hydrothermal synthesis of metal nanoparticles using glycerol as a reducing agent. *J. Supercrit. Fluids*, 90, 53–59, 2014.
46. Lee, G.H., Moon, H., Kim, H., Lee, G.H., Kwon, W., Yoo, S., Myung, D., Yun, S.H., Bao, Z., Hahn, S.K., Multifunctional materials for implantable and wearable photonic healthcare devices. *Nat. Rev. Mater.*, 5, 2, 149–165, 2020.
47. Ladpli, P., Nardari, R., Rewari, R., Liu, H., Slater, M., Kepler, K., Wang, Y., Kopsaftopoulos, F., Chang, F.K., Multifunctional energy storage composites: design, fabrication, and experimental characterization, in: *Energy Sustainability*, vol. 50237, p. V002T01A004, American Society of Mechanical Engineers, New York, 2016, June.

48. Shirshova, N., Qian, H., Shaffer, M.S., Steinke, J.H., Greenhalgh, E.S., Curtis, P.T., Kucernak, A., Bismarck, A., Structural composite supercapacitors, in: *Composites Part A: Applied Science and Manufacturing*, vol. 46, pp. 96–107, 2013.
49. Staaf, L.G.H., Lundgren, P., Enoksson, P., Present and future supercapacitor carbon electrode materials for improved energy storage used in intelligent wireless sensor systems. *Nano Energy*, 9, 128–141, 2014.
50. Friend, C., Smart materials: the emerging technology. *Mater. World*, 4, 16–18, 1996.
51. Halonen, N., Pálvölgyi, P.S., Bassani, A., Fiorentini, C., Nair, R., Spigno, G., Kordas, K., Bio-based smart materials for food packaging and sensors–a review. *Front. Mater.*, 7, 82, 2020.
52. Arrieta, M.P., Peponi, L., López, D., López, J., Kenny, J.M., An overview of nanoparticles role in the improvement of barrier properties of bioplastics for food packaging applications. *Food packaging*, 1, 391–424, 2017.
53. Liu, C., Luan, P., Li, Q., Cheng, Z., Xiang, P., Liu, D., Hou, Y., Yang, Y., Zhu, H., Biopolymers derived from trees as sustainable multifunctional materials: a review. *Adv. Mater.*, 33, 28, 2001654, 2021.
54. Huang, S., Liu, X., Chang, C., Wang, Y., Recent developments and prospective food-related applications of cellulose nanocrystals: A review. *Cellulose*, 27, 2991–3011, 2020.
55. Ali, A. and Andriyana, A., Properties of multifunctional composite materials based on nanomaterials: A review. *RSC Adv.*, 10, 28, 16390–16403, 2020.
56. Salonitis, K., Pandremenos, J., Paralikas, J., Chryssolouris, G., Multifunctional materials used in automotive industry: A critical review, in: *Engineering Against Fracture: Proceedings of the 1st Conference*, Springer Netherlands, pp. 59–70, 2009.
57. Carle, D. and Blount, G., The suitability of aluminium as an alternative material for car bodies. *Mater. Des.*, 20, 267–272, 1999.
58. Garces, J.M., Moll, D.J., Bicerano, J., Fibiger, R., McLeod, D.G., Polymeric nanocomposites for automotive applications. *Adv. Mater.*, 12, 23, 1835–1839, 2000.
59. Liu, H., Ding, W., Lei, S., Tian, X., Zhou, F., Selective adsorption of CH4/N2 on Ni-based MOF/SBA-15 composite materials. *Nanomaterials*, 9, 2, 149, 2019.
60. Arefi, M., Zamani, M.H., Kiani, M., Size-dependent free vibration analysis of three-layered exponentially graded nanoplate with piezomagnetic face-sheets resting on Pasternak's foundation. *J. Intell. Mater. Syst. Struct.*, 29, 5, 774–786, 2018.
61. Gibson, R.F., A review of recent research on mechanics of multifunctional composite materials and structures. *Compos. Struct.*, 92, 12, 2793–2810, 2010.
62. Xiao, Y., Qiao, W., Fukuda, H., Hatta, H., The effect of embedded devices on structural integrity of composite laminates. *Compos. Struct.*, 153, 21–29, 2016.
63. Behl, M., Razzaq, M.Y., Lendlein, A., Multifunctional shape-memory polymers. *Adv. Mater.*, 22, 31, 3388–3410, 2010.

9

Polymer Emulsions, Surface, and Interface

Bharti N. Naik[1], Subhalaxmi Pradhan[2*] and Chandu S. Madankar[1†]

[1]*Department of Oils, Oleochemicals and Surfactants Technology Institute of Chemical Technology, Matunga, Mumbai, India*
[2]*Division of Chemistry, School of Basic Sciences, Galgotias University, Greater Noida, Uttar Pradesh, India*

Abstract
Polymeric emulsions play a pivotal role in diverse industrial applications. Comprising polymer particles dispersed in a continuous liquid phase, these emulsions offer unique properties, including stability, flexibility, and controlled release. The synthesis of polymeric emulsions involves the emulsification of monomers followed by polymerization, yielding tailored materials with tunable characteristics. These materials find applications in coatings, adhesives, textiles, and biomedical fields, showcasing their adaptability. Advances in emulsion polymerization techniques enable the precise control of particle size and morphology, influencing the final product's performance. As sustainable alternatives gain importance, the development of eco-friendly polymeric emulsions using bio-based monomers reflects ongoing research trends. Understanding the synthesis, properties, and applications of polymeric emulsions is crucial for optimizing their performance in various industries.

Keywords: Emulsion, polymer emulsion, surface, interface, adhesives, textile

9.1 Introduction

Emulsion polymerization is a highly significant process in the production of various polymers used in numerous industrial applications. The process involves the polymerization of monomers dispersed in water, with the aid of surfactants and initiators to create stable emulsions. Some key reasons why emulsion polymerization is favored in various industries include environmental benefits, safety, ease of handling, product performance, versatility, and economic benefits [1].

Emulsion polymerization's versatility allows for the production of a wide range of polymer types with tailored properties, enabling their use in diverse industrial applications. Its ability to create stable, finely dispersed polymer particles in water-based systems contributes to the popularity of these materials in various industries [4]. This is the process of dispersing hydrophobic monomers in water using emulsifiers or surfactants. Emulsifiers are molecules

**Corresponding author*: subhalaxmipradhan@gmail.com
†*Corresponding author*: chandumadankar@gmail.com

Divya Bajpai Tripathy, Anjali Gupta and Arvind Kumar Jain (eds.) Multifunctional Materials: Engineering and Biological Applications, (225–244) © 2025 Scrivener Publishing LLC

that have both hydrophilic (water-attracting) and hydrophobic (water-repelling) parts. They stabilize the system by forming a layer around the monomer droplets, preventing them from coalescing. Once the monomer droplets are dispersed in water, the initiation step involves the activation of free radicals that will start the polymerization reaction. Initiators in emulsion polymerization act to generate free radicals through two primary mechanisms: thermal decomposition and redox (oxidation-reduction) reactions [3]. In emulsion polymerization, initiators can be either water-soluble or oil-soluble. Water-soluble initiators, like sodium persulfate (NaPS), can generate radicals in the aqueous phase, while oil-soluble initiators, such as 2,2′-azobisisobutyronitrile (AIBN), function in the hydrophobic regions of the emulsion. The activated free radicals generated by the initiators attack the double bonds in the monomer molecules, leading to chain growth as monomers add to the growing polymer chains. Polymerization continues until chain termination occurs, which can happen through various mechanisms like recombination of radicals or combination with certain functional groups, resulting in the end of the polymerization process [2].

The selection of the emulsion type depends on the intended application and desired properties. The ratio of oil to water, the nature and concentration of surfactants or stabilizers, the method of emulsion preparation (such as mixing techniques, shear rates, and temperatures), and particle sizes within a given sample influence the type of emulsion formed [1]. It is a crucial parameter in emulsion polymerization as it influences the properties of the resulting latex, affecting characteristics such as stability, viscosity, film formation, and mechanical properties [1]. Controlled radical polymerization methods have revolutionized the synthesis of polymers by enabling the production of materials with well-defined characteristics in terms of molar mass, composition, and architecture. This stands in contrast to traditional free radical polymerization methods, which often result in polymers with broad molecular weight distributions and less precise compositions [5]. The present chapter focused on the classification and properties of emulsions, the role of emulsions on surface chemistry, the synthesis and characterization of polymeric emulsions, and their real-life applications.

9.2 Emulsion, Types of Emulsions, and Properties

An emulsion is a combination of two immiscible liquids, usually consisting of one liquid spread as minute droplets inside another. The dispersed phase (the liquid that creates the droplets) and the continuous phase (the liquid in which the droplets are dispersed) are the two basic components of an emulsion. The inclusion of emulsifying agents, which might be surfactants or other chemicals that lessen the surface tension between the two liquids, maintains the stability of an emulsion.

9.2.1 Classification of Oil Emulsions

The three basic forms of emulsions are water as dispersed phase in oil as continuous phase; Water-in-oil (W/O) emulsion, similarly oil-in-water (O/W), and complex emulsions [6, 7].

Table 9.1 Various states of stability in water-in-oil mixtures [12].

Sr. no.	State of emulsion	Water content (%)	Appearance
1	Stable	60–80	Brown
2	Mesostable	-	Black/brown
3	Unstable	<10	-
4	Entrained water	10–30	Black

9.2.1.1 Water-in-Oil Emulsions (W/O)

A water-in-oil emulsion consists of water-insoluble materials present in water as a dispersed phase [8]. Over 95% of crude oil emulsions produced in the oil sector are of the W/O type [9]. The three critical ingredients necessary for a stable water-in-oil emulsion are solvent, surfactant, and water [10]. Research indicates that stability is the most vital property of an emulsion. Resins and asphaltenes present in compounds act as natural surfactants, playing an important role in maintaining the stability of an emulsion [11]. Classification of water-in-oil mixtures into four states: stable, mesostable, unstable, and entrained water with properties is depicted in Table 9.1.

9.2.1.2 Oil-in-Water Emulsions (O/W)

In this type of emulsion, oil is the dispersed phase and the water is the dispersion medium or continuous phase. Dealing with emulsions in the petroleum sector should be done in a very appropriate manner to avoid massive losses [13]. However, water-in-oil emulsions are more common. The term reversal emulsion is used for oil-in-water emulsion.

9.2.1.3 Multiple Emulsions

In some situations, water-in-oil-in-water (W/O/W) and oil-in-water-in-oil (O/W/O) can be discovered. Multiple emulsions are very complex in nature as small droplets of oil are trapped in big droplets of oil and are dispersed in water (continuous phase), in the case of W/O/W emulsions. This complex mixture is stabilized by a mix of hydrophilic and hydrophobic surfactants. These emulsions need at least two emulsifiers in the system, one with a low and the other with a high HLB [14–17].

9.2.2 Properties of Emulsion

Emulsions exhibit several properties that distinguish them from other types of mixtures. The key properties of emulsions include:

- Immiscibility of Components: Emulsions involve two immiscible liquids, typically oil and water. These liquids would not normally dissolve in each other.

- Dispersed and Continuous Phases: Emulsions consist of a dispersed phase (small droplets of one liquid) and a continuous phase (the medium in which the droplets are dispersed).
- Stability: Stability is a critical property of emulsions. Without the use of emulsifying agents, the droplets of the dispersed phase would tend to coalesce and separate from the continuous phase over time.
- Emulsifying Agents: Emulsifying agents, such as surfactants, are substances that help stabilize emulsions by reducing the surface tension between the immiscible liquids. They prevent droplet coalescence and aid in the formation and maintenance of the emulsion.
- Particle Size Distribution: Emulsions have a characteristic distribution of droplet sizes. The size of the droplets can influence the appearance, texture, and stability of the emulsion.
- Opacity or Translucency: The presence of small droplets in an emulsion can affect its appearance. O/W emulsions are generally translucent, while W/O emulsions can be more opaque.
- Rheological Properties: Emulsions can exhibit specific flow properties, such as viscosity and thixotropy, which can impact their performance in various applications, such as food processing or cosmetics.
- pH Sensitivity: The stability of some emulsions can be influenced by the pH of the system. Changes in pH may lead to the breakdown of the emulsion.
- Temperature Sensitivity: Temperature can affect the stability of emulsions. Some emulsions may undergo phase separation or changes in viscosity with variations in temperature.
- Application-Specific Properties: Emulsions are widely used in various industries, and their properties can be tailored for specific applications. For example, food emulsions may be designed for texture, appearance, and flavor stability, while cosmetic emulsions may focus on skin feel and stability.

9.2.3 Types of Emulsion Polymerization

9.2.3.1 Miniemulsion Polymerization (Nanoemulsion)

There is a substantial difference between the nucleation of particles, growth processes, and dynamics of miniemulsion, microemulsion, and traditional emulsion polymerizations. This is mostly due to the various starting circumstances (i.e., the conditions immediately preceding the commencement of polymerization), which have a significant impact on the following particle nucleation and growth processes. The most commonly acknowledged particle nucleation processes in traditional emulsion polymerization are micellar nucleation [18–23] and homogeneous nucleation [24–26]. Due to low surface area, monomer droplets do not participate in particle nucleation. However, when subjected to high-pressure homogenization, monomer droplets containing insoluble hydrocarbon compounds become the loci for particle nucleation. This is due to an increase in surface area due to homogenization and particle nuclei are generated to capture the radicals within the continuous aqueous phase. The monomer droplets contain lyophilic, less molecular weight components, which act as costabilizers and result in the slowing of monomer droplets being converted to bigger droplets. Due to this stable miniemulsion precipitations are formed, which are then polymerized by free radicals. This is a special kind of technology that has been labeled miniemulsion polymerization [27–32]. There are several new miniemulsion polymerizations that are constantly being explored for commercial applications.

9.2.3.2 Microemulsion Polymerization

The transparent reaction system composed of microemulsion drops is thermally stable and has an enormously large oil-water intersurface area. The system consisting of an anionic surfactant in conjunction with a cosurfactant is the most common.

The combination of amphipathic cosurfactant with the absorbed layer of negatively charged surfactant results in relatively stable polymer particles, which in turn reduce interfacial tension and increase membrane flexibility. The resulting polymer molecular weight is relatively high, with few polymer chains per particle [33–37]. This cannot be accomplished using emulsion or miniemulsion polymerization. Furthermore, the formation and growth of particle nuclei also differ significantly.

Research interest in polymerization techniques has gained attention since the 80's due to its applications in fine latex, flocculants, conducting polymers, the biomedical field of colloidal materials, photochemistry, and other applications [35].

There are several research data available on conventional emulsion polymerization [38–56], miniemulsion [32, 57–61], and microemulsion [34–37, 62] polymerization which gained more attention these days.

9.2.3.3 Inverse Emulsion Polymerization

It entails dispersion followed by polymerization of hydrophilic monomers in a nonaqueous continuous phase, generally in an aqueous solution. The emulsification arrangements are largely based on the steric stabilization mechanisms which differ significantly from those used in more traditional oil-in-water emulsion polymerization methods. This is simply due to the ineffectiveness of the electrostatic stabilization mechanism unstabilising inverse emulsion polymerization.

Bis(2 - ethylhexyl) sulfosuccinate is a unique negatively charged surfactant that is soluble in water and oil. This compound is applied for the stabilization of inverse emulsion or microemulsion polymerization techniques. In this water-in-oil kind of colloidal system, both oil-soluble and water-soluble initiators were utilized to begin the free radical chain polymerization. There are several research reported on the processes and dynamics of inverse emulsion polymerization [63–66].

Except for the inverse microemulsion polymerization of acrylamide, the field of inverse emulsion polymerization has received little attention. These acrylamide-based compounds are most commonly used as polymeric flocculants in water treatment. The high molecular weight of the resulting polymer and the formation of colloid which aids in quick polymer dissolution in water are two significant benefits of this polymerization technique.

9.3 Role of Emulsion in Surface Chemistry

Emulsions formed without the use of surfactants might initially appear mixed but are generally unstable and tend to separate into distinct layers based on the densities of the involved phases. The addition of surfactants or other stabilizing agents is indeed crucial to maintaining the stability of the emulsion over extended periods. Overall, emulsions play a significant role in various industrial sectors due to their ability to combine immiscible components for specific purposes while maintaining stability through appropriate formulation and stabilization techniques [67].

Surface chemistry involves studying the behavior and interactions that occur at the interfaces between different phases, such as between solids, liquids, and gasses. Adsorption, catalysis, and colloids are key aspects of surface chemistry that play vital roles in various fields. Adsorption is a crucial phenomenon wherein molecules or atoms from a gas or liquid phase adhere to the surface of a solid or liquid, forming a thin film. This accumulation of molecules at the surface, rather than penetrating into the bulk of the material, characterizes adsorption. The substance or molecules that concentrate or accumulate at the surface are referred to as the adsorbate. The material or surface on which this accumulation occurs is known as the adsorbent. Various factors influence adsorption, including the nature of the adsorbate and adsorbent, surface area, temperature, pressure, and the presence of other substances in the system [68]. The interfacial tension at an emulsion droplet interface does not necessarily remain constant at every point along the interface. Interfacial tension refers to the force acting per unit length between two immiscible phases at their interface. In an emulsion, the interfacial tension between the dispersed phase (such as oil) and the continuous phase (like water) can vary due to factors such as the nature of the phases involved, the presence of surfactants or stabilizers, temperature, and other environmental conditions [69]. General aspects of emulsions, such as the role of surfactants in stabilizing them, the tendency for immiscible phases to separate over time without stabilization, and their various industrial applications, remain applicable across different size classifications [68, 69].

The molar heat of adsorption is an important thermodynamic property that provides insight into the strength of the interaction between the adsorbate molecules and the adsorbent surface. It influences the efficiency of adsorption processes and can vary depending on factors such as the nature of the adsorbate and adsorbent, surface properties, temperature, and pressure [68].

Surface modification techniques play a vital role in modifying the properties of polymers by focusing on their surface characteristics without changing their inherent chemical composition. Some important polymeric techniques are given in Table 9.2. Physical methods such as injection molding, extrusion, lamination, and template techniques are employed to modify the surface of polymers [70].

Emulsions are inherently thermodynamically unstable and prone to eventual separation due to the immiscibility of their components. The properties of emulsions can change over time, leading to instability. Thermodynamic stability, which assesses the equilibrium between the phases, and kinetic stability, which deals with the rate of change in emulsion properties over time, is crucial for predicting and managing emulsion behavior. The mechanisms behind emulsion stability are explored by considering both thermodynamic factors such as free energy changes, phase equilibria, and kinetic aspects such as droplet

Table 9.2 Some important polymer emulsion techniques [69].

Sr. no.	Polymer emulsion techniques
1	Microemulsion
2	Miniemulsion
3	Spontaneous emulsion
4	Surfactant alcohol mixture
5	Swelling and polymerization

coalescence or breakdown over time. This comprehensive understanding helps in designing emulsions with desired properties for various applications and in developing strategies to enhance their stability and shelf life [67].

9.4 Polymeric Emulsion, Types, and Their Functions

Polymeric emulsions are materials made up of polymers dispersed in a liquid medium, typically water. They find applications in various industries like paints, coatings, adhesives, cosmetics, and more due to their versatile properties [69]. There are several types of polymeric emulsions, each serving different functions.

9.4.1 Acrylic Emulsion

The emulsion polymerization of acrylic monomers has gained popularity in various industries due to its economical and environmentally friendly nature. Acrylic latexes produced through this method are utilized in coatings, adhesives, paper finishes, and other applications. Typically, stable latexes are created with low solid content (around 30–40%) to prevent particle agglomeration, which can cause colloidal instability. However, there is a growing interest in developing high-solid-content latexes with around 60% polymer content. These high-solid-content latexes offer several advantages, including reduced transportation costs, shorter preparation times, and quicker film curing times. Despite these benefits, achieving high-solid-content latexes has posed challenges [71].

9.4.2 Styrene-Butadiene Emulsion

At low levels of SBR admixture, the polymer's ability to enhance adhesion between the two-phase materials (like OPC–CAC–G mortar) is limited. The viscosity and deformability of the polymer might not be sufficient to create a significant bridging effect between the materials, resulting in lower bond strength. Conversely, when the polymer dosage is too high, an overly dense film of polymer can form. This dense film acts as a barrier preventing effective contact between the hydration products and the concrete matrix. As a result, the mechanical engagement force weakens due to this hindrance, adversely impacting bond strength. Excessive amounts of polymer admixture can significantly retard the hydration reaction of cement. This retarding effect interferes with the formation of the required C-S-H (calcium silicate hydrate) gel and disrupts the bridging between these gels, affecting the microstructure and compactness of hydration products. This disruption can further contribute to reduced bond strength. Additionally, the polymer, when present in excessive amounts, might fail to act as a rigid support. This lack of proper support affects the overall strength of the material [71].

9.4.3 Vinyl Acetate Emulsions

VAc (Vinyl acetate) is a crucial monomer in the production of waterborne dispersions. It finds extensive use in applications such as adhesives, interior decorative coatings, paper coatings, and re-dispersible powders. Its ability to form stable dispersions in water makes it valuable in various formulations where water-based systems are preferred. These alkoxysilane monomers, VTMS (Vinyl trimethyl silane), and VTES (Vinyl triethyl silane) serve as coupling agents, adhesion promoters, and crosslinking agents. They possess reactive silane groups that can chemically bond with surfaces, enhancing adhesion and promoting crosslinking in various materials [72]. There are several applications of VAc emulsions such as adhesives, coatings, and concrete adhesives. The addition of vinyl alkoxysilane monomers to VAc-based adhesives can improve adhesion properties by enhancing bonding to different surfaces [72]. In coatings, the incorporation of these alkoxysilane monomers with VAc can enhance the coating's adhesion to substrates, improve durability, and provide water and chemical resistance [72]. In the realm of concrete, the combination of VAc and alkoxysilane monomers has been explored as additives. They can potentially improve concrete performance by enhancing adhesion between cementitious materials and aggregates, promoting better bonding, and offering increased durability [72].

9.4.4 Polyurethane Emulsion

Polyurethane is indeed a widely used material in various industries due to its excellent properties such as handle ability, functional versatility, and relatively low cost compared to other materials.

- Utilization of Toxic Solvents: The preparation process of synthetic polyurethane often involves the use of toxic solvents. These solvents are used in various stages of production, including polymer synthesis, processing, and application. Exposure to these solvents can pose health risks to workers involved in the manufacturing process and may also have adverse environmental impacts if not handled properly [73].
- Petrochemical-Based Raw Materials: Most synthetic polyurethanes are derived from petrochemical-based raw materials. The reliance on fossil fuels not only contributes to the depletion of non-renewable resources but also raises concerns regarding carbon emissions and environmental sustainability [73].
- Health and Environmental Hazards: The use of toxic solvents and petrochemical-based raw materials in polyurethane production can lead to

environmental pollution, air and water contamination, and potential health hazards for workers, consumers, and ecosystems [73].

9.4.5 Epoxy Emulsions

Epoxy coatings are widely recognized for their effectiveness in protecting metallic surfaces against corrosion and wear. They offer various advantages, including:

- Excellent Mechanical Properties: Epoxy coatings exhibit good mechanical properties such as hardness, toughness, and abrasion resistance, making them suitable for protecting surfaces in demanding environments.
- High Chemical Resistance: They possess excellent resistance to a wide range of chemicals, which enhances their suitability for use in harsh industrial settings where exposure to corrosive substances is common.
- Strong Adhesion: Epoxy coatings form strong bonds with metallic substrates, ensuring robust adhesion and long-term protection against corrosion and degradation.
- Electrical Insulation: They have excellent electrical insulating properties, making them suitable for applications where electrical insulation is required.

Research and development in this area aim to further improve the performance, durability, and application characteristics of waterborne epoxy coatings to meet industry standards while minimizing environmental impact [74].

9.4.6 Functions of Polymeric Emulsions

- Binding: Polymeric emulsions play a major role during binding and adhesion and it has good water resistance, flexibility, and other enhanced properties. Emulsions bind together components in various formulations, providing structural integrity and cohesion [76]. They enhance the adhesion of coatings, paints, or adhesives to substrates, improving their performance and durability [75]. Polymeric emulsions can impart water resistance to coatings and adhesives, protecting surfaces from moisture damage [76]. Many emulsions offer flexibility, which is crucial in applications where materials need to expand, contract, or withstand movement without cracking or breaking [75]. Some emulsions provide resistance to chemicals, making coatings or adhesives suitable for environments where exposure to chemicals is expected [76]. They can enhance properties like UV resistance, hardness, gloss, or matte finish in coatings and paints [75]. Understanding the types and functions of polymeric emulsions helps industries choose the most suitable emulsion for their specific application needs, ensuring the desired performance and quality of their products [75].

9.5 Preparation Method and Characterization of Polymer Emulsions

The preparation of polymer emulsions involves the emulsification of a monomer or preformed polymer in water with the aid of surfactants and other stabilizing agents. Using water as the dispersion medium in making functional polymers through the emulsion polymerization method offers several advantages, especially in addressing environmental concerns related to volatile organic compounds [69]. The materials required for the preparation of polymer emulsions are monomers or preformed polymers, water, surfactants/emulsifiers, stabilizers, initiators (for polymerization if using monomers), reaction vessel, stirrer/agitator, and heat source (if needed). The mentioned materials are frequently used in the formulation of various solutions, coatings, pharmaceuticals, and other applications due to their specific properties and functionalities [78].

9.5.1 Methods of Preparation

Polymeric nanoparticles (PNPs) have emerged as versatile and valuable tools for various applications due to their unique molecular designs, which can be integrated into nanospheres or nanocapsules. These nanoparticles are employed in diverse fields such as medicine, electronics, conducting materials, photonics, environmental technology, sensor systems, and pollution control [77].

a) **Polymerization Reaction or Nanoparticle Formation from Macromolecules**
This method involves synthesizing nanoparticles through polymerization reactions or transforming macromolecules into nanoparticles. Various polymerization techniques like emulsion polymerization, mini-emulsion, or micro-emulsion polymerization are used to create PNPs [77].
b) **Ionic Gelation Method**
In this technique, nanoparticles are formed through ionic interactions between oppositely charged molecules, leading to the formation of stable polymeric structures. For instance, polyelectrolyte complexes formed via interactions between polymers with opposite charges can be employed to create nanoparticles [77].

Table 9.3 Examples of some natural and synthetic polymers.

Sr. no.	Natural polymer	Synthetic polymer
1	Chitosan	Polylactides
2	Gelatin	Polyglycolides
3	Sodiun alginate	PLGA (polylactide-co-glycolide)
4	Albumin	Polyunhydride

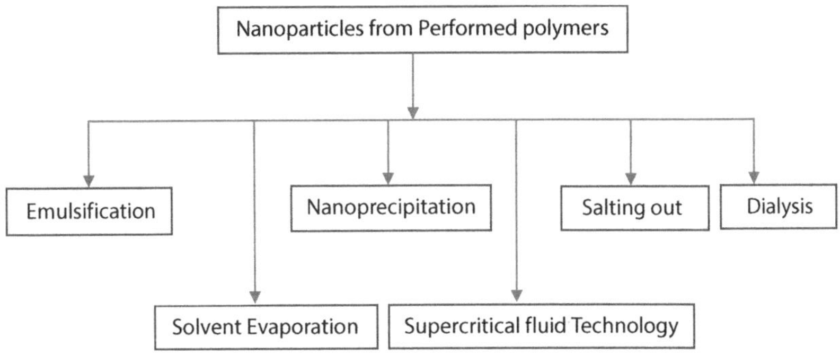

Figure 9.1 Various polymerization techniques.

9.5.2 Classification of Polymeric Emulsions

Polymeric emulsions are classified into different types such as Natural polymeric emulsions and synthetic polymeric emulsions and the example is given in Table 9.3. Different polymerization technique is represented in Figure 9.1.

9.5.3 Characterization of Polymer Emulsions

- Particle Size Analysis: Determine the size distribution of the polymer particles using techniques like dynamic light scattering (DLS), laser diffraction, or microscopy [79].
- Stability Testing: Assess the stability of the emulsion over time by monitoring changes in particle size, appearance, and phase separation [67].
- Emulsion Viscosity: The viscosity of an emulsion is a critical parameter that determines its flow behavior and stability. When examining the viscosity of an emulsion concerning shear rate, it is essential to understand how the emulsion responds to applied forces [78].
- Chemical Composition Analysis: Employ techniques such as spectroscopy (FTIR, NMR), chromatography (GC-MS, HPLC), or elemental analysis to identify and quantify the components within the emulsion [80].
- Zeta Potential Measurement: Determine the surface charge of the polymer particles to evaluate their stability and potential interactions [81].
- Rheological Analysis: Evaluate the emulsion's mechanical properties, such as elasticity and deformation behavior, using a rheometer [82].
- Morphological Characterization: Utilize electron microscopy (SEM or TEM) to visualize the morphology and structure of the polymer particles in the emulsion. This method allows for a detailed understanding of particle size distribution and morphology at the nanoscale. TEM coupled with image analysis software provides quantitative data that can be used to describe and characterize the particles present in the sample, offering valuable insights for various research and industrial applications in fields such as materials science, nanotechnology, pharmaceuticals, and more [79].

9.6 Surface and Interface Characterization of Polymer Emulsion

Surface and interface characterization of polymer emulsions involves analyzing and understanding the properties, structures, and interactions occurring at the surface and interfaces of these complex systems. Several techniques are employed to study these aspects:

a) Surface Tension and Contact Angle Measurement:
Surface tension measurements provide insights into the surface energy of the emulsion, affecting its wetting behavior [78]. Contact angle measurements help to understand the interaction between the emulsion droplets and various substrates, indicating surface hydrophobicity or hydrophilicity [78].
b) Atomic Force Microscopy (AFM):
AFM allows for high-resolution imaging of surfaces at the nanoscale, providing topographical information and insights into surface roughness, particle size, and morphology [81].
c) Scanning Electron Microscopy (SEM):
SEM enables imaging of the emulsion's surface and interfaces at higher magnifications, providing detailed structural information about particles, their distribution, and surface morphology [83].
d) Transmission Electron Microscopy (TEM):
TEM, especially in the case of polymer nanoparticles within the emulsion, offers high-resolution imaging of nanoscale structures, providing details about particle size, shape, and aggregation [78].
e) X-ray Photoelectron Spectroscopy (XPS):
XPS analyzes the elemental composition and chemical states of the emulsion's surface layers, providing information about surface functional groups and chemical bonding [83].
f) Fourier Transform Infrared Spectroscopy (FTIR):
FTIR analyzes molecular vibrations, identifying functional groups present at the emulsion's surface and interfaces, aiding in chemical characterization [83].
g) Dynamic Light Scattering (DLS):
DLS measures particle size distribution and polydispersity of the emulsion, helping characterize the colloidal stability and aggregation behavior [79].
h) Zeta Potential Analysis:
Zeta potential measurements evaluate the surface charge of particles within the emulsion, indicating stability and interparticle interactions [81].
i) Electron Spectroscopy for Chemical Analysis (ESCA):
ESCA provides a detailed chemical analysis of the surface, identifying elements and compounds present at the interface of the emulsion [84].

By employing these surface and interface characterization techniques, researchers gain comprehensive insights into the physical, chemical, and structural properties of polymer emulsions. This information is crucial for optimizing formulations, understanding stability, controlling properties like adhesion or barrier properties, and tailoring emulsions for specific applications in industries ranging from coatings and paints to pharmaceuticals and cosmetics [78].

9.7 Applications of Polymeric Emulsions

Emulsions play a crucial role in various industries, including food, pharmaceuticals, cosmetics, and paints. In addition to the basic types of emulsions, there are also multiple variations and special cases of emulsions and concentrated emulsions. The stability of emulsions can be influenced by factors like temperature, pH, and the presence of stabilizers.

Nanotechnology is used in the development of innovative materials as a useful tool in many disciplines of research. Material manipulation at the nanoscale scale, on the other hand, allows for the creation of new products with unparalleled performance and increased attributes [85]. In general, there are three primary matrix types in composite systems: polymer, metal, and ceramic, with diverse additives in various forms such as lamina, fillers, fiber (short and continuous), flake, and particles. Nanocomposites are characterized in this sector as having at least one component at a nanometric scale [86].

Polymers are far more complicated than metals or ceramics among the commonly used matrices, but they are also much cheaper and easier to process. Furthermore, polymers have a reduced modulus and strength when used at lower temperatures [87]. As a result, these materials' properties with the absence of necessary attributes must be improved. Nanofillers are used as filler components in PNCs (polymeric nanocomposites) and are classified according to their chemical origin, physical structure, and particle morphologies. These particles are represented by geometries and nanodimension as linear, two-dimensional, and layered [86]. Several researchers have studied the application of PNCs in various domains, but the primary issue is selecting the PNC-generating procedure based on the final anticipated features [88–94].

Emulsion-based coatings and adhesives are becoming more popular as people become more aware of the health and safety risks associated with solvent-based polymer production methods. However, as compared to solvent-based procedures, emulsion-based approaches sometimes require more research to attain equivalent or greater application performance. The use of nanoparticles in emulsion-based coatings and adhesives is a viable method of improving performance.

Gloss, adhesion, hardness, stability toward moisture and heat, and flexibility are typical requirements for coatings [95]. Adhesives must be capable of binding to rough surfaces and then sustain a minimal amount of stress when debonded [96]. Because water is used as a synthesis medium rather than organic solvents, emulsion-based paints and adhesives are becoming more economically relevant. Although they are less harmful to the environment, emulsion-based goods frequently require property upgrading by hazardous procedures.

Emulsion-based coatings form a soft film on curing; therefore, solvent-based coatings have better gloss and hardness due to the formation of cross-links [97]. Solvent-based pressure-sensitive adhesives (PSAs) create a continuous gel network more easily [98].

Nano-ZnO particles have high luminous transmittance and intense UV absorption. Nano-ZnO particles are included in polymers to increase heat resistance as well as mechanical and optical qualities for coating applications [99]. The UV shielding characteristics of nanocomposites containing 60 nm ZnO particles have superior properties.

Titanium dioxide (TiO_2) pigments are commonly used to give opacity and whiteness in paints, inks, and paper due to effective reflection or refraction of light compared to other pigments [100]. Strong UV-absorbing property increases the longevity of TiO_2-based exterior paints or coatings. Furthermore, they are naturally harmless and may be used in a wide range of goods with no health or safety issues because of their great hardness (hardness on the Mohs scale is 9). Nano-Al_2O_3 particles are employed in numerous commercial applications such as abrasives, insulators, and coatings. They are generally used to increase the scratch and abrasion resistance of coatings [101, 102].

9.8 Conclusion

The study and understanding of polymer emulsions have significant implications across various industries, offering versatile solutions for the development of products ranging from paints and coatings to pharmaceuticals and personal care items. Polymer emulsions serve as crucial components in numerous everyday products, including paints, adhesives, coatings, cosmetics, pharmaceuticals, and food items. Their versatility makes them indispensable across diverse industries. Understanding and controlling the critical parameters like phase ratio, concentration of emulsifiers, agitation, and temperature are pivotal for ensuring the longevity and performance of emulsion-based products. Emulsion preparation methods, such as polymerization reactions or ionic gelation, play a crucial role in determining the characteristics and properties of polymer nanoparticles, whether in the form of nano-capsules or nanospheres. Employing advanced characterization techniques like microscopy (TEM, SEM), spectroscopy (FTIR, XPS), rheology analysis, and surface tension measurements provide in-depth insights into the structural, morphological, and chemical properties of polymer emulsions. Surfactants play a pivotal role in stabilizing polymer emulsions by forming a protective layer around droplets, preventing coalescence, and ensuring long-term stability. The industrial relevance of polymer emulsions lies in their ability to enhance product performance, offer controlled release of active ingredients, improve coating and adhesion properties, and facilitate the development of innovative materials for various applications.

References

1. Fresco-Cala, B. and Cardenas, S., Advanced polymeric solids containing nano-and microparticles prepared via emulsion-based polymerization approaches. A review. *Anal. Chim. Acta*, 1208, 339669, 2022.
2. Chern, C.-S., *Principles and Applications of Emulsion Polymerization*, John Wiley & Sons, New Jersey USA, 2008.
3. El-hoshoudy, A.N.M.B., Emulsion polymerization mechanism, in: *Recent Research in Polymerization*, p. 1, 2018.
4. Zhang, S. *et al.*, Measurement methods of particle size distribution in emulsion polymerization. *Chin. J. Chem. Eng.*, 39, 1–15, 2021.
5. Herrera-Ordonez, J., Controversies on the mechanism and kinetics of emulsion polymerization: An updated critical review. *Adv. Colloid Interface Sci.*, 320, 103005, 2023.

6. Hoshyargar, V., Marjani, A., Fadaei, F., Shirazian, S., Prediction of flow behavior of crude oil-in-water emulsion through the pipe by using rheological properties. *Orient. J. Chem.*, 28, 109–113, 2012.
7. Auflem, I.H., *Influence of Asphaltene Aggregation and Pressure on Crude Oil Emulsion Stability*, Norwegian University of Science and Technology, Trondheim, Norway, 2002.
8. Lei J, Gao Y, Hou X, Sheng Z, Zhang C, Du F. A simple and effective strategy to enhance the stability and solid-liquid interfacial interaction of an emulsion by the interfacial dilational rheological properties. *Soft Matter.*, 24, 16, 24, 5650–5658, 2020.
9. Ali, M.F. and Alqam, M.H., Role of asphaltenes, resins and other solids in the stabilization of water in oil emulsions and its effects on oil production in Saudi oil fields. *Fuel*, 79, 1309–1316, 2000.
10. B.B. Filippini, J. John, A. Deborah, *Water-in-Oil Emulsion*, 2004, US Patent 2004/0176263 A1.
11. Fridjonsson, E.O., Graham, B.F., Akhfash, B.F., May, E.F., Johns, M.L., Optimized Droplet Sizing of Water-in-Crude Oil Emulsions Using Nuclear Magnetic Resonance. *Energy Fuels*, 28, 1756–1764, 2014.
12. Fingas, M. and Fieldhouse, B., Formation of water-in oil emulsions and application to oil spill modelling. *J. Hazard. Mater.*, 107, 37–50, 2004.
13. Abdel-Raouf, M.E.-S. (ed.), Factors affecting the stability of crude oil emulsions, in: *Crude Oil Emuls. Stab. Charact*, pp. 183–204, 2012.
14. Khan, A.Y., Talegaonkar, S., Iqbal, Z., Ahmed, F.J., Khar, R.K., Multiple emulsions: an overview. *Curr. Drug Deliv.*, 3, 429–443, 2006.
15. Muschiolik, G., Multiple emulsions for food use. *Curr. Opin. Colloid Interface Sci.*, 12, 213–220, 2007.
16. Pal, R., Rheology of simple and multiple emulsions. *Curr. Opin. Colloid Interface Sci.*, 16, 41–60, 2011.
17. Jiao, J. and Burgess, D.J., Rheology and stability of water-in-oil-in-water multiple emulsions containing Span 83 and Tween 80. *AAPS PharmSci.*, 5, 12, 2003.
18. Harkins, W.D., A general theory of the reaction Ioci in emulsion polymerization. *J. Chem. Phys.*, 13, 381, 1945.
19. Fortuny, M., Graillat, C., McKenna, T.F., Araújo, P.H., Pinto, J.C. Modeling the nucleation stage during batch emulsion polymerization. *Aiche Journal*, 51, 2521–2533, 2005.
20. Harkins, W.D., A general theory of the mechanism of emulsion polymerization. *J. Chem. Phys.*, 13, 381, 1947.
21. Smith, W.V., The Kinetics of styrene emulsion polymerization. *J. Am. Chem. Soc.*, 70, 3695, 1948.
22. Smith, W.V. and Ewart, R.H., The Kinetics of emulsion polymerization. *J. Chem. Phys.*, 16, 592, 1948.
23. Smith, W.V., Chain initiation in styrene emulsion polymerization. *J. Am. Chem. Soc.*, 71, 4077, 1949.
24. Priest, W.J., Partice growth in the aqueous polymerization of vinyl acetate. *J. Phys. Chem.*, 56, 1977, 1952.
25. Roe, C.P., Surface chemistry aspects of emulsion polymerization. *Ind. Eng. Chem.*, 60, 20, 1968.
26. Fitch, R.M. and Tsai, C.H., Particle formation in polymer colloids, III: Prediction of the number of particles by a homogeneous nucleation theory, in: *Polymer Colloids*, R.M. Fitch (Ed.), Plenum Press, New York, 73, New York, 1971.
27. Fitch, R.M. and Tsai, C.H., Homogeneous nucleation of polymer colloids. IV: The role of soluble oligomeric radicals, in: *Polymer Colloids*, R.M. Fitch (Ed.), p. 103, Plenum Press, New York, 1971.
28. Fitch, R.M., The homogenours nucleation of polymer colloids. *Br. Polym. J.*, 5, 467, 1973.

29. Ugelstad, J., El - Aasser, M.S., Vanderhoff, J.W., Initiation of polymerization in monomer droplets. *J. Polym. Sci.: Polym. Lett. Ed.*, 11, 503, 1973.
30. Ugelstad, J., Hansen, F.K., Lange, S., Emulsion polymerization of styrene with sodium hexadecyl sulphate/hexadecanol mixtures as emulsifiers. Initiation in monomer droplets. *Die Makromol. Chem.*, 175, 507, 1974.
31. Durbin, D.P., El-Aasser, M.S., Poehlein, G.W., Vanderhoff, J.W., Influence of monomer preemulsification on formation of particles from monomer drops in emulsion polymerization. *J. App. Polym. Sci.*, 24, 3, 703–7, 1979.
32. Chamberlain, B.J., Napper, D.H., Gilbert, R.G., Polymerization within styrene emulsion droplets. *J. Chem. Soc., Faraday Trans.*, 1, 78, 591, 1982.
33. Choi, Y.T., El-Aasser, M.S., Sudol, E.D., Vanderhoff, J.W., Polymerization of styrene miniemulsions. *J. Polym. Sci.: Polym. Chem. Ed.*, 23, 2973, 1985.
34. Capek, I. and Chern, C.S., Radical polymerization in direct mini-emulsion systems. *New Poly. Tech. Syn. Meth.*, 155, 101–65, 2001.
35. Mark, H.F., Bikales, N.M., Overberger, C.G., Mengers, G., *Encyclopedia of Polymer Science and Engineering*, p. 718, Wiley, New York, 1987.
36. Paleos, C.M. (Ed.), *Polymerization in Organized Media, Gordon and Breach*, p. 215, Chapter 4, Kluwer Academic Publishers, 1992.
37. Kumar, P. and Mittal, K.L. (Eds.), *Handbook of Microemulsion Science and Technology*, p. 679, Marcel Dekker, New York, 1999.
38. Capek, I., Radical polymerization of polar unsaturated monomers in direct microemulsion systems. *Adv. Colloid Interface Sci.*, 80, 85, 1999.
39. Poehlein, G.W. and Dougherty, D.J., Continuous emulsion polymerization. *Rubber Chem. Technol.*, 50, 601, 1977.
40. Li, B. and Brooks, B.W., Semi-batch Processes for Emulsion Polymerisation. *Polym. Int.*, 29, 41, 1992.
41. Šňupárek, J., Principles and limits of polymer latex tailoring. *Prog. Org. Coat.*, 29, 225, 1996.
42. Gao, J. and Penlidis, A., Mathematical modeling and computer simulator/database for emulsion polymerizations. *Prog. Polym. Sci.*, 27, 403, 2002.
43. Hubbard, A. (Ed.), *Encyclopedia of Surface and Colloid Science*, pp. 4220–41, Marcel Dekker, New York, 2002.
44. Nomura, M., Tobita, H., Suzuki, K., Emulsion polymerization: Kinetic and mechanistic aspects. *Adv. Polym. Sci.*, 175, 1, 2005.
45. Alexander, E. and Napper, D.H., Emulsion polymerization, in: *Progress in Polymer Science*, D.P. Jensen (Ed.), p. 145, Pergamon Press, Oxford, 1971.
46. Barrett, K.E., *Dispersion Polymerization in Organic Media*, Wiley, New York, 1975.
47. Ugelstad, J. and Hansen, F.K., Kinetics and mechanism of emulsion polymerization. *Rubber Chem. Technol.*, 49, 536, 1976.
48. Vanderhoff, J.W., Mechanism of emulsion polymerization, in: *J. Polym. Sci. Polym. Symp*, vol. 72, pp. 161–98, Wiley Subscription Services, Inc., New York, 1985.
49. Wang, Q.U., Fu, S., Yu, T., Emulsion polymerization. *Prog. Polym. Sci.*, 19, 703, 1994.
50. Aslamazova, T.R., Emulsifier-free latexes and polymers on their base. *Prog. Org. Coat.*, 25, 109, 1995.
51. Nagai, K., Radical polymerization and potential applications of surface-active monomers. *Trends Polym. Sci.*, 4, 122, 1996.
52. Guyot, A., Recent advances and challenges in the synthesis of polymer colloids. *Colloid Surface A: Physicochem. Eng. Aspects*, 153, 11, 1999.
53. Capek, I., Radical polymerization of polyoxyethylene macromonomers in disperse systems. *Adv. Polym. Sci.*, 145, 1, 1999.

54. Kawaguchi, H., Functional polymer microspheres. *Prog. Polym. Sci.*, 25, 1171, 2000.
55. Capek, I., On the role of oil-soluble initiators in the radical polymerization of micellar systems. *Adv. Colloid Interface Sci.*, 91, 295, 2001.
56. Capek, I., Sterically and electrosterically stabilized emulsion polymerization. Kinetics and preparation. *Adv. Colloid Interface Sci.*, 99, 77, 2002.
57. Cunningham, M.F., Living/controlled radical polymerizations in dispersed phase systems. *Prog. Polym. Sci.*, 27, 1039, 2002.
58. El - Aasser, M.S. and Miller, C.M., Preparation of latexes using miniemulsions, in: *Polymeric Dispersions — Principles and Applications*, J.M. Asua (Ed.), pp. 109–126, Kluwer, Dordrecht, 1977.
59. Sudol, E.D. and El – Aasser, M.S., Miniemulsion polymerization, in: *Emulsion Polymerization and Emulsion Polymers*, P.A. Lovell and M.S. El-Aasser (Eds.), pp. 699–722, Wiley, Chichester, 1977.
60. Antonietti, M. and Landfester, K., Polyreactions in miniemulsions. *Prog. Polym. Sci.*, 27, 689, 2002.
61. Asua, J.M., Miniemulsion polymerization. *Prog. Polym. Sci.*, 27, 1283, 2002.
62. Chern, C.S., Microemulsion Polymerization, in: *Encyclopedia of Polymer Science and Technology*, Wiley - Interscience, New York, 2003.
63. Barton, J. and Capek, I., *Radical Polymerization in Disperse Systems*, Ellis Horwood, New York, 1994.
64. Vanderhoff, J.W., Bradford, E.B., Tarkowski, H.L., Shaffer, J.B., Wiley, R.M., Inverse emulsion polymerization. *Adv. Chem. Ser.*, 34, 32, 1962.
65. Candau, F., Polymer Association Structures: Microemulsions and Liquid Crystals, in: *ACS Symposium Series*, M. El Nokaly (Ed.), vol. 384, p. 47, ACS Publication, 1989.
66. Candau, F., Scientific Methods for the Study of Polymer Colloids and Their Applications, in: *NATO ASI Series*, F. Candau and R.H. Ottewill (Eds.), p. 73, D. Reidel, Dordrecht, 1990.
67. Nour, A.H., Emulsion types, stability mechanisms and rheology: A review. *Int. J. Innov. Res. Sci. Stud. (IJIRSS)*, 1, 1, 11–17, 2018.
68. Adamson, A.W. and Gast, A.P., *Physical chemistry of surfaces*, vol. 150, Interscience publishers, New York, 1967.
69. Neelambaram, P., Shankar, A., Sykam, K., Kumar, DB.R., Chakrabarty, A., Narayan, R., Siloxane-based high solid acrylic latex by mini-emulsion polymerization for coatings with improved water resistance. *Prog. Org. Coat.*, 171, 107011, 2022.
70. Nemani, S.K., Annavarapu, R.K., Mohammadian, B., Raiyan, A., Heil, J., Haque, M.A., Abdelaal, A., Sojoudi, H., Surface Modification: Surface Modification of Polymers: Methods and Applications (Adv. Mater. Interfaces 24/2018). *Adv. Mater. Interfaces*, 5, 24, 1870121, 2018.
71. Shi, C. *et al.*, Interfacial bonding properties of styrene-butadiene rubber and ethylene vinyl acetate emulsion-modified OPC–CAC–G repair mortar. *Constr. Build. Mater.*, 370, 130747, 2023.
72. Barquero, A. *et al.*, Understanding the emulsion copolymerization kinetics of vinyl acetate and vinyl silanes. *Polym. Chem.*, 11.13, 2390–2398, 2020.
73. Wu, Z., Zhang, X., Kang, S., Liu, Y., Bushra, R., Guo, J., Zhu, W., Khan, M.R., Jin, Y., Song, J., Preparation of biodegradable recycled fiber composite film using lignin-based polyurethane emulsion as strength agent. *Ind. Crops Prod.*, 197, 116512, 2023.
74. Soleimani, M., Bagheri, E., Mosaddegh, P., Rabiee, T., Fakhar, A., Sadeghi, M., Stable waterborne epoxy emulsions and the effect of silica nanoparticles on their coatings properties. *Prog. Org. Coat.*, 156, 106250, 2021.
75. Zeng, G., Aladejana, J.T., Li, K., Xue, Q., Zhou, Y., Luo, J., Dong, Y., Li, X., Li, J., A tough bioadhesive inspired by pearl layer and arthropod cuticle structure with desirable water resistance, flame-retardancy, and antibacterial property. *Int. J. Biol. Macromol.*, 253, 127669, 2023.

76. Wang, N., Ma, Z., Ma, L., Zhang, Y., Zhang, K., Ban, Q., Wang, X., Synergistic modification of structural and functional characteristics of whey protein isolate by soybean isoflavones non-covalent binding and succinylation treatment: A focus on emulsion stability. *Food Hydrocoll.*, 144, 108994, 2023.
77. Sahoo, S., Gopalan, A., Ramesh, S., Nirmala, P., Ramkumar, G., Shifani, S.A., Subbiah, R., Lalvani, J., II, Preparation of polymeric nanomaterials using emulsion polymerization. *Adv. Mater. Sci. Eng.*, 2021, 1–9, 2021.
78. Kaji, M., Fujiwara, S., Sakai, K., Sakai, H., Characterization of O/W emulsions prepared by PEG-diisostearate amphiphilic random copolymer. *J. Oleo Sci.*, 66, 10, 1121–1128, 2017.
79. Calderó, G., Jose Garcia-Celma, M., Solans, C., Formation of polymeric nano-emulsions by a low-energy method and their use for nanoparticle preparation. *J. Colloid Interface Sci.*, 353, 2, 406–411, 2011.
80. Cong, R., Hollis, C., Bautista, J., Hill, T., Bailey, K., Joseph, K., Use graphene coated silica core shell particles to provide rapid, precise, and equivalent chemical composition distribution analysis for polyolefin materials by high temperature thermal gradient interaction chromatography. *J. Chromatogr. A*, 1709, 464393, 2023.
81. Mohammadi-Jam, S., Waters, K.E., Greenwood, R.W., A review of zeta potential measurements using electroacoustics. *Adv. Colloid Interface Sci.*, 309, 102778, 2022.
82. Tadros, T.F., Emulsion formation, stability, and rheology. *Emulsion Formation Stability*, 1–75, Berkshire, UK, 2013.
83. Ma, H., Wang, R., Xi, Z., Gao, H., Fryda, H., Wang, P., Regulation and inhibition of early whitening of calcium aluminate cement/hemihydrate gypsum decorative mortar by polymer emulsion. *Constr. Build. Mater.*, 393, 132023, 2023.
84. Foucher, A.C. and Stach, E.A., Identifying dynamic restructuring effects in nanocatalysts by combining *in situ* scanning transmission electron microscopy and *in situ* X-ray absorption spectroscopy: A review. *Catal. Today*, 428, 114417, 2023.
85. Li, S., *Fabrication of nanostructured materials for energy applications*, Royal Institute of Technology, Stockholm, 2008, PhD thesis.
86. Barton, J., Niemczyk, A., Czaja, K., Korach, L., Polymer composites, biocomposites and nanocomposites. Production, composition, properties and application fields. *CHEMIK*, 68, 4, 280–7, 2014.
87. Akca, E. and Gursel, A., A review on the matrix toughness of thermoplastic materials. *Period Eng. Nat. Sci.*, 3, 2, 2303–4521, 2015.
88. Ansari, M.O., Khan, M.M., Ansari, S.A., Amal, I., Lee, J., Cho, M.H., Enhanced thermoelectric performance and ammonia sensing properties of sulfonated polyaniline/graphene thin films. *Mater. Lett.*, 11, 4, 159–62, 2014.
89. Ansari, M.O., Khan, M.M., Ansari, S.A., Lee, J., Cho, M.H., Enhanced thermoelectric behaviour and visible light activity of Ag-TiO_2/polyaniline nanocomposite synthesized by biogenic-chemical route. *RSC Adv.*, 4, 6, 23713–9, 2014.
90. Ansari, M.O., Khan, M.M., Ansari, S.A., Raju, K., Cho, M.H., Enhanced thermal stability under DC electrical conductivity retention and visible light activity of Ag/TiO_2/polyaniline nanocomposite film. *ACS Appl. Mater. Interfaces*, 6, 11, 8124–33, 2014.
91. Ansari, M.O., Khan, M.M., Ansari, S.A., Cho, M.H., DC electrical conductivity retention and electrical compensation of polyaniline by TiO_2 at higher loading percentages in polyaniline/TiO_2 nanocomposites. *Electron. Mater. Lett.*, 11, 4, 559–64, 2015.
92. Ansari, M.O., Polythiophene nanocomposites for photodegradation applications: past, present and future. *J. Saudi Chem. Soc.*, 19, 494–504, 2015.

93. Ansari, M.O., Khan, M.M., Ansari, S.A., Cho, M.H., Electrically conductive polyaniline sensitized defective-TiO_2 for improved visible light photocatalytic and photoelectrochemical performance: a synergistic effect. *New J. Chem.*, 39, 8381–8, 2015.
94. Yang, C., Wei, H., Guan, L., Guo, J., Wang, Y., Yan, X. *et al.*, Polymer nanocomposites for energy storage, energy saving, and anticorrosion. *J. Mater. Chem. A*, 3, 14929–41, 2015.
95. Zhou, S. and Wu, L., *Functional Polymer Coatings: Principles, Methods, and Applications*, L. Wu and J. Baghdachi (Eds.), pp. 1–71, John Wiley & Sons, Hoboken, NJ, 2015.
96. Satas, D., *Handbook of Pressure Sensitive Adhesive Technology*, 2nd ed, Van Nostrand Reinhold, New York, 1989.
97. Gumfekar, S.P., Kunte, K.J., Ramjee, L., Kate, K.H., Sonawane, S.H., Synthesis of $CaCO_3$–P (MMA–BA) nanocomposite and its application in water based alkyd emulsion coating. *Prog. Org. Coat.*, 72, 632, 2011.
98. Tobing, S.D. and Klein, A., Molecular parameters and their relation to the adhesive performance of emulsion acrylic pressure-sensitive adhesives. II. Effect of crosslinking. *J. Appl. Polym. Sci.*, 79, 2558, 2001.
99. Xiong, M., Gu, G., You, B., Wu, L., Preparation and characterization of poly (styrene butylacrylate) latex/nano-ZnO nanocomposites. *J. Appl. Polym. Sci.*, 90, 1923–31, 2003.
100. Allen, N.S., Edge, M., Verran, J., Stratton, J., Maltby, J., Bygott, C., Photocatalytic titania based surfaces: environmental benefits. *Polym. Degrad. Stab.*, 93, 1632, 2008.
101. Sow, C., Riedl, B., Blanchet, P., UV-waterborne polyurethane-acrylate nanocomposite coatings containing alumina and silica nanoparticles for wood: mechanical, optical, and thermal properties assessment. *J. Coat. Technol. Res.*, 8, 211, 2011.
102. Kaboorani, A. and Riedl, B., Nano-aluminum oxide as a reinforcing material for thermoplastic adhesives. *J. Ind. Eng. Chem.*, 18, 1076, 2012.

10

A Comprehensive Review on Advancement in Nano Polymer System for Drug Targeting

Debashish Paramanick[1,2*], Deepika Modi[1], Farheen[1] and K. Nagarani[2]

[1]*School of Medical & Allied Science, KR Mangalam University, Gurgram, Haryana, India*
[2]*School of Medical & Allied Science, Galgotias University, Greater Noida, Uttar Pradesh, India*

Abstract

Nanotechnology is linked to a variety of scientific and technological areas, such as medical, biomedical, and technical sciences, as well as environmental and process engineering. In the course of research, it is essential, in the medical sciences and other fields of science and engineering, to update and restructure the landscape of problems in nanotechnological applications. Nanotechnology has now strongly established itself in the field of drug delivery. A new technique of drug delivery systems is provided by the latest novel drug delivery systems (NDDSs). Nanomedicine and nano delivery systems seem to be a comparatively new but rapidly advancing science where nanomaterials are being used to be seen as a screening method or to deliver therapeutic agents controlled to specific targeted sites. Nanotechnology provides a lot of benefits for site-specific and goal delivery of specific therapies to treat severe biological disorders. A number of the challenges in the treatment of different diseases have been initiated by nanomaterials (chemotherapeutic agents, biological agents, immunotherapy agents, etc.). Nano-based drug carrier via a deep analysis of nanomaterial targets identification and applications in the enhancement of efficacy of both new and old drugs (e.g., natural products, etc. The modulation of the dosage forms of pharmacological medicines but also specific tissue targeting. Until the last few, the main target of NDDS was encapsulation with anti-cancer molecules like paclitaxel, vincristine, and doxorubicin, which remained one of the most successful groups of NDDS for liposomes and polymer-drug conjugates. Nanocarriers are colloidal nano-scale structures capable of transporting anticancer agents, as well as small molecular weight capsules or macromolecules such as genes or proteins, away from normal tissues and into tumors as an indirect method of oriented treatment. Nanocarriers have many times greater cytotoxicity in this tumor, as opposed to loosening tablets, with reduced toxicity to calm the body. This chapter describes the Smart drug delivery systems based on nanotechnology with different types of nanoparticles, i.e., polymeric, carbon nanotubes, quantum dots, silver nanoparticles, and many others, to demonstrate a broad variety of useful properties.

Keywords: Nanotechnology, drug targeting, drug delivery, precision medicine, controlled release, advanced therapeutics

**Corresponding author*: debashishparamanick02@gmail.com

Divya Bajpai Tripathy, Anjali Gupta and Arvind Kumar Jain (eds.) Multifunctional Materials: Engineering and Biological Applications, (245–272) © 2025 Scrivener Publishing LLC

10.1 Introduction

Nanotechnology has established a profitable study area for researchers focused on drug delivery. The Greek term "dwarf" served as the inspiration for the prefix "nano." A nanometer (nm) is equivalent to 10 m to 9 m in size. Nanoparticles (NPs) differ from regular materials due to their increased surface area and the presence of quantum phenomena [1]. The National Institutes of Health (NIH) in the United States defines nanomedicine as precise medical treatments at the molecular level aimed at curing diseases and repairing injured tissues like bone, muscle, and nerves [2]. Nanocarriers investigated for drug delivery systems include liposomes, solid lipid NPs, dendrimers, polymers, silicon or carbon compounds, and magnetic NPs (as demonstrated in Figure 10.1) [3–5]. In 1974, Norio Taniguchi, an engineer at Tokyo University, first used the term "nanotechnology." Nanotechnology, rediscovered in the early 1990s, is roughly defined as the creation of materials and systems with a size restriction of less than 100 nm. The size of the 100 nm border is variable and depends on the specific purpose it is being utilized for. In medicine, sizes smaller than a micrometer are referred to as nanometrics. This scale is utilized for controlling drug distribution within the body and interactions between implants and tissues. In the past ten years, nanotechnology has captured the interest of experts from several fields [6, 7]. Nanostructures are increasingly popular in the field of medicine because they can be utilized to deliver medications more precisely to target tissues through controlled release, either by encapsulating drugs or attaching medicinal chemicals [8]. Nanomedicine is a field that utilizes nanoscience knowledge and technology to address medical biology, disease prevention, and therapy [9]. The field encompasses nanorobots, nanosensors, actuated materials, and nano-dimensional materials, such as sensory devices, used within live cells. A therapeutic method utilizing NPs for imaging has been created. Nanotechnology is employed in the production of pharmaceuticals, cosmetics, nutraceuticals, and therapeutic agents [10]. Although these products

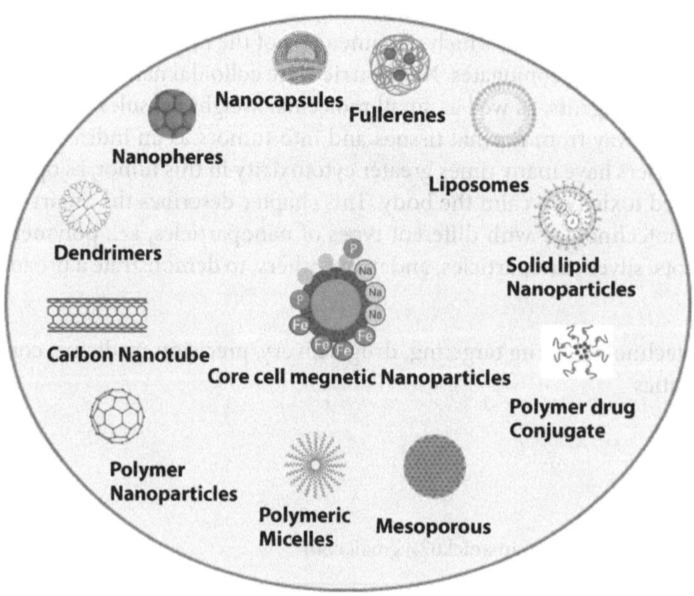

Figure 10.1 Diagrammatic presentation of nanoformulations.

may not consistently display quantum behavior, they can interact with biological molecules in a distinctive manner when compared to larger versions. Nanosystems have an advantage over conventional delivery methods because they may be targeted to specific places in the body. Nanomedicine is a branch of nanotechnology. The nanoconjugates possess distinctive properties for diagnosis and drug delivery [11]. Nanomaterials and their surrounding environments can undergo many interactions involving electrical, chemical, thermal, and mechanical properties. This chapter emphasizes targeted medicine delivery systems and novel therapeutics utilizing nanotechnology. An examination of the latest advancements in nano-DDS technology and how they can be used in cancer treatment, diagnosis, gene delivery, and other areas is also included. Nanomedicine has the potential to significantly impact the treatment of human diseases and improve human physiology in the future.

10.2 Targeted Drug Delivery

The creation of targeted medicine delivery is the next logical step after NP modification. Due to increased epithelial connections, NPs can penetrate inflamed or injured tissue. Passively or actively, this infiltration can occur [12]. Active targeting occurs when the drug carrier system is coupled to a tissue- or cell-specific ligand, while passive targeting occurs when an NP enters the target organ via leaky junctions [13]. It should be possible for an NP drug delivery system to reach pathological tissues and recognize, bind, and distribute its drug load there while minimizing or avoiding drug-induced tissue harm.

As a result, covering NP surfaces with specific targeted ligands is the most prevalent technique. A variety of targeted ligands have been proposed [13, 14]. Due to their relative simplicity of synthesis, stability, and control of conjugation chemistry, small organic compounds are the most often used targeting agents. The selectivity and affinity of these targeted ligands may not be as anticipated. Due to its strong affinity for streptavidin, biotin (vitamin H) is commonly utilized for NP conjugation [15]. It has been found that folic acid (vitamin B9) has a high affinity for the endogenous folate receptor, which has led to its investigation as a potential cancer treatment [16]. Others, such as short peptides and antibodies as well as small compounds that are task specific, have been developed as well [14]. Liposomes are another useful medication delivery tool. Lipid monomers can be tailored for physicochemical features like size and charge by designing specific lipid monomers, which can also comprise surface-targeting ligands as stated above [17]. Since the liposomal makeup is identical to that of the targeted cell membrane, a greater lipid-lipid exchange occurs with this system as well. In this way, lipophilic medicines are transported faster from the liposomal lipid layer to the cell membrane [17–19].

10.3 Designing Nano-Based Drug Delivery

Nanoscale pharmaceutics provide numerous obstacles. NPs can be classified into three categories: one-dimensional (used in chemical and biological sensors), two-dimensional (such as carbon nanotubes), and three-dimensional (like dendrimers). Nanomaterials can be created using both bottom-up and top-down processes [20]. Nano-drug delivery systems need to be carefully modified considering the surface charge, hydrophobicity, and drug release characteristics. NPs' interaction with the biological environment depends on their surface charge and intensity. Zeta potential, an indirect measure of surface charge, governs the stability of a substance [21]. Other techniques for assessing surface hydrophobicity include probe adsorption, contact angle measurements, and X-ray photon correlation spectroscopy, in addition to hydrophobic interaction chromatography and biphasic partitioning. Loading capacity is assessed by quantifying the percentage of medicine that binds to the polymer [22]. Medication loading assays are utilized to ascertain the duration of medication release from a test subject [23].

Biomedical sciences' pharmaceutical technology is a growing field. In pharmaceutical nanotechnology, nanomaterials and devices including drug delivery, diagnostics, imaging, and biosensors are used to apply nanotechnology to the pharmacy field. A molecular-level diagnosis and therapy of disease has been improved by pharmaceutical nanotechnology. Cancer, diabetes mellitus, neurological illnesses, as well as detecting germs and viruses associated with infections, can all be combated by pharmaceutical nanotechnology. In the next 10 years, the market is projected to be inundated with nanotechnology-based pharmaceuticals. Intelligent and smart drug delivery systems based on nanotechnology are predicted to become the most important and powerful alternative to traditional dosage forms [24]. Based on the drug and the polymer's physicochemical nature, the preparation technique for NPs and the choice of matrix materials are determined by a number of factors. Biocompatibility and toxicity of the final product, degree of biodegradability, desired drug release profile, inherent qualities of the drug (aqueous solubility and stability), size of NPs needed, and surface features (charge and permeability) [25]. Hydrophilic polymers, ionic gelation, coacervation, or polymerization of monomers can be used to create nanoparticulate systems. For the creation of NPs, further technologies include supercritical fluid technology and particle replication in nonwetting templates [26]. When grafting nanoparticulate drug delivery, systemic distribution is the most important factor to consider as it must undergo opsonization. NP engineering is therefore focused on clearance and biodistribution. The kidneys or extravazation are responsible for the removal of NPs from the body (10 nm particle). Cells of the reticuloendothelial system remove particles smaller than 10 nm. NPs 100 nm in size are less likely to be filtered by the liver and remain in the blood for a longer period of time after ingestion. A tumor's or its surroundings' accumulation rate of NPs is also influenced by the size of the particles [23]. The use of colloidal drug delivery mechanisms such as liposomes, micelles, or NPs in cancer therapy has been extensively studied. As a result of their tiny size and lower toxicity, as well as their regulated time release of the medication and alteration of pharmacokinetics and biological distribution, drug delivery systems are extremely successful. As a result of tumor cell resistance to various anticancer medications, chemotherapy is typically ineffective in curing cancer. In most cases, resistance develops when cancer cells begin expressing a protein, known as p-glycoprotein that is capable of pumping anticancer drugs out of a cell as quickly as they cross through the cell's outer membrane. New research shows that NPs may be able to get anticancer drugs into cells without triggering the p-glycoprotein pump [23].

These NPs were tested in human colorectal cancers resistant to paclitaxel to see if they had any efficacy *in vivo*. Human colon cancer cell lines were treated with paclitaxel encapsulated in wax NPs (HCT-15). It is possible to circumvent paclitaxel's insolubility issues by conjugating it with albumin. Abraxane, an injectable nano-suspension containing paclitaxel coupled to biocompatible proteins such as albumin, is approved for the treatment of breast cancer. In prior formulations of paclitaxel, the solvent Cremophor-EL produces acute hypersensitivity responses. Premedication with steroids and antihistamines is required before paclitaxel is administered. The medicine is then slowly infused over a period of many hours. Paclitaxel was bound to albumin, resulting in a greater dose of medication being delivered in a shorter period of time. Solvent-related toxicities are also eliminated due to the absence of solvents in the formulation. Abraxane's response rate in a Phase III clinical trial was roughly double that of Taxol, a solvent-containing medication [27, 28].

10.4 Targeting Strategies

To ensure successful drug delivery, nanocarriers need to meet two key conditions. As a first step, medications should be able to reach the desired tumor sites after delivery with little volume and activity loss in blood circulation. Two: Drugs should not harm healthy tissue, but should solely kill tumor cells. Two ways are available to meet these requirements: passive and active medication targeting [29, 30].

A combination of active and passive therapies that target cancer cells with drugs. Tumor cells are disordered, with endothelial cells revealing leaky vasculature via which nanocarriers escape. On the other hand, targeted NPs in tumor tissues are endocytosed by receptor-mediated endocytosis even when they are non-targeted (Figure 10.2 indicating endocytosis of dendrimers) or passively targeted (not depicted in the diagram) [31].

10.4.1 Passive Targeting

Nano drugs can accumulate in tumor tissues due to passive targeting, which takes advantage of tumor vasculature's particular pathophysiological properties. Tumor arteries are typically extremely disordered, dilated, and porous. This leads to larger gap junctions between endothelial cells and a weakened lymphatic system. According to the EPR effect, "leaky vascularization" permits macromolecules to migrate into the tumor zone up to 400 nm in diameter

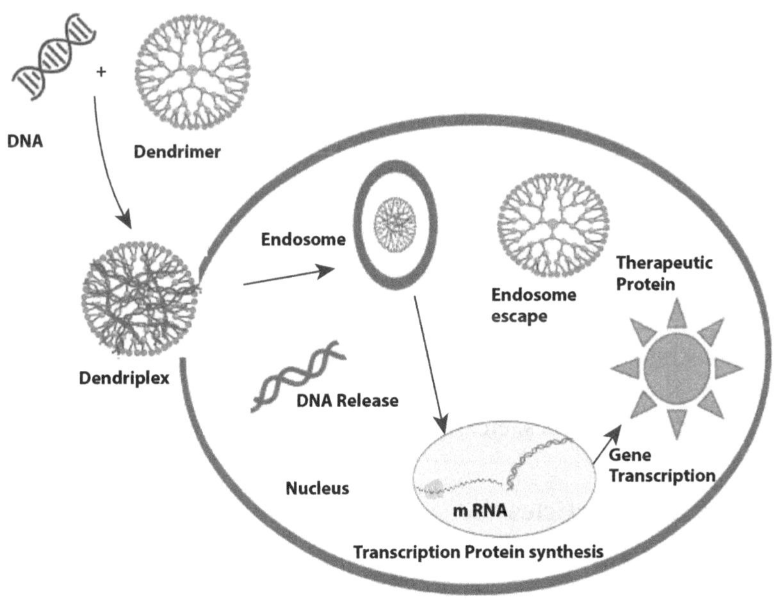

Figure 10.2 Schematic diagram for a possible route in the use of dendrimers as gene delivery vectors.

[32, 33]. In the early days of passive drug targeting, liposomes were one of the first nanoscale technologies to be used. Those that are more sophisticated have a synthetic polymer coating on them that shields them from immunological destruction. Passive targeting is further supported by the EPR effect, the milieu around tumor tissue, which is different from that of healthy cells [34]. Fast-growing tumor cells have a high metabolic rate, which means they demand more oxygen and nutrients. It is as a result of this stimulation of glycolysis that an acidic environment is created [35]. A liposome intended to be stable at physiological pH 7.4 but disintegrate in acidic conditions has been developed to take advantage of this fact [36]. They have a number of drawbacks while being the basis for therapeutic therapy. As a result of the random nature of this method, it is not always possible to target all cells within a tumor. In addition, many cancers do not display an EPR effect, and vascular permeability may not be uniformly distributed throughout a tumor [37].

10.4.2 Active Targeting

This can be done by attaching affinity ligands (antibodies, peptides, or small molecules) to the surface of the nanocarriers via a number of chemical conjugation methods. Through the production of receptors or epitopes on the cell surface, nanocarriers will recognize and bind to target cells through ligand-receptor interactions [37]. Those receptors must be substantially expressed on tumor cells, but not on normal cells, to obtain high specificity. In addition, the receptors should be expressed uniformly and not lost into the bloodstream [38]. It is also possible for targeting conjugates to be internalized through receptor-mediated endocytosis following binding to target cells. To construct an endosome, targeting conjugates must first bind to their receptors, and then the plasma membrane must close around the complex to form an endosome. Acidic pH or enzymes could release medications from the newly created endosome. Nano drugs now licensed for therapeutic use are extremely simple and lack active targeting or triggered drug release components, despite the intrigue of active targeting [39]. Furthermore, nano drugs that are currently being tested in clinical trials do not have any unique targeting capability. Targeted drug delivery can only be completely explored if specific diseases are targeted, if therapeutic drug qualities, as well as their site and mode of action, are ideal for targeting and if the delivery vehicles are optimal for product development [40].

10.5 Types of Nano Drug Delivery Systems

Different writers classify nano-drug delivery technologies in different ways. There are many pharmacological nanotechnology-based systems, which can be referred to as nanomedicines, such as polymeric NPs and magnetic NPs, dendrimers, nanosuspensions, nanocrystals, quantum dots, micelles, liposomes, etc. (as shown in Figure 10.3) [4].

10.5.1 Biopolymeric Nanoparticles

In drug delivery systems, a variety of biopolymeric materials are used to deliver the drugs. Materials and their qualities are described in the next section [8].

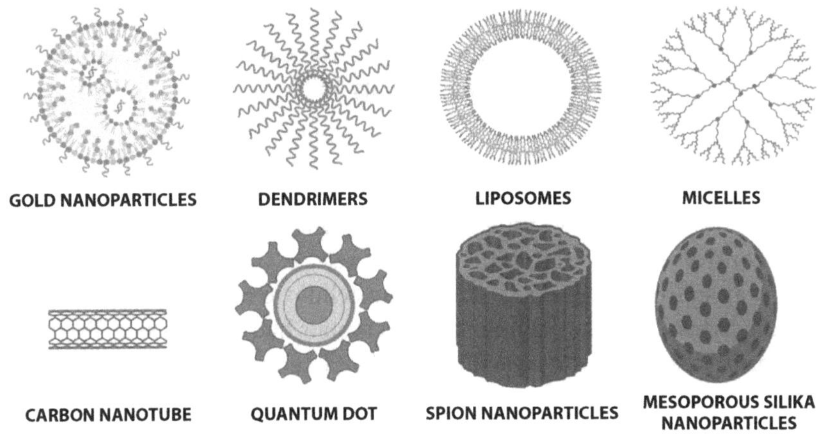

Figure 10.3 Schematic representation of the eight nanocarriers used in smart drug delivery systems.

10.5.1.1 Chitosan

Chitosan has muco-adhesive qualities and can be used to act at epithelial junctions that are very tight and constrictive. In addition to buccal, intestinal, nasal, ocular, and pulmonary epithelia, chitosan-based NPs are frequently used for sustained drug release systems [41–45]. Ceftazidime was delivered to the eye using an isotonic hydroxypropyl methylcellulose (HPMC) solution with chitosan, sodium tripolyphosphate, and Hyaluronic Acid NPs at a concentration of 0.75% (w/w). NPs in contact with mucin at varied mass ratios were used to calculate the rheological synergism parameter [46].

When chitosan NPs were brought in contact with mucin, the viscosity was reduced to a minimum. Due to their mucoadhesion, the NPs interacted well with the eye's mucosa and allowed the antibiotic to be released more slowly, extending its shelf life in the eye. When evaluated on two cell lines, Te NPs showed no cytotoxicity (ARPE-19 and HEK 239T). Because the antibacterial activity of the NPs was preserved, they were a suitable formulation for the administration of ocular medicines with improved mucoadhesive qualities, the researchers concluded [8].

To administer medications into the oral cavity, Pistone *et al.* [47] produced NPs of chitosan, alginate, and pectin [48]. NPs' solubility in saliva and their cytotoxicity potential in oral cell lines were used to determine the formulations' biocompatibility [47]. It took at least 2 hours for alginate NPs to stabilize in artificial saliva. Under all test settings, chitosan NPs were revealed to be the most cytotoxic (concentration and time). A possible explanation of the observed cytotoxicity is Zn2+ (a cross-linking agent) [8].

There were advantages and disadvantages to each formulation in terms of oral absorption, requiring further research. Additionally, Liu *et al.* developed NPs of carboxymethyl chitosan to breach the blood-brain barrier (BBB), increasing the amount of medication in the brain and improving treatment efficacy, therefore minimizing systemic drug exposure. Encapsulation efficiency was 80% and drug loading was 35% for the NPs with a diameter of 218.762.41 nm. The concentrations of CBZ remained higher than they had been in the past (P,0.05). The brain is over 240 minutes [49].

Through oral administration of hyaluronic acid-coated chitosan NPs, Jain and Jain examined the efflux of 5-FU into the stomach. Throughout the stomach-to-colon transit, 5-FU was released in quantities that were protected from excretion during stomach and small intestine transit, according to release assays. For colon cancer, for example, a high local concentration of medications might lengthen the exposure time, which would boost the antitumor efficacy and reduce systemic toxicity [50].

10.5.1.2 Cellulose

It is widely used in drug delivery systems to modify the medicines' solubility and gelation, which in turn controls their release [50]. The oral release of repaglinide (an anti-hyperglycemic—RPG) was studied by Elseoud *et al.* using cellulose nanocrystals and chitosan NPs. RPG-containing hybrid chitosan-cellulose nanocrystals had a mean size distribution of 197 nm. These hybrid NPs and nanocrystalline oxidized cellulose with RPG have an average diameter of 251–310 nm [51].

Due to the presence of hydrogen bonds between cellulose nanocrystals and the drug, the NPs formed with oxidized nanocrystals had a lesser release than those made with native nanocrystals. It is based on the conjugation of calcium alginate beads with 5-fuoroacyl (5-FU) loaded carboxymethylcellulose (CMC). Swelling and muco-adhesion were larger in beads with lower CMC fractions in the simulated colonic environment. There was a 90% release of 5-FU contained in the beads when intestinal enzymes were present [52].

A study by Hansen *et al.* looked at four cellulose derivatives for use in medication release into the nasal mucosa. An additional excipient was tested in conjunction with these cellulose derivatives. Acyclovir was used as the drug model in this method. Also examined were the polymers' ciliary beat frequency (CBF) and their infusion via the tissue system of the nasal cavity. When cellulose derivatives were combined with polymer graft copolymer, thermally induced viscosity increased. When acyclovir was coupled with cationic hydroxyethylcellulose, its penetration into the nasal mucosa increased. CBF found that none of the cellulose derivatives had any adverse effects on nasal mucosal tissues and cells [53].

10.5.2 Dendrimers

A new type of highly branched, monodisperse polymeric macromolecules called dendrimers. In addition to having capabilities similar to proteins, these molecules can transport antibodies or hormones to their target sites [54]. After adding carbon and other atoms to nitrogen, dendrimers take on their spherical shape through a sequence of chemical processes. They consist of three components [55]:

- ➢ initiator core.
- ➢ interior layers composed of repeating units, radically attached to the interior core.
- ➢ exterior (terminal functionality) attached to the outermost interior generations.

Mechanism of Drug Delivery Through Dendrimers:

They act as the prodrug, in which the drug molecule wraps or attaches itself to the dendrimer. The central core is composed of amines, which may be substituted with sugar in some instances. There are several and identical reaction sites in all core compounds.

Numerous amines are present on the surface of all full generations; multiple acids are present on the surface of all half generations [56, 57].

10.5.3 Nanosuspensions

Nanosuspensions are colloidal dispersions of nanosized drug particles stabilized by surfactants made by milling or high-pressure homogenization [58]. For the preparation of the same, techniques such as wet milling, high-pressure homogenizer, emulsion-solvent evaporation, melt emulsification process, and supercritical fluid are utilized. Stabilizers, organic solvents, and other additives such as buffers, salts, polyols, osmogent, and cryoprotectants can be used to make nanosuspension. A particle with a diameter bigger than 1 mm can be obtained using traditional milling and precipitation methods [59].

Advantages of Nanosuspensions:
Because the particle size is tiny, IV use is possible, and the solid crystal phase allows for large loading [60]. For substances with a high log P value, a high melting point, and a high dosage, the nanosuspension formulation technique is ideal. Passive targeting is one of the additional benefits of nanosuspension. The nanosuspension's formulation is crucial. Ostwald ripening and agglomeration are prevented by the stabilizers employed in nanosuspension. The stabilizer controls the physical stability and *in vivo* behavior of nanosuspension. When microemulsion is utilized to make nanosuspension, co-surfactants are used [59, 61]. The commercial manufacturing of nanosuspensions mostly involves medium milling (nanocrystals) and high-pressure homogenization (Dissociate in aqueous media and Nanopure in nonaqueous media). Nano-edge [62] is a combination of precipitation and high-pressure homogenization. Because nanosuspensions are thermodynamically unstable, they must be solidified into tablets or pilates [63].

10.5.4 Nanocrystals

Nanocrystals are groupings of atoms that unite into a "cluster" and are less than 1 mm in size. Typical diameters range between 10 and 400 nm. They have physical and chemical properties that fall midway between bulk solids and molecules. As the size of the particle shrinks, the effective surface area grows, increasing solubility and bioavailability [64].

Top-down and bottom-up techniques can be used to create nanocrystals. Bottom-up approaches include pearl/ball milling, high-pressure homogenization, microfluidizer technology, piston gap homogenization, and spray drying, whereas top-down techniques include high-pressure homogenization, microfluidizer technology, piston gap homogenization, and spray drying [64–66]. They are crystalline, as the name implies, but they can also be partially or fully amorphous, depending on the manufacturing procedure. Nanocrystals are made using a mix of technologies such as nano edge technology (microprecipitation and high shear pressures) and nano pure XP technology [66].

Nano sonication can also make nanocrystals. To make nanosized crystals, high-speed stirring creates a nanosuspension, which is then followed by wet milling, high-pressure homogenization, nano crystallization, and spray drying. Nanocrystals of nelfinavir mesylate were created by Nayak *et al.* To optimize the pharmacokinetic characteristics of the nanocrystal production, they adopted a ball milling process (solubility and bioavailability). Using mannitol as a cryoprotectant, the produced nanocrystals were lyophilized. The improved formulation's particle size and zeta potential were determined to be 740.12 + 79.21 nm and 23.31 + 1.10 mV, respectively, with a polydispersity index of 0.20 + 0.07. When compared to raw crystals, scanning electron microscopy (SEM) and differential scanning calorimetry (DSC) research revealed that the nanocrystals had a modest drop in crystallinity [67].

10.5.5 Polymeric NPs

As colloidal particles in the 10–1000 nm range, polymeric NPs (polymer NPs) can take the form of either spheres or branching core-shell structures. Biodegradable synthetic polymers, such as polylactide–polyglycolide copolymers, polyacrylates, and polycaprolactones, as well as natural polymers, such as albumin, gelatin, alginate, gellan and collagen, were used to make them [68]. Three drug-incorporation mechanisms can be used to classify polymeric nanocarriers. Polymeric carriers that use covalent chemistry for direct drug conjugation are among the first (e.g., linear polymers). Hydrophobic interactions between medicines and nanocarriers fall into the second category (e.g., polymeric micelles from amphiphilic block copolymers). Hydrogels, which are water-filled depots for hydrophilic drug encapsulation, are included in the third group of polymeric nanocarriers.

10.5.6 Polymer-Drug Conjugates (Prodrugs)

Since the first polymer-drug conjugate was described in the 1970s [29, 69, 70], many more have been produced. Macromolecular polymers can be added to medications to increase their blood circulation time. PEG can be used to preserve the activity of protein or peptide medicines, which are easily digested by the human body (PEGylation). l-asparaginase, for example, has a plasma half-life of up to 357 hours when PEGylated [71]. Natural l-asparaginase has a half-life of 20 hours without PEG. To improve cancer therapy pharmacokinetics, tiny molecular anticancer medicines can also be PEGylated. There are clinical trials underway for cancer therapy with PEG camptothecin (PROTHECAN®) [72, 73].

10.6 Characterization of Nano-Drug Delivery System

NP size is typically measured using photon correlation spectroscopy (PCS) or dynamic light scattering (DLS). On the basis of Brownian motion, photon correlation spectroscopy primarily determines the hydrodynamic diameter of NPs. Additionally, molecular weight and crystallinity should be evaluated, as these factors might also affect drug release and degradation [74]. NPs also have a zeta potential, which is used to calculate surface charge density. The surface morphology of NPs is studied using SEM and atomic force microscopy (AFM). As well as DSC and TGA, X-ray crystallography can be used to study the thermal profile of NPs. Drug encapsulation efficiency and physical features such as size, shape, stability of size, drug loading, and drug release should also be assessed [75]. The surface charge of NPs affects various features of polymer dispersion, such as physical stability and redispersibility, as well as their *in vivo* performance [23]. A general overview of the characterization of the nanosystem is given in [22, 75–79].

10.7 Challenges of Nanotechnology for Drug Delivery

The application of nanotechnology to medication delivery has made dramatic and effective development, as indicated by the fact that several nano drugs are now on the market, but there are still a number of major obstacles to overcome in this field.

10.7.1 Biological Understanding

The mechanisms underlying intracellular absorption, trafficking, and the fate of nanomaterials in complex biological networks must be better understood before bio nanotechnologies may be used to humans. Current delivery systems are plagued by a number of serious problems (e.g., rapid clearance by the immune system, low targeting efficiency, and difficulty in crossing biological barriers). A comprehensive grasp of the basic science of NP transport will allow medication delivery to be controlled and manipulated [80, 81].

10.7.2 Safety Concern

Nanotoxicology is a field that has arisen in tandem with nanomedicine. According to Nanotoxicology, interactions between nanomaterials and biological systems have the potential to have harmful consequences [82]. Preliminary nanotoxicity studies have led to the hypothesis that nanomaterials may contribute to the production of free radicals [83]. Neurological injury [84] and unwanted penetration through the epidermis or other physiological barriers into parts of the body that are more sensitive to toxic effects [85]. Nanomaterials' toxicity is thought to be affected by a variety of mechanisms, including physicochemical qualities, physical attributes, and environmental circumstances. The nucleus, mitochondria, and plasma membranes are damaged as a result of the cytotoxicity of naked quantum dots [86]. According to another investigation, the release of free Cd2+ ions from cadmium-containing quantum dots is hazardous [87]. N-acetylcystein was attached to their surfaces, which lowered their toxicities [88].

Inhaled, swallowed and skin-absorbed NPs can permeate cells and even reach the cell nucleus, where they may come into close contact with genetic material if they are small enough. It is important to consider the patient population and the complete manufacturing and disposal process when evaluating nanomaterial toxicity. It is necessary to establish standards and testing methodologies that can serve as benchmarks for the development of innovative classes of materials in order to address safety concerns [89].

10.7.3 Manufacturing Issue

Large-scale production of nanomaterials for uniform and repeatable manufacturing and commercialization is a difficulty for nano-drug delivery. The preparation process and high cost of materials used in a variety of nano-drug delivery systems may make them unsuitable for large-scale manufacture. Nanomaterials with low concentrations, agglomeration, and the chemical process are among the problems of scaling up. A laboratory-scale modification of nanomaterial size or composition is much easier than a large-scale modification. When working with nanomaterials, the biomedical community should reassess the level of control needed. To identify nanomaterials, a statistical technique may be used rather than precise control of physical dimensions [29].

10.7.4 Economic and Financial Barriers

Nanomedicine can potentially be hampered by economic and financial obstacles. Because of their costs and complexity, nanoproducts are expected to face even bigger obstacles in the deployment of personalized medicine in general [90].

Contrary to popular belief, nano-drug delivery technologies are still in their infancy. Large pharmaceutical corporations are able to give startup companies the financial resources and experience needed to achieve regulatory and commercial success since nano drugs and medical devices have substantial development costs [29].

10.8 Evaluation of Nanotechnology for Industrial Applications

Consumers benefit from nanotechnology in food-related applications [91]. Nanotechnology has a number of applications in the food industry, including:

- In this application, food products with nanostructures or nanotextures are produced that have a better taste and include less fats and emulsifiers that are stable. Canola oil, ice cream, etc., are only a few examples. In addition, microbiological pathogens are destroyed during this operation. In this form of application, materials cannot be absorbed into the circulation since the food nanostructures are easily assimilated in the gastrointestinal tract.
- These applications include the masking of the flavor of various food items using liposomes or biopolymer nano-encapsulated compounds. Like fish oils. Processing and refining particular substances. As a result of the increased availability of supplements and other additives in biological systems, the distribution of compounds to tissues can be altered, and the pharmacokinetics of those substances can be affected [91, 92].
- Due to its organic nature, no additional fats or emulsifiers are required to enhance the taste. The wide surface area and organic nature of these foods mean that they are needed in smaller quantities, and they have a greater absorption and bioavailability rate.
- Packaging and temperature/moisture stability of food items are improved using plastic polymers containing NPs. Microbial development is also prevented. However, consumers may be exposed to NPs and may eat them as a result [91, 93].
- Nanofiltration: Porous silica and regenerated cellulose membranes are used to purify water and remove undesirable elements from food. However, these NPs may end up in filtered products, posing health hazards.
- These are employed in the production of fertilizers, insecticides, and biocides, among other things, because of their greater efficacy and reduced usage of solvents in the field. People who work with hazardous substances may be exposed to health concerns as a result of this.
- Food labeling nanosensors: Apps that monitor food conditions during transit and storage are included in this sort of application. Such packaging will release preservatives only when they're actually required.

❖ Nanotechnology has a number of advantages, but it also carries a number of potential hazards. Despite their small size, man-made nanomaterials can have serious toxicological effects on the body. Cells in humans could be infected with them [94]. Engineered nanomaterials represent a risk to the environment and the population, which is a cause for concern. Nanomedicine manufacture requires knowledge of nanoproduct toxicity and biokinetics. Human and atmospheric exposure, pharmacokinetics, effective risk detection, and persistence or accumulation of nanomaterials in cells and subcellular structures are all required for a hazard assessment to be made [95].

10.9 Application of Nanoparticle Technology

10.9.1 Cancer Therapy

As a result of chemotherapeutic drugs' non-specificity, the treatment's adverse effects are severe, affecting the whole body. Cancer is a biological process that is extremely complex and might be called a sickness of many diseases, according to some experts [96]. To now, the major goal of chemotherapy is to eliminate all rapidly dividing cells. Patients must also deal with life-altering side effects as a result of this medication, including the death of rapidly growing cells in the hair follicles and intestinal epithelium [97]. Because NPs have been developed, chemotherapy has gained a new channel of attack. Toxic effects on normal tissues and organs can be minimized with NPs that are carefully constructed [98, 99]. Detailed description of a smart drug delivery system for cancer treatment based on nanoparticles are shown in Figure 10.4.

Their absorption, distribution, metabolism, and elimination (ADME) profiles depend on many structural features [100, 101]. Dendrimers are highly branched macromolecules

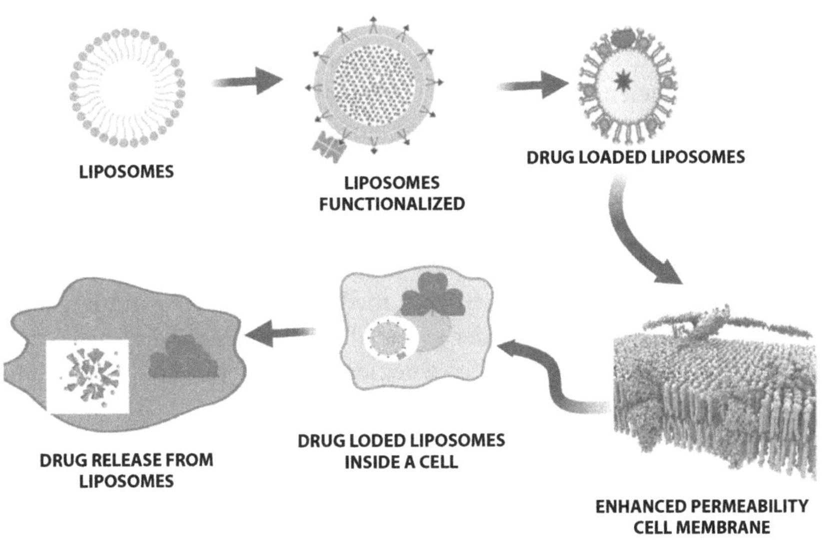

Figure 10.4 Step-wise illustration of nanoparticle-based smart drug delivery system for cancer therapy.

with multiple functional groups available for the attachment of drugs, targeting agents, and imaging agents. Methotrexate has been successfully localized, imaged, and delivered *in vitro* using a polyfunctional dendrimer system [102]. Surface derivatization using PEGylation, acetylation, glycosylation, and different amino acids can increase the therapeutic index of cytotoxic medicines [103, 104].

When it came to the safe and effective delivery of drug to the tumour site, liposomes are wonderful candidates. Liposomes have exceptional properties, including the ability to contain hydrophilic and hydrophobic bioactives and the ability to modify their surface and composition. In addition to the various NPs that have shown promise in cancer treatment, carbon nanotubes are the most recent. A single-walled carbon nanotube (SWCNT) and a multi-walled carbon nanotube (MWCNT) are two types of carbon nanotubes (MWCNTs) [105, 106]. The hollow interior of carbon nanotubes is extremely hydrophobic, making it easy to load them with water-insoluble medicines. Surface functionalization is possible due to the vast surface area, and can be done particularly for a cancer receptor or contrast agent [107].

10.9.2 Diagnostic Testing

Academic research has focused on the diagnostic use of NPs, which has not yet been approved for clinical usage [47]. Failings of fluorescent markers in diagnostic tests, such as fading after one usage, color matching, and dye use restrictions owing to bleeding, have forced researchers to turn to fluorescent NPs as a solution [48].

These NPs, which can be employed for both treatment and diagnosis [108], have recently attracted a lot of attention. Drug conjugates, dendrimers, surfactant aggregates (micelles and vesicles), core-shell particles, and carbon nanotubes are all examples of NPs that use this method. As long as the drug and imaging agent are combined into one smart formulation, it is possible to monitor both drug action and NP location at the target site [109].

10.9.3 HIV and AIDS Treatment

An individual's immune system is almost completely destroyed when infected with the human immunodeficiency virus (HIV). When this disease was first being treated, it required patients to take 30 to 40 medications a day [110]. In the past decade, pharmaceutical advances have reduced the number of pills taken daily to just a few [111]. Antiretroviral (ARV) medications can be delivered intracellularly as well as to the brain using polymeric NPs, according to research [112]. HIV infections can also be prevented using this technique in conjunction with vaccines [113].

As a result of nanotechnology, antiretroviral medications are now more readily available and more widely accepted [114]. In order to be effective, antiretroviral medicines must be able to pass the mucosal epithelial barrier whether administered orally or by other routes other than parenteral administration (suppository and patches, etc.). HIV infects and multiplies mostly in lymphoid tissue. Many studies have shown that NPs containing antiretroviral medicines can target monocytes and macrophages *in vitro* [115, 116], a prime example of the superiority and efficacy of NP systems in the delivery of sustained and targeted drugs [117].

10.9.4 Nutraceutical Delivery

Nutraceuticals are food-derived, standardized components that have apparent health advantages, such as antioxidants [118]. In addition to providing additional health advantages and reducing the risk of certain chronic conditions, they are often consumed as a complement to various allopathic treatments. For orally taken nutraceuticals, food matrix interactions, water solubility, degradation and metabolism, and epithelial permeability all affect bioavailability [119].

Curcumin is one of many nutraceuticals that has anti-inflammatory, antioxidative, antiapoptotic, and antiangiogenic properties (diferuloylmethane). In order to overcome this issue, many approaches have been used, including liposomes, phospholipid vesicles, and polymer-based nano-formulations [120, 121]. Curcumin has a 9-fold increased oral bioavailability when combined with piperine (an absorption enhancer) [122]. Comparing Theracurmin NPs to curcumin powder, another study found that Theracurmin had a 40-fold larger area under the curve (AUC) in rats and a 27-fold higher AUC in healthy human volunteers, as well as inhibiting effects on alcohol intoxication in healthy human participants [123].

10.9.5 Vaccines

In addition, NPs are being created for use in oral vaccination. It has been shown that chitosan NPs containing DNA can provide immunological protection against a variety of allergens. Anaphylaxis produced by peanut allergen was reduced in mice that were given chitosan NPs that contained the allergen gene [124, 125]. In addition, chitosan NPs harboring allergen genes from house dust mites were employed to immunize mice orally. NPs administered orally activated the gene in the stomach and small intestinal epithelial cells, inhibiting a Th2-mediated IgE response [126, 127].

10.9.6 Gene Delivery

For the prevention of bacterial and viral infections, NPs were produced. The Peyer's Patches in mice were able to take up poly (lactic-co-glycolic acid) (PLGA) particles carrying Staphylococcal enterotoxin B and produce toxin-specific antibodies and secretory IgA-specific antibodies in saliva, gut wash fluids, and bronchial-alveolar wash fluids [128, 129].

An immunological response was not elicited by oral administration of soluble enterotoxin. pH-dependent micro-particules were produced in another study that could be targeted to either the small or large intestine. When employed as an alternative to intra-colorectal injection, it provides good protection against a viral infection challenge [124, 130].

10.9.7 Brain Targeting

Protected by the BBB, which regulates its homeostasis, the nervous system is one of the body's most delicate microenvironments Most new medications for the central nervous system cannot be developed because of the BBB. This very complex structure strictly controls the transport of ions and macromolecules from the blood to the brain, protecting the brain from accidents and sickness. Enzymatic activity and active efflux transport

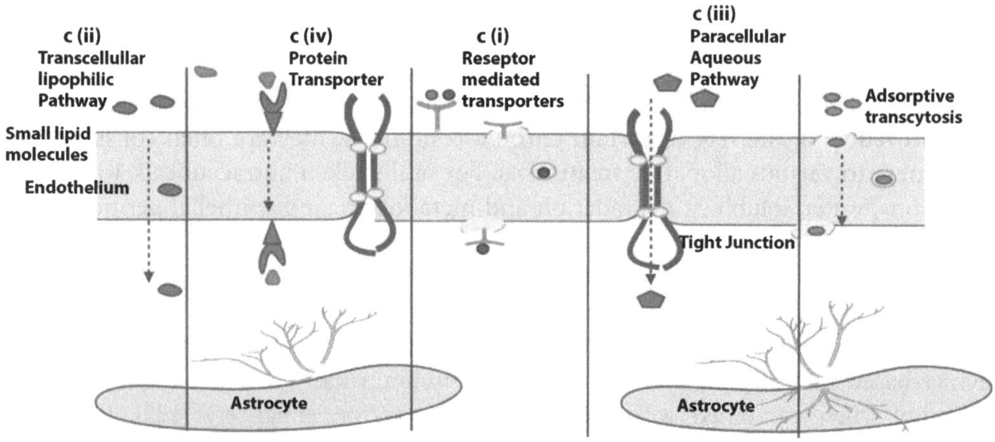

Figure 10.5 Transport mechanism for brain drug delivery.

mechanisms characterize the BBB, which is composed of relatively impermeable endothelial cells with tight connections (as depicted in Figure 10.5). Water-soluble compounds are successfully prevented from entering the CNS via blood circulation, while lipid-soluble molecules can be reduced in the brain by the action of enzymes or efflux pumps. A number of neurological illnesses cannot be treated because the BBB only allows the selective transfer of chemicals that are vital to brain function. Multiple techniques are being developed to improve medicine delivery across the BBB as a result. This can be achieved by using a chimeric construct that is able to undergo receptor-mediated transcytosis (Figure 10.5c, inset i). Examples include polysorbate 80/LDL, transferrin receptor binding antibody (such as OX26), lactoferrin, cell-penetrating amino acids, and melano transferrin [23, 131–135].

Transcellular diffusion is the movement of solute particles via endothelial cells (Figure 10.5c, inset ii). BBB may be transported via this channel only if the chemicals are tiny, lipid-soluble, hydrophilic, and non-ionized [134]. Transcellular diffusion is also driven by a negative concentration gradient, just as paracellular diffusion is (Figure 10.5c, inset iii). Glucose transporter GLUT-1 and long amino acid transporter (LAT) are two examples of transporter proteins that may transport chemicals across the BBB via an active transport mechanism (Figure 10.5c, inset iv) [135]. Glucose or amino acids first bind to transporter proteins on the blood side of the BBB in order to begin the process. The transfer of glucose or amino acids into the brain is then caused by the conformational shift of transporter proteins [136].

10.9.8 Anthrax Vaccine Uses Nanoparticles to Produce Immunity

Spore-forming Bacillus anthracis is responsible for the anthrax infection. Lethal factor and edoema factor are the two enzyme components responsible for B. anthracis' harmful effects. Current anthrax vaccines contain antibodies against PA, which confers protective immunity against B. anthracis infection. mice and guinea pigs tested with a more effective and easier-to-administer anthrax vaccine than the current vaccination showed great efficacy in testing, claim scientists at the University of Michigan Medical School in Infection and Immunity, which was published in August. Multiple shots are required, followed by annual boosters, even though the vaccination is safe and effective. "Nanoemulsion" is a suspension of water, soybean oil, alcohol, and surfactant that is emulsified to generate droplets of 200–300 nm in size. To guard against anthrax, the oil particles are small enough to transport a crucial protein into the nasal membranes. As a result, the immune system is primed to fight off infection as soon as it comes into contact with the entire organism. If the droplets were lined up side by side, it would take roughly 265 to equal the breadth of a human hair. In addition, the oil particles are small enough to allow immune system cells to react to the anthrax protein and initiate a protective immunological response when they enter the nasal membranes. When the immune system confronts the full microbe, it is primed to kill it off quickly. Researchers claim the nanoemulsion anthrax vaccine has another advantage besides eliminating the need for needles: It is easier to store and use in areas without refrigeration. In the event of a future attack in which a terrorist spreads anthrax bacteria, a vaccination that is both effective and easy to give would be a vital tool for health officials. After being exposed to anthrax, people might be given a nasal nanoemulsion-based anthrax vaccine and antibiotics if it proves successful in humans, the researchers said. After exposure to certain diseases, vaccinations are used to increase the immune response [23].

10.10 Future of Nanomedicine and Drug Delivery System

Nanomedicine is currently one of the most exciting fields of research, according to the National Science Foundation. In the previous two decades, a great deal of research in this field has resulted in the filing of 1500 patents and the completion of dozens of clinical trials. Unsurprisingly, cancer is a disease where nonmedical technology has been most helpful in both diagnosing and treating it, as described above. It's safe to say that the application of nanomedicine and nano-drug delivery systems will continue to be a research and development focus for decades to come, thanks to the use of NPs for precise medication delivery to cancer/tumor cells, without affecting the normal cell's physiology. Some of the NPs shown in this communication are nm-sized, while others are submicrometer-sized (over 100 nm). As a next step, researchers should focus on developing materials with improved homogeneity, drug loading and release capacity, etc [137]. This review also examines the progress made in the use of metal NPs for diagnostic purposes. Despite the widespread acceptance of nanomedicine and nano-drug delivery systems, its real impact on the healthcare system, including in cancer therapy/diagnosis, is still quite limited at this point. As a relatively new branch of science, barely two decades of research have been conducted on the subject, and many crucial fundamental qualities are still unknown, this is understandable. In the future, research will focus on identifying the core indicators of sick tissues, including important biological markers that allow for absolute targeting without disrupting the normal cell function. It is hoped that as our knowledge of diseases at the molecular level increases, new pathways for nanomedicine would open up [138]. As a result, nanomedicine applications will advance in the future as we better understand illness molecular markers. Further study is needed to expand the use of nanomedicine beyond what has been detailed in this review using the known nanoprobes and nanotheragnostics products. Yet to be perfected are concepts such as the controlled release of specific medications at beleaguered sites and technologies for assessing these events, as well as the effects of drugs on tissues/cells and theoretical mathematical models of prediction. Biomaterials and formulation research appear to be the focus of many nanomedicine studies. It is the results of animal studies and transdisciplinary research that will provide valuable information for medication therapeutic and diagnosis studies. These types of studies need a significant amount of time and research resources. Due to an increasing global trend toward more accurate medications and diagnostics methods, nanomedicine and nano-drug delivery technology have a promising future. The basic idea of developing nanorobots (and nanodevices) that work in tissue diagnosis and repair mechanisms with full external control mechanisms has generated a lot of excitement among researchers. New nanomaterials must therefore be subjected to a thorough impact analysis to determine their potential acute or chronic harmful effects on humans and the environment. Affordability of nanomedicines is another field of research that requires additional exploration. Regulation of nanomedicines, as described in the previous section will continue to evolve in tandem with nanomedicine applications' advancements [138].

Conclusion

Drug delivery and human therapies will be dominated by nanotechnology. Despite the fact that medication delivery systems are still in their infancy, they have a bright future. Researchers in physics, chemistry, biochemistry, and other fields can use nanotechnology, which is still in its infancy, to design systems that may eventually rival the sophistication and precision of biological structures. In the future, nanotechnology could revolutionize how we treat diseases through medicine delivery. However, there are still many obstacles to overcome before this sector can produce clinically effective treatments. Design and testing of novel approaches for influencing how NPs interact with the body are now preventing these technologies from being translated into therapeutics. It is still necessary to characterize molecular targets precisely and to ensure that these molecules are expressed only in the targeted organs, in order to prevent effects on healthy tissues. The fate of the medications, once they reach the nucleus and other sensitive organelles of the cell, must also be understood. Due to the fact that nanosystems boost medication delivery efficiency, dose calibration may be necessary. Despite this, the future's still bright and wide open, with many possibilities.

References

1. Biswas, A., Islam, M., Choudhury, Z.S., Mostafa, A., Kadir, M.F., Nanotechnology based approaches in cancer therapeutics. *Adv. Nat. Sci. Nanosci. Nanotechnol.*, 5, 4, 043001, 2014.
2. Saha, M., Nanomedicine: Promising Tiny Machine for the Healthcare in Future-A Review. *Oman Med. J.* [Internet], 24, 4, 242, 2009 [cited 2024 Aug 12, Available from: /pmc/articles/PMC3243873/.

3. Thassu, D., Pathak, Y., Deleers, M., Nanoparticulate drug-delivery systems: An overview, in: *Nanoparticulate Drug Deliv Syst* [Internet], pp. 1–31, 2007 Mar 30 [cited 2024 Aug 12]; Available from: https://www.taylorfrancis.com/chapters/edit/10.1201/9781420008449-1/nanoparticulate-drug-delivery-systems-overview-deepak-thassu-yashwant-pathak-michel-deleers.
4. Chavda, V.P., Nanobased Nano Drug Delivery: A Comprehensive Review, in: *Appl. Target Nano Drugs Deliv. Syst. Nanosci. Nanotechnol. Drug Deliv.*, pp. 69–92, 2018 Jan 1.
5. Wilczewska, A.Z., Niemirowicz, K., Markiewicz, K.H., Car, H., Nanoparticles as drug delivery systems. *Pharmacol. Rep.*, 64, 5, 1020–1037, 2012.
6. Bath, J. and Turberfield, A.J., DNA nanomachines. *Nat. Nanotechnol.*, 2, 5, 275–84, 2007, [Internet]. [cited 2024 Aug 12]; Available from: https://www.nature.com/articles/nnano.2007.104.
7. Sharma, D., Sharma, N., Pathak, M., Agrawala, P.K., Basu, M., Ojha, H., Nanotechnology-based drug delivery systems: challenges and opportunities, in: *Drug targeting and stimuli sensitive drug delivery systems*, pp. 39–79, 2018
8. Patra, J.K., Das, G., Fraceto, L.F., Campos, E.V.R., Rodriguez-Torres, M.D.P., Acosta-Torres, L.S., *et al.*, Nano based drug delivery systems: recent developments and future prospects. *J. Nanobiotechnol.*, 16, 1, 1–33, [Internet]. 2018 Sep 19 [cited 2024 Aug 12]; Available from: https://link.springer.com/articles/10.1186/s12951-018-0392-8.
9. Lam, P.L., Wong, W.Y., Bian, Z., Chui, C.H., Gambari, R., Recent Advances in Green Nanoparticulate Systems for Drug Delivery: Efficient Delivery and Safety Concern. *Nanomedicine*, 12, 4, 357–85, [Internet]. 2017 Feb 1 [cited 2024 Aug 12]; Available from: https://www.tandfonline.com/doi/abs/10.2217/nnm-2016-0305.
10. Haba, Y., Kojima, C., Harada, A., Ura, T., Horinaka, H., Kono, K., Preparation of poly(ethylene glycol)-modified poly(amido amine) dendrimers encapsulating gold nanoparticles and their heat-generating ability. *Langmuir* [Internet], 23, 10, 5243–6, 2007 May 8 [cited 2024 Aug 12]; Available from: https://pubs.acs.org/doi/abs/10.1021/la0700826.
11. Chhikara, B.S., Kumar, R., Rathi, B., Krishnamoorthy, S., Kumar, A., Prospects of Applied Nanomedicine: potential clinical and (bio) medical interventions via nanoscale research advances. *J. Mater. Nanosci.*, 3, 2, 50–56, 2016.
12. Varshosaz, J. and Farzan, M., Nanoparticles for targeted delivery of therapeutics and small interfering RNAs in hepatocellular carcinoma. *World J. Gastroenterol.*, 21, 42, 12022, 2015.
13. Liu, R., Kay, B.K., Jiang, S., Chen, S., Nanoparticle Delivery: Targeting and Nonspecific Binding. *MRS Bull.* [Internet], 34, 6, 432–40, 2009 [cited 2024 Aug 13]; Available from: https://www.cambridge.org/core/journals/mrs-bulletin/article/abs/nanoparticle-delivery-targeting-and-nonspecific-binding/68181B365D3DB58A419E08C82BF2800E.
14. Friedman, A.D., Claypool, S.E., Liu, R., The smart targeting of nanoparticles. *Curr. Pharm. Des.*, 19, 35, 6315–6329, 2013.
15. Pramanik, A.K., Siddikuzzaman, Palanimuthu, D., Somasundaram, K., Samuelson, A.G., Biotin Decorated Gold Nanoparticles for Targeted Delivery of a Smart-Linked Anticancer Active Copper Complex: *In Vitro* and *in Vivo* Studies. *Bioconjug. Chem.*, 27, 12, 2874–85, 2016 Dec 21.
16. Zhao, X., Li, H., Lee, R.J., Targeted drug delivery via folate receptors. *Expert Opin. Drug Discov.*, 5, 3, 309–19, 2008 Mar.
17. Kelly, C., Jefferies, C., Cryan, S.-A., Targeted Liposomal Drug Delivery to Monocytes and Macrophages. *J. Drug Deliv.* [Internet], 2011, 1, 727241, 2011 Jan 1 [cited 2024 Aug 13]; Available from: https://onlinelibrary.wiley.com/doi/full/10.1155/2011/727241.
18. Singh, R. and Lillard, J.W., Nanoparticle-based targeted drug delivery. *Exp. Mol. Pathol.*, 86, 215–23, 2009.

19. Pattni, B.S., Chupin, V.V., Torchilin, V.P., New Developments in Liposomal Drug Delivery. *Chem. Rev.* [Internet], 115, 19, 10938–66, 2015 Oct 14 [cited 2024 Aug 13]; Available from: https://pubs.acs.org/doi/full/10.1021/acs.chemrev.5b00046.
20. Farokhzad, O.C. and Langer, R., Nanomedicine: Developing smarter therapeutic and diagnostic modalities. *Adv. Drug Deliv. Rev.*, 58, 14, 1456–9, 2006 Dec 1.
21. Otsuka, H., Nagasaki, Y., Kataoka, K., PEGylated nanoparticles for biological and pharmaceutical applications. *Adv. Drug Deliv. Rev.* [Internet], 55, 3, 403–19, 2003 Feb 24 [cited 2024 Aug 12]; Available from: https://pubmed.ncbi.nlm.nih.gov/12628324/.
22. Kreuter, J., Physicochemical characterization of polyacrylic nanoparticles. *Int. J. Pharm.*, 14, 1, 43–58, 1983.
23. Bhatia, S., *Natural polymer drug delivery systems: Nanoparticles, plants, and algae*, pp. 1–225, Springer, Cham, 2016 Jan 1.
24. Rana, V. and Sharma, R., Recent Advances in Development of Nano Drug Delivery, in: *Appl Target Nano Drugs Deliv Syst Nanosci Nanotechnol Drug Deliv.*, pp. 93–131, 2019 Jan 1.
25. Tice, T.R. and Gilley, R.M., Preparation of injectable controlled-release microcapsules by a solvent-evaporation process. *J. Control. Release*, 2, C, 343–52, 1985 Nov 1.
26. Allemann, E., Gurny, R., Doelker, E., Drug-loaded nanoparticles : preparation methods and drug targeting issues. *Eur. J. Pharm. Biopharm.*, 39, 5, 173–91, 1993.
27. Arap, W., Pasqualini, R., Ruoslahti, E., Cancer treatment by targeted drug delivery to tumor vasculature in a mouse model. *Science*, 279, 5349, 377–380, 1998.
28. Koziara, J.M., Lockman, P.R., Allen, D.D., Mumper, R.J., Paclitaxel nanoparticles for the potential treatment of brain tumors. *J. Control. Release*, 99, 2, 259–69, 2004 Sep 30.
29. Bamrungsap, S., Zhao, Z., Chen, T., Wang, L., Li, C., Fu, T., *et al.*, Nanotechnology in Therapeutics: A Focus on Nanoparticles as A Drug Delivery System. *Nanomedicine* [Internet], 7, 8, 1253–71, 2012 Aug [cited 2024 Aug 12] Available from: https://www.tandfonline.com/doi/abs/10.2217/nnm.12.87.
30. Sinha, R., Kim, G.J., Nie, S., Shin, D.M., Nanotechnology in cancer therapeutics: bioconjugated nanoparticles for drug delivery. *Mol. Cancer Ther.* [Internet], 5, 8, 1909–17, 2006 Aug 1 [cited 2024 Aug 12]; Available from: /mct/article/5/8/1909/234929/Nanotechnology-in-cancer-therapeutics.
31. Smith, A.M., Duan, H., Mohs, A.M., Nie, S., Bioconjugated quantum dots for *in vivo* molecular and cellular imaging. *Adv. Drug Deliv. Rev.*, 60, 11, 1226–40, 2008 Aug 17.
32. Matsumura, Y., A new concept for macromolecular therapeutics in cancer chemotherapy: mechanism of tumoritropic accumulation of proteins and the antitumor agent smancs. *Cancer Res.*, 6, 6397–6392, 1986.
33. Kubik, T., Bogunia-Kubik, K., Sugis, M., Nanotechnology on duty in medical applications. *Pharm. Biotechnol.*, 6, 17–33, 2005 [cited 2024 Aug 12]; Available from: https://www.ingentaconnect.com/content/ben/cpb/2005/00000006/00000001/art00004.
34. Pelicano, H., Martin, D., Xu, R., Oncogene, P.H., Pelicano, H., Martin, D.S., Xu, R.H., Huang, P., Glycolysis inhibition for anticancer treatment. *Oncogene*, 25, 34, 4633–4646, 2006.
35. Yatvin, M.B., Kreutz, W., Horwitz, B.A., Shinitzky, M., pH-sensitive liposomes: Possible clinical implications. *Science (80-)*, 210, 4475, 1253–5, 1980.
36. Jain, R.K., Barriers to drug delivery in solid tumors. *Sci. Am.* [Internet], 271, 1, 58–65, 1994 [cited 2024 Aug 12]; Available from: https://www.jstor.org/stable/24942767.
37. Sudimack, J. and Lee, R.J., Targeted drug delivery via the folate receptor. *Adv. Drug Deliv. Rev.*, 41, 2, 147-162, 2000.
38. Wu, Y., Sefah, K., Liu, H., DNA aptamer–micelle as an efficient detection/delivery vehicle toward cancer cells. *Proc. Natl. Acad. Sci.* [Internet], 107, 1, 5–10, 2010 [cited 2024 Aug 13]; Available from: https://www.pnas.org/doi/abs/10.1073/pnas.0909611107.

39. Leamon, C.P. and Reddy, J.A., Folate-targeted chemotherapy. *Adv. Drug Deliv. Rev.*, 56, 8, 1127–41, 2004 Apr 29.
40. Farokhzad, O.C. and Langer, R., Impact of nanotechnology on drug delivery. *ACS Nano* [Internet], 3, 1, 16–20, 2009 Jan [cited 2024 Aug 13]; Available from: https://pubs.acs.org/doi/abs/10.1021/nn900002m.
41. Portero, A., Remuñán-López, C., Criado, M.T., Alonso, M.J., Reacetylated chitosan micropheres for controlled delivery of anti-microbial agents to the gastric mucosa. *J. Microencapsul.*, 19, 6, 797–809, 2002 Nov.
42. Artursson, P., Lindmark, T., Davis, S.S., Illum, L., Effect of chitosan on the permeability of monolayers of intestinal epithelial cells (Caco-2). *Pharm. Res.*, 11, 9, 1358–61, 1994.
43. Fernández-Urrusuno, R., Calvo, P., Remuñán-López, C., Vila-Jato, J.L., Alonso, M.J., Enhancement of nasal absorption of insulin using chitosan nanoparticles. *Pharm. Res.*, 16, 10, 1576–81, 1999.
44. De Campos, A.M., Sánchez, A., Alonso, M.J., Chitosan nanoparticles: a new vehicle for the improvement of the delivery of drugs to the ocular surface. Application to cyclosporin A. *Int. J. Pharm.*, 224, 1–2, 159–168, 2001.
45. Al-Qadi, S., Grenha, A., Carrión-Recio, D., Al-Qadi, S., Grenha, A., Carrión-Recio, D., Seijo, B., Remuñán-López, C., Microencapsulated chitosan nanoparticles for pulmonary protein delivery: *in vivo* evaluation of insulin-loaded formulations. *J. Control. Release*, 157, 3, 383–390, 2012.
46. Pist, S., Goycoolea, F.M., Young, A., Smistad, G., Hiorth, M., Formulation of polysaccharide-based nanoparticles for local administration into the oral cavity. *Eur. J. Pharm. Sci.*, 96, 381–389, 2017.
47. Kolluru, L.P., Rizvi, S.A.A., D'Souza, M., D'Souza, M.J., Formulation development of albumin based theragnostic nanoparticles as a potential delivery system for tumor targeting. *J. Drug Targeting*, 21, 1, 77–86, 2013 Jan.
48. Wolfbeis, O.S., An overview of nanoparticles commonly used in fluorescent bioimaging. *Chem. Soc. Rev.* [Internet], 44, 14, 4743–68, 2015 Jul 7 [cited 2024 Aug 13]; Available from: https://pubs.rsc.org/en/content/articlehtml/2015/cs/c4cs00392f.
49. Liu, S., Yang, S., Hoasian, P.C., Intranasal administration of carbamazepine-loaded carboxymethyl chitosan nanoparticles for drug delivery to the brain. *J. Pharm. Sci.*, 13, 1, 72–81, 2018.
50. Jain, A. and S.J., Optimization of chitosan nanoparticles for colon tumors using experimental design methodology. *Artif. Cells Nanomed. Biotechnol.*, Taylor Fr [Internet], 44, 8, 1917–26, 2016 Nov 16 [cited 2024 Aug 13]; Available from: https://www.tandfonline.com/doi/abs/10.3109/21691401.2015.1111236.
51. Abo-Elseoud, W.S., Hassan, M.L., Sabaa, M.W., Basha, M., Hassan, E.A., Fadel, S.M., Chitosan nanoparticles/cellulose nanocrystals nanocomposites as a carrier system for the controlled release of repaglinide. *Int. J. Biol. Macromol.*, 111, 604–613, 2018.
52. Agarwal, T., Narayana, S.N.G.H., Pal, K., Pramanik, K., Giri, S., Banerjee, I., Calcium alginate-carboxymethyl cellulose beads for colon-targeted drug delivery. *Int. J. Biol. Macromol.*, 75, 409–417, 2015.
53. Hansen, K., Kim, G., Desai, K.G.H., Patel, H., Olsen, K.F., Curtis-Fisk, J., *et al.*, Feasibility Investigation of Cellulose Polymers for Mucoadhesive Nasal Drug Delivery Applications. *Mol. Pharm.*, 12, 8, 2732–41, 2015 Aug 3.
54. Tripathy, S. and Das, M.K., Dendrimers and their applications as novel drug delivery carriers. *J. Appl. Pharm. Sci.*, 3, 09, 142–9, 2013•japsonline.com [Internet]. 2013 [cited 2024 Aug 13]; Available from: https://japsonline.com/admin/php/uploads/1064_pdf.pdf.

55. Hawker, C., A new convergent approach to monodisperse dendritic macromolecules. *J. Chem. Soc., Chem. Commun.*, 15, 1010–1013, 1990.
56. Miller, L.L., Duan, R.G., Tully, D.C., Tomalia, D.A., Electrically conducting dendrimers. *J. Am. Chem. Soc.*, 119, 5, 1005–10, 1997 Feb 5.
57. Priya, P., Sivabalan, M., Jeyapragash, R., Dendrimer: a novel polymer. *Int. J. Res. Pharm. Chem.* [Internet], 3, 2, 495–501, 2013 [cited 2024 Aug 13]; Available from: https://scholar.google.com/scholar?hl=en&as_sdt=0%2C5&q=Dendrimer%3A+a+novel+polymer.+2013&btnG=#d=gs_cit&t=1723542107904&u=%2Fscholar%3Fq%3Dinfo%3AXwQBqf6hIhMJ%3Ascholar.google.com%2F%26output%3Dcite%26scirp%3D0%26hl%3Den.
58. Shegokar, R. and Müller, R.H., Nanocrystals: industrially feasible multifunctional formulation technology for poorly soluble actives. *Int. J. Pharm.* Elsevier [Internet], 399, 1–2, 129–39, 2010 [cited 2024 Aug 13]; Available from: https://www.sciencedirect.com/science/article/pii/S0378517310005715.
59. Patil, S.A., Rane, B.R., Bakliwal, S.R., Pawar, S.P., Nano suspension: At a glance. *Int. J. Pharm. Sci.* [Internet], 3, 947–60, 2011 [cited 2024 Aug 13]; Available from: https://scholar.google.com/scholar?cluster=17861571631002364444&hl=en&oi=scholarr#d=gs_cit&t=1723542616729&u=%2Fscholar%3Fq%3Dinfo%3AnOqb83m0yRgJ%3Ascholar.google.com%2F%26output%3Dcite%26scirp%3D0%26scfhb%3D1%26hl%3Den.
60. Bhatia, S., *Natural polymer drug delivery systems: Nanoparticles, plants, and algae*, pp. 33–93, Springer, Cham, 2016 Jan 1. https://doi.org/10.1007/978-3-319-41129-3.
61. Mudgil, M., Gupta, N., Nagpal, M., Pawar, P., Nanotechnology: Nanotechnology: a new approach for ocular drug delivery system. *Int. J. Pharm. Pharm. Sci.*, 4, 2, 105–12, 2012.
62. Pu, X., Sun, J., Li, M., He, Z., Formulation of Nanosuspensions as a New Approach for the Delivery of Poorly Soluble Drugs. *Curr. Nanosci.*, 5, 4, 417–27, 2012 Jun 18.
63. Gupta, R.B. and Kompella, U.B. (eds.), *Nanoparticle technology for drug delivery*, vol. 53, pp. 53–84, Taylor & Francis, New York, 2006.
64. Brunner, E., Reaktionsgeschwindigkeit in heterogenen Systemen. *Z. Phys. Chem.*, 47U, 1, 56–102, 1904 Jan 1.
65. De Waard, H., Hinrichs, W.L.J., Frijlink, H.W., A novel bottom-up process to produce drug nanocrystals: controlled crystallization during freeze-drying. *J. Control. Release*, 128, 2, 179–183, 2008.
66. Keck, C.M. and Müller, R.H., Drug nanocrystals of poorly soluble drugs produced by high pressure homogenisation. *Eur. J. Pharm. Biopharm.*, 62, 1, 3–16, 2006.
67. Nayak, U.Y., Naresh, D.N., Rajan, O.P., Preparation and characterization of Nelfinavir Mesylate nanocrystals by ball milling. *Int. J. Res. Pharm. Sci.* [Internet], 4, 4, 586–92, 2013 [cited 2024 Aug 13]; Available from: https://www.researchgate.net/publication/286315449_Preparation_and_characterization_of_Nelfinavir_Mesylate_nanocrystals_by_ball_milling.
68. Modi, D., Nirmal, J., Warsi, M.H., Bhatia, M., Hasan, N., Kesharwani, P., *et al.*, Formulation and development of tacrolimus-gellan gum nanoformulation for treatment of dry eye disease. *Colloids Surf. B Biointerfaces*, 211, 112255, 2022 Mar 1.
69. Gros, L., Ringsdorf, H., Schupp, H., Polymeric Antitumor Agents on a Molecular and on a Cellular Level? *Angew. Chem. Int. Ed. Engl.*, 20, 4, 305–25, 1981.
70. Ringsdorf, H., Structure and properties of pharmacologically active polymers. *J. Polym. Sci. Polym. Symp.*, 51, 135–53, 1975.
71. Ho, D.H., Brown, N.S., Yen, A., Holmes, R., Keating, M., Abuchowski, A., Newman, R.A., Krakoff, I.H., Clinical pharmacology of polyethylene glycol-L-asparaginase. *Drug Metab. Dispos.*, 14, 3, 349–352, 1986.
72. Greenwald, R.B., Choe, Y.H., McGuire, J., Conover, C.D., Effective drug delivery by PEGylated drug conjugates. *Adv. Drug Deliv. Rev.*, 55, 2, 4493–4495, 2011.

73. Zou, J., Jafr, G., Themistou, E., Yap, Y., Wintrob, Z.A.P., Alexandridis, P., Ceacareanu, A.C., Cheng, C., pH-Sensitive brush polymer-drug conjugates by ring-opening metathesis copolymerization. *Chem. Commun.*, 47, 15, 4493–4495, 2011.
74. Jain, K. and Jain, K., *The handbook of nanomedicine* . vol. 404, Humana Press, Totowa, 2008.
75. Joudeh, N. and Linke, D., Nanoparticle classification, physicochemical properties, characterization, and applications: a comprehensive review for biologists. *J. Nanobiotechnol.*, 20, 1, 262, 2022.
76. Vyas, S.P. and K.R.K., Targeted and Controlled Drug Delivery: Novel Carrier Systems. *Int. J. Pharm.*, 267, 1–2, 157, 2003 Nov.
77. Nogueira De Assis, D., Carla, V., Mosqueira, F., Mário, J., Vilela, C., Andrade, M.S., *et al.*, Release profiles and morphological characterization by atomic force microscopy and photon correlation spectroscopy of 99mTechnetium-fluconazole nanocapsules. *Int. J. Pharm.* [Internet], 349, 152–60, 2008 [cited 2024 Aug 13]; Available from: https://www.sciencedirect.com/science/article/pii/S037851730700662X.
78. Molpeceres, J., Aberturas, M.R., Guzman, M., Biodegrable nanoparticles as a delivery system for cyclosporine: Preparation and characterization. *J. Microencapsulation*, 17, 5, 599–614, 2000.
79. Zur Mühlen, A., Zur Mühlen, E., Niehus, H., Mehnert, W., Atomic force microscopy studies of Solid Lipid Nanoparticles. *Pharm. Res.*, 13, 9, 1411–6, 1996.
80. Stolnik, S., Illum, L., Davis, S.S., Long circulating microparticulate drug carriers. *Adv. Drug Deliv. Rev.*, 16, 195–214, 2012•Elsevier [Internet]. 1995 [cited 2024 Aug 13]; Available from: https://www.sciencedirect.com/science/article/pii/S0169409X1200289X.
81. Dobrovolskaia, M.A. and McNeil, S.E., Immunological properties of engineered nanomaterials. *Nat. Nanotechnol.*, 2, 8, 469–478, 2007.
82. Jiang, W., Kim, B., Rutka, J., Nanoparticle-mediated cellular response is size-dependent. *Nat. Nanotechnol.*, 3, 145–50, 2008 [cited 2024 Aug 13]; Available from: https://www.nature.com/articles/nnano.2008.30.
83. Shvedova, A., Castranova, V., Kisin, E., Schwegler-Berry, D., Murray, A., Gandelsman, V., *et al.*, Exposure to carbon nanotube material: Assessment of nanotube cytotoxicity using human keratinocyte cells. *J. Toxicol. Environ. Health Part A*, 66, 20, 1909–26, 2003.
84. Thrall, L., Study links TiO2 nanoparticles with potential for brain-cell damage. *Environ. Sci. Technol.*, 40, 14, 4326–7, 2006 Jul 15.
85. Oberdörster, G. and E.O., Nanotoxicology: an emerging discipline evolving from studies of ultrafine particles. *Environ. Health Perspect.*, 113, 7, 823–39, 2005 Jul [cited 2024 Aug 13]; Available from: https://ehp.niehs.nih.gov/doi/abs/10.1289/ehp.7339.
86. Lovrić, J., Bazzi, H.S., Cuie, Y., Fortin, G.R.A., Winnik, F.M., Maysinger, D., Differences in subcellular distribution and toxicity of green and red emitting CdTe quantum dots. *J. Mol. Med.*, 83, 5, 377–85, 2005 May.
87. Kirchner, C., Liedl, T., Kudera, S., Cytotoxicity of colloidal CdSe and CdSe/ZnS nanoparticles. *Nano Lett. ACS Publ.* [Internet], 5, 2, 331–8, 2005 Feb [cited 2024 Aug 13]; Available from: https://pubs.acs.org/doi/abs/10.1021/nl047996m.
88. Choi, A.O., Cho, S.J., Desbarats, J., Lovrić, J., Maysinger, D., Quantum dot-induced cell death involves Fas upregulation and lipid peroxidation in human neuroblastoma cells. *J. Nanobiotechnol.* Springer [Internet], 5, 1–3, 2007 [cited 2024 Aug 13]; Available from: https://link.springer.com/article/10.1186/1477-3155-5-1.
89. De Jong, W.H. and Borm, P.J., Drug delivery and nanoparticles: applications and hazards. *Int. J. Nanomed., Taylor Fr* [Internet], 3, 2, 133–49, 2008 [cited 2024 Aug 13]; Available from: https://www.tandfonline.com/doi/abs/10.2147/ijn.s596.

90. Graffignini., M.J., Corporate strategies for nanotech companies and investors in new economic times. *Nanotechnol. Law Bus.* [Internet], 6, 251, 2009 [cited 2024 Aug 13]; Available from: https://heinonline.org/hol-cgi-bin/get_pdf.cgi?handle=hein.journals/nantechlb6&section=25.
91. Chaudhry, Q. and Castle, L., Food applications of nanotechnologies: an overview of opportunities and challenges for developing countries. *Trends Food Sci. Technol.* Elsevier [Internet], 22, 11, 595–603, 2011 [cited 2024 Aug 13]; Available from: https://www.sciencedirect.com/science/article/pii/S0924224411000021.
92. Knezevic, I. and Griffiths, E., Biosimilars - Global issues, national solutions. *Biologicals*, 39, 5, 252–5, 2011 Sep.
93. Kickhoefer, V.A., Han, M., Raval-Fernandes, S., Poderycki, M.J., Moniz, R.J., Vaccari, D., *et al.*, Targeting vault nanoparticles to specific cell surface receptors. *ACS Nano*, 3, 1, 27–36, 2009 Jan.
94. Piccinno, F., Gottschalk, F., Seeger, S., Nowack, B., Industrial production quantities and uses of ten engineered nanomaterials in Europe and the world. *J. Nanopart. Res.* [Internet], 14, 9, 1–11, 2012 Sep 1 [cited 2024 Aug 13]; Available from: https://link.springer.com/article/10.1007/s11051-012-1109-9.
95. Oberdörster, G., Safety assessment for nanotechnology and nanomedicine: concepts of nanotoxicology. *J. Intern. Med.* [Internet], 267, 1, 89–105, 2010 Jan 1 [cited 2024 Aug 13]; Available from: https://onlinelibrary.wiley.com/doi/full/10.1111/j.1365-2796.2009.02187.x.
96. Hanahan, D. and Weinberg, R.A., Hallmarks of cancer: The next generation. *Cell* [Internet], 144, 5, 646–74, 2011 Mar 4 [cited 2024 Aug 13]; Available from: http://www.cell.com/article/S0092867411001279/fulltext.
97. Baudino, T.A., Targeted cancer therapy: the next generation of cancer treatment. *Curr. Drug Discov. Technol.* [Internet], 12, 1, 3–20, 2015 [cited 2024 Aug 13]; Available from: https://www.ingentaconnect.com/content/ben/cddt/2015/00000012/00000001/art00003.
98. Shen, B., Ma, Y., Yu, S., Ji, C., Smart Multifunctional Magnetic Nanoparticle-Based Drug Delivery System for Cancer Thermo-Chemotherapy and Intracellular Imaging. *ACS Appl. Mater. Interfaces*, 8, 37, 24502–8, 2016 Sep 21.
99. Huang, B., Abraham, W.D., Zheng, Y., Bustamante López, S.C., Luo, S.S., Irvine, D.J., Active targeting of chemotherapy to disseminated tumors using nanoparticle-carrying T cells. *Sci. Transl. Med.*, 7, 291, ra94, 2015 Jun 10.
100. Somani, S. and Dufès, C., Applications of dendrimers for brain delivery and cancer therapy. *Nanomedicine*, 9, 15, 2403–14, 2014 Oct 1.
101. Kaminskas, L.M., Boyd, B.J., Porter, C.J.H., Dendrimer pharmacokinetics: The effect of size, structure and surface characteristics on ADME properties. *Nanomedicine*, 6, 6, 1063–84, 2011 Aug.
102. Quintana, A., Raczka, E., Piehler, L., Lee, I., Myc, A., Majoros, I., *et al.*, Design and function of a dendrimer-based therapeutic nanodevice targeted to tumor cells through the folate receptor. *Pharm. Res.*, 19, 9, 1310–6, 2002 Sep 1.
103. Baker Jr, J.R., Dendrimer-based nanoparticles for cancer therapy. *ASH Educ. Program.*, 2009, 1, 708–719, 2009.
104. Cheng, Y., Zhao, L., Li, Y., Xu, T., Design of biocompatible dendrimers for cancer diagnosis and therapy: current status and future perspectives. *Chem. Soc. Rev.* [Internet], 40, 5, 2673–703, 2011 Apr 18 [cited 2024 Aug 13]; Available from: https://pubs.rsc.org/en/content/articlehtml/2011/cs/c0cs00097c.
105. Rastogi, V., Yadav, P., Bhattacharya, S.S., Mishra, A.K., Verma, N., Verma, A., *et al.*, Carbon Nanotubes: An Emerging Drug Carrier for Targeting Cancer Cells. *J. Drug Deliv.* [Internet], 2014, 1, 670815, 2014 Jan 1 [cited 2024 Aug 13] Available from: https://onlinelibrary.wiley.com/doi/full/10.1155/2014/670815.

106. Sanginario, A., Miccoli, B., Demarchi, D., Carbon Nanotubes as an Effective Opportunity for Cancer Diagnosis and Treatment. *Biosens*, 7, 1, 9, 2017 Feb 15 [cited 2024 Aug 13]; Available from: https://www.mdpi.com/2079-6374/7/1/9/htm.
107. Dinesh, B., Bianco, A., Ménard-Moyon, C., Designing multimodal carbon nanotubes by covalent multi-functionalization. *Nanoscale* [Internet], 8, 44, 18596–611, 2016 Nov 10 [cited 2024 Aug 13]; Available from: https://pubs.rsc.org/en/content/articlehtml/2016/nr/c6nr06728j.
108. Janib, S.M., Moses, A.S., MacKay, J.A., Imaging and drug delivery using theranostic nanoparticles. *Adv. Drug Deliv. Rev.*, 62, 11, 1052–63, 2010 Aug 30.
109. Bhojani, M.S., Van Dort, M., Rehemtulla, A., Ross, B.D., Targeted imaging and therapy of brain cancer using theranostic nanoparticles. *Mol. Pharm.* [Internet], 7, 6, 1921–9, 2010 Dec 6 [cited 2024 Aug 13]; Available from: https://pubs.acs.org/doi/abs/10.1021/mp100298r.
110. Moss, J.A., HIV/AIDS Review. *Radiol. Technol.* [Internet], 84, 3, 247–67, 2013 Jan 1 [cited 2024 Aug 13]; Available from: http://www.radiologictechnology.org/content/84/3/247.full.
111. Bartlett, J. and Moore, R.D., Improving HIV therapy. *Sci. Am.* [Internet], 279, 1, 84–7, 1998 [cited 2024 Aug 13]; Available from: https://www.jstor.org/stable/26057856.
112. Mamo, T., Moseman, E.A., Kolishetti, N., Salvador-Morales, C., Shi, J., Kuritzkes, D.R., *et al.*, Emerging nanotechnology approaches for HIV/AIDS treatment and prevention. *Nanomedicine*, 5, 2, 269–85, 2010.
113. Khalil, N.M., Carraro, E., Cótica, L.F., Mainardes, R.M., Potential of polymeric nanoparticles in AIDS treatment and prevention. *Expert Opin. Drug Deliv.* [Internet], 8, 1, 95–112, 2011 Jan [cited 2024 Aug 13]; Available from: https://www.tandfonline.com/doi/abs/10.1517/17425247.2011.543673.
114. Dev Jayant, R., Nanotechnology for the Treatment of NeuroAIDS. *J. Nanomed. Res.*, 3, 1, 00047, 2016.
115. Shah, L.K. and Amiji, M.M., Intracellular delivery of saquinavir in biodegradable polymeric nanoparticles for HIV/AIDS. *Pharm. Res.* [Internet], 23, 11, 2638–45, 2006 Nov 13 [cited 2024 Aug 13]; Available from: https://link.springer.com/article/10.1007/s11095-006-9101-7.
116. Mallipeddi, R. and Rohan, L.C., Progress in antiretroviral drug delivery using nanotechnology. *Int. J. Nanomed.*, 533–547, 2010.
117. Destache, C.J., Belgum, T., Christensen, K., Shibata, A., Sharma, A., Dash, A., Combination antiretroviral drugs in PLGA nanoparticle for HIV-1. *BMC Infect. Dis.* [Internet], 9, 1, 1–8, 2009 Dec 9 [cited 2024 Aug 13]; Available from: https://link.springer.com/articles/10.1186/1471-2334-9-198.
118. Aggarwal, B.B., Van Kuiken, M.E., Iyer, L.H., Harikumar, K.B., Sung, B., Molecular targets of nutraceuticals derived from dietary spices: potential role in suppression of inflammation and tumorigenesis. *Exp. Biol. Med.*, 234, 8, 825-849, 2009.
119. Mcclements, D.J., Nanoscale Nutrient Delivery Systems for Food Applications: Improving Bioactive Dispersibility, Stability, and Bioavailability. *J. Food Sci.* [Internet], 80, 7, N1602–11, 2015 Jul 1 [cited 2024 Aug 13]; Available from: https://onlinelibrary.wiley.com/doi/full/10.1111/1750-3841.12919.
120. Mohanty, C. and Sahoo, S.K., The *in vitro* stability and *in vivo* pharmacokinetics of curcumin prepared as an aqueous nanoparticulate formulation. *Biomaterials*, 31, 25, 6597–611, 2010 Sep 1.
121. Carvalho D de, M., Takeuchi, K.P., Geraldine, R.M., de Moura, C.J., Torres, M.C.L., Production, solubility and antioxidant activity of curcumin nanosuspension. *Food Sci. Technol.* [Internet], 35, 1, 115–9, 2015 [cited 2024 Aug 13]; Available from: https://www.scielo.br/j/cta/a/MngxC7yjbmBGnY4PGwQkKkh/.
122. Shaikh, J., Ankola, D.D., Beniwal, V., Singh, D., Kumar, M.N.V.R., Nanoparticle encapsulation improves oral bioavailability of curcumin by at least 9-fold when compared to curcumin

administered with piperine as absorption enhancer. *Eur. J. Pharm. Sci.*, 37, 3–4, 223–30, 2009 Jun 28.

123. Sasaki, H., Sunagawa, Y., Takahashi, K., Imaizumi, A., Fukuda, H., Hashimoto, T., *et al.*, Innovative Preparation of Curcumin for Improved Oral Bioavailability. *Biol. Pharm. Bull.*, 34, 5, 660–5, 2011 May 1.
124. Pridgen, E.M., Alexis, F., Farokhzad, O.C., Polymeric nanoparticle drug delivery technologies for oral delivery applications. *Expert Opin. Drug Deliv.* [Internet], 12, 9, 1459–73, 2015 Sep 2 [cited 2024 Aug 13]; Available from: https://www.tandfonline.com/doi/abs/10.1517/17425247.2015.1018175.
125. Roy, K., Mao, H.Q., Huang, S.K., Leong, K.W., Oral gene delivery with chitosan–DNA nanoparticles generates immunologic protection in a murine model of peanut allergy. *Nat. Med.*, 5, 4, 387–91, 1999 Apr [cited 2024 Aug 13]; Available from: https://www.nature.com/articles/nm0499_387.
126. Chew, J.L., Wolfowicz, C.B., Mao, H.Q., Leong, K.W., Chua, K.Y., Chitosan nanoparticles containing plasmid DNA encoding house dust mite allergen, Der p 1 for oral vaccination in mice. *Vaccine*, 21, 21–22, 2720–9, 2003 Jun 20.
127. Li, G., Liu, Z., Liao, B., Zhong, N., Induction of Th1-Type Immune Response by Chitosan Nanoparticles Containing Plasmid DNA Encoding House Dust Mite Allergen Der p 2 for Oral Vaccination in Mice. *Cell. Mol. Immunol.*, 6, 1, 45–50, 2009 [cited 2024 Aug 13]; Available from: https://www.nature.com/articles/cmi20096.
128. Eldridge, J.H., Hammond, C.J., Meulbroek, J.A., Staas, J.K., Gilley, R.M., Tice, T.R., Controlled vaccine release in the gut-associated lymphoid tissues. I. Orally administered biodegradable microspheres target the peyer's patches. *J. Control. Release*, 11, 1–3, 205–14, 1990 Jan 1.
129. Eldridge, J.H., Meulbroek, J.A., Staas, J.K., Tice, T.R., Gilley, R.M., Vaccine-Containing Biodegradable Microspheres Specifically Enter the Gut-Associated Lymphoid Tissue Following Oral Administration and Induce a Disseminated Mucosal Immune Response. *Adv. Exp. Med. Biol.* [Internet], 251, 191–202, 1989 [cited 2024 Aug 13]; Available from: https://link.springer.com/chapter/10.1007/978-1-4757-2046-4_18.
130. Zhu, Q., Talton, J., Zhang, G., Cunningham, T., Wang, Z., Waters, R.C., *et al.*, Large intestine–targeted, nanoparticle-releasing oral vaccine to control genitorectal viral infection. *Nat. Med.*, 18, 8, 1291–6, 2012 Jul 15 [cited 2024 Aug 13]; Available from: https://www.nature.com/articles/nm.2866.
131. Kreuter, J., Influence of the surface properties on nanoparticle-mediated transport of drugs to the brain. *J. Nanosci. Nanotechnol.*, 4, 5, 484–8, 2004 May.
132. Pardridge, W.M., Drug and gene targeting to the brain with molecular trojan horses. *Nat. Rev. Drug Discov.*, 1, 2, 131–9, 2002 Feb [cited 2024 Aug 13]; Available from: https://www.nature.com/articles/nrd725.
133. Ji, B., Maeda, J., Higuchi, M., Inoue, K., Akita, H., Harashima, H., *et al.*, Pharmacokinetics and brain uptake of lactoferrin in rats. *Life Sci.*, 78, 8, 851–5, 2006 Jan 18.
134. Scherrmann, J.M. and Temsamani, J., The use of Pep: Trans vectors for the delivery of drugs into the central nervous system. *Int. Congr. Ser.*, 1277, 199–211, 2005 Apr 1.
135. Gabathuler, R., Arthur, G., Kennard, M., Chen, Q., Tsai, S., Yang, J., *et al.*, Development of a potential protein vector (NeuroTrans) to deliver drugs across the blood–brain barrier. *Int. Congr. Ser.*, 1277, 171–84, 2005 Apr 1.
136. Taylor, R., Hare, J.P., Abdul-Sada, A.K., Kroto, H.W., Isolation, separation and characterisation of the fullerenes C 60 and C 70 : the third form of carbon. *J. Chem. Soc. Chem. Commun.* [Internet], 0, 20, 1423–5, 1990 Jan 1 [cited 2024 Aug 13]; Available from: https://pubs.rsc.org/en/content/articlelanding/1990/c3/c39900001423.

137. Petros, R.A. and Desimone, J.M., Strategies in the design of nanoparticles for therapeutic applications. *Nat. Rev. Drug Discov.*, 9, 8, 615–27, 2010 Jul 9 [cited 2024 Aug 13]; Available from: https://www.nature.com/articles/nrd2591.
138. Pandit, A. and Zeugolis, D.I., Twenty-Five Years of Nano-Bio-Materials: Have We Revolutionized Healthcare? *Nanomedicine* [Internet], 11, 9, 985–7, 2016 May 1 [cited 2024 Aug 13]; Available from: https://www.tandfonline.com/doi/abs/10.2217/nnm.16.42.

11

Multifunctional Materials in Engineering and Processing Engineering of Multifunctional Materials

Akash Kumar[1]*, Srasti Yadav[1] and Shelly Kujur[2]

[1]*Division of Chemistry, School of Basic Sciences, Galgotias University, Greater Noida, Uttar Pradesh, India*

[2]*Applied Sciences and Humanities, Galgotias College of Engineering and Technology, Greater Noida, Uttar Pradesh, India*

Abstract

Integrating diverse functions in a single material system is a fundamental task, especially if those functions appear to eliminate each other. There are numerous contemporary applications that enhance the quality of life and address significant global concerns, made possible by an understanding of function-structure links and the development of expertise in the system approach for multifunctionality. In this way, the developments can be shortened and products based on multifunctional materials can become more and more individualized and adaptive. Multifunctional materials are considered so as to meet the specific requirements through customized properties. Multifunctional materials that can respond to external stimuli are referred to as smart materials since they mimic the behavior of natural materials in this way. Furthermore, the creation of materials with processes like those found in nature is now possible because of the development of biomimetics in material science: constructing whole things from molecules. This chapter focuses on Multifunctional materials in engineering and processing Engineering reported in the literature and the processing means developed.

***Keywords*:** Multifunctional materials, nanomaterials, characterization, application

11.1 Introduction

This is the "golden age of materials," wherein new materials are being developed on the spot to meet the growing demand from people for more capable mobility. These materials are chosen for their multifunctionality because, although they frequently draw inspiration from biological analogs, they are engineered to carry out several tasks by carefully combining several functional characteristics. Materials with desired electrical, magnetic, optical, thermal, or other characteristics that meet previously unachievable performance

**Corresponding author*: kumarakash@galgotiasuniversity.edu.in

Divya Bajpai Tripathy, Anjali Gupta and Arvind Kumar Jain (eds.) Multifunctional Materials: Engineering and Biological Applications, (273–294) © 2025 Scrivener Publishing LLC

parameters are referred to as multifunctional materials [1]. These characteristics are coupled with one another or with certain mechanical characteristics like strength, ductility, and stiffness. The shape and capabilities of the products created, as well as the design philosophy itself, will be profoundly altered by the introduction of such materials into manufacturing. Since composite materials may attain multifunctionality by carefully choosing several components to mix to make the material, they can be regarded as multifunctional materials [2]. On the other hand, nature provides the best multifunctional materials; almost all biological material systems are composites, which are usually endowed with a superior set of properties since they need to be able to perform a variety of functions well for example, bones are naturally occurring organic/inorganic composites made of collagen and hydroxyapatite that have various important functions like producing blood, moving muscles, and supporting the body [3]. Moreover, a wide range of multifunctional materials is demonstrated by biology, including material systems that combine structural capacity with fully integrated sensing, actuation, and healing capabilities [4]. As a result, a material's multifunctionality might be obtained by modeling natural structure formations by developing the material from the micro- to macroscales (a process known as the "bottom-up approach") [5]. A novel alternative strategy in material synthesis is computationally driven material development. It is predicated on multiscale material and process modeling that spans length scales and a wide range of time. Different models atomistic ones were first employed. The development of this field will make it possible to build at the atomic and molecular level with several functions by defining certain attributes [6]. It might be argued that the process of creating a unique multifunctional material is expensive and carries a considerable risk [7]. As a result, the industry's and consumers' approval of this kind of material will determine how financially viable such an endeavor may be. Furthermore, the processing methods that are now accessible restrict the use of these materials. The high processing costs are still a result of costly and time-consuming procedures. Thus, a great deal of study has to be done on the environmentally friendly and mass-processing of these materials in addition to the current work on the creation of new materials [8]. Nowadays, it is thought that ecoefficiency in particular is necessary for a new technology to be widely accepted by businesses, governments, and the general public. The current trend is producing more goods and services with less waste and pollution while utilizing fewer resources [9]. In terms of mass processing, the commercialization of goods based on multifunctional materials can only be possible with technologies that can support large production volumes at much reduced costs [10]. The recognition of these cutting-edge materials' distinct properties and the creative flexibility they afford designers is another important component in their widespread adoption [11]. The design of the parts, subassemblies, and ultimately the product itself must be fundamentally changed to use these materials to mass-produce commodities in the real world [12]. The goals of the contemporary design philosophy are not those of composite materials with particular multifunctionalities, but rather those of homogeneous materials. Structural composite materials, smart or intelligent materials, and nanostructured materials are the three main categories of multifunctional materials that have been identified and are under consideration [13]. The present chapter tries to outline the development of multifunctional materials and go over the major challenges and processing issues.

11.2 Synthesis and Fabrication of Multifunctional Materials

Materials can be categorized as building materials (such as insulation for heat insulation), nuclear materials, aerospace materials, refractory materials for high temperatures, and biomaterials based on their intended purposes [14]. Materials' structural integrity can be assessed by microscopy or spectroscopic analysis. The structure of the materials can be divided into four categories: macrostructure (large-scale structure), microstructure, mesostructured, and hierarchical. The materials can be classed into four categories: 0D, 1D, 2D, and 3D. structure. Two common approaches are used to synthesize nanomaterials: top-down and bottom-up (Figure 11.1). The two strategies seek to either increase particle size through bottom-up methods or decrease it through top-down methods. In order to accomplish this goal, physical, chemical, and biological techniques can be applied. Bulk or large-scale materials are used in the top-down method. It uses both chemical and physical techniques to reduce particle size [15]. Nanomaterial synthesis is cutting-edge. There is not a clear division between different approaches [16, 17].

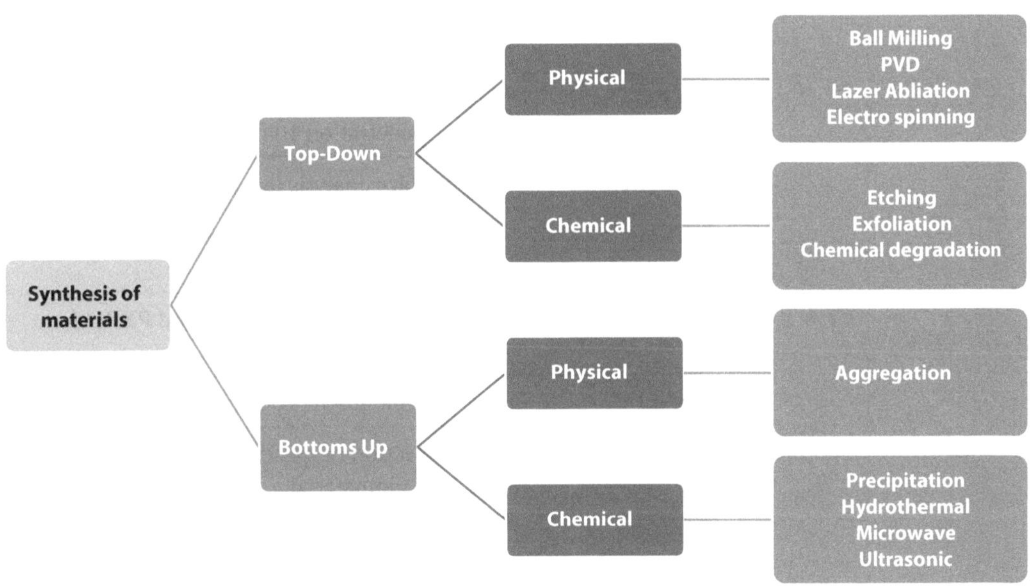

Figure 11.1 Synthesis and fabrication routes of multifunctional materials.

11.3 Characterization Techniques for Multifunctional Materials

The materials should be characterized after synthesis in order to verify the target materials' characteristics and ensure a successful synthesis. A multitude of analytical techniques are employed in the characterization of materials. Based on their principles, applications, and the data gleaned from their research, the characterization approaches can be categorized in Table 11.1 [16].

Table 11.1 Some common characterization techniques for multifunctional materials.

Details	Type	Technique	Remark
Composition	Qualitative	Elemental analysis	Analysis of C, H, N, S, O, X (F, Cl, Br, I)
		Mass spectrometry	Information about the molecular weight and m/z value.
	Quantitative	AAF	Element in solution measured in ppm
		ICP/AAS	composition of inorganic-based nanomaterials
		XRF	Identify the oxidation state and the ratio of the multi-valence element
		XPS	For surface examination with a depth of penetration 10–100 Å
		EDAX	Surface analysis technique
		TGA	Differentiate between components according to temperature
Structure	Crystal	XRD	SXRD for larger crystals and PXRD for powder or small crystals
		SAED	Cell parameters determination
	Specifics	EXAFS	Detection of specific elements in a nanomaterial
		EELS	TEM-based analysis
		XPS	Peak deconvolution can be used to determine the local structure of each element in the sample.
	Linkage	FT-IR	Identify bonds
		NMR	Solid and Liquid

(*Continued*)

Table 11.1 Some common characterization techniques for multifunctional materials. (*Continued*)

Details	Type	Technique	Remark
Morphology	Light electron	Optical light microscopy	magnification 1000×
		TEM	For particle size and lattice fringe determination
		SEM	Particle size greater than 25 nm
Size	Probe: Solid	Scanning probe microscopy	Atomic force microscopy (AFM), STM, scanning tunneling microscopy (STM), and scanning probe electrochemistry (SPE)).
		SEM	
		TEM	
		AFM	Thickness of 2D material
	Liquid	Dynamic light scattering	Light scatters in all directions
		Laser diffraction spectroscopy	Examine the laser beam's diffraction as a result of its interaction with any particles.
		Nanoparticle tracking analysis	Offer a way for observing and analyzing particles in liquids moving with Brownian motion
Optical properties	Absorption	UV-Vis	Analyzing the electrical transition that occurs in nanomaterials.
		Diffuse reflectance spectroscopy	Determine the optical band gap for solid-state nanoparticles
	Emission	Fluorescence	Study the charge transfer characterization
		Photoluminescence	Study the particles with long-lifetime emission

11.4 Structure-Property Relationships in Multifunctional Materials

The relationship of structure-property in multifunctional materials refers to the intricate connections between the arrangement or composition of the material's structure and its various physical, mechanical, chemical, and functional properties. Understanding this relationship is important for designing and engineering the materials with personalized properties to suit precise applications. Here are structure-property relationships that work in multifunctional materials [17].

11.4.1 Structure and Composition

The composition and structure of materials play a vital role in defining their properties. For instance, in a composite material, the arrangement of different phases or constituents at the microscale influences properties like strength, stiffness, conductivity, and thermal stability [18].

11.4.2 Crystal Structure

In crystalline materials, the atomic arrangement within the crystal lattice dictates properties such as electrical conductivity, optical behavior, and mechanical strength. Defects or modifications in the crystal structure can lead to changes in these properties [19].

11.4.3 Interfaces and Boundaries

Interfaces between different phases or grains within a material can significantly impact its properties. Grain boundaries, for example, can affect mechanical strength, electrical conductivity, and diffusion rates in polycrystalline materials [20].

11.4.4 Processing Methods

The techniques used to process or manufacture a material can influence its microstructure and, consequently, its properties. Factors such as temperature, pressure, and processing time can affect grain size, porosity, and phase distribution, thereby altering material properties [21].

11.4.5 Phase Transitions

Phase transitions, such as melting, solidification, and phase transformations, can introduce changes in material properties. For example, the transition from an amorphous to a crystalline structure can affect mechanical strength and transparency.

11.4.6 Doping and Alloying

Introducing dopants or alloying elements into a material can modify its electronic, optical, and mechanical properties. This modification occurs through changes in the material's band structure, lattice parameters, or defect concentration [22].

11.4.7 Nanostructuring

Manipulating the material at the nanoscale can lead to unique properties due to quantum effects, increased surface area, and enhanced mechanical properties. Nanostructured materials often exhibit improved strength, catalytic activity, and electrical conductivity compared to their bulk counterparts [23].

11.4.8 Functionalization

Adding functional groups or molecules to the surface of a material can impart specific functionalities such as chemical reactivity, adsorption capacity, or biocompatibility. Functionalization strategies are commonly employed in materials for applications such as sensors, membranes, and biomedical devices [24].

By understanding and controlling these structure-property relationships, researchers and engineers can design multifunctional materials with tailored properties to meet the demands of diverse applications, including electronics, energy storage, healthcare, and environmental remediation [25].

11.5 Processing of Multifunctional Materials [26–28]

11.5.1 Processing Techniques for Multifunctional Materials

The techniques for processing multifunctional materials include different approaches for constructing the materials exhibiting many functionalities like thermal, electrical, mechanical, and optical properties. These materials got uses in various fields such as biomedical, electronics, energy aerospace, and many more. There are some processing techniques Here are some common processing techniques which have been used for multifunctional materials.

11.5.1.1 Material Manufacture

Layer-by-Layer Deposition: Altering the way that various materials are deposited in layers to produce composites with specific characteristics.

Melt Blending: Combining several elements in a molten state, then solidifying the mixture to create a composite.

Polymerization In Situ: Concurrent monomer polymerization and functional filler dispersion to produce composites with improved characteristics.

11.5.1.2 Synthesis of Nanomaterials

Chemical Vapor Deposition (CVD): Development of nanowires (NWs) or thin films on a reactant through vapor phase chemical reactions.

Sol-Gel Process: Inorganic material formation from a predecessor solution via condensation and hydrolysis processes.

Electrospinning: Method of creating nanofibers from polymer melts or solutions using electrostatic forces.

11.5.1.3 *Hybrid Material Integration*

Hybridization: Combining materials with various macro- or nanoscale characteristics to produce synergistic effects.

Interfacial Engineering: Interactions between various components are optimized to improve functionality and compatibility.

11.5.1.4 *Post-Processing Methodologies/Techniques*

Heat Treatment: Enhancing a material's mechanical strength, electrical conductivity, or crystallinity by annealing or sintering.

Functionalization of Surface: Functional molecules or coatings are applied to a material's surface to change its characteristics.

Mechanical Processing: Materials can be tailored in terms of shape and surface properties by grinding, milling, or polishing.

It is possible to design multifunctional materials with particular qualities and functions appropriate for a range of applications by customizing and combining these processing techniques. Furthermore, new developments in materials science and engineering keep broadening the spectrum of processing techniques accessible for the creation of multifunctional materials.

11.5.2 Microstructural Evolution During Processing of Multifunctional Materials

11.5.2.1 *Mechanical Milling*

High-energy ball milling is used in mechanical milling to cause alloying, fracturing, and mechanical deformation in powders. Grain refinement, phase transition, and defect creation are all parts of the microstructural evolution that produce fine-grained materials with improved mechanical characteristics.

11.5.2.2 *Hot Isostatic Pressing (HIP)*

Powders or porous materials are combined with high pressure and temperature using HIP. The process of microstructural evolution results in increased mechanical strength and structural homogeneity through densification, grain growth, and porosity elimination.

11.5.2.3 *Extrusion and Sintering*

Ceramic, metal, and polymer-based materials are frequently shaped and densified by extrusion followed by sintering. During sintering, microstructural evolution includes grain growth, necking, and particle rearrangement, resulting in the formation of dense, interconnected structures with customized porosity and grain size distributions.

11.5.2.4 Additive Manufacturing (3D Printing)

Selective laser melting (SLM) and stereolithography are two examples of additive manufacturing processes that construct components layer by layer from digital designs. During laser or UV exposure, solidification, phase transformation, and grain growth occur, and processing parameters can be controlled to affect the material's properties and microstructure.

11.5.3 Effect of Processing Parameters on Properties of Multifunctional Materials

11.5.3.1 Temperature and Pressure

Temperature and pressure throughout the manufacturing process are examples of processing factors that can have a big impact on the material's microstructure and phase composition. For instance, in the case of composite materials, changes in temperature and pressure during sintering or curing can affect how the phases of reinforcement are distributed, resulting in differences in the materials' mechanical, thermal, and electrical characteristics.

11.5.3.2 Chemical Composition and Stoichiometry

In multifunctional materials, exact control over stoichiometry and chemical composition is essential. Changes in composition can affect attributes like conductivity, magnetism, or optical behavior by causing variations in the creation of phases, crystal structures, and defect concentrations.

11.5.4 Mechanical and Structural Properties of Multifunctional Materials

High-energy ball milling is used in mechanical milling to cause alloying, fracturing, and mechanical deformation in powders. Grain refinement, phase transition, and defect creation are all parts of the microstructural evolution that produce fine-grained materials with improved mechanical characteristics. Materials can be tailored in terms of shape and surface properties by grinding, milling, or polishing. It is possible to design multifunctional materials with particular qualities and functions appropriate for a range of applications by customizing and combining these processing techniques. Furthermore, new developments in materials science and engineering keep broadening the spectrum of processing techniques accessible for the creation of multifunctional materials.

11.5.5 Structural Properties of Multifunctional Materials

The composition and structure of materials play a vital role in defining their properties. For instance, in a composite material, the arrangement of different phases or constituents at the microscale influences properties like strength, stiffness, conductivity, and thermal stability.

11.6 Multifunctional Composites and Nanocomposites

Materials having innovative functionality and higher performance achieved through sophisticated nano-processing processes are referred to as multifunctional nanocomposites. In this domain, the principal objective is to create functional nanocomposites that can harvest energy by adapting to changes in temperature, light, and vibration. Likewise, highly ductile cementitious materials that exhibit self-healing properties.

11.6.1 Metal-Based Nanomaterials

These nanoparticles include gold and silver, which have the unusual capacity to change light into heat and have been shown to be highly effective in the treatment of tumors. Treatment options for the tumors include photothermal therapy, which includes causing heat damage to the tissues and cells that contain the nanoparticles, or heat-triggered drug release, in which pharmaceuticals are housed inside the nanoparticle. Core-shell magnetic nanoparticles were created by Lee *et al.* (2013) for photothermal treatment [29]. These magnetic nanoparticles experience variations in thermal energy due to fluctuations in their magnetic characteristics when exposed to an alternating current or magnetic field. The thermal energy was measured with a high-radiofrequency heating apparatus. Because radio frequencies penetrate much deeper than magnetic fields, they are typically used to cause fluctuations in magnetic nanoparticles. Because they can be identified using magnetic resonance imaging (MRI), the magnetic nanoparticles have a second function: they let researchers track how the particles travel throughout the human body when a patient receives treatment. Abdelhamid *et al.* reported that $AgFeO_2$ has a strong antibacterial effect on a variety of bacterial species [30]. Fruits and vegetables were tested for freshness using graphene-enhanced Raman spectroscopy (GERS) with Ag NPs. Surface-assisted laser desorption-ionization mass spectrometry (SALDI-MS) can be utilized to micro-extract proteins and other analytes using silver nanoparticles as a surface. Chitosan can also be added to it in order to separate and detect bio thiols [31].

11.6.2 Sensors

These multifunctional nanomaterials are also useful for developing effective, multifunctional sensors. In a variety of fields, including robotics, energy harvesting, *in vitro* diagnostics, displays, and innovative therapeutics, flexible mechanical and electrical sensing devices have shown considerable promise. The development of e-skins, wearable and implantable electronics, and sophisticated sensors with the capacity for self-healing and self-power has been greatly aided by multifunctional materials. Graphene, hybrid composites, organic/inorganic matrix arrays, NWs or nanotube assemblies, and other very flexible, ultrasensitive, and transparent multifunctional materials have all been used to make these astounding advancements. Fe_3O_4@C@CdTe core/shell microspheres with multifunctional magnetic and luminous properties were designed by Wang *et al.* and can probe Cu^{2+} ions in aqueous solutions [32]. When tested in water, the resultant nanosensor shows a higher sensitivity to Cu^{2+} ions than to other competing metal ions. Additionally, the core/shell Fe_3O_4 microspheres' high magnetic characteristic aids in the collection and separation of Cu^{2+} ions using a standard magnet.

11.6.3 Self-Healing Materials

Materials that are remodeling and self-healing Scientific attention is piqued by the idea of sophisticated structures with the capacity for remodeling or self-healing [33]. Material phases that self-generate in reaction to damage have been proven to incorporate biologically inspired self-healing capabilities. Through *in situ* polymerization and the release of microencapsulated phases, healing is accomplished. As an alternative, consider microvascular material architectures, wherein vascular circulatory systems are integrated into materials to act as a conduit for the transfer of functionality. It is now feasible to continuously heal and repair vascular materials. Nevertheless, the microvascular network can also be used for many other functions, such as self-heating or cooling, self-diagnosis and assessment, and self-morphing structures healing agents and self-healing with it [34]. Highly cross-linked polymeric materials containing multiple furans and multiple maleimides produced by the Diels-Alde reaction make up the former category. When the temperature rises over 120°C, the "intermonomer" connections break, but they reattach when the temperature drops. The procedure can be utilized to repair the broken polymer segments and is completely reversible. The latter group includes an agent that is liquid at least at the temperature necessary for healing. Usually, the healing ingredient is embedded and contained in the matrix of the composites. Due to the capillary effect, the healing agent is released into the crack planes as soon as the capsules are destroyed, sealing the cracks. There are two types of healing agents: single-component and two-component. The presence of air is necessary for the curing of single-component agents like polyvinyl acetate and cyanoacrylate. They are therefore unsuitable for deep-seated fracture healing within the composite matrix. The two components of the agent are a hardener and a polymerizable resin. In order for the fractured portion to be bound, polymerization is triggered when they combine. With the use of capsules with delicate walls, the drug is implanted within a composite matrix.

11.7 Electrical and Thermal Properties of Multifunctional Materials

11.7.1 Electrical Conductivity of Multifunctional Materials

Materials that display several functions at once, such as electrical conductivity, mechanical strength, thermal characteristics, etc., are referred to as multifunctional materials. These materials' electrical conductivity can vary greatly based on their composition, structure, and intended use (Table 11.2). The following are typical instances of materials with multiple uses and different electrical conductivity characteristics such as carbon nanotubes (CNTs), graphene, conductive polymers, metal-organic frameworks (MOFs), hybrid perovskites, composite materials, and conductive inks and coatings. Multifunctional materials' electrical conductivity may be adapted to suit needs in a variety of applications, from electronics and energy storage to sensing and structural materials, by using a variety of synthesis procedures, doping strategies, and structural alterations [35, 36].

Table 11.2 Electrical property of some multifunctional materials.

No.	Material	Formula	Form	Piezoelectric constant (pm/V or pC/V)	References
1	Ammonium dihydrogen phosphate (ADP)	$NH_4H_2PO_4$	Single crystal	d36=48	[37]
2	Barium titanate	$BaTiO_3$	Single crystal	d15=587	[38]
3	Barium titanate	$BaTiO_3$	Polycrystalline ceramic	d15=270	[39]
4	Lead zirconate titanate (PZT)	$PbZr_{0.6}Ti_{0.40}O_3$	Polycrystalline ceramic	d33=117	[40]
5	Lead lanthanum zirconate titanate (PLZT)	$Pb_{0.925}La_{0.5}Zr_{0.56}Ti_{0.44}O_3$	Polycrystalline ceramic	d33=545	[41]
6	Polyvinylidene fluoride	$(CH_2CF_2)_n$	Polycrystalline ceramic	d31=28	[42]
7	Potassium dihydrogen phosphate (KDP)	KH_2PO_4	Single crystal	d36=21	[43]
8	Quartz	SiO_2	Single crystal	d11=2.3	[44]
9	Zinc oxide	ZnO	Single crystal	d33=12	[45]

11.7.2 Thermal Conductivity of Multifunctional Materials

Multifunctional materials' thermal conductivity must be understood and controlled in order to advance a wide range of technological applications. To fully realize the promise of these materials in tackling modern issues and promoting innovation across sectors, multidisciplinary research activities must continue.

The thermal conductivity is denoted by k and quantifies a material's capacity to conduct the heat. It is ruled by Fourier's Law of Heat Conduction, which states that the rate of heat transfer (Q) through a material is directly proportional to the temperature gradient (∇T) and the cross-sectional area (A) perpendicular to the direction of heat flow.

$$Q = -k \cdot A \cdot \nabla T$$

The inverse relationship between thermal conductivity and thermal resistance (R) is often utilized, where $R = 1/k$ [35].

Factors Influencing Thermal Conductivity

There are several factors that influence the thermal conductivity of multifunctional materials such as material composition, microstructure, and temperature.

11.8 Optical and Magnetic Properties of Multifunctional Materials

11.8.1 Optical Properties of Multifunctional Materials

The optical characteristics of multifunctional materials are important for many applications, such as biomedical imaging, photonics, optoelectronics, and sensing. The following are some crucial elements of multifunctional materials' optical characteristics:

11.8.1.1 Transparency and Opacity

The degree of transparency or opacity shown by multifunctional materials can vary based on their composition, structure, and manufacturing process. While opaque materials are used in shielding applications and photovoltaic devices, transparent materials are necessary for applications like windows, displays, and optical lenses.

11.8.1.2 Optical Absorption and Transmission

The intrinsic optical characteristics of multifunctional materials as well as the existence of impurities or imperfections control how much light they absorb and transmit. Photodetectors, solar cells, and optical filters use materials with specific absorption spectra, whereas optical fibers and imaging applications favor materials with excellent transmission.

11.8.1.3 Photoluminescence and Fluorescence

Certain materials with several uses display fluorescence or photoluminescence, which is the ability to absorb light at one wavelength and release it at a different one. These materials are used in bioimaging, displays, sensors, and light-emitting diodes (LEDs).

11.8.1.4 Bandgap Engineering

It is possible to modify a material's optical characteristics, such as wavelengths of emission and absorption, by adjusting its bandgap. This is accomplished by methods including alloying, doping, heterostructure engineering, and quantum confinement. It is essential for maximizing the efficiency of optoelectronic devices, such as LEDs and solar cells.

11.8.1.5 Plasmonic and Metamaterial Effects

Nanoscale light-matter interactions can be precisely controlled thanks to plasmonic and metamaterial structures. Multifunctional materials can display unusual optical phenomena including surface plasmon resonance, negative refractive index, and perfect absorption by manipulating the geometry, composition, and arrangement of nanostructures. This opens up new possibilities for sensing, imaging, and energy harvesting.

11.8.1.6 *Chiral and Optical Activity*

Chiral multifunctional materials spin the incident light's plane of polarization, a property known as optical activity. Chiral sensors, optical switches, and metamaterials with customized chirality all make use of this characteristic.

11.8.1.7 *Biocompatibility and Bioimaging*

Multifunctional materials having optical characteristics compatible with biological tissues are crucial for diagnosis, treatment, and imaging in biomedical applications. Because of their special optical qualities, materials including CNTs, quantum dots, and upconversion nanoparticles are designed for photothermal treatment, drug delivery, and bioimaging.

Comprehending and using the optical characteristics of multifunctional materials facilitates the creation of cutting-edge technologies with uses in a variety of industries, including energy and telecommunications, healthcare, and environmental monitoring. Further investigation in this field is expected to open up new avenues for the development of optical materials and devices with improved functionality and performance [46].

11.8.2 Magnetic Properties of Multifunctional Materials

The capacity of multifunctional materials to display magnetic behavior in addition to their other capabilities is referred to as their magnetic characteristics. These qualities have a wide range of applications in industries including electronics, energy, healthcare, and more, so understanding them is essential. The main features of the magnetic characteristics of multifunctional materials are explained here.

11.8.2.1 *Types of Magnetic Property*

Ferromagnetism - Materials exhibiting spontaneous magnetization, even in the absence of an external magnetic field, and maintaining a magnetic moment such as nickel, iron, and cobalt.

Antiferromagnetism - There is zero net magnetization when adjacent atomic magnetic moments line up in opposition to one another. Materials that are antiferromagnetic include chromium oxide (Cr_2O_3) and manganese oxide (MnO).

Ferrimagnetism - Comparable to ferromagnetism, however, even in the absence of an external field, uneven magnetic moments result in a net magnetization. Ferrimagnetic materials consist of certain rare-earth compounds and magnetite (Fe_3O_4).

11.8.2.2 *Factors Influencing Magnetic Properties*

Crystal Structure - The magnetic behavior is determined by the arrangement of atoms in the crystal lattice. Different magnetic characteristics, such as ferromagnetism, antiferromagnetism, or ferrimagnetism, can result from different crystal structures.

Magnetic Anisotropy - Magnetic anisotropy is the directional dependency of a material's magnetic characteristics. The crystal structure or outside variables like stress or temperature cause this anisotropy.

Doping and Defects - A material's magnetic characteristics can be altered by adding flaws or dopants. For instance, doping a material with transition metal ions can modify the magnetic interactions inside it, changing the material's magnetic properties [46].

11.9 Applications of Multifunctional Materials

11.9.1 Energy Applications of Multifunctional Materials

The multifunctional materials play a vital role in allowing and attaining sustainable and renewable energy paths of the future. Rapid industrialization and the development of a wide range of consumer goods with ever-increasing power consumption are the outcomes of readily available and sensibly valued energy. Alternatives to green energy are now central to meet this rising demand. Materials science research is advancing the development of renewable energy generation, distribution, transmission, electrical and chemical energy storage, energy efficiency, and enhanced energy management systems—all of which are necessary for a sustainable future. Improving sustainable methods beyond Li-ion batteries is not the only difficulty facing the advanced materials sector, it's also critical to consider how these energy systems might be incorporated into mechanical load routes. The future products will only be able to achieve effective material utilization and sub-system mass and volume minimization via the use of multifunctionality technology. There are some significant energy applications of multifunctional materials such as energy storage devices (battery electrodes and supercapacitors), energy conversion devices (photovoltaic cells and thermoelectric generators), energy harvesting and sensing (piezoelectric materials and triboelectric materials), and energy infrastructure and grid integration (high-performance conductors and energy storage integration). In conclusion, multifunctional materials are essential to the advancement of many energy-related activities, such as infrastructure development, energy conversion, storage, harvesting, and sensing. Accelerating the shift to a resilient and sustainable energy future will need sustained research and development in this area [47].

11.9.2 Biomedical Applications of Multifunctional Materials

Implant materials for the human body are currently being explored, which are multifunctional biomaterials that can release, modify, or deploy therapeutic capability. Multifunctional shape-shifting biomaterials are being explored for use in implants that may be used in minimally invasive surgical procedures and for health technologies that assist senior citizens with daily tasks like housework. In the future, multifunctional biomaterials will follow nature's lead and go beyond simple remodeling of the surrounding environment. In addition to their structural function, these biomaterials may also have sensory properties like audition (hearing), somatosensation (touch), olfaction (smell), and nociception (pain).

These materials, which are next-generation multifunctional implant materials, will significantly increase the possibilities of patient well-being and age-appropriate healthcare [48].

11.9.3 Electronics Applications of Multifunctional Materials

In vitro diagnostics, energy harvesting, robotics, displays, and upgraded therapeutics are just a few of the many areas where flexible mechanical and electrical sensing devices have shown significant promise. For e-skins, wearable and implantable technology, and sophisticated sensors with the ability to be self-sufficient and self-healing, multifunctional materials have proved crucial. Utilizing organic/inorganic matrix arrays, hybrid composites, graphene, NWs or nanotube assemblies, and other extremely flexible, ultrasensitive, and transparent multifunctional materials, amazing advancements have been made. The difficulty is in converting this technology into novel products (from healthcare to robots) that have the potential to completely transform how people interact with society [21].

11.10 Future Directions in Multifunctional Materials

To address new difficulties and progress technology across several disciplines, future approaches in multifunctional materials will require investigating novel material combinations, production processes, and applications. Following are some important topics of study and development for the future [49].

11.10.1 Tailored Properties for Specific Applications

Create materials that are multifunctional and have features that are specifically suited for applications including environmental cleanup, biomedical devices, energy storage, and sensing. Also, create previously unheard-of combinations of attributes, such as high strength, conductivity, and biocompatibility, and investigate novel material combinations and nanostructures.

11.10.2 Smart and Adaptive Materials

Investigate the creation of intelligent, adaptable, multifunctional materials that can dynamically adjust their characteristics and behavior in response to external inputs like light, temperature, or mechanical force. Look into materials that can self-regulate, meaning they can change their characteristics on their own to suit the needs of the user or the environment [50].

11.10.3 Advanced Fabrication Techniques

Explore cutting-edge production methods to accurately control material structure and characteristics at the nanoscale, such as molecular self-assembly, additive manufacturing, and nanolithography. Examine scalable and economical manufacturing techniques to create multipurpose materials and products on a commercial basis [51].

11.10.4 Sustainable and Eco-Friendly Materials

Create multipurpose materials with little environmental effect by utilizing eco-friendly and sustainable raw ingredients and production techniques. Inspect bioinspired methods for creating multipurpose materials with qualities derived from natural systems, such as biodegradability, self-healing, and self-cleaning [52].

11.10.5 Integration of Multiple Functionalities

Examine methods for combining many functions into a single material or gadget in order to produce synergistic effects and improve performance. Examine methods for managing the interactions between various features in order to maximize the behavior of the system as a whole [53].

11.11 Emerging Trends and Developments in Multifunctional Materials

The capacity of multifunctional materials to display several features at once has led to a great deal of study and development in this area in recent years. Here are a few recent advancements and trends in this area [54].

11.11.1 Nanotechnology Integration

New opportunities have been made possible by the incorporation of nanotechnology into multifunctional materials. Multifunctionality can be enabled by nanomaterials by endowing the basic material with special qualities like increased mechanical strength, electrical conductivity, and thermal stability [55].

11.11.2 Smart Materials

The creation of smart materials, which can react to external stimuli like temperature, light, pH, or magnetic field, is the result of advancements in materials science. These materials are adaptable and can vary in shape, color, or conductivity, which makes them useful for a variety of applications such as actuators, sensors, and flexible constructions [56].

11.11.3 Biocompatible and Bioinspired Materials

Creating multipurpose materials that work with biological systems is a priority as interest in biomedical applications grows. Bioinspired materials are being investigated for medical implants, medication delivery, tissue engineering, and other uses. They are designed to resemble the structures and functions found in nature [57].

11.11.4 Energy Harvesting and Storage

In response to the growing need for efficient energy storage devices and renewable energy sources, multifunctional materials are being developed for energy harvesting and storage applications. Particularly interesting are materials that can both convert and store energy at the same time, such as hybrid materials for electrochemical energy storage or piezoelectric materials for mechanical energy harvesting [58].

11.11.5 Additive Manufacturing (3D Printing)

Complex multifunctional structures with customized features may now be designed and fabricated thanks to additive manufacturing techniques like 3D printing. This lowers material waste and manufacturing costs while facilitating quick prototype and material customization for particular applications.

11.11.6 Environmental Sustainability

The development of multipurpose materials with little environmental impact and no effect on ecosystems is becoming more and more important. This covers the use of recyclable or biodegradable materials, environmentally friendly production techniques, and renewable feedstocks [59].

11.11.7 Cross-Disciplinary Collaborations

Collaborations between researchers in the fields of materials science, chemistry, physics, biology, engineering, and computer science are becoming more and more common in multifunctional materials research [60]. By integrating a variety of views and areas of expertise to tackle difficult problems, this interdisciplinary approach promotes creativity.

Overall, the area of multifunctional materials is developing quickly due to advances in research, technology, and social demands [61]. Combining many functions into one material has the potential to transform a number of sectors and provide solutions for urgent global issues [40].

11.12 Conclusion

In conclusion, we have briefly elaborated on the multifunctional materials in engineering and processing engineering. The introduction of multifunctional materials to bulk manufacture will modify significantly the way products are designed. This introduction is barred primarily by the lack of cost-effective processing approaches. A detailed description of the novel materials that will regularly be the focus of the industrial research community has been attempted. To fulfil industrial, technological, and financial goals, some specific directions for future research projects and strategic advances have been provided.

References

1. Lendlein, A. and Trask, R.S., Multifunctional materials: concepts, function-structure relationships, knowledge-based design, translational materials research. *Multifunct. Mater.*, 1, 010201, 2018, https://doi.org/10.1088/2399-7532/aada7b.
2. Arabolla Rodríguez, R., Avila Santos, M., Aliev, A.E., Walton, R.I., Tavera Carrasco, L.A., Pérez Cappe, E.L., González Montiel, M., Pérez Reyes, E.O., Della Santina Mohallem, N., Kashtiban, R.J., Mosqueda Laffita, Y., Leyva Insunza, C., Shearing, P.R., Brett, D.J.L., P and Fe doping, a strategy to develop light and magnetic responsive multifunctional materials: The case of LiMn2O4. *J. Alloys Compd.*, 978, 172837, 2024, https://doi.org/10.1016/j.jallcom.2023.172837.
3. Sahay, R., Reddy, V.J., Ramakrishna, S., Synthesis and applications of multifunctional composite nanomaterials. *Int. J. Mech. Mater. Eng.*, 9, 25, 2014. https://doi.org/10.1186/s40712-014-0025-4.
4. Bonardd, S., Nandi, M., Hernández García, J.I., Maiti, B., Abramov, A., Díaz Díaz, D., Self-Healing Polymeric Soft Actuators. *Chem. Rev.*, 123, 736–810, 2023, https://doi.org/10.1021/acs.chemrev.2c00418.
5. Müller, M. and de Pablo, J.J., Computational Approaches for the Dynamics of Structure Formation in Self-Assembling Polymeric Materials. *Annu. Rev. Mater. Res.*, 43, 1–34, 2013, https://doi.org/10.1146/annurev-matsci-071312-121618.
6. Solga, A., Cerman, Z., Striffler, B.F., Spaeth, M., Barthlott, W., The dream of staying clean: Lotus and biomimetic surfaces. *Bioinspir. Biomim.*, 2, S126, 2007, https://doi.org/10.1088/1748-3182/2/4/S02.
7. Wang, S., Gao, M., Yao, Z., Liu, Y., Wu, M., Li, Z., Liu, Y., Sun, W., Pan, H., A nanoconfined-LiBH4 system using a unique multifunctional porous scaffold of carbon wrapped ultrafine Fe3O4 skeleton for reversible hydrogen storage with high capacity. *Chem. Eng. J.*, 428, 131056, 2022, https://doi.org/10.1016/j.cej.2021.131056.
8. He, A., Li, S., Ma, J., Yang, Z., Environmental Friendly Polymer Materials for Sustainable Development. *Int. J. Polym. Sci.*, 2014, e107028, 2014, https://doi.org/10.1155/2014/107028.
9. Braungart, M., McDonough, W., Bollinger, A., Cradle-to-cradle design: creating healthy emissions – a strategy for eco-effective product and system design. *J. Clean. Prod.*, 15, 1337–1348, 2007, https://doi.org/10.1016/j.jclepro.2006.08.003.
10. Verma, J. and Khanna, A.S., Digital advancements in smart materials design and multifunctional coating manufacturing. *Phys. Open*, 14, 100133, 2023, https://doi.org/10.1016/j.physo.2022.100133.
11. El-Husseiny, H.M., Mady, E.A., Hamabe, L., Abugomaa, A., Shimada, K., Yoshida, T., Tanaka, T., Yokoi, A., Elbadawy, M., Tanaka, R., Smart/stimuli-responsive hydrogels: Cutting-edge platforms for tissue engineering and other biomedical applications. *Mater. Today Bio*, 13, 100186, 2022, https://doi.org/10.1016/j.mtbio.2021.100186.
12. Munir, M., Sinambela, E.A., Halizah, S.N., Khayru, R.K., Mendrika, V., Review of Vocational Education Curriculum in the Fourth Industrial Revolution and Contribution to Rural Development. *J. Soc. Sci. Stud. (JOS3)*, 2, 5–8, 2022, https://doi.org/10.56348/jos3.v2i1.20.
13. Salonitis, K., Pandremenos, J., Paralikas, J., Chryssolouris, G., Multifunctional materials: engineering applications and processing challenges. *Int. J. Adv. Manuf. Technol.*, 49, 803–826, 2010, https://doi.org/10.1007/s00170-009-2428-6.
14. Sengupta, P. and Manna, I., Advanced High-Temperature Structural Materials in Petrochemical, Metallurgical, Power, and Aerospace Sectors—An Overview, in: *Future Landscape of Structural Materials in India*, D. Bhattacharjee and S. Chakrabarti (Eds.), pp. 79–131, Springer Nature, Singapore, 2022, https://doi.org/10.1007/978-981-16-8523-1_5.

15. Abid, N., Khan, A.M., Shujait, S., Chaudhary, K., Ikram, M., Imran, M., Haider, J., Khan, M., Khan, Q., Maqbool, M., Synthesis of nanomaterials using various top-down and bottom-up approaches, influencing factors, advantages, and disadvantages: A review. *Adv. Colloid Interface Sci.*, 300, 102597, 2022, https://doi.org/10.1016/j.cis.2021.102597.
16. Abdelhamid, H.N., *An Introductory Review on Advanced Multifunctional Materials*, 2023, https://doi.org/10.26434/chemrxiv-2023-w78g3-v3.
17. Díez, A.G., Rincón-Iglesias, M., Lanceros-Méndez, S., Reguera, J., Lizundia, E., Multicomponent magnetic nanoparticle engineering: the role of structure-property relationship in advanced applications. *Mater. Today Chem.*, 26, 101220, 2022, https://doi.org/10.1016/j.mtchem.2022.101220.
18. Eivazzadeh-Keihan, R., Sadat, Z., Lalebeigi, F., Naderi, N., Panahi, L., Ganjali, F., Mahdian, S., Saadatidizaji, Z., Mahdavi, M., Chidar, E., Soleimani, E., Effects of mechanical properties of carbon-based nanocomposites on scaffolds for tissue engineering applications: A comprehensive review. *Nanoscale Adv.*, 6, 337–366, 2024.
19. Khan, G.R., Crystallographic, structural and compositional parameters of Cu–ZnO nanocrystallites. *Appl. Phys. A*, 126, 311, 2020, https://doi.org/10.1007/s00339-020-03480-y.
20. Meyers, M.A., Mishra, A., Benson, D.J., Mechanical properties of nanocrystalline materials. *Prog. Mater Sci.*, 51, 427–556, 2006, https://doi.org/10.1016/j.pmatsci.2005.08.003.
21. Ortega, N., Kumar, A., Scott, J.F., Katiyar, R.S., Multifunctional magnetoelectric materials for device applications. *J. Phys.: Condens. Matter*, 27, 504002, 2015, https://doi.org/10.1088/0953-8984/27/50/504002.
22. Wang, X., Zhao, B., Kan, W., Xie, Y., Pan, K., Review on Low-Cost Counter Electrode Materials for Dye-Sensitized Solar Cells: Effective Strategy to Improve Photovoltaic Performance. *Adv. Mater. Interfaces*, 9, 2101229, 2022, https://doi.org/10.1002/admi.202101229.
23. Tong, Y., Liu, J., Wang, L., Su, B.-J., Wu, K.-H., Juang, J.-Y., Hou, F., Yin, L., Dou, S.X., Liu, J., Liang, J., Carbon-Shielded Single-Atom Alloy Material Family for Multi-Functional Electrocatalysis. *Adv. Funct. Mater.*, 32, 2205654, 2022, https://doi.org/10.1002/adfm.202205654.
24. Hu, X., Wang, T., Li, F., Mao, X., Surface modifications of biomaterials in different applied fields. *RSC Adv.*, 13, 20495–20511, 2023, https://doi.org/10.1039/D3RA02248J.
25. Uniyal, S., Choudhary, K., Sachdev, S., Kumar, S., Nano-bio fusion: Advancing biomedical applications and biosensing with functional nanomaterials. *Opt. Laser Technol.*, 168, 109938, 2024, https://doi.org/10.1016/j.optlastec.2023.109938.
26. Clare, A.T., Chalker, P.R., Davies, S., Sutcliffe, C.J., Tsopanos, S., Selective laser sintering of barium titanate–polymer composite films. *J. Mater. Sci.*, 43, 3197–3202, 2008, https://doi.org/10.1007/s10853-007-2396-x.
27. Chryssolouris, G., Georgoulias, K., Michalos, G., Production Systems Flexibility: Theory and Practice. *IFAC Proc.*, 45, 15–21, 2012, https://doi.org/10.3182/20120523-3-RO-2023.00442.
28. Kruth, J.-P., Levy, G., Klocke, F., Childs, T.H.C., Consolidation phenomena in laser and powder-bed based layered manufacturing. *CIRP Ann.*, 56, 730–759, 2007, https://doi.org/10.1016/j.cirp.2007.10.004.
29. Lee, S.-M., Kim, H.J., Ha, Y.-J., Park, Y.N., Lee, S.-K., Park, Y.-B., Yoo, K.-H., Targeted Chemo-Photothermal Treatments of Rheumatoid Arthritis Using Gold Half-Shell Multifunctional Nanoparticles, *ACS Nano.*, 7, 50–57, 2013, https://doi.org/10.1021/nn301215q.
30. Abdelhamid, H.N., Talib, A., Wu, H.-F., Facile synthesis of water soluble silver ferrite (AgFeO2) nanoparticles and their biological application as antibacterial agents. *RSC Adv.*, 5, 34594–34602, 2015, https://doi.org/10.1039/C4RA14461A.
31. Abdelhamid, H.N. and Wu, H.-F., Facile synthesis of nano silver ferrite (AgFeO2) modified with chitosan applied for biothiol separation. *Mater. Sci. Engineering: C*, 45, 438–445, 2014, https://doi.org/10.1016/j.msec.2014.08.071.

32. Wang, H., Sun, L., Li, Y., Fei, X., Sun, M., Zhang, C., Li, Y., Yang, Q., Layer-by-Layer Assembled Fe3O4@C@CdTe Core/Shell Microspheres as Separable Luminescent Probe for Sensitive Sensing of Cu2+ Ions. *Langmuir*, 27, 11609–11615, 2011, https://doi.org/10.1021/la202295b.
33. Plaisted, T.A., Amirkhizi, A.V., Arbelaez, D., Nemat-Nasser, S.C., Nemat-Nasser, S., Self-healing structural composites with electromagnetic functionality, in: *Smart Structures and Materials 2003: Industrial and Commercial Applications of Smart Structures Technologies*, pp. 372–381, 5054, SPIE, 2003, https://doi.org/10.1117/12.483894.
34. Kanu, N.J., Gupta, E., Vates, U.K., Singh, G.K., Self-healing composites: A state-of-the-art review. *Composites, Part A*, 121, 474–486, 2019, https://doi.org/10.1016/j.compositesa.2019.04.012.
35. Zhang, X., Zhao, J., Liu, K., Li, G., Zhao, D., Zhao, Z., Wan, J., Yang, X., Bai, R., Wang, Y., Zhang, W., Yan, X., Monolithic dynamic covalent polymer aerogels via ambient pressure drying with super-flexibility, weldability, closed-loop recyclability, and multifunctionality, 2021, https://doi.org/10.26434/chemrxiv-2021-gxp40.
36. Polla, D.L. and Francis, L.F., Processing and Characterization of Piezoelectric Materials and Integration into Microelectromechanical Systems. *Annu. Rev. Mater. Sci.*, 28, 563–597, 1998, https://doi.org/10.1146/annurev.matsci.28.1.563.
37. Dhanalakshmi, M., Parthiban, S., Swaminathan, M., Crystal growth and optical properties of ammonium dihydrogen phosphate (ADP) crystals doped with Fe(II) and Fe(III): Oxidation number effects. *Mater. Today Proc.*, 29, 1119–1124, 2020, https://doi.org/10.1016/j.matpr.2020.05.200.
38. Wada, S., Suzuki, S., Noma, T., Suzuki, T., Osada, M., Kakihana, M., Park, S.-E., Cross, L.E., Shrout, T.R., Enhanced Piezoelectric Property of Barium Titanate Single Crystals with Engineered Domain Configurations. *Jpn. J. Appl. Phys.*, 38, 5505, 1999, https://doi.org/10.1143/JJAP.38.5505.
39. Esmaeili, S., Ehsani, M.H., Toghraie, D., Saber-Samandari, S., The investigation of piezoelectric features of lead-free barium titanate ceramic materials at different initial temperatures in the tetragonal phase using molecular dynamics simulation. *Results Phys.*, 63, 107851, 2024, https://doi.org/10.1016/j.rinp.2024.107851.
40. Tiwari, B., Babu, T., Choudhary, R.N.P., Piezoelectric lead zirconate titanate as an energy material: A review study. *Mater. Today Proc.*, 43, 407–412, 2021, https://doi.org/10.1016/j.matpr.2020.11.692.
41. Lee, Y., Choi, H., Jang, J., Fabrication and characterization of lead lanthanum zirconate titanate ceramic-based fully transparent piezoelectric loudspeakers for next-generation electronic devices. *Sens. Actuators A: Phys.*, 382, 116087, 2025. https://doi.org/10.1016/j.sna.2024.116087.
42. Fukada, E. and Furukawa, T., Piezoelectricity and ferroelectricity in polyvinylidene fluoride. *Ultrasonics*, 19, 31–39, 1981, https://doi.org/10.1016/0041-624X(81)90030-5.
43. Milek, J.T. and Neuberger, M., Potassium Dihydrogen Phosphate (KDP), In: *Linear Electrooptic Modular Materials*, pp. 177–222, Springer US, Boston, MA, 1972, https://doi.org/10.1007/978-1-4684-6168-8_13.
44. Suter, J.J., Piezoelectric properties of quartz crystal resonators below 10 K. *Cryogenics*, 30, 547–548, 1990, https://doi.org/10.1016/0011-2275(90)90058-K.
45. Bhadwal, N., Ben Mrad, R., Behdinan, K., Review of Zinc Oxide Piezoelectric Nanogenerators: Piezoelectric Properties, Composite Structures and Power Output. *Sensors*, 23, 3859, 2023, https://doi.org/10.3390/s23083859.
46. Madhan, K. and Murugaraj, R., Enrichment of optical, electrical, and magnetic properties of Li_{+}, La_{3+} doped $BaTiO_3$ perovskite multifunctional ceramics. *Appl. Phys. A*, 126, 97, 2020, https://doi.org/10.1007/s00339-020-3285-2.

47. Wang, H., Liang, X., Wang, J., Jiao, S., Xue, D., Multifunctional inorganic nanomaterials for energy applications. *Nanoscale*, 12, 14-42, 2020, https://pubs.rsc.org/en/content/articlehtml/2020/nr/c9nr07008g.
48. Kim, D., Shin, K., Kwon, S.G., Hyeon, T., Synthesis and Biomedical Applications of Multifunctional Nanoparticles. *Adv. Mater.*, 30, 1802309 2018, https://doi.org/10.1002/adma.201802309.
49. Archer, R.J., Becher-Nienhaus, B., Dunderdale, G.J., Hozumi, A., Recent Progress and Future Directions of Multifunctional (Super)Wetting Smooth/Structured Surfaces and Coatings. *Adv. Funct. Mater.*, 30, 1907772, 2020, https://doi.org/10.1002/adfm.201907772.
50. Qader, İ.N., Kök, M., Dagdelen, F., Aydoğdu, Y., A review of smart materials: researches and applications. *ECJSE*, 6, 755–788, 2019. https://doi.org/10.31202/ecjse.562177.
51. Harish, V., Ansari, M.M., Tewari, D., Yadav, A.B., Sharma, N., Bawarig, S., García-Betancourt, M.-L., Karatutlu, A., Bechelany, M., Barhoum, A., Cutting-edge advances in tailoring size, shape, and functionality of nanoparticles and nanostructures: A review. *J. Taiwan Inst. Chem. Eng.*, 149, 105010, 2023, https://doi.org/10.1016/j.jtice.2023.105010.
52. Elfaleh, I., Abbassi, F., Habibi, M., Ahmad, F., Guedri, M., Nasri, M., Garnier, C., A comprehensive review of natural fibers and their composites: An eco-friendly alternative to conventional materials. *Results Eng.*, 19, 101271, 2023, https://doi.org/10.1016/j.rineng.2023.101271.
53. Jose, S., Sariga, Varghese, A., Design and structural characteristics of conducting polymer-metal organic framework composites for energy storage devices. *Synth. Met.*, 297, 117421, 2023, https://doi.org/10.1016/j.synthmet.2023.117421.
54. González, C., Vilatela, J.J., Molina-Aldareguía, J.M., Lopes, C.S., LLorca, J., Structural composites for multifunctional applications: Current challenges and future trends. *Prog. Mater Sci.*, 89, 194–251, 2017, https://doi.org/10.1016/j.pmatsci.2017.04.005.
55. Malik, S., Muhammad, K., Waheed, Y., Nanotechnology: A Revolution in Modern Industry, *Molecules,* 28, 661, 2023, https://doi.org/10.3390/molecules28020661.
56. Roy, R., Bhakta, D., Roy, S., Applications and Types of Smart Materials, in: *Modeling, Characterization, and Processing of Smart Materials*, pp. 266–296, IGI Global, 2023, https://doi.org/10.4018/978-1-6684-9224-6.ch013.
57. Harun-Ur-Rashid, M., Jahan, I., Foyez, T., Imran, A.B., Bio-Inspired Nanomaterials for Micro/Nanodevices: A New Era in Biomedical Applications. *Micromachines*, 14, 1786, 2023, https://doi.org/10.3390/mi14091786.
58. Agarwal, R., Singh, S., Shalan, A.E., Sustainable Energy Storage Devices and Device Design for Sensors and Actuators Applications, in: *Sustainable Energy Storage in the Scope of Circular Economy*, pp. 225–290, John Wiley & Sons, Ltd, 2023. https://doi.org/10.1002/9781119817741.ch10.
59. Tsoutsos, T., Frantzeskaki, N., Gekas, V., Environmental impacts from the solar energy technologies. *Energy Policy,* 33, 289–296, 2005. https://doi.org/10.1016/S0301-4215(03)00241-6.
60. Dzogbewu, T.C., Amoah, N., Afrifa Jnr, S., Fianko, S.K., de Beer, D.J., Multi-material additive manufacturing of electronics components: A bibliometric analysis. *Results Eng.*, 19, 101318, 2023. https://doi.org/10.1016/j.rineng.2023.101318.
61. Luo, L., Zhang, F., Wang, L., Liu, Y., Leng, J., Recent Advances in Shape Memory Polymers: Multifunctional Materials, Multiscale Structures, and Applications. *Adv. Funct. Mater.*, 2312036. https://doi.org/10.1002/adfm.202312036.

12

Multifunction Materials Optoelectronic

Amit Kumar Srivastava*, Prashant Kumar Mishra, Aakash Mathur, Gurupada Maity and Yogendra Kumar

Division of Physics School of Basic Sciences, Galgotias University, Greater Noida, Uttar Pradesh, India

Abstract

The development of high-performance multifunctional materials technologies for more efficient mobility; as structural energy storage systems; to promote better lifestyles and improve healthcare by monitoring and remodeling according to our personal well-being; to act as an intelligent system for active sensing and self-diagnosis; to morph and adapt in response to the local environment; and to improve product sustainability from cradle-to-grave via new assembly and disassembly processes, has become an extremely active research area of great importance. In recent years, many multifunctional materials in the form of thin films, two-dimensional materials, nanostructures, and nanocomposites have been put forth. They can function astonishingly well because of their unique properties, whether electronic, magnetic, optical, thermal, or elsewhere. Multifunctional materials with glaring merits have radically different energy band structures and electrical properties, covering superconductors, metals, semi-metals, semiconductors, and insulators and displaying outstanding applications. Especially, these materials are very promising with respect to their integration into optoelectronic devices such as conductive additives, transparent conductive electrodes, and charge transfer layers. Optoelectronic devices are those electronic devices that operate on both light and electrical currents. This can include electrically driven light sources such as laser diodes and light-emitting diodes, components for converting light to an electrical current such as solar and photovoltaic cells, and devices that can electronically control the propagation of light. Multifunctional materials can be crucial in the creation of such optoelectronic devices because they have the ability to repeatedly and reversibly change some of their functions, characteristics, or behaviors over time in response to external conditions. The silicon-based technologies now used in these optoelectronic devices are slowly reaching their performance limits. Therefore, it is necessary to gain insight into new materials having interdisciplinary properties. Due to the enormous variety of material combinations and consequent qualities and functions, they should also be far more suited to the application than present unifunctional materials. Beyond system performance, these composites may also save costs, at least in the post-production stage, which is something to keep in mind. Also, such an interdisciplinary field can be extended to computational engineering approaches for virtual material design and to advanced fabrication schemes taking advantage of digitalization. In this way, development cycles can be shortened and products based on multifunctional materials can become more and more adaptive and individualized.

In this chapter, we summarized the unique optical and electronic properties of multifunctional materials as well as recent advances in their optoelectronic device applications. First, the superb and

**Corresponding author*: akrsri@gmail.com

Divya Bajpai Tripathy, Anjali Gupta and Arvind Kumar Jain (eds.) Multifunctional Materials: Engineering and Biological Applications, (295–334) © 2025 Scrivener Publishing LLC

programmable optical and electrical features of multifunctional materials are introduced. Second, a summary of the fundamental ideas, structures, and components utilized in optoelectronic device applications, including light-emitting diodes (LEDs), solar cells, photodetectors, and displays, is given, along with some recent developments in this area. Last but not least, there have been some significant issues raised that require more research in terms of the practical optoelectronic device applications of multifunctional materials.

***Keywords*:** Multifunction materials, optoelectronic device, light-emitting diodes, solar cell, photodiode act

12.1 Multifunction Materials Optoelectronic

In recent years, many multifunctional materials in the form of thin films, two-dimensional materials, nanostructures, and nanocomposites have been put forth. They can function astonishingly well because of their unique properties, whether electronic, magnetic, optical, thermal, or elsewhere. Multifunctional materials with glaring merits have radically different energy band structures and electrical properties, covering superconductors, metals, semimetals, semiconductors, and insulators and displaying outstanding applications. These materials are especially promising with respect to their integration into optoelectronic devices such as conductive additives, transparent conductive electrodes, and charge transfer layers. Optoelectronic devices are those electronic devices that operate on both light and electrical currents. This can include electrically driven light sources such as laser diodes and light-emitting diodes (LEDs), components for converting light to an electrical current such as solar and photovoltaic (PV) cells, and devices that can electronically control the propagation of light. Multifunctional materials can be crucial in the creation of such optoelectronic devices because they have the ability to repeatedly and reversibly change some of their functions, characteristics, or behaviors over time in response to external conditions. The silicon (Si)-based technologies now used in these optoelectronic devices are slowly reaching their performance limits. Therefore, it is necessary to gain insight into new materials having interdisciplinary properties. Due to the enormous variety of material combinations and consequent qualities and functions, they should also be far more suited to the application than present unifunctional materials. Beyond system performance, these composites may also save costs, at least in the post-production stage, which is something to keep in mind. Also, such interdisciplinary fields can be extended to computational engineering approaches for virtual material design and to advanced fabrication schemes taking advantage of digitalization. In this way, development cycles can be shortened and products based on multifunctional materials can become more and more adaptive and individualized.

These days, the material science research community is focusing on creating new structures and materials that can be effectively employed in a multitude of approaches, including energy harvesting, healthcare, transportation, optoelectronics and photonics, sensing, memory applications, and many more by taking advantage of their various properties [1]. Certain electrical, optical, magnetic, thermal, and other properties of these materials can be adjusted to meet the needs of various applications. Comparing multifunctional materials to their non-functional counterparts, multifunctional materials have multiple of these

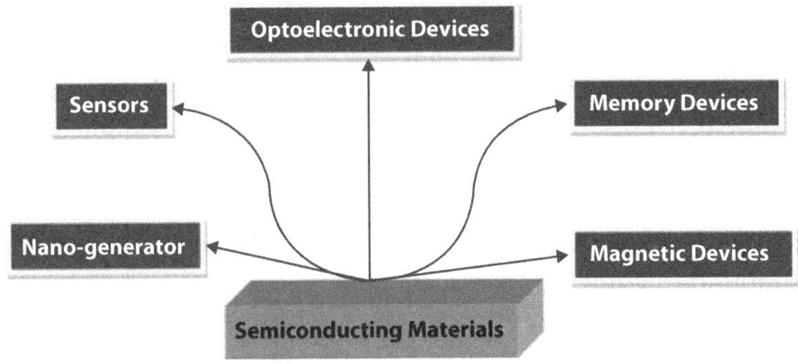

Figure 12.1 Schematic of applications of multifunctional materials.

features present together. Its extensive range of capabilities is made possible by these materials' ability to carry out structural, non-structural, or both types of functions.

Multifunctional materials can be found in their natural condition or can be created through various ways to gain multifunctional qualities. The materials are superior to others because of their effective and adaptable qualities, reduced size, inexpensive cost, lightweight nature, and low power consumption [2]. Figure 12.1 depicts a schematic of how multifunctional materials are applied.

12.1.1 Overview of Optoelectronic Materials

A branch of photonics science called optoelectronics studies and applies electrical devices that interact with light and systems in which photons and electrons coexist. Optoelectronics operate fundamentally on the basis of the optical impact on the electronic reaction in particular optically sensitive semiconductor substances. Since the electrical conductivity of semiconducting materials varies across orders of magnitude as a result of temperature variations, optical irradiation, and the amount of impurities contained, they are naturally chosen. Optoelectronic devices are those that provide that kind of interaction; as Figure 12.2 illustrates, these devices typically encompass the transformation of energy from optical to electrical or vice versa.

The category of materials known as semiconductors is well known to have reasonable bandgaps for energy (0.2 eV to 4.0 eV) and electrical resistivities (10^{-3} Ω cm to 10^{6} Ω cm) that lie between the metal and insulator categories. One of the most important properties of semiconductors is the energy gap (E_g) that exists between the valence and conduction bands. This characteristic essentially determines whether the semiconductor absorbs, transmits, or emits light at a given wavelength.

Si and germanium (Ge) are the two most traditional and thoroughly researched semiconductors among the numerous types. Given that they are both classified as elemental

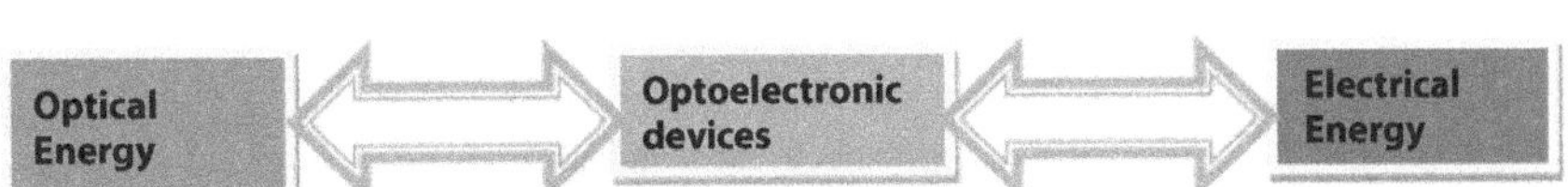

Figure 12.2 An optoelectronic device's schematic representation of energy conversion.

semiconductors since they are made up of a single species of their respective atoms and are members of the 4th group of the periodic table.

Although these elemental semiconductors (such as Si and Ge) have many applications, their usage in optoelectronics is fairly limited because of some significant disadvantages. Considering that Si and Ge are indirect-type bandgap semiconductors, their light emission and absorption are both subpar. Elemental semiconductor-based optoelectronic devices are not energy-efficient when used as a whole. Direct bandgap compounds are presently a partially viable solution to the limitations associated with the use of essential semiconductors in optoelectronic devices or alloy semiconductors, which are composed of elements from distinct periodic table groups, such as III-V, IV-IV, and II-VI. Due to their easy synthesis and numerous desired features, like improved device performance, these materials are highly appealing for use in optoelectronic devices.

GaAs is one of the most developed compound semiconductors in terms of both material and device technology. GaAs is best suited for microwave devices (such as Gunn diodes), high-speed devices (because of their comparatively higher carrier mobility), and effective optoelectronic devices [such as LEDs, lasers, and photodetectors (PDs)]. In addition to GaAs, group III-V alloy semiconductors are also appealing for various applications, such as LED (such as GaN and GaP), Gunn diode (such as InP), light detector (such as InSb), and laser (such as AlGaAs, InGaAs, and InGaAsP), employed in optical fiber communication networks (at wavelength few micrometers). Group II-VI alloy semiconductors find extraordinary applications in the creation of solar cells, PDs (CdSe, PbTe, and HgCdTe), fluorescent materials (ZnS), and solar cell development (ZnO, CdTe, and CdS).

The fundamental mechanism of operation in all optoelectronic devices is the interaction of electromagnetic waves or optical radiations with target materials. Figure 12.3 illustrates the various optical processes that can happen when light rays strike a substance, including transmission, absorption, and reflection.

The generation and annihilation of electron-hole pairs, and hence the overall effectiveness of optoelectronic devices, are largely dependent on the optical absorption mechanism and magnitude of these processes. One can increase the amount of the incoming optical signal that is absorbed by minimizing both the transmission loss and reflection to the lowest possible degree.

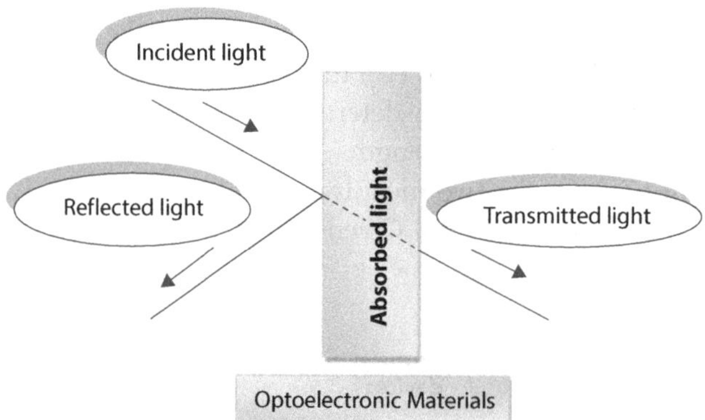

Figure 12.3 Various kinds of light interaction with optoelectronic substances.

When light is absorbed, incoming photons provide the filled valence band electrons the energy they need to jump into the highest energy band (conduction band 'C. B.') and create holes in the lowest energy band (valence band 'V. B.') in the process. This energy is at least equal to the energy of the bandgap. In the reverse procedure, when the excited electrons recombine with holes utilizing energy in the form of the radiative or non-radiative process. The essential mechanism by which LEDs, lasers, and other display systems, among other light sources, work is the release of light, which causes excited electrons to move from the conduction band to the valence band.

Through appropriate light absorption, the optoelectronic devices produce extra electrons and holes (in pair form). The carrier production rate and device efficiency are determined by the degree of absorption of light. The photons, or energy quanta of radiation, clash with semiconductor electrons in the simplest process of light absorption. As illustrated in Figure 12.4, if the lowest energy of the incoming photon is $h\nu \geq E_g$, the valence band electron will be driven into the conduction band by optical absorption, creating a hole in the valence band. The most frequent process in optical detectors is the phenomenon of optical energy being converted (via the absorption process) into electron-hole pairs. Energy and momentum are maintained throughout the absorption process because it essentially involves an electron and a photon colliding or being scattered. Given that photon momentum is essentially zero compared to electron momentum.

In order to maintain momentum conservation, it is preferred that the transition in k-space (or momentum space) is vertical or the same. Only materials with direct bandgaps, like GaAs, InP, and GaAlAs, are capable of this. Alternatively, Si, Ge, and GaP as indirect semiconductors, where the lowest energy value of C. B. and the highest energy value of V. B. are not located in the same value of k in k-space [Figure 12.4(b)], the creation or annihilation of phonons, or lattice vibration energy quanta, always aids in the inter-band electron transition.

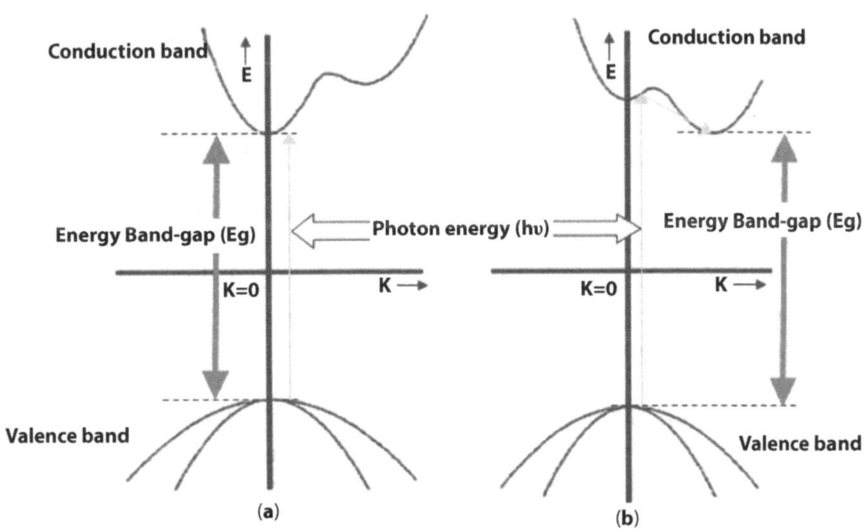

Figure 12.4 Electron transition of semiconductor with (a) direct bandgap and (b) indirect bandgap.

12.1.2 Introduction to Multifunctional Materials

When we talk about multifunctional materials in the context of optical materials, we are talking about sophisticated materials that combine several functionalities with optical qualities, which makes them desirable and adaptable for a variety of applications. These materials offer additional properties like mechanical strength, flexibility, or electrical conductivity in addition to their ability to control light in different ways. Here is a summary of multifunctional materials in the realm of optical materials. Light interacts with optical materials, and they produce many properties like absorption, transparency, reflectivity, and refractivity and develop optical multifunction materials by adding additional functionalities, such as mechanical strength, flexibility, or other structural features, to enhance the optical properties. When producing optical components like lenses or cameras, for example, materials must be subjected to mechanical stress. Most optical materials have high electrical resistivity; reducing their electrical resistivity is a feature of multifunctional optical materials that increases their applications in optoelectronic devices, such as transparent conductive materials that are necessary for touch screens, solar panels, and electronic displays. Multifunctional optical materials allow for dynamic changes in their optical properties in response to external stimuli. Applications where the material's optical properties may be actively adjusted, such as smart windows, tunable lenses, and adaptive optics, take advantage of this responsiveness. Multifunctional optical materials are used in medical imaging, therapy, and diagnostics in the field of biophotonics. These materials could combine biocompatibility and optical qualities, enabling the creation of cutting-edge biomedical instruments and sensors. Multifunctional optical materials are used to help develop various photonic devices, including quantum optics and optical communications. The creation of compact, effective photonic devices is made possible by these materials. The application of nanotechnology facilitates the development of multifunctional optical materials. These multifunctional structures are often combined with nanomaterials with unique optical properties, such as quantum dots (QDs), to achieve exceptional optical functions.

Multifunctional materials for optical applications are used in interdisciplinary areas such as engineering, materials science, and optics. Innovative technologies in fields like electronics, photonics, medicine, and other fields can achieve new heights with the capacity to combine optical properties with other functionality

12.1.3 Application of Multifunctional Materials in Optoelectronics

In optoelectronics, devices that interact with both light and electrical impulses are studied and used, and multifunctional materials are essential to this field's advancement. These materials are engineered to demonstrate a blend of optical and electrical properties, facilitating the creation of novel devices with improved capabilities. Important optoelectronic applications for multifunctional materials include the following:

- Transparent Conductive Films: In applications such as touch screens, flexible displays, and smart windows, where optical purity and electrical conductivity are crucial, multifunctional materials possessing high optical transparency in the visible range and high electrical conductivity are indispensable [3–5].

- Photodetectors: These multipurpose materials are used in the fabrication of PDs, which convert light impulses into electrical signals. These materials integrate electronic functionality with optical sensitivity to improve the efficiency and sensitivity of PDs used in sensors, cameras, and communication systems [6, 7].
- Light-Emitting Diodes (LEDs): LEDs are based on the electroluminescence theory and are made of efficient materials that serve multiple purposes. This method results in LEDs designed for lighting and display applications having improved color purity, brightness, and energy efficiency [8, 9].
- Solar Cells: The development of solar cell technology depends on materials with many functions. These materials show promise for increasing mechanical flexibility, charge carrier mobility, and light absorption, all of which lead to more efficiency [10, 11].
- Optical Modulators and Switches: Optical modulators and switches that regulate the phase or intensity of light signals are built using materials with several functions. These devices are crucial parts of optical communication networks and data transmission systems because they combine electronic control with optical features [12, 13].
- Flexible and Stretchable Electronics: The development of stretchable and flexible electronics makes use of multifunctional materials with mechanical flexibility and optical-electronic capabilities. These materials allow conformal electronic devices, wearable sensors, and bendable screens to be made [14, 15].
- Optical Sensors: Multifunctional materials are used in the fabrication of optical sensors, which find application in industrial sensing, healthcare, and environmental monitoring, among other fields [16, 17].

Multifunctional materials are generally used in optoelectronics to enable the creation of devices with improved functionality, performance, and flexibility.

12.2 Multifunctional Materials for Light-Emitting Diodes (LEDs)

LED is a semiconductor device based on the optoelectronic concept that emits light depending on the applied voltage across the diode, causing electrons and holes to recombine. Large angles of view, strong contrast between colors, minimal power use, color tunability, and self-luminescence are just a few benefits of several colors LEDs, which are considered the next generation of displays used in smartphones, laptops, televisions (TVs), and virtual reality technology. Furthermore, a wider variety of uses in developing optoelectronic devices, including foldable color displays, and smart watches are made possible by the exceptional flexibility and thin thickness of several color LEDs. Moreover, a clever and interesting method of visualizing the electronic output signal of sensors to a range of visible colors is to integrate multi-color LEDs with sensors. This method has great potential for use in a variety of applications, including electronic skin sensors and real-time ECG monitors.

12.2.1 Basic Concept of LEDs

Solid-state light-emitting devices are called LEDs. It produces less heat and lasts longer than traditional light bulbs. Additionally, the all-solid-state, small light source may be created, giving it exceptional impact resistance and significantly increasing the design options available to customers.

One of the earliest and most basic optoelectronic devices, Since its discovery in the early 1900s, the LED has been used extensively in science and industry for a wide range of applications, including vehicle tail lights, sensors, optical communication networks, display systems, and logic circuits. Both LEDs and lasers are essentially electroluminescence devices, meaning that their light emission is fueled by the creation of excess carriers through the application of an electric field or the infusion of current.

The development of effective LEDs has been centered around alloy materials including III-V and II-VI semiconductors since the mid-1950s. Since these materials have a direct bandgap by nature, they promote radiative light emission over everything else, which results in higher efficiency. Furthermore, semiconducting materials emit light across the broad spectrum, ranging from the visible to the ultraviolet (UV) and infrared (IR) regions. The techniques involved in the growth of semiconductors like molecular beam epitaxy and metal-organic chemical vapor deposition along with bandgap engineering using compound semiconductors have made it possible to create a wide range of solid-state light sources, such as LEDs and lasers that are active in the visible, UV, and IR regions of the spectrum, as shown in Figure 12.5. It is widely recognized that in terms of light source quality, semiconductor laser diodes are a more glamorous light source than LEDs in terms of technology. But the subsequent device (the laser diode), which requires more sophisticated drive and cooling electronics, also demands special care during manufacture. A number of fabrication and operational factors affect how well-efficient LEDs are designed and run. In actuality, different LEDs can be built for different purposes, including basic p-n junction LEDs and the most effective category of LEDs, such as double heterostructure (DH) LEDs, surface-emitting LEDs, edge-emitting LEDs, resonant-cavity LEDs, white LEDs (WLEDs), etc.

Like a normal diode, the LED consists of a chip of semiconducting material doped with impurities to create a p-n junction. Figure 12.6 describes the internal components of LED.

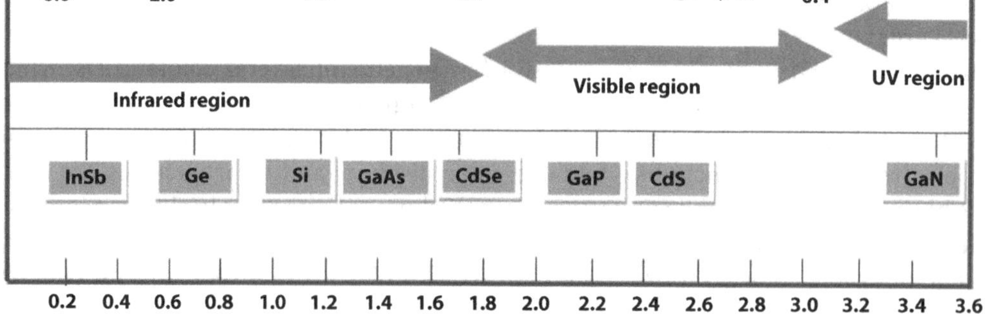

Figure 12.5 Bandgap energy and emission wavelength of the most practical semiconductors.

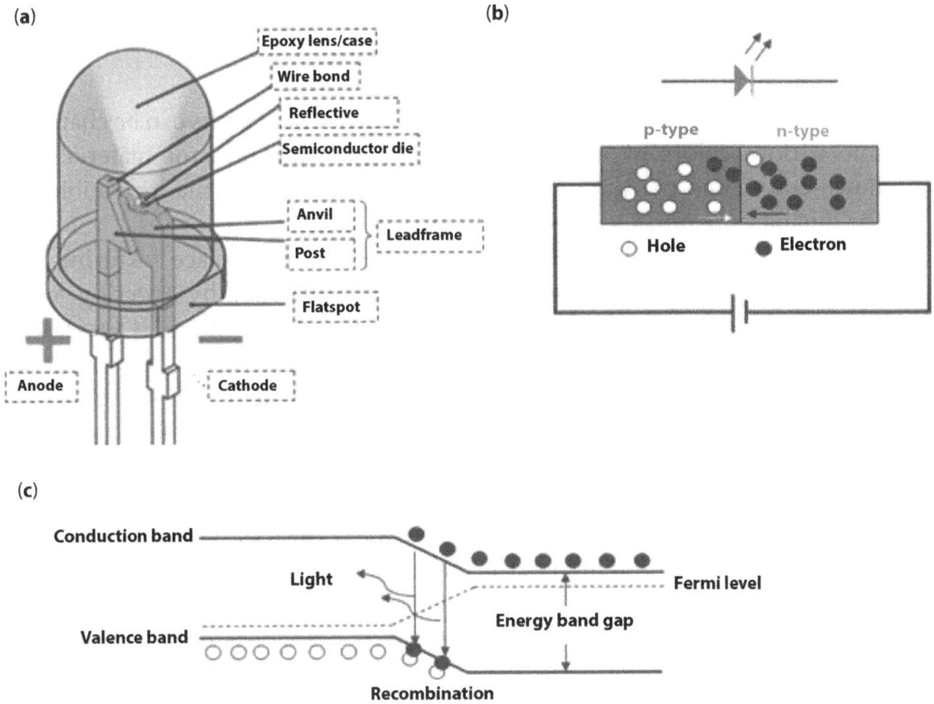

Figure 12.6 LED schematic (a) internal structure, (b) circuit diagram with symbol, and (c) band diagram.

The electrons and holes, or charge carriers, enter the junction from electrodes that vary. The wavelength of the light emitted depends on the energy bandgap of the semiconductor voltages. An electron releases energy in the form of a photon as depicted in the figure when it collides with a hole, dropping into a lower energy level. forming the p-n junction. Since Si and Ge are indirect band gap materials, the electrons and holes recombine in these materials through a non-radiative transition that does not result in visual emission. The LED's materials have a straight band gap that matches energy found in visible, near-UV, and near-IR light. Gallium arsenide-based red and IR devices marked the beginning of LED development. The n-type substrate is typically used to construct LEDs, and an electrode is affixed to the p-type layer that is deposited on its surface. Although less often, P-type substrates can also exist. A lot of commercial LEDs employ sapphire substrates, particularly GaN/InGaN LEDs. The majority of the materials used to make LEDs have extremely high refractive indices. This indicates that at the material/air surface interaction, a significant amount of light will be reflected back into the material. Light extraction in LEDs is therefore a crucial component of LED manufacturing and is the focus of extensive research and development.

12.2.2 Multifunctional Materials for Improved Efficiency and Color-Tuning of LEDs

LEDs can be made more efficient and capable of color tweaking with the use of multifunctional materials. LEDs are now widely used in many different applications, including displays, communication devices, and lighting. The following multipurpose materials and techniques help to enhance LED color adjustment and efficiency:

- *Quantum Dots (QDs):* The blue or UV light that an LED emits can be changed into various colors by using these nanoscale semiconductor materials as down-conversion phosphors. The exact control of color is made possible by the tunable emission spectra of QDs [18, 19].
- *Nanostructured Surfaces:* Utilizing nanostructures or surface texturing on LED surfaces might increase light extraction efficiency by lowering total internal reflection. By doing this, the LED's overall brightness is increased [11, 20].
- *Photonic Crystals:* Enhancing light extraction and efficiency, photonic crystal integration with LEDs can help regulate the directionality of light emission [21, 22].
- *Graphene:* Excellent thermal conductivity of graphene-based materials contributes to effective heat dissipation from LED devices. Doing this, LEDs maintain prolonged and high performance [23–25].
- *Tunable Filters:* LEDs can be dynamically color-tuned by adding tunable filters or color conversion layers, which lets users change the hue or color temperature [9, 26].
- *Multiple Emission Centers:* It is possible to produce WLEDs with better color rendering index and color temperature flexibility by using materials with many emission centers [27].
- *Nanostructured Light Extraction Layers:* The LED structure can be made more efficient overall and improve light extraction by using nanostructures like nanowires or nanorods [28, 29].

Combining these multifunctional materials and techniques can greatly improve the efficiency, color quality, and adaptability of LEDs for a variety of applications. Research on innovative materials and design approaches is still ongoing in order to advance LED technology.

12.2.3 Emerging Materials for High-Performance LEDs

The creation of high-performance LEDs relies on constant developments in materials research.

Scientists are investigating a range of novel materials to improve LED efficiency, brightness, and color quality. Key developing materials for high-performance LEDs include the following:

- *Perovskite Nanocrystals (PNCs):* The efficiency and color purity of LEDs have been demonstrated to be enhanced by materials based on perovskites. Suitable for usage in LEDs, PNCs have good optical characteristics and tunable emission spectra [30, 31].
- *Organic Light-Emitting Diodes (OLEDs):* Organic materials are used as the emissive layer in OLEDs. Flexibility, tiny form factor, and the possibility

of low-cost manufacturing are among their benefits. The goal of current research is to optimize OLEDs for lighting applications. OLED technology is widely employed in display applications [32].
- *Graphene:* Graphene is being investigated for usage in transparent electrodes and as a heat-spreading material in LED devices due to its exceptional electrical and thermal conductivity [33, 34].
- *Transition Metal Dichalcogenides (TMDs):* TMDs, such as tungsten diselenide (WSe2) and molybdenum disulfide (MoS2), are semiconducting 2D materials that have the potential to be used in LEDs [35, 36].
- *Quantum Dots (QDs):* Semiconductor nanocrystals with size-dependently adjustable emission spectra are known as QDs. They are used in LEDs to produce a large color gamut with excellent color efficiency and purity [18, 37].
- *Nanowires and Nanorods:* LED light extraction efficiency can be increased by using nanostructures like nanowires and nanorods, which will improve overall performance [38, 39].
- *Gallium Nitride (GaN) and Indium Gallium Nitride (InGaN):* The fundamental component of blue and WLEDs is III-nitride material. The goal of ongoing research is to maximize the performance of these materials in order to get increased brightness and efficiency [40, 41].
- *Metasurfaces:* The directionality of light emission from LEDs can be controlled using metasurfaces made of nanostructured materials, improving light extraction and efficiency [42, 43].

These innovative materials show the variety of research paths being investigated to improve LED performance—including efficiency, brightness, color quality, and application adaptability. It is expected that continued progress in these areas will help upcoming generations of high-performance LED technology.

12.3 Multifunctional Materials for Solar Cells

Developing multifunctional materials is essential to creating solar cells that are both effective and adaptable. These components are made to fulfill several functions inside the framework of the solar cell, resulting in improved efficiency, economy, and sustainability.

- Perovskite materials have garnered a lot of attention recently due to their exceptional efficiency and low cost of production [43, 44].
- Transparent conductive oxides are widely used in solar cells as transparent electrodes. Multifunctional TCOs can be made to work as anti-reflective coatings and provide better electrical conductivity and transparency [45, 46].
- Multifunctional coatings applied to solar cell surfaces can reduce light reflection, increase light trapping, and repel dust and grime. These characteristics reduce maintenance requirements and enhance long-term performance [10, 47].

These versatile materials could be used to create solar cells with high efficiency in converting solar radiation into electrical power, as well as other properties like flexibility, transparency, and scalability for a range of applications, according to research. Material science advancements drive innovation in solar cell technology.

12.3.1 Basic Concepts of Solar Cells

The operating principle of solar cells is the PV effect, which is the generation of a potential difference at the interface of two different materials in response to electromagnetic radiation. The PV effect is closely associated with the photoelectric effect, which is the emission of electrons from a substance that has absorbed light over a threshold frequency that varies depending on the material. In 1905, Albert Einstein discovered that this effect could be explained if light were composed of distinct energy particles known as photons. Such a photon's energy (E_p) is provided by

$$E_p = h\upsilon$$

where υ is the light's frequency and h is Planck's constant. In 1921, Einstein won the Nobel Prize in Physics for his explanation of the photoelectric phenomenon.

As seen in Figure 12.7(a), when a photon is absorbed by a material, its energy is used to excite an electron from an initial energy level E_i to a higher energy level E_f. Only when the electron energy levels Ei and E_f are present to the point where their difference, $h\upsilon = E_f - E_i$, equals the photon energy, can photons be absorbed. In a perfect semiconductor, electrons can occupy energy levels above the conduction band edge (E_C) and below the valence band edge (E_V). Between those two bands, there are no allowed energy levels that electrons could work with. The energy difference, denoted as $Eg = E_C - E_V$, is therefore known as the bandgap. A photon with energy less than E_g will not be absorbed by a perfect semiconductor; instead, it will pass through the substance unaffected.

The valence and conduction band structures of a true semiconductor are not flat; rather, they change in response to an electron's momentum within the semiconductor, which is represented by the so-called k-vector. This suggests that the energy of an electron depends on its momentum due to the periodic structure of the semiconductor crystal. If the conduction band minimum and the valence band maximum occur at the same k-vector, then

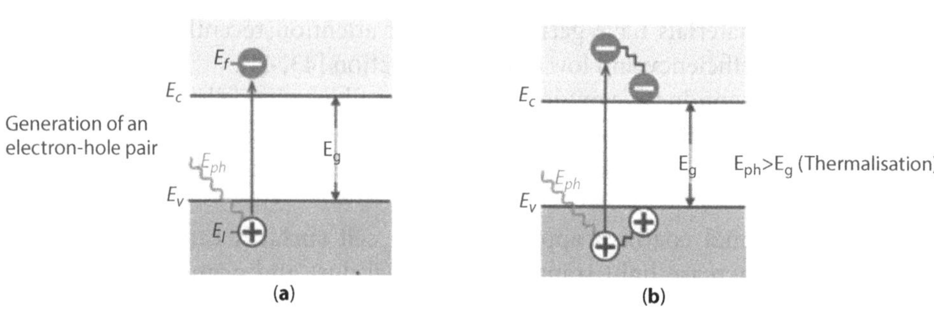

Figure 12.7 Photon energy ($E_{Ph} = h\upsilon$) and bandgap (E_g) are used to illustrate how a photon is absorbed in a semiconductor (a) $E_{Ph} = E_g$ (b) $E_{Ph} > E_g$.

an electron can be excited from the valence to the conduction band without affecting its momentum. This kind of semiconductor is known as a direct bandgap material. Any material having an indirect bandgap is incapable of having its electron excited without leading to a change in momentum. Only through momentum exchange—that is, by taking on or imparting momentum to the crystal lattice's vibrations—can the electron modify its momentum relative to the crystal. The absorbing semiconductor, also known as the absorber, can be significantly thinner in a direct bandgap material because the absorption coefficient is substantially higher than in an indirect bandgap material [24]. A void is generated at E_i if an electron is stimulated from E_i to E_f. This emptiness is known as a hole and functions similarly to an elementary charged particle. Thus, as shown in Figure 12.8, the absorption of a photon results in the formation of an electron-hole pair. The chemical energy of the electron-hole pair is obtained from the photon's radiative energy.

Thermodynamics sets a maximum conversion efficiency limit for radiative energy to chemical energy. It is demonstrated that in non-concentrated and highly concentrated sunlight, respectively, this thermodynamic limit is between 67% and 86% [25].

As seen in Figure 12.8, the electron-hole pair will typically recombine, or for the electron to return to its initial energy level Ei. Next, the energy will be converted to additional electron-hole pair, or lattice vibrations (non-radiative recombination) or released as photons (radiative recombination). From energy stored to deposited, the thin layer of the membrane are on the both side of absorber such that the flow of electrons and holes are on opposite sides of the membrane [25]. The stored energy is to be used for working in a circuit. Most solar cells have these membranes consisting of n- and p-type materials.

A solar cell's construction before recombining the electron and hole pairs must reach the membranes, and the transit time of the charge carriers to the membranes must be less than their lifetime. This requirement limits the thickness of the absorber. In order for the charge carriers to function in an external circuit, they are ultimately removed from the

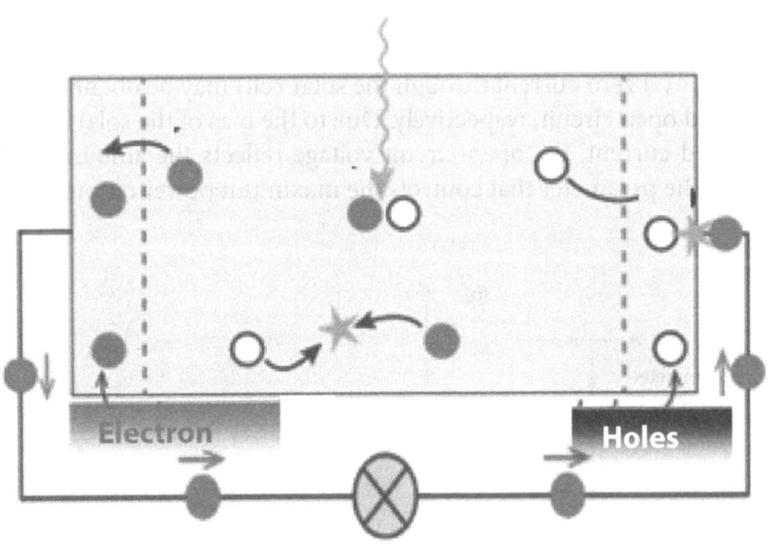

Figure 12.8 Based on a very basic description of a solar cell, an electron-hole pair that is isolated by membranes is created when a photon is absorbed.

solar cells using electrical connections (Figure 12.8). In the end, the chemical energy of the electron-hole pairs is converted to electric energy. Once the electrons have completed their circuit, they will recombine with holes at a metal-absorber interface (Figure 12.8).

As shown in Figure 12.7(b), the two main loss processes in single-junction solar cells are (i) unable to convert the photons to electricity, which is below the bandgap, and (ii) the thermalization of photons with energies is above the bandgap. In the conversion process, these two mechanisms alone result in a loss of almost half of the incident solar energy [26]. Consequently, a single-junction solar cell's maximum energy conversion efficiency is far less than the thermodynamic limit. Shockley and Queisser computed this single bandgap limit for the first time in 1961 [27].

The solar cell diode's I-V curves in the absence of light and the presence of light are superposed to form a curve. The diode law changes when a cell is illuminated because it increases the typical "dark" currents in the diode.

$$I = I_o\left[exp\frac{qV}{nKT} - 1\right] - I_L \tag{12.1}$$

where I_o = saturation (or leakage) current in absence of light, q = Electric charge on carrier, V = voltage (applied across the diode), n = ideality factor, k = Boltzmann's constant, T = temperature, and I_L = photo generated current.

Figure 12.9 depicts a typical circuit for obtaining the current-voltage curve. Numerous solar cell properties, including the efficiency, fill factor (FF), open-circuit voltage (V_{OC}), and short-circuit current (I_{SC}), can be calculated from these features. These factors determine a solar panel's rating. When the voltage across a solar cell is zero, or when the solar cell is short circuited, the current flowing through it is known as the short-circuit current. The short-circuit current is due to light-generated carriers being produced and gathered. At the most moderate resistive loss mechanisms, the light-generated current and the short-circuit current of an ideal solar cell are equal.

As a result, the maximum current "I_{SC}" (at zero voltage across the solar cell) and maximum voltage "V_{OC}" (at zero current through the solar cell) may be obtained from solar cell at short circuit and open circuit, respectively. Due to the bias of the solar cell junction with the light-generated current, the open-circuit voltage reflects the amount of forward bias on the solar cell. The parameter that controls the maximum power output from a solar cell

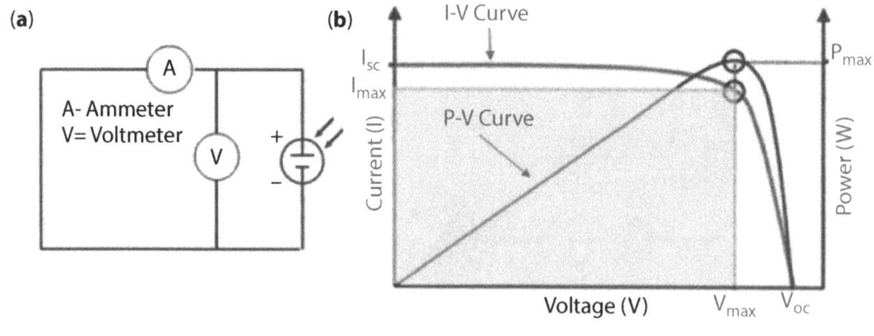

Figure 12.9 Solar cell schematic (a) circuit diagram with symbol and (b) current voltage curve.

along with Voc and Isc is called the "fill factor," or simply "FF." The relationship between the solar cell's maximum power and the product of Voc and Isc is known as the FF.

$$FF = \frac{P_{max}}{V_{OC}I_{sC}} = \frac{V_{max}I_{max}}{V_{OC}I_{SC}} \tag{12.2}$$

The area of the largest rectangle that will fit inside the current voltage curve, as illustrated in Figure 12.9(b), serves as the FF's visual depiction of the "squareness" of the solar cell. Efficiency is the most commonly used parameter to compare the performance of various solar cells. Efficiency is defined as the solar cell's energy output divided by its input. Efficiency is influenced by the solar cell's temperature as well as the type and strength of the incident sunlight. It also shows how well the solar cell is working.

12.3.2 Multifunctional Materials for Enhanced Absorption and Conversion Efficiency of Solar Cells

The utilization of materials with multiple functions is essential for improving the absorption and conversion efficiency of solar cells. These materials are made to have a variety of uses, which can help them overcome different obstacles in the field of solar cell technology. Multifunctional materials contribute significantly to solar cell efficiency in the following ways:

- Multifunctional materials are engineered to exhibit enhanced light absorption across a broad spectrum of wavelengths. To do this, one can incorporate plasmonic materials, QDs, or additional photo-absorbing nanostructures [11, 48].
- Light Trapping and Scattering utilizing nanostructured materials, light can be absorbed and distributed inside the solar cell, extending the photons' path through the active material. Both the possibility of photon absorption and the overall efficiency of the solar cell increase [49, 50].
- Self-cleaning coatings can help ensure optimal light absorption by preventing dust and grime buildup on the surface of solar cells. Furthermore, anti-reflection coatings lessen sunlight reflection, increasing the amount of light that reaches the solar cell [51, 52].
- Additionally, multifunctional materials can be engineered to improve solar cells' stability and robustness, increasing their ability to withstand environmental elements including heat, moisture, and UV rays [53].
- Tandem or multi-junction solar cell architectures frequently use multifunctional materials. A wider spectrum of sunlight can be absorbed by these designs, increasing total efficiency [54, 55].

Scholars continue to explore and develop new multifunctional materials to address the evolving problems in solar cell technology and improve the overall efficiency of solar energy conversion systems.

12.3.3 Emerging Materials for High-Performance Solar Cells

The study of solar cells is a dynamic topic, and scientists are always looking into new materials to improve solar cell performance. The following cutting-edge substances have potential for use in high-performance solar cells:

- Perovskite solar cells have attracted a lot of interest because of their impressive efficiency advancement. These materials have high absorption coefficients and inexpensive production costs. They are generally based on organic-inorganic lead halide perovskites. The goal of ongoing research is to increase scalability and stability [53, 56].
- Tandem solar cells enable more effective light absorption because of their numerous layers with distinct bandgaps. Tandem structures with increased overall efficiency can be created by combining materials such as perovskites with conventional Si or other cutting-edge materials [57, 58].
- Organic semiconductors are flexible and inexpensive to process. Scientists are attempting to increase the stability and efficiency of OPV materials. Particularly non-fullerene acceptors are drawing interest due to their potential to increase organic solar cells' efficiency [59, 60].
- Size-dependent absorbance and programmable bandgaps are features of semiconductor nanocrystals known as colloidal QDs. Researchers are looking into using these materials in solar cells to obtain effective and tunable absorption since they can be precisely controlled in terms of their properties throughout the synthesis process [61, 62].
- Transparent conducting electrodes in solar cells require TCOs. To increase overall device performance, new TCO materials are being developed, such as metal oxide nanoparticles, which have improved conductivity, transparency, and flexibility [63, 64].
- Metal-organic frameworks (MOF's) are materials that are highly surface-area and porous. Because of MOFs' special structural and electrical characteristics, researchers are looking into incorporating them into solar cells for uses including light harvesting and carrier transport [65, 66].
- In solar cells, plasmonic nanoparticles—like those made of gold or silver—can improve light scattering and absorption. Scholars are examining methods to include these nanomaterials to enhance the solar energy conversion efficiency [67, 68].

These substances just serve as a sample of the wide variety of novel materials being investigated to improve solar cell technology. In an effort to get these materials closer to commercial viability, ongoing research intends to address issues like stability, scalability, and cost-effectiveness.

12.4 Multifunctional Materials for Photodetectors

Modern optoelectronic devices are not complete without PDs, which translate optical information into electrical signals and are crucial to many conventional applications including communication, imaging, and sensing [69–71]. Furthermore, due to the rapid development of emerging artificial intelligence technologies like robotics and autonomous driving, people have higher expectations for the miniaturization, lightness, integration, and multifunctionality of PDs in order to detect more optical information, such as light's intensity, spectrum, polarization, and incidence angle [72, 73]. Even though the field of PDs has seen a great deal of research and successful outcomes, the majority of this work has focused on light intensity detection, and there is still a dearth of research on multifunctional PDs for the detection of light polarization, spectrum, and incidence angle [74, 75]. A multifunctional PD is one that can simultaneously detect light polarization, spectrum, and incident angle information in addition to light intensity in a single device [69, 76]. Traditional multifunctional detectors are implemented physically using gratings, waveplates, large optical lenses, and multiple PDs. This leads to a large device size, complicated structure, and high cost, and makes it difficult to integrate flexible multifunctional detectors in the future. In order to further minimize the dimensions and weight of the apparatus, self-powered PDs must also be created [77].

12.4.1 Basic Concepts of Photodetectors

An optical receiver's job is to retrieve the data sent over the lightwave system and transform the optical signal back into electrical form. A PD, which uses the photoelectric effect to turn light into electricity, is its primary component. A PD needs the same specifications as an optical source. It ought to be highly reliable, quiet, responsive, sensitive, and affordable. Its dimensions ought to match those of the fiber core. PDs composed of semiconductor materials are best suited to meet these requirements.

Optical absorption is the basic principle by which photodetection works. This lesson covers fundamental ideas that are used to characterize all PDs, including responsivity, quantum efficiency, rise time, and bandwidth.

Examine the semiconductor slab schematically depicted in the accompanying image (Figure 12.10). Every time a photon enters the semiconductor and is absorbed, an electron-hole pair is created if the energy hυ of the incident photons is greater than the bandgap energy.

An applied voltage creates an electric field that sweeps electrons and holes across the semiconductor, causing an electric current to flow. In other words, the photocurrent I_p and the incident optical power P_{in} are exactly proportional.

$$I_P = R_d\, P_{in} \tag{12.3}$$

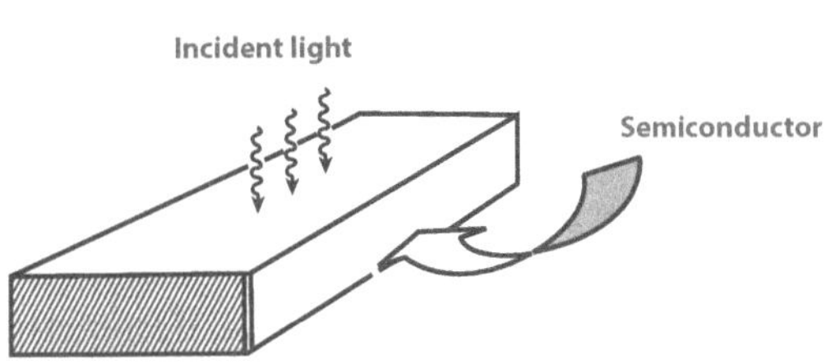

Figure 12.10 A photo detector made of a semiconductor slab.

Since greater values of R_d result in greater current being produced at a given input power, the constant R_d—which is expressed in ampere/watts (A/W)—is referred to as the PD's responsivity.

The quantum efficiency, denoted as η, is a fundamental quantity that can be used to express the responsivity Rd. It is defined as

$$\eta = \frac{Electron\ generation\ rate}{Photon\ incident\ rate} = \frac{\frac{I_p}{q}}{\frac{P_{in}}{h\upsilon}} = \frac{h\upsilon}{q} R_d \tag{12.4}$$

Thus, the responsivity R_d is provided by

$$R_d = \frac{nq}{h\upsilon} \approx \frac{n\lambda}{1.24} \tag{12.5}$$

A PD's responsivity rises with wavelength λ due to the fact that it can detect more photons with the same optical power. It is not anticipated that this linear dependence on λ will last indefinitely since, eventually, photon energy drops too low to produce electrons. This occurs in semiconductors for hv < E_g, where E_g is the bandgap. At that point, the quantum efficiency η becomes zero.

The absorption coefficient α allows the dependency of η on λ to be accessed. Assuming that the semiconductor slab's facets in the aforementioned figure have an antireflection coating, $P_{tr} = e^{-\alpha W} P_{in}$ represents the power transmitted through the slab with width W. One way to express the absorbed power is as

$$P_{abs} = P_{in} - P_{tr} = \left\lfloor 1 - e^{-\alpha W} \right\rfloor P_{in} \tag{12.6}$$

Since every absorbed photon produces a pair of electrons and holes, the quantum efficiency η may be found using

$$\eta = \frac{P_{abs}}{P_{in}} = 1 - e^{-\alpha W} \tag{12.7}$$

As anticipated, η drops to zero at α = 0. Conversely, if αW >> 1, then η approaches 1.

The wavelength dependence of α for a number of semiconductor materials that are frequently used to create PDs for lightwave systems is depicted in the accompanying picture (Figure 12.11).

Because that material can only be utilized as a PD for $\lambda < \lambda_c$, the wavelength λc at which α becomes zero is known as the cutoff wavelength. This image illustrates how PDs can be made from indirect-bandgap semiconductors, such Si and Ge, despite the fact that their absorption edges are not as sharp as those of direct-bandgap materials. For most semiconductors,

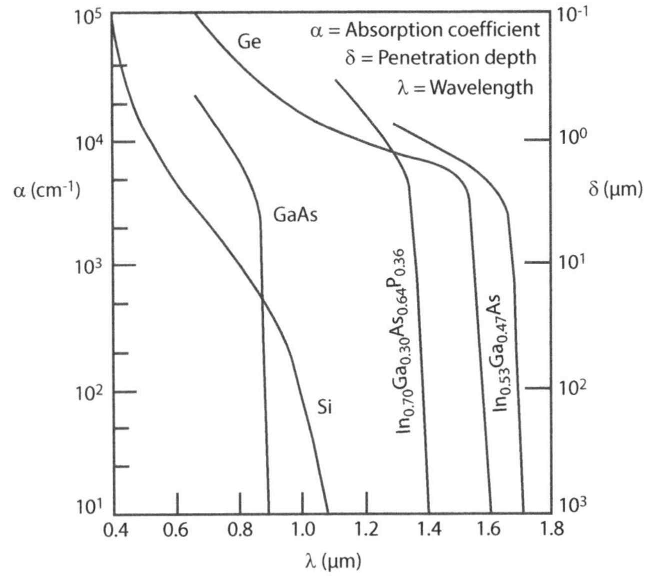

Figure 12.11 Variation of absorption coefficient, penetration depth with wavelength.

large values of α (~ 10^4 cm^{-1}) may be obtained, and for W ~ 10 μm, η can reach 100%. This characteristic demonstrates the photodetection effectiveness of semiconductors.

12.4.2 Multifunctional Materials for Improved Sensitivity and Response Time of Photodetectors

The sensitivity and reaction time of PDs can be greatly increased by multifunctional materials, allowing them to detect light more quickly and precisely. A few types of versatile materials that are commonly used to improve these characteristics in PDs are as follows:

- Due to their tunable bandgaps, QDs offer precisely regulated absorption spectra. PDs can be made more sensitive by using QDs, even at wavelengths that are not in the visible spectrum. Furthermore, by facilitating efficient charge separation and transfer, QDs can speed up response times [78, 79].
- Nanostructures with plasmonic properties can enhance light-matter interactions and focus electromagnetic fields at the nanoscale. Plasmonic components can increase the sensitivity of PDs by raising their absorption cross-section [80, 81].
- Quantized energy levels and improved light absorption are produced by semiconductor quantum well structures, which confine charge carriers in two dimensions. The spectrum response and PDs' sensitivity can be tailored by varying the quantum wells' composition and thickness [82, 83].

These multifunctional materials can be used in PDs to improve sensitivity, widen spectral coverage, increase quantum efficiency, and speed up response times. Applications in communication, sensing, imaging, and PVs will be made possible by this.

12.5 Multifunctional Materials for Optical Sensors

Optical sensors are now widely used in modern technology, with uses in consumer electronics, industrial automation, medical diagnostics, and environmental monitoring, among other areas. Optical sensors use the way light interacts with materials to identify changes in the biological, chemical, or physical aspects of their surroundings. Because of their high sensitivity, selectivity, and precision, optical sensors are especially beneficial for finding analyte concentrations at low levels in intricate matrices.

12.5.1 Basic Concept of Optical Sensors

The basis of optical sensing is the interaction of light with matter. A light source's ability to absorb, reflect, or scatter light can provide details about an object's composition, shape, size, and distance from another source. It is made up of a light receiver (light detector) and a light emitter (light source). From transmitter to receiver, light passes via a medium (typically air, although it might also include other components). The quantity that has to be measured affects (modulates) the transmission route. The measurand can have an impact on the light's many qualities. There are several kinds of receivers (photoresistor, PV cell, photodiode, phototransistor, etc.), emitters (light bulb, gas laser, laser diode, LED), transmitters, and potential add-ons that might be used to complete an optical sensor system. It is obvious that the designer has a great deal of flexibility in selecting the ideal solution for the intended application due to the abundance of options. A new generation of optical sensors was made possible by the advancement of lasers and fiber optics in the 20th century. Based on their operating principles, settings, and applications, optical sensors may be divided into various groups. These sensors are more sensitive and accurate than previous types of sensors, and they can measure a larger variety of physical values. Typical varieties of optical sensors include fiber optic sensors, photonic sensors, surface plasmon sensors, and micro electromechanical sensors.

The components of all-optical sensors are the same, regardless of how they operate or what their goals are: light sources, PDs, and optical components to direct the light between. The measuring item's inherent characteristics control the signal on the PD and, consequently, the sensor reading will be anywhere along this light path. Thus, knowledge of the many optical components that may be employed, as well as the potential kinds of lighting and detecting hardware, is necessary to understand the fundamental concepts behind the operation of different types of optical sensors. The arrangement of the optical sensor system is shown in Figure 12.12.

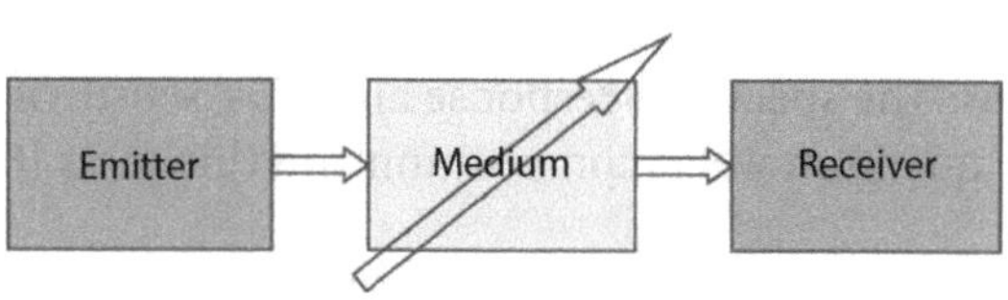

Figure 12.12 The fundamental arrangement of an optical sensor system.

a) Light Sources
An optical sensor's light source is its fundamental component. It serves as the "medium" for the transfer of information, but it may also function as a part of the sensing circuit by modulating the light that is released, for instance. Characteristics of a light source include its power consumption, emission spectrum, radiant intensity, coherence level, longevity, and any other factors that are important for that particular application [84].

b) Photodetectors
An optoelectronic device called a PD transforms incident light or other electromagnetic radiation in the visible, IR, and UV spectrums into electrical impulses [85]. A photomultiplier tube (PMT) is a highly sensitive detector that is commonly used to amplify and detect low-intensity optical signals. It is particularly useful in applications where high sensitivity and low-light detection are essential. The external photoelectric effect serves as the foundation for the photomultiplier's operation. This involves the removal of photoelectrons from the surface of the metal upon light of threshold frequency. To liberate electrons from the surface, minimum energy corresponding to minimum frequency must be equivalent to a so-called work function, W. This is described by Einstein's equation:

$$E_{KE} = h.v - W$$

As the photocurrent can be in minimal value depending on the incident photons, this minimal photocurrent can be enhanced using PMT. The wavelength-dependent quantum efficiency, or $\eta(\lambda)$, which represents the ratio of emitted photoelectrons to incident photons, is what defines PMTs.

Compared to photomultipliers, photodiodes are more simpler to handle and operate since they do not require a high voltage or a certain type of detector housing [86]. Using the action of photons on the charge carriers in the depletion zone at the pn-junction of a semiconductor diode, photodiodes produce electron-hole pairs when photons are absorbed. A reverse voltage that is applied causes the charge carriers to drift in the direction of the external electrodes, producing a photocurrent that is proportionate to the intensity of the light. Most photodiodes are composed of Ge or Si.

c) Optical Elements:
A vast array of optical components may be integrated into an optical sensor. Mirrors and lenses, dispersive components including optical fibers, filters, prisms and gratings, and optical modulators are a few examples of them [87]. In order to direct light, create and modify pictures, control beam forms, restrict and choose wavelength ranges, and modulate light, optical components are needed.

12.5.2 Multifunctional Materials for Improved Sensitivity and Selectivity of Optical Sensors

Multifunctional materials play a crucial role in enhancing the sensitivity and selectivity of optical sensors. These materials often integrate multiple functionalities, such as optical, electrical, and chemical properties, to enable more versatile and efficient sensing capabilities.

a) Plasmonic Nanomaterials: These metallic nanomaterials are important components, being stimulated by incident light, to enhance the efficiency of optical sensors [88]. These metallic nanostructures produce surface plasmon resonances in response to incident light, which in turn causes plasmon—a unified oscillation of free electrons—to dominate. Strong absorption, scattering, and near-field amplification at the plasmonic-metal nanoparticles' native frequency are caused by a large number of these electrons engaging in surface plasmon resonance [89]. Because of this resonance coupling, plasmonic nanostructures have remarkable absorption and scattering profiles at these particular frequencies, which have led to novel uses for them. These materials exhibit localized surface plasmon resonances, leading to increased electromagnetic field enhancements near the sensor surface, which in turn improves detection sensitivity [90]. Many plasmonic sensors have been designed and developed during the past ten years for a variety of analytes, including neurotransmitters and explosive chemicals [91, 92]. Plasmonic sensors exhibit performance and efficiency that is entirely on par with or even better than traditional detection techniques. Immobilized nanoparticles deposited on substrates, which can trace analytes on-site with the use of a portable image analysis device (like smartphones), will primarily shape the future of plasmonic sensors. Because these sensor arrays use distinct ligands for each unique analyte, they also enable multi-analyte determination.
b) Quantum Dots: QDs, which are luminous semiconductor nanocrystals, are especially appealing for biological sensing and imaging applications. A great deal of chemical sensing methods and bioassays rely on fluorescence. In this context, QDs' distinct optical characteristics are far superior to those of conventional molecular fluorophores. Numerous issues with traditional luminescence sensors can be resolved and a whole new range of applications can be realized thanks to their remarkable photo-stability, narrow fluorescence emission, broad absorption spectrum, relatively high quantum yields, and ability to tune their luminescence characteristics by controlling particle size. Graphene QDs, or GQDs, have become a very promising sensing material to be included with the created sensors. GQDs, functionalized GQDs, and their composites have been demonstrated in a number of recent research to improve metal ion optical detection [93]. Carbon QDs (CQDs) have exceptional photoluminescence capabilities that make them very promising for use in sensing applications. Charge transfer and dipole-dipole interaction are used to examine the mechanism of interaction between CQDs and chemical analytes. These interactions alter their fluorescence characteristics, boosting fluorescence signals and enhancing the performance of optical sensors [94].
c) Functionalized Nanomaterials: Selective detection of target analytes can be made possible by surface functionalizing nanomaterials with certain molecules or ligands. This is especially helpful in biological and chemical sensing applications because sensor selectivity is increased by the interaction between the functionalized surface and the target analyte [95]. The scientific community has become interested in functional nanomaterials due to their unique combination of advantageous chemical and physical properties, including

superior heat and electrical conductivity, superior optical properties, chemical stability, and high mechanical strength [96].

12.5.3 Emerging Materials for High-Performance Optical Sensors

A number of recently developed materials have demonstrated significant potential for high-end optical uses. Artificially constructed materials that are used to manipulate and shape the flow of electromagnetic waves—or maybe any other kind of physical wave—are known as metamaterials. Metamaterials can enable features like negative refraction and super-resolution imaging, which might transform optical equipment. famed as meta-atoms, these artificial materials are made up of a variety of nanostructures that serve as the atoms and molecules in conventional materials. In contrast to normal materials, metamaterials' characteristics are dictated by the unique architectures of their meta-atoms rather than the inherent qualities of their chemical ingredients [97, 98]. By means of effective interactions between incident electromagnetic waves and meta-atoms, metamaterials are able to display remarkable physical characteristics and may manipulate light fields. Some of these features are even unattainable in naturally occurring materials [99]. Topological insulators are surface-conductive materials with insulating bulk properties. Strong second harmonic production and photocurrent effects have been shown in optical investigations of these materials, which may be helpful for detectors and lasers [100]. The family of topologically intriguing materials, which includes superconductors, semimetals, and insulators, has expanded during the past fifteen years. 2D TMDs are important research materials for optical sensors. These materials are finding niche uses in optoelectronics and next-generation electronics, which depend on the final thickness of atoms. When bulk materials are reduced to monolayers, 2D TMDs show distinctive electrical and optical features that develop from surface effects and quantum confinement that occur during the shift from an indirect to a direct bandgap. Together with their enormous exciton binding energy and variable bandgap, TMDs are a viable option for a range of optoelectronic devices, such as solar cells, photo-detectors, LEDs, and photo-transistors [101, 102]. Perovskite materials, due to the flexible substitution of many ions in its system, the perovskite family has a large number of members. Unlike metal chalcogenides, PNCs do not require surface passivation in order to maintain high quantum yields. They are also very resistant to defects. It's interesting to note that perovskites frequently only have defect structures and trap states in their valence and/or conduction bands, not in the bandgap's mid-states. These qualities basically improve them and greatly benefit their sensing applications. Photonic devices heavily rely on nonlinear optical materials. The field of nonlinear optics has developed and been associated with several fields and swift advancements in fiber optics and optical communications technology. Nonlinear optics encompasses a wide range of activities, from the basic research of light-matter interaction to the development of components and systems for a wide range of scientific, military, and medical uses [103]. Within the PhC, electromagnetic waves scatter. At some wavelengths, destructive interference occurs, resulting in the formation of a photonic bandgap (PBG), which is analogous to the energy bandgap of electron waves in a semiconductor. The possibility of generating a PBG makes it feasible to control how light propagates. PhCs are used in a number of notable applications, including self-collimation, negative refraction, optical diodes, and light bending [104].

It is important to note that the development and adoption of these emerging materials for high-performance optical applications depend on ongoing research, technological advancements, and the ability to scale up their production for practical use. Researchers and engineers continue to explore and optimize these materials for various applications to unlock their full potential.

12.6 Multifunctional Materials for Display Technologies

12.6.1 Basic Concept of Display Technologies

The twentieth century has seen significant improvements in the design and production of display systems. It has been a significant journey from Braun's 1897 invention of the cathode-ray-tube (CRT) to the present-day status of flat panel displays. Over this time, technology has developed steadily, occasionally replacing earlier technologies and other times complementing them. This has led to the development of numerous useful technologies, including CRT, LCD, PDP, field emitter displays (FEDs), OLEDs, and many more. Drive electronics have also seen similar innovations. Vacuum tubes, the technology available at the time, were used to build the electronics at first [105]. Discrete semiconductor devices were used in the middle of the 1950s, and semiconductor integrated circuits were used in the 1970s [105–107]. The development of very large-scale integration (VLSI) and the easy availability of digital signal processing (DSP) chips, and semiconductor electronics have reached an unprecedented level of integration and function versatility today. Engineers' goal of creating a device that would require "no space," react "instantaneously," cost "nothing," and carry out any instruction has nearly been realized. The paradox of the situation is that there is a growing demand for ever-growing display screens, yet drive electronics are becoming smaller.

Present Devices: The driver electronics and display screens are not independent components; rather, they are a part of an information system. It is now commonplace to link a display to the system it serves, such as avionic displays, gaming displays, healthcare displays, military displays, TV displays, and so forth, depending on the situation. Based on the data content, the operating environment, and—above all—the requirements for the man-machine interface, each of these categories of displays is linked to a particular application segment and has specific needs. Depending on the application, each of these classes of displays has different market prices and volumes. For instance, computer and TV displays have become commodities that are sold for hundreds of thousands at low prices, but avionic displays are exceptionally expensive, come in small quantities, and have extremely strict specifications. Currently, TVs for entertainment, computer monitors, and mobile devices for communication are the top-selling displays [108].

In the following subsections, features distinctive to TV, computer, and mobile phone display systems are explained as examples, either by frequency or amplitude modulation, or by using different digital modulation techniques, and then transmitted. There is a limited amount of bandwidth required for this process. As a result, a lot of work has historically been put into the research and design of the TV signal. For instance, the bandwidth allotted for color TV at the time of its proposal was nearly identical to that of a monochrome

TV, even though it carried three equivalent monochrome signals. In addition, the color signal needed to work with TV receivers that were already in use, monochrome models. Researchers had to come across sophisticated solutions, which they did. The switch from low-definition to high-definition TV presented similar challenges in terms of compatibility and bandwidth conservation. In parallel, a lot of work has been done on bandwidth compression methods to lower the quantity of bandwidth needed for transmission or storage while maintaining the audio or video signal's quality.

12.6.1.1 Display System: Computer Monitor

The foundation of the ongoing IT revolution is the computer. The display connected to a computer system is commonly referred to as a computer display monitor, and it has emerged as the preferred output port for computer-human communication. There is typically no long-distance transmission required in the case of a computer system between the computer hardware that powers the display and the display itself. Thus, there is no need for the signal between the system and the monitor to be modulated or demodulated. In this instance, the computer's signals are intrinsically digital, but the CRT-based display systems are intrinsically analog. The interfaces to convert digital to analog format to power the display were built into the monitor to address this issue (the same procedure is employed for observing a DTV signal on an analog TV). Nevertheless, these interfaces are not needed given the availability of matrix-like flat panels, particularly those that are LCD based.

Most display systems are made to handle both digital and analog video signals in different formats. The design of the exhibit device must match these input signals. The main video processing card handles the required signal processing. Internal memories are needed to store one or two frames, as required by signal processing algorithms. The enormous advancements in memory technology have reduced the requirement for high-density memories to be as large or expensive, which has allowed display resolutions to rise.

12.6.1.2 Display System: Cell Phone

In the past, the caller identity and, subsequently, short message service (SMS) communications were displayed on a basic monochrome alphanumeric display found on cell phones. Today's cell phones have advanced from getting such basic displays to having multi-colored, higher-resolution screens that can display moving images. This is required when new features are added to cell phones in response to user requests and manufacturers' desire to give their goods an advantage over rival models. Nowadays, a high-end cell phone's display is usually a 2" diagonal with resolution (320 x 240), offering at least 6 bits of color. It can display images in video format from a DVD or the internet. The battery should be charged at a high mean interval. Thus, the display should keep the amount of power used to the unconditional minimum. However, it must have sufficient brightness for sunlight readability, requiring a higher input power that clashes with the requirement for low power consumption. Higher resolutions will be demanded over time because of the more complex operations it will be expected to perform. OLEDs [109] are expected to meet many of these requirements.

12.6.1.3 Computer-Assisted Visualization

Along with advancements in electronics for consumers, there has emerged a niche market for display technologies focused on "visualization," which is demanding and crucial. It is significant to highlight that the meaning of "display" varies greatly between the communities of "developers and users of visualization tools" and "designers of TV, computer monitors, and mobile phones." For the earlier display technologies, drive electronics, CRT, LCD, PDP, OLED, active matrix, passive matrix, and similar terms would be involved. For those who use visualization, this would imply scientific or information visualization, both of which require handling substantial amounts of raw data and converting it into a visual format at the end. It is done to take advantage of people's remarkable ability to use visual cues to identify patterns and create relationships.

Scientific visualization places a strong focus on tangible objects, such as engineering machine parts, medical pictures of human organs, and so forth. The creator of a scientific visualization engine needs to be an expert in geometric modeling, graphics and image processing, software engineering, and the direct or indirect implementation of 3-D displays using 2-D projection techniques. The presentation of information involves managing multidimensional data, such as web pages, conferences, archives, and EOLSS materials. The inability to see small details and the entire picture of data on small display screens makes it difficult to create useful illustrations in this situation.

12.6.1.4 Performance Requirements and Specifications for Display Screens

An individual "sees" the image on an engineer-designed display screen through his or her eyes. To satisfy the viewer's perceptual needs, the display screen designer should be able to make full use of the state-of-the-art material and computational capabilities. With the invention of CRT for black-and-white TV in the 1930s, an attempt was made to comprehend and apply man-machine interaction. Displays were produced and sold using the technologies, materials, and processes that were available at the time, as well as our understanding of how people see. Although there has been an increase in research on technologies and visual perception since the 1930s, display screen design challenges persist. The creativity of those early scientists and display engineers who created working display screens with "so little knowledge" as was available at the time is astounding considering this. Good physical shape, weight, and volume were understood, as was the significance of high luminosity, high resolution, excellent contrast, large viewing angle, quick response, appropriate power consumption, and low cost. However, as time passed and technology advanced for a variety of reasons, the unit's volume for a specific screen size ended up being the deciding factor in this industry's evolution. The dimensions of a display unit's face (display screen) and depth determine its volume. The screen is equally wide and tall. However, it became customary to mention the form-factor, or width-to-height ratio, along with the diagonal's length instead of two lengths. Another crucial factor is the display unit's depth. Individuals would prefer a display unit with even lower performance if it had a smaller depth, according to experience. The last 20 years have seen a rise in flat-panel display sales due to user preference. We only discuss flat-panel displays in this section because CRT displays are widely understood.

12.6.2 Multifunctional Materials for Improved Color Purity and Brightness of Displays

The exhibits change people's reading preferences and cut down on paper consumption. Every industry and every person needs a monitor. Display applications are used in billboards, TV screens, cell phones, home appliances, and automobile dashboards. Therefore, research continuously improves display technical specifications. CRTs, which were used in the first display screens, had two main disadvantages: their small size and high consumption of electricity. Lightweight and comfortable and thin, liquid-crystal displays (LCDs) have primarily taken the place of CRTs. However, since LCD screens are not fitted for backlighting, the emission of a full-color spectrum via a color filter is required. LEDs have recently been introduced as an excellent alternative for these issues of LCD and have emerged as the preferred option for the industry as well. Three different types of illumination are used in LED backlights; the majority of displays right now use WLEDs. Utilizing edge-lit backlights helps create a thin, light appearance. Hot spots frequently appear on low-quality light guide plates, which can lead to issues with low light uniformity and inefficient light extraction. Direct-lit backlights, the second kind of light source, are more advantageous than edge-lit backlights when it comes to contrast, luminosity, and affordability because they employ WLEDs. Three primary color sources as red, green, and blue (RGB). LEDs are used but the light source's parameters like attenuation rates, color shift, cost effectiveness, and a high technical threshold were major issues.

12.6.3 Emerging Materials for High-Performance Displays

QDs and core-shell QDs (CSQDs) display technology is used in QLEDs (QDs LEDs). QDs exhibit narrow emission and wide absorption. Through QD films, which offer excellent color saturation performance, backlight LEDs in blue produce outstanding RGB light. But according to, this technology is categorized as a non-self-luminous light source. Su *et al.* [125] studied the application of transparent QLEDs in RGB that are constructed on a flexible material substrate and integrated vertically using UV glue.

For transparent RGB QLEDs, they reported high quantum efficiency proving that individually controllable RGB QLEDs are feasible. Mini-LED technology is another ancestor of micro-LED technology. The compact size of mini-LEDs is an advantage. To optimize the light field type of mini-LEDs—which can be used as a backlight source for displays and lighting—Ye *et al.* [124] suggested a modified package structure. They recently worked on RGB micro-LEDs that are incredibly small and are entirely made of inorganic semiconductor materials. These LEDs have the advantages of high brightness, high reliability, quick response times, and photoelectric conversion efficiency due to their material properties. These emerging materials not only cover the simple oxides like ZnO, ZnS, Ga_2O_3, and CuO, but they also cover QDs of (I-III-VI) groups.

12.7 Multifunctional Materials for Optical Communications

12.7.1 Basic Concept of Optical Communication

An optical communication system is one in which data is sent and received using light as a carrier. This kind of communication has been around for as long as humans have, even in prehistoric times. However, recent technological developments have made it possible to use this quick media extensively for contemporary worldwide communication. Light would naturally be the ideal carrier of a communication signal because nothing can travel faster than light and because it is immune to electromagnetic interference. As a result, optical communication has proven to be a desirable data transmission method. The channel that transmits light messages from one place to another, optical fibers, is another alluring feature of optical communication systems. Because it is composed of glass, it is a strong, flexible, small-cross sectional area, and reasonably priced medium that can compete with coaxial cables and conventional wires in many communication-related fields. It also adds a negligible distortion factor to the data being transmitted. Fiber optical communication technologies have a large capacity [110] and very low transmission loss as a result.

We now have access to a communication medium that crosses oceans between the continents that make up the global communication network known as "The internet," owing to optical communication systems. It is reasonable to conclude that optical communication systems have had a profound impact on our daily lives and the way that we communicate with one another. This emphasizes how crucial it is to conduct additional research and development to maximize its potential. Any digital or analog signal can be converted by the transmitter into an intensity-based light signal (amplitude modulation) or a discrete light pulse (pulse modulation). Since their emission wavelengths coincide with the low-loss fiber's wavelength region of attenuation, light-producing diodes are typically utilized for this. Since a laser-producing diode is more efficient than a non-laser light-emitting diode, such as LEDs, it is most frequently employed. The signal is received by a receiver at the end of the optical fiber after traveling through it. There, it is amplified and demodulated to return as an electrical signal that can be decoded and understood by the electronic equipment. High-capacity fiber is essential, but in order to keep up with its pace, quick optical transmitters and receivers are also necessary.

(a) Fiber Cable

Low-loss silica fiber was suggested as a transmission medium for optical communications in the late 1960s. The lowest theoretical limit of silica fiber, a loss of 0.2 dB/km, was achieved by research and development of this medium by the late 1970s. This made it possible for low-loss silica to function better in a fiber that may be as thin as a human hair than in any other metal-based medium that was available at the time. Since interest in the potential of such systems has grown significantly over time, this accomplishment made it possible to design systems that use optical communication. The concept of total internal reflection is essential to the transmitting principle of fiber optics. In situations where a light beam is traveling from one medium to another and the angle between it and the second medium is small enough, the light beam will entirely reflect off of the second medium and stay in the first medium. To ensure complete internal reflection, a fiber cable is composed of a low refractive index covering (cladding) surrounding a high refractive index core. In order to shield the fiber cable from external interference and stop noise from entering and departing, an additional coating is applied all around it.

(b) Optical Amplification

For long-distance optical communication, optical amplifications are necessary to boost a weak signal and increase its maximum fiber length before it becomes significantly lost and unintelligible. These gadgets are essential for today's international communication networks, like the Internet. Optical-electrical-optical repeaters (OEO), which convert an optical signal back into an electrical signal before regenerating it with more power, were replaced by optical amplifiers. This is due to the high cost of OEO as well as their challenging creation and application. Furthermore, although OEO is restricted to a single signal modulation and bandwidth, a single optical amplifier can operate with multiple of them. Raman amplifiers, semiconductor optical amplifiers, and erbium-doped waveguide amplifiers are only a few of the several methods of optical amplification.

(c) Transmitter
Devices known as optical transmitters take an electrical signal and use light waves to transform it into an optical signal. Since light-producing semiconductor devices are the most potent and efficient ones on the market, using one makes sense. Instead of producing incoherent light as LEDs do, laser diodes create coherent light, making them particularly more effective for uses like optical transmitters. Though LEDs were more widely used despite their drawbacks because of their simplicity, this began to change with the development of vertical-cavity surface-emitting lasers (VCSELs), which combine the affordability of LEDs with the efficiency of earlier laser diodes. As a result, laser diode technology has surpassed that of LEDs in terms of the application of optical transmitters.

(d) Receiver
Receivers are devices that create the electrical signal that was there before it became optical by converting the optical signal into an electrical one. These gadgets make use of parts known as photosensors or PDs. Similar to optical transmitters in optical communication systems, semiconductor elements are also used by optical receivers to perform their functions. To generate an electrical signal that is correctly converted, the semiconductor is connected to a limiting amplifier and a transimpedance amplifier. A local oscillator laser can also be used in optical systems that use coherent light signals to aid in the conversion process.

(e) Wavelength-Division Multiplexing
The technique known as wavelength-division multiplexing (WDM) expands the bandwidth that optical modulators and detectors can handle. With this method, different light wavelengths can be carried simultaneously by a single carrier, which can subsequently be demultiplexed at the receiver [111]. This makes it feasible to gain extra bandwidth and to perform wavelength-division duplexing, or bidirectional communication within a single optical fiber. Optical filters, such as grating, prism, and interference filters, can be used to multiplex and demultiplex light signals. WDM gearbox systems come in one-way and two-way versions. The only devices needed for one-way transmission are a multiplexer at the transmitter and a demultiplexer at the receiver. Both need to be present at the transmitter and the receiver during a two-way broadcast.

12.7.2 Multifunctional Materials for Improved Transmission and Modulation of Optical Signals

In the past decade, emerging applications like telepresence and high-throughput e-science demand a significant portion of network bandwidth. To meet this demand, researchers have developed optical communication systems for transport backbone networks, metropolitan areas, and access networks, improving signal processing functions like amplification, filtering, and dispersion compensation. Multifunctional materials play a crucial role in optimizing optical signal modulation and transmission in optical communication systems. These materials are designed with properties that enhance the performance of components such as fibers, modulators, and detectors. Material advances have enabled increases in transmission and modulation capabilities in high-performance optical communication. The interplay of 2D materials, plasmonics, Si photonics, chalcogenide glasses, perovskites, QDs, and photonic crystal fibers is driving a constant evolution of optical communication technologies. By incorporating these multifunctional materials into the design of optical communication components, researchers and engineers can advance optical communication technologies by increasing the effectiveness, speed, and dependability of optical data transmission and modulation.

12.7.3 Emerging Materials for High-Performance Optical Communication

The following discusses multifunctional materials and their functions in high-performance optical communication. Other thermally drawable materials have been suggested and shown to be able to be used in place of silica glass to construct optical fibers that can be used for a variety of purposes. First among these materials are polymers, which include ZEONEX, PDMS, hydrogels, thermoplastic cyclic olefin copolymer (TOPASfi), and polymethyl methacrylate (PMMA). Either microstructure is added, or the core is doped with additional materials to raise the refractive index, in order to achieve light guidance. Polymer-based fiber is particularly biocompatible and flexible, which makes it a viable option for use in medical applications. Soft glass, such as chalcogenide and telluride glass, 2D materials, plasmonics, semiconductors, Si photonics, perovskites, QDs, and photonic crystals, is another material type that is frequently used to create optical fiber. This large group of materials has a lower drawing temperature than regular silica glass. The temperature at which soft-glass fiber is typically drawn varies in the hundreds of degrees Celsius based on the composition of the various materials, while silica-glass fiber is pulled at a temperature between 1900 and 2100 °C based on the different preform structures or volumes. Owing to the materials' flexibility, step-index, gradient-index, and other microstructures can be used to create the fiber [16]. In a similar vein, photonic crystals can be employed in optical fibers to produce PBGs, which boost the transmission of particular wavelengths while lowering signal loss and selectively permitting certain wavelengths to pass through. The optical signals are modulated and altered by semiconductors with nonlinear optical characteristics. Compact devices for signal modulation and transmission enhancement can be developed thanks to plasmonic materials' ability to confine and manipulate light at the nanoscale. Optical modulators can be equipped with 2D materials such as graphene, which is a single layer of carbon atoms organized in a hexagonal lattice, to regulate the phase and intensity of light signals. Devices based on graphene provide low power consumption and high-speed modulation. Beyond graphene, 2D materials with the potential for better modulation and detection in optical systems include TMDs [112]. Si photonics makes it possible to integrate optical components onto a chip, which makes it easier to create small, effective signal transmission and modulation devices. With continuous advancements in effective modulators, detectors, and light sources built on Si chips, Si photonics continues to be a focal point. By using this method, optical communication systems operate more efficiently overall and integrate components more optimally [113]. Perovskite materials have emerged as a promising option for optical modulators and detectors for communication applications in the last ten years because of their good optical characteristics (such as tunable band gap and high carrier mobility), excellent electric characteristics (such as long carrier diffusion length), and, most importantly, a straightforward and affordable synthesis process carried out in non-vacuum conditions [114]. However, QDs have optical characteristics that can be adjusted, which makes them useful for various optical communication components including modulators and light-emitting devices [115].

12.8 Multifunctional Materials for Future Optoelectronics

12.8.1 Multifunctional Materials for Emerging Optoelectronic Applications

However, their use in large group screens causes technical maintenance issues in addition to challenges with mass transfer. It is difficult and time-consuming to replace intermittent LEDs when black spots or hue shifts show up on the screen. Furthermore, because of their material absorption characteristics, red LED chips are more likely than blue or green LED chips to produce excessive heat, which lowers their EQE and photoelectric conversion efficiency. Moreover, these CSQDs have significantly fewer surface defects and traps due to the overgrowth of multiple layers, such as ZnS, ZnO, InP, and others. Based on band structure and its formation of the CQDs, the outer shell confinement of carriers is possible to prevent surface bonds from dangling with respect to bare QDs. Of all the I-III-VI core/shell QD varieties, a few examples of CQDs, like CuGaS2/ZnS (CGS/ZnS), have been suggested to have significant application potential. PVs, LSCs, and LEDs are superior due to their larger PLQY, large Stokes shift, small self-absorption loss, and high photostability. Recently several sustainable QDs have been developed and put into use, including Si QDs, CQDs, and I-III-VI QDs. I-III-VI QDs have shown great promise in a range of optoelectronic devices due to their immediate composition-tunable transparency as TCO materials. Notably, by creating suitable wide-band-gap outer shell materials like ZnS, ZnSe, and ZnO, the various surface imperfections of I-III-VI QDs can also be adequately passivated. As such, they are unable to satisfy display requirements. Reports suggest a wide color gamut of 85.4% (Rec. 2020) and 114.4% (National Television System Committee, NTSC) using directional control over RGB micro-LEDs produced sapphire substrate. The results mentioned above suggest that the most effective way to achieve full color right now is to combine blue micro-LEDs with QDs. Micro-LEDs are advantageous due to their high contrast, low use of electricity, extended lifespan, and quick response time [116–118]. To enhance color purity, modern displays also include a color filter, which absorbs most of the wavelength band and only permits certain wavelengths to pass. As a result, the efficiency of light extraction decreases substantially. The improvement of micro-LED hybrid QDs' color purity has not received a lot of focus in research. A monitor's ability to show a wider color gamut is limited by low color purity, which additionally contributes to low color saturation. Improving color purity is therefore essential. The excited blue light passes through to the red and green QDs layer, and the associated colors of red and green are emitted after being absorbed by the QDs.

High color saturation and conversion efficiency are rendered possible by this indicated solution, that satisfies display requirements.

12.8.2 Challenges and Opportunities in the Field of Multifunctional Materials for Optoelectronic Applications

Emerging as possibilities to move beyond Si-based electronics are 2D materials and their HSs. As fascinating and unconventional 2D-physics and promising optoelectrical real estate are revealed, they are developing into a strong contender for transparent, flexible, low-power optoelectronics and quantum technology of the future. However, due to the lack of logical and scalable methods for the fabrication of HSs other than mechanical transfer techniques, the progress in this direction is limited to mostly proof-of-concept studies. Hetero-integration of 2D devices and scalable synthesis of 2D HSs are essential to advancing the potential into practical applications. Many attempts are being made in this direction, with the CVD as well as MOCVD methods being regarded as the most capable for marketing. However, before 2D technology can be applied in the real world, several important issues must be addressed and a firm grasp of growth kinetics must be established. This article methodically discusses growth mechanisms and comprehensive future targets to highlight the difficult problems. Entering the 2D paradigm requires large-scale, electronic-grade quality 2D HSs with high management over interface width, layer-by-layer number, doping, structure, twist angle, and dependability via a universal growth strategy. In addition to the synthesis problem, a significant challenge that requires thorough examination to witness remarkable potential uses of 2D HSs at the ultimate thinness restricts is the devices' stability in the environment. Two-dimensional heterostructures have been observed recently, offering a platform beyond CMOS technology. Integrating logic and memory as a basis for logic-in-memory in addition to von Neumann architecture and neuromorphic computing has received a great deal of attention (Christensen *et al.* [119]; LeCun *et al.* [120]). Phase-change memory, resistive random-access memory, and spintronic memory—which offer both memory and logic units—are some of the new technologies that have emerged (Kaspar *et al.* [121]; Xia and Yang [122]) [123].

12.9 Conclusion and Future Directions

12.9.1 Summary of the Key Concepts and Findings

With their striking advantages, multifunctional materials span superconductors, metals, semi-metals, semiconductors, and insulators and exhibit remarkable applications. They also have drastically varied energy band topologies and electrical properties. In terms of their incorporation into optoelectronic devices, such as charge transfer layers, transparent conductive electrodes, and conductive additives, these materials are particularly promising. Due to the enormous variety of material combinations and consequent qualities and functions, they should also be far more suited to the application than present unifunctional materials. Beyond system performance, these composites may also save costs, at least in the post-production stage, which is something to keep in mind. This multidisciplinary subject can also be expanded to include innovative fabrication techniques that leverage digitalization and computational engineering methods for virtual material design. Development cycles can be reduced in this way, and goods made of multifunctional materials can become increasingly customized and flexible.

We outlined the special optical and electronic characteristics of multifunctional materials in this chapter, along with the most recent developments in their use in optoelectronic devices. First, the exceptional optical and electrical properties of multifunctional materials are described. These properties are programmable. Second, along with some recent advancements in this field, a synopsis of the basic concepts, structures, and parts used in optoelectronic device applications is provided. These applications include LEDs, solar cells, PDs, optical sensors, display technology, and optical communications. Finally, some important questions about the usefulness of multifunctional materials in optoelectronic device applications have been brought up, requiring further investigation.

12.9.2 Future Directions and Challenges in the Field of Multifunctional Materials for Optoelectronics

Optoelectronics, in our opinion, will be a crucial technical advance in the ensuing decades. In order to enable previously unattainable technological breakthroughs, we are gathering data on materials and equipment. However, we must respect the environment and the resources left for the next generation. Future technology design must take circular lifecycles, self-powering devices, and renewable materials into consideration. Optoelectronics of the future will be dependent on the latest developments in novel materials. While Si and other inorganic materials enjoyed success in the previous century, organic and hybrid organic-inorganic materials will increasingly find their way into optoelectronics. We do not think that organic will inevitably be less expensive than inorganic, nor that it will certainly replace inorganic. We do discover that using hybrid and organic materials will lead to new possibilities. The application of hybrid halide perovskite in optoelectronics is among the most recent pieces of proof. Using earthly, non-toxic components that can be recycled with low-cost and clean processing is the main driving force in the exploration and selection of novel materials. Delivering high-performing devices with minimal environmental impact—that is, by using non-toxic materials and environmentally friendly production techniques—is the paradigm for optoelectronics of the future. The research and development of new technologies is now significantly more difficult than it was in the past due to this problem. It appears that a multidisciplinary strategy combining knowledge from process engineering and basic science is required. The field may experience a notable advancement if machine learning is applied to particular jobs. The screening of new materials that meet the above-mentioned circular lifespan and renewable criteria is one of the most productive uses. In this case, optoelectronics may find and synthesize a new class of materials much more quickly by applying machine learning. As an illustration, the recent surge in interest in halide perovskites showed that the vast library of thousands of possible optoelectronic materials had previously gone unnoticed. It is practically impossible to investigate every potential use for halide perovskites using a straightforward experimental methodology. To take full advantage of this new class of materials, machine learning applications appear to be practically required.

References

1. Hongwei, T., Gang, L., Run-Wei, L., Multifunctional Optoelectronic Device Based on Resistive Switching Effects, in: *Recent Development in Optoelectronic Devices*, S. Ruby (Ed.), p. Ch. 3, IntechOpen, Rijeka, 2018.
2. Siengchin, S., A review on lightweight materials for defence applications: Present and future developments. *Def. Technol.*, 24, 1–17, 2023.
3. Sotelo-Vazquez, C. *et al.*, Multifunctional P-Doped TiO2 Films: A New Approach to Self-Cleaning, Transparent Conducting Oxide Materials. *Chem. Mater.*, 27, 3234–3242, 2015.
4. Duy, L.T. *et al.*, Multifunctional Nanohybrid of Alumina and Indium Oxide Prepared Using the Atomic Layer Deposition Technique. *ACS Appl. Mater. Interfaces*, 13, 59115–59125, 2021.
5. Kidambi, P.R. *et al.*, Multifunctional oxides for integrated manufacturing of efficient graphene electrodes for organic electronics. *Appl. Phys. Lett.*, 106, 063304, 2015.
6. Zhao, Y. *et al.*, Multifunctional Perovskite Photodetectors: From Molecular-Scale Crystal Structure Design to Micro/Nano-scale Morphology Manipulation. *Nano Micro Lett.*, 15, 187, 2023.
7. Chen, H. *et al.*, New concept ultraviolet photodetectors. *Mater. Today*, 18, 493–502, 2015.
8. Liu, D. *et al.*, Multifunctional applications of triazine/carbazole hybrid thermally activated delayed fluorescence emitters in organic light emitting diodes. *J. Mater. Chem. C*, 7, 12470–12481, 2019.
9. Ma, S. *et al.*, Multi-Color Light-Emitting Diodes. *Coatings*, 13, 182, 2023.
10. Mozumder, M.S. *et al.*, Recent developments in multifunctional coatings for solar panel applications: A review. *Sol. Energy Mater. Sol. Cells*, 189, 75–102, 2019.
11. Wu, C. *et al.*, Multifunctional nanostructured materials for next generation photovoltaics. *Nano Energy*, 70, 104480, 2020.
12. Chen, H. *et al.*, All-optical modulation with 2D layered materials: status and prospects. *Nanophotonics*, 9, 2107–2124, 2020.
13. He, J. *et al.*, Design of a Multi-Functional Integrated Optical Switch Based on Phase Change Materials. *Photonics*, 9, 320, 2022.
14. Wu, W., Stretchable electronics: functional materials, fabrication strategies and applications. *Sci. Technol. Adv. Mater.*, 20, 187–224, 2019.
15. Hwan, J., Kim, L.S.E., Kim, B.H., Applications of flexible and stretchable three-dimensional structures for soft electronics. *Soft Sci.*, 3, 16, 2023.
16. Liu, Z. *et al.*, Multifunctional Smart Optical Fibers: Materials, Fabrication, and Sensing Applications. *Photonics*, 6, 48, 2019.
17. Xie, M. *et al.*, Flexible Multifunctional Sensors for Wearable and Robotic Applications. *Adv. Mater. Technol.*, 4, 1800626, 2019.
18. Mohamed, W.A.A. *et al.*, Quantum dots synthetization and future prospect applications. *Nanotechnol. Rev.*, 10, 1926–1940, 2021.
19. Amini, P. *et al.*, High Throughput Quantum Dot Based LEDs, in: *Energy Efficiency Improvements in Smart Grid Components*, M.E. Moustafa (Ed.), p. Ch. 11, IntechOpen, Rijeka, 2015.
20. Lozano, G. *et al.*, Metallic nanostructures for efficient LED lighting. *Light Sci. Appl.*, 5, e16080–e16080, 2016.
21. Yue, Y. and Gong, J.P., Tunable one-dimensional photonic crystals from soft materials. *J. Photochem. Photobiol., C*, 23, 45–67, 2015.
22. Abd-Elnaiem, A.M. *et al.*, Recent Progress in the Fabrication of Photonic Crystals Based on Porous Anodic Materials. *Energies*, 16, 4032, 2023.
23. Yap, P.L. *et al.*, Tuning the Multifunctional Surface Chemistry of Reduced Graphene Oxide via Combined Elemental Doping and Chemical Modifications. *ACS Omega*, 4, 19787–19798, 2019.

24. Chi, H. *et al.*, Recent advances in graphene based photoresponsive materials. *Prog. Nat. Sci.: Mater. Int.*, 29, 603–611, 2019.
25. Junaid, M. *et al.*, A Review on Graphene-Based Light Emitting Functional Devices. *Molecules*, 25, 4217, 2020.
26. Weng, Y. *et al.*, Hybrid Device of Blue GaN Light-Emitting Diodes and Organic Light-Emitting Diodes with Color Tunability for Smart Lighting Sources. *ACS Omega*, 7, 5502–5509, 2022.
27. Tian, J., Xie, J., Zhuang, W., Recent Advances in Multi-Site Luminescent Materials: Design, Identification and Regulation. *Materials*, 16, 2179, 2023.
28. Bae, E.-J. *et al.*, Enhanced Light Extraction from Organic Light-Emitting Diodes with Micro-Nano Hybrid Structure. *Nanomaterials*, 12, 1266, 2022.
29. Zhang, W. *et al.*, Nanostructured inorganic electrochromic materials for light applications. *Nanophotonics*, 10, 825–850, 2021.
30. Gao, Y. *et al.*, High-Performance Perovskite Light-Emitting Diodes Enabled by Passivating Defect and Constructing Dual Energy-Transfer Pathway through Functional Perovskite Nanocrystals. *Adv. Mater.*, 34, 2207445, 2022.
31. Kim, Y.-H. and Tae-Woo, L., Engineering Colloidal Perovskite Nanocrystals and Devices for Efficient and Large-Area Light-Emitting Diodes. *Acc. Mater. Res.*, 4, 655–667, 2023.
32. Ràfols-Ribé, J. *et al.*, High-performance organic light-emitting diodes comprising ultrastable glass layers. *Sci. Adv.*, 4, eaar8332, 2018.
33. Ababay Ketema, W. and Worku, A.D., Recent advances of graphene-based materials for emerging technologies. *Results Chem.*, 5, 100971, 2023.
34. Ahmad, F. *et al.*, Advances in graphene-based electrode materials for high-performance supercapacitors: A review. *J. Energy Storage*, 72, 108731, 2023.
35. Pu, J. and Takenobu, T., Monolayer Transition Metal Dichalcogenides as Light Sources. *Adv. Mater.*, 30, 1707627, 2018.
36. Wang, C., Yang, F., Gao, Y., The highly-efficient light-emitting diodes based on transition metal dichalcogenides: from architecture to performance. *Nanoscale Adv.*, 2, 4323–4340, 2020.
37. Lin, Q. *et al.*, Flexible Quantum Dot Light-Emitting Device for Emerging Multifunctional and Smart Applications. *Adv. Mater.*, 35, 2210385, 2023.
38. Cameron, D. *et al.*, Core–Shell Nanorods as Ultraviolet Light-Emitting Diodes. *Nano Lett.*, 23, 1451–1458, 2023.
39. Panfil, Y.E., Oded, M., Banin, U., Colloidal Quantum Nanostructures: Emerging Materials for Display Applications. *Angew. Chem. Int. Ed.*, 57, 4274–4295, 2018.
40. Meyaard, D.S., Huang, Kuo, Shen, *et al.*, 9 - Efficiency droop in gallium indium nitride (GaInN./gallium nitride (GaN. LEDs, in: *Nitride Semiconductor Light-Emitting Diodes*, J. Huang, H.-C. Kuo, and S.-C. Shen, (Eds.), pp. 279–300, LEDs., Woodhead Publishing, Cambridge, United Kingdom, 2014.
41. Kim, H.-S. *et al.*, Unusual strategies for using indium gallium nitride grown on silicon 111. for solid-state lighting. *Proc. Natl. Acad. Sci.*, 108, 10072–10077, 2011.
42. Zhang, G. *et al.*, Metasurface-tuned light-matter interactions for high-performance photodetectors. *Fundam. Res.*, 4, 224–233, 2024.
43. Ali, A., Mitra, A., Aïssa, B., Metamaterials and Metasurfaces: A Review from the Perspectives of Materials, Mechanisms and Advanced Metadevices. *Nanomaterials*, 12, 1027, 2022.
44. Cai, Y. *et al.*, Multifunctional Enhancement for Highly Stable and Efficient Perovskite Solar Cells. *Adv. Funct. Mater.*, 31, 2005776, 2021.
45. Maurya, S.K., Recent Progress in Transparent Conductive Materials for Photovoltaics. *Energies*, 15, 8698, 2022.
46. Contreras, M.A. *et al.*, Replacement of Transparent Conductive Oxides by Single-Wall Carbon Nanotubes in Cu(In,Ga.Se2-Based Solar Cells. *J. Phys. Chem. C*, 111, 14045–14048, 2007.

47. Hossain, M.J. *et al.*, Self-assembled multifunctional nanostructures for surface passivation and photon management in silicon photovoltaics. *Nanophotonics*, 10, 4611–4621, 2021.
48. Zhang, X. *et al.*, Semitransparent quantum dot solar cell. *Nano Energy*, 22, 70–78, 2016.
49. Peter Amalathas, A. and Alkaisi, M.M., Nanostructures for Light Trapping in Thin Film Solar Cells. *Micromachines*, 10, 619, 2019.
50. Selvapriya, R. *et al.*, Reaction induced multifunctional TiO2 rod/particle nanostructured materials for screen printed dye sensitized solar cells. *Ceram. Int.*, 47, 8094–8104, 2021.
51. Syafiq, A. *et al.*, Application of transparent self-cleaning coating for photovoltaic panel: a review. *Curr. Opin. Chem. Eng.*, 36, 100801, 2022.
52. Wu, P. *et al.*, Transparent Self-Cleaning Coatings: A Review. *Coatings*, 13, 1270, 2023.
53. Afre, R.A. and Pugliese, D., Perovskite Solar Cells: A Review of the Latest Advances in Materials, Fabrication Techniques, and Stability Enhancement Strategies. *Micromachines*, 15, 192, 2024.
54. Zheng, Z. *et al.*, Tandem Organic Solar Cell with 20.2% Efficiency. *Joule*, 6, 171–184, 2022.
55. Ašmontas, S. and Mujahid, M., Recent Progress in Perovskite Tandem Solar Cells. *Nanomaterials*, 13, 1886, 2023.
56. Fu, Q. *et al.*, Multifunctional Two-Dimensional Polymers for Perovskite Solar Cells with Efficiency Exceeding 24%. *ACS Energy Lett.*, 7, 1128–1136, 2022.
57. Brinkmann, K.O. *et al.*, Perovskite–organic tandem solar cells. *Nat. Rev. Mater.*, 9, 202–217, 2024.
58. Anaya, M. *et al.*, ABX3 Perovskites for Tandem Solar Cells. *Joule*, 1, 769–793, 2017.
59. Tufail, M.K. *et al.*, Designing efficient materials for high-performance organic solar cells: Detailed chemical space exploration, machine learning and virtual screening. *Chem. Phys. Lett.*, 834, 140974, 2024.
60. Basha, B. *et al.*, Designing of novel organic semiconductors materials for organic solar cells: A machine learning assisted proficient pipeline. *Inorg. Chem. Commun.*, 153, 110818, 2023.
61. Akhil, S.,. and Balakrishna, R.G., CuBiSe2 Quantum Dots as Ecofriendly Photosensitizers for Solar Cells. *ACS Sustain. Chem. Eng.*, 10, 13176–13184, 2022.
62. Khan, I., Saeed, K., Khan, I., Nanoparticles: Properties, applications and toxicities. *Arabian J. Chem.*, 12, 908–931, 2019.
63. Shen, J.-J., Recently-explored top electrode materials for transparent organic solar cells. *Synth. Met.*, 271, 116582, 2021.
64. Aepuru, R. *et al.*, Chapter 19 - Transparent conducting electrode materials for solar cell technologies, in: *Green Sustainable Process for Chemical and Environmental Engineering and Science*, R.B. Inamuddin, A.M. Asiri, M.M. Rahman (Eds.), pp. 377–407, Elsevier, Amsterdam, Netherlands, 2021.
65. Chen, B. *et al.*, Emerging applications of metal-organic frameworks and derivatives in solar cells: Recent advances and challenges. *Mater. Sci. Eng.: R: Rep.*, 152, 100714, 2023.
66. Heo, D.Y. *et al.*, Metal-Organic Framework Materials for Perovskite Solar Cells. *Polymers*, 12, 2061, 2020.
67. Subramanyam, P. *et al.*, Emerging materials for plasmon-assisted photoelectrochemical water splitting. *J. Photochem. Photobiol. C*, 51, 100472, 2022.
68. Arinze, E.S. *et al.*, Plasmonic Nanoparticle Enhancement of Solution-Processed Solar Cells: Practical Limits and Opportunities. *ACS Photonics*, 3, 158–173, 2016.
69. Kelley, S.O. *et al.*, Advancing the speed, sensitivity and accuracy of biomolecular detection using multi-length-scale engineering. *Nat. Nanotechnol.*, 9, 969–980, 2014.
70. Casalino, M. *et al.*, State-of-the-art all-silicon sub-bandgap photodetectors at telecom and datacom wavelengths. *Laser Photonics Rev.*, 10, 895–921, 2016.
71. Chen, H. *et al.*, Nanostructured Photodetectors: From Ultraviolet to Terahertz. *Adv. Mater.*, 28, 403–433, 2016.

72. Xue, J. *et al.*, Narrowband Perovskite Photodetector-Based Image Array for Potential Application in Artificial Vision. *Nano Lett.*, 18, 7628–7634, 2018.
73. El Hamdani, S., Benamar, N., Younis, M., Pedestrian Support in Intelligent Transportation Systems: Challenges, Solutions and Open issues. *Transp. Res. Part C: Emerging Technol.*, 121, 102856, 2020.
74. Chen, G. *et al.*, Air-Stable Highly Crystalline Formamidinium Perovskite 1D Structures for Ultrasensitive Photodetectors. *Adv. Funct. Mater.*, 30, 1908894, 2020.
75. Xu, W. *et al.*, *2D Perovskite Single Crystals for Photodetectors: From Macro- to Microscale*, physica status solidi RRL. *Rapid Res. Lett.*, 15, 2100183, 2021.
76. Meng, J., Cadusch, J.J., Crozier, K.B., Detector-Only Spectrometer Based on Structurally Colored Silicon Nanowires and a Reconstruction Algorithm. *Nano Lett.*, 20, 320–328, 2020.
77. Perumal Veeramalai, C. *et al.*, Lead–halide perovskites for next-generation self-powered photodetectors: a comprehensive review. *Photonics Res.*, 9, 968–991, 2021.
78. Seo, K.-H. *et al.*, Improving of Sensitivity of PbS Quantum Dot Based SWIR Photodetector Using P3HT. *Materials*, 14, 1488, 2021.
79. Zhao, X. *et al.*, Lead Chalcogenide Colloidal Quantum Dots for Infrared Photodetectors. *Materials*, 16, 5790, 2023.
80. Lin, K.-T., Lin, H., Jia, B., Plasmonic nanostructures in photodetection, energy conversion and beyond. *Nanophotonics*, 9, 3135–3163, 2020.
81. Yan, N. *et al.*, Plasmonic Enhanced Nanocrystal Infrared Photodetectors. *Materials*, 16, 3216, 2023.
82. Talib, M. *et al.*, Improvements in the Performance of a Visible–NIR Photodetector Using Horizontally Aligned TiS3 Nanoribbons. *ACS Omega*, 4, 6180–6191, 2019.
83. Tan, C.L. and Hooman, M., Emerging technologies for high performance infrared detectors. *Nanophotonics*, 7, 169–197, 2018.
84. Light Sources. *Opt. Sens.*, 7–26, 2010, https://doi.org/10.1002/9783527629435.ch2.
85. Photodetectors. *Opt. Sens.*, 27–36, 2010, https://doi.org/10.1002/9783527629435.ch3.
86. Argueta-Diaz, V., Light sources and detectors, in: *Optical Sensors*, pp. 2-1-2-15, IOP Publishing, Bristol, United Kingdom, 2023.
87. Optical Elements. *Opt. Sens.*, 37–54, 2010, https://doi.org/10.1002/9783527629435.ch4.
88. Amirjani, A. and Rahbarimehr, E., Recent advances in functionalization of plasmonic nanostructures for optical sensing. *Microchim. Acta*, 188, 57, 2021.
89. Cottancin, E., Del Fatti, N., Halté, V., Optical, Structural and Magneto-Optical Properties of Metal Clusters and Nanoparticles, in: *Nanoalloys: Synthesis, Structure and Properties*, D. Alloyeau, C. Mottet, C. Ricolleau (Eds.), pp. 331–368, Springer London, London, 2012.
90. Amirjani, A., Koochak, N.N., Fatmehsari Haghshenas, D., Investigating the Shape and Size-Dependent Optical Properties of Silver Nanostructures Using UV–vis Spectroscopy. *J. Chem. Educ.*, 96, 2584–2589, 2019.
91. Wang, J. *et al.*, Highly Selective Rational Design of Peptide-Based Surface Plasmon Resonance Sensor for Direct Determination of 2,4,6-trinitrotoluene TNT. Explosive. *Sens. Actuators, B*, 264, 279–284, 2018.
92. Milkani, E., Lambert, C.R., McGimpsey, W.G., Direct detection of acetylcholinesterase inhibitor binding with an enzyme-based surface plasmon resonance sensor. *Anal. Biochem.*, 408, 212–219, 2011.
93. Anas, N.A.A. *et al.*, Development of Graphene Quantum Dots-Based Optical Sensorfor Toxic Metal Ion Detection. *Sensors*, 19, 3850, 2019.
94. Nazri, N.A.A. *et al.*, Carbon quantum dots for optical sensor applications: A review. *Opt. Laser Technol.*, 139, 106928, 2021.

95. Patil, M.D. *et al.*, Functional Nanomaterials for Sensing Devices, in: *Smart Nanostructure Materials and Sensor Technology*, R.K. Sonker, K. Singh, R. Sonkawade (Eds.), pp. 77–95, Springer Nature Singapore, Singapore, 2022.
96. Lakshmi, D. *et al.*, Functionalized Nanomaterials, Classification, Properties, and Functionalization Techniques, in: *Functionalized Nanomaterials Based Supercapacitor: Design, Performance and Industrial Applications*, C.M. Hussain and M.B. Ahamed (Eds.), pp. 65–92, Springer Nature Singapore, Singapore, 2024.
97. Cai, W. and Shalaev, V., Introduction, in: *Optical Metamaterials: Fundamentals and Applications*, W. Cai and V. Shalaev (Eds.), pp. 1–10, Springer New York, New York, NY, 2010.
98. Ziolkowski, R.W. and Engheta, N., Introduction, History, and Selected Topics in Fundamental Theories of Metamaterials, in: *Metamaterials*, pp. 1–41, 2006.
99. Smith, D.R., Pendry, J.B., Wiltshire, M.C.K., Metamaterials and Negative Refractive Index, vol. 305, pp. 788–792, 2004.
100. Anirban, A., 15 years of topological insulators. *Nat. Rev. Phys.*, 5, 267–267, 2023.
101. Manzeli, S. *et al.*, 2D transition metal dichalcogenides. *Nat. Rev. Mater.*, 2, 17033, 2017.
102. Ahamed, M., II, Shakeel, N., Anwar, N., Asiri, Structure and Fundamental Properties of Black Phosphorus, in: *Black Phosphorus: Synthesis, Properties and Applications*, R.B. Inamuddin, and A.M. (Eds.), pp. 139–156, Springer International Publishing, Cham, 2020.
103. Semin, S. *et al.*, Nonlinear Optical Properties and Applications of Fluorenone Molecular Materials, vol. 9, p. 2100327, 2021.
104. Butt, M.A., Khonina, S.N., Kazanskiy, N.L., Recent advances in photonic crystal optical devices: A review. *Opt. Laser Technol.*, 142, 107265, 2021.
105. Fernández, M.R., Casanova, E.Z., Alonso, I.G., Review of Display Technologies Focusing on Power Consumption. *Sustainability*, 7, 10854–10875, 2015.
106. Huang, Y. *et al.*, Mini-LED, Micro-LED and OLED displays: present status and future perspectives. *Light Sci. Appl.*, 9, 105, 2020.
107. Anwar, A.R. *et al.*, Recent Progress in Micro-LED-Based Display Technologies. *Laser Photonics Rev.*, 16, 2100427, 2022.
108. Zhan, T. *et al.*, Multifocal displays: review and prospect. *PhotoniX*, 1, 10, 2020.
109. Zhang, S. *et al.*, Research Progress of Micro-LED Display Technology. *Crystals*, 13, 1001, 2023.
110. Essiambre, R.J. *et al.*, Capacity Limits of Optical Fiber Networks. *J. Lightwave Technol.*, 28, 662–701, 2010.
111. Iga, K., Surface-emitting laser-its birth and generation of new optoelectronics field. *IEEE J. Sel. Top. Quantum Electron.*, 6, 1201–1215, 2000.
112. Novoselov, K.S. *et al.*, Electric Field Effect in Atomically Thin Carbon Films. *Science*, 306, 666–669, 2004.
113. Soref, R., The Past, Present, and Future of Silicon Photonics. *IEEE J. Sel. Top. Quantum Electron.*, 12, 1678–1687, 2006.
114. Kim, H.-S. *et al.*, Lead Iodide Perovskite Sensitized All-Solid-State Submicron Thin Film Mesoscopic Solar Cell with Efficiency Exceeding 9%. *Sci. Rep.*, 2, 591, 2012.
115. Brus, L.E., Electron–electron and electron-hole interactions in small semiconductor crystallites: The size dependence of the lowest excited electronic state. *J. Chem. Phys.*, 80, 4403–4409, 1984.
116. Ye, Z.T. and Wu, J.-Y., Use of Recycling-Reflection Color-Purity Enhancement Film to Improve Color Purity of Full-Color Micro-LEDs. *Nanoscale Res. Lett.*, 17, 1, 2022.
117. Chen, H.-W. *et al.*, Liquid crystal display and organic light-emitting diode display: present status and future perspectives. *Light Sci. Appl.*, 7, 17168–17168, 2018.
118. Desroches, L.-B. and Ganeshalingam, M., The dynamics of incremental costs of efficient television display technologies. *Technol. Forecast. Soc. Change*, 90, 562–574, 2015.

119. Christensen, D.V. *et al.*, 2022 roadmap on neuromorphic computing and engineering. *Neuromorph. Comput. Eng.*, 2, 022501, 2022.
120. LeCun, Y., Yoshua, B., Hinton, G., Deep learning. *Nature*, 521, 436–444, 2015.
121. Kaspar, C. *et al.*, The rise of intelligent matter. *Nature*, 594, 345–355, 2021.
122. Xia, Q. and Yang, J.J., Memristive crossbar arrays for brain-inspired computing. *Nat. Mater.*, 18, 309–323, 2019.
123. Chakraborty, S.K. *et al.*, Challenges and opportunities in 2D heterostructures for electronic and optoelectronic devices. *iScience*, 25, 2022.
124. Ye *et al.*, CSP-LEDs combined with light guide without reflective matrix for antiglare design. *IEEE Access*, 8, 156718–156726, 2020.
125. Su *et al.*, Flexible and tandem quantum-dot light-emitting diodes with individually addressable red/green/blue emission. *Npj Flex Electron*, 5, 8, 2021.

13

Analytical Tools for Multifunctional Materials

Javed Khan and Shikha Yadav*

Department of Pharmacy, School of Medical and Allied Sciences, Galgotias University, Greater Noida, Uttar Pradesh, India

Abstract

Recent years have seen a rise in interest in multifunctional materials due to the unusual combination of electrical, mechanical, thermal, magnetic, and optical capabilities they offer. In a wide range of fields, such as storing energy, electronics, sensing, and catalysis, these materials show significant promise. Understanding the structure-property correlations of multifunctional materials is crucial to their design and optimization, but this requires sophisticated analytical methods. In order to better understand the morphology, crystal structure, and composition of multifunctional materials, the article first discusses structural characterization techniques like X-ray diffraction (XRD), scanning electron microscopy (SEM), transmission electron microscopy (TEM), or atomic force microscopy (AFM). Spectroscopic methods are next discussed in detail; these include Fourier-transform infrared spectroscopy (FTIR), Raman spectroscopy, ultraviolet-visible spectroscopy, and X-ray photoelectron spectroscopy (XPS), all of which provide insight into chemical bonding, electronic structure, and optical features of a substance. In addition, the study emphasizes the usefulness of thermal analysis techniques including differential scanning calorimetry (DSC), thermogravimetric analysis (TGA), or thermal conductivity measurements in characterizing the thermal behavior and stability of multifunctional materials. The importance of electrical characterization techniques, including impedance spectroscopy or Hall effect measurements, in shedding light on the underlying principles of charge transport and electrical behavior is also highlighted.

***Keywords*:** Multifunctional materials, structural characterization, spectroscopy techniques, thermal analysis, eelectrical characterization, microscopy methods, raman spectroscopy, x-ray photoelectron spectroscopy (XPS)

13.1 Introduction

The purpose of detecting targets and calculating their quantities, an essential area of analytical research, is the sensitive or selective detection of both chemical and biological analytes. Numerous scientific and technological sectors, including analytical chemistry, life sciences, material science, biomedical diagnostics, drug development, food security, personal care, and monitoring the environment, including the Internet of Things, have made extensive use of sensing [1, 2]. The development of different instrumental techniques that have obtained ultrasensitive detection of both chemical and biological analytes, such as spectroscopic

**Corresponding author*: shikha.yadav@galgotiasuniversity.edu.in

Divya Bajpai Tripathy, Anjali Gupta and Arvind Kumar Jain (eds.) Multifunctional Materials: Engineering and Biological Applications, (335–364) © 2025 Scrivener Publishing LLC

Table 13.1 Various characterization techniques are compared.

Technique	Advantages	Limitations	Performed studies
SEM	Magnification of up to 1 million.	Samples need to be conductive. Not accurate for measurements less than 10 nm.	Surface topography. Pore size. Particle size. Membrane Thickness.
TEM	Higher resolution than SEM.	Images do not show topography data. Staining the sample may be required.	Nanoparticles. Nanofilms. Nanopores. Membrane thickness.
AFM	No sample preparation. Similar resolution to TEM. Measurements of mechanical properties.	Requires more processing time. Lower depth of field.	Nano-profiling. Surface roughness. Pore size distribution. Membrane stiffness.
FTIR	Detection of a variety of compounds. High sensitivity of parts per million (ppm). Fast analysis time (in seconds).	Cannot analyze aqueous samples. Cannot detect molecules of two identical atoms.	Functional groups. Polymer-solvent compatibility. Polymer-filler interaction. Miscibility of polymer blends. Membrane degradation.
Raman	No sample preparation. More sensitive to functional groups with better intensity peaks.	Release of fluorescent light of some samples may cause background noise. Polar molecules have lower Raman signals.	Functional groups. Crystal structure. Polymer chain orientation. Polymer blends. Membrane fouling.
XPS	High sensitivity. Quantification of ions. Measurements of thin films of nanometers.	Solid samples. Cannot detect hydrogen and helium. Peaks overlap for some elements. Long processing time.	Elemental composition. Functional groups. Formula of chemical compounds. Membrane degradation.
NMR	Less background interference. Detection of polar molecules.	Liquid samples. Paramagnetic elements have less NMR signal.	Functional groups. Polymer-blend miscibility. Membrane decomposition.
XRD	No sample preparation. Detection of a wide range of crystalline compounds.	Heavy elements are less sensitive to XRD. Less accuracy for small crystals. Peaks overlap for some compounds.	Membrane purity. Compounds chemical formula. Crystal structure. Polymer chain distance.

approaches [3], detection devices coupled in the chromatography [4], mass spectroscopy, electrochemical techniques [5], and others [6], has historically contributed significantly to advances in chemical detection. There is a pressing need for the invention of portable, miniaturized devices that can perform accurate chemical analyses at a low cost and in real time [7, 8]. This holds true despite the fact that instrumental analytical procedures are generating capacities that are increasing at a dizzying rate in terms of both complexity and sensitivity. Materials chemistry has always been an integral part of analytical chemistry [9]. The discovery of new materials has stimulated the creation of innovative analytical methods and hardware capable of detecting the analytes at the single-molecular level [10]. The ability to measure minute changes in a material's physical and chemical makeup has aided in the development of materials that may induce detecting responses or the transduction for electrical or spectroscopic signals by modifying their optical as well as electronic characteristics, such as changes in impedance, which color, fluorescence, Raman emission, and surface plasmon resonance [11]. Stimuli-responsive materials have been demonstrated to be useful for the detection of numerous analytes, including gasses, pH, ATP, microRNA, synthetic pesticides, proteins, and microorganisms. Materials can function as analyte-responsive or signal-transduction components as well as important components that help with chemical detection. Examples of these components include a secondary selective layer, an adsorption module, or a scaffold that can be coupled with traditional instrumental techniques to improve the detection performance. Materials have been utilized, for instance, as traditional absorbing substances in a solid-phase extraction technique to pre-treat the samples as well as stationary phases in electrochromatography or gas and liquid chromatography to separate targets with high efficiency [12, 13]. Several different types of materials, including zeolites, conjugated polymers, carbon nanotubes, graphene [14, 15], and metal-organic frameworks (MOFs), have emerged as a result of advancements in chemistry and material science over the past century and have significantly aided the development of chemical detection. Because of their one-of-a-kind optical, electrical, or catalytic properties, these materials have been integrated into novel detection platforms that have found a wide variety of successful applications in the field of chemical detection. Graphene has been utilized in the development of very sensitive, portable, self-powered detectors in the detection of various ions and biological compounds [16, 17] due to its remarkable electrical, chemical, optical, and mechanical characteristics. Conjugated polymers were developed as a reliable substance for detecting explosives [18]. By incorporating carbon nanotubes into smart and portable devices, non-line-of-sight gas detection can be accomplished wirelessly and without the need for a clear line-of-sight [19, 20]. Chemical and biological detection is poised to advance thanks to the discovery of novel chemically specific materials that can incorporate new features while retaining the beneficial properties of these established materials, such as flexible accessibility through from the bottom-up self-assembly and porous scaffolds that offer a large area for analyte uptake Table 13.1.

13.2 Spectroscopy Technique

Although spectroscopy has been used for some time in biomedicine, the latest technology developments for basic and translational research in addition to clinical application have been nothing short of amazing. This tendency is shown in a quick summary of the growing field of FTIR-based diagnosis. Colonic tissue has been used as an example to examine the diagnosis potential of the FTIR spectroscopy method [21] ever since the technology was discovered as one that could distinguish between normal and pathological colonic tissues. Colonic tissues are useful because their well-defined architecture changes in response to various disorders. The infrared micro spectrometer's [22] capacity to recognize patterns for data analysis, and the creation of techniques to compare spectral data from sequential slices of normal and pathological tissues [23] helped spread the use of this paradigm. During normal settings, IR spectra of colonic tissue, like cervical tissues, include specific signatures from carbohydrates (glycogens) in the form of a triad. However, during aberrant metabolism associated with sickness or malignancies, these signatures are diminished or altered [24]. These markers served as a foundation for establishing tissue diagnosis and tracking signal changes related to glycogen depletion. The pathology was determined based on these results, which were independently confirmed by an experienced pathologist analyzing sequential sections [25]. Over time, scientists realized that signals to glycogen/carbohydrates, nucleic acids, lipids, and others were just as effective in detecting cancer in biopsies as they were in detecting a wide range of other abnormalities and malignancies at the cellular and tissue levels [26]. This discovery gave rise to the concept of shared biomarkers. The expanded and reevaluated instrumentation's potential utility became crucial as it opened the door to investments at clinical levels. Automated data analysis using multivariate computational approaches, such as focal plane area (FPA) sensors, cluster analysis, and others, has made the apparatus more objective and increased its translational potential. Research methods often analyze morphological information, crystal structure, groups with functions, or chemical make-up. Electron microscopy techniques such as scanning and transmission electron microscopy (TEM) as well as atomic force microscopy or AFM, are commonly used to study membrane morphology. Traditional X-ray diffraction (XRD) uses rays of light from an X-ray source to learn more about a crystal's form, size, and composition. Small-angle X-ray scattering (SAXS) and wide-angle X-ray scatter (WAXS) can add particle and pore size distribution information to crystallography data. Functional groups are identified using Fourier-transform infrared radiation (FTIR), Raman, and NMR spectroscopy. Energy-dispersion X-ray (EDS), fluorescence X-ray (XRF), and XPS are common technologies for chemical analysis. The research presented here discusses each technology's pros and cons.

13.2.1 UV-Vis Spectroscopy

UV-visible spectroscopy studies matter-radiation interactions. UV is the electromagnetic (EM) spectrum's 10–380 nm area. UVA is 320–380 nm, UVB is 280–320, and UVC is 100–280 nm. Vacuum ultraviolet (VUV) is also used to refer to the 10–200 nm region, which is exclusively measured in vacuum. 380–750 nm is the Vis area. UV-Vis spectroscopy, also known as "electronic spectroscopy," involves the activation of atoms' outermost electrons, which create molecules [27]. Transmittance, reflectance, and photoluminescence modes are used for UV-Vis measurements. Photoluminescence measurements can be taken without a reference material, but transmittance and reflectance must be. This spectroscopic technique is used *in situ* with portable instruments and non-invasive methods in art conservation. Except for glass items or stained-glass windows, the analyzed artworks have opaque appearances that need reflectance mode measurement. Reflectance spectrum upon bulk

Table 13.2 UV-Vis spectroscopy's most typical detectors and light sources.

Detectors	Sensitivity	Radiation source	Emission range
Photomultiplier tube	<190–850 nm	Tungsten-Halogen filament lamps With the addition of iodine, a higher filament temperature is obtained	320–3200 nm
Silicon photodiodes	200–1100 nm	Deuterium lamp	185–400 nm
Silicon charge-coupled devices (CCD)	200–1100 nm	Xenon arc lamp	270–1000 nm

samples or weight tone dyes and pigments are usually harder to interpret than transmittance spectrum due to factors like surface roughness as well as texture, particle dimension or distribution, packing weight, compositional homogeneity, and paint layer thickness. At the National Gallery in the United Kingdom in the 1930s and 1940s, Rawlins brought Vis spectroscopic measurements to cultural heritage [28]. In the 1930s, the Fogg Art Museum in Cambridge, Massachusetts, used vis reflectance spectroscopy to characterize paints and pigments [29]. Reflectance spectra in the Vis of Duccio di Boninsegna painting Maestà were acquired by Manlio Santini using a method he devised at the Istituto Centrale del Restauro (currently Istituto Superiore per la Conservazione ed il Restauro) in Rome after World War II [30]. Improved Vis reflectance spectroscopic technologies were also developed in paint composition laboratories or by paint producers at the same time [31]. This method was perfected at the National Gallery Conservation Laboratory in London [32]. The non-invasive technique of fiber optic reflection spectroscopy (FORS) has been utilized and improved in this field since the early 1980s, when fiber optical equipment became available, and beginning with the pioneering research of Mauro Bacci at the Centre di Ricerca sulle Onde electromagnetism at the Consiglio Nazionale Delle Ricerche (which became the Institute di Fisica Applicata "Nello Carrara" in [33, 34]). Following the research of E. René de la Rie [35, 36], fluorescence spectroscopy has emerged as a useful analytical technique for the study of cultural artifacts. Micro-spectrofluorimetry [37] is used to analyze organic colorants or binders in art restoration by combining fluorescence data with confocal microscopes. To analyze dyes and lake colors with a chemical fluorescence fingerprint, Maria Joo Melo and colleagues created microspectrofluorimetry using a reference database. When it comes to collecting data quickly and accurately, micro-spectrofluorimetry is hard to beat. In addition to being used in the lab, it may be used "*in situ*" for transportable devices Table 13.2 [38].

13.2.2 FTIR Spectroscopy

FTIR spectroscopy determines a solid, liquid, and gas's emission or absorption of the infrared spectrum. Unlike a dispersive spectrometer, which calculates power over a restricted frequency range, the FTIR spectrometer captures high-resolution data over a large spectrum range (between 4000 to 400 cm−1). FTIR and strong perceptible (UV-Vis) spectroscopy measures how much light a material absorbs at each frequency. The most direct method, "dispersive spectroscopy," involves focusing a monochromatic laser beam on a sample, detecting the quantity of absorbed light, or revising it for each wavelength. The Fourier transform (FT) spectroscopy is less intuitive for similar data. This approach may focus a bar, and array, containing several wavelengths of light on the specimen and measure how much of the beam is absorbed. Another data point is obtained by changing the wave's frequency. This procedure is repeated several times in a short time, and a computer gets the information. A broad light source that covers the frequency range to be estimated creates an interferogram. A Michelson interferometer (described in the next section) with a motor-driven mirror array catches the light. The interferometer occasionally blocks, mediates, impedes, and transmits each column's light frequency as this mirror moves. The column leaving an interferometer has a varied range at each second and mirror location because the frequencies are modified at various rates [39]. Infrared light's tissue deep penetration limits FTIR spectroscopy to the biochemical study of tissues that span a few dozen per micrometers. Conventional FTIR spectroscopy, which operates in transmission mode using no angle of contact between the emitter and sample, cannot repeat spacer thickness with liquid samples [40]. This constraint has been overcome through ATR-FTIR. Reflections are measured by ATR-FTIR. IR radiation enters a high-refractive-index crystal (typically 45°) and undergoes complete internal reflection before leaving the crystal and being directed into an IR detector. ATR-FTIR can examine high-absorbing, thick substances that do not transmit IR radiation because it analyzes reflected light. This approach directly measures gas, fluidic, or thin-film solid-state materials [41, 42]. Furthermore, microscopic Micro-FTIR uses an IR spectrometer and visible light microscope to detect condensed-phase molecules [43]. It measures thin film samples. The microscope focuses the infrared laser beam on the sample, although the measurement largely originates from the target's focal point. Even a tiny displacement in the radiation source and sample could influence the results. Micro-FTIR allows local measurement of a sample point, while traditional FTIR delivers average knowledge from a homogenized sample. FTIR can also detect pathogenic changes in molecules and materials, resulting in differing spectra between individuals and healthy controls. Pathology alters a sample's infrared spectrum by modifying its intensity or peak frequencies. Chemical changes in the molecules' composition might weaken connections,

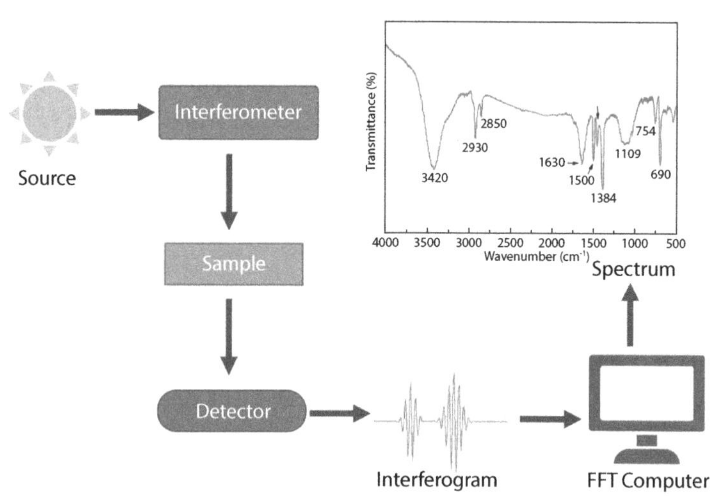

Figure 13.1 FTIR spectroscopy elements for functional group identification.

decrease mass, or shift vibrations that stretch due to temperature changes, changing the bands' vibrational frequencies Figure 13.1 [44].

13.2.3 Raman Spectroscopy

Raman spectroscopy is a technique for studying molecular structures using the Raman scattering impact, which was discovered by C.V. Raman in 1928. However, the typical Raman spectroscopy's reduced sensitivity and application scope are caused by the comparatively weak Raman scattering mechanism, with an incidence intensity of 1010 times. To this end, it is essential that the enhancing impact of material surfaces be used in Raman investigations. For example, Fleischmann *et al.* [45] evaluated the intensity of a Raman scattering signal from pyridine molecules adsorbed in the surface over roughed silver electrodes in 1974 and found that it was amplified by a factor of 106. Later, Van Duyne confirmed this phenomenon, which he labeled a surface amplification effect [46]. With resonance, Surface-enhanced Raman spectra (SERS) can achieve an enhancement factor (EF) of up to 1010, paving the way for the creation of a powerful analytical tool capable of detecting analytes at low concentrations. Because of advances in SERS, the poor sensitivity of the Raman spectroscopy method can be increased, and structural information that is difficult to identify with traditional Raman spectroscopy can now be collected [47]. According to the EM theory, SERS EF is heavily influenced by substrate properties [48, 49]. In SERS analysis, analytes are adsorbed onto an active substrate and then irradiated with monochromatic light from a laser. A Raman spectrometer can examine the scattered light as a result. Applications in the real world necessitate picking a substrate that does its job well, delivering adequate improvement, reproducibility, and durability. Among the several metals available, silver and gold are the most typical choices for developing the substrates. Recently, as knowledge of nanomaterials and nanotechnology has increased, so has interest in building optimal SERS substrates to produce hot spots. Findings from this study suggest that Raman signals can be greatly amplified by boosting the local field around hot spots between multiple nanoparticles. SERS's enhanced detection efficiency has led to its widespread usage in the identification of many analytes [50, 51], including DNA fragments, illicit substances, and proteins. Traditional methods of DNA analysis are notoriously time-consuming and costly due to their reliance on intricate analytical techniques. With its ultrahigh sensitivity and in-depth fingerprint information, SERS has recently emerged as a potential technology for the fast detection and structural analysis of DNA. Given the importance of diagnostic tests in this regard, research into the creation of a highly sensitive and precise analytical tool capable of measuring numerous protein markers concurrently has received a great deal of attention [52]. In the meanwhile, SERS provides a viable method for multiplexed tests. When used in tandem with chemometric techniques, SERS's capacity to resolve, analyze, and classify complicated spectrum data is particularly attractive. In addition to the apparent benefits in drug recognition, SERS also shows considerable promise for the safety of drugs and medication guidance.

13.2.4 X-Ray Photoelectron Spectroscopy

This section is an element of a broader set of resources meant to familiarize readers with the fundamentals of X-ray photoelectron spectroscopy (XPS), one of the more extensively used surface-sensitive techniques. The frequent misunderstanding of characterization data in materials science has been recently documented in studies. Although materials analysis methods like XPS are becoming more accessible and user-friendly, the number of professionals in these fields has remained stable. Therefore, many first-time XPS users try to interpret information on their own and incorrectly. The purpose of this paper is to give a suitable introduction to XPS by synthesizing the authors' experience working with potential or novice users and the information found in various books as well as other publications [53–55]. Those who are new to XPS are urged to seek advice from more seasoned users after reading this paper as a starting point. Researchers can benefit greatly from the guidance of the facility staff when it comes to the gathering and interpretation of the XPS data, as XPS devices are frequently housed in shared user facilities. To help researchers recognize and prevent some possible misinterpretation that emerges in the literature, this study presents key principles needed to comprehend and apply XPS to a broad range of research in science and technology areas. Concerns or subjects that are outside the purview of this study but have been identified as sources of confusion and mistakes in the literature are highlighted in Caution Boxes. Quantification [56, 57], electron path lengths [58], background signals [59], and peak fitting [60] are all discussed further in other works in this series, as are instrument calibration [60] and consistent terminology [61]. The analytical method of XPS involves bombarding the surface of a material with X-rays and then measuring the kinetic energy of the electrons that are released from the substance. Surface sensitivity and the capacity to reveal chemical state data from the substances in the sample are the two main properties that make this technique so effective as an analytical tool. XPS has been used to investigate the surface chemistry for nearly every known substance, from plastics and textiles to dirt and semiconductors, with the exception of hydrogen and helium. It is the surfaces of materials that are responsible for their interactions with one another. Because surfaces or surface contamination determine properties like wettability, adherence, corrosion, transfer of charge, and catalysis, understanding and analyzing surfaces are crucial [62].

13.2.5 NMR Spectroscopy

Nuclear magnetic resonance (NMR) Spectroscopy has developed into a strong, interdisciplinary technology since its development in the 1940s. Since 1944, nine prize laureates have invented resonance technologies for recording atomic nuclei's magnetic characteristics. Protons showed the NMR phenomenon soon afterward. After years of growth, FT NMR was introduced in the 1960s, which was followed by the assessment of non-invasive experimental and medical imaging in the early 1980s. At the same time, atomic-resolution 3D protein structure elucidation in aqueous solution was developed [63]. Superconducting materials and reliable electronic equipment also contributed to this accomplishment. NMR has been employed in physics, chemistry, and biology in practically limitless ways. Studying biological systems requires three sorts of applications: macromolecule structure and function, metabolism, and *in vivo* representations of anatomical structure and functional (physiological) state [64]. Lactate, pyruvate, alanine, or creatine in a solution with red blood cells produced 1H signals in 1977 [65]. In pharmacology, toxicology, or inborn metabolic diseases, 1H NMR of bodily fluids provides a lot of metabolism information [66, 67]. Since these early projects, HR-MAS NMR on biofluids, cell extracts, or soft tissues has been employed to research various diseases and potentially hazardous procedures

[68, 69]. This chapter introduces NMR for metabolic profiling. It focuses on the essentials or prerequisites of NMR's effective use as a metabolites profiling instrument rather than reviewing all its applications in this field. To achieve high throughput and reliable findings, metabolite profiling NMR hardware requirements are analyzed and contrasted, as well as robotics and automated workflow control. Sample setup and handling are crucial for meaningful comparisons between hundreds of specimens and thousands of specimen's characteristics in a metabolite profile investigation. Thus, all aspects recognized to influence biofluid sample properties are covered and extensively explored. Urine samples must be handled and prepared carefully to control the vast range of variables that affect the 1H NMR spectra. Homeostasis reduces sample variability in blood serum, plasma, and cerebrospinal fluid (CSF). NMR Experiments and Processing summarizes the information needed to acquire high-quality one-dimensional (1D) or two-dimensional (2D) biofluid NMR spectra. Finally, the situation for data pre-processing, which is the intermediary step between collecting NMR raw spectrum or utilizing uni- and multivariate analysis and modeling approaches, is examined. To simplify, strengthen, and improve metabolomics analysis and modeling, this is crucial.

13.3 Microscopy Technique

Microscopy, as a phrase, is commonly used to describe the tools that allow us to view with our own eyes what is normally invisible to the naked sight [70]. Microscopy is one of the most widely used characterization methods, allowing for the analysis of the surface morphology or microstructure of a variety of micro/nano-vehicles [71, 72]. Using optical microscopic methods, we may examine the agglomeration propensity of systems for delivery, which presumably affects particle size and, in turn, bioactive efficiency. The use of simple optical microscopy is still widely accepted as a valid method for analyzing delivery vehicles, managing product properties, and gauging bioactive stability. However, standard imaging techniques have a low resolution, which makes 3D assessments more difficult. To overcome

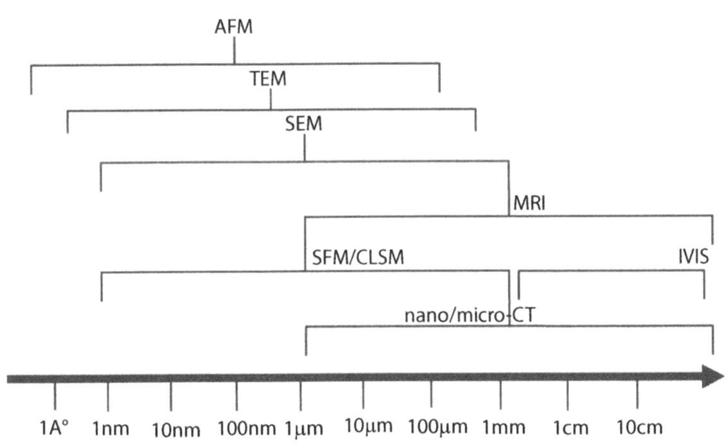

Figure 13.2 Resolution varies widely between optical microscopy, SEM, TEM, and AFM.

the low-resolution quality of optical microscopy, advanced microscopy methods are used to create high-magnification micrographs; in this vein, CLSM, scanning electron microscopy (SEM), TEM, and AFM are widely used for the evaluation of bioactive-loaded delivery systems. Microscopy can image delivery vehicles to reveal particle size and shape, dispersity, collection tendencies, mechanical properties, and even the fate of biological agents within *in vitro*, *ex vivo*, and *in vivo* settings, as shown in Figure 13.2 [73, 74].

13.3.1 SEM

SEM, as a powerful surface imaging instrument, can be used to get useful data about the specimen's visual appearance or particle size, both of which are related to the product's (bio) functionality. Accelerated electron beams are used to create a variety of surface signals on solid-state materials, allowing for imaging with SEM. Normal scanning electron microscopes have a very narrow beam because the electron gun's lighted beam is focused by a condenser lens with an aperture that effectively filters out high-angle electrons. One at a time, an objective lens brings the beam's focus to the specimen. When a beam hits a specimen's surface, it can generate signals including high-energy electron reflection, secondary electron radiation, and X-ray emission. These signals could be picked up by various detectors and analyzed to reveal details about the specimen's molecular structure and chemical composition. Because of the high accuracy and resolving power of SEM images, a wide variety of delivery systems engineered through different methods like spray drying/freeze drying [75], electrical spraying [76], electrostatic atomizing [77], and liposomal entrapment [78] can be evaluated for their structural/morphological properties. Bioactive-loaded nano/micro-delivery structures require this. This form of microscopy is valuable due to its flexibility and excellent resolution. SEM's high magnification and depth of focus make it appropriate for analyzing 0.1–1000 m structures [79]. Despite its advantages, SEM lacks the resolution of other imaging methods and is not particularly useful for characterizing the interior structure of a material. Additionally, artifacts and morphological alterations distinct from the initial state of the specimen may be produced due to the necessity of a rather laborious preparation of the specimen (staining, freeze-drying, metal coating) and vacuum operation. Low-vacuum SEM and environmental SEM (ESEM) are two examples of cutting-edge SEM methods that have evolved to address these challenges; they eliminate the need for specimen preparation inherent to traditional SEM methods and allow for the observation of substances in their natural environment [80]. However, the resolutions achieved with this more lenient strategy are far poorer. The ESEM is not the only effective imaging method that does not need dehydration; the FIB-SEM, HR-SEM, or cryo-SEM are also viable alternatives.

13.3.2 TEM

Next to SEM and AFM, TEM is the most commonly used technique to conduct configurational evaluation of nanovehicles [81]. TEM is a potential tool for advancing food/pharmaceutical technology. Preparing specimens for TEM using techniques including freeze-fracturing, negative staining, or vitrification by plunge-freezing reveals fundamental facts about the delivery system at length scales of 1–100 nm. With the right preparation, this method can be used to successfully characterize both the solid substances and the dehydrated forms of nanovehicles used in delivery systems. In this respect, preparing specimens for classical approaches to TEM is less difficult but more time-consuming and costly [82]. In addition, TEM methods can be utilized to properly characterize both structural aspects and the chemical makeup of various samples. TEM resolution has been increased to 0.05 nm. TEM resolution is 0.07 nm. TEM resolution depends on sample thickness and electron beam acceleration voltage. Increasing electron beam acceleration voltage should theoretically improve resolution. Like other biological specimens, electron microscopy examinations of food components are best carried out at a voltage of > 100 kV to prevent electron damage. A TEM machine generates images by passing an electron beam of high voltage (80–300 kV) through an object and observing the resulting interaction with the specimen's internal structure [83, 84]. Thermionic (tungsten filaments or the lanthanum hexaboride crystal) or field-emission (guns and thin tungsten needle) sources are employed to generate conventional electron beams. The produced electron beam is accelerated by a voltage of 40–400 kV applied to an anode. The accelerated light then travels via electrostatic or EM lenses before being focused on the sample. Parts in the specimen are electron-transparent, while other regions might deflect or soak up the accelerated beam. The electrons are collected and then magnified by objective lenses to create the final image of the material [85]. The final product can be viewed on a fluorescent screen with a phosphor coating. Connecting this setup to a CCD camera's sensor yields an image that may be shown digitally. New detectors that directly capture electrons, improved CCD cameras, and cutting-edge optical precise imaging systems improve nanoscale sample pictures [86].

13.3.3 AFM

AFM, a form of scanning probe microscopy, characterizes bioactive-loaded micro/nanocarriers. Scanners, sample holders, force sensors, control electronics, and computers are involved. The x-y piezoelectric elements for lateral rotation or the vertical piezoelectric element for probe tip movement are induced by the control electronics. They digitize and upload the cantilever's horizontal deflections. Force sensors assess the specimen-probe tip force during the experiment. Control electronics cause the z-axis piezoelectric components to vertically adjust the probe tip to maintain this force. The tip scans the sample surface using the x-y piezoelectric components. To maintain tip-sample distance, the z-axis of the piezoelectric rotates up and down, indicating sample topography. 10-11–10-6 N is the usual force between the material and the probe tip. Nanoscale specimens can be imaged nondestructively by measuring tiny stresses. AFM has developed several operational modes for various uses in recent decades, which can be found in earlier papers [87]. Because of its great lateral and force resolutions, AFM can topographically visualize bioactive delivery systems. This allows high-resolution imaging of biological systems' 3D structures at nanoscale levels in a number of modalities with no specimen change. AFM can assess soft/liquid specimens through tapping mode, analyze nanocarriers' inherent structure by analyzing their adhesion/elastic behavior, and offer useful information above topographical imaging or material identification. More specifically, innovations within AFM visualization have provided nanoscience with a wide range of special opportunities to assess surface characteristics (e.g., Young's modulus or friction) as well as single-molecule characteristics (such as flexibility, dimension, size, cross-sectional packaging, quantity, and even mass/length value) through different AFM methods [88, 89]. AFM requires no staining or drying and is non-invasive compared to SEM and TEM. Micro/nanoemulsions as well as other liquid-based delivery systems require the specimen to be examined in a liquid circumference. Due to specimen deposition on the support surface, AFM nano-imaging of transport vehicles may alter the particles' native morphology or size. Nanostructural bodies may also be damaged by tip-surface interactions. The biggest drawback of AFM measurements may be the tip's larger dimensions compared to the specimens [90].

13.3.4 Confocal Microscopy

Confocal microscopes have a small resolution advantage over "widefield" (epifluorescence) microscopes, but their main advantage is optical sectioning, which produces high-contrast images. Sharp characteristics from the field of view and fuzzy features from outside the focus are superimposed in widefield microscope images. The focus plane of a confocal microscope is sharp because it blocks the latter. Using a high-resolution objective lens, a confocal microscope is an instrument that may create optical sections finer than 1 μm without slicing the sample. It can precisely quantify the levels and location of fluorescent molecules, which is important for assigning molecules to specific physiological compartments or analyzing their colocalization. A single confocal image (or "slice") may be sufficient for measurement if it reflects the sample's whole thickness, but one can also collect several confocal pictures while changing the focus to produce a three-dimensional database (or "z-stack") to reassemble and measure the sample volume. One or more carefully selected hole apertures are fundamental to all confocal microscopes. The traditional confocal laser-scanning microscopy (CLSM) is shown in Figure 13.2. Focusing a laser beam into a material excites fluorescent molecules within the cone of light. The objective lens collects fluorescence from excited fluorescent molecules and focuses it via a pinhole. The pinhole prevents fluorescence emission from either side of the focus plane from reaching the detector. A spinning disc confocal microscope (SD), uses hundreds of focused beams from a disc of pinholes to illuminate the material. The pinholes gather fluorescence, which is monitored by a digital camera to parallelize numerous confocal light paths. To image the specimen's complete field of vision, the disc spins. SDs are optimized for speed, so they may not be ideal for applications that prioritize image quality. The pros and cons associated with various confocal modalities are explored in "Choosing the right microscope" [91].

13.3.5 Fluorescence Microscopy

All medical research uses fluorescence microscopy. Confocal, two-photon, and wide-field fluorescence microscopy have advanced biology. Fluorescence microscopy images are naturally noisy because microscopic detectors like photomultiplier tubes (PMTs) and charged coupled device (CCD) cameras collect only 102 photons per pixel, compared to 105 in photography [92]. Poisson noise, not Gaussian noise, quantized the optical signal during fluorescence microscopy [93, 94]. One way to do this is to increase the stimulation laser and lamp power, but this is limited by the amount of light a biological sample may absorb in and, essentially, by the fluorescence both saturated state rate, which means that the signal from fluorescence will cease to expand when the stimulation strength is too high. Stretching the imaging period (pixel dwell time, exposure time, number of lines, and frame averages) may damage the sample, but it produces clear images. Dynamic or real-time imaging requires millisecond-accurate image capture, making extended imaging times challenging. Thus, a fluorescence microscope with an image-denoising algorithm is essential to biological science. The performance of a denoising algorithm can be measured and benchmarked with the use of a high-quality denoising dataset. A new emphasis on denoising with real noisy pictures, such as those captured by smartphones and digital single-lens reflex cameras (DSLRs), has prompted a shift in the focus of image-denoising methods and datasets away from Gaussian noise and toward these more realistic types of noise [94, 95]. A reliable Poisson noisy-dominated blurring dataset containing actual fluorescence microscope pictures is lacking, however. This research aims to address such a lack. To be more specific, we create a Fluorescence Microscopy, as well as denoising (FMD) dataset with 12,000 noisy microscopy pictures taken with three popular imaging techniques (confocal, two-photon, or wide-field) as well as three representative biological samples (cells, zebrafish, and mouse brain tissues). We use picture averaging to generate ground-truth and noisy images at five noise levels with our high-quality commercial microscopy. With and without ground truth, the dataset is used to compare classical denoising approaches to deep learning models. Benchmark code and FMD dataset are open-source. This dataset employs noisy fluorescence microscope photos for Poisson-Gaussian denoising.

13.4 Thermal Analysis Technique

The common denominator among the numerous methods of thermal analysis (TA) is the measurement of a material's response to heat or cold (or, in some cases, to being held isothermally). In an effort to categorize materials based on their measurable physical attributes, temperature has been used to establish a correlation between the two [96]. Though FDM has been around for a while and is considered an advanced technology [97, 98], it still has a lot of potential in the printing industry because of the wide range of materials it can work with and the fact that early printer machines were only capable of working with a small subset of those materials. TA is widely recognized as one of the key studies as well as quality control techniques in the creation and production of polymeric materials, and at this stage, academia as well as industry need additional data for improved 3D printing technology. In addition to assessing physical qualities, TA is also used to better understand materials' thermal or mechanical histories, characterize and design manufacturing processes, and predict how long various materials will last in service. When evaluating a material's thermal performance, two of TA's most commonly used characteristics are the onset dissolution temp (also known as the starting decomposition temperature) and the temperature at which decomposition begins at 5% mass loss (T5%). T5% is generally considered more dependable than temperature since it is highly dependent on the slope of the lowering region in thermogravimetric (TG) curve, which is often observed when more than one stage of deterioration is recognized. As composites for 3D printing increase in the biological, mechanical, and electrical sectors, molecular-level encouragement can considerably decrease physical aging in the glassy state. DSC, DTA, and DMA can compute the material's transition temperature (Tg), another relevant measure. Melting behaviors is the most critical element, however glass transition and additives can also affect print quality. In addition, the issue of security is frequently disregarded. Working with a melt, 3D printers produce volatile compounds, thus the lab housing them needs to be prepared with safety precautions. Although 3D printers are now affordable for the masses, not all of the spaces in which they are used are optimal for their intended purposes. The importance of safeguards is frequently overlooked. Using TA (in conjunction with other methods) can be a huge help in this respect.

13.4.1 DSC

The statistical technique in differential scanning calorimetry (DSC) can be used to great effect when attempting to determine the nature of a polymer's physical properties. Glass transition as well as additional effects that demonstrate either change to heat capacity and a latent heat can be characterized using DSC, as can melting, crystallization, and mesomorphic temperatures of transition and the related enthalpy and entropy changes. Calorimetry stands out as a unique technique. Although interpretation can occasionally be difficult due to their simplicity or universality, the energy parameters derived from calorimetry (heat capacities CP or the integral along temperature T entropy H) have a clear physical meaning. In DSC, there are normally two sample positions: one for the sample of interest and one for the reference sample, which can either be an empty crucible or one filled with an inert substance. There are numerous monographs that cover the operation of a DSC in great depth; as such, we will not repeat that material here. This applies to temperature-regulated DSC (TMDSC), as well as data treatment and calibration. The many gadgets can be categorized in various ways. Concerning accuracy, one must wonder if the instrument measures each and every bit of heat retained or released in the sample. Therefore, this will be the standard by which DSC devices are categorized [99].

13.4.2 Thermogravimetric Analysis (TGA)

Analyzing how a material's chemical, physical, or structural properties change as a result of heating or cooling is what TA is all about. Most chemical processes, physical qualities, and structural transformations are, in principle, sensitive to temperature. By extension, every experimental characterization of a material where the temperature is altered serves as a form of TA. Unfortunately, the use of this word is now restricted to a narrow set of procedures concerned with TG or calorimetric effects [100]. Differential TA (DTA), thermogravimetry (TG), and its variants (DTG), or DSC (differential scanners calorimetry) are now widely recognized as the principal methodologies linked with TA. Thermal analyses are the embodiment of other methods used to compute thermal conductivity, particular heat, or thermal diffusivity. The evaluation of material qualities can make use of a variety of methods, either individually or in combination. One of the most accurate thermal studies is thermogravimetric (TG/DTG), which is commonly used to characterize the thermal durability of natural fiber polymer composites by examining the dependence of their composition and structure on thermal degradation behavior. Analytically, TG/DTG measures the weight change that takes place as a specimen is heated to ascertain its thermal stability and the proportion of volatile components. TG/DTG, on the other hand, is an experimental method in which the sample weight is tracked across time and temperature. Usually, a steady rate of heating is used to warm the sample. Usually done in air and an inert environment (such as helium and argon), this procedure involves recording the weight as an indication of rising temperature. The TG/DTG device's thermobalance tracks temperature-induced shifts in instrument mass. Having gas circulate through the experiment helps create a stable medium by removing moisture. Because of variations in gas density, buoyancy corrections must be applied to TGA readings. If stability modifications are not done, the specimen will appear to gain mass as the heating experiment progresses. Taking a buoyancy-corrected control sample is a common practice in TG/DTG analysis [101].

13.4.3 Thermal Conductivity Measurements

Over the years, many techniques have been developed to calculate the thermal conductivity of powders and large samples of solids. The thermal conductivity in metal hydride (MH) beds may be evaluated with a number of different techniques. Generally speaking, there are two types of steady-state methods or transient methods that are used to quantify thermal conductivity. Due to the nature of heat conduction, there is a finite set of materials that may be used with any given technique. After all points in the sample have reached equilibrium temperature, the thermal conductivity may be calculated using Fourier's law from measurements of the temperature gradient or heat flow. Two subcategories of steady-state approaches exist, based on how the heat flux is measured: comparative and absolute. The heat flow of the sample is inferred from the temperature distribution of the reference sample, which is placed in contact with the specimen's test material and has a specified thermal conductivity in comparison procedures. The heat flux is often calculated from the electrical power used in absolute methods when the sample is heated using a constant-power heat source. We opted to employ the radial heat flux technique, the compared cutting bar technique, the guarded heat flow meter method, or the guarded hotplate method to determine the effective thermal conductivity (ETC) of MH beds after studying the existing literature on steady-state approaches. The use of transient techniques and other non-steady-state approaches to measurement is becoming increasingly common. Transient methods are used to monitor the temperature across time since the heat source is intermittent or produces heat in pulses. Therefore, there are two types of ephemeral strategies: periodic heat flux strategies and transitory temperature flux strategies [102]. Hot wire, flow meter, thermal probe, transient plane sources (TPS), and laser flash techniques are common transient methods used to calculate the ETCs of MH beds.

13.4.4 DMA

Dynamic mechanical analysis is a crucial and efficient method for figuring out the morphology or viscoelastic characteristics of crystalline polymers or composite materials in relation to primary concessions and other important parameters such as crosslinking density, liquid fragility [103], dynamic/complex viscosity storage/loss compliance, creep compliance/stress-relaxation the modulus, and the non-Arrhenius change in connection periods with temperature [104]. Young's modulus is directly correlated with storage modulus (E0) or dynamic modulus. It is commonly used to describe the firmness (or lack thereof) of a substance. In this context, E0 refers to the capacity of a substance to temporarily store the energy that is applied to it. The loss modulus (E00), also known as the dynamic lost modulus, is a measure of a material's viscous response to an applied force and is commonly used to characterize the material's propensity to dissipate that force [105]. The dynamic loss coefficient is sensitive to variations in molecular mobility, transition, relaxation process, morphology, and structural heterogeneity, and is thus typically related to the concept of "internal friction." The component derived from the plot is known as the modulus of complexity (shear modulus), marked by (E). High Tan d values suggest that the material has a large amount of non-elastic strain component, whereas low values indicate that the material is very elastic. Since the mobility of the chain molecules within the fiber/matrix interface is reduced as bonding between the two increases, the damping factor is reduced. The Tan d (E00/E0) measurement for a system is consequently bigger when there is less loss of energy with respect to its storage capacity. Molecular motions and viscoelasticity play a role in the damping factor, as do defects such as displacements, grain boundaries, phase borders, and interfaces [106].

13.4.5 Thermo-Optical Analysis

Thermo-optics uses absorbed light to measure sample temperature rise in laser spectroscopy. Transmission methods are inferior to thermo-optical absorbance measurements, which use both laser beams to interact inside a material. Compared to simple transmission measurements, increased sensitivity or precision enhances outcomes. Pump light beams heat samples in thermo-optical spectrophotometries. Since most materials' reflective index changes with temperature, the heated sample diverts the second probe beam. A periodical thermal optical element controls probe beam intensity from a modulated pump laser. Profile distortion is measured by position- and power-sensitive detectors and demodulated using a lock-in amplifier that signals average, or box-car average. The experiment's probe volume is determined by the pump and probe beams' overlap zone, which could be both collinear and crossing. In picoliter probe volumes, crossed-beam instruments exhibit high mass sensitivity and high concentration sensitivity. Localized absorbance at liquid-solid and gas-solid interfaces is detected using surface-sensitive techniques. Many thermo-optical terminologies exist. Analogies with standard optical elements characterize these methods: A cylinder lens in a crossed-beam thermal lens, a prism used in thermally displacement spectroscopy, a grating in heat refraction, a mirror with an angle in thermally driven stretching, or a simple change in optical path length in interferometry [107–109]. Thermally induced beam disturbance can only be approximated using standard optical components because the thermo-optical component is aberrant. Thermo-optical elements change with time. Thermo-optical spectrophotometrics are more sensitive and precise than transmission measurements. Thermo-optical methods are sensitive because heat transfer is linear. Pump laser power increases sample temperature correspondingly. An endless thermo-optical signal can be generated from a modest absorption utilizing a high-power pump laser. Solvent absorbance determines signal size. A redesigned pump beam's time-varying temperature rise makes thermo-optical approaches exact. Electronics demodulation methods include lock-in amplifiers as well as box-car averaging, or signal averages using regression analysis to improve measurement precision. Absorbance detection limits for solution background absorbance are a few parts per thousand using a steady probe laser using these demodulation methods [110].

13.5 Mechanical Testing Technique

There is no substitute for thorough mechanical or physical testing when it comes to polymers and composites, whether it be for the purposes of product development and evaluation, quality control, satisfying application performance criteria, or manufacturing. Many different industries, including the aircraft, automobile, customer, medical, and defense sectors, require mechanical or physical examinations to ensure the material fulfills their standards [111]. Mechanical assessment of polymer composites can be used to determine mechanical properties like strength and stiffness, which can then be applied to the design of a composite's structure. The most common standardized mechanical testing for polymer composites includes tensile (tension), flexion (bending), impacting (shock), shear (bending), or compressive (open and closed holes), while common physical tests include the absorption of water, density, empty content, robustness, or scratch resistance. Researchers also conducted the compression process, shear, and interlaminar strength tests to determine relevant parameters for a failure criterion model including final strength and failing predictions for composite samples [112].

13.5.1 Tensile Tests

Uniaxial tensile assessments are frequently used in both academics and industry to get an understanding of the strength and ductility of a material because of its low cost and ease of application. According to the norms, the specimen geometry must be such that the stress and strain are uniform over the length of the gauge and in the cross-section of the specimen. Similar to a uniaxial test, the stress tensor has just one non-zero component because there is only a single source of external load. The yield power, point of yielding elongation, tension strength, modulus of Young, elongation on break, and true (or Cauchy) tension against correct (or logarithmic) strain distribution of a material can all be determined in this method. Metals' thermomechanical characteristics at high temperatures are best assessed by tensile testing. Some examples of commonly applied standards include ASTM E21-20 and ISO 6892-2. Merklein or Lechler, who both worked in the automotive sector and focused on the hot stamping process, undertook experiments to evaluate the thermomechanical characteristics of Ultra Heavy Strength Steel. Steel's conduct at high temperatures is of interest to civil engineers because of their frequent use of the material in building construction. For this purpose, Chen and Young employed steady and transient tests to investigate the behavior of high-temperature cold-formed steel. With the goal of minimizing the amount of springback, there has been a lot of research into the thermo-mechanical properties of these compounds [113]. Finally, and in preparation for the next section, it is possible to achieve multi-axial stress levels with just one loaded direction, with proper sample shape or boundary conditions. Tensile tests that can be utilized to investigate the planar strain situation, work best with specimens whose breadth is much larger than their axial length. It is safe to presume that the diagonal distortion is negligible or close to zero. The stress-strain state is no longer uniaxial due to the induction of positive tension stress along the path of translation by the plane of strain condition. The Yld2003 model is calibrated using the plane strain test by Aretz *et al.* [114] and an inverse approach is linked to a plane strain test by Ha *et al.* [115, 116].

13.5.2 Compression Testing

Another typical technique for characterization is compression testing, which, in contrast to stretching, compresses the material. The stress on the sample is determined by measuring the force applied, just as in tensile testing. The stress and strain on the sample are measured in order to generate stress-strain curves. For compressive loading, two flat platens are used in place of hands. Compression tests can be conducted either in a controlled environment or in the wild. Fluid within the sample is redistributed vertically by placing it in an impermeable well and then compressing it with a porosity platen. When a sample is compressed between a solid, non-porous platen and a solid, impermeable base, the fluid flow is channeled in a radial direction. Extensive research into soft tissues relies on compression testing. The nonlinear elastic behavior of skin and subcutaneous tissue was measured by Wu *et al.* using unconfined compression tests. Composite materials obtained from pigs' feet were used in their experimentation. To determine friction coefficients in unrestricted compression testing in pig brain tissue, Rashid *et al.* [117] employed a hybrid experimental-computational approach. Brain samples were compressed between metal platens at different strain rates, either in a connected (no slip) or lubricated (pure slip) situation. The collected stress data was utilized to estimate the coefficients of friction in numerical simulations. Confined compression experiments were used by Yu *et al.* to quantify spinal cord gray or white matter mechanical characteristics. The anisotropy of gray or white matter samples was examined in transverse and axial directions. Compression testing in a moist environment with physiologically realistic loads offers many benefits. Compression testing improves mechanical property measurements for several load-bearing tissues in the body. Compression testing, like tensile testing, uses stress-strain curves to directly evaluate characteristics. Sample preparation is one of the technique's drawbacks. Samples must be cut to exact specifications and have smooth planes. Friction between the platen and specimen can cause stress-strain readings to overestimate [118].

13.5.3 Flexure

Researchers have studied the print's reaction to bending stresses by analyzing the flexural characteristics of additively made CFRPs. Flexural testing of short carbon dioxide fiber-reinforced 3D polymer compound composites was performed by Ning *et al.* and the results were compared to those of non-reinforced polymer specimens. The specimen's short CFRPs with 5 wt.% carbon fiber had 11.82%, 21.86%, and 16.82% higher flexural stress, flexible robustness, and modulus than natural thermoplastic material. Yan *et al.* conducted research to determine how changing the short carbon fiber content of additively made short CFRPs affected their flexural capabilities. The results revealed that the flexural modulus was improved by 93.4%, 129.4%, and 243.4%, and the flexural strengths were enhanced by 44.5%, 83.3%, and 114%, when the weight content of short fibers was 30%, 40%, and 50%, respectively. Flexural qualities of continuous CFRPs have been found to be superior to those of short CFRPs. For additively made nylon-reinforced continuous carbon fiber composites, evaluations have revealed flexural strengths of 772.6 MPa and moduli of 85.3 GPa, which pioneered a fresh approach to making thermosetting polymers with continuous carbon fiber reinforcement. To conduct a compression test, scientists have looked into using additively made sandwich panels having lattice core structures. These panels were made in two distinct configurations. The process begins with differential plug and bond, then transitions to continuous lattice fabricating (CLF). Lattice core architectures for these panels were also compared for their base areas using triangular, quadratic, or hexagonal geometries. This variant resulted in the creation of six sandwich panels. The findings confirmed that extremely lightweight sandwich panels may be manufactured using this innovative additive manufacturing (AM) approach. The flexural response of these panels has been demonstrated to be superior to that of conventional sandwich panels [119].

13.5.4 Hardness

For practical purposes, hardness is important. Surface hardness controls wear and tear in components and semi-finished products. The safety and effectiveness of technical systems or constructions depend on the materials' hardness. Because of hard materials' central role in the manufacturing value chain, multinational research efforts are constantly developing unique hard materials. Electroless nickel is a tribology-based hard coating. The substrate influences surface testing because the electroless gold deposit is too thin. Thus, microhardness tests use modest loads to assess coating hardness. The amount of phosphorus and boron in an electroless Ni-P or Ni-B coating determines its hardness. However, the hardness of electroless Ni-B increases with an increase in boron material, while that of electroless Ni-P increases with a decrease in phosphorus content. The coating's hardness initially increases with increasing phosphorus material up to 8%, then decreases. The as-deposited coating reached 910 HV0.1 in hardness. Electroless coating hardness increases with the temperature of annealing up to a point. Precipitation of silver phosphide crystallites with appropriate thermal treatments strengthens supersaturated Ni–P alloys. Phosphides block dislocation movement, making the material harder. Grain coarsening causes surface brittleness and dislocation propagation, reducing Ni–P film hardness. As-deposited, borohydride-reduced electroless silver deposits are robust and resistant to mechanical wear. Electroless Ni-B coatings can be used as a substitute for gold in electronics and are more durable than tool steel and have durable chromium coatings [120]. Ni-B coating hardness can be optimized to approximately 1400 HV0.1 through parameter manipulation of the deposition process [121]. An aluminum alloy with a nickel-boron coating that is 800 HV0.1. While no diffusion takes place at the interface, heat treatments conducted in a neutral atmosphere (95% Ar + 5%H2) cause the Ni, Ni2B, or Ni3B phases to crystallize, hence increasing the hardness of the coatings. It is 1300 HV0.1. The hardness profiles obtained from nano-indentation testing on treated and untreated materials show that the coating is homogeneous in hardness [122]. After that, certain samples of Ni-B coating were nitride in a vacuum with pure nitrogen. Nitrides brought the hardness up to 1500 HV0.1. Researchers Krishnaveni *et al.* [123] looked into the correlation between heat treatment temperature and the micro-hardness of electroless Ni-B coatings, and they discovered two peaks at 350 C and 450 C. Past 450 C, coatings become more malleable due to the aggregation of Ni3B particles.

13.5.5 Tribological

Tribology comes from the Greek word tribos, which means to rub or slide. Tribology is "the science of friction, wear, and lubrication between interacting surfaces that are in relative motion, regardless of the existence of a medium that separates them," according to its definition. Since its inception, tribology has been widely used in numerous industries, including those dealing with automobiles [124], engineering, polymer technology, nanotechnologies, medicine, or cosmetics. Its importance in culinary uses has just recently been recognized. Tribometers measure the amount of wear, tear, or lubrication experienced by two moving surfaces that are rubbing against one another. In the beginning, tribometers were created to determine the effectiveness of lubricants in reducing wear and to identify the surface qualities of designed materials. Tribology is useful for assessing the tongue, upper palate, or tooth surface, all of which are involved in perceiving the food's complex orosensory qualities. To properly use a tribometer, one must have a firm grasp of the instrument's theory, workings, and operation [125].

13.6 Electrical and Magnetic Techniques

As a powerful and related technique for manipulating magnetic particles, magnetic techniques of different kinds are currently used extensively in the field of processing minerals with tremendous success thanks to their efficiency, ease of use, inexpensive costs of operation, and minimal impact on the environment. In order to put this supremacy into practice, many people have been working hard and coming up with novel approaches. A better understanding of the theoretical or operational principles of magnetized techniques is necessary because, despite its tremendous achievements in the field of processing minerals, the mineral industry is currently facing new difficulties that must be taken into account in the design of new magnetized techniques for the environment and sustainable requirements. From prototype to mass personalized production, AM for electrical machinery offers practically endless possibilities. The use of AM in the production of electrical devices is an exciting development because of the many advantages it has over more traditional manufacturing methods. These include the ability to realize 3D designs (which take into account electrical, magnetic, mechanical, and thermal factors), the use of recyclable constructions, and the maximization of material utilization. Many other types of electrical machinery and their components have been manufactured using the conventional, foundational AM procedures detailed in the vast majority of these books. Selective laser melting (SLM), fusion deposition modeling (FDM), and binder jets printing (BJP) technologies have all been demonstrated to have the best potential for the production of electrical devices among all AM techniques [126].

13.6.1 Conductivity Measurements

The electrical properties of materials utilized in the electronics industry, such as conductivity and sheet resistance, are of paramount importance. Integrated electronic systems manufacture places a premium on semiconductors, but other electronic materials such as conductors and superconducting materials are also crucial. Sheet resistance is often used to characterize the electrical characteristics of thin films, while conductivity is utilized to characterize the electrical characteristics of bulk homogeneous materials like metals and semiconductors (virgin wafers). Several different types of electrical structures, such as epitaxial layers, metallization layers, and 2D structures, can exist in thin films. For conductivity and resistance of sheets measurement of virgin wafers, which are epitaxial layers, and the metallization to the semiconductor sector, contact direct current (dc) techniques are still routinely employed, but contactless approaches are becoming increasingly significant. This is associated with novel materials being developed for usage in the electronics sector, such as graphene, GaN, and SiC, and the advancements that have occurred in materials technology.

Some conducting materials, such as GaN and SiC, are not amenable to measurement with the commonplace four-point probe technique. In order to quickly and accurately measure such substances, contactless measuring techniques have been developed [127].

13.6.2 Dielectric Spectroscopy

Dielectric spectroscopy's non-destructive measurement capabilities have made it a valuable tool for studying biological suspensions of cells and tissues [128]. Dielectric spectroscopy has expanded from its traditional use in the laboratory to include bioprocess controls in industry and medical diagnostics. The focus of this article is on (a) theoretical investigation of the electrical behavior of the biological cells, (b) dielectric methods for the monitoring of cells during physiological circumstances, and (c) single-cell techniques for characterizing individual cells. Dielectric spectroscopy has been used to investigate many different types of biological cells, and the resulting spectra have been analyzed using theories derived from different electrical models of cells. The advent of modern-day personal computers has allowed us to deal with more intricate and accurate models of cells, such as those that incorporate internal organelles or are ellipsoidal in shape rather than spherical. Instruments able to make fast, automated observations across a broad frequency range have made it possible to investigate dynamic dielectric behavior in time-dependent events. Cell sedimentation, cell collection, cell division and development in cultures [129, 130], organ degeneration [131], and single frog embryogenesis [132] are some of the phenomena that have been examined thus far. There has been a recent uptick in the use of single-cell analysis in the biomedical and health sciences. When compared to the "suspension" method, which characterizes cell populations, the dielectric spectroscopy method for characterizing individual cells offers numerous advantages. Preparing a homogenous population of cells, as is necessary for the suspension method, can be avoided with the single-cell approach.

13.6.3 Magnetic Susceptibility Measurements

Since magnetic characteristics change noticeably during a phase transition, measuring magnetic susceptibility is an intriguing method for pinpointing the boundaries of phases in magnetic systems. The method involves using a magnetic field and a pendulum to hang the alloy. The sample is then subjected to heating or cooling during the phase transition, at which point its inherent magnetism causes it to be deflected from its original location. When the pendulum is returned to its starting position, the magnetic field is adjusted by adjusting the applied current. Several works on the basics of NMR spectroscopy have covered the various factors that influence the chemical change of a magnet nucleus in sufficient detail. Magnetic susceptibility is a key consideration. The bulk sensitivity of the environment in which the resonant nuclei are located determines the location of a line on the proton resonance spectrum of a molecule. The addition of the paramagnetic species was found to alter the chemical shift of a certain kind of proton in a solvent. In this context, we will refer to the volume susceptibility variation as. The theoretical formulation for water-based solutions of paramagnetic compounds explains the variation in magnetic resonance retention of protons between the two solutions. A capillary containing about 2% t-butyl alcohol with water is placed within a nuclear EM resonance tube and spun while a reference substance composed of about 2% t-butyl ethanol is added. (Usually, t-butyl alcohol would not have much of an impact on the dissolved chemical's reactivity.) The volumetric susceptibilities of the two solutions will cause the methylation proton in the t-butyl molecules of alcohol to resonate at different frequencies, resulting in two distinct resonance lines. Alternatives to t-butyl alcohol include acetone and dioxan; cyclohexane and tetramethyl silane are good choices for non-aqueous solutions. Another option is to utilize the organic solvent's own resonance line as a standard, presuming no interaction with the solute [133].

13.6.4 Magnetostriction Measurements

A magnetic field changes the form of a magnetostrictive material. Most ferromagnetic materials show magnetostriction. These materials can be used for sensing and actuation since the phenomena involve the bilateral transfer of energy between magnet and elastic states. They are also used in many vibration control systems for mechanical equipment. Magnetostriction actuators' low driving voltages make them beneficial in medical applications and simplify amplifier construction. Transformers' buzzing sound comes from magnetostrictive materials vibrating at twice the frequency of an alternating magnetic field. If a magnetostrictive material is mechanically stressed, its permeability to magnets will vary due to an opposite magnetostrictive effect. The density of the magnetic flux pattern will change due to the change in the magnetic permeability if the material is additionally subjected to a magnetic field that is alternating from a coil using an alternating current. Depending on the magnetic permeability of the material, the alternating magnetic field induces an emf of varying polarity in an additional "pick up" coil. Magnetostrictive transducers rely on this transformation to transfer electrical energy into mechanical motion. It is usual practice to manufacture rods of modern magnetically restrictive substances like Terfenol-D with residual magnetic fields that are generally horizontal to the rod axis. Having the external magnetic field aligned with the axis of the rod has no effect on the orientation of the domains within the substance. So, they are irrelevant to magnetostriction and do not make it worse. As a result, the greatest magnetostriction produced by a rod having randomly aligned domains is only around two-fifths of that. This means that all domains should be aligned at right angles to the rod's axis. This is approximal in manufacturing. A mechanical pre-load is necessary since the rods are not initially aligned along their axis. The impact of pre-stress on the dynamic performance of Terfenol-D transducers has been investigated [134]. Magnetostriction measurement techniques can be either direct or indirect, depending on whether the stress is sensed directly or inferred from a strain-dependent characteristic. As opposed to indirect methods, which can only measure saturated magnetostriction sat [135], direct methods may measure magnetostrictive stress as an effect of the applied field.

13.6.5 Hall Effect Measurements

The Hall effect is often measured using dc magnetic fields. Extremely thin semiconducting materials, such as molybdenum disulfide (MoS2), present formidable hurdles for studies of the dc Hall effect in electrical devices. We present the findings of Hall effect measurements performed using ac magnetic pulls and lock-in monitoring of the Hall potential of field effect transistors using an ultrathin MoS2 channel. There are benefits to studying the ac hall impact rather than using dc measurements. The ac Hall effect data stood in for the gate voltage to provide estimates of carrier concentration and Hall mobility. They employed a field of magnetic strength two orders of magnitude lower than previous research on MoS2 equipment, which used dc magnetic fields. As a promising material for nanoelectronics or nano optoelectronic devices, MoS2, a prototype semi-conducting transition-metal dichalcogenide, has garnered attention. In particular, MoS2-channeled field-effect transistors (FETs) have been extensively explored. Monolayer MoS2 FETs exhibit high electron mobility. Measurement of basic material properties including carrier concentration and mobility of carriers is essential for studying MoS2 devices and other 2D materials. Hall effect measurement data are used to estimate carrier concentration. Hall effect measurements on FETs with fundamentally thin MoS2 channels are challenging to perform, hence just a few articles have reported their results [136]. The extremely low bias current for totally thin devices, significant voltage noise due to an excessive amount of localized state [137], and high contact resistance related to Schottky barriers on the contact point junctions [138] confound this measurement. Localized states create voltage distortion due to carrier movement or number variations. Hall effect measurements of MoS2 FETs were previously shown for devices with relatively high carrier densities created by top-gating through small dielectric and ionic-liquid gating and devices with relatively low contact resistance, which is obtained by vacuum annealing to nearly prevent the Schottky barrier's effect or by using material connections [139, 140].

13.7 Conclusion

This provides a comprehensive overview of the analytical tools used for characterizing multifunctional materials. These materials, with their diverse combination of properties, hold immense promise for a wide range of applications. However, understanding the structure-property relationships and optimizing their performance requires the utilization of advanced analytical techniques. The review covered a variety of analytical methods, including structural characterization techniques such as XRD, SEM, TEM, and AFM, which enable the visualization and characterization of the morphology and composition of multifunctional materials. Spectroscopic techniques like FTIR, Raman spectroscopy, UV-Vis spectroscopy, and XPS provide valuable insights into their chemical bonding, electronic structure, and optical properties. TA techniques, including DSC, TGA, and thermal conductivity measurements, contribute to understanding the thermal behavior and stability of multifunctional materials. Electrical characterization methods, such as impedance spectroscopy and Hall effect measurements.

References

1. Broza, Y.Y., Zhou, X., Yuan, M., Qu, D., Zheng, Y., Vishinkin, R., Khatib, M., Wu, W., Haick, H., Disease detection with molecular biomarkers: from chemistry of body fluids to nature-inspired chemical sensors. *Chem. Rev.*, 119, 22, 11761–817, 2019 Nov 15.
2. Mitsubayashi, K., Niwa, O., Ueno, Y. (Eds.), *Chemical, gas, and biosensors for internet of things and related applications*, Elsevier, 2019 Jun 14.
3. Haddad, P.R. and Jackson, P.E., *Ion chromatography*, Elsevier, London, 1990 Jul 16.

4. Guetens, G., De Boeck, G., Highley, M.S., Wood, M., Maes, R.A., Eggermont, A.A., Hanauske, A., De Bruijn, E.A., Tjaden, U.R., Hyphenated techniques in anticancer drug monitoring: II. Liquid chromatography–mass spectrometry and capillary electrophoresis–mass spectrometry. *J. Chromatogr. A*, 976, 1-2, 239–47, 2002 Nov 8.
5. Bard, A.J., Faulkner, L.R., White, H.S., *Electrochemical methods: fundamentals and applications*, John Wiley & Sons, New Jersey, 2022 May 31.
6. Skoog, D.A., Holler, F.J., Crouch, S.R., *Principles of instrumental analysis*, Cengage learning, Boston, Massachusetts, 2017 Jan 27.
7. Diamond, D., Coyle, S., Scarmagnani, S., Hayes, J., Wireless sensor networks and chemo-/bio-sensing. *Chem. Rev.*, 108, 2, 652–79, 2008 Feb 13.
8. Meng, Z., Stolz, R.M., Mendecki, L., Mirica, K.A., Electrically-transduced chemical sensors based on two-dimensional nanomaterials. *Chem. Rev.*, 119, 1, 478–598, 2019 Jan 3.
9. Di Nardo, F., Chiarello, M., Cavalera, S., Baggiani, C., Anfossi, L., Ten years of lateral flow immunoassay technique applications: Trends, challenges and future perspectives. *Sensors*, 21, 15, 5185, 2021 Jul 30.
10. Schedin, F., Geim, A.K., Morozov, S.V., Hill, E.W., Blake, P., Katsnelson, M.I., Novoselov, K.S., Detection of individual gas molecules adsorbed on graphene. *Nat. Mater.*, 6, 9, 652–5, 2007 Sep.
11. Banica, F.G., *Chemical sensors and biosensors: fundamentals and applications*, John Wiley & Sons, Chichester, UK, 2012 Aug 15.
12. Sinha Roy, K., Goud, D.R., Mazumder, A., Chandra, B., Purohit, A.K., Palit, M., Dubey, D.K., Triazine-Based Covalent Organic Framework: A Promising Sorbent for Efficient Elimination of the Hydrocarbon Backgrounds of Organic Sample for GC–MS and 1H NMR Analysis of Chemical Weapons Convention Related Compounds. *ACS Appl. Mater. Interfaces*, 11, 17, 16027–39, 2019 Apr 9.
13. Negreira, N., de Alda, M.L., Barceló, D., On-line solid phase extraction–liquid chromatography–tandem mass spectrometry for the determination of 17 cytostatics and metabolites in waste, surface and ground water samples. *J. Chromatogr. A*, 1280, 64–74, 2013 Mar 8.
14. Shirakawa, H., Louis, E.J., MacDiarmid, A.G., Chiang, C.K., Heeger, A.J., The electronic structures of-polyacetylene. *J. Chem. Soc., Chem. Comm.*, 16, 578, 1977.
15. Ferrari, A.C., Meyer, J.C., Scardaci, V., Casiraghi, C., Lazzeri, M., Mauri, F., Piscanec, S., Jiang, D., Novoselov, K.S., Roth, S., Geim, A.K., Raman spectrum of graphene and graphene layers. *Phys. Rev. Lett.*, 97, 18, 187401, 2006 Oct 30.
16. Pumera, M., Ambrosi, A., Bonanni, A., Chng, E.L., Poh, H.L., Graphene for electrochemical sensing and biosensing. *TrAC, Trends Anal. Chem.*, 29, 9, 954–65, 2010 Oct 1.
17. Singh, E., Meyyappan, M., Nalwa, H.S., Flexible graphene-based wearable gas and chemical sensors. *ACS Appl. Mater. Interfaces*, 9, 40, 34544–86, 2017 Oct 11.
18. McQuade, D.T., Pullen, A.E., Swager, T.M., Conjugated polymer-based chemical sensors. *Chem. Rev.*, 100, 7, 2537–74, 2000 Jul 12.
19. Mirica, K.A., Weis, J.G., Schnorr, J.M., Esser, B., Swager, T.M., Mechanical drawing of gas sensors on paper. *Angew. Chem. Int. Ed.*, 51, 43, 10740–5, 2012 Oct 22.
20. Azzarelli, J.M., Mirica, K.A., Ravnsbæk, J.B., Swager, T.M., Wireless gas detection with a smartphone via rf communication. *Proc. Natl. Acad. Sci.*, 111, 51, 18162–6, 2014 Dec 23.
21. Salman, A., Argov, S., Ramesh, J., Goldstein, J., Sinelnikov, I., Guterman, H., Mordechai, S., FT-IR microscopic characterization of normal and malignant human colonic tissues. *Cell. Mol. Biol.*, 47, 22, 159–66, 2001 Jan 1.
22. Lasch, P. and Naumann, D., Spatial resolution in infrared microspectroscopic imaging of tissues. *Biochim. Biophys. Acta (BBA)-Biomembranes*, 1758, 7, 814–29, 2006 Jul 1.

23. Argov, S., Ramesh, J., Salman, A., Sinelnikov, I., Goldstein, J., Guterman, H., Mordechai, S., Diagnostic potential of Fourier-transform infrared microspectroscopy and advanced computational methods in colon cancer patients. *J. Biomed. Opt.*, 7, 2, 248–54, 2002 Apr 1.
24. Chiriboga, L., Xie, P., Yee, H., Zarou, D., Zakim, D., Diem, M., Infrared spectroscopy of human tissue. IV. Detection of dysplastic and neoplastic changes of human cervical tissue via infrared microscopy. *Cell. Mol. Biol. (Noisy-le-Grand, France)*, 44, 1, 219–29, 1998 Feb 1.
25. Argov, S., Sahu, R.K., Bernshtain, E., Salman, A., Shohat, G., Zelig, U., Mordechai, S., Inflamatory bowel diseases as an intermediate stage between normal and cancer: A FTIR-microspectroscopy approach. *Biopolymers: Original Res. Biomol.*, 75, 5, 384–92, 2004 Dec 5.
26. Sahu, R.K., Argov, S., Salman, A., Huleihel, M., Grossman, N., Hammody, Z., Kapelushnik, J., Mordechai, S., Characteristic absorbance of nucleic acids in the Mid-IR region as possible common biomarkers for diagnosis of malignancy. *Technol. Cancer Res. Treat.*, 3, 6, 629–38, 2004 Dec.
27. Picollo, M., Aceto, M., Vitorino, T., UV-Vis spectroscopy. *Phys. Sci. Rev.*, 4, 4, 20180008, 2018 Nov 14.
28. Rawlins, F.I., Studies in the colorimetry of paintings: A note in conclusion. *Tech. Stud. Field Fine Arts*, 10, 230–31, 1942.
29. Barnes, N.F., A spectrophotometric study of artists' pigments. *Tech. Stud. Field Fine Arts*, 7, 3, 120–38, 1939 Jan.
30. Santini, M., Applicazione del metodo spettrofotometrico nelle determinazioni colorimetriche sui dipinti. *Boll. dell'Istituto centrale del restauro*, 23, 95–129, 1955.
31. Duncan, D.R., The identification and estimation of pigments in pigmented compositions by reflectance spectrophotometry. *J. Oil Colour Chem. Assoc.*, 12, 300–24, 1962.
32. Bacci, M., Picollo, M., Trumpy, G., Tsukada, M., Kunzelman, D., Non-invasive identification of white pigments on 20th-century oil paintings by using fiber optic reflectance spectroscopy. *J. Am. Inst. Conserv.*, 46, 1, 27–37, 2007 Jan 1.
33. Peruzzi, G., Ciccola, A., Bosi, A., Serafini, I., Negozio, M., Hamza, N.M., Moricca, C., Sadori, L., Favero, G., Nigro, V., Postorino, P., Applying Gel-Supported Liquid Extraction to Tutankhamun's Textiles for the Identification of Ancient Colorants: A Case Study. *Gels*, 9, 7, 514, 2023 Jun 25.
34. Dupuis, G., Elias, M., Simonot, L., Pigment identification by fiber-optics diffuse reflectance spectroscopy. *Appl. Spectrosc.*, 56, 10, 1329–36, 2002 Oct 1.
35. Lanteri, L., Agresti, G., Pelosi, C., A new practical approach for 3D documentation in ultraviolet fluorescence and infrared reflectography of polychromatic sculptures as fundamental step in restoration. *Heritage*, 2, 1, 207–15, 2019 Jan 15.
36. Romani, A., Clementi, C., Miliani, C., Favaro, G., Fluorescence spectroscopy: a powerful technique for the noninvasive characterization of artwork. *Acc. Chem. Res.*, 43, 6, 837–46, 2010 Jun 15.
37. Claro, A., Melo, M.J., de Melo, J.S., van den Berg, K.J., Burnstock, A., Montague, M., Newman, R., Identification of red colorants in van Gogh paintings and ancient Andean textiles by microspectrofluorimetry. *J. Cult. Heritage*, 11, 1, 27–34, 2010 Jan 1.
38. Melo, M.J. and Claro, A., Bright light: microspectrofluorimetry for the characterization of lake pigments and dyes in works of art. *Acc. Chem. Res.*, 43, 6, 857–66, 2010 Jun 15.
39. Chai, J., Zhang, K., Xue, Y., Liu, W., Chen, T., Lu, Y., Zhao, G., Review of MEMS based Fourier transform spectrometers. *Micromachines*, 11, 2, 214, 2020 Feb 20.
40. Grdadolnik, J.,., ATR-FTIR spectroscopy: Its advantage and limitations. *Acta Chim. Slov.*, 49, 3, 631–42, 2002 Jan 1.
41. Jelle, B.P., Nilsen, T.N., Hovde, P.J., Gustavsen, A., Accelerated climate aging of building materials and their characterization by Fourier transform infrared radiation analysis. *J. Build. Phys.*, 36, 1, 99–112, 2012 Jul.

42. La Russa, M.F., Ruffolo, S.A., Barone, G., Crisci, G.M., Mazzoleni, P., Pezzino, A., The use of FTIR and micro-FTIR spectroscopy: an example of application to cultural heritage. *Int. J. Spectro.*, 2009, 187, 2009.
43. Liu, Y., Yang, Z., Desyaterik, Y., Gassman, P.L., Wang, H., Laskin, A., Hygroscopic behavior of substrate-deposited particles studied by micro-FT-IR spectroscopy and complementary methods of particle analysis. *Anal. Chem.*, 80, 3, 633–42, 2008 Feb 1.
44. Coates, J., Interpretation of infrared spectra, a practical approach.
45. Fleischmann, M., Hendra, P.J., McQuillan, A.J., Raman spectra of pyridine adsorbed at a silver electrode. *Chem. Phys. Lett.*, 26, 2, 163–6, 1974 May 15.
46. Albrecht, M.G. and Creighton, J.A., Anomalously intense Raman spectra of pyridine at a silver electrode. *J. Am. Chem. Soc.*, 99, 15, 5215–7, 1977.
47. Zhang, X., Zheng, Y., Liu, X., Lu, W., Dai, J., Lei, D.Y., MacFarlane, D.R., Hierarchical porous plasmonic metamaterials for reproducible ultrasensitive surface-enhanced Raman spectroscopy. *Adv. Mater.*, 27, 6, 1090–6, 2015.
48. Jensen, T., Kelly, L., Lazarides, A., Schatz, G.C., Electrodynamics of noble metal nanoparticles and nanoparticle clusters. *J. Cluster Sci.*, 10, 295–317, 1999 Jun.
49. Otto, A., The 'chemical'(electronic) contribution to surface-enhanced Raman scattering, in: *Journal of Raman Spectroscopy: An International Journal for Original Work in all Aspects of Raman Spectroscopy, Including Higher Order Processes, and also Brillouin and Rayleigh Scattering*, vol. 36, pp. 497–509, 2005 Jun.
50. Qian, Y., Fan, T., Yao, Y., Shi, X., Liao, X., Zhou, F., Gao, F., Label-free and Raman dyes-free surface-enhanced Raman spectroscopy for detection of DNA. *Sens. Actuators, B*, 254, 483–9, 2018 Jan 1.
51. Pu, H., Xie, X., Sun, D.W., Wei, Q., Jiang, Y., Double strand DNA functionalized Au@ Ag Nps for ultrasensitive detection of 17β-estradiol using surface-enhanced raman spectroscopy. *Talanta*, 195, 419–25, 2019 Apr 1.
52. Lee, W.Y., Weinberg, O.K., Evans, A.G., Pinkus, G.S., Loss of full-length GATA1 expression in megakaryocytes is a sensitive and specific immunohistochemical marker for the diagnosis of myeloid proliferative disorder related to down syndrome. *Am. J. Clin. Pathol.*, 149, 4, 300–9, 2018 Mar 7.
53. Watts, J.F. and Wolstenholme, J., *An introduction to surface analysis by XPS and AES*, John Wiley & Sons, Hoboken, New Jersey, 2019 Nov 4.
54. Briggs, D. and Grant, J.T., XPS: Basic principles, spectral features and qualitative analysis, in: *Surface analysis by Auger and X-ray photoelectron spectroscopy*, pp. 31–56, 2003.
55. Watts, J.F., An introduction to surface analysis by XPS and AES. *Chapter*, 5, 149–56, 2003.
56. Briggs, D. and Riviere, J.C., *Practical Surface Analysis*, D. Briggs and M.P. Seah (Eds.) Journal of Vacuum Science & Technology A, manjusaka.de, 2020.
57. McHugh, J.A. and Czanderna, A.W., Secondary ion mass spectrometry, in: *Methods of Surface Analysis*, Wolsky, S.P. and Czanderna, A.W. (Eds.), 1975.
58. Engelhard, M.H., Baer, D.R., Herrera-Gomez, A., Sherwood, P., Introductory guide to backgrounds in XPS spectra and their impact on determining peak intensities. *J. Vac. Sci. Technol. A*, 38, 6, 38, 2020 Dec 1.
59. Wolstenholme, J., Procedure which allows the performance and calibration of an XPS instrument to be checked rapidly and frequently. *J. Vac. Sci. Technol. A*, 38, 4, 4, 2020 Jul 1.
60. Baer, D.R. and Shard, A.G., Role of consistent terminology in XPS reproducibility. *J. Vac. Sci. Technol. A*, 38, 3, 3, 2020 May 1.
61. Steinhardt, R. and Serfass, E., X-ray photoelectron spectrometer for chemical analysis. *Anal. Chem.*, 23, 11, 1585–90, 1951 Nov 1.

62. Siegbahn, K., Nordling, C., Fahlman, A., Nordberg, R., Hamrin, K., Hedman, J., Johansson, G., Bergmark, T., Karlsson, S.E., Lindgren, I., Lindberg, B., *Nova Acta Regiae Soc. Sci. Upsaliensis*, Catalytic Science Series, worldscientific.com, 2023.
63. Wüthrich, K., NMR with proteins and nucleic acids. *Europhys. News*, 17, 1, 11–3, 1986.
64. Gadian, D.G., NMR and its applications to living systems. *Application of NMR to the Study of Cells and Body Fluids*, 5, 19, 1039–1057, 1995.
65. Brown, F.F., Campbell, I.D., Kuchel, P.W., Human erythrocyte metabolism studies by 1H spin echo NMR. *FEBS Lett.*, 82, 1, 12–6, 1977 Oct 1.
66. Bales, J.R., Higham, D.P., Howe, I., Nicholson, J.K., Sadler, P.J., Use of high-resolution proton nuclear magnetic resonance spectroscopy for rapid multi-component analysis of urine. *Clin. Chem.*, 30, 3, 426–32, 1984 Mar 1.
67. Ross, A., Schlotterbeck, G., Dieterle, F., Senn, H., NMR spectroscopy techniques for application to metabonomics. *Handb. metabonomics metabolomics.*, 1, 55–112, 2007 Jan.
68. Nicholson, J.K., Connelly, J., Lindon, J.C., Holmes, E., Metabonomics: a platform for studying drug toxicity and gene function. *Nat. Rev. Drug Discovery*, 1, 2, 153–61, 2002 Feb 1.
69. Fernie, A.R., Trethewey, R.N., Krotzky, A.J., Willmitzer, L., Metabolite profiling: from diagnostics to systems biology. *Nat. Rev. Mol. Cell Biol.*, 5, 9, 763–9, 2004 Sep 1.
70. Shur, J. and Price, R., Advanced microscopy techniques to assess solid-state properties of inhalation medicines. *Adv. Drug Delivery Rev.*, 64, 4, 369–82, 2012 Mar 30.
71. Klang, V. and Valenta, C., Lecithin-based nanoemulsions. *J. Drug Delivery Sci. Technol.*, 21, 1, 55–76, 2011 Jan 1.
72. Klang, V., Valenta, C., Matsko, N.B., Electron microscopy of pharmaceutical systems. *Micron*, 44, 45–74, 2013 Jan 1.
73. Sennour, M. and Esnouf, C., Contribution of advanced microscopy techniques to nano-precipitates characterization: case of AlN precipitation in low-carbon steel. *Acta Mater.*, 51, 4, 943–57, 2003 Feb 25.
74. Turner, Y.T., Roberts, C.J., Davies, M.C., Scanning probe microscopy in the field of drug delivery. *Adv. Drug Delivery Rev.*, 59, 14, 1453–73, 2007 Dec 10.
75. Moayyedi, M., Eskandari, M.H., Rad, A.H., Ziaee, E., Khodaparast, M.H., Golmakani, M.T., Effect of drying methods (electrospraying, freeze drying and spray drying) on survival and viability of microencapsulated Lactobacillus rhamnosus ATCC 7469. *J. Funct. Foods*, 40, 391–9, 2018 Jan 1.
76. Rostamabadi, H., Mahoonak, A.S., Allafchian, A., Ghorbani, M., Fabrication of β-carotene loaded glucuronoxylan-based nanostructures through electrohydrodynamic processing. *Int. J. Biol. Macromol.*, 139, 773–84, 2019 Oct 15.
77. Bruni, G.P., de Oliveira, J.P., Gómez-Mascaraque, L.G., Fabra, M.J., Martins, V.G., da Rosa Zavareze, E., López-Rubio, A., Electrospun β-carotene–loaded SPI: PVA fiber mats produced by emulsion-electrospinning as bioactive coatings for food packaging. *Food Packag. Shelf Life*, 23, 100426, 2020 Mar 1.
78. Sarabandi, K. and Jafari, S.M., Effect of chitosan coating on the properties of nanoliposomes loaded with flaxseed-peptide fractions: Stability during spray-drying. *Food Chem.*, 310, 125951, 2020 Apr 25.
79. Srinivasan, C., Mullen, T.J., Hohman, J.N., Anderson, M.E., Dameron, A.A., Andrews, A.M., Dickey, E.C., Horn, M.W., Weiss, P.S., Scanning electron microscopy of nanoscale chemical patterns. *ACS Nano*, 1, 3, 191–201, 2007 Oct 31.
80. Wight, S.A. and Zeissler, C.J., Environmental scanning electron microscope imaging examples related to particle analysis. *Microsc. Res. Tech.*, 25, 5-6, 393–7, 1993 Aug.
81. Rubin, M.A., Jasiuk, I., Taylor, J., Rubin, J., Ganey, T., Apkarian, R.P., TEM analysis of the nanostructure of normal and osteoporotic human trabecular bone. *Bone*, 33, 3, 270–82, 2003 Sep 1.

82. Danino, D., Cryo-TEM of soft molecular assemblies. *Curr. Opin. colloid Interface Sci.*, 17, 6, 316–29, 2012 Dec 1.
83. Lorenz, K. and Kulp, K., Cereal-and Root Starch Modification by Heat-Moisture Treatment. *I. Physico-Chemical Properties. Starch-Stärke*, 34, 2, 50–4, 1982.
84. Tiede, K., Tear, S.P., David, H., Boxall, A.B., Imaging of engineered nanoparticles and their aggregates under fully liquid conditions in environmental matrices. *Water Res.*, 43, 13, 3335–43, 2009.
85. Kuntsche, J., Horst, J.C., Bunjes, H., Cryogenic transmission electron microscopy (cryo-TEM) for studying the morphology of colloidal drug delivery systems. *Int. J. Pharm.*, 417, 1-2, 120–37, 2011 Sep 30.
86. Koning, R.I. and Koster, A.J., Cryo-electron tomography in biology and medicine. *Ann. Anatomy-Anatomischer Anz.*, 191, 5, 427–45, 2009 Jan 1.
87. Garcıa, R. and Perez, R., Dynamic atomic force microscopy methods. *Surf. Sci. Rep.*, 47, 6-8, 197–301, 2002 Sep 1.
88. Ruggeri, F.S., Habchi, J., Cerreta, A., Dietler, G., AFM-based single molecule techniques: unraveling the amyloid pathogenic species. *Curr. Pharm. Des.*, 22, 26, 3950–70, 2016 Jul 1.
89. Wang, C., Cheng, T., Zhang, D., Pan, X., Electrochemical properties of humic acid and its novel applications: A tip of the iceberg. *Sci. Total Environ.*, 863, 160755, 2023 Mar 10.
90. Benne, N., Leboux, R.J., Glandrup, M., van Duijn, J., Vigario, F.L., Neustrup, M.A., Romeijn, S., Galli, F., Kuiper, J., Jiskoot, W., Slütter, B., Atomic force microscopy measurements of anionic liposomes reveal the effect of liposomal rigidity on antigen-specific regulatory T cell responses. *J. Control. Release*, 318, 246–55, 2020 Feb 1.
91. Jonkman, J., Brown, C.M., Wright, G.D., Anderson, K.I., North, A.J., Tutorial: guidance for quantitative confocal microscopy. *Nat. Protoc.*, 15, 5, 1585–611, 2020 May.
92. Morris, P.A., Aspden, R.S., Bell, J.E., Boyd, R.W., Padgett, M.J., Imaging with a small number of photons. *Nat. Commun.*, 6, 1, 5913, 2015 Jan 5.
93. Nam, S., Hwang, Y., Matsushita, Y., Kim, S.J., A holistic approach to cross-channel image noise modeling and its application to image denoising, in: *Proceedings of the IEEE conference on computer vision and pattern recognition*, pp. 1683–1691, 2016.
94. Pawley, J. (Ed.), *Handbook of biological confocal microscopy*, Springer Science & Business Media, New York, 2006 Jun 2.
95. Plotz, T. and Roth, S., Benchmarking denoising algorithms with real photographs. *Proceedings of the IEEE conference on computer vision and pattern recognition*, pp. 1586–1595, 2017.
96. Prime, R.B., Bair, H.E., Vyazovkin, S., Gallagher, P.K., Riga, A., Thermogravimetric analysis (TGA). *Therm. Anal. polymers: Fundam. applications.*, 3, 241–317, 2009 Apr.
97. Mohamed, O.A., Masood, S.H., Bhowmik, J.L., Optimization of fused deposition modeling process parameters: a review of current research and future prospects. *Adv. Manuf.*, 3, 42–53, 2015 Mar.
98. Carneiro, O.S., Silva, A.F., Gomes, R., Fused deposition modeling with polypropylene. *Mater. Des.*, 83, 768–76, 2015.
99. Bershteĭn, V.A. and Egorov, V.M., Differential scanning calorimetry of polymers: physics, chemistry, analysis, technology. *DSC study of main relaxations in the poly(imide-dimethylsiloxane) block copolymers.*, 37, Taylor and Francis Online homepage 1994 Jan.
100. Monteiro, S.N., Calado, V., Rodriguez, R.J., Margem, F.M., Thermogravimetric behavior of natural fibers reinforced polymer composites—An overview. *Mater. Sci. Engineering: A*, 557, 17–28, 2012 Nov 15.
101. Principles of thermal analysis and calorimetry. *R. Soc. Chem.*, 5, 100333, 2019 Mar 15.
102. dos Santos, W.N., Mummery, P., Wallwork, A., Thermal diffusivity of polymers by the laser flash technique. *Polym. Test.*, 24, 5, 628–34, 2005 Aug 1.

103. Baumann, T.F., Worsley, M.A., Han, T.Y., Satcher Jr., J.H., High surface area carbon aerogel monoliths with hierarchical porosity. *J. Non-Cryst. Solids*, 354, 29, 3513–5, 2008 Jul 1.
104. Bera, M. and Maji, P.K., Effect of structural disparity of graphene-based materials on thermo-mechanical and surface properties of thermoplastic polyurethane nanocomposites. *Polymer*, 119, 118–33, 2017 Jun 16.
105. Jawaid, M., Khalil, H.A., Hassan, A., Dungani, R., Hadiyane, A., Effect of jute fibre loading on tensile and dynamic mechanical properties of oil palm epoxy composites. *Composites, Part B*, 45, 1, 619–24, 2013 Feb 1.
106. Zhang, Z., Wang, P., Wu, J., Dynamic mechanical properties of EVA polymer-modified cement paste at early age. *Physics Procedia*, 25, 305–10, 2012 Jan 1.
107. Harris, J.M. and Dovichi, N.J., Thermal lens calorimetry. *Anal. Chem.*, 52, 6, 695A–706A, 1980 May 1.
108. Whinnery, J.R., Laser measurement of optical absorption in liquids. *Acc. Chem. Res.*, 7, 7, 225–31, 1974 Jul 1.
109. Pelletier, M.J., Thorsheim, H.R., Harris, J.M., Laser-induced thermal diffraction for calorimetric absorption measurements. *Anal. Chem.*, 54, 2, 239–42, 1982 Feb 1.
110. Da Costa, G., Real-time recording of light patterns in heavy hydrocarbons: a theoretical analysis. *Appl. Opt.*, 19, 20, 3523–8, 1980 Oct 15.
111. Stone, J., Measurements of the absorption of light in low-loss liquids. *JOSA*, 62, 3, 327–33, 1972 Mar 1.
112. Woodruff, S.D. and Yeung, E.S., Refractive index and absorption detector for liquid chromatography based on Fabry-Perot interferometry. *Anal. Chem.*, 54, 7, 1174–8, 1982 Jun 1.
113. Saba, N., Paridah, M.T., Jawaid, M., Mechanical properties of kenaf fibre reinforced polymer composite: A review. *Constr. Build. Mater.*, 76, 87–96, 2015 Feb 1.
114. Grèze, R., Manach, P.Y., Laurent, H., Thuillier, S., Menezes, L.F., Influence of the temperature on residual stresses and springback effect in an aluminium alloy. *Int. J. Mech. Sci.*, 52, 9, 1094–100, 2010 Sep 1.
115. Coër, J., Bernard, C., Laurent, H., Andrade-Campos, A., Thuillier, S., The effect of temperature on anisotropy properties of an aluminium alloy. *Exp. Mech.*, 51, 1185–95, 2011 Sep.
116. Aretz, H., Hopperstad, O.S., Lademo, O.G., Yield function calibration for orthotropic sheet metals based on uniaxial and plane strain tensile tests. *J. Mater. Process. Technol.*, 186, 1-3, 221–35, 2007 May 7.
117. Griffin, M., Premakumar, Y., Seifalian, A., Butler, P.E., Szarko, M., Biomechanical characterization of human soft tissues using indentation and tensile testing. *JoVE (Journal Visualized Experiments)*, 13, 118, e54872, 2016 Dec.
118. Vickers, N.J., Animal communication: when i'm calling you, will you answer too? *Curr. Biol.*, 27, 14, R713–5, 2017 Jul 24.
119. Striemann, P., Eichenhofer, M., Schupp, D., Niedermeier, M., Huelsbusch, D., Walther, F., Compression testing of additively manufactured continuous carbon fiber-reinforced sandwich structures. *Mater. Test.*, 60, 9, 801–8, 2018 Sep 30.
120. Srinivasan, K.N., Meenakshi, R., Santhi, A., Thangavelu, P.R., John, S., Studies on development of electroless Ni–B bath for corrosion resistance and wear resistance applications. *Surf. Eng.*, 26, 3, 153–8, 2010 Apr 1.
121. Oraon, B., Majumdar, G., Ghosh, B., Improving hardness of electroless Ni–B coatings using optimized deposition conditions and annealing. *Mater. Des.*, 29, 7, 1412–8, 2008 Jan 1.
122. Vitry, V., Delaunois, F., Dumortier, C., Mechanical properties and scratch test resistance of nickel–boron coated aluminium alloy after heat treatments. *Surf. Coat. Technol.*, 202, 14, 3316–24, 2008 Apr 15.

123. Krishnaveni, K., Narayanan, T.S., Seshadri, S.K., Electroless Ni–B coatings: preparation and evaluation of hardness and wear resistance. *Surf. Coat. Technol.*, 190, 1, 115–21, 2005 Jan 3.
124. Shafi, W.K., Raina, A., Haq, M.I., Khajuria, A., Applications of industrial tribology. *Int. Res. J. Eng. Technol.*, 5, 1, 1285–9, 2018.
125. Dresselhuis, D.M., De Hoog, E.H., Stuart, M.C., Van Aken, G.A., Application of oral tissue in tribological measurements in an emulsion perception context. *Food Hydrocolloids*, 22, 2, 323–35, 2008 Mar 1.
126. Chen, L., Qian, Z., Wen, S., Huang, S., High-gradient magnetic separation of ultrafine particles with rod matrix. *Miner. Process. Extr. Metall. Rev.*, 34, 5, 340–7, 2013 Aug 1.
127. Lamichhane, T.N., Sethuraman, L., Dalagan, A., Wang, H., Keller, J., Paranthaman, M.P., Additive manufacturing of soft magnets for electrical machines—A review. *Mater. Today Phys.*, 15, 100255, 2020 Dec 1.
128. Valdes, L.B., Resistivity measurements on germanium for transistors. *Proc. IRE*, 42, 2, 420–7, 1954 Feb.
129. Harris, C.M., Todd, R.W., Bungard, S.J., Lovitt, R.W., Morris, J.G., Kell, D.B., Dielectric permittivity of microbial suspensions at radio frequencies: a novel method for the real-time estimation of microbial biomass. *Enzyme Microb. Technol.*, 9, 3, 181–6, 1987 Mar 1.
130. Asami, K., Gheorghiu, E., Yonezawa, T., Real-time monitoring of yeast cell division by dielectric spectroscopy. *Biophys. J.*, 76, 6, 3345–8, 1999 Jun 1.
131. Raicu, V., Saibara, T., Enzan, H., Irimajiri, A., Dielectric properties of rat liver in vivo: analysis by modeling hepatocytes in the tissue architecture. *Bioelectrochem. Bioenerg.*, 47, 2, 333–42, 1998 Dec 1.
132. Asami, K., Characterization of biological cells by dielectric spectroscopy. *J. Non-Cryst. Solids*, 305, 1-3, 268–77, 2002 Jul 1.
133. Brown, S.D. and Betley, T.A., Peters JC A Low-Spin d5 Iron Imide: Nitrene Capture by Low-Coordinate Iron (I) Provides the 4-Coordinate Fe (III) Complex [PhB (CH2PPh2) 3] Fe: N-p-tolyl. *J. Am. Chem. Soc.*, 125, 2, 322–3, 2003 Jan 15.
134. Hristoforou, E. and Ktena, A., Magnetostriction and magnetostrictive materials for sensing applications. *J. Magn. Magn. Mater.*, 316, 2, 372–8, 2007 Sep 1.
135. Calkins, F.T., Dapino, M.J., Flatau, A.B., Effect of prestress on the dynamic performace of a Terfenol-D transducer, in: *Smart structures and materials 1997: smart structures and integrated systems*, vol. 3041, SPIE, pp. 293–304, 1997 Jun 6.
136. Cui, X., Lee, G.H., Kim, Y.D., Arefe, G., Huang, P.Y., Lee, C.H., Chenet, D.A., Zhang, X., Wang, L., Ye, F., Pizzocchero, F., Multi-terminal transport measurements of MoS2 using a van der Waals heterostructure device platform. *Nat. Nanotechnol.*, 10, 6, 534–40, 2015 Jun.
137. Zhu, W., Low, T., Lee, Y.H., Wang, H., Farmer, D.B., Kong, J., Xia, F., Avouris, P., Electronic transport and device prospects of monolayer molybdenum disulphide grown by chemical vapour deposition. *Nat. Commun.*, 5, 1, 3087, 2014 Jan 17.
138. Allain, A., Kang, J., Banerjee, K., Kis, A., Electrical contacts to two-dimensional semiconductors. *Nat. Mater.*, 14, 12, 1195–205, 2015 Dec.
139. Sangwan, V.K., Arnold, H.N., Jariwala, D., Marks, T.J., Lauhon, L.J., Hersam, M.C., Low-frequency electronic noise in single-layer MoS2 transistors. *Nano Lett.*, 13, 9, 4351–5, 2013 Sep 11.
140. Martinez, I., Ribeiro, M., Andres, P., Hueso, L.E., Casanova, F., Aliev, F.G., Photodoping-Driven Crossover in the Low-Frequency Noise of MoS 2 Transistors. *Phys. Rev. Appl.*, 7, 3, 034034, 2017 Mar 31.

14

Novel Study on Different Polysaccharides and Its Application in Solar Cell

Ashlesha P. Kawale[1]*, Nishant Shekhar[1], Arti Srivastava[1], Navin Pradhan[1], Pravat K. Swain[2] and S.Y. Bodkhe[3]

[1]*Department of Chemistry, Guru Ghasidas Vishwavidyalaya Central University, Bilaspur, Chhattisgarh, India*
[2]*Department of Chemistry, Dr. J N College, Rasalpur, Balasore, Odisha, India*
[3]*National Environmental Engineering Research Institute, Nagpur, Maharashtra, India*

Abstract
Scientific and commercial sectors have shown a great deal of interest in dye-sensitized solar cells (DSSCs), especially in those that use different types of nanoparticle coatings. While fluid electrolytes are commonly used in DSSCs, their instability and tendency to vaporize have hindered commercialization efforts. Additionally, electron recombination at semiconductor-liquid electrolyte interfaces has been identified as a factor reducing DSSC performance. This issue worsens when the photoanode is exposed to a vaporized electrolyte solution, altering charge density at the semiconductor-electrolyte junction and initiating photo corrosion on the photoanode. The discovery of ionic conductivity in polymers has marked a significant advancement in DSSC technology. Electrolytes combining polymers with salt offer a promising avenue for addressing these challenges. Polysaccharides, in particular, show potential for enhancing DSSC performance and stability. They can also help lower the overall cost of these gadgets. Moreover, they can substitute synthetic polymers, which eases environmental concerns. With the purpose of enhancing the function and stability of DSSCs, the current chapter attempts to present a thorough synopsis of the main polysaccharides. These include chitosan, cellulose, starch, xanthan, carboxymethyl cellulose, carrageenan, alginate, and gellan gum. By exploring these polysaccharides in detail, the chapter seeks to elucidate their roles and potential applications in advancing DSSC technology.

Keywords: Polysaccharides, DSSC, biopolymer, solar cells, electrolytes

14.1 Introduction

Earth is free of solar energy. Even under severely demanding conditions, the sun emits twelve thousand terawatts (TW) of radiation onto the earth's surface, well beyond human needs. It is 10 thousand times larger than the global population's energy usage. About 10% of solar cell efficiency would satisfy the current energy needs if 0.1% earth's surface is covered [1–3].

**Corresponding author*: ashleshakawale1996@gmail.com

Divya Bajpai Tripathy, Anjali Gupta and Arvind Kumar Jain (eds.) Multifunctional Materials: Engineering and Biological Applications, (365–392) © 2025 Scrivener Publishing LLC

All roads for using solar energy, capitalize on the well-designed steps of capturing, converting, and storing. The most effective methods for converting and conserving solar energy are photovoltaic (PV) cells. PV cells, also known as photo electrochemical solar cells, are capable of directly converting the free radiation produced by the sun (photon) into electrical power. Numerous electrical equipment, including those in the residential, automotive, agricultural, and electricity sectors, among others, can be powered by direct or conserved electrical energy in the form of current. The discovery of solar cells dates back centuries ago. The immediate transformation of sunlight into high-quality electrical energy is one of the most alluring features of PV technology among all renewable energy sources [4]. A potentially useful method of converting solar energy involves a DSSC, which functions similarly to photosynthesis in plants. Notably, DSSCs have a number of benefits, such as being simple to assemble, having cheap material prices, being easy to process using simple equipment, using eco-friendly ingredients, and being able to create adaptable solar cells. Electrolyte materials are essential to the operation of DSSCs because they facilitate the transfer of charge carriers from the photoanode to the counter electrode (CE). Electrolyte systems serve as the foundation for a multitude of electrochemical devices, such as fuel cells, batteries, supercapacitors, and electrode-containing sensitized solar cells. Electrolytes are ionic conductors that typically have conductivities between 10^2 and 10^5 S/cm. Unlike in external circuits where charge motion is governed by electrons, in electrochemical cells, the transfer of charge is facilitated by the motion of cations and anions present in the electrolyte.

Ionic liquids, diluted salt solutions (usually with concentrations of 1-2 molar), and strong acid/base solutions are examples of conventional electrolytes. These electrolytes provide an adequate supply of cations and anions to ensure optimal cell performance. However, they often exist in liquid form, exhibit high volatility, possess caustic properties, and tend to degrade over time due to evaporation, thereby affecting cell performance [5]. They significantly affect the durability and effectiveness of the DSSCs' light-to-electric conversion processes. A unit called a solar cell uses light energy from the sun to create electricity. While the world's supply of fossil fuels, which is sufficient for its current energy needs, is running out, the need for energy is increasing daily.

The environmental threat posed by non-biodegradable plastics derived from petroleum has become a significant global concern, as these materials persist in the environment without undergoing microbial degradation. In response to escalating fossil fuel prices and the environmental impact of synthetic polymers, researchers have increasingly turned their attention to biodegradable alternatives. Through time, robust biopolymeric systems with improved mechanical and electrical characteristics have been created by synthesizing polysaccharides like the ones listed above. Unlike their synthetic counterparts, materials derived from biopolymers are biodegradable, making them a more environmentally friendly choice. Bio-polymer electrolytes (BEs) are a type of high molecular weight natural polymer that includes dissolved salts, serving as effective ion conductors. These materials have a lot of benefits, such as high energy density, solvent-free composition, leaking-free nature, lightweight construction, ease of production, and high ionic conductivity values (10^6 to 10^4 S/cm). These properties are particularly useful for devices that store electricity such as supercapacitors, batteries, fuel cells, and DSSCs [6].

14.2 Generation of Photovoltaic Cell

According to the kind of material used, the application, the maximum conversion rate, and the cost of generating PV electricity, solar cells may be categorized into three main generations:

14.2.1 First-Generation Solar Cell

- Mono crystalline solar cells
- Poly crystalline silicon (Si) on low-cost substrate
- Hybrid Si cells

The 1st generation of PVs employs materials with a high degree of purity and few structural flaws, including single crystals. First-generation PV devices have so far achieved 13% power-conversion efficiency [4]. However, the enormous expense of producing big, ultra-pure Si crystals at high and sustained temperatures necessitates the search for less expensive materials as their substitutes in the solar industry. The Figure 14.1 shows an example of first-generation solar cell.

14.2.2 Second-Generation Solar Cells

- Copper indium diselenide cells
- Cadmium telluride cells
- Gallium arsenide cells
- Amorphous Si cells (shown in Figure 14.2)

These are typically referred to as thin film solar cells that use semiconductors like cadmium telluride (CdTe), copper indium gallium selenide (CIGS), or amorphous Si and have a thickness of less than one micron. These solar cells are less expensive than Si solar panels.

Figure 14.1 Crystalline silicon solar cell.

Figure 14.2 Amorphous silicon solar cell.

Devices of the second generation are built using low-energy, intensive preparatory methods including electroplating and vapor deposition. Thin films in semiconductors allow for the use of fewer materials, which also lowers the cost of production. The maximum power conversion is lower than with crystalline Si solar cells, despite the cheaper cost. Additionally, certain concerns were raised since this type of solar cells included dangerous hazardous chemicals.

14.2.3 Third-Generation Solar Cells

- Dye-sensitized solar cells (DSSCs)
- Copper zinc tin sulfide solar cells
- Quantum dot solar cells
- Polymer solar cells
- Organic solar cells
- Perovskite solar cells

Third-generation solar cells are very efficient in converting light into energy with low production costs. Solar energy may be used in a way that makes it one of the most efficient and affordable sources of energy in the future by utilizing thin film technology.

14.3 Advantages of Solar Cells

- Because the dye, semiconductor, and electrolyte's surface charges are separated individually, recombination is less likely.
- Renewable Energy Source: Because solar energy comes from the sun, which will continue to shine for billions of years, it is abundant and unbounded.
- Eco-friendly: Greenhouse gas emissions from the generation of solar energy are negligible, rendering it a sustainable and eco-friendly substitute for fossil

fuels. In addition to lowering air and water pollution, it helps slow down climate change.

- Lower Energy Expenses: By using free sunshine to create power for residences, companies, and communities, solar panels may drastically cut or even completely eliminate energy expenses after they are installed.
- Minimal Maintenance Requirements: Since solar energy systems do not have any moving components and often need little maintenance, they offer cheap operating costs. Over the course of the system's life, this lowers operational expenses.
- Energy Independence: By lowering dependency on imported fossil fuels and centralized energy systems, solar power fosters resilience to changes in energy prices and supply interruptions.
- Scalability and Modularity: Solar panels may be customized to satisfy a variety of energy requirements, from large-scale utility projects to small-scale residential installations. Because they are modular, it will be simple to extend or improve them as energy needs alter.
- Employment Creation: The solar sector contributes to economic growth and job possibilities in local areas by creating jobs in manufacturing, set up, upkeep, and research and development.
- Technological Developments: As solar cell technology continues to evolve, solar energy becomes more competitive with traditional energy sources due to improvements in efficiency, durability, and price.

14.4 Disadvantage of All-Generation Solar Cells

- PV panels need one of the most expensive and complex production processes since they are made of single-cell Si crystals.
- The same materials experience charge separation, which encourages a high charge carrier recombination rate.
- Materials with the highest levels of purity are needed to produce PV panels.
- Air conditioning was required to maintain the cold in the mainframes.
- Intermittency: The production of solar energy is reliant on sunshine, which changes during the day and is absent at night. It may be difficult to continuously satisfy energy demand due to this intermittency, particularly in the absence of suitable energy storage options.
- Weather Dependency: The weather, including clouds, rain, and even shade from surrounding objects, can have an impact on solar panels. The predictability and dependability of solar energy output may be impacted by this weather unpredictability.
- Land Use: The installation of solar power plants requires a sizable amount of land, especially for large-scale utility projects. Conflicts over land usage may develop in places where there is a shortage of land or when ecological or agricultural issues are of the utmost importance.

- Challenges with Energy Storage: It is still difficult to store extra energy produced during bright days for usage in gloomy or nocturnal situations. Although battery backup technologies are advancing, they may not yet be economically feasible for all applications and might raise the total expense of a solar energy system.
- Resource Extraction: Raw resources including Si, cadmium, and rare earth metals must be extracted and processed in order to make solar panels. If not handled properly, the mining and processing of these resources may have negative effects on the environment and society.
- Effects on the Life Cycle: Although solar panels provide clean energy when they are in use, there are environmental issues with their production, shipping, installation, and disposal at the end of their useful lives. Maintaining appropriate recycling and disposal procedures is crucial to reducing environmental effects throughout the product's lifespan.

For the purpose of developing highly effective solar energy technology, much research has been done. Producing solar cells, including Si-based PV cells and thin-film, polymer, and DSSC, is the most promising approach. A lot of interest has been shown in DSSCs, one of the most intriguing advancements in third-generation solar cell technology. Over five articles are published per day in this discipline, which is expanding quickly. The benefits of DSSCs are simple construction, great performance, and low-cost photo-conversion. Since the first study by Kalyanasundaram K. [7], the ability of DSSCs to function even under diffuse light characteristics has attracted attention on a global scale. These cells have an impressive 13% conversion efficiency. As a result, the field of DSSCs is the focus of this work.

14.5 Dye-Sensitized Solar Cell

The solar radiation received from the sun annually is estimated to be approximately ten times greater than current global energy demands. To utilize this vast solar energy supply, a variety of PV technologies have been created, including organic, inorganic, and hybrid cells. With no negative environmental effects, these PV cells provide a clean, sustainable energy source by transforming sunlight into electricity. In 1954, Bell Laboratories designed the first practical PV cell using diffused Si p-n junction technology, achieving an efficiency of 6%. Even though Si-based solar cells now have 25% efficiency, the need for very pure Si, the use of hazardous chemicals during manufacture, and the expensive cost of production prevent them from being used widely. This prompted a search for less expensive and more ecologically friendly solar cell substitutes. In 1991, O'Regan and Graetzel introduced a groundbreaking PV cell inspired by plant photosynthesis processes, achieving efficiencies ranging from 7.1% to 7.9%. These cells, often referred to as DSSCs, have attracted a lot of interest due to their inexpensive production costs and several advantages. The photoanode has a significant impact on DSSC performance and is essential for attaining high efficiency in converting energy. The creation of DSSCs with improved performance is made possible by features in nanostructured photoanodes, such as large surface areas, minimal electron recombination, and a large electron transfer efficacy. Currently, much of the research in the field is focused on developing innovative photoanodes for DSSCs. Factors such as semiconductor material, synthesis methods, morphology, and structure are vital for controlling the PV performance of DSSCs. Notably, in the areas of building-integrated and indoor light-harvesting applications, DSSCs have recently achieved efficiencies exceeding 14%. This summary aims to provide an overview of the advancements and potential of DSSCs, highlighting their significant role in the quest for efficient and sustainable solar energy technologies [8], which are quickly becoming the top PV technology. The electrolyte is crucial in these devices since it affects their efficiency and durability. In addition to collecting electrons that reach the counter-electrode, the redox pair is dissolved in the organic solvent and is crucial to the production of the oxidized dye that acts as the sensitizer [8]. Indeed, the electrolyte plays a crucial role in DSSCs by enabling both rapid regeneration of the sensitizer and a delayed recombination process with electrons in the titanium dioxide (TiO_2) conduction band. However, iodine-based redox pairs exhibit several unfavorable characteristics.

14.6 Component of DSSC

The development of effective and useful DSSCs, such as roll-to-roll or sheet-like structured designs, has been the emphasis of late. Most of these DSSCs typically consist of these five main components:

- Transparent conducting electrode
- Photoanode (semiconductor)
- Dyes (sensitizer)
- Electrolyte
- Counter electrode

In a DSSC, these parts cooperate to absorb light, create electron-hole pairs, move charge carriers, and finally produce electricity. Due to its many advantages—including low production costs, simplicity of fabrication, and good low-light performance—DSSCs are a viable option for a range of solar energy harvesting applications.

14.6.1 Transparent Conducting Electrode

Transparent glass is covered with a conductive tin oxide doped with either fluorine (FTO) or indium (ITO) to serve as the substrate for the anode and cathode (Figure 14.3). layer to allow for the passage of light and electrons. The DSSC has a clear glass substrate layer with a conductive layer on the front and back sides. Both the entry of sunlight into the cell and the conductivity of electrons into the external circuit depend on this layer. SnO_2:F is the most used transparent conductive electrode (TCE) in DSSCs, primarily due to its outstanding heat resistance, high conductivity, and superb optical transparency capabilities for visible light. Other transparent conductive oxide (TCO) substrates, such as ITO, which comprises poisonous, scarce, and costly materials, perform worse than FTO.

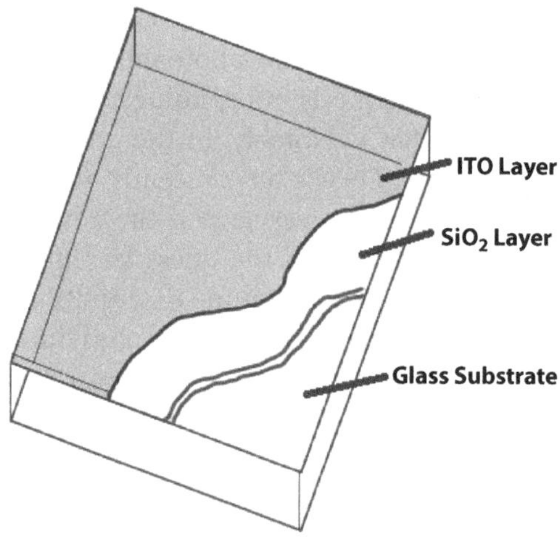

Figure 14.3 ITO glass plates.

14.6.2 Photoanode (Semiconductor)

On the conductive glass substrate (photo electrode), a nanocrystalline wide band gap semiconductor was deposited to give the required substantial surface area for the adsorption of sensitizers (dye molecules). Due to its advantages over other semiconducting photoanodes like ZnO, SnO_2, and others, titanium dioxide was chosen as the semiconductor. It is reasonably priced, extensively available, highly photosensitive, and structurally robust both in solutions and when exposed to sunlight. On the other hand, TiO_2 absorbs ultraviolet (UV) light due to its high band gap semiconductor (3.2 eV). UV radiation does not make up the complete solar spectrum. Dye compounds are employed to absorb visible light. Monocrystalline anatase-coated TiO_2 particles with an average particle size of 10-15 nm have been found to be the best choice for efficient device operation because they readily form mesoscopic films that maximize the dye load per unit area and also permit electrolyte infiltration through their pores, slowing the recombination events.

14.6.3 Dye (Sensitizer)

In order to use the light energy from the sun, the sensitizer or dye that is attached to the photoanode is crucial. Important design criteria must be met in order to create dyes for DSSCs that are more effective. The sensitizing dyes must firmly stick to the surface of the photocatalyst (TiO_2) in order to effectively inject electrons into the conduction band of the material. For the oxidized dye to be efficiently charged and regenerable, its lowest unoccupied molecular orbital (LUMO) must be higher than the TiO_2 conduction band and its highest occupied molecular orbital (HOMO) must be lower than the hole-transport material. Lastly, the dye's light-harvesting ability in the visible and/or nearby infrared (IR) regions needs to be substantial for its absorption process to be as successful as feasible [9].

- The dye's absorption spectra should encompass the entirety of the visible and near-infrared portions of the solar spectrum.

- An excited dye's charged phase energy has to be greater than the semiconductor's conduction band's leading edge in order for electrons to move between it and the conducting band of the semiconductor with efficiency. Redox capacity should be high enough in its ground state to allow for quick regeneration using either a hole conductor or an electron donation via the electrolyte.
- They must, for instance, have anchoring groups that can adhere to and inject electrons into the TiO_2 surface.
- Under continuous light irradiation, they must remain stable in interaction with the various species comprising the electrolyte.

A minimum of one billion redox revolutions under illumination, or about two decades of exposure to sunlight at air mass (AM) 1.5, is needed for long-term photo and thermal consistency investigations.

Due to their prolonged excitation lifespan, robust visual absorption, and effective metal-to-ligand charge transfer, transition metal coordinated compounds were used as sensitizers in early DSSC designs [8]. The power conversion efficiency of DSSCs utilizing N719 dye as the active ingredient has surpassed that of other dyes.

14.6.4 Electrolyte

One of the most vital components of the DSSC is the electrolyte. Solvents, ion conductors, and additives make up the bulk of liquid electrolytes. The ionic conductor can dissolve in the presence of organic solvents, which also regulate the migration of electrolytes into semiconductors and CEs. As a result, the solvent has to be low viscous, high in dielectric constant, and high in conductivity. During DSSC function, it is in charge of internal charge carrier transit amid electrodes and continuous dye and self-regeneration. The power-to-light efficiency (PCE) and long-term stability of the cell are highly valued by the electrolytes. Iodide/triiodide (I/I3) is the most common redox pair, although other options include Br/Br_3, sulfur/polysulfide (Sn2/nS2), and metal complexes such cobalt complexes (CoII/CoIII). The enhancement of electrolytes, electrode surfaces, and mesoporous films is the key area of research in the current study.

14.6.5 Counter Electrode

The back of the DSSC features Figure 14.4 a second glass substrate covered in a thin layer of platinum microcrystals, which acts as a catalyst for the reduction of oxidized redox couples. Collecting charges from the materials used in hole transport is the CE's key function. Technically, platinum is the most recommended catalyst for producing effective devices because of its electrocatalytic activity for triiodide reduction and higher stability. Graphene, carbon black, carbon nanotubes, metal alloys, conducting polymers, and their composites are some examples of carbon cathodes that are being researched as alternatives to platinum because of their high cost. Different researchers have worked on all of the fundamental parts of the DSSC, and several enhancements have been proposed.

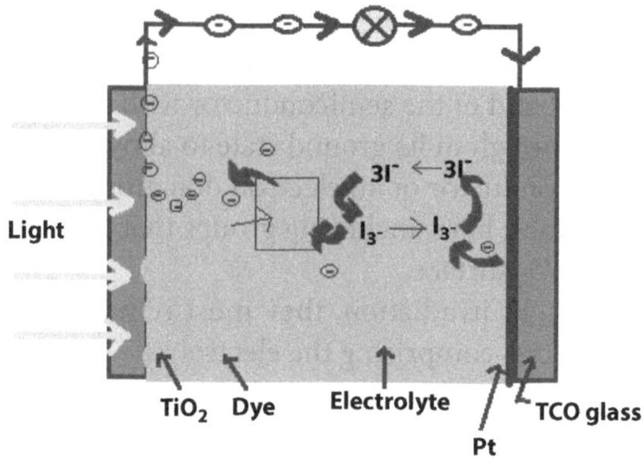

Figure 14.4 Working principle of DSSC.

14.7 Operating Principle of Dye-Sensitized Solar Cell

Its similarity to the mechanism of photosynthesis in plants leads to the comparison that it frequently bears with artificial photosynthesis. TCO substrates not only provide support for the components of the cell but also allow solar radiation to flow through solar materials, effectively collecting and transmitting produced electrons to the CE with little energy loss. Excitation of dye molecules by solar radiation results in the formation of excitons, or electron-hole pairs, as they transition from their ground state in the HOMO to their excited state in the LUMO. Thereafter, the excited dye oxidizes and holes are left in the oxidized dye as the liberated electrons diffuse to the conduction band of the semiconductor metal oxide (SMO)/dye link. Redox mediators—like the triiodide-based I/I3 redox pair, where I is oxidized to I3 by releasing electrons gathered by the oxidized dye—then return the oxidized dye to its neutral state. Finally, the oxidized mediators, acting as messenger ions, are transported to the CE and oxidized.

Excitable electrons from the dye molecules go through the TCO, SMO, and CE as a result of the sun's radiation-stimulating dye molecules. A redox pair in the electrolyte or hole transporter provides electrons that the oxidized dye absorbs, reducing it simultaneously. By gathering electrons from the CE, the oxidized redox pair is then restored. Making electricity is made easier by the movement of electrons from the TCO (SMO) to the TCO (CE).

14.8 Excitation Process

The photo-sensitized dye molecule atop the metal oxide semiconductor absorbs photons from sunlight, which excites the dye's electrons from the HOMO to the LUMO state and releases photogenerated electrons.

$$S + \text{Photon}\,(hv) \rightarrow S*$$

1. **Injection Process**
When an electron is introduced into a semiconductor nanostructured sheet with a large band gap, the dye molecule undergoes oxidation. Now that they are separated, the electrons remain within the conduction band and the holes continue to be in the oxidized dye.

$$S* + TiO_2 \rightarrow e^{-}(TiO_2) + S^{+}$$

2. **Energy Generation**
By diffusing through the TiO_2 layer and onto the TCO substrates, the electrons infused into the semiconductor conduction band finally flow into the external circuit, producing a current. In the process, electrons move between TiO_2 nanoparticles and are then collected and sent to a load, where they generate electrical energy and carry out tasks.

$$e^{-}(TiO_2) + C.E \rightarrow TiO_2 + e^{-}(C.E) + \text{electrical energy}$$

3. **Regeneration of Dye**
By accepting ground-state electrons through redox mediators—typically iodide/triiodide (I-/I3) in the electrolyte—the oxidized dye is returned to its neutral state. The CE is electrically connected to the CE. I-/I3-redox electrolytes are used as intermediates between the TiO_2 photoanodes and the CE, facilitating electron exchange. The I ion redox mediator must provide electrons to repair the oxidized dye molecules since it is oxidized to I3- (triiodide ions).

$$S^{+} + (3/2)I^{-} \rightarrow S + (1/2)I_3^{-}$$

4. **Electron Recapture Reaction**
By substituting an electron from the external load for the one supplied internally, the I3- reduces back to the I-ion. Dispersing charge-compensating cations at the junction between the photoanode and electrolyte improves the flow of electrons in the semiconductor in the conducting band of the photoanode. As a result, there are no long-term changes brought about by electricity production in DSSC.

$$(1/2)I_3^{-} + e^{-}_{(C.E)} \rightarrow (3/2)I^{-} + C.E$$

5. **Electron Recombination**
There is no discernible photocurrent, yet recombination with the titania layer electrons takes place in the lack of a redox mediator to rapidly intercept and decrease the oxidized dye (S+). Recombination is a process that is undesirable yet is constant.

$$S^+ + e^-(TiO_2) \rightarrow S \text{ (Recombination)}$$

14.9 Roll of Polysaccharides in Dye-Sensitized Solar Cells

The best efficiency solar cells are made of inorganic metals, such as Si, as compared to polymer and organic solar cells. However, these solar cells are exceedingly harmful to the environment, not renewable, not biodegradable, and are very expensive. As an alternative to Si solar cells, DSSCs are appropriate. Although DSSCs do not have the same efficiency as Si solar cells, they are nonetheless environmentally friendly, flexible, biodegradable, and reasonably priced. The study is ongoing, and over time, DSSC effectiveness increases.

In this study, we will examine the action of polysaccharides in solar cells and how they contribute to DSSC's increased efficiency. Long chains of monosaccharides joined by glycosidic bonds are known as polysaccharides. Polysaccharides are compounds that are present in many types of living things, including plants. The polysaccharides utilized in solar cells include cellulose, starch, chitosan, xanthan, agarose, amylopectin, and carboxymethyl cellulose, among others. These all play a significant part in DSSC, whether it is through the use of electrodes, electrode binders, electrolytes, or body designs. Recently, a lot of research has been conducted on organic solar cells, and the efficiency of power conversion has increased by a value ranging from 3.5% to 10.6%.In addition to their capacity to bind, polysaccharides also possess the qualities of viscosity, ionic conductivity, and thin film formation. Polysaccharides are among the several biopolymers most frequently employed for the DSSC's electrolyte production. We will look at each polysaccharide's function in a solar cell in turn, as well as how solar cells get more efficient.

14.9.1 Chitosan

The shells of crabs and prawns may be used to make chitosan, which is the deacetylated form of chitin. N-acetyl-D-glucosamine (2-acetamido-2-deoxy-D-gluconopyranose) is connected to other molecules in a linear chain to form chitin by a (1,4) linkage. With a highly concentrated sodium hydroxide solution, chitin is alkaline N-DE acetylated into chitosan at a high temperature and for a significant period of time. Chitosan's molecular formula is $C_6H_{11}O_4N$, or poly[(1,4)-2 amino-2 deoxy-D-glucopyranose] Figure 14.5 [10].

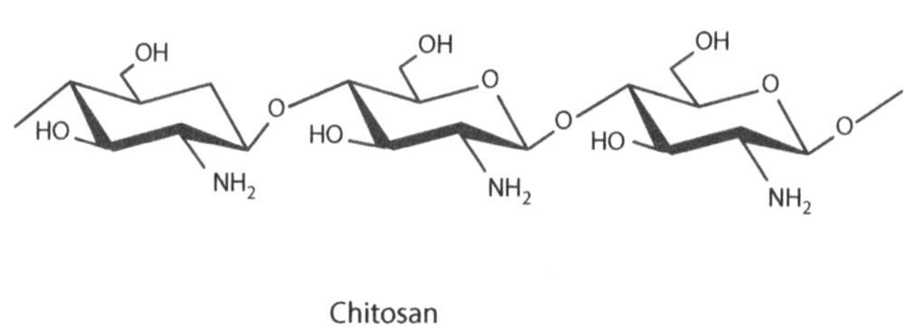

Figure 14.5 Structure of chitosan.

14.9.2 Preparation and Characterization of Chitosan-Based TiO_2 Electrode for Dye-Sensitized Solar Cells

Inexpensive and ecologically friendly, DSSCs have these qualities. This makes it a suitable substitute for the solar cells that are now based on Si semiconductors. Si solar cells with a 15–16% efficiency are employed in commercial applications. Because exceptionally pure crystalline minerals are required to make the solar cells, these panels are exceedingly costly. Environmental contamination is a byproduct of the harmful elements used in the manufacture of Si. Polycrystalline Si and other thin-film solar cells are relatively less costly, but they have a lower efficiency. Although less expensive, amorphous Si solar cells only last three to four years [9].

Numerous researchers are changing various parts of the DSSC to improve performance and longevity such as an electrode, an electrolyte, a change in the DSSC's construction material, etc. We will discuss the change to the electrode portion of the DSSC in the next section. The issue with mesoporous films containing TiO_2 nanoparticles was that less dye could be loaded into the pores if the film's thickness was reduced and its pore size increased. Low surface area is also a result of this. As a result, efficiency declines. To enhance a film's surface area, there are several methods. This method reduces the pore size and enhances porosity by tightly binding the TiO_2 nanoparticle to a binder material, such as chitosan. This leads to a greater dye loading because of the huge surface area and enhanced film thickness. The performance of the DSSC can be impacted by a variety of events, including transport property.

En Mei Jin's research suggests changing the light electrode to improve the effectiveness of DSSC. In this experiment, four distinct chitosan solution compositions were made, and they were combined with TiO_2 paste to create chitosan-based TiO_2 paste. This paste was molded onto FTO glass, and the DSSC was created.

Instrument:

- the use of field emission scanning electron microscopy (FE-SEM) to examine the morphology of TiO_2 film [11].
- Measures of electron transit and recombination times include intensity-modulated photocurrent spec. (IMPS) and intensity-modulated photo voltage spec [11].

By using FE-SEM, IMPS, and IMVS to analyze the surface structure of photo electrodes Figure 14.6 with four distinct compositions comprising 1.5wt %(D1), 2.0wt %(D2), 2.5wt %(D3), and 3.0wt %(D4), the following results are shown: Various pore sizes exist in the TiO_2 film. The D2-based TiO_2 film is thick and has the lowest pore diameter [10]. This led to an increase in porosity and surface area. More dye was loaded into the film as a result, along with an increased diffusion coefficient, high current density, and quick electron transit time. So, DSSC's efficiency went increased. The binder for electrodes is made of chitosan. With various chitosan binder solutions as the foundation for the photo electrode, DSSCs were created. Apart from their elevated diffusion coefficient and diffusion length, the D2-based DSSCs demonstrated a reduced electron recombination time and a faster electron transit time. Effectiveness is key.

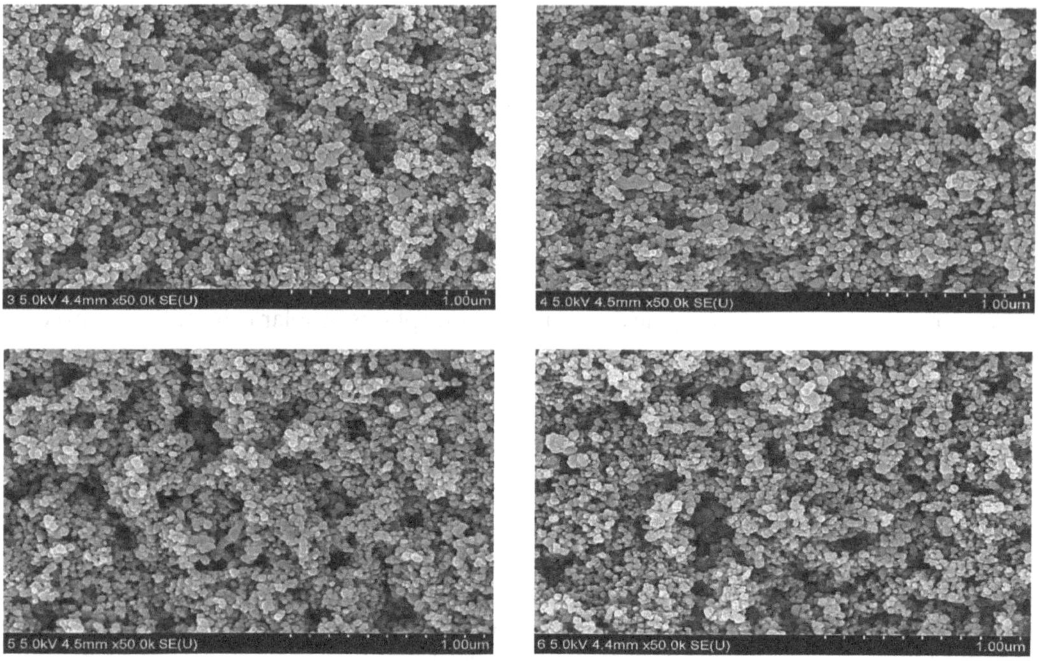

Figure 14.6 Field emission scanning electron images of the surface morphologies of chitosan-based TiO_2 films [11].

14.9.3 Cellulose

Cellulose, the most prevalent natural polymer on the planet, provides a resource that is biocompatible, renewable, economical, and ecologically benign. It is produced directly from natural plants like wood, cotton, and hemp as well as from microorganisms like bacteria, algae, and fungi. The most prevalent of these, wood-origin cellulose, is utilized most frequently as wood pulp. Chemically, cellulose is made up of repetitive linear chains of -D-glucopyranose units that are bonded together by glycosidic bonds at the 1,4 positions Figure 14.7. Numerous intra- and intermolecular hydrogen bonds occur, resulting in a variety of cellulose structural configurations [12].

CNF-Templated Mesoporous Structure as Solar Cell Electrode:
The solar cell's creation has revolutionized the energy-generating industry by providing a substantial substitute for conventional forms of energy like Si solar cells. Materials typically used in solar cell manufacturing, such as ceramics, metals, and plastics, pose environmental challenges. However, the utilization of cellulose nanomaterials presents a

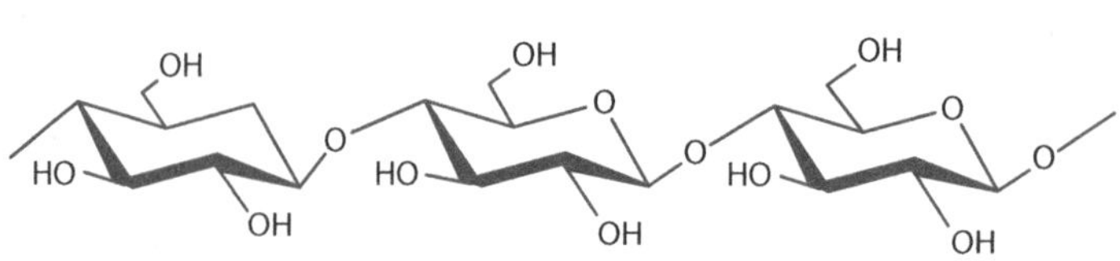

Figure 14.7 Structure of cellulose.

promising solution to this issue. Cellulose, characterized by numerous intra- and intermolecular hydrogen bonds, offers diverse configurations of structures. Nanoscale cellulose structures, such as nanocrystals or nanofibrils, have special qualities including high elasticity, low thermal expansion, complex surface chemistry, transparency, and anisotropy. Because of these characteristics, nanocellulose is perfect for use in mechanical energy harvesters, energy storage systems, PV devices, and catalytic components. Mesoporous structures based on nanocellulose, as well as flexible thin films, fibers, and networks, are being developed and are finding growing applications in many energy-related technologies. This shift toward nanocellulose materials holds promise for mitigating environmental impacts associated with conventional solar cell materials and advancing sustainable energy solutions [11]. The percentage of environmentally friendly components in such energy-related devices is greatly increased by the usage of nanocellulose. The capacity of cellulose to create thin films allows for modification of the DSSC photoanode. It can hold the particle in place by forming a hydrogen bond between the oxygen in the TiO_2 nanoparticle and the hydroxyl group of the cellulose. This positive interaction causes the film's pores to become smaller and have a higher surface area. This may facilitate quick and free ion transportation.

Cellulose nanofiber (CNF) network is employed as a scaffold for the construction of solar cell electrodes in this research project with the same name offered by Xudong Wang [11]. TiO_2 material that is both photoactive and conductive is applied to the CNF template. This creates a void-like nanofiber network with significant surface area and high porosity. CNFs' hydrophilic properties aid in the creation of a homogeneous coating with tens of nm in thickness. When it comes to electron transport, CNF-generated cross-linked TiO2 networks are more favorable than mesoporous TiO2 nanoparticle films [10]. CNF-Templated TiO2 nanofibers offer structural advantages for the transfer of charge and photon interaction. High photocurrent density and increased incident photon-to-current efficiency in the conversion are the outcomes of this, as indicated by the bigger surface area and improved dye molecule loading [10]. In contrast to DSSCs based on nanoparticles of the same thickness, the nanofiber-based devices exhibited a higher open circuit voltage but a somewhat lower photocurrent density. The 6 m-thick fibrous electrode's electron recombination lifespan was about three times longer than that of the light electrode based on nanoparticles [10]. In addition, TiO_2 hollow fibers with the same film thickness showed 2–3 times shorter electron collecting periods than TiO_2 nanoparticle networks [10]. Due to its ability to make thin, flexible films, cellulose nanomaterial is employed in solar devices. With a unique mix of chemical, structural, dielectric, and optical characteristics that are uncommon in other polymer systems, cellulose is a polymer with many special qualities and benefits. Thus, the application of cellulose nanomaterial as a useful component in contemporary energy gadgets has the potential to usher in a new era of material innovation.

14.9.4 Starch

(C6H10O5)n is the starch molecule's fundamental chemical formula. Glucose monomers linked together in alpha 1,4 links make up the polysaccharide known as starch Figure 14.8. The linear polymer amylose is the most basic kind of starch; the branching type is called amylopectin.

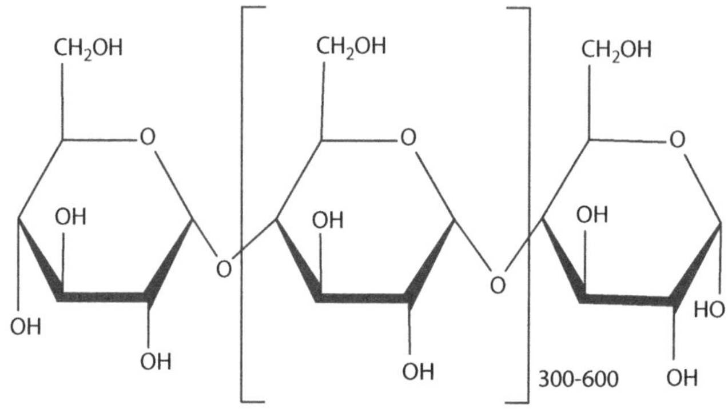

Figure 14.8 Structure of starch.

Effect of Cross-Linking on the Performance of Starch-Based Biopolymer as Gel Electrolyte for Dye-Sensitized Solar Cell:

Moving on to another area of the DSSC in this section will allow us to improve and adjust its functionality. Despite having advantages like high ion diffusion and strong ionic conductivity, liquid electrolytes have drawbacks including corrosion of electrodes and electrolyte leakage. These things influence the DSSC's effectiveness. Therefore, some gelling substance is needed, such as starch or gellan gum. Gel polymer electrolyte (GPE) has the advantage of solving all the problems mentioned previously. It possesses both the properties of solid-like ionic conductivity and liquid-like viscosity.

The liquid electrolyte in this work is changed into a GPE. Starch is an accessible biopolymer. When utilizing a starch-based biopolymer electrolyte, there is a risk of producing a starch-iodine complex because iodide/triiodide is the redox pair that is typically used in DSSCs. When iodine particles are incorporated within the cage-like structure of amylopectin and starch and I2 interact in a condition of ambient moisture, a starch-iodine complex can develop. This compound impedes the electrolyte's capacity to contain free iodine, which has a negative impact on the recombination process and electrolyte conductivity. To circumvent the formation of the starch-iodine complex, a cross-linking approach is employed. Cross-linking restricts iodine molecules from entering the starch polymer chain, thereby making iodine accessible for ionic transport. For cross-linking, three distinct techniques are used. Citric acid is used as an addition because of its multi-carboxylic structure, which enables it to connect to the hydroxyl groups in starch.

In the study conducted by Pavithra Nagaraj and Asija Sasidharan, a novel biopolymer electrolyte system was developed. In this system, raw potato starches that have been cross-linked with citric acid are combined with an iodide/iodine redox pair Figure 14.9. Glycerol is utilized as the plasticizing agent in the mixture [12]. To examine the structural changes in cross-linked starch employed as electrolytes, Fourier transform infrared spectroscopy, or FT-IR, was done. We looked at how cross-linking affected the crystallinity of starch using X-ray diffraction spectroscopy (XRD) and different scanning calorimetry (DSC) methods. The cross-linked polymer gel electrolytes' conductivity was determined by electrochemical impedance spectroscopy (EIS), and PV measurements were then done for DSSCs that were manufactured with the electrolytes utilized. Citric acid and starch interact to generate cross-links, as demonstrated by the process in question. Citric acid will create ester groups

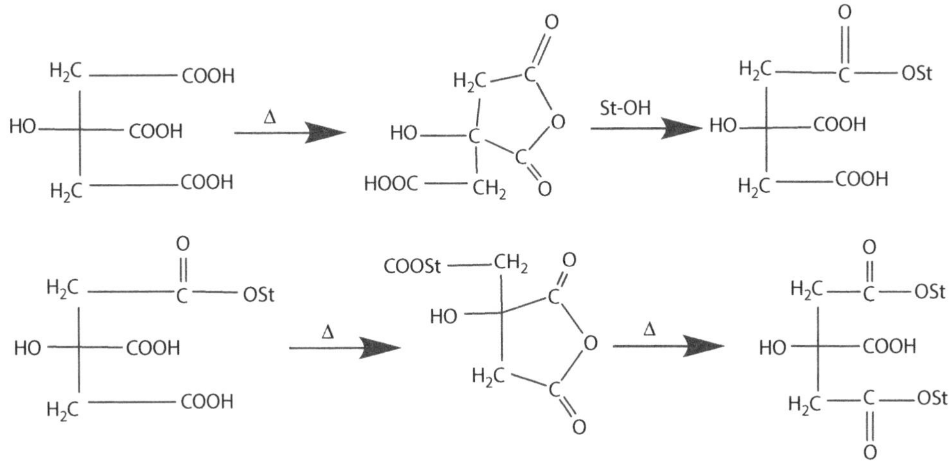

Figure 14.9 Reaction of citric acid with starch [13].

with the -OH groups on starch molecules, which will prevent the iodine molecule from passing through. Citric acid will also enter the polymer chain during cross-linking [13]. The starch-iodine combination is so prevented from forming. Along with making the starch less permeable to water, cross-linking also makes the electrolyte more stable [13].

The three primary parameters that define a DSSC's economy are its fill factor (FF), short circuit current (Jsc), and open circuit voltage (Voc). Jsc is primarily dependent on ionic mobility in the electrolyte, FF is determined by the uniformity and pore-filling capacity of the electrolyte layer, and Voc is dependent on electron lifetime and recombine resistance in the TiO_2 conduction band. With an increase in starch content, the electron lifespan lengthens [11]. When compared to DSSCs using gel electrolytes, liquid electrolytes showed better efficiency. In comparison to liquid electrolytes, GPEs demonstrated higher photocurrent conversion efficiency and Voc. This could be the result of the cross-linked starch inhibiting the charge recombination reaction [13]. Cross-linking starch polymer is employed as an electrolyte in gels. According to X-ray diffraction, cross-linking techniques have an impact on the amount and quality of starch's crystalline structure. Cross-linking also increased the electron lifespan, recombination resistance, and photo-conversion efficiency of DSSC.

14.9.5 Xanthan

A characteristic microbial polysaccharide called xanthan is generated by particular bacterial species like Xanthomonas campestris [14].

Hydrogel Electrolyte Based on Xanthan Gum: Green Route Toward Stable Dye-Sensitized Solar Cells:

GPE is used in place of liquid electrolytes because it has fewer drawbacks than liquid electrolytes, including leakage, vaporization, and electrode corrosion. It is difficult to manufacture an electrolyte that should be harmless and nonflammable. Polymers like xanthan can be used to address this problem. It bonds well with water. This quality aids in the formation of hydrogel

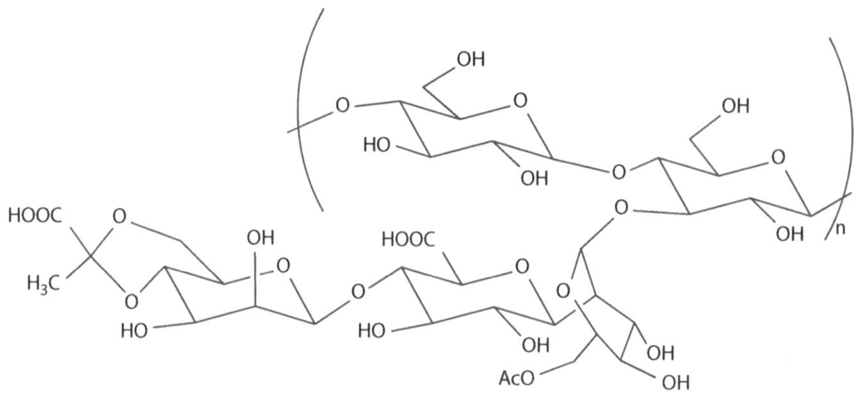

Figure 14.10 Structure of xanthan.

electrolytes. GPE has an issue with its capacity to fill pores. The high viscosity of GPE makes it more difficult for the material to flow than the liquid electrolyte, which causes it to have problems filling the pore space of TiO_2 nanoparticles. A cheap, water-soluble polysaccharide known as xanthan gum (XG) is produced by the Xanthomonas campestris bacterium and it has a pentasaccharide repeating unit as its main structural component, which is made up of glucose, mannose, and glucuronic acids Figure 14.10 [17]. This structure may be formed into a three-dimensional hydrogel network with thixotropic properties [17]. Thixotropic is a characteristic of xanthan. It indicates that it possesses the ability to behave like a liquid electrolyte. It behaves more like a liquid electrolyte and loses some of its viscosity when it is shacked, which increases its capacity to fill pores. This characteristic should enable the XG-based hydrogel to efficiently enter the TiO_2 photoanode's mesopores as a liquid, resolving the pore-filling problems associated with conventional more viscous polymer-based electrolytes.

In contrast to the general trend, the performance of xanthan consisting of GPEs actually improves over time. With time, hydrogel electrolyte's effectiveness grew [15, 16]. GPEs have a lower efficiency than liquid electrolytes, although, with time, their efficiency has grown. In fact, a jellified electrolyte kept the XG form from "freezing" and preventing leakage from the cell [13]. Environmental benefits also come from the hydrogel-forming capability. The most noteworthy discovery of this work is the substantial increase in stability that can be obtained by gelling the electrolyte with XG polymer. This suggests that XG polymer is a viable option for a low-cost, bio-derived jellifying agent for DSSCs. This eliminates the need for oil-derived polymers, critical raw materials (CRMs), and volatile and hazardous organic solvents, offering an economical and ecologically beneficial substitute. The research effectively suggested hydrogel electrolytes that are 100% water-based and derived from biosourced XG. Through electrochemical investigation, it was observed that the efficiency of DSSCs containing the hydrogel improved over time.

14.9.6 Carboxy Methyl Cellulose

CMC gum is made from cellulose by substituting carboxymethyl groups for part of the hydroxyl groups in the glucopyranose monomer, which makes up the cellulose backbone Figure 14.11. The sodium version of CMC that is often used is sodium carboxymethyl cellulose.

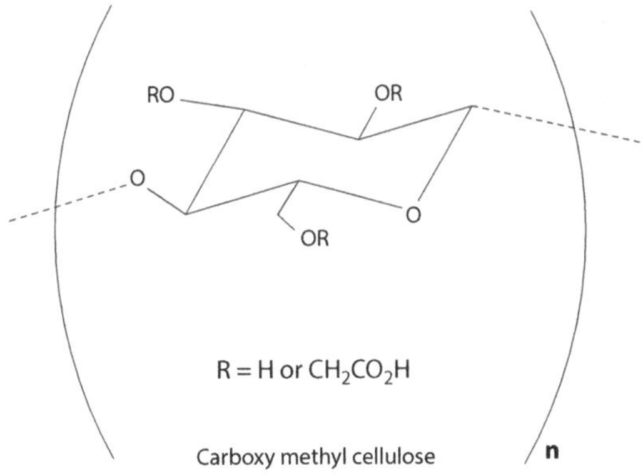

Figure 14.11 Carboxy methyl cellulose.

Electrical Properties of Plasticized Sodium-Carboxy Methylcellulose (NaCMC)-Based Polysulfide Solid Polymer Electrolyte:

Sensitized solar cells have surpassed Si-based solar cells because of their consistently superior performance to manufacturing cost ratio. Low-cost and widely accessible materials are also employed in low-cost sensitized solar cells. Sensitized solar cells therefore provide hope for the future of energy generation. Once more, the issue with solid polymer electrolyte (SPE) will be covered in this section. Numerous factors in this situation have an impact on the DSSC's performance. The first thing to take into account is the SPE's crystalline structure. The ions are grouped in a crystalline structure in a constant and predictable manner. Ion mobility is unaffected by this. The ion-to-ion attraction force is another element. Ions are bound together by the electrostatic force of attraction in crystalline form. It also results in ions being less mobile. Thus, to enhance the SPE's performance, the crystalline component must be eliminated. By dissolving the regular, established arrangement of the ions and lowering the columbic force, this can be accomplished. By increasing the amorphous nature of the electrolyte, in essence. The problems have been studied extensively with SPEs. Many polymers, such as poly (ethylene oxide), poly (methyl methacrylate), and poly (vinyl alcohol), have been studied by researchers [18]. Sodium-Carboxymethylcellulose (NaCMC), an abundantly accessible and ecologically benign cellulose derivative, was employed as the host polymer in this study suggested by N.N.S. Baharun and four others. Polymer electrolytes based on NaCMC have demonstrated strong electrolyte performance [19]. Chemical additives can be added to polymer electrolytes to plasticize them and enhance their performance. Ethyl carbonate (EC), propylene carbonate (PC), dimethyl carbonate (DMC), diethyl carbonate (DEC), and many other plasticizers can increase the activeness of the polymer electrolyte. In order to improve the amorphousness of the thin film, ethylene carbonate (EC) is added to NaCMC in this study [19]. The electrochemical analysis has been completed with positive results. Beyond 40 wt% EC, the conductivity decreased [19]. It may be inferred that the jointed plasticizer's viscosity and ionic conductivity are strongly connected. The molecular weight of EC is low (88.06 g mol-1), its dielectric constant value is high (= 89.6), and it has a high donor count [20]. Ion mobility can be increased by adding EC since it can reduce the local viscosity around the ions that carry charges. Furthermore, salt dissociation may result from EC's high dielectric constant qualities, which may lessen the Columbic force between the cations and anions of the salts. In this instance, the polysulfide ions in the SPE were more numerous due to the EC's involvement in the NaCMC film because the EC's functional group was used to complexate the polysulfide ions. Ionic conductivity increases as a result of this. However, EC concentrations over 40% weighted produced ion agglomeration, which lowers the density of moving ions. Therefore, with greater EC concentration, the conductivity dropped. An efficient method of producing a plasticized polysulfide polymer electrolyte was solution casting. The ionic conductivity of NaCMC-based polysulfide film was enhanced by adding ethylene carbonate (EC), which increased the amorphous character of the ions and resulted in high ion mobility.

14.9.7 Carrageenan

Carrageenan is a red seaweed (Rhodophyceae)-derived anionic sulfated linear polysaccharide with a linear backbone made of alternating (1 3)- and (1 4)-linked (1 D-galactopyranose links Figure 14.12 [12].

A Novel Biopolymer Gel Electrolyte System for DSSC Applications:
Carrageenan is one of the biopolymers that has been widely utilized in many commercial applications, such as gelling, thickening, and stabilizing agents, particularly in food goods [21]. Additionally, carrageenan is utilized in industrial applications, medicines, and medical treatments in addition to cosmetics [22]. Due to solvent leakage and evaporation, using liquid electrolytes makes it difficult for DSSC to be commercialized and maintain long-term stability. The inclusion of a gel-polymer electrolyte system, which allows it to capture the solvent in its polymer form, is one of the problems that may be solved. Special polysaccharides, a class of naturally occurring polymers, were investigated for DSSC applications because of their ability to gel. Agarose, chitosan, and carrageenan are only a few of the few polysaccharides that have been employed as polymer electrolytes in DSSC applications, with efficiency levels as high as 7%. A specific type of red seaweed known as carrageenan is a linear sulfated polysaccharide. Due to the development of double helical structures and subsequent cation-specific aggregation, this polysaccharide gels at room temperature and is stable. A. Ahmad's study involved the utilization of a new I-/I3- redox couple-containing gel biopolymer electrolyte for DSSC applications, which was made utilizing -carrageenan in dimethyl sulfoxide (DMSO), an eco-friendly solvent [23]. Hot-melt material that was 25 m thick was sandwiched between the electrodes to create a DSSC. Pre-drilled openings in the counter-electrode allowed the hot carrageenan electrolyte to be introduced in between. By employing IV measurements and impedance spectroscopy, the DSSCs were characterized. The addition of carrageenan resulted in an increase and a decrease in the open-circuit

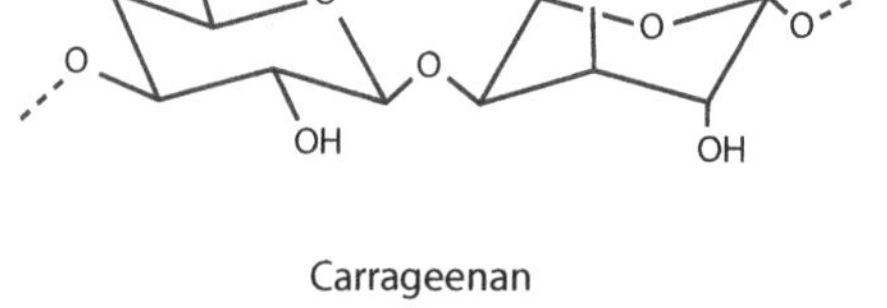

Figure 14.12 Structure of carrageenan.

voltage and short-circuit current of the generated DSSCs [23]. One possible explanation for the overall improvement of DSSCs based on carrageenan polymer electrolyte might be the increase in short-circuit current. The development of a unique biopolymer gel-electrolyte system using -carrageenan and a redox pair in dimethyl sulfoxide. Due to enhanced electron recombination, the addition of -carrageenan increased the DSSCs' short-circuit current but lowered their open-circuit voltage.

14.9.8 Alginate

Brown algae have a large amount of alginic acid, commonly known as algin, which is a hydrophilic polymer that when hydrated produces a sticky gum. Alginates Figure 14.13 are the names given to the salts of metals like sodium and calcium.

Photoactive Titanium Dioxide–Doped Sodium Alginate Film for Dye-Sensitized Solar Cell Application:

Because they are cheaper, more flexible, renewable, and environmentally benign than inorganic solar cells, polymer solar cells are now preferred over them. Its efficiency is just slightly less than that of an inorganic solar cell. It is, however, constantly updated. Low-cost PV solar cells are made possible by the liquid electrolyte, which interacts directly with the I/I3 redox pair and allows for flawless dye regeneration, resulting in very high solar-to-electrical conversion efficiencies (7–11%) [24]. The stability and long-term function of the cell are affected by solvent evaporation or leakage, though. To enable commercial use of these devices, the liquid electrolyte must be replaced with a solid charge-transport material that provides hermetic closure and stability, lowers design constraints, and gives the cell a variety of shape options.

A nanocomposite consisting of TiO_2, natural dye (N-Dye), and sodium alginate (Na-Alg) was thus created and characterized by Uddin M. J. and three other authors in order to address the problems associated with polymer electrolytes. This nanocomposite was employed as a solid electrolyte based on polymers. In this work, N-Dye and TiO_2 were chosen as the donor and acceptor ligands, respectively, with the intention of improving the

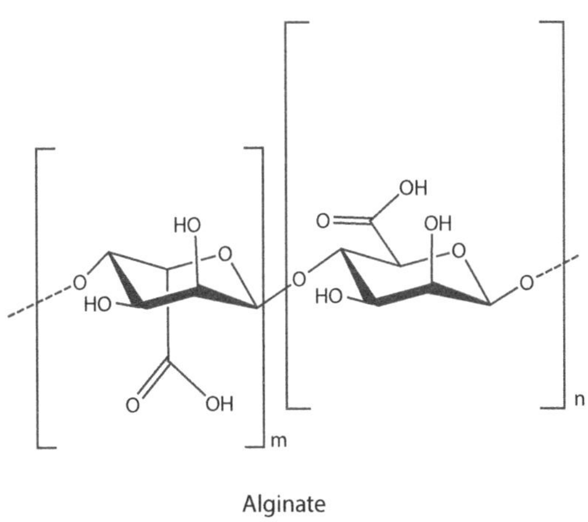

Figure 14.13 Structure of alginate.

solid electrolyte's stability and performance in preparation for its potential application in DSSCs [25]. The primary objective of this study is to investigate the electrical, optical, and physio-mechanical properties of a nanocomposite of Na, Al, and TiO_2 using N-Dye. After being prepared, a thin coating of N-dye containing Na-Alg/TiO_2 was cast onto the glass. Results from analyses of all glass films with various TiO_2 compositions were obtained.

Temperature Dependence of Conductivity:
The temperature climbed from the morning till midday as the intensity of the sun grew. Consequently, conductivity also rose. We can comprehend how distinct wt% Na-Alg/TiO_2 nanocomposites that include N-dye behave in terms of conductivity. The graph clearly shows that conductivity increases fast at low temperatures while the temperature rises. Later, the temperature rose much further. When temperatures are high, the curve becomes linear [25]. The nanocomposite with 8%wt of TiO_2 has the highest conductivity of all TiO_2 weight percentages. Charge carriers are moving and in greater quantities as a result of this. As the temperature increased, the charge carrier bonded both the polymer and the TiO_2 particle broke free.

Mechanical Properties:
Tensile strength increased in proportion to the weight percentage of TiO_2 content. The main process involves the creation of a hydrogen connection between the hydroxyl and carbonyl groups in TiO_2 and alginate [25].

14.9.9 Gellan Gum

An anionic polymer that is soluble in water, gellan gum is produced by the bacteria Sphingomonas elodea. When first discovered, it was thought to be a replacement for a gelling agent. A tetrasaccharide makes up the repeating unit of the polymer. It contains one residue of each L-rhamnose and D-glucuronic acid, as well as two residues of D-glucose Figure 14.14.

Solid Gellan Gum Polymer Electrolyte for Energy Application:
Polysaccharides are crucial in creating flexible, affordable, and renewable solar cells. Gellan gum is the final polymer in the sequence of rolls of polysaccharides in solar cells. Similar to other polysaccharide gellan gum that was utilized to change the DSSC's electrolyte

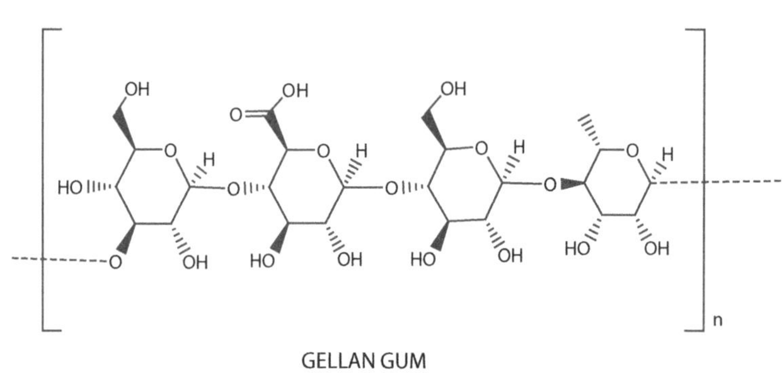

Figure 14.14 Structure of gellan gum.

component. SPE may be utilized to solve the flaws in liquid electrolytes, such as leakage, evaporation, corrosion of the electrode, and others. Compared to liquid electrolytes, SPEs provide a number of benefits, including ideal form, no leaks, mechanical strength, and design freedom. As a result, we are free to choose any mechanical design. In addition to its benefits, it has certain drawbacks, such as a low ionic conductivity. SPEs' ionic conductivity is influenced by [26]:

- Crystalline nature of material
- Cation and anion motion
- Ion pair formation

GPEs can be used to resolve each of these issues. Both the characteristics of a solid and a liquid electrolyte are present. In this section, we will learn about an entirely novel sort of bio polymer electrolyte system that has the conductivity of a solid and the viscosity of a liquid. The gel biopolymer electrolyte containing potassium iodide-doped phytagel biopolymer was created by Rahul Singh and B. Bhattacharya's research [26]. One glucuronic acid, one rhamnose, and two glucose units were used as the bacterial substrate in the production of the chemical phytogetagen. Doping of potassium iodide (KI) in SPE also improves the same features, in a similar way to how doping in semiconductors boosts electrical, optical, and structural characteristics. In addition, the quantity of ions (mobile ions) rises when KI is added. However, there should be a cap on how much KI may be added; otherwise, ion aggregation (grouping) from high ion concentration results in reduced ionic mobility.

Impedance Spectroscopy:
Since there are more mobile charge carriers, ionic conductivity increases as KI concentration climbs. Additionally, beyond a certain point, continued growth in KI results in a reduction in ionic conductivity. This may be described as the multiple ion creation process (aggregation). The reduced dissociation is what causes the lowest conductivity. Decreased mobility can be attributed to either increased ion association or triplet ion formation [27].

Polarized Microscopy:
It illustrates how the Phytagel/gellan gum and KI-doped biopolymer film's amorphous structure is encouraged by the uniform distribution of salt (KI) in the POM image. Doping with KI led to an increase in the film's amorphous region. In the biopolymer-salt combination, this amorphous region facilitates the movement of ions. Higher KI concentrations led to an increase in the charge carrier number of (K +ve and I-ve) ions, which raised the conductivity values [28]. A solid gel electrolyte doped with KI and based on gellan gum/phytagel has been produced and characterized using various techniques. This demonstrates that doping KI with more charge carriers (cations/anions) increases overall conductivity whereas charge pair creation processes cause conductivity to decrease.

14.10 Results and Discussion

The third generation of solar cells, known as DSSC, are made with low-cost, environmentally acceptable raw ingredients and by using straightforward, low-cost production techniques.

The device's active component, which regulates the flow of charges between electrodes, is still crucial to the overall efficiency of the cell. Since liquid electrolytes have high energy conversion efficiency, they have been the most frequently employed electrolytes in DSSCs to date. The electrolyte and electrodes have high charge transfer resistance and low ionic mobility, which lowers the cell's efficiency in converting electrical energy into chemical energy. Additionally, the fact that the majority of polymer electrolytes are not biodegradable has raised environmental problems. The effective introduction of biopolymers as an affordable, ecologically friendly, and clean substitute electrolyte for DSSCs is now taking place [29]. The exponential rise in the total number of research publications published in a given year suggests that interest in biopolymer electrolytes has increased significantly since they were effectively used in DSSCs. A biopolymer-based DSSC has so far been reported to have the greatest efficiency, 9.61% [30]. It more nearly mimics the DSSC (10%) based on synthetic polymers [31, 32]. The efficiency of biopolymer-based DSSCs is not as high as that of other solar cells on the marketplace, such as Si-based solar cells (26%) [33], perovskite solar cells (25.2%) [34], fluid-electrolyte solar cells (14.2%) [35], and ruthenium dye-based solar cells (14.3%) [36]. This is because the technology surrounding biopolymer-based DSSCs is still in its early stages. Yet, studies on biopolymer electrolytes are promising and might result in flexible manufacturing, long-term stability, low cost, high efficiency, and ecological sustainability [37].

14.11 Future Prospects

The utilization of biopolymers presents a promising, eco-friendly approach for manufacturing DSSCs. In recent years, there has been a growing focus on developing biopolymer-based electrolytes, encompassing their design, effectiveness, and exploration of suitable biopolymer candidates. Table 14.1 shows factors affecting performance of biopolymer-based DSSC and advantages of biopolymer-based electrolyte. Nonetheless, there are still issues to be

Table 14.1 Collective information about factors affecting performance of biopolymer-based DSSC and advantages of biopolymer-based electrolyte.

Factors affecting the role of the biopolymer-based DSSC	**Benefits of electrolytes based on biopolymers over traditional electrolytes**
Biopolymer structure, blending, and modification	Modifiable chemical structure
Physical properties of biopolymer	Long-term stable photovoltaic performance
Plasticization	Economical
Solvent selection	Sustainable
Iodide salt types	Mechanical flexibility
Salt concentration	Low-cost fabrication
Ionic liquids	Environment friendly
Nanoparticles	Simple

resolved, namely with regard to mass manufacturing, stability, conductivity restrictions, and efficiency in comparison to traditional solar cell technologies. Biopolymers offer unique chemical structures that make them suitable for electrolyte production in solar energy harvesting systems. The abundance of hydroxyl groups along the polymer chain backbones provides ample opportunities for modification, thereby enhancing the electrical conductivity of these materials. Various modification techniques such as grafting, crosslinking, or nanoparticle incorporation can be employed to tailor their properties. For instance, without altering the polymer chain structure, simple modification techniques can include highly conductive functional groups like acyl, carboxylic, or phthaloyl groups to increase conductivity and performance. However, it's worth noting that modified biopolymer-based electrolytes may exhibit reduced long-term stability compared to unmodified counterparts. Thus, further research is necessary to address issues related to longevity, compatibility, and stability in comparison to other electrolyte components.

14.12 Conclusion

The third generation of solar cells, known as DSSC, is made with low-cost, environmentally acceptable raw ingredients and by using straightforward, low-cost production techniques. Still essential to the cell's overall effectiveness is the active part of the apparatus, which controls the passage of charges between electrodes. Liquid electrolytes are the most often used electrolytes in DSSCs to date because of their high energy conversion efficiency. However, the design and endurance of the cells are limited by the need for good sealing to avoid leaks and the evaporation of organic solvents. Additionally, they are in charge of the CE's corrosion brought on by the electrolyte. A major step toward overcoming crucial concerns about the stability and operation of these solar energy conversion devices is the inclusion of DSSCs. Polysaccharides such as carrageenan, chitosan, cellulose, starch, xanthan, carboxymethyl cellulose, alginate, and gellan gum provide a variety of advantages in the context of DSSC technology, including increased performance, stability, and environmental benefits. These solar cells might be made more effective, stable, and ecologically friendly by incorporating polysaccharides into them, which presents a promising way to get around persistent problems in the sector. Future breakthroughs in solar energy technology might be sparked by continued research and development in this field, which would also make it easier for people to use it more widely. A detailed evaluation of the early research leads to the prediction that, due to their unique properties, biopolymer electrolytes would dominate the electrolyte use in DSSCs in the next years.

Acknowledgment

The authors thank to Guru Ghasidas Vishwavidyalaya Central University, Bilaspur, CG-495009, India. The authors also thank Dr. J N College, P.O.: Rasalpur, Balasore-756021, Odisha, India and Berhampur Degree College, Raj Berhampur, Balasore-756058, Odisha, India for providing the necessary facilities.

References

1. Grimes, C.A., Varghese, O.K., Ranjan, S., Hydrogen: The Solar Generation of Hydrogen by Water Photo electrolysis. *Springer Sci. Bus. Media*, 2007. https://doi.org/10.1007/978-0-387-68238-9.
2. Yu, K. and Chen, J., Enhancing solar cell efficiencies through1-D nanostructure. *Nanoscale Res. Letter*, 4, 1–10, 2009. DOI 10.1007/s11671-008-9200-y.
3. Conibeer, G., Third generation photovoltaics. *Mater. Today*, 10, 42–50 2007. https://doi.org/10.1016/S1369-7021(07)70278-X.
4. Gong, J., Sumathy, K., Qiao, Q., Zhou, Z., Review on dye-sensitized solar cells (DSSCs): Advanced techniques and research trends. *Renew. Sustain. Energy Rev.*, 68, 234–246, 2017. http://dx.doi.org/10.1016/j.rser.2016.09.097.
5. Konwar, S., Singh, D., Strzalkowski, K., Masri, M.N.B., Yahya, M.Z.A., Markus, D., Savilov, S.V., Singh, P.K., Stable and efficient dye-sensitized solar cells and supercapacitors developed using ionic liquid-doped biopolymer electrolytes. *Molecules*, 28, 5099, 2023. https://doi.org/10.3390/molecules28135099.
6. Ghazali, N.M. and Samsudin, A.S., Progress on biopolymer as an application in electrolytes system: A review study. *Mater. Today: Proc.*, 49, 3668–3678 2022. https://doi.org/10.1016/j.matpr.2021.09.473.
7. Kalyanasundaram, K., Photochemical and photoelectrochemical approaches to energy conversion. *Fundam. Sci.*, 2010. 978-2-940222-36-0.
8. O'Regan, B. and Grätzel, M., A low-cost, high-efficiency solar cell based on dye-sensitized colloidal TiO_2 films. *Nature*, 353, 737–740, 1991. https://doi.org/10.1038/353737a0.
9. Sugathan, V., John, E., Sudhakar, K., Recent improvements in dye sensitized solar cells. A review. *Renew. Sustain. Energy Rev.*, 52, 54–64, 2015. http://dx.doi.org/10.1016/j.rser.2015.07.076.
10. Kabir, E., Kumar, P., Kumar, S., Adelodun, A.A., Kim, K.H., Solar energy: Potential and future prospects. *Renew. Sustain. Energy Rev.*, 82, 894–900, 2018. https://doi.org/10.1016/j.rser.2017.09.094
11. Jin, E.M., Park, K.H., Yun, J.J., Hong, C.K., Hwang, M.J., Park, B.K., Kim, K.W., Gu, H.A., Photovoltaic properties of TiO_2 photoeletrode prepared by using liquid PEG-EEM binder. *Surf. Rev. Lett.*, 17, 52–20, 2010. https://doi.org/10.1142/S0218625X10013576.
12. Li, Q., Wang, S., Jin, X., Huang, C., Xiang, Z., The application of polysaccharides and their derivatives in pigment, barrier, and functional paper coatings. *Polymers*, 2020. https://doi.org/10.3390/polym12081837.
13. Zia, K.M., Tabasum, S., Masif, M., Sultan, N., Aslam, N., Noreen, A., Zuber, M., A review on synthesis, properties and applications of natural polymer based carrageenan blends and composites. *Int. J. Biol. Macromol.*, 2016. http://dx.doi.org/doi:10.1016/j.ijbiomac.2016.11.095.
14. Wang, X., Yao, C., Wang, F., Li, Z., Cellulose-based nanomaterials for energy applications. *Nano Micro Small*, 2017. https://doi.org/10.1002/smll.201702240.
15. Riaz, T., Iqbal, M.W., Jiang, B., Chen, J., A review of the enzymatic, physical, and, chemical modification techniques of xanthan gum. *Int. J. Biol. Macromol.*, 2021. https://doi.org/10.1016/j.ijbiomac.2021.06.196.
16. Nagaraj, P., Sasidharan, A., David, V., Sambandam, A., Effect of cross-linking on the performances of starch-based bioplolymer as gel electrolyte for dye-sensitized solar cell applications. *Polymers*, 9, 667, 2017. https://doi.org/10.3390/polym9120667.
17. Maa, X., Chang, P.R., Yu, J., Stumborg, M., Properties of biodegradable citric acid-modified granular starch/thermoplastic pea starch composites. *Carbohydr. Polym.* 75, 1–8, 2009. https://doi.org/10.1016/j.carbpol.2008.05.020.
18. Aitola, K., Domanski, K., Baena, J.P.C., Sveinbjornsson, K., Saliba, M., Abate, A., Gratzel, M., Kauppinen, E., Johansson, E.M.J., Tress, W., Hagfeldt, A., Boschloo, G., High tempearature-stable

perovskite solar cell based on low-cost carbon nanotube hole contact. *Adv. Mater.*, 29, 1606398 2017. https://doi.org/10.1002/adma.201606398.

19. Cho, W., Kim, Y.R., Song, D., Choi, H.W., Kang, Y.S., High-efficiency solid-state polymer electrolyte dye-sensitized solar cells with a bi-functional porous layer. *J. Mater. Chem. A*, 2, 17746–17750 2014. https://doi.org/10.1039/C4TA04064C.
20. Baharun, N.N.S., Mingsukang, M.A., Buraidah, M.H., Woo, H.J., Arof, A.K., Electrical properties of plasticized sodium-carboxymethylcellulose (NaCMC) based polysulfide solid polymer electrolyte. *Conferences*, 2018. https://doi.org/10.1109/ICTON.2018.8473830.
21. Saha, D. and Bhattacharya, S., Hydrocolloids as thickening and gelling agents in food: a critical review. *J. Food Sci. Technol.*, 47, 587–597 2010. https://doi.org/10.1007/s13197-010-0162-6.
22. Popa, E.G., Gomes, M.E., Reis, R.L., Cell delivery systems using alginate–carrageenan hydrogel beads and fibers for regenerative medicine applications. *Biomacromolecules*, 12, 3952–3961 2011. https://doi.org/10.1021/bm200965x.
23. Jafirin, S., Ahmad, I., Ahmad, A., Carboxymethyl cellulose from kenaf reinforced composite polymer electrolytes based 49% poly(methyl methacrylate)-grafted natural rubber. *Malays. J. Anal. Sci.*, 18, 376–384, 2014.
24. Barbé, C.J., Arendse, F., Compte, P., Jirousek, M., Lenzmann, F., Shklover, V., Grätzel, M., Nanocrystalline titanium oxide electrodes for photovoltaic applications. *J. Am. Ceram. Soc.*, 80, 3157–3171, 2005. https://doi.org/10.1111/j.1151-2916.1997.tb03245.x.
25. Uddin, M.J., Islam, J.M.M., Rahman, M.A., Khan, M.A., Development of photoactive titanium dioxide doped sodium alginate film for dye sensitized solar cell application. *Int. J. Thin Film Sci. Technol.*, 6, 135–138, 2017. http://dx.doi.org/10.18576/ijtfst/060307.
26. Singh, R., Bhattacharya, B., Rhee, H.W., Singh, P.K., New biodegradable polymer electrolyte for dye sensitized solar cell. *Int. J. Electrochem. Sci.*, 9, 2620–2630, 2014. https://doi.org/10.1016/S1452-3981(23)07952-X.
27. Bhattacharya, B., Lee, J.Y., Geng, J., Jung, H.T., Park, J.K., Effect of cation size on solid polymer electrolyte based dye-sensitized solar cells. *Langmuir*, 25, 3276–3281, 2009. https://doi.org/10.1021/la8029177.
28. Ramesh, S., Shanti, R., Morris, E., Studies on the plasticization efficiency of deep eutectic solvent in suppressing the crystallinity of corn starch based polymer electrolytes. *Carbohydr. Polym.*, 87, 701–706, 2012. https://doi.org/10.1016/j.carbpol.2011.08.047.
29. Hasan, M.M., Islam, M.D., Rashid, T.U., Biopolymer-based electrolytes for dye-sensitized solar cells: A critical review. *Energy Fuel*, 34, 15634–15671, 2020. https://doi.org/10.1021/acs.energyfuels.0c03396.
30. Buraidah, M.H., Shah, S., Teo, L.P., Chowdhury, F., II, Careem, M.A., Albinsson, I., Mellander, B.E., Arof, A.K., High efficient dye sensitized solar cells using phthaloylchitosan based gel polymer electrolytes. *Electrochim. Acta*, 245, 846–853, 2017. https://doi.org/10.1016/j.electacta.2017.06.011.
31. Freitas, J.N.D., Nogueira, A.F., Paoli, M.A.D., New insights into dye-sensitized solar cells with polymer electrolytes. *J. Mater. Chem.*, 19, 5279–5294, 2009. https://doi.org/10.1039/B900928K.
32. Aghazada, S. and Nazeeruddin, M.K., Ruthenium complexes as sensitizers in dye-sensitized solar cells. *Inorganics.*, 6, 2018. http://dx.doi.org/10.3390/inorganics6020052.
33. Yoshikawa, K., Kawasaki, H., Yoshida, W., Irie, T., Konishi, K., Nakano, K., Uto, T., Adachi, D., Kanematsu, M., Uzu, H., Yamamoto, K., Silicon heterojumction solar cell with interdigitated back contacts for a photoconversion efficiency over 26%. *Nat. Energy*, 2, 17032, 2017. https://doi.org/10.1038/nenergy.2017.32.
34. Yang, X., Wang, W., Ran, R., Zhou, W., Shao, Z., Recent advances in $Cs_2AgBiBr_6$-based halide double perovskites as lead-free and inorganic light absorbers for perovskite solar cells. *Energy Fuels*, 34, 10513–10528, 2020. https://doi.org/10.1021/acs.energyfuels.0c02236.

35. Ji, J.M., Zhou, H., Eom, Y.K., Kim, H.K., 14.2% Efficiency dye-sensitized solar cells by co-sensitizing novel Thieno[3,2-b]indole-based organic dyes with a promising porphyrin sensitizer. *Adv. Energy Mater.*, 10, 2000124, 2020. https://doi.org/10.1002/aenm.202000124.
36. Tomar, N., Agrawal, A., Dhaka, V.S., Surolia, P.K., Ruthenium complexes based dye sensitized solar cells: Fundamentals and research trends. *Sol. Energy*, 207, 59–76, 2020. https://doi.org/10.1016/j.solener.2020.06.060.
37. Pal, A.K. and Potter, H.C., Advances in solar energy: solar cells and their applications. *Adv. Solar Energy Res.*, 75–127, 2018. https://doi.org/10.1007/978-981-13-3302-6_4.

15

Multifunctional Biopolymers: Types, Preparation, and Industrial Applications

Surabhi Pandey, Sweekriti Choudhry and Anurag Singh*

Department of Food Technology, Harcourt Butler Technical University, Kanpur, Uttar Pradesh, India

Abstract

Researchers and professionals from a wide range of disciplines are very interested in biopolymers since they are a leading functional material that may be used in high-value applications. Understanding the fundamentals and practical applications of biopolymers is crucial for addressing several intricate issues related to health and well-being. This requires interdisciplinary study. Much work has gone toward substituting biodegradable materials, especially those produced from natural resources for synthetic polymers to lessen the impact on the environment and the reliance on fossil fuels. To fulfill the demands of ever-growing applications, several varieties of natural or biopolymers have been produced in this area. Because of their special qualities, these biopolymers are already employed in the food industry and are starting to be used increasingly in the pharmaceutical and medical sectors.

Keywords: Biopolymers, biodegradability, natural, starch, cellulose, applications

15.1 Introduction

The advantageous qualities of polymers such as their stability, hardness, and simplicity of manufacture have made them an indispensable part of contemporary life. Although synthetic plastics are affordable and widely available, their unregulated use has already severely damaged the ecosystem and is already posing a threat to all life on Earth. Finding alternatives is desperately needed since the growing use of synthetic plastics in everyday goods poses a severe threat to humans. Bio-based, renewable, and biodegradable polymers will likely take the place of manufactured polymers determined from petrochemicals. Biopolymers are natural compounds that are shown in normally happening sources. These days, biopolymers are a hot theme due to their potential applications within the nourishment, pharmaceutical, material, and restorative segments as well as their capacity to address the developing issue of natural contamination. The term "biopolymer" starts from the Greek concepts "bio" and "polymer," which signify nature and living beings, separately [1]. The biopolymers are found to be biocompatible and biodegradable, which makes them valuable for a run of employment. Food-grade movies, emulsions, materials for nourishment industry bundling, and

**Corresponding author*: anurag.niftem@gmail.com

Divya Bajpai Tripathy, Anjali Gupta and Arvind Kumar Jain (eds.) Multifunctional Materials: Engineering and Biological Applications, (393–418) © 2025 Scrivener Publishing LLC

materials for medicine transport, restorative inserts (counting organs), tissue frameworks, wound recuperating, and dressing materials are among them [2].

The foremost predominant macromolecules are called biopolymers, and they include huge non-polymeric particles like lipids and macrocycles. Additionally, lipids, proteins, carbohydrates, and nucleic acids are among them [2, 3]. Plastics, engineered filaments, and test materials such as carbon nanotubes are cases of engineered macromolecules [2, 4]. In expansion to rehashing units of these compounds, the atomic spines of nucleic acids, saccharides, and amino acids can have an assortment of chemical side chains connected to them that advance their capacities. Polylactic corrosive (PLA) and polyhydroxyalkanoates are two cases of biopolymers that are made by conventional chemical strategies in organisms or hereditarily altered life forms [polyhydroxyalkanoates (PHAs)]. These are made up of carbs from cellulose and proteins from collagen or drain. Microorganisms can be hereditarily adjusted to create biopolymers with particular highlights valuable for high-end therapeutic applications counting tissue designing and medicate conveyance.

Biopolymers have pulled in a part of intrigued in an extent of applications that call for maintainable and biodegradable arrangements. Treatment of illness still to a great extent depends on the advancement of drug conveyance frameworks to extend the action of bioactive chemicals, and a colossal advance has been made in this range. In this occurrence, engineered, common, and semi-synthetic polymers are commonly utilized in the advancement of sedate conveyance methods [5]. The far-reaching utilization of synthetic and chemical-based polymers within the nourishment and therapeutic businesses raises a few natural concerns. Developing natural mindfulness of supportability, contamination controls, and civil strong squander administration is driving the improvement of biopolymer-based bundling materials [6]. Utilizing biopolymers brings down carbon dioxide (CO_2) outflows, dependence on petroleum-based assets, and squanders from municipal strong waste [7]. The most objective of this audit is to offer a few understanding of biopolymers and the assortment of employment they have over numerous businesses.

15.2 Sources of Biopolymers

In many situations where biodegradable and sustainable solutions are required, biopolymers have attracted a lot of interest [2]. Biopolymers, which include huge non-polymeric molecules like lipids and macrocycles, are the most common macromolecules. They also include nucleic acids, proteins, carbohydrates, and lipids [3]. Based on where they come from, biopolymers are categorized, as mentioned in the Table 15.1.

15.2.1 Cellulose

One of nature's most prevalent biopolymers, cellulose was the first thermoplastic polymer found in plants and is an essential part of plant fiber materials. This polysaccharide has a linear structure made up of 100–1000 β-linked glucose units that repeat (1–4). In this case, the glucose unit is called D-Glucose, and it consists of around 100,000 repeating units of hydroxyl (OH) and CH_2OH groups arranged in the same plane. Cellulose demonstrates biocompatibility and is renewable and biodegradable, which makes its derivatives extremely beneficial for a wide range of uses. These uses include textiles, coatings, laminates, optical

Table 15.1 Various sources of biopolymer materials.

Biopolymers	Sources	Structure	Reference
Chitin	Corals, horseshoe worms, lamp shells, sponges, squid, cuttlefish, and clams are examples of aquatic species	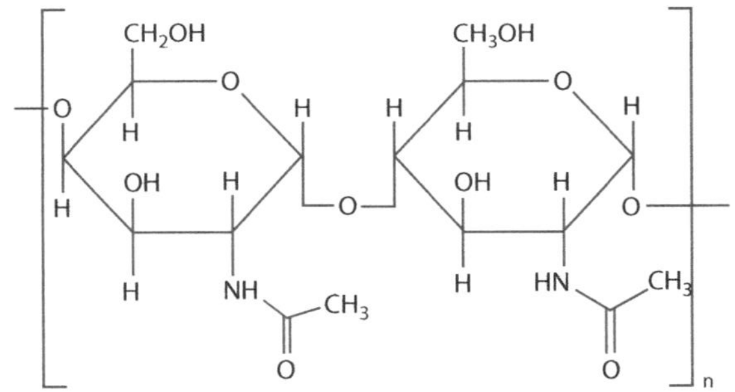	[38]
Gelatin	Pigs, cattle, fish, and poultry	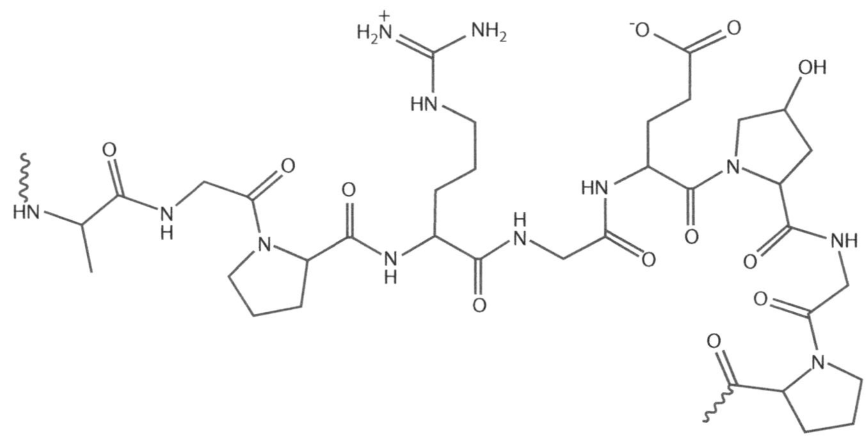	[39]

(*Continued*)

Table 15.1 Various sources of biopolymer materials. (*Continued*)

Biopolymers	Sources	Structure	Reference
Chitosan	Fungi, mollusks, algae, crustaceans, and insects	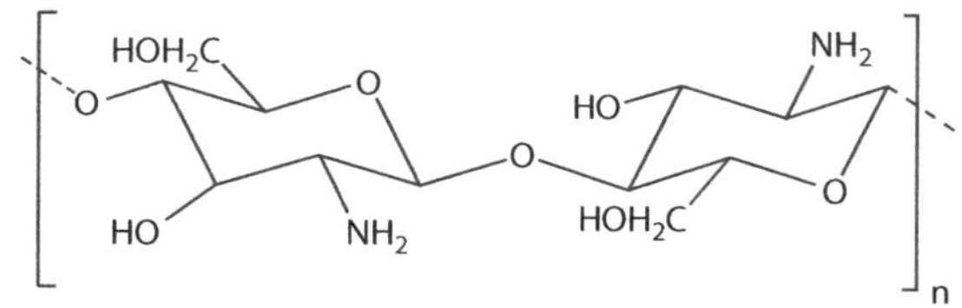	[40]
Collagen	Rat tail tendon, porcine, and bovine skin-derived collagen type I, alternatives such as human placenta and human skin-derived collagen type I collagen type III isolated from bovine placenta, human placenta, human skin, chicken skin, and rat skin	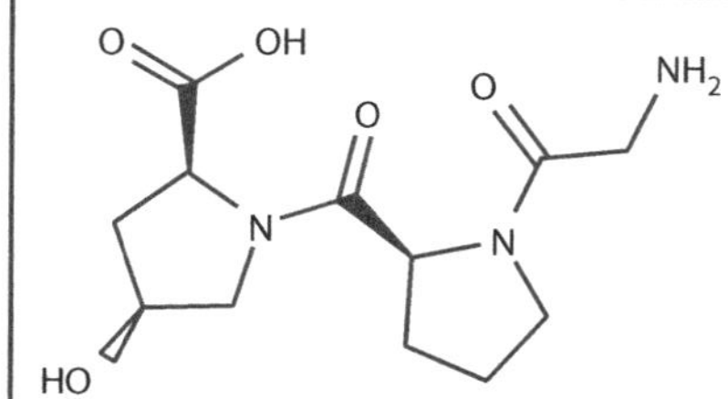	[41]

(*Continued*)

Table 15.1 Various sources of biopolymer materials. (*Continued*)

Biopolymers	Sources	Structure	Reference
Cellulose	Agricultural trashes, such as seaweed, rice husk, and sugarcane bagasse. Plant sources like wood, bamboo, sugarbeet, and banana rachis	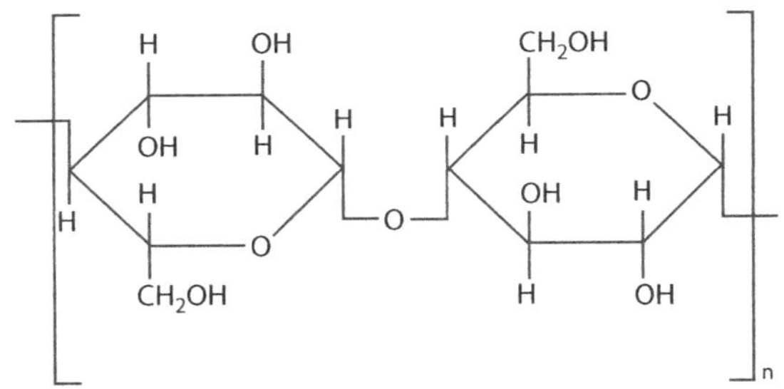	[42]
Pectin	Citrus fruits,apple pomace, cocoa husk sunflower heads, sugar beet, pumpkin, watermelon, pears, and potato pulp	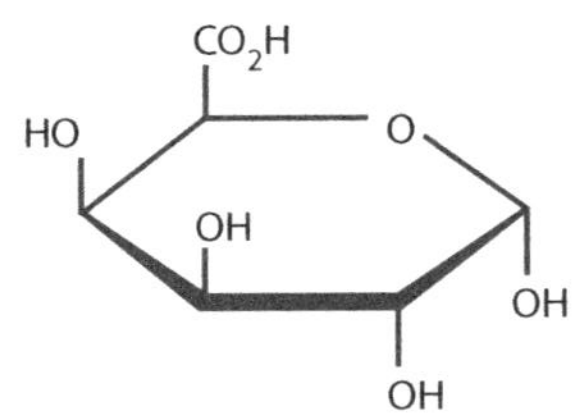	[43]
Alginate	Seawood	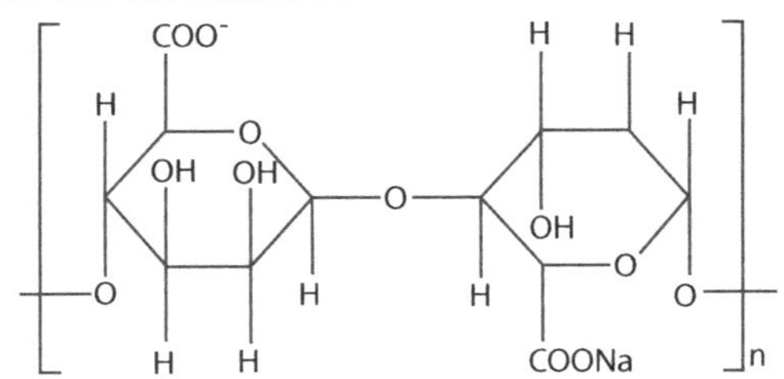	[44]

(*Continued*)

Table 15.1 Various sources of biopolymer materials. (*Continued*)

Biopolymers	Sources	Structure	Reference
Xanthan gum	*Xanthomonas campestris, Rhone Poulenc, Mero-Rousselot-Santia, and Sanofi-Elf.*	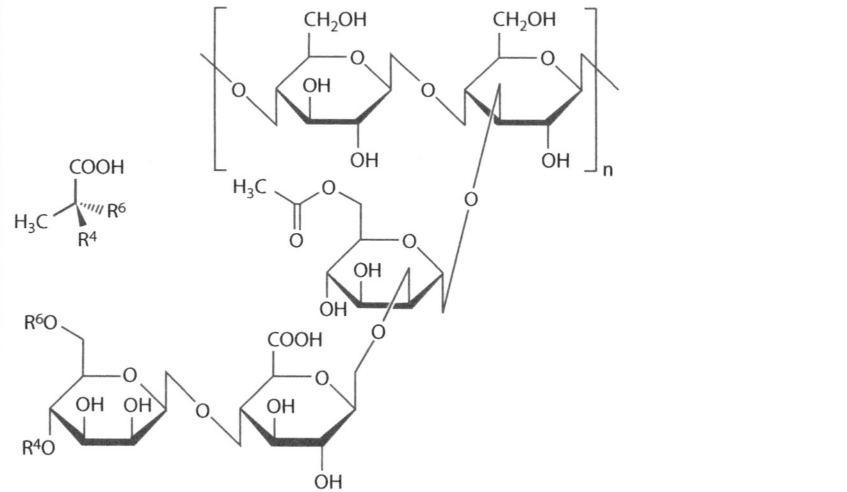	[45]
Starch	Potatoes, maize, cassava, rice, sorghum, banana, wheat, and yams	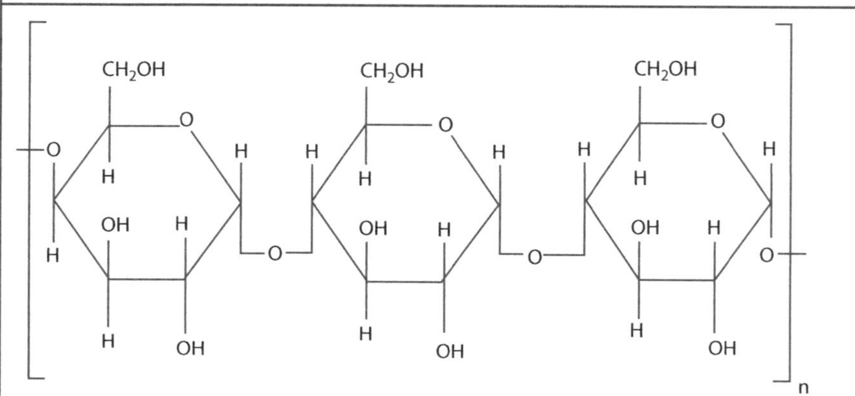	[46]

(*Continued*)

Table 15.1 Various sources of biopolymer materials. (*Continued*)

Biopolymers	Sources	Structure	Reference
Lignin	Vegetable peel, pomace, molasses, beet pulp & distillate	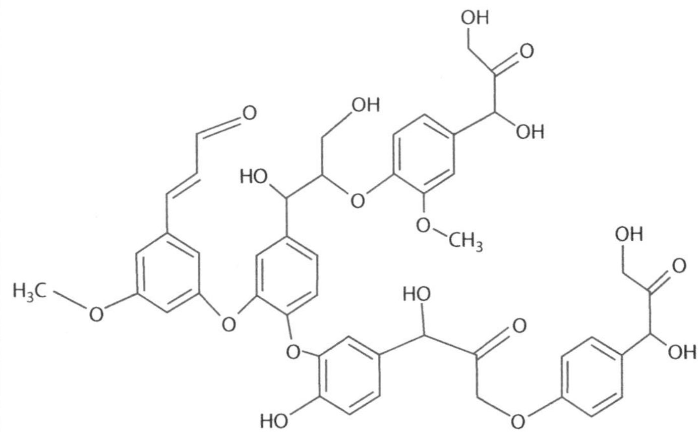	[47]
Polyhydroxyalkanoates (PHAs)	Whey hydrolysates, cheese residues, broken grains, and malt dust	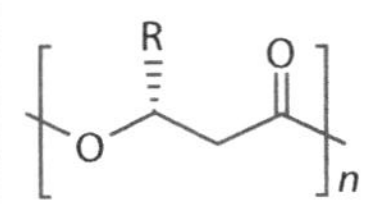	[47]
Polycaprolactone	Polycondensation of ε -caprolactone	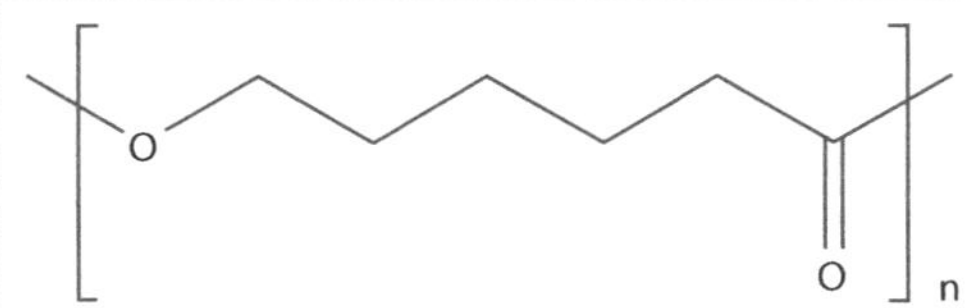	[48]
Polylactic acid (PLA)	Vegetable peel, pomace, molasses, beet pulp & distillate	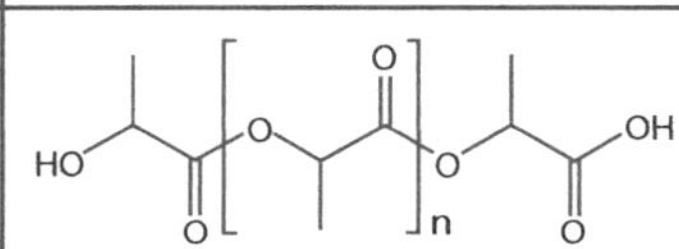	[47]

films, medicines, immobilization of proteins and antibodies, and the creation of cellulose composites using synthetic polymers and other biopolymers [8, 9]. Numerous industries utilize cellulose extensively, but the paper, textile, and pharmaceutical industries use it the most. Cellulose demonstrates its potential to stick to biological surfaces by acting as a mucoadhesive and bioadhesive in medication delivery systems [10]. This trait is particularly beneficial for increasing the effectiveness of medication administration. Furthermore, cellulose is used in the food and pharmaceutical sectors as a coating material, thickening agent, and stabilizing agent, all of which improve the formulation and caliber of different goods. Because of its natural qualities and ability to coexist with biological systems, cellulose is incredibly versatile and important for the creation of useful and sustainable materials in a variety of sectors. Its vital importance in several scientific and industrial pursuits is shown in the ongoing study of cellulose derivatives and uses [11].

15.2.2 Starch

While starch, a naturally occurring biopolymer, and cellulose have similar structures, their internal linking is different. According to a study, pyranose, an open-chain D-glucose molecule, is the basic building block of starch. In essence, starch is a homopolymer made up of D-glucopyranose repeating units [12]. Starch is sourced from a variety of foods, including potatoes, cassava, rice, corn, and tapioca [13, 14]. It is often used to alter the physical characteristics of food items. It is an adaptable substance that finds use in many different sectors as an adhesive, thickening agent, and moisture-retentive material [15]. D-glucopyranose in starch has an open chain structure that makes it versatile for use in a variety of applications, particularly in the food sector. The creation of customized biopolymers is made possible by the variety of starches found in various plant sources. The use of starch to alter food properties emphasizes how crucial it is to improve texture, viscosity, and general quality in food applications. Furthermore, starch's versatility in a variety of commercial and industrial contexts is demonstrated by its use as a thickening agent, adhesive, and moisture-retention substance. According to a study, the investigation of starch-based biopolymers is still a major topic of study and application, helping to create materials that are adaptable and sustainable for use in a variety of applications.

15.2.3 Gelatin

Raw collagen from the meat industry is the source of gelatin, a naturally occurring biopolymer [16]. Raw collagen from different animal sources, such as skin, bones, and connective tissues, is hydrolyzed to produce gelatin [17]. To be more precise, gelatin is often made from type-I collagen and does not include cysteine. Gelatin has a wide range of uses due to its flexibility, especially in the biomedical field. Gelatin is used in bone and tissue engineering because of its biocompatibility and capacity to promote cellular proliferation and regeneration. It can be used as a scaffold for tissue restoration in wound dressings. Due to its characteristics, gelatin may be used in medication delivery and gene transfection processes, where it can serve as a carrier for medicinal substances. Beyond its use in biomedicine, gelatin is used to improve the condition of hair and hasten the healing process following sports-related injuries. Gelatin and collagen have long been used in the food sector to coat sausages and other meat products [18]. Meat products are better preserved and of higher quality thanks to this use. The fact that gelatin comes from collagen emphasizes its natural source, and because of its special qualities, it is used in a variety of sectors. Gelatin is a vital and adaptable biomaterial in the biomedical and food-related sectors, and its uses are continually being discovered by the field's continuing research and development efforts.

15.2.4 Chitosan

The biggest structural polysaccharide based on nitrogen, chitosan, is made up of repeated modified glucose subunits [19]. Chitosan is a versatile biomaterial with diverse uses. It is derived from chitin, a polymer found in the exoskeleton of crustaceans, fungal cell walls, and several other animal sources [20]. Chitosan has extraordinary features. Because chitosan has a special structure that allows it to break down into regular body components, it contains reactive groups that attach to both microbial and human cells. Its safety profile and biocompatibility are enhanced by this property. Chitosan is primarily used extensively in the biomedical sector. Chitosan is frequently utilized in the chromatography area to produce chromatography columns for isolating lectins, demonstrating its value in the processes of separation and purification [21]. Furthermore, chitosan has exhibited noteworthy biological actions, such as stimulating the peritoneum and inhibiting the formation of tumors in mice [12]. Its biological potential is further expanded by its capacity to promote nonspecific host resistance against infections [22]. Because it speeds up the healing process, chitosan is essential in applications involving wound healing. Its value in many medical and biotechnological contexts stems from this feature as well as its capacity to immobilize enzymes and cells [22]. In addition, chitosan has a number of advantageous qualities, including the capacity to produce gels, hydrophilicity, biodegradability, nontoxicity, and physiological inertness [23]. Chitosan has found uses in the food business because of these beneficial qualities; biosensors are one such use. Chitosan is a material that has great promise for a wide range of scientific and commercial applications due to its unique properties and biocompatibility. Research is still ongoing to find new applications for chitosan that are sustainable and creative.

15.2.5 Polycaprolactone

The synthetic polyester biopolymer known as Poly(ε-caprolactone) (PCL) is created by polymerizing caprolactone with a catalyst, usually stannous octate, in a ring-opening manner. This adaptable polymer is used in many different applications, demonstrating its useful qualities and versatility. One noteworthy usage for PCL is as an additive in starch, a naturally occurring biopolymer. According to a study, this addition improves impact strength, enhances biodegradability, and lowers costs. The judicious combination of PCL and starch results in materials with enhanced overall performance. PCL is widely used because of a number of special qualities that it possesses. It is affordable and provides a cost-effective solution for a range of uses. Moreover, PCL has a high degree of hardness, which makes it appropriate in situations where longevity is essential. Its attractiveness is increased by its biocompatibility, particularly in biomedical applications [24]. Compared to other biopolymers, PCL degrades more slowly, which is one of its unique qualities [25]. When it comes to medication delivery applications, this feature is very helpful since it allows for the regulated and prolonged release of medicinal substances. Because PCL is hydrophobic, it has strong chemical resistance to biological fluids, which increases its usefulness for a variety of medicinal applications [26]. It is noteworthy that PCL blends well with other synthetic polymers, opening up possibilities for the creation of materials with specific qualities. As a biomaterial, PCL is used in tissue engineering because it can break down its ester linkage and make it more compatible with biological systems [27].

15.2.6 Polyvinyl Alcohol (PVA)

This is a thermoplastic biopolymer that is produced by hydrolyzing polyvinyl acetate, which is its precursor. It is an environmentally favorable choice since biological microbes help it decompose [28, 29]. PVA is a common and adaptable biopolymer due to its diverse range of characteristics and uses. PVA's high solubility in water, which is linked to its higher crystallinity, is one of its distinguishing qualities [30]. Its versatility in a range of applications is facilitated by its solubility. PVA is used to make a variety of polymer end products, including food packaging, surgical threads, and liquors [31, 32]. Its chemical bonding capacity to material surfaces adds to its flexibility, enabling easy retention on the water's surface [33]. This characteristic creates opportunities for use in stabilization and surface modification, for example. PVA is widely used in food packaging because it is biodegradable and nontoxic, and because it complies with safety regulations due to its biocompatibility [34]. Furthermore, PVA has become more well-known in the biomedical field, where it is used in applications including wound dressing, medication delivery systems, contact lenses, and heart surgery. It is an important substance for various medicinal applications because of its compatibility with biological systems.

15.2.7 Protein

Successful applications in several research have shown that proteins are useful in the production of films and coatings [35]. One important milk protein that may make up more than 80% of the total protein in milk is casein, which has been used specifically in the creation of films. The characteristics of casein-derived films include their water-insoluble nature and opacity. Nonetheless, after being submerged for a whole day, the mass of the film increases dramatically (up to 90%), demonstrating their ability to absorb water. Casein-based films are robust and adaptable because they retain strong mechanical qualities despite their capacity to absorb water [36]. Apart from casein, films made from zein and gluten proteins are also produced by adding alcohol-soluble proteins to a product and letting it form a thin layer that dries later. Under ideal circumstances, wheat gluten-based films have a glossy look and are water resistant. Even though these films are insoluble in water, when immersed, they exhibit water absorption. Conversely, food products having a short shelf life can be protected by zein-based films. Zein is a protein that comes from corn that helps to prolong the shelf life of perishable goods by preserving them [37].

15.3 Methods of Biopolymer Processing

The sort, quality, and amount of dissolvable utilized as well as the sort of preparation utilized to make the ultimate structure, which is able to decide how the components are associated,

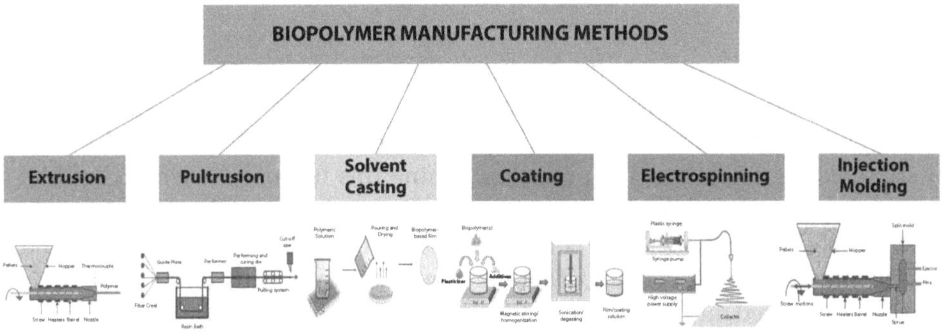

Figure 15.1 Manufacturing methods of biopolymers.

are extra variables that influence how biopolymers work in expansion to their structure and substance. The multi-step handle of creating biopolymers requires information and an exhaustive understanding of the behavior of the components all through preparation. Biopolymers can be handled into an assortment of items, such as bundling movies, covered paper, movies, plates, glasses, and cutlery things, depending on the method utilized (cast movies, blow molding, coextruded films). The fundamental forms in handling any biopolymer incorporate dissolving the combination of biopolymers and after that, depending on the fabric to be made, blow molding, expulsion, and casting (Figure 15.1). Different fabricating strategies of biopolymers incorporate

15.3.1 Extrusion

It is noteworthy that the most used technique for creating composite materials is extrusion. By passing the material to be extruded through a die, this process guarantees a uniform cross-sectional area of the finished product. An essential part of this process is the extruder, which is a cylindrical chamber containing a hopper. The material is first introduced into the hopper, where it is subsequently transported through the cylindrical chamber's three main portions. The content goes through a number of crucial processes in these stages. To guarantee that the components are evenly distributed throughout the composite, it is first carefully mixed. The next phase involves compressing the material, which helps the particles align and improves the uniformity of the composite. The treated material is then pushed through the die to take on the required shape. Fiber-reinforced thermoplastic composites may be made using this extrusion method in addition to traditional materials. These composites benefit from the integration of fibers into the matrix material during the extrusion process, which improves mechanical attributes like strength and durability. As previously mentioned, the extrusion process offers a flexible and effective way to create composite materials with desired qualities and uniform cross-sectional profiles. According to Allen [48], the process of producing composite materials involves the use of extruders, which is indicative of the method's popularity and efficacy across many sectors. This method is particularly useful for forming continuous profiles or forms, which makes it an important tool for fabricating composite materials for a variety of uses in construction and industry.

15.3.2 Pultrusion

A continuous manufacturing method called pultrusion is used to create fiber-reinforced polymer (FRP) composites with uniform cross-sectional forms and high strengths. In fact, the word "pultrusion" itself is a compound of the words "pull" and "extrusion," which accurately captures the essence of the procedure. Using this technique, reinforcing fibers are continuously pulled or tensioned via a resin solution. Following a thorough impregnation of these fibers with a polymer matrix, the composite material is molded and cured to take on its ultimate form [49]. A supply of reinforcing fibers, frequently in the form of rovings, mats, or textiles, is fed into a resin bath to start the pultrusion process. The resin surrounds and impregnates the fibers, creating a wetted profile. The resin is usually a liquid thermosetting polymer. The wetted fibers are next run through a shaping die, which establishes the final product's cross-sectional geometry. Following their release from the die, the wetted fibers go through a curing process which typically involves heat during which the polymer matrix changes chemically in order to harden and create a link with the reinforcing fibers. Pultrusion has the benefit of producing homogeneous, continuous profiles with high stiffness and strength. This makes it the perfect technique for producing parts like beams, tubes, and rods that have uniform cross-sectional forms. The method is very successful in producing FRP materials with customized mechanical characteristics at a reasonable price. Pultruded composites are used in a variety of industries where strong, lightweight materials are required, such as infrastructure, automotive, aircraft, and construction.

15.3.3 Solvent Casting Method

The biopolymer is combined with an aqueous or hydroalcoholic mixture in the solvent casting technique, also known as the solution function or wet processing method. A study says that the use of a solvent that enables the polymer to be suspended in a film-forming solution is the basis of this procedure, which is then followed by solvent evaporation and polymer chain reformulation. Alcohol, water, or other organic solvents are commonly used to dissolve the selected polymer. Sometimes the best results come from heating the suspension polymer solution or altering its pH. The polymer–solvent mixture is poured onto a mold, drum, or flat surface and left to dry for a predefined period of time [50, 51].

15.3.4 Coating Method

Applying an arrangement coating based on a polymer or component combination is the coating handle. This arrangement is connected to a fabric known as a substrate. The solvent-casting strategy and the coating preparation have assorted applications. It is conceivable to apply the coating for down-to-earth, corrective, or both objectives [52]. The coating strategy has a few variations. The foremost common strategy, plunging, is immersing the question within the coating arrangement. It is broadly utilized within the nourishment industry, for case, when coating natural products and vegetables, to act as a boundary against dampness and gasses [53, 54]. The coating is connected straightforwardly to the product's surface amid the brushing preparation employing a brush or other brushing apparatuses [55]. Various beads of the coating arrangement are splashed onto the ultimate product during the showering handle. High-pressure splash weapons, discuss atomization frameworks, or water-powered shower spouts are commonly utilized to encourage the method [56]. An atomizer spout associated with a source of tall electrical potential is utilized to shower the coating arrangement electrically [57].

15.3.5 Electrospinning Method

With exact control over their diameter and qualities, charged polymer strands may be produced through the use of modern technology called electrospinning. A strong electrical voltage is first created between two electrodes to start the electrospinning process. For charged polymer strands to form, this high voltage is essential. The polymer solution is made, usually by dissolving it in a solvent. After that, a nozzle is used to feed this solution at a preset flow rate. To get the required properties in the final fibers, the flow rate is meticulously regulated. The polymer solution runs through the nozzle, which is exposed to a high electrical voltage. The polymer solution receives a charge from this electric field, which causes charged strands to form. The strands stretch and thin as a result of surface tension being overcome by the electrostatic repulsion between the charged components in the polymer solution. The fibers' diameter may be precisely controlled thanks to this method. The resultant fibers may be designed and used in a variety of ways since their sizes generally range from two to several micrometers [58]. Either a high-voltage electric field alone or mechanical pressure combined with gravity can be used to force the polymer solution through the nozzle. The polymer solution can flow more smoothly through the nozzle when it is pulled through by gravitational or mechanical pressure. The evaporation of the solvent is one of the crucial phases in the electrospinning process. For the fibers to solidify and adopt a stable form, the solvent present in the polymer solution must evaporate. The polymer chains unite to form solid fibers when the solvent evaporates [59]. These fibers retain the charged qualities that were imparted during the electrospinning process, which adds to their special qualities. Applications for electrospinning techniques may be found in biotechnology, materials science, tissue engineering, and other areas. The resultant nanofibers find usage in many different applications, including scaffolds for tissue regeneration, filtration materials, drug delivery systems, and sensors.

15.3.6 Three-Dimensional Printing Method

Additive manufacturing, or three-dimensional (3D) printing, is a cutting-edge technology that makes it possible to create items layer by layer. Through the use of a print head and nozzle, material is deposited into a substrate in this process, allowing for the creation of complex, pre-designed shapes. Choosing the material for 3D printing is the first stage in the process. Depending on the particular 3D printing process being utilized, this material may take the shape of resins, powders, or filaments. A substrate is created with a predefined geometry. To guarantee accuracy in the finished product, this geometry is frequently developed using computer-aided design (CAD) software [60]. Layer-by-layer material deposition is the first step in the 3D printing process. With the help of a nozzle, the print head moves precisely, releasing the chosen material onto the substrate in accordance with the shape that was pre-planned. By layering one layer of material on top of the other, the item is progressively constructed in three dimensions. The process of construction never ends until every layer of the item is completed. The finished result might have rich features and complicated architecture thanks to this layering technique. The material may be put in a molten or semi-liquid condition and solidify quickly upon deposition, depending on the particular 3D printing process. A curing phase can be required, depending on the kind of 3D printing method and the material utilized [61]. The printed structures are taken out of the support material when the 3D printing procedure is finished and any necessary curing time has passed. If any support structures are utilized, they are removed, leaving the finished 3D-printed item. Post-processing techniques, such as polishing, surface finishing, or other treatments, can be used to enhance the printed object's appearance and functionality.

15.3.7 Injection Molding

A popular manufacturing technique is injection molding, especially for producing goods made of biopolymers. To mold thermoplastic polymers into the appropriate shapes, this technique entails a number of processes and factors. A thermoplastic polymer is first inserted into an injection molding machine's cylindrical chamber to start the operation. The temperature of the polymer material is raised to the point where it melts and flows readily. The end of the heated chamber is attached to a mold, which is usually constructed of metal. The chamber of the mold is made to fit the intended final product's form. After that, the molten material is injected into the mold while being held to certain temperatures and pressures. To guarantee correct mold filling and the creation of the intended product, these parameters are closely regulated. Using a ram and stamp combination, the injected material is hydraulically forced into the mold. The mold is placed in a cold environment once the material has been injected and formed inside of it. This quick cooling aids in the molten material's solidification, transforming it into a form that keeps the mold's shape. The substance that has hardened in the cold mold is processed further. The material is compressed and melted by an injection molding machine's revolving and reciprocating screw. By taking this extra step, you can be confident that any residual flaws are fixed and the finished product is consistent. It is frequently used in the production of a broad range of commodities, including consumer goods, packaging materials, and complex automobile components. In large-scale production, the technique is renowned for its efficacy, speed, and affordability [49].

15.4 Life Cycle Assessment of Biopolymers

Life cycle assessment (LCA) may be a device for evaluating a material's natural execution and supportability. A careful approach called LCA is utilized to survey how an item or process's entire life cycle influences the environment. By taking into consideration the stages of generation, utilization, and transfer, assessment LCA helps in assessing the maintainability of biopolymers [62]. A biopolymer's life cycle and natural impact are closely related, and strategies such as life cycle assessment (LCA) are frequently used to assess this. The capacity of biopolymers to break down normally could be a vital component of their supportability. Since both biopolymers and biomass are broken down into carbon and water by the activity of chemicals and microorganisms, their biodegradation forms are comparable. A metric known as EdK is utilized to survey the potential biodegradability of biodegradable polymers quantitatively. The biodegradation rate, or the amount of CO_2 transmitted amid the debasement handle partitioned by the sample's hypothetical CO_2 concentration, is utilized to calculate this metric. Put something else, it quantifies the degree to which a biodegradable polymer breaks down into CO_2, consequently uncovering its environmental noteworthiness. Tests in controlled settings are vital to assess biodegradability. Bioreactors are filled with genuine soil samples to form an environment that mirrors the characteristic world. The rates of biodegradation for reference materials are measured over a two-week period. With EdK values of 100 and 0, respectively, starch and polyethylene are regularly utilized as reference materials in this context. Whereas polyethylene could be a manufactured polymer and symbolizes non-biodegradable materials, starch, a normal polymer created from biomass, speaks to adding up to biodegradability. The method of deciding EdK values follows to ISO 14852, which calls for the explanatory parameter of advanced CO_2 location. This standardized approach ensures comparability and consistency when assessing the biodegradability of different biopolymers.

15.5 Applications of Biopolymers

The food business has seen significant advancements in biotechnologies and applied engineering. Many food science developments, from genetic engineering to improved preservatives and complex materials for creative materials, food quality control, and packaging, are focused on the worldwide food scarcity that an ever-increasing population faces. Food research is significantly impacted by new products in the biopolymer arena, according to recent studies. These materials are now being investigated for potential use in innovative and distinctive food products that include qualities including improved nutritional value, bioresponsiveness, advanced detection, and a variety of biodegradable options. By promoting an ecologically sensitive approach to food engineering, their application is revolutionizing the manufacturing, design, and packaging of food [63].

15.5.1 Active Packaging

Agreeing with Hassan *et al.* [101] and Janjarasskul and Suppakul [102], dynamic bundling is any bundling method in which dynamic chemicals are intentionally included in the pressing fabric or the bundle headspace in arrange to extend the soundness and security of the stuffed merchandise. The establishment of dynamic bundling depends on specific active fixings implanted within the biopolymer framework, such as probiotics, cancer prevention agents, dampness safeguards, antimicrobials, and phytochemicals, as well as the characteristics of the biopolymers that make up the bundling fabric, as mentioned in Table 15.2. Agreeing with Majid *et al.* [103], the bundling strategy can discharge the desired dynamic fixings whereas moreover blocking gasses, dampness, and other undesirable substances. Based on the parts they play, dynamic bundling frameworks can be broadly separated into two bunches:

(i) emitters, which discharge dynamic fixings with craved properties to make strides in the bundle, and (ii) foragers, which reduce the number of undesirable substances like ethylene, CO_2, oxygen, dampness, and off-odors within the bundling environment [64]. Useful components, such as basic oils and substances with antibacterial, antifungal, and antioxidant properties, have progressed the dynamic movies in this way. The foremost promising dynamic

Table 15.2 Biopolymers used for active packaging.

Biopolymer	Active agent	Finding	Reference
Polyvinyl alcohol (co-ethylene)	Gallic acid and umbelliferone	The characteristics of films in terms of antioxidant, heat, and mechanical (tensile strength, Young's, and elastic moduli) significantly enhanced.	[66]
Zein/gelatin and polyethylene	Oregano essential oil	considerable fruit weight loss and bacterial growth inhibition (lowest total plate counts of 4.9 and 4.3 log CFU/g for longan and strawberry, respectively).	[67]
Polyvinyl alcohol and gelatin	Amaranthus leaf extract (ALE)	Compared to neat-film (3 day), fish and poultry packed in ALE with active film had a much longer shelf life—up to 12 days.	[68]
Zein and chitosan	α-tocopherol	When mushrooms were stored at 4°C for 12 days, zein-chitosan-tocopherol (ZCT) films enhanced their physicochemical characteristics and enzyme activity compared to the control group.	[69]
Sodium alginate and chitosan	Cinnamon essential oil (CEO)	Enhanced mechanical, functional, and physical qualities.	[70]
Chitosan	Apricot kernel oil	*Bacillus subtilis* and *Escherichia coli* growth was successfully prevented by film since no viable colony was seen.	[71]
Zein	Pomegranate peel extract (PPE)	increased the overall phenolic content (~0.05 mg GA/g) relative to the control (~0.002 mg GA/g), delayed fat oxidation, and kept the Himalayan cheese's sensory quality.	[72]
Zein and gelatin	Tea polyphenol (TP)	decrease in kiwi fruit browning, weight loss, and bacterial degradation.	[73]
Gelatin	Grapefruit seed extract and titanium dioxide	The film's tensile strength (63.4 MPa), stiffness (9.6%), water contact angle (59.3o), and antioxidant activity (DPPH, 7.5–31% and ABTS, 29–57%) were all improved by the addition of active components.	[74]
Chitosan	Thyme essential oil	Avocados with the coating had a lower incidence of C. gloeosporioides.	[75]

fixings to be included in dynamic bundling movies are antimicrobial and antioxidant compounds, which can expand item rack life, protect quality, and make strides in item security by ending the improvement and spread of nourishment pathogens [65].

15.5.2 Fruits and Vegetable Industry

Ponders have appeared that an assortment of factors contribute to the misfortune and debasement of natural products and vegetables after collection. Insufficient care and consideration within the bundling and promoting forms can moreover lead to post-harvest misfortunes. Natural products and vegetables ruin rapidly after collection due to an assortment of variables, such as microscopic organisms, scratches, bruises, bothers and mice, winged creatures, and other creatures, as well as lacking bundling, poor capacity buckets, and the utilization of wooden cases [76]. Specialists say that one of the most straightforward strategies to keep veggies from softening is to pack them in eatable wrapping, as mentioned in Table 15.3. Citric acid is one of the foremost critical consumable coating materials that's habitually utilized in bundling, concurring to Jin *et al.* [104]. Eatable bundling materials made of bio-derived polymers such as carrageenan, chitosan, and alginate were utilized [77]. To protect the postharvest longan natural products, a coating of 1.29% (w/v) chitosan with 0.42% glycerol and 1.49% (w/v) carrageenan with 0.03% glycerol was connected. The reaction surface technique (RSM) test uncovered moo quality changes and amount of misfortunes [22].

Table 15.3 Biopolymers used for packaging of different food commodities.

Polymer	**Food**	**Activity**	**Reference**
Cellulose	Cabbage	The film demonstrated strong antibacterial efficacy against Aeromonas hydrophila.	[79]
Cellulose	Tomato	Extended the shelf life.	[79]
Polylactide	Berries	Extends shelf life up to 15 days.	[80]
Chitosan	Mangoes	Shelf life increased up to 20 days.	[81]
Carboxymethyl cellulose	Soft cheese	Lengthen the white soft cheese's shelf life efficient against yeast, mold, and bacterial counts.	[82]
Cellulose	Cheese	Increased shelf life and freshness of cheese.	[83]
Hydroxypropyl methylcellulose	Walnuts	Prevent oxidation of walnuts during storage.	[84]
Carboxymethyl cellulose	Chicken breast fillets	Meat's longer shelf life and lower microbiological count were observed.	[85]
Bacterial cellulose	Meat	Extended shelf life in storage conditions.	[86]

15.5.3 Meat Industry

The production, processing, preserving, and packaging sectors make up the meat industry's chain of activity. Meat's shelf life is reduced by microbial contamination and protein oxidation, which also raises the possibility of foodborne illnesses. As a result, the packing supplies and wmethods have a direct effect on the meat's quality and the meal's safety. Meat that has been actively packed in edible films retains its juice content, slows down the rate of rancidity, and guards against moisture loss in cold storage situations [78]. In recent times, several inventive packaging technologies have been created, such as packaging made of edible, naturally degradable, and nanoparticle materials. The primary benefits of this type of packaging for beef products, according to Fang *et al.* [105], are increased economic value, extended shelf life, and quality assurance. A film was created to stop cooked pig sausages from oxidizing using two naturally occurring antioxidants: chitosan and a tea polyphenol solution [78]. To extend the shelf life of natural semi-finished hog meat, polysaccharides including agar, gelatin, maize starch, and citrus pectin were included into a film-forming coating and stored for two days at a temperature of –1 to +1°C [78].

15.5.4 Dairy Industry

Fermented milk, butter, cheese, yogurt, and similar functional foods make up the majority of dairy products. Dairy products are high in protein, fats, vitamin A, and minerals including phosphorus, magnesium, and calcium. They also make up a sizable portion of the food that the body needs to continue receiving its necessary nutrients. The food industry uses a variety of biopolymers to keep milk products aesthetically pleasing while guarding against contamination and rancidity (aroma alterations) [78]. Zein-oleic acid (Z-OA) and Z-OA-xanthan gum (XG) coatings were applied to Minas Padrao cheeses, and the cheeses' ability to be preserved for up to 56 days was examined. When compared to cheese samples that were unpackaged and plastic-packed, the coated samples showed the retention of physicochemical properties such as protein, ash, chlorides, and acidity. When compared to unpackaged samples, the biodegradable covering reduced moisture loss in the cheese and avoided early microbial contamination [87].

15.5.5 Bakery and Confectionery Industry

Starch is the essential fixing in bread and treats, particularly gluten-free assortments, because it gives the ultimate item with structure and surface. Vegetables, cereal grains, tubers, roots, and the tissue of natural products and vegetables are wealthy sources of starch, a polysaccharide comprising a long chain of glucose atoms [88]. Be that as it may, starches are mostly utilized as tidying and gelling specialists within the confectionery trade. Beneath an assortment of processing conditions, local starches can improve their useful properties and adaptability. Within the heating and confectionery businesses, starch subordinates were utilized, among other things, to duplicate fat in chocolate fillings, substitute gluten in gluten-free items, and keep the bread from staying together.

15.5.6 Medical Industry

The qualities of biopolymers incorporate biocompatibility, non-toxicity, and biodegradability. They have incredible potential and their utilization in embedded gadgets is developing rapidly [89]. Making utilize of sedate conveyance methods to boost the power of bioactive chemicals could be a significant approach to treating infections, and critical advances have been made in this space. In this setting, medicate conveyance strategies are regularly created utilizing engineered, semi-synthetic, and normal polymers [90]. Biopolymers have a few therapeutic employments, counting suturing, sticking, attachment, covering, impediment,

Table 15.4 Biopolymers used for different medical applications.

Biopolymer	Medical application	Reference
Chitosan	Implants used in cardiology as heart valves, ophthalmology as contact lenses, and neurosurgery as nerve regeneration.	[95]
Collagen	Cardiovascular implants, bone marrow, and scaffolds.	[94, 96]
Polyhydroxyalkanoates (PHAs)	Implants in the nerves, vascular cartilage, and esophagus.	
Gelatin	Heart transplants, orthopedic bone substitutes, and dermatology's use of 3D biometrics.	
Hyaluronic acid	Otolaryngology for the cartilage and vocal fold tissue development.	
PHB	Surgical implants and as cell culture scaffolds.	[97]
Chitosan	Wound dressing material for wound closure, healing, and regeneration.	[79]
Chitin	Types of films and fibers to treat the wounds.	[98]
Collagen, chitosan, and hyaluronic acid	Dermis, cartilage, skin regeneration, vascular engineering, and soft tissue restoration.	[99]
Gelatin composites reinforced with chitosan and hyaluron	Scaffold in tissue regeneration, hard tissue regeneration in orthopedics.	[99]
Alginate	Cell transplantation and drug delivery regenerative medicine.	[100]
Fibrin	Growth of tumors, the healing of wounds, and blood clotting Hemostatic agent, sealant, and surgical glue.	[101]
Keratin	Cornea tissue engineering and skin regeneration.	[102]

confinement, contact hindrance, cell expansion, tissue direction, and controlled sedate conveyance [91, 92]. Synthetic platforms and biopolymers are utilized to make an environment that underpins tissue mending and cell development. The constituents of movies, Pickering emulsions, hydrogels, nanogels, nanofibers, associated permeable platforms, and 3D-printed frameworks are collagen, keratin, gelatin, sericin, and fibroin [93, 88]. Table 15.4 records a number of occurrences of how biopolymers are utilized in medication.

15.6 Conclusion and Future Prospectives

The industrial interest in biopolymers has been steadily increasing over time. Future biopolymer manufacturers will have a huge demand for new materials. While biomaterials and biopolymers hold great promise for the production of sustainable materials, certain obstacles must be overcome before further progress can be made. The higher cost of these materials compared to traditional materials limits their use in several sectors. Gaining expertise in biopolymers and biomaterials is crucial for creating materials with an eye toward the future. Enhancement of the physical properties of biomaterials is also necessary to meet the requirements of industrial uses. Research on this topic is meant to contribute to the development of more sustainable materials in the future as well as now. The investigation and use of biopolymers are the result of interdisciplinary cooperation motivated by the need to find sustainable substitutes for conventional polymers. Biopolymers have a wide range of uses in the food, pharmaceutical, and medical industries, which highlights their potential to revolutionize these sectors and usher in a more ecologically friendly and sustainable future.

References

1. Ezeoha, S.L., Production of Biodegradable Plastic Packaging Film from Cassava Starch. *IOSR J. Eng.*, 3, 10, 14–20, 2013. https://doi.org/10.9790/3021-031051420.
2. Baranwal, J., Barse, B., Fais, A., Delogu, G.L., & Kumar, A. Biopolymer: A Sustainable Material for Food and Medical Applications. *Polymers*, 14, 5, 1–22, 2022. https://doi.org/10.3390/polym14050983.
3. Stryer, L., Berg, J.M., Tymoczko, J.L., *Biochemistry And Study Guide*, 5th ed, W.H. Freeman, New York, NY, USA, 2002.
4. Gullapalli, S. and Wong, M.S., Nanotechnology: A Guide to Nano-Objects. *Chem. Eng. Prog.*, 107, 5, 28–32, 2011.
5. Ruso, J.M. and Messina, P.V., *Biopolymers for Medical Applications*, CRC Press, Boca Raton, FL, USA, 2017, https://doi.org/10.1201/9781315368863.
6. Siracusa, V., Rocculi, P., Romani, S., Rosa, M.D., Biodegradable polymers for food packaging: a review. *Trends Food Sci. Technol.*, 19, 12, 634–643, 2008. https://doi.org/10.1016/j.tifs.2008.07.003.
7. Peelman, N., Ragaert, P., De Meulenaer, B., Adons, D., Peeters, R., Cardon, L., Van Impe, F., Devlieghere, F., Application of bioplastics for food packaging. *Trends Food Sci. Technol.*, 32, 2, 128–141, 2013. https://doi.org/10.1016/j.tifs.2013.06.003.
8. Liu, Y., Ahmed, S., Sameen, D.E., Wang, Y., Lu, R., Dai, J., Li, S., Qin, W., A review of cellulose and its derivatives in biopolymer-based for food packaging application. *Trends Food Sci. Technol.*, 112, 532–546, 2021. doi:10.1016/j.tifs.2021.04.016.

9. Jin, K., Tang, Y., Liu, J., Wang, J., Ye, C., Nanofibrillated cellulose as coating agent for food packaging paper. *Int. J. Biol. Macromol.*, 168, 331–338, 2021. https://doi.org/10.1016/j.ijbiomac.2020.12.066.
10. Javanbakht, S. and Shaabani, A., Carboxymethyl cellulose-based oral delivery systems. *Int. J. Biol. Macromol.*, 133, 21–29, 2019. https://doi.org/10.1016/j.ijbiomac.2019.04.079.
11. Saha, A., Tyagi, S., Gupta, R.K., Tyagi, Y.K., Natural gums of plant origin as edible coatings for food industry applications. *Crit. Rev. Biotechnol.*, 37, 8, 959–973, 2017. https://doi.org/10.1080/07388551.2017.1286449.
12. Wang, C., Fernández de Ávila, B.E., Mundaca-Uribe, R., Lopez-Ramirez, M.A., Ramírez-Herrera, D.E., Shukla, S., Steinmetz, N.F., Wang, J., Active Delivery of VLPs Promotes Anti-Tumor Activity in a Mouse Ovarian Tumor Model. *Small*, 16, 20, 1–26, 2020. https://doi.org/10.1002/smll.201907150.
13. Tosh, B., Thermal analysis of cellulose esters prepared from different molecular weight fractions of high α-cellulose pulp. *Indian J. Chem. Technol.*, 18, 6, 451–457, 2011.
14. Kaur, B., Ariffin, F., Bhat, R., Karim, A., A. Progress in starch modification in the last decade. *Food Hydrocolloids*, 26, 2, 398–404, 2012. https://doi.org/10.1016/j.foodhyd.2011.02.016.
15. Dara, P.K., Raghavankutty, M., Sebastian, N., Chatterjee, N.S., Mathew, S., Ravishankar, C.N., Anandan, R., Rheological, Physico-chemical, and Surface-Active Properties of Gelatin Extracted from Bigeye Tuna (Thunnus obesus) Skin Waste. *J. Aquat. Food Prod. Technol.*, 29, 5, 428–444, 2020. https://doi.org/10.1080/10498850.2020.1749745.
16. Ahmed, M.A., Al-Kahtani, H.A., Jaswir, I., AbuTarboush, H., Ismail, E.A., Extraction and characterization of gelatin from camel skin (potential halal gelatin) and production of gelatin nanoparticles. *Saudi J. Biol. Sci.*, 27, 6, 1596–1601, 2020. https://doi.org/10.1016/j.sjbs.2020.03.022.
17. Pellicer, E., Nikolic, D., Sort, J., Bar., M.D., Zivic, F., Grujovic, N., Grujic, R., Pelemis, S., Advances in applications of industrial biomaterials. *Adv. Appl. Ind. Biomater.*, vol. 1, 1–214, 2017. https://doi.org/10.1007/978-3-319-62767-0.
18. Moon, S., II, Lee, C.W., Taniguchi, I., Miyamoto, M., Kimura, Y., Melt/solid polycondensation of L-lactic acid: An alternative route to poly(L-lactic acid) with high molecular weight. *Polymer*, 42, 11, 5059–5062, 2001. https://doi.org/10.1016/S0032-3861(00)00889-2.
19. Arancibia, M.Y., López-Caballero, M.E., Gómez-Guillén, M.C., Fernández-García, M., Fernández-Martín, F., Montero, P., Antimicrobial and rheological properties of chitosan as affected by extracting conditions and humidity exposure. *Lwt*, 60, 2, 802–810, 2015. https://doi.org/10.1016/j.lwt.2014.10.019.
20. Abdella, A.A., Ulber, R., Zayed, A., Chitosan-toluidine blue beads for purification of fucoidans. *Carbohydr. Polym.*, 231, 115686, 2020. https://doi.org/10.1016/j.carbpol.2019.115686.
21. Lin, M.G., Lasekan, O., Saari, N., Khairunniza-Bejo, S., Effect of edible coatings based on chitosan and carrageenan on longan (Dimocarpus longan) fruits after. *CyTA J. Food*, 16, 1, 490–497, 2018. https://doi.org/10.1080/19476337.2017.1414078.
22. Chaudhary, S., Kumar, S., Kumar, V., Sharma, R., Chitosan nanoemulsions as advanced edible coatings for fruits and vegetables: Composition, fabrication and developments in last decade. *Int. J. Biol. Macromol.*, 152, 154–170, 2020. https://doi.org/10.1016/j.ijbiomac.2020.02.276.
23. Vert, M., Polyme`res de fermentation, L'Act. *Chim.*, 1, 79, 2002.
24. Hsiao, C.J., Lin, J.F., Wen, H.Y., Lin, Y.M., Yang, C.H., Huang, K.S., Shaw, J.F., Enhancement of the stability of chlorophyll using chlorophyll-encapsulated polycaprolactone microparticles based on droplet microfluidics. *Food Chem.*, 306, June 2019, 125300, 2020. https://doi.org/10.1016/j.foodchem.2019.125300.
25. Zolfagharlou Kouhi, M., Behzad, T., Ghasemi-Mobarakeh, L., Allafchian, A., Moazzami Goudarzi, Z., Enayati, M.S., Proceeding toward the development of poly(ε caprolactone)/

cellulose microfibrils electrospun biocomposites using a novel ternary solvent system. *J. Text. Inst.*, 111, 2, 249–259, 2020. https://doi.org/10.1080/00405000.2019.1633216.
26. Mondal, D., Griffith, M., Venkatraman, S.S., Polycaprolactone-based biomaterials for tissue engineering and drug delivery: Current scenario and challenges. *Int. J. Polym. Mater. Polym. Biomater.*, 65, 5, 255–265, 2016. https://doi.org/10.1080/00914037.2015.1103241.
27. Okada, M., Chemical syntheses of biodegradable polymers. *Prog. Polym. Sci. (Oxford)*, 27, 1, 87–133, 2002. https://doi.org/10.1016/S0079-6700(01)00039-9.
28. Jacobsen, S., Plastcising PLA. *Polym. Eng. Sci.*, *39*, 1303–1310, 1999.
29. Lv, C., Liu, D., Tian, H., Xiang, A., Non-isothermal crystallization kinetics of polyvinyl alcohol plasticized with glycerol and pentaerythritol. *J. Polym. Res.*, 27, 66, 1–7, 2020. https://doi.org/10.1007/s10965-020-2038-4.
30. Maharana, T., Mohanty, B., Negi, Y.S., Melt–solid polycondensation of lactic acid and its biodegradability. *Prog. Polym. Sci.*, *34*, 1, 99–124, 2009. ISSN 0079-6700, 2009. https://doi.org/10.1016/j.progpolymsci.2008.10.001.
31. Zeng, J.B., Li, Y.D., Zhu, Q.Y., Yang, K.K., Wang, X.L., Wang, Y.Z., A novel biodegradable multiblock poly(ester urethane) containing poly(l-lactic acid) and poly(butylene succinate) blocks. *Polymer*, 50, 5, 1178–1186, 2009. https://doi.org/10.1016/j.polymer.2009.01.001.
32. Farmahini-Farahani, M., Xiao, H., Zhao, Y., Poly lactic acid nanocomposites containing modified nanoclay with synergistic barrier to water vapor for coated paper. *J. Appl. Polym. Sci.*, 131, 20, 1–6, 2014. https://doi.org/10.1002/app.40952.
33. Jayasekara, R., Harding, I., Bowater, I., Lonergan, G., Biodegradability of a selected range of polymers and polymer blends and standard methods for assessment of biodegradation. *J. Polym. Environ.*, 13, 3, 231–251, 2005. https://doi.org/10.1007/s10924-005-4758-2.
34. Qiu, C., Li, X., Ji, N., Qin, Y., Sun, Q., Xiong, L., Rheological properties and microstructure characterization of normal and waxy corn starch dry heated with soy protein isolate. *Food Hydrocoll.*, 48, 1–7, 2015. https://doi.org/10.1016/j.foodhyd.2015.01.030.
35. Sadek, C., Pauchard, L., Schuck, P., Fallourd, Y., Pradeau, N., Le Floch-Fou.r., C., Jeantet, R., Mechanical properties of milk protein skin layers after drying: Understanding the mechanisms of particle formation from whey protein isolate and native phosphocaseinate. *Food Hydrocoll.*, 48, 8–16, 2015. https://doi.org/10.1016/j.foodhyd.2015.01.014.
36. Wang, Q. and Padua, G.W., Properties of zein films coated with drying oils. *J. Agric. Food. Chem.*, 53, 9, 3444–3448, 2005. https://doi.org/10.1021/jf047994n.
37. Satitsri, S. and Muanprasat, C., Chitin and Chitosan Derivatives as Biomaterial Resources for Biological and Biomedical Applications. *Molecules*, 25, 24, 1–25, 2020. https://doi.org/10.3390/MOLECULES25245961.
38. Noor, N.Q.I.M., Razali, R.S., Ismail, N.K., Ramli, R.A., Razali, U.H.M., Bahauddin, A.R., Zaharudin, N., Rozzamri, A., Bakar, J., Shaarani, S.M., Application of Green Technology in Gelatin Extraction: A Review. *Processes*, *9*, 2227, 2021. https://doi.org/10.3390/pr9122227.
39. Aranaz, I., Alcántara, A.R., Civera, M.C., Arias, C., Elorza, B., Caballero, A.H., Acosta, N., Chitosan: An overview of its properties and applications. *Polymers*, 13, 19, 1–27, 2021. https://doi.org/10.3390/polym13193256.
40. Davison-Kotler, E., Marshall, W.S., García-Gareta, E., Sources of collagen for biomaterials in skin wound healing. *Bioengineering*, 6, 3, 1–15, 2019. https://doi.org/10.3390/bioengineering6030056.
41. Farooq, A., Patoary, M.K., Zhang, M., Mussana, H., Li, M., Naeem, M.A., Mushtaq, M., Farooq, A., Liu, L., Cellulose from sources to nanocellulose and an overview of synthesis and properties of nanocellulose/zinc oxide nanocomposite materials. *Int. J. Biol. Macromol.*, 154, 1050–1073, 2020. https://doi.org/10.1016/j.ijbiomac.2020.03.163.

42. Chandel, V., Biswas, D., Roy, S., Vaidya, D., Verma, A., Gupta, A., Current Advancements in Pectin: Extraction, Properties and Multifunctional Applications. *Foods*, 11, 17, 1–30, 2022. https://doi.org/10.3390/foods11172683.
43. Lee, K.Y. and Mooney, D.J., Alginate: Properties and biomedical applications. *Prog. Polym. Sci. (Oxford)*, 37, 1, 106–126, 2012. https://doi.org/10.1016/j.progpolymsci.2011.06.003.
44. Habibi, H. and Khosravi-Darani, K., Effective variables on production and structure of xanthan gum and its food applications: A review. *Biocatal. Agric. Biotechnol.*, 10, May 2016, 130–140, 2017. https://doi.org/10.1016/j.bcab.2017.02.013.
45. Petri, D.F.S., Xanthan gum: A versatile biopolymer for biomedical and technological applications. *J. Appl. Polym. Sci.*, 132, 23, 2015. https://doi.org/10.1002/app.42035.
46. Carvalho, A.J.F., Starch: Major sources, properties and applications as thermoplastic materials. *Monomers, Polym. Compos. Renew. Resour.*, 1, 321–342, 2008. https://doi.org/10.1016/B978-0-08-045316-3.00015-6.
47. Espinoza, S.M., Patil, H., II, San Martin Martinez, E., Casa.as Pimentel, R., Ige, P.P., Poly-ε-caprolactone (PCL), a promising polymer for pharmaceutical and biomedical applications: Focus on nanomedicine in cancer. *Int. J. Polym. Mater. Polym. Biomater.*, 69, 2, 85–126, 2020. https://doi.org/10.1080/00914037.2018.1539990.
48. Allen., K.A., Processing of Thermoplastic Composites, in: *Polymer and Composites Processing*, M. Kessler (Ed.), Online-Iowa State University, USA, 2012.
49. Pechová, V., Gajdziok, J., Muselík, J., Vetchý, D., Development of Orodispersible Films Containing Benzydamine Hydrochloride Using a Modified Solvent Casting Method. *AAPS PharmSciTech*, *19*, 6, 2509–2518, 2018. https://doi.org/10.1208/s12249-018-1088-y.
50. Flórez, M., Cazón, P., Vázquez, M., Selected Biopolymers' Processing and Their Applications: A Review. *Polymers*, 15, 3, 2023. https://doi.org/10.3390/polym15030641.
51. Kausar, A., Polymer coating technology for high performance applications: Fundamentals and advances. *J. Macromol. Science, Part A: Pure Appl. Chem.*, 55, 5, 440–448, 2018. https://doi.org/10.1080/10601325.2018.1453266.
52. Cisneros-Zevallos, L. and Krochta, J.M., Dependence of coating thickness on viscosity of coating solution applied to fruits and vegetables by dipping method. *J. Food Sci.*, 68, 2, 503–510, 2003. https://doi.org/10.1111/j.1365-2621.2003.tb05702.x.
53. Parreidt, T.S., Schott, M., Schmid, M., Müller, K., Effect of presence and concentration of plasticizers, vegetable oils, and surfactants on the properties of sodium-alginate-based edible coatings. *Int. J. Mol. Sci.*, *19*, 3, 1–21, 2018. https://doi.org/10.3390/ijms19030742.
54. Md Nor, S. and Ding, P., Trends and advances in edible biopolymer coating for tropical fruit: A review. *Food Res. Int.*, 134, October 2019, 109208, 2020. https://doi.org/10.1016/j.foodres.2020.109208.
55. Andrade, R.D., Skurtys, O., Osorio, F.A., Atomizing Spray Systems for Application of Edible Coatings. *Compr. Rev. Food Sci. Food Saf.*, 11, 3, 323–337, 2012. https://doi.org/10.1111/j.1541-4337.2012.00186.x.
56. Khan, M.K., II, Nazir, A., Maan, A.A., Electrospraying: a Novel Technique for Efficient Coating of Foods. *Food Eng. Rev.*, 9, 2, 112–119, 2017. https://doi.org/10.1007/s12393-016-9150-6.
57. Bhardwaj, N. and Kundu, S.C., Electrospinning: A fascinating fiber fabrication technique. *Biotechnol. Adv.*, 28, 3, 325–347, 2010. https://doi.org/10.1016/j.biotechadv.2010.01.004.
58. Zhao, P., Gu, H., Mi, H., Rao, C., Fu, J., Turng, L.-S., Fabrication of Scaffolds in Tissue Engineering: A Review. *Front. Mech. Eng.*, 13, 107–119, 2018.
59. Valino, A.D., Dizon, J.R.C., Espera, A.H., Chen, Q., Messman, J., Advincula, R.C., Advances in 3D printing of thermoplastic polymer composites and nanocomposites. *Prog. Polym. Sci.*, 98, 101162, 2019. https://doi.org/10.1016/j.progpolymsci.2019.101162.

60. Stansbury, J.W. and Idacavage, M.J., 3D printing with polymers: Challenges among expanding options and opportunities, in: *Dental Materials*, vol. 32, pp. 54–64, Elsevier Inc, USA, 2016, https://doi.org/10.1016/j.dental.2015.09.018.
61. Smith, G., Schenck, R., Lynd, L., Environmental Life Cycle Assessment of Polylactic Acid and Polyethylene Terephthalate Bottles. *J. Ind. Ecol.*, *27*, 1, 194–206, 2023.
62. Laine, C., Harlin, A., Hartman, J., Hyvärinen, S., Kammiovirta, K., Krogerus, B., Pajari, H., Rautkoski, H., Setälä, H., Sievänen, J., Uotila, J., Vähä-Nissi, M., Hydroxyalkylated xylans - Their synthesis and application in coatings for packaging and paper. *Ind. Crops Prod.*, 44, 692–704, 2013. https://doi.org/10.1016/j.indcrop.2012.08.033.
63. Wyrwa, J. and Barska, A., Innovations in the food packaging market: active packaging. European. *Food Res. Technol.*, 243, 10, 1681–1692, 2017. https://doi.org/10.1007/s00217-017-2878-2.
64. Jideani, V.A. and Vogt, K., Antimicrobial Packaging for Extending the Shelf Life of Bread—A Review. *Crit. Rev. Food Sci. Nutr.*, 56, 8, 1313–1324, 2016. https://doi.org/10.1080/10408398.2013.768198.
65. Luzi, F., Puglia, D., Dominici, F., Fortunati, E., Giovanale, G., Balestra, G.M., Torre, L., Effect of gallic acid and umbelliferone on thermal, mechanical, antioxidant and antimicrobial properties of poly (vinyl alcohol-co-ethylene) films. *Polym. Degrad. Stab.*, 152, 162–176, 2018. https://doi.org/10.1016/j.polymdegradstab.2018.04.015.
66. Cai, J., Lu, W., kan, Q., Chen, X., Cao, Y., Xiao, J., Volatile composition changes of fruits in a biopolymer-coated polyethylene active packaging: Effects of modified atmosphere and packaging-shaped bacterial community. *Food Res. Int.*, 152, 110843, 2022. ISSN 0963-9969, https://doi.org/10.1016/j.foodres.2021.110843.
67. Kanatt, S.R., Development of active/intelligent food packaging film containing Amaranthus leaf extract for shelf life extension of chicken/fish during chilled storage. *Food Packag. Shelf Life*, 24, February 2019, 100506, 2020. https://doi.org/10.1016/j.fpsl.2020.100506.
68. Zhang, L., Liu, Z., Sun, Y., Wang, X., Li, L., Combined antioxidant and sensory effects of active chitosan/zein film containing α-tocopherol on Agaricus bisporus. *Food Packag. Shelf Life*, 24, September 2019, 100470, 2020. https://doi.org/10.1016/j.fpsl.2020.100470.
69. Zhang, W., Shu, C., Chen, Q., Cao, J., Jiang, W., The multi-layer film system improved the release and retention properties of cinnamon essential oil and its application as coating in inhibition to penicillium expansion of apple fruit. *Food Chem.*, 299, June, 125109, 2019. https://doi.org/10.1016/j.foodchem.2019.125109.
70. Priyadarshi, R., Sauraj, Kumar, B., Deeba, F., Kulshreshtha, A., Negi, Y.S., Chitosan films incorporated with Apricot (Prunus armeniaca) kernel essential oil as active food packaging material. *Food Hydrocolloids*, 85, 158–166, 2018. https://doi.org/10.1016/j.foodhyd.2018.07.003.
71. Mushtaq, M., Gani, A., Gani, A., Punoo, H.A., Masoodi, F., A. Use of pomegranate peel extract incorporated zein film with improved properties for prolonged shelf life of fresh Himalayan cheese (Kalari/kradi). *Innovative Food Sci. Emerging Technol.*, 48, 2017, 25–32, 2018. https://doi.org/10.1016/j.ifset.2018.04.020.
72. Xia, C., Wang, W., Wang, L., Liu, H., Xiao, J., Multilayer zein/gelatin films with tunable water barrier property and prolonged antioxidant activity. *Food Packag. Shelf Life*, 19, November 2018, 76–85, 2019. https://doi.org/10.1016/j.fpsl.2018.12.004.
73. Riahi, Z., Priyadarshi, R., Rhim, J.W., Bagheri, R., 106314, Gelatin-based functional films integrated with grapefruit seed extract and TiO2 for active food packaging applications. *Food Hydrocoll.*, 112, 1–10, July 2020, 2021. https://doi.org/10.1016/j.foodhyd.2020.106314.
74. Correa-Pacheco, Z.N., Bautista-Baños, S., Valle-Marquina, MA., Hernández-López, M., The Effect of Nanostructured Chitosan and Chitosan-thyme Essential Oil Coatings on Colletotrichum gloeosporioides Growth in vitro and on cv Hass Avocado and Fruit Quality. *J. Phytopathol.*, 165, 297–305, 2017. https://doi.org/10.1111/jph.12562.

75. Liu, Y., Ahmed, S., Sameen, D.E., Wang, Y., Lu, R., Dai, J., Li, S., Qin, W., A review of cellulose and its derivatives in biopolymer-based for food packaging application. *Trends Food Sci. Technol.*, 112, April, 532–546, 2021. https://doi.org/10.1016/j.tifs.2021.04.016.
76. Abdul Khalil, H.P.S., Banerjee, A., Saurabh, C.K., Tye, Y.Y., Suriani, A.B., Mohamed, A., Karim, A.A., Rizal, S., Paridah, M.T., Biodegradable Films for Fruits and Vegetables Packaging Application: Preparation and Properties. *Food Eng. Rev.*, 10, 3, 139–153, 2018. https://doi.org/10.1007/s12393-018-9180-3.
77. Suresh, S., Pushparaj, C., Subramani, R., Recent development in preparation of food packaging films using biopolymers. *Food Res.*, 5, 6, 12–22, 2021. https://doi.org/10.26656/FR.2017.5(6).082.
78. Singh, M. and Sahareen, T., Investigation of cellulosic packets impregnated with silver nanoparticles for enhancing shelf-life of vegetables. *LWT*, 86, 116–122, 2017. https://doi.org/10.1016/j.lwt.2017.07.056.
79. Bandyopadhyay, S., Saha, N., Brodnjak, U.V., Sáha, P., Bacterial cellulose and guar gum based modified PVP-CMC hydrogel films: Characterized for packaging fresh berries. *Food Packag. Shelf Life*, 22, 1–9, September, 2019. https://doi.org/10.1016/j.fpsl.2019.100402.
80. Dey, D., Dharini, V., Selvam, S.P., Sadiku, E.R., Kumar, M.M., Jayaramudu, J., Gupta, U.N., Physical, antifungal, and biodegradable properties of cellulose nanocrystals and chitosan nanoparticles for food packaging application. *Mater. Today: Proc.*, 38, xxxx, 860–869, 2021. https://doi.org/10.1016/j.matpr.2020.04.885.
81. Youssef, A.M., El-Sayed, S.M., El-Sayed, H.S., Salama, H.H., Dufresne, A., Enhancement of Egyptian soft white cheese shelf life using a novel chitosan/carboxymethyl cellulose/zinc oxide bionanocomposite film. *Carbohydr. Polym.*, 151, 9–19, 2016. https://doi.org/10.1016/j.carbpol.2016.05.023.
82. Al-Moghazy, M., Mahmoud, M., Nada, A.A., Fabrication of cellulose-based adhesive composite as an active packaging material to extend the shelf life of cheese. *Int. J. Biol. Macromol.*, 160, 264–275, 2020. https://doi.org/10.1016/j.ijbiomac.2020.05.217.
83. Aydogdu, A., Sumnu, G., Sahin, S., Fabrication of gallic acid loaded Hydroxypropyl methylcellulose nanofibers by electrospinning technique as active packaging material. *Carbohydr. Polym.*, 208, 241–250, 2019. https://doi.org/10.1016/j.carbpol.2018.12.065.
84. Nouri Ala, M.A. and Shahbazi, Y., The effects of novel bioactive carboxymethyl cellulose coatings on food-borne pathogenic bacteria and shelf life extension of fresh and sauced chicken breast fillets. *Lwt*, 111, May, 602–611, 2019. https://doi.org/10.1016/j.lwt.2019.05.092.
85. Shafipour Yordshahi, A., Moradi, M., Tajik, H., Molaei, R., Design and preparation of antimicrobial meat wrapping nanopaper with bacterial cellulose and postbiotics of lactic acid bacteria. *Int. J. Food Microbiol.*, 321, November 2019, 108561, 2020. https://doi.org/10.1016/j.ijfoodmicro.2020.108561.
86. Youssef, A.M., Assem, F.M., El-Sayed, S.M., Salama, H., Salam, A.E.-.M.H., Utilization of Edible Films and Coatings as Packaging Materials for Preservation of Cheeses. *J. Packag. Technol. Res.*, 1, 2, 87–99, 2017. https://doi.org/10.1007/s41783-017-0012-3.
87. Chen, S., Chen, Y., Wang, Z., Chen, H., Fan, D., Renewable bio-based adhesive fabricated from a novel biopolymer and soy protein. *RSC Adv.*, 11, 19, 11724–11731, 2021. https://doi.org/10.1039/d1ra00766a.
88. Rebelo, R., Fernandes, M., Fangueiro, R., Biopolymers in Medical Implants: A Brief Review. *Procedia Eng.*, 200, 236–243, 2017. https://doi.org/10.1016/j.proeng.2017.07.034.
89. Cisneros, C.G., Bloemen, V., Mignon, A., Synthetic, natural, and semisynthetic polymer carriers for controlled nitric oxide release in dermal applications: A review. *Polymers*, 13, 5, 1–26, 2021. https://doi.org/10.3390/polym13050760.

90. Chen, Y., Hung, S.-T., Chou, E., Wu, H.-S., Review of Polyhydroxyalkanoates Materials and other Biopolymers for Medical Applications. *Mini-Rev. Org. Chem.*, 15, 105–121, 2018.
91. Tsai, C.H., Wang, P.Y., Lin, I.C., Huang, H., Liu, G.S., Tseng, C.L., Ocular drug delivery: Role of degradable polymeric nanocarriers for ophthalmic application. *Int. J. Mol. Sci.*, 19, 9, 1–20, 2018. https://doi.org/10.3390/ijms19092830.
92. Nagarajan, S., Abessolo Ondo, D., Gassara, S., Bechelany, M., Balme, S., Miele, P., Kalkura, N., Pochat-Bohatier, C., Porous Gelatin Membrane Obtained from Pickering Emulsions Stabilized by Graphene Oxide. *Langmuir*, 34, 4, 1542–1549, 2018. https://doi.org/10.1021/acs.langmuir.7b03426.
93. Jammalamadaka, U. and Tappa, K., Recent advances in biomaterials for 3D printing and tissue engineering. *J. Funct. Biomater.*, 9, 1, 1–14, 2018. https://doi.org/10.3390/jfb9010022.
94. Udayakumar, G.P., Muthusamy, S., Selvaganesh, B., Sivarajasekar, N., Rambabu, K., Banat, F., Sivamani, S., Sivakumar, N., Hosseini-Bandegharaei, A., Show, P.L., Biopolymers and composites: Properties, characterization and their applications in food, medical and pharmaceutical industries. *J. Environ. Chem. Eng.*, 9, 4, 105322, 2021. https://doi.org/10.1016/j.jece.2021.105322.
95. Bonartsev, A.P., Bonartseva, G.A., Shaitan, K.V., Kirpichnikov, M.P., Poly(3-hydroxybutyrate) and poly(3-hydroxybutyrate)-based biopolymer systems. Biochemistry (Moscow) Supplement Series B. *Biomed. Chem.*, 5, 1, 10–21, 2011. https://doi.org/10.1134/S1990750811010045.
96. Joseph, S.M., Krishnamoorthy, S., Paranthaman, R., Moses, J.A., Anandharamakrishnan, C., A review on source-specific chemistry, functionality, and applications of chitin and chitosan. *Carbohydr. Polym. Technol. Appl.*, 2, January, 100036, 2021. https://doi.org/10.1016/j.carpta.2021.100036.
97. Di Martino, A., Sittinger, M., Risbud, M.V., Chitosan: A versatile biopolymer for orthopaedic tissue-engineering. *Biomaterials*, 26, 30, 5983–5990, 2005. https://doi.org/10.1016/j.biomaterials.2005.03.016.
98. Álvarez-Castillo, E., Aguilar, J.M., Bengoechea, C., María Luisa López-Castejón, A.G., Rheology and Water Absorption Properties of Alginate – Soy. *Polymers*, 13, 1807, 2021.
99. Nelson, D.W. and Gilbert, R.J., Extracellular Matrix-Mimetic Hydrogels for Treating Neural Tissue Injury: A Focus on Fibrin, Hyaluronic Acid, and Elastin-Like Polypeptide Hydrogels. *Adv. Healthcare Mater.*, 10, 22, 1–20, 2021. https://doi.org/10.1002/adhm.202101329.
100. Sellappan, L.K., Anandhavelu, S., Doble, M., Perumal, G., Jeon, J.H., Vikraman, D., Kim, H.S., Biopolymer film fabrication for skin mimetic tissue regenerative wound dressing applications. *Int. J. Polym. Mater. Polym. Biomater.*, 71, 3, 196–207, 2022. https://doi.org/10.1080/00914037.2020.1817019.
101. Hassan, B., Chatha, S.A.S., Hussain, A., II, Zia, K.M., Akhtar, N., Recent advances on polysaccharides, lipids and protein based edible films and coatings: A review. *Int. J. Biol. Macromol.*, 109, 1095–1107, 2018. https://doi.org/10.1016/j.ijbiomac.2017.11.097.
102. Janjarasskul, T. and Suppakul, P., Active and intelligent packaging: The indication of quality and safety. *Crit. Rev. Food Sci. Nutr.*, 58, 5, 808–831, 2018. https://doi.org/10.1080/10408398.2016.1225278.
103. Majid, I., Ahmad Nayik, G., Mohammad Dar, S., Nanda, V., Novel food packaging technologies: Innovations and future prospective. *J. Saudi Soc. Agric. Sci.*, 17, 4, 454–462, 2018. https://doi.org/10.1016/j.jssas.2016.11.003.
104. Jin, K., Tang, Y., Liu, J., Wang, J., Ye, C., Nanofibrillated cellulose as coating agent for food packaging paper. *Int. J. Biol. Macromol.*, 168, 331–338, 2021. https://doi.org/10.1016/j.ijbiomac.2020.12.066.
105. Fang, Z., Zhao, Y., Warner, R.D., Johnson, S.K., Active and intelligent packaging in meat industry. *Trends Food Sci. Technol.*, 61, 2, 60–71, 2017. https://doi.org/10.1016/j.tifs.2017.01.002.

16

Nano-Pesticides, Nano-Herbicides and Nano-Fertilizers: Future Perspective

Priyanka Chhabra[1]*, Akshara Johari[2], Divya Bajpai Tripathy[3] and Anjali Gupta[3]

[1]*Centre for Medical Biotechnology, Amity Institute of Biotechnology, Amity University, Greater Noida, Uttar Pradesh, India*
[2]*Division of Forensic Science, School of Basic and Applied Sciences, Galgotias University, Greater Noida, Uttar Pradesh, India*
[3]*Division of Chemistry, School of Basic Sciences, Galgotias University, Greater Noida, Uttar Pradesh, India*

Abstract
Agriculture is an essential part of the national economy in emerging countries. Rising food production rates have a major effect on the country's economic growth. The amount of fertilizer and insecticides applied determines the rate at which food is produced in the country. Food security, climate change, soil health, and water availability are all issues that directly affect food production and agricultural expansion. As a result, regulating agricultural challenging circumstances is essential in order to improve the supply of food for the increasing global population in a sustainable way. Nanotechnology, an emerging technology, offers promise in a variety of applications including solar, electronics, optical science, and medicine. To satisfy food production objectives and battle different emerging agricultural concerns, scientists are developing nanotechnology solutions.

***Keywords*:** Nanotechnology, agriculture, nano-pesticides, nano-herbicides, nano-fertilizers

16.1 Introduction

Nanotechnology is the application of chemistry, physics, biology, medicine, and engineering to manipulate the structure of matter on the nanoscale scale. This allows for the creation of materials with unique qualities such as a huge surface area, a specific point of action, and a sluggish reaction time. "Nano" is derived from the Greek word "nanos," which means "dwarf" (little). "Nanotechnology is mostly about working with one atom, one molecule, or ions to distinguish, merge, and alter the shape of materials" [1]. Whether they are made or found in nature, nanoparticles (NPs) have at least one dimension that is between 1 and 100 nm [1]. The most important things about NPs are their shape, size, hydrophobicity, solubility-release of toxic species, surface area or roughness, surface species contaminations

**Corresponding author*: pchhabra188@gmail.com

Divya Bajpai Tripathy, Anjali Gupta and Arvind Kumar Jain (eds.) Multifunctional Materials: Engineering and Biological Applications, (419–440) © 2025 Scrivener Publishing LLC

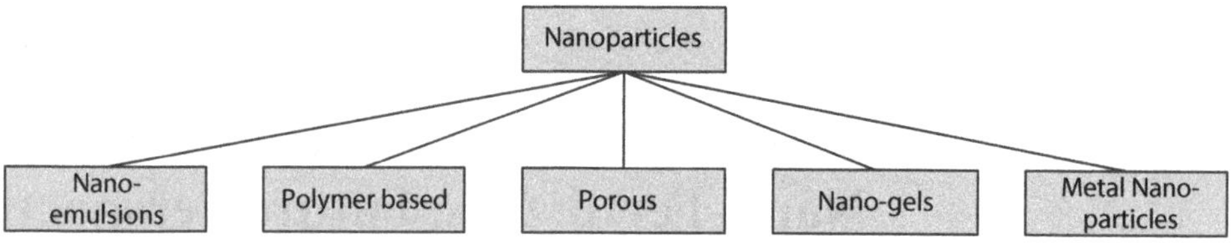

Figure 16.1 Types of nanoparticles [2].

or adsorption, when they were made or their history, reactive oxygen species (ROS) O2/ H2O, their ability to produce ROS, structure, composition, competitive binding sites with receptors, dispersion, and aggregation. Nanotechnology applications in agriculture, particularly crop production, have recently received a lot of attention, with the primary goal of making better use of resources by developing pesticide delivery systems or sensors (for example, for water or nutrient status). Most of the so-called nano-pesticides and nano-fertilizers suggested to date include reformulation of registered active ingredients (AIs), with the goal of achieving superior performance relative to the existing AIs and mitigating the primary shortcomings of present agrochemical products [2, 3]. Targeted delivery of the AI to the pest and/or greater fertilizer's usefulness are among the tactics aimed to help maintain or even boost yields with drastically decreased application rates while minimizing adverse effects of agriculture on ecosystems and human health [4]. It has been argued that nano-agrochemicals may be superior to traditional goods, and there are high hopes for the sector's nanotechnology uses. In spite of the fact that more than a decade's worth of research has been produced, no comprehensive comparison has yet been conducted [4]. One of the primary environmental issues facing the world today is the rising buildup of dangerous and non-recyclable waste products brought on by the unchecked use of chemicals.

Expanding the use of biotechnology tools and techniques is one strategy to tackle this problem since they can safeguard beneficial biota, boost agricultural output, and lessen the detrimental effects of pesticides on agroecosystems and the entire biosphere [5, 6]. It takes a lot of different pesticides to manage weeds, pests, and crop diseases in intensive farming. As a result, nanotechnology is beginning to show promise for resolving these issues in agricultural and food production. Recently, research on agrochemical industry applications of nanotechnology has intensified, particularly in the creation and design of novel plant-protection products [6]. As a result, nano-based plant-protection solutions may and will be crucial to agriculture in the future.

16.2 Nanotechnology and Its Importance in Agriculture

Nanotechnology is important in many fields of science, including food technology, crop improvement (such as genetically modified crops), seed technology, precision farming (location-based management), nano-fertilizers to balance crop nutrition, plant disease diagnosis, weed control, water management, biosensors, and pest control [7]. As it stands now, controlled environment agriculture (CEA) technology is a great way to bring nanotechnology into agriculture and use it there [8]. Precision farming has been a goal for a long time because it is the best way to use fertilizers, pesticides, herbicides, and other inputs to get the most out of a crop while using the least amount of them [9]. Precision farming uses computers, global positioning system (GPS), geological information systems (GIS), and remote sensing devices for tracking extremely localized environmental factors [9, 10]. This lets farmers find out if crops are growing as well as they can, or if there are problems with the crops or the ecological environment and where they are [10]. The use of precision agriculture can also help reuse farm waste and reduce pollution in the environment. One of the most important things that nanotechnology-enabled devices will do is make it easier for autonomous sensors to be linked to a global positioning system (GPS) for real-time analysis [11]. These nano-sensors could be put all over the field, where they could watch how the soil is doing and how the crops are growing. When these two technologies are combined in sensors, they will make more sensitive equipment that can react faster to changes in the environment [11, 12]. For example: (a) Nano-sensors that use carbon nanotubes or nano-cantilevers can trap and measure small molecules [11]. b) NPs or nano-surfaces can be made to send an electrical or chemical signal when bacteria or other pests or disease-causing organisms are nearby [12]. c) Other nanosensors work by initiating an enzymatic response or by employing nanoengineered branching molecules known as dendrites as probes to bind to certain chemicals and proteins [11]. They are distinct from bulk materials in that they have certain physio-chemical characteristics. Because of their large surface area to volume ratio and therefore increased molecular reactivity, they have applications in the chemical, optical, electrical, mechanical, and biological areas [13]. Nanotechnology has grown into a fast-developing technical subject with a huge potential economic impact as a result of the extraordinary properties of NPs [14]. Although certain nanoscale materials have been in use for years (such as window and sunglass lenses), others are still in the research and development stages (such as sunscreen, cosmetics, and textiles). It is quite remarkable how nanotechnology advancements are being applied to the preservation of the environment and human health. Engineered nanomaterials (ENMs) have exceptional physiochemical properties that make them appropriate for a wide range of applications in many research domains including biology, energy, cosmetics, medicines, and many others [15]. During their production operations, ENMs may convert waste or go through recycling procedures. As a result of their rapidly expanding applications, ENMs have prompted serious concerns about their potential consequences on both living (biotic) and nonliving (abiotic) ecosystem components [16]. In addition to playing a crucial role in the bioaccumulation and transportation of ENMs in food chains at various trophic levels, plants are essential for the maintenance of ecological systems. Encountering/interfering with internal metabolic structures and pathways is a simple process for ENMs to enter plant cells [17]. On the other hand, analogous to the negative effects of biotechnology on the development of resistance or genetic modification, nanotechnology has detrimental effects on the biological and chemical environment. Before using nanotechnology in agricultural techniques, an environmental impact study must also be conducted. In this article, we examine the use of nanotechnology tools in agriculture, including nano-fertilizer, nano pesticide, nanosensor, and nano herbicide, as well as their effect as growth promoters, nutritional supplements, and pesticides, and their environmental impact evaluation [18]. Also mentioned is the advantage of nanotechnology over conventional agricultural operations.

16.3 Functions of Nanomaterials in Agriculture

Undoubtedly, nanotechnology is helping to revitalize the agriculture and food-producing businesses. In comparison to traditional fertilizers [19], NPs have shown to be a potential option for the production of nano-fertilizers. Therefore, using nano-fertilizers in agriculture can cut down on the overuse of chemical fertilizers, reducing environmental pollution. The use of NPs was encouraged due to their advantages in improving nutritional qualities and plant resistance to stressors [20]. Utilizing carriers that can entrap, encapsulate, absorb, or attach active chemicals, agricultural nano-formulations were created. Undoubtedly, nanotechnology is helping to revitalize agriculture and food-producing businesses [21]. In comparison to traditional fertilizers, NPs have shown to be a potential option for the production of nano-fertilizers. Therefore, using nano-fertilizers in agriculture can cut down on the overuse of chemical fertilizers, reducing environmental pollution [22]. The use of NPs was encouraged due to their advantages in improving nutritional qualities and plant resistance to stressors. Utilizing carriers that can entrap, encapsulate, absorb, or attach active chemicals, agricultural nano-formulations were created [23]. By giving plants resilience to diverse challenges, nano-priming is a cutting-edge seed priming method that helps to increase seed germination, seed growth, and yield. When compared to all other seed priming techniques, nano-priming is noticeably more successful. NPs offer a significant part in seed priming by enhancing surface flexibility and electron exchange abilities, which are linked to many components of plant cells and tissues [24]. Inducing the creation of nanopores in the shoot, which aids in the uptake of water, activating ROS/antioxidant processes in the seeds, creating hydroxyl radicals to weaken cell walls, and acting as an inducer for the fast hydrolysis of starch are all effects of nano-priming. It also facilitates the distribution of H2O2, or ROS, across biological membranes and increases the production of aquaporin genes, which are involved in water intake as shown in Figure 16.2. By stimulating amylase, nano-priming causes starch breakdown, which in turn stimulates seed germination [25].

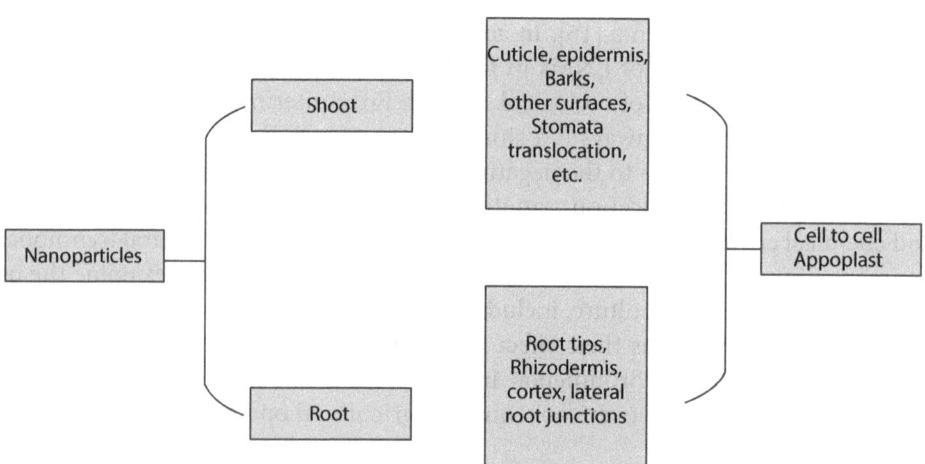

Figure 16.2 Illustration of nanoparticles and its transportation in plants [24].

16.3.1 Crop Protections

Crops are attacked by a variety of adversaries, including weeds (monocots, dicots, and parasitic weeds), pests (insects, mites, mice, birds, and mammals), and phytopathogens (bacteria, fungi, and viruses). There are various types of attackers that can be deployed, including stand reducers, photosynthetic rate reducers, leaf senescence accelerators, light stealers, assimilate sappers, and tissue consumers. For instance, Helicoverpa armigera consumes more than 150 plant species worldwide, including tomatoes, chickpeas, pigeon peas, and others. A polyphagous pest, fruitflies (Bactrocera dorsalis Hendel) eat a variety of fruits and vegetables, including guava, apple, mango, orange, and banana [26, 27]. Additionally, insects serve as carriers of several plant diseases that can seriously harm crops [28]. In order to be effective, nano-pesticides "either include very small particles of pesticidal active substances or other small designed structures." Nano-pesticides can improve agricultural formulations' wettability, dispersion, and ability to stop undesirable pesticide movement (i.e., by reducing organic solvent runoff). Nanomaterials (NMs) and bio-composites possess the stiffness, permeability, crystallinity, thermal stability, solubility, and biodegradability needed to formulate nano-pesticides [29]. Additionally, nano-pesticides have a large specific surface area, increasing their affinity to the target. Some of the nano-pesticide delivery methods that have recently been considered for plant protection include nanoemulsions, nanoencapsulates, nanocontainers, and nanocages [29, 30]. Basically, the nano-formulations ought to break down rapidly in the soil and more gradually in plants, leaving residual amounts in meals that are below regulatory norms. As a nano-pesticide, sodium dodecyl sulfate (SDS) modified photocatalytic TiO2/Ag NMs were mixed with dimethomorph (DMM), a pesticide frequently utilized in agriculture. In trials on vegetable seedlings (of cabbage and cucumber), the pesticide's efficiency was improved while its dispersivity and soil breakdown rates were raised. The SDS modification of the NMs significantly improved DMM absorption [30, 31]. The chitosan and alginate microencapsulation method was employed to create the SDS-modified Ag/TiO2 imidacloprid nano-formulation. It was examined on soybean plants that had been replanted in soil with a pH of 6.2 and a dry matter content of 3.1%. The formulation residues in the soil and plants dissolved more quickly over the first eight days, and by 20 days, they were negligible to undetectable. In the aforementioned applications, SDS was employed to speed up the photo-degradation of soil-borne NPs. As an alternative, highly photodegradable Ag/TiO2 particles (5e7 nm) were created and produced utilizing polyoxyethylene laurel ether (POL). They examined the effectiveness of POL and SDS-produced Ag/TiO2 NPs for the breakdown of the herbicide 2,4-D under visible and UV light. Due to the absence of any national or international institutional frameworks or norms relating to the use of NMs in agriculture, none of the aforementioned outcomes can be regarded as being secured. Pesticide toxicity or biosafety is yet another important issue in agricultural productivity [19]. Applications of nano-pesticides raise concerns about the long-term impact of pesticides on human health and the environment. Nanoemulsions have the potential to be more effective pesticide delivery methods because of their greater kinetic stability, smaller size, reduced viscosity, and optical transparency [18, 32]. At the nanoscale, formulation stability is also a crucial factor. Polymer stabilizers including poly(acrylic acid)-b-poly(butyl acrylate (PAA-b-PBA), polyvinylpyrrolidone (PVP), and polyvinyl alcohol were effectively used to manufacture a stable nano-pesticide (bifenthrin) (PVOH). Bifenthrin particles measuring 60e200 nm were created using a flash nano-precipitation process. Commercial application of such approaches must take into account the polymers' long-term durability. Although not intended for agricultural use, a nano-permethrin formulation, stabilised with natural plant surfactants and free of synthetic polymers, proved to be an effective larvicide. It is necessary to conduct more research into the usage of natural stabilizers in nano-pesticide formulations for agricultural plant protection [20]. The creation of NPs that may be employed as a coating or protective layer to permit the delayed release of conventional pesticides and fertilizers could be another area of advanced study. For encapsulating "agrochemicals such as fertilizers, plant growth boosters, and insecticides," nano-clay materials primarily offer interacting surfaces with a high aspect ratio. In the aforementioned applications, SDS was employed to speed up the photo-degradation of soil-borne NPs [33]. As an alternative, highly photodegradable Ag/TiO_2 particles (5e7 nm) were created and produced utilizing POL. They examined the effectiveness of POL and SDS-produced Ag/TiO_2 NPs for the breakdown of the herbicide 2,4-D under visible and UV light. Due to the absence of any national or international institutional frameworks or norms relating to the use of NMs in agriculture, none of the aforementioned outcomes can be regarded as being secured. Pesticide toxicity or biosafety is yet another important issue in agricultural productivity [34]. Applications of nano-pesticides raise the uncertainties around the long-term consequences of pesticides on human health and the environment. Nanoemulsions have the potential to be more effective pesticide delivery systems because of their improved kinetic stability, smaller size, low viscosity, and optical transparency. At the nanoscale, formulation stability is also a crucial factor. Polymer stabilizers including Poly(acrylic acid)-b-Poly(butyl acrylate (PAA-b-PBA), PVP, and Polyvinyl alcohol were effectively used to manufacture a stable nano-pesticide (bifenthrin) (PVOH). Bifenthrin particles measuring 60e200 nm were created using a flash nano-precipitation process. Commercial application of such approaches must take into account the polymers' long-term durability. reported formulation of artificial polymer-free nano-permethrin as a successful larvicide that was stabilized by plant-extractable natural surfactants, while not being for agricultural use. It is necessary to conduct more research into the usage of natural stabilizers in nano-pesticide formulations for agricultural plant protection [35]. The creation of NPs that may be employed as a coating or protective layer to permit the delayed release of conventional pesticides and fertilizers could be another area of advanced study. For encapsulating "agrochemicals such as fertilizers, plant growth boosters, and insecticides," nano-clay materials primarily offer interacting surfaces with a high aspect ratio.

16.3.2 Crop Growth

To encourage their usage in agricultural applications, several researchers have recently investigated the impact of NMs on plant germination and growth [29]. Nanoscience is an emerging scientific innovation platform that includes the creation of techniques for a variety of low-cost nanotech applications aimed at improving seed germination, plant growth development, and environmental adaption. Seed germination is a critical stage of a plant's life cycle, promoting seedling development, survival, and population dynamics. However, several factors influence seed germination, including the climate, genetic makeup, moisture availability, and soil fertility. Numerous studies in this area have shown that the use of NMs improves germination, as well as plant growth and development. For example, MWCNTs promote seed germination in a wide range of crop species, including tomato, corn, soybean, barley, wheat, maize, peanut, and garlic. Furthermore, it was demonstrated that Fe/SiO2 NMs have tremendous potential for improving seed germination in barley and maize. In a similar vein, nano SiO2, TiO2, and Zeolite applications promote seed germination in crop plants. Despite much research on the benefits of NMs, the fundamental processes by which they can expedite germination remain unknown. A few studies have demonstrated that NMs can puncture the seed coat, stimulate water absorption and consumption, activate the enzymatic system, and therefore improve germination and seedling growth. However, the process of water absorption caused by NMs inside the seed is not well known. Multi-walled carbon nanotubes (MWCNTs) are a type of NM that has received a lot of attention in both basic research and technological development due to their unique nanostructures and exceptional properties like high electrical conductivity, large and special area, and significant thermal stability. NMs such as ZnO, TiO2, MWCNTs, FeO, ZnFeCu-oxide, and hydroxyfullerenes have been demonstrated to increase crop quality and accelerate growth in a variety of crop species, including mustard, onion, spinach, tomato, potato, and wheat. According to studies conducted in water and soil, the long-term availability of all doped nutrients to the plant over the entire crop period of cultivation is critical for promoting germination, growth, flowering, and fruiting. For example, urea fertilizer coated with hydroxyapatite NM slowly and uniformly releases nitrogen over up to 60 days, whereas traditional bulk fertilizer releases nitrogen unevenly and only within 30 days, reducing plant nutrient efficiency and negatively impacting crop growth. In contrast, the analysis in various studies reveals inconsistent results about the favorable effects of NMs on seed germination and crop growth, including differences could be caused by a number of NM characteristics, including as size, shape, surface coating, and electrical properties, as well as dosage and administration method.

16.3.3 Soil Enhancement

Ground and structural strengthening, as well as environmental sustainability, are critical concerns that conventional soil restoration technologies are currently facing as a result of industrialization and urbanization. Traditional soil strength improvement products such as cement and chemical grouts (sodium silicate, acrylate, and epoxy) have limited in-field treatment capabilities and cause significant disruption. In addition, they contaminate the environment and are expensive to use, so it is critical to investigate new, unique materials that will address these difficulties. Pore water pressure has a considerable impact on soil strength, especially when subjected to dynamic loading because the soil is a three-phase medium composed of soil grains, water, and air between the grains. NMs are utilized to improve soil strength at the macro level by reinforcing the soil skeleton and changing the pore fluid, as explained in NMs used in soil strength enhancement. The micronization process progresses as the grain size decreases from gravel to sand, fine particle, ultrafine particle, and finally NP, in comparison to the grain size of natural rock and soil mass. We witness a decrease in particle size as well as changes in soil composition and structure since the type and characteristics of most minerals vary with particle size, such as large quartz and feldspar to sheet mica and clay. For instance, through the incorporation of mineral particles with a wide surface area within clay liners and bases, soils exhibit outstanding mechanical (such as high ductility) and chemical (such as species-selective diffusion) features on a macroscopic scale. On the nanoscale, NPs are smaller than clay particles. When a particle's size is equal to or less than the wavelength of light waves or the De Broglie wave, its periodic boundary condition is shattered, and the surface atomic density declines. As a result, the physical properties of the substance change. Meanwhile, particularly through fine soil that is not under high pressure, the NPs can readily scatter in the pore space between the soil grains. The rheological suspension qualities of the NPs have an impact on their capacity to increase soil strength because some NPs are injected into the soil in the form of NP-water suspension. Long induction times are possible for the NP suspension, during which the viscosity is rather low and behaves Newtonianly. With time, the suspension's viscosity will quickly rise, and the fluid will eventually begin to gel at a regulated gelling moment. For the following reasons, the solid-like gel can increase soil tensile strength. First, in a suspension of water, NPs like layered silicate can create a colloid [21]. The soil's pores are filled with the NP gel, which changes the fluid in the pores from a fluid state with no shear resistance to a colloidal one with shear resistance. Additionally, the gel can delay the process of pore pressure production, which will lessen the danger of liquefaction under dynamic loading. In addition, the forming gel has a high viscosity and can strengthen the bonding between sand matrixes, increasing the soil mass's shear strength. Additionally, hydraulic conductivity decreases, and the gel in the sand pores appears heavier and more concentrated at the particle contacts. To summarize, the microscopic properties of NMs, such as size, microstructure, suspension properties, surface effect, and rheology of NP suspension, alter soil composition, structure, and particle interactions, thereby improving soil strength [22].

16.3.4 Stress Tolerance

Abiotic stress, which reduces average yields for the majority of major crop plants by more than 50%, is the main factor in agricultural production worldwide, which is continually impacted by a variety of unfavorable environmental circumstances. The severity or extent of abiotic stress influences how the plant responds to it. For optimum abiotic environmental conditions to occur, the plant must grow. Abiotic stress is defined as any variation from the ideal environmental parameters that is inadequate due to physical or chemical conditions. Abiotic stresses include temperature (heat, cold, and freezing), irradiation, drought, waterlogging, heavy metals, a lack of nutrients, and salt. Numerous abiotic variables, such as salinity, soil alkalinity, heavy metal toxicity, extremely high or low temperatures, and so on, have a negative impact on plant growth and development, as well as the production of agro-economic crops. Numerous acidic soil conditions have a negative impact on soil nutrients, causing nutritional shortages in plants and impairing their normal physiological ability for growth and development. The primary cause of ROS generation is abiotic stress. Under normal conditions, the synthesis and removal of ROS are balanced. However, under abiotic stress circumstances, an increase in ROS production disrupts the ROS balance, resulting in phytotoxicity by impairing protein structure and function. ROS are created by cell organelles such as mitochondria, peroxisomes, and chloroplasts in plants. Overreduction of the electron transport chain in mitochondria produces oxygen radicals and hydrogen peroxide. Chloroplasts are the principal source of O2 and H2O2 production. These superoxides are transformed into hydrogen peroxide, either naturally or via the enzyme superoxide dismutase. O_2 radicals are formed in the peroxisomal matrix when xanthine and hypoxanthine are converted to uric acid in the presence of the enzyme xanthine oxidase.

They damage biomolecules such as proteins, lipids, carbohydrates, and DNA, resulting in cell death. Changes in the environment can disrupt a plant's metabolic balance, causing it to house metabolic and genetic activity within the cell. Plants retain the defense mechanisms they ingested to deal with abiotic stress situations. These pathways cause the plant's metabolic systems to be reprogrammed in order to support the biophysical and chemical developments caused by abiotic stress. The advanced stage of the stress response includes the regulation of proteins involved in cellular damage prevention as well as the synthesis of stress-related genes. Secondary metabolites help plants battle abiotic stress by maintaining cell integrity, insulating the photosystem from ROS, aiding signal transmission, and assisting in polyamine production. The plant cell wall acts as a physical barrier to stress perception and actively contributes to plant adaptation in both biotic and abiotic stress conditions. Salinity is a major abiotic stressor. It hinders food production and worsens the rising need for agricultural products for food. Salinity is a major concern in the scientific\community to attain sustainable crop production. Since the bulk of the important agricultural plant species fall under the lycophyte classification, salt stress is one of the most serious environmental stresses that can seriously impair crop yield. A variety of biochemical and physiological processes related to plant development and production are negatively impacted by salinity stress. Cold stress has actual impacts such as permeability distortion and ion leakage from the membrane, both of which are damaging to the plant's germination and development. Certain species, on the other hand, are more vulnerable to cold stress than others; these plants have lesser membrane thickness than sensitive plants. Heat stress is defined as a high temperature at such a high level for so long that it permanently damages plant growth and development. Heat stress promotes the production of ROS and causes oxidative stress, which causes membrane ion leakage and membrane lipid degeneration, followed by protein degradation. Because heavy metal stress increases toxicity and stunts plant development, it poses a severe danger to crops globally. Plants under heavy metal stress develop less because of a lack of critical nutrients caused by disruptions in the absorption of important supplements as well as the inhibition of enzyme activity. NPs help plants defend themselves against abiotic stress, which reduces the harm it causes. NPs' very tiny size makes it easy for them to regulate ion channels, which supports plant development and seed germination. In addition, their enormous surface areas support high absorption rates and targeted delivery of chemicals. On the other hand, NPs contribute to the generation of ROS and are hazardous to plants. The increased ROS concentration caused by NP exposure may be connected to the amplification of stress signals, which cause plants to activate their effective defensive systems. Thus, the signaling substance's initiation encourages gene expression, which leads to better stress resistance. It is stated that more molecular and cellular study is needed to fully understand how NPs work in plant systems. It is also critical to determine whether NPs act as stress promoters or stress inhibitors.

16.3.5 Precision Farming

The rapidly developing field of study known as nanoscience and technology has completely transformed numerous technical specialties, including the electrical, pharmaceutical, power, and automobile sectors. This technology has enormous potential in the realms of medicine and plants, including biosensors, tissue synthesis, smart medication release, and nano-fertilizers. In comparison to their bulk counterparts, NPs have larger surface-to-volume fractions and unique properties that make them particularly suitable for utilization [13]. A promising method to boost plant development and productivity is the collaboration of nanotechnologies with agricultural and food research. With the aid of this technology, farmers can make efficient and accurate use of NPs together with other scarce resources like water and pricey synthetic fertilizers. By reducing the use of energy resources and waste, precision farming aids in the sustainability and environmental friendliness of agriculture [15].

16.4 Focused Nano-Agromaterials

16.4.1 Nano-Fertilizers

Nutrient fertilizers known as nano-fertilizers are made up entirely or in part of formulations that are nanostructured and may be given to plants, enabling the effective absorption or controlled release of active components. Due to the low plant absorption efficiency of conventional bulk fertilizers, greater dosages must be used. This goal can be achieved sustainably and accurately through nanotechnology, which is why researchers are working to develop various metal and metal oxide NPs for application in agriculture and plant science.

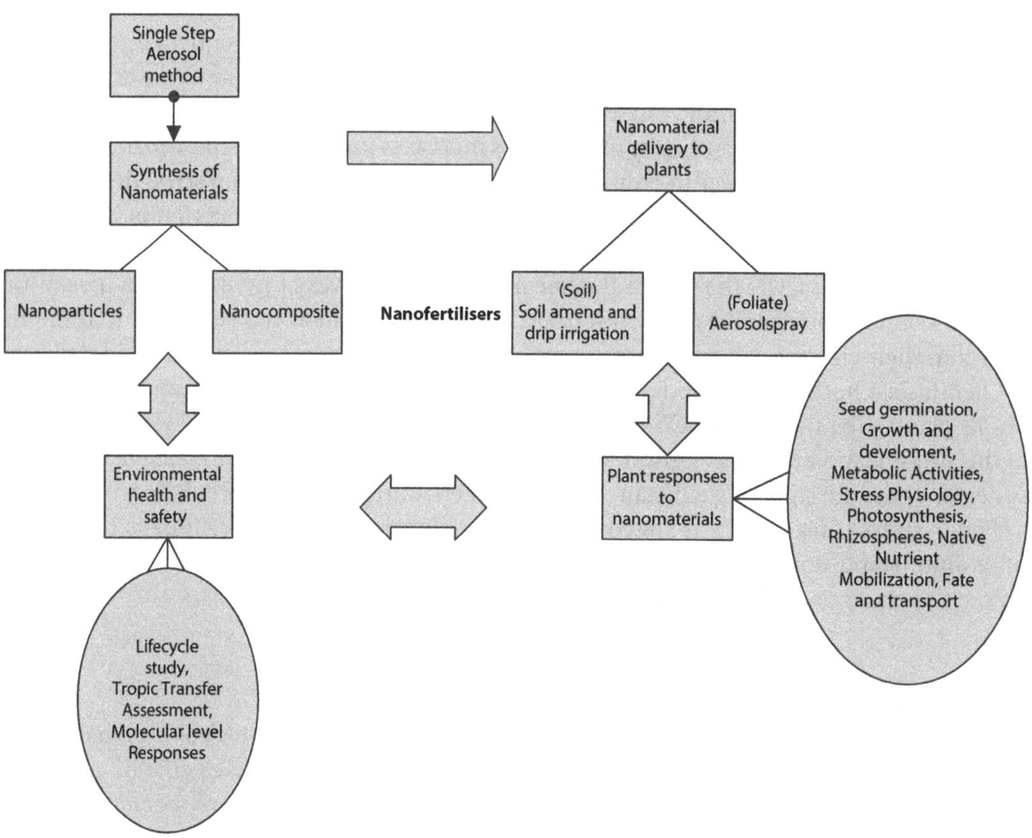

Figure 16.3 A systematic investigation on synthesis and implications of nano-fertilizers [16].

The toxicity and biocompatibility of nano-fertilizers must be determined, as well as the environmental health and safety implications of the technology. Figure 16.3 contains details about the varieties of NPs, their nanoscale characteristics, methods of NP administration, plants experimented on, and the responses observed. This is because the impact of nanoscale materials on plants and the subsequent reactions of these plants are influenced by a multitude of factors, including soil composition, environmental conditions, and the specific properties of the NPs at the nanoscale level.

We go through particular instances of NPs being employed as nutrients or fertilizers in the following subsections. Information on the type of NPs, nanoscale properties, mode of NP delivery, tested plants, and studied responses because the effects of nanoscale materials and corresponding plant responses depend on various factors related to soil, environment, and nanoscale properties. Chemical fertilizers supply plants with the nutrients necessary for maximum development and yield; yet, existing agricultural techniques cannot meet the rising demand for food without relying heavily on fertilizers. Limited nutrient usage efficiency and environmental limits connected with the use of chemical fertilizers continue to be a significant obstacle to agriculture's progress toward sustainable sustainability. Several solutions, including the use of precision fertilization, divided or targeted application, fertigation, and nano-fertilizers, have been proposed to boost fertilizer usage efficiency. One of the most promising ways to greatly increase global horticulture production to satisfy the rising food demands of the population while also being sustainable in the face of climate change is through the use of nanotechnology to the creation of new types of fertilizers. Nano-fertilizers, when used properly, may provide slow, steady nutrition to plants while also boosting productivity, reducing wasteful fertilizer runoff, and protecting the environment. Nano-fertilizer uses in agriculture may provide a possibility to attain global food production sustainability. The industry is under great food production pressure since nutritional inadequacies in human populations are mostly caused by the use of less nutritious food and a low dietary intake of fruits and vegetables.

16.4.1.1 Macronutrients Nano-Fertilizers

Macronutrients such as nitrogen (N), phosphorus (P), potassium (K), magnesium (Mg), sulfur (S), and calcium (Ca) are being combined with NMs to deliver an exact amount of nutrients to crops. This approach reduces the need for large quantities of these nutrients and decreases the costs associated with purchasing and transporting them. These macronutrient nano fertilizers are made up of one or more nutrients that have been encapsulated in a particular NM. Therefore, it is urgently necessary to conduct research from a practical perspective to produce new fertilizers with high nutrition efficiency and environmentally benign to replace the traditional macronutrient fertilizers. Studies on mesoporous silica NPs, hydroxyapatite, and urea-modified zeolites as nano-fertilizers for slow or controlled nitrogen release have demonstrated promising results. A biosafe nano-fertilizer, serving as a phosphorus source, was developed as a nanostructured water-phosphorite suspension (with particle sizes between 60–120 nm). This innovation utilized ultrasonic material dispersion to convert raw phosphorite from the Syundyukovskoe deposits in Tatarstan into the inaugural phosphatic nano-fertilizer. The application of this nano-fertilizer significantly enhanced morphometric indicators, fresh yield, fruit yield, and the overall quality of production in the examined plant species, multiplying the outcomes several times over.

16.4.1.2 Micronutrients Nano-Fertilizers

Micronutrients, required by plants in minuscule or trace quantities, play vital roles in supporting essential metabolic processes. Zinc (Zn), in particular, serves as a critical structural element or regulatory cofactor for numerous enzymes and proteins, making it indispensable for plant growth. This nutrient is key in safeguarding plants against harmful pathogens and is essential for carbohydrate synthesis, protein metabolism, and the regulation of auxin, a plant hormone. Boron (B), on the other hand, is engaged not only in the production of plant cell walls and their lignification but also in plant development and a variety of other physiological activities. As a result, it is critical to apply the appropriate quantities of Zn and B to horticultural crops in order to achieve maximum yields with high quality. Iron (Fe) is an essential element that, despite being needed in only trace quantities, is vital for the normal growth and development of plants. Insufficient or excessive amounts of iron can disrupt critical metabolic and physiological functions within plants, leading to reduced crop yields. Manganese (Mn) is crucial for both the metabolic and physiological activities in plants, and it enhances the plant's ability to tolerate various environmental stresses through its role as a cofactor in numerous enzymes. Additionally, it is necessary for photosynthesis, adenosine triphosphate (ATP) synthesis, chlorophyll, fatty acids, and protein production, as well as the synthesis of secondary metabolites such as lignin and flavonoids. MnO and FeO NPs, on the other hand, were not only less hazardous than their ionic counterparts, but they also increased the development of lettuce seedlings by 12% to 54%. Other micronutrient nano-fertilizers have been tried in a variety of crops as well.

16.4.1.3 Nano-Biofertilizers

Biofertilizers are mixtures or substances comprising one or more types of microorganisms. These organisms enhance soil fertility by activities such as fixing nitrogen from the atmosphere, making phosphorus more soluble, or boosting plant growth by producing substances that promote development. Consequently, nano biofertilizers might be characterized as the incorporation of biofertilizers with nanostructures or NPs to enhance plant development. Controlling the release of biofertilizers in the soil and extending the shelf life of formulations are key to achieving this objective. The interaction between NPs and microorganisms, the shelf life of biofertilizers, and their distribution are some of the most essential factors in the development of nano biofertilizers. It was discovered that the interaction between gold NPs and plant growth-boosting rhizobacteria had beneficial impacts. However, the shelf life of biofertilizers is a limiting issue in these formulations, and NMs can be used to increase it. Utilizing nanoformulations can improve the biofertilizers' resistance to desiccation, heat, and ultraviolet degradation. For instance, polymeric nanoparticle coatings can be utilized to build formulations that are resistant to desiccation, hence extending the lifespan of these items. In addition, NPs can be employed to enhance the transport of biofertilizers to soil and plants. Trials involving the addition of hydrophobic silica NPs to a water-in-oil emulsion demonstrated an increase in product distribution and a prolongation of shelf life by reducing desiccation. However, the manufacture of nano biofertilizers is hampered by the fact that nanoscale constructions are often smaller than cells. In this context, filters consisting of radially aligned carbon nanotube walls that can absorb Escherichia coli might be employed as a potential method to gather other microbes from fermentation operations and transfer them to plants. Therefore, nano biofertilizers are capable of overcoming some limits of biofertilizers, although this technology still requires more study and development.

The application of nutrients is essential for maintaining soil fertility and enhancing crop yield and quality. Globally, precise nitrogen management of horticulture crops is a significant concern due to the reliance on chemical fertilizers. Not only are traditional fertilizers expensive for the grower, but they may also be damaging to individuals and the environment. This has prompted the quest for ecologically benign fertilizers, especially ones with high nutrient-use efficiency, and nanotechnology is emerging as a possible option. Nano fertilizers offer significant benefits for nutrient management due to their enhanced efficiency in nutrient utilization. Nutrients, either alone or in combination, are attached to nano-scale adsorbents, resulting in a slower release of nutrients compared to conventional fertilizers. This method not only improves fertilizer-use efficiency but also reduces nutrient leaking into groundwater. In addition, nano-fertilizers may be used to increase a plant's resistance to abiotic stress, and when combined with microorganisms (so-called nano-biofertilizers), they give significant additional advantages.

16.4.2 Nano-Pesticides

In commercial agriculture, pesticide usage is a common practice, and the development of novel, effective, and target-specific pesticides is ongoing. Consequently, a huge number of pesticides are evaluated annually. Very little (0.1%) of the sprayed pesticides reach their intended targets, while the vast majority (99.9%) poison the environment. The widespread use of pesticides poses significant risks to the food chain and human health, including the development of resistance among weeds, insects, and pathogens. It's important to note, however, that without pesticides, the loss of human lives could increase a thousandfold compared to deaths caused by pesticide exposure. While biopesticides offer a reduction in the harmful effects associated with synthetic pesticides, their effectiveness is often hindered by slow action and dependency on environmental conditions. Nano-pesticides, on the other hand, have the capacity to overcome these limitations. Due to the slow degradation and controlled release capabilities of NPs, nano-pesticides can offer prolonged protection against pests. Consequently, nano-pesticides play a crucial role in achieving efficient and sustainable pest management, potentially reducing the need for synthetic pesticides and minimizing environmental damage. Nano-pesticides differ from conventional pesticides in their mode of action, enhancing their effectiveness. Nano-sized particles may be transported in both dissolved and colloidal phases, and this sort of mechanism explains why their behavior differs from that of ordinary solutes of the same particles. AI solubility may promote mobility and breakdown by soil-dwelling microbes. Because nanoparticle-based pesticides improve the solubility of AI, they are thought to have a lower environmental effect than traditional pesticides. Overall, nano-pesticides save energy and water because they are used less often and in smaller amounts than traditional pesticides. They also make pesticides work better and increase crop yields and crop productivity by reducing waste and labor costs and increasing crop yields. However, nano-pesticides may cause health problems for a number of reasons listed by the Environmental Protection Agency (EPA) of the United States. These include:

i) Nano-pesticides can be absorbed through the skin because they are so small and can pass through cell membranes
ii) Nano-pesticides can be inhaled because they can go deep into the lungs and move to the brain by crossing the blood-brain barrier
iii) the durability and reactive potential of some NM. One of the most challenging areas of the pesticide business is the nano-formulation of conventional pesticides with polymers or metal NPs.

By manipulating the outer shell of the nano-capsule, nanoencapsulation of pesticides is advantageous in a regulated and gradual release of the AI, which releases low doses over a prolonged time period and prevents excess run-off of undesirable pesticides. Another benefit of nanocarriers in plant protection is site-specific distribution and active component stability. Similar to this, a nanoemulsion of oil or water increases the solubility and effectiveness of a pesticide against many pests. It is essential to think about and discuss the molecular and physicochemical properties of a pesticide before employing it to protect crops, regardless of the formulation type (conventional or nano). Different schools of thought have recently voiced their opinions on the advantages and disadvantages of moving from a pesticide's conventional form to its nanoform. Indeed, nano-pesticides must be tested for some of the most fundamental molecular and physicochemical properties that affect efficacy, stability, and environmental and/or human safety. The following subsections guide readers through some of the essential qualities of a pesticide that determine whether it is suitable for nano-formulation or not, as well as how these elements may be employed in the creation of nano-formulations.

16.4.3 Nano-Herbicides

Getting rid of unwanted weeds is a severe hazard while farming. These weeds' flooding growth can be inhibited by several chemical herbicides; however, most of these chemical-based herbicides have been shown to have persistent toxicity even in small amounts. As a result, nano-herbicides are used as a safe technique to avert such a catastrophe. Chemical stability, solubility, bioavailability, photodecomposition, and soil sorption have all been documented with nano-herbicides. Recent research shows that combining NPs with active chemicals such as atrazine with poly (epsilon-caprolactone) NPs as the carrier improves herbicidal action. Additionally, it was possible to diminish the sorption of herbicides by encapsulating them in sodium triphosphate and chitosan NPs, which minimized their toxicity in comparison to the single AI and decreased environmental risk. Chitosan NPs have recently been cross-linked with diuron disulfide links to regulate the release of herbicides dependent on glutathione levels. The toxicity was decreased and plant growth was successfully encouraged as a consequence. The chemical stability, solubility, bioavailability, photodecomposition, and soil sorption of nano-herbicides have all been described. Recent research shows that formulations of NPs combined with active chemicals, like atrazine with poly (epsilon-caprolactone) NPs as the carrier, can boost herbicidal efficacy by reducing the mobility of herbicide in the soil and genotoxicity. In agriculture, weeds are viewed as a severe issue since they dramatically lower crop vigor and output. By employing nanotechnology to apply nano-herbicides in an environmentally benign method that doesn't leave any residual toxicity in the soil and environment, weed problems can be solved. The chemical stability, solubility, bioavailability, photodecomposition, and soil sorption of nano-herbicides have all been described. Recent research shows that formulations of NPs combined with active chemicals, like atrazine with poly (epsilon-caprolactone) NPs as the carrier, can boost herbicidal efficacy by reducing the mobility of herbicide in the soil and genotoxicity. Herbicides could be released under regulated conditions to increase their effectiveness and reduce their adverse effects.

16.5 Methods for Synthesis

The two basic categories of NM synthesis techniques are (1) top-down approach and (2) bottom-up approach, as shown in Figure 16.4 and Figure 16.5.

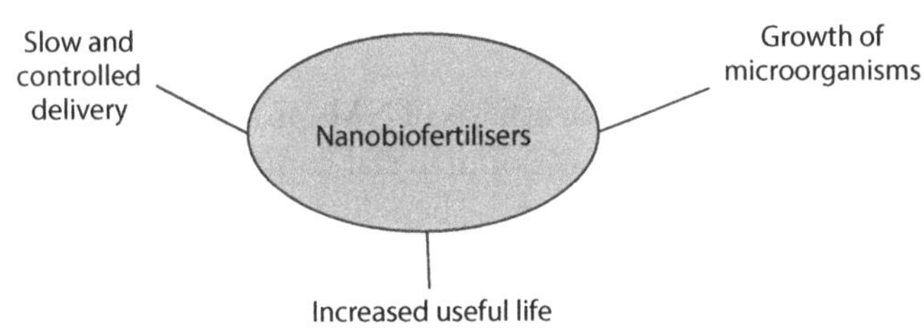

Figure 16.4 Effect of nano-biofertilizers on plants [24].

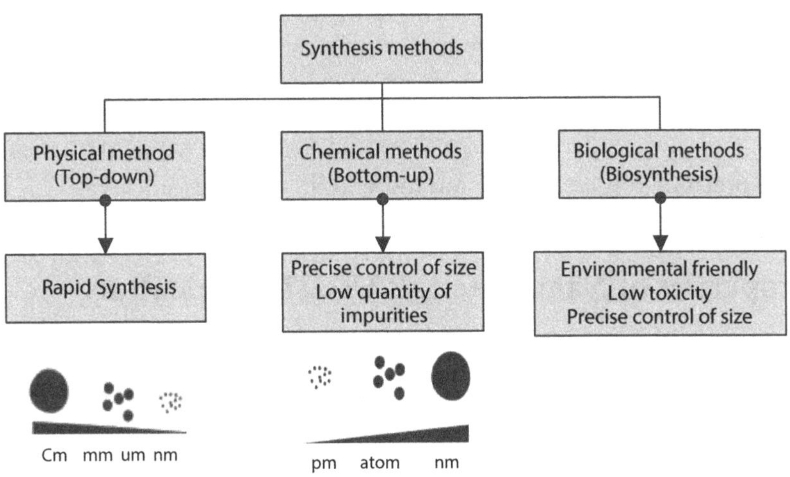

Figure 16.5 Schematic illustration of synthesis techniques used for nanomatreials [29].

16.5.1 Top-Down Synthesis

In the top-down approach, nanoscale structures or particles are produced by deconstructing bulk material. These synthesis techniques expand upon existing methods used to generate particles smaller than one micron in diameter. Top-down processes involve either elimination or segmenting the bulk material, or adapting large-scale production techniques to achieve the desired structures with specific characteristics [29]. They are intrinsically simpler. The top-down strategy's main drawback is the surface structure's flaws. NM synthesis frequently employs chemical vapor deposition (CVD). The CVD technique has several uses, including (1) coating surfaces with thin films, (2) synthesizing pure bulky materials and powder, and (3) creating composite materials. There are several gas and surface reactions that take place throughout the incredibly complex CVD process [31]. Typically, a number of variables, including substrate temperature, reactor pressure, the composition of the NMs, and gas-phase chemistry, affect how NMs are synthesized. The gradual reduction of bulk materials to tiny sizes is necessary for the production of NMs. This method is the foundation for several operations, including milling, grinding, CVD, physical vapor deposition, and other breakdown methods. This method is used to generate a number of NMs, including CNTs, CNFs, carbon spheres, graphene, carbon NPs, and spherical magnetite. The synthesis of NMs based on carbon is frequently carried out using the CVD method.

16.5.2 Bottom-Up Method

The converse is true for the bottom-up synthesis of NMs, in which very simple chemicals are used to create NPs [35]. The effectiveness of material components with self-assembly processes that produce nanostructure is one of the benefits of bottom-up techniques. Examples of the bottom-up strategy in action include the creation of quantum dots during epitaxial growth and the creation of NPs from colloidal suspensions. This method is the foundation for many processes, including spinning, biological synthesis, sol-gel synthesis, and green synthesis. The alternative approach is known as "bottom-up," which can potentially generate less waste and thus be more economical. In this method, materials are built atom by atom, molecule by molecule, or cluster by cluster from the ground up. This approach is recognized as a bottom-up strategy. Many of these techniques are currently under investigation or are just beginning to be utilized in the production of nano-powders for commercial applications [36]. Some of the well-known bottom-up strategies documented for the manufacture of luminous NPs include the organometallic chemical pathway, sol-gel process, reverse-micelle method, template-assisted sol-gel synthesis, hydrothermal method, electrodeposition, colloidal precipitation technique, and more [37].

16.6 Properties of Nanomaterials Used in Agriculture

In addition to chemical and physical processes, NPs may be created biologically, using the biosynthesis approach. There are a variety of natural sources for this function, including plants, fungi, and bacteria. The advantage of this method is that it allows for more control over particle toxicity and size [38]. For each application, the best strategy will necessitate a synthesis capable of creating mass-scale particles with regulated physicochemical characteristics, resulting in a homogenous and target-specific nano-formulation. Scientists are under immense pressure to develop novel technologies that not only match the productivity needs of growers but also fit the economic budgets of both producers and the production industry [39]. Nanotechnology may provide a solution to these difficulties by developing high-performance NMs. As previously said, the goal of these revolutionary fertilizers is to manage the release of nutrient AIs at a very gradual rate in line with or proportionate to crop development. The primary goal of employing these nano-fertilizers is to improve nutrient usage efficiency, which will lead to precision agriculture. Various NPs, their oxides, and nano-formulations of traditional nutrients have been turned into beneficial inputs in the form of nano-fertilizers [40]. When sprayed at a specified concentration on different crops, the conversion of these NPs has shown encouraging. These NPs have different properties, as shown in Table 16.1, which are flexible for various applications.

Organic and inorganic NMs are used to create NPs. Furthermore, their synthesis differs in terms of the physical or chemical procedures used. Metal oxides such as zinc oxide titanium oxide, and magnesium oxide are examples of inorganic NMs. Organic NMs, on the other hand, include lipids, polymers, and carbon nanotubules. NPs frequently exhibit distinct physical and chemical characteristics as shown in Table 16.1. For example, the

Table 16.1 Properties of nanomaterials and their effectiveness [38].

S. no.	Properties	Examples
1	Catalytic	Enhanced catalytic performance via increased surface area to volume ratio.
2	Electrical	Enhanced electrical conductivity in ceramic and magnetic nano-composites, coupled with elevated electrical resistance in metal.
3	Magnetic	Elevated magnetic coercivity achieved until reaching a critical grain size, followed by superparamagnetic characteristics.
4	Mechanical	Improved hardness and toughness of metals and alloys, ductility and super-plasticity of ceramic.
5	Optical	Enhanced strength and resilience in metals and alloys, alongside increased ductility and super-plasticity in ceramics.
6	Sterical	Shifts in the spectral characteristics of optical absorption and fluorescence, alongside improved quantum efficiency in semiconductor crystals.
7	Biological	Heightened selectivity, utilizing hollow spheres for targeted drug delivery and precise release. Improved permeation across biological barriers (such as membranes and the blood-brain barrier) and enhanced biocompatibility.

electrical, optical, and chemical characteristics of NPs may differ significantly from those of the bulk components. Materials act extremely differently at the nanoscale than at larger dimensions, and it is still very difficult to anticipate the physical and chemical properties of such small particles [41].

16.7 Researches and Advancements

The development of "smart delivery systems" for nano-fertilizers and nano-pesticides also paves the way for better fertilizer formulation. These systems increase nutrient absorption, reduce nutrient loss in plant cells, and deliver targeted and regulated nano-pesticide concentrations. The dissemination of NMs into environmental and soil ecosystems stems from progress in agricultural nanotechnology. The interaction of NMs with the soil, influenced by their composition and the existence of organic or inorganic soil constituents, can lead to physical, chemical, or biological modifications. Such alterations or agglomerations can significantly affect the stability, reactivity, toxicity, and target specificity of the NMs Accessing the destiny of NMs in the soil necessitates thorough research [38]. Through the ingestion of plant-derived goods supplied through delivery systems or processed meals, NMs may enter human food chains. The development and commercialization of NM-enabled technologies in agriculture are hindered by a number of issues, despite the fact that they have created several potential to support the expansion of sustainable agriculture [36]. The most significant technological obstacles to utilizing NMs to their full potential are the uncertainties surrounding plant-NM interactions and NM uptake, the limited routes available for delivering NMs into plant architecture, the lack of knowledge regarding the fate of NMs in the environment, and the potential health risks associated with NMs entering the food chain. The fundamental processes of nano-enabled solutions are now mostly focused on phenomenological and theoretical observations, both of which require more investigation [42, 43]. It is important to understand the biophysics of how NMs interact with the leaf cuticle and chloroplast cells in the context of nano-agrochemicals for improving plant function. It is important to thoroughly examine the characteristics of NMs, such as size, dosage, exposure period, surface chemistry, structures, immunological response, accumulation, retention time, and other consequences. More in-depth studies must be conducted in order to identify the many processes generating the beneficial benefits. Such knowledge will make it possible to create more intelligent NM designs that will enhance photosynthesis and reduce plant stress. With innovative techniques like the delivery of nano fertilizers and micronutrients, agrochemical encapsulated nanocarrier systems, and nano pesticide delivery, including green pesticides and biopesticides, nanotechnology may support the agro-food sector [44]. These technologies may not be especially feasible and struggle to meet the ongoing demands of food delivery and production. The market currently lacks a significant presence of commercial products, indicating an urgent need for action. Additionally, the impact of NMs on human health, safety, and environmental consequences remains uncertain. To minimize risks associated with their use, an in-depth analysis of the soil's physical and chemical characteristics in the areas where NPs are applied is economic. Research on human and environmental toxins can help us comprehend the complicated interactions between the agroecosystem, nanoscale agrochemicals, exposure levels, and people. To evaluate the mechanistic use of NMs and their agro-ecological toxicity, a thorough examination is required.

16.8 Future Perspective

According to research so far, nanotechnology by itself is unlikely to result in a novel mode of action for plant protection or nutrition, but its incorporation into the science of formulation will undoubtedly enable the design of more complex solutions that may aid in lessening the environmental effect of contemporary agriculture, and both benefit human health and help ensure global food security. Agriculture urgently needs innovation to fulfill the rising food demand while minimizing its environmental effect. The creation of innovative products that are competitive and have the potential to increase the sustainability of agriculture in the future requires major research investment. Hopefully, appropriate nanotechnology applications may play a significant part in achieving this objective. Under the present climate change scenario, there is a significant deal of interest in maximizing the production capacity of agricultural crops due to the potential advantages of nano-fertilizers [43]. Reduced leaching and volatilization caused by the use of traditional fertilizers are the main economic advantages of using nano-fertilizers. To create effective, versatile, stable, economical, and environmentally friendly NMs, collaborative research among institutions studying various applications of NMs is essential. This would also make it easier to complete the picture regarding the function, course, behavior, and assessment of NMs' ecotoxicity. To create integrated remediation solutions, the role of NMs in bioremediation should also be investigated. Additionally, the majority of studies on nano-assisted agriculture rely on controlled laboratory tests, and there is little information on their use in the field [44]. Considerable advancements in nanoparticle applications have been observed in pharmaceutical and biomedical domains. Limited documentation exists on the effective deployment and innovation of sustainable NPs within the agricultural sector. Consequently, the integration of nanotechnology could significantly improve agricultural methodologies and yield multifaceted benefits. Nevertheless, the toxicity of NPs poses a significant hurdle. In response, a range of strategic measures are being formulated to mitigate these adverse effects [45].

References

1. Jalil, S.U. and Ansari, M., II, Role of nanomaterials in weed control and plant diseases management, 421-434, in: *Nanomaterials for Agriculture and Forestry Applications*, Elsevier, Netherlands, 2020.
2. Mandal, A. and Ray Banerjee, E., Introduction to Nanoscience, Nanotechnology and Nanoparticles, in: *Nanomaterials and Biomedicine*, pp. 1–39, Springer, Singapore, 2020.
3. Singh, R.P., Application of nanomaterials toward development of nanobiosensors and their utility in agriculture, in: *Nanotechnology*, pp. 293–303, Springer, Singapore, 2017.
4. Dwivedi, S., Saquib, Q., Al-Khedhairy, A.A., Musarrat, J., Understanding the role of nanomaterials in agriculture, in: *Microbial Inoculants in Sustainable Agricultural Productivity*, pp. 271–288, Springer, New Delhi, 2016.
5. Sarkar, M.R., Rashid, M.H.O., Rahman, A., Kafi, M.A., Hosen, M., II, Rahman, M.S., Khan, M.N., Recent advances in nanomaterials based sustainable agriculture: An overview. *Environ. Nanotechnol. Monit. Manage.*, *18*, 100687, 2022.
6. He, X., Deng, H., Hwang, H.M., The current application of nanotechnology in food and agriculture. *J. Food Drug Anal.*, *27*, 1, 1–21, 2019.

7. Chhipa, H., Applications of nanotechnology in agriculture, in: *Methods in Microbiology*, vol. 46, pp. 115–142, Academic Press, London, UK, 2019.
8. Khot, L.R., Sankaran, S., Maja, J.M., Ehsani, R., Schuster, E.W., Applications of nanomaterials in agricultural production and crop protection: a review. *Crop Prot.*, *35*, 64–70, 2012.
9. Baker, S., Volova, T., Prudnikova, S.V., Satish, S., Prasad, N., Nanoagroparticles emerging trends and future prospect in modern agriculture system. *Environ. Toxicol. Pharmacol.*, *53*, 10–17, 2017.
10. Qureshi, A., Singh, D.K., Dwivedi, S., Nano-fertilizers: a novel way for enhancing nutrient use efficiency and crop productivity. *Int. J. Curr. Microbiol. App. Sci.*, *7*, 2, 3325–3335, 2018.
11. Abd El-Azeim, M.M., Sherif, M.A., Hussien, M.S., Tantawy, I.A.A., Bashandy, S.O., Impacts of nano- and non-nanofertilizers on potato quality and productivity. *Acta Ecol. Sin.*, *40*, 5, 388–397, 2020.
12. Yaseen, R., Ahmed, A., II, Omer, A.M., Agha, M.K., Emam, T.M., Nano-fertilizers: Bio-fabrication, application and biosafety. *Nov. Res. Microbiol. J.*, *4*, 884–900, 2020.
13. Nuruzzaman, M.D., Rahman, M.M., Liu, Y., Naidu, R., Nanoencapsulation, nano-guard for pesticides: a new window for safe application. *J. Agric. Food. Chem.*, *64*, 7, 1447–1483, 2016.
14. Sarkar, N., Chaudhary, S., Kaushik, M., Nano-fertilizers and Nano-pesticides as Promoters of Plant Growth in Agriculture, in: *Plant-Microbes-Engineered Nano-particles (PM-ENPs) Nexus in Agro-Ecosystems*, pp. 153–163, Springer, Cham, 2021.
15. Kumar, S., Nehra, M., Dilbaghi, N., Marrazza, G., Hassan, A.A., Kim, K.H., Nano-based smart pesticide formulations: Emerging opportunities for agriculture. *J. Control. Release*, *294*, 131–153, 2019.
16. Sachdev, S. and Ahmad, S., Role of Nanomaterials in Regulating Oxidative Stress in Plants, in: *Nanobiotechnology*, pp. 305–326, Springer, Cham, 2021.
17. Rana, R.A., Siddiqui, M.N., Skalicky, M., Brestic, M., Hossain, A., Kayesh, E., Islam, T., Prospects of nanotechnology in improving the productivity and quality of horticultural crops. *Horticulturae*, *7*, 10, 332, 2021.
18. Solanki, P., Bhargava, A., Chhipa, H., Jain, N., Panwar, J., Nano-fertilizers and their smart delivery system, in: *Nanotechnologies in Food and Agriculture*, pp. 81–101, Springer, Cham, 2015.
19. Abid, N., Khan, A.M., Shujait, S., Chaudhary, K., Ikram, M., Imran, M., Maqbool, M., Synthesis of nanomaterials using various top-down and bottom-up approaches, influencing factors, advantages, and disadvantages: A review. *Adv. Colloid Interface Sci.*, 102597, 300, 2021.
20. Khan, F.A., Synthesis of nanomaterials: methods & technology, in: *Applications of Nanomaterials in Human Health*, pp. 15–21, Springer, Singapore, 2020.
21. Mahesh, K.V., Singh, S.K., Gulati, M., A comparative study of top-down and bottom-up approaches for the preparation of nanosuspensions of glipizide. *Powder Technol.*, *256*, 436–449, 2014.
22. Huang, Y. and Wang, L., Experimental studies on nanomaterials for soil improvement: a review. *Environ. Earth Sci.*, *75*, 6, 1–10, 2016.
23. Nikolaidis, P., Analysis of green methods to synthesize nanomaterials. *Green Synth. Nanomaterials Bioenergy Appl.*, 125–144, 2020.
24. Saleh, T.A., Development and synthesis of nanoparticles and nanoadsorbents, in: *Interface Science and Technology*, vol. 34, pp. 127–165, Elsevier, Netherlands, 2022.
25. Biswas, A., Bayer, I.S., Biris, A.S., Wang, T., Dervishi, E., Faupel, F., Advances in top–down and bottom–up surface nanofabrication: Techniques, applications & future prospects. *Adv. Colloid Interface Sci.*, *170*, 1-2, 2–27, 2012.
26. Duhan, J.S., Kumar, R., Kumar, N., Kaur, P., Nehra, K., Duhan, S., Nanotechnology: The new perspective in precision agriculture. *Biotechnol. Rep.*, *15*, 11–23, 2017.

27. Singh, H., Sharma, A., Bhardwaj, SK., Arya, S.K., Bhardwaj, N., Khatri, M., Recent advances in the applications of nano-agrochemicals for sustainable agricultural development. *Environ. Sci.: Process. Impacts*, 23, 2, 213–239, 2021.
28. Khan, I., Saeed, K., Khan, I., Nanoparticles: Properties, applications and toxicities. *Arabian J. Chem.*, *12*, 7, 908–931, 2019.
29. Abid, N., Khan, A.M., Shujait, S., Chaudhary, K., Ikram, M., Imran, M., Maqbool, M., Synthesis of nanomaterials using various top-down and bottom-up approaches, influencing factors, advantages, and disadvantages: A review. *Adv. Colloid Interface Sci.*, *300*, 102597, 2022.
30. Li, N., Zhao, P., Astruc, D., Anisotropic gold nanoparticles: synthesis, properties, applications, and toxicity. *Angew. Chem. Int. Ed.*, *53*, 7, 1756–1789, 2014.
31. Oktaviani, O., Nanoparticles: Properties, applications and toxicities. *Jurnal. Latihan*, *1*, 2, 11–20, 2021.
32. Rahimi, H.R. and Doostmohammadi, M., Nanoparticle synthesis, applications, and toxicity, in: *Applications of Nanobiotechnology*, pp. 1–16, IntechOpen, London, UK, 2019.
33. Jadhav, A.S., Urkude, R.P., Jangam, S.S., Current review on nanoparticles: properties, application and toxicity, 2021.
34. Fiol, D.F., Terrile, M.C., Frik, J., Mesas, F.A., Álvarez, V.A., Casalongué, C.A., Nanotechnology in plants: Recent advances and challenges. *J. Chem. Technol. Biotechnol.*, *96*, 8, 2095–2108, 2021.
35. Sangeetha, J., Hospet, R., Thangadurai, D., Adetunji, C.O., Islam, S., Pujari, N., Al-Tawaha, A.R.M.S., Nanopesticides, nanoherbicides, and nanofertilizers: the greener aspects of agrochemical synthesis using nanotools and nanoprocesses toward sustainable agriculture, in: *Handbook of Nanomaterials and Nanocomposites for Energy and Environmental Applications*, pp. 1663–1677, 2021.
36. Sangeetha, J., Thangadurai, D., Hospet, R., Purushotham, P., Karekalammanavar, G., Mundaragi, A.C., Harish, E.R., Agricultural nanotechnology: concepts, benefits, and risks, in: *Nanotechnology*, pp. 1–17, Springer, Singapore, 2017.
37. Baker, S., Satish, S., Prasad, N., Chouhan, R.S., *Ind. Appl. Nanomater.*, 341–363, 2019.
38. Sau, S., Sarkar, S., Mitra, M., Gantait, S., Recent trends in agro-technology, post-harvest management and molecular characterisation of pomegranate. *J. Hortic. Sci. Biotechnol.*, *96*, 4, 409–427, 2021.
39. Usman, M., Farooq, M., Wakeel, A., Nawaz, A., Cheema, S.A., ur Rehman, H.., Sanaullah, M., Nanotechnology in agriculture: Current status, challenges and future opportunities. *Sci. Total Environ.*, *721*, 137778, 2020.
40. Bernela, M., Rani, R., Malik, P., Mukherjee, T.K., Nanofertilizers: applications and future prospects, in: *Nanotechnology*, pp. 289–332, Jenny Stanford Publishing, Singapore, 2021.
41. Gahukar, R.T. and Das, R.K., Plant-derived nanopesticides for agricultural pest control: challenges and prospects. *Nanotechnol. Environ. Eng.*, 5, 1, 1–9, 2020.
42. Prajapati, D., Pal, A., Dimkpa, C., Singh, U., Devi, K.A., Choudhary, J.L., Saharan, V., Chitosan nanomaterials: A prelim of next-generation fertilizers; existing and future prospects. *Carbohydr. Polym.*, 119356, 288, 2022.
43. Chhipa, H., Nanofertilizers and nanopesticides for agriculture. *Environ. Chem. Lett.*, *15*, 1, 15–22, 2017.
44. Zulfiqar, F., Navarro, M., Ashraf, M., Akram, N.A., Munné-Bosch, S., Nanofertilizer use for sustainable agriculture: Advantages and limitations. *Plant Sci.*, *289*, 110270, 2019.
45. Joshi, H., Choudhary, P., Mundra, S.L., Future prospects of nanotechnology in agriculture. *Int. J. Chem. Stud.*, *7*, 2, 957–963, 2019.

17

Nano-Surfactants: Types, Synthesis, Properties, and Potential Applications

Divya Bajpai Tripathy[1]*, Sonali Kesarwani[1], Anjali Gupta[1] and Priyanka Chhabra[2]

[1]School of Basic and Applied Science, Galgotias University, Greater Noida, Uttar Pradesh, India
[2]Center of Medical Biotechnology, Amity Institute of Biotechnology, Amity University, Noida, Uttar Pradesh, India

Abstract
Nano-surfactants belong to a class of surfactants that have recently received remarkable recognition because of their exclusive characteristics and latent applications. Nano-surfactants refer to surfactant molecules or particles that have been engineered or modified at the nanoscale. Surfactants are compounds that reduce the surface tension between two substances, such as a liquid and a solid or a liquid and a gas. These materials typically consist of a hydrophobic tail and a hydrophilic head, which allows them to self-assemble into nanoparticles with a high surface area. Similarly, nano-surfactants are nanoscale molecules that have two different moieties with different polarities. For example, active plasmonic magnetic nano-surfactants, synthesized through acid-triggered treatment of gold-ferric oxide dumbbell-shaped nanocrystals. Other examples include asymmetric fullerene nano-surfactants, and charged zirconium phosphate nano-surfactants. Nano-surfactants have a wide range of applications in various fields, including drug delivery, cosmetics, the food industry, and environmental remediation. In drug delivery, nano-surfactants are used to enhance the solubility and bioavailability of drugs and can be functionalized with targeting ligands for specific delivery to cells or tissues. They can also be modified to release drugs in response to specific stimuli, which improves drug efficacy while reducing toxicity. In cosmetics, nano-surfactants are used in skin care products and sunscreens to improve product texture and provide UV protection. In the food industry, they are used as emulsifiers to improve the stability and texture of food products. In environmental remediation, nano-surfactants are used to remove pollutants from soil and water by increasing their solubility and facilitating their removal. DosSurf™ 100 is commercially available as a nano-surfactant with DASOS, a USA-based company with a claim to reduce the contact angle with crude oils, and can be commercially used in nano-surfactants enhanced oil recovery. The unique properties of nano-surfactants make them promising materials for various applications. In spite of gaining interest from the researchers, a good review article is unavailable in this field. The current article reviews some of the important aspects of nano-surfactants that will help the researchers get good knowledge on the topic.

***Keywords*:** Nano-surfactants, Janus symmetry, enhance oil recovery, nanoparticles

**Corresponding author*: divyabaj@gmail.com

Divya Bajpai Tripathy, Anjali Gupta and Arvind Kumar Jain (eds.) Multifunctional Materials: Engineering and Biological Applications, (441–460) © 2025 Scrivener Publishing LLC

17.1 Introduction

The concept of nano-surfactants (NSs) is based on the combination of multifunctional approaches of nanoparticles (NPs) with the specific amphiphilic properties of surfactant molecules to design a molecule with exceptional surface-active properties and enhanced absorption and multifunctionalities. NSs are nanosized amphiphilic molecules with Janus symmetry and have the ability to adsorb onto the surface and alter their properties [1]. A nanomaterial type with amphiphilic characteristics, it is similar to that of any surfactant, with a hydrophilic head group and hydrophobic tail [2]. Hydrophilic silicon oxide NPs with hydrophobically altered moiety through exploiting silane coupling agents can be one of the best-known examples of such kind [3]. These altered nanosized amphiphiles were exploited to create a Pickering emulsion with exceptional stability over their conventional counterparts. Janus particles contributed as another class of NSs, acknowledged as two distinct sides of structures and/or chemical configurations. Furthermore, asymmetric design imparts exclusive amphiphilicity to Janus particles in comparison to their homogeneous counterparts [4]. Janus balance that is because of their Janus symmetry helps to control their amphiphilicity [5]. However, it is not easy to create NSs with precise syllable structures with decidedly active surfaces. Jasus NSs have been reported with different morphologies like spherical, dumbbell-shaped, and match-stick shape. The morphology of these surfactants decides their packing density and symmetry properties. Similar to those of surfactant molecules, Jasus particles-based NSs also have the ability to form aggregates at specific concentrations, similar to the micelles, and show strong absorption at the interfaces [6, 7]. These aggregates have the ability to greatly alter the fluid dynamics and their stratification behaviors. Such specific properties based on their special structural characteristics make them suitable for a wide range of applications such as enhanced oil recovery (EOR) in the oil industry, targeted drug delivery in the biomedical field, corrosion inhibitors in desalination plants, activity enhancers for pesticides in the agriculture field, and in cosmetics [8, 9]. Still, this field of research explored their applicability in various sectors of science and engineering like electronics, electrical, and polymer industries. Since now, very less articles are available on this field specifically review articles that can provide comprehensive knowledge on the related field. The scope of this review paper would include a comprehensive examination of the current state of knowledge in the field, focusing on recent advancements, key findings, and emerging trends. The review would provide a critical analysis and synthesis, classification, mechanism, potential applications, and future research prospectives of NSs aiming to offer a valuable resource for researchers, scientists, and practitioners in the field of NSs.

17.2 History of Nano-Surfactants

A literature review in the field of NSs revealed that this term was first used in 2011 in the *"Proceedings of the Technical Workshops for the Hydraulic Fracturing Study: Chemical & Analytical Methods"* for a composition where surfactants were present in the range from 500 to 1,000 ppm and account only for 0.05% - 0.1% of the total fluid volume [10]. Then, in 2014, this word reappeared in the title *"Investigation of Nano-Surf Nano Actor Compositions for Inhibition of Salt Deposit in Oil and Gas Production*" [11]. In both the above work, NSs words were used for the surfactant's compositions consisting of NPs to enhance their surface activities and are not validated by the current definition of NSs. Although the same approach continued till now where the NPs containing surfactant composition were termed NSs but on contrary, in the same 2011, Janus NPs with amphiphilic properties have also been investigated and termed NPs [12] and completely changed the perspectives of researchers to think about NSs. Till now, the term "nano-surfactants" was coined for the nanosized molecules with polar and nonpolar moieties on the same molecules, the "nano-surfactant" word is used to develop amphiphilic molecules that are able to create micelles-like aggregates alike to the surfactant molecules as well as NPs in surfactants solutions.

17.3 Types of Nano-Surfactants

On the basis of their origin, NSs may be broadly classified into the following two groups:17

17.3.1 Nano-Surfactants Type 1 (Nanoparticles in Surfactant Moiety)

The term "nano-surfactants" has been coined for the amphiphilic NPs characterized with NPs of Janus symmetry in which two groups of different polarity (lipophilic and hydrophilic) are linked together just like a surfactant molecule that has a hydrophilic polar head and a hydrophobic tail. These NSs have abilities similar to surfactants in order to molecular self-assembly and sensitivity toward surface properties such as interfacial tension (IFT), wettability, foaming, and emulsion stability [13, 14]. Nano surfactants are covering the size range of 100–1000 nm [15]. On the basis of their morphology, NSs may be spherical or dumbbell-shaped or may have the shape of match sticks. In spherical types of NSs, hydrophobic groups may contribute to the core with hydrophilic parts oriented outwards to make the surface of the sphere or vice versa depending upon the medium in which they are present. Dumbbell-shaped NPs are made up of two lopes just like the dumbbell in which one lobe is hydrophilic and another one is hydrophobic type. Matchstick-type NSs are a new class recently explored that has a polar head and hydrophobic tail. Furthermore, when adjudicated through the adsorption at the interfaces of emulsions, such molecules formed into free-standing hierarchical structures, comprising capsules, inverse capsules, and 2-D sheets [1].

17.3.2 Nano-Surfactants Type 2 (Formulations with Nanoparticles in Surfactant Solutions)

In some recent research, the "nano-surfactant" term has also been used for formulations made up of nano-sized particles with natural conventional surfactants [17–19]. Incorporating NPs into surfactant solutions generally tends to stabilize the metal oxides with sizes ranging from 1 to 100 nm [20–23] and due to this, these types have attracted interest as potential candidates for EOR-surfactant formulations [24, 25]. Such amended properties attributed to the compositions claim their superiority over conventional bulk material counterparts. These formulations were found to have great chemical stability, particularly in the formulations of suspensions or emulsions.

17.4 Synthesis of Nano-Surfactants

Saeedi *et al.* (2014) produced nanosized biosurfactants using autochthonous Pseudomonas aeruginosa MM1011 and examined their stability for one year at room temperature [15]. This report was the first report available on the stability of nano rhamnolipid exploiting Pseudomonas aeruginosa. Dae-Yoon Kim *et al.* (2017) [13] synthesized C60 NS (C60NS). In this research, the author proposed an amphiphilic and asymmetric NS that has both the LCs (liquid crystals) favoring organic functions as well as LC-repelling NPs. The LC-favoring group is directly tethered to the LC-repelling group. The modification of this NS was performed by attaching alkyl chain and cyanobiphenyl mesogens to the nano building block of C60, resulting in the desired product (C60NS), which was studied for its application in the fabrication of optoelectronic devices. Lin *et al.* (2018) [26] designed a two-dimensional ultrathin-charged Zirconium phosphate nanoplatelets as NS and studied their applications in order to disperse the reduced graphene oxide (RGO) agglomerates in liquid phase or in chiral nematic LCs.

Emadi *et al.* (2017) [27] synthesized nano silica particles-based NS formulations. In this work, they used different wt% of Cedr extraction (1, 2, 3, 4, 5, 6, 7, 8, 10) in 100,000 ppm of brine along with different concentrations of nano-silica (500, 1000, 1500, 2000, 2500 ppm). These surfactants were studied for their various surface-active properties such as CMC and IFT. These surfactants were also evaluated for their application in EOR.

Mohajeri and Hemmati (2015) [28] synthesized Zr NPs -based NSs. In order to synthesize these surfactants, they first prepared ZrO_2 NPs using the sol-gel method [16]. In this synthesis, Zirconium oxychloride ($ZrOCl_2$) has been used as a source of Zr. NS was synthesized by mixing 2.1 g of citric acid, 1.2 g of succinic acid, 0.5 g of CTAB (Cetyl Trimethyl Ammonium Bromide), 10 ml of ethoxylatednonyl phenol, and 20 ml of ethylene oxide with 3.2 g of $ZrOCl_2$. This surfactant was evaluated on the basis of its surface-active properties like IFT, CMC, and wettability. A comparison of their performance properties has also been made with commercially available cationic (CTAB) and anionic (SDS) surfactants in terms of their rheological properties along with their applications specifically in EOR.

Gu *et al.* (2022) [29] synthesized and explained the self-assembly of Janus symmetrical inorganic NSs of matchstick-shaped with precise surface amphiphilicity. In this work, matchstick-shaped Janus NSs were synthesized that resembled the surfactant molecules and explored their self-assembly. Great extent of amphiphilicity was attained via ligand-exchange reaction in hard–soft acid–base having high selectivity on the matchsticks surface of NSs with CdS stems and Ag_2S heads. Synthesized NSs extemporaneously gathered into varied ordered structures like cylindrical, lamellar, wrinkled, curved, and micellar structures depend upon the IFT and vertical asymmetry held through their surface ligands and geometry.

Lin *et al.* (2018) studied the dispersibility and agglomeration of deoxidized graphene oxide in chiral nematic LCs through charged 2D NSs. In this work, a novel approach to constructing 3D ordered chiral architecture containing RGO as well as using the anisotropic characteristics of RGO to create functional soft matter was proposed which involved the exploitation of a new two-dimensional charged zirconium phosphate (CZ) nanoplatelet NS to liquid phase containing ordered soft matter and dispersed RGO [25].

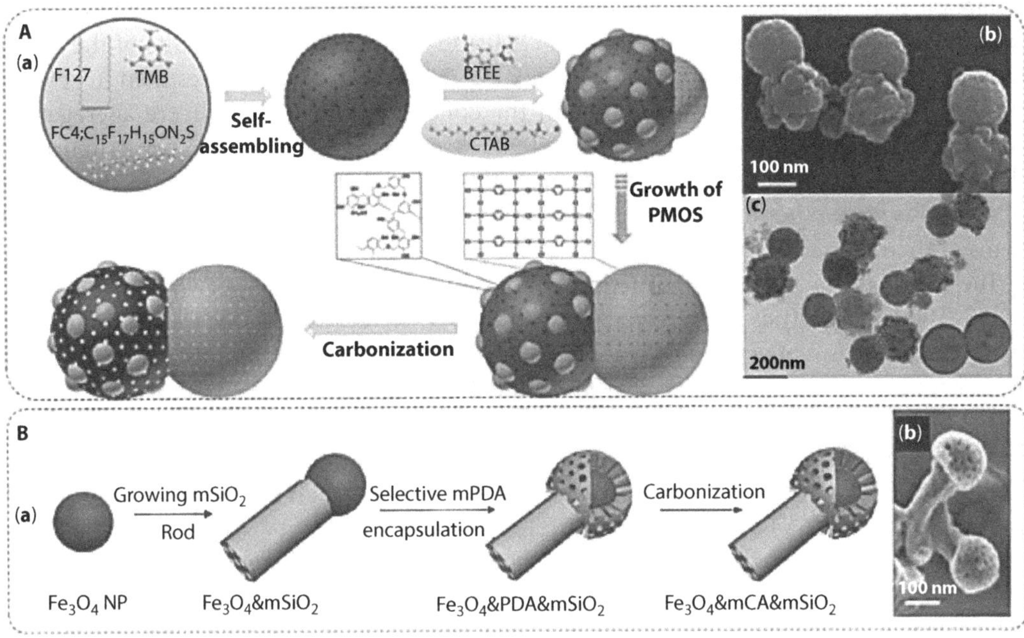

Figure 17.1 (Aa) Synthesis of Janus different Janus NSs, (Ab) SEM, and (Ac) TEM. (Ba) Fabrication scheme and (Bb) SEM image of iron-oxide based Janus NPs (used with copyright permission) [31].

Yang *et al.* (2017) effectively created the dumbbell-shaped amphiphilic Janus particles, with one nonpolar and one polar sphere, and exploited them as interface-active solid catalysts for stabilizing the Pickering emulsions. The study revealed that the Janus catalyst displayed improved catalytic ability than conventional Pt/C catalysts in aqueous hydrogenation reactions [30].

Zhao *et al.* (2021) produced the e dual-mesoporous Janus NPs, with one pure 1-D mesoporous silica nanorod strictly connected with a mesoporous magnetic iron-oxide nanosphere. Adjustments in their hydrophilic and hydrophobic volume ratio provided them with amphiphilic characteristics and made them potential emulsion stabilizers (Figure 17.1) [31].

Khowdiary *et al.* (2022) synthesized novel NSs and evaluated them as effective corrosion inhibitors and biocide metal complex NP surfactants via reacting sulfonamide with selenious acid and characterized them through elemental analysis, FTIR spectra, ^{1}H-, and ^{13}-NMR spectroscopic techniques to confirm the purity of the synthesized compounds [32].

17.5 Characterization

Characterization of NSs can be done to study their morphology, aggregation behavior, IFT, and wettability. In addition, studies of core properties are also required to understand the micellar size.

Kim *et al.* (2017) [13] investigated the core properties of C_{60}NS that influenced the automatic orientations of molecules as well as their interactions between anisotropic media, surfactant materials, and solid substrates through polarized optical microscopy (POM). In order to study the interactions of NSs POM study was performed. Results of the POM study revealed that doping of 0.1 wt% loads of C_{60}NS into the LCs cell resulted in several micrometer-sized particles. An increase in the concentration showed severe aggregation that occurred due to the coalescence of C60NS, which also disturbed the construction of 2D uniform layers and subsequently formed 3D micro-particles. Remarkably, multifolds LC domains were randomly oriented. This different behavior of C_{60}NS can be explained in terms of their concentration.

The X-ray diffraction (XRD) method was used to study the aggregation behavior of C_{60}NS. The XRD analysis of uniform dispersion of CZ-RGO suspensions was made to confirm the completion of crystallinity of bulk zirconium phosphate nanocrystals. In the XRD of NS crystals, strong peaks at 11.9°, 19.72°, and 25° confirmed the crystallinity.

The SEM analysis was used to detect the morphology of CZ-RGO and Zeta potential was used to confirm the complete neutralization of negatively charged protons exfoliated CZ nanoplatelets on zirconium phosphate nanocrystals through strong base TBAOH. At this stage, the value of zeta potential was found to be -51.38 mV. Through zeta potential, the dispersion stability of CZ-RGO suspensions can also be analysed, whereas the UV-Vis-NIR spectroscopy is used to analyze the dispersion stability of RGO in CZ-RGO suspension.

IFT is an important parameter used to characterize the NS formulations. The presence of NSs reduces the IFT when present in the different interfaces specifically when checked for oil and water interfaces. This characteristic of NSs makes them suitable for surfactant EOR. In order to test enhance oil recovery NMR measurements. NMR measurements were conducted before waterflooding and after waterflooding, and NS was injected, followed by chase waterflooding, and the spatial oil saturations were assessed via NMR data.

The wettability of NS formulations was also investigated via NMR studies. In order of a qualitative assessment of wettability, the core samples were flushed with 4 PVs of fresh crude oil before NMR measurements. The T_2 distributions of samples saturated with oil at S_{swirr} before and after 4 weeks of aging were recorded. Results revealed that T_2 distributions of samples that have more oil-wet surfaces shifted to shorter T_2 time as compared to the samples that have less oil-wet surfaces.

NS flooding tests combined with NMR were used to evaluate the NS potential of oil mobilization. This test was used to study the effect of soaking NS under different wettability conditions. In addition to that, the oil mobilization by NS injection was also evaluated using different injection rates of chase sea water and remaining oil saturation after primary water flooding using different injection rates of chase water and remaining oil saturation.

17.6 Properties of Nano-Surfactants

NS molecules have the characteristics of NPs along with the properties of surfactants. They are amphiphilic molecules of nanosized, thus displaying the synergetic approach of both the surfactants as well as of NPs. These peculiar characteristics make them potential candidates to be exploited in various industrial sectors like lubricants, EOR, and printing inks.

17.6.1 Molecular Self-Assembly

The surface properties of nanomaterials play a crucial role in their molecular self-assembly. To achieve the desired chemical and physical properties of nanomaterials, the molecular packing structure on their surface must be carefully considered. Kim and her co-workers (2018) [32] proposed a novel NS that could automatically construct a macroscopic surface alignment layer for LC molecules. The C60NS NS, designed and synthesized, was composed of mesogenic cyanobiphenyl moieties and a fullerene nano atom, along with flexible alkyl chains [33]. Its amphiphilic nature, derived from the LC-favoring and LC-repelling groups in its asymmetric structure, results in them self-assembling into monolayered protrusions on the surface when introduced into an anisotropic LC medium. This self-assembly results in phase separation, and the monolayered protrusions amplify and transfer the surface‘s molecular orientational order to the bulk, leading to automatic vertical molecular alignment on the macroscopic scale. Gu *et al.* (2020) [29] examined the molecular self-assembly of a Janus NS by optimizing the length of the nonpolar stem and revealed that shorter hydrophobic tails revealed a weaker relative absorbance that increased with increase in the hydrophobic tail.

17.6.2 Surface Hydrophobicity and Interfacial Tension

IFT and vertical geometry of NSs can be altered to the required value via controlling the length of hydrophobic moiety and the surface hydrophobicity of the nonpolar stem of the NS tension. Modulation of self-assembled vertical lamellar phase into curved structures like cylindrical and micellar form, acquire the free energy change based on the identical relationship.

In assemblies, the controlled length of the hydrophobic stem has been represented as "d," and the extent of nonpolarity of the NSs is defined as "γ." A systematic decrease in the length of the stem decreases the d, which results in a self-assembly with curvature "R" of small radii. Decrease in the radii, resulted in the phase transition of the aggregates from vertical lamellae to wrinkler and interconnected cylindrical symmetry. Reduction in the nonpolarity due to the surface of shorter hydrocarbon ligands reduces the "γ" that controlled its compatibility with the polar solvent phase. Earlier studies, done on the monounsaturated C_{18} hydrophobic group also endorsed this approach by revealing that NSs with C_{18} precursors showed 1.2, 1.1-, and 1.05-times larger surface tension in polar solvents than their C_{12}, C_{14}, and C_{16} counterparts, respectively. A decrease in the value of R due to the reduction in "γ" explicates the phase transition of the surfactants from the lamellar structure to micellar and cylindrical forms. In addition to the above explanation, the values of exposed area "Δa" were found augmented in the sequence of curved, wrinkled, cylindrical, and micellar structures with a minimum of 0.2 for curved and a maximum of 1.0 for wrinkled assemblies [30].

Very recently, Yang *et al.* (2023) determined the equilibrium surface tension of Carbon Dots solution using a surface tensiometer via the Wilhelmy plate technique. In this study, a spinning drop interfacial tensiometer was exploited to assess the IFT between CDs aqueous solution and crude oil whereas hydrodynamic diameters and zeta potentials of Carbon Dots solutions were done through Zetasizer Nano ZS. The results of this study revealed that carbon dots displayed the same surface tension pattern as their conventional counterparts.

Increase in the concentration of CD NSs decrease the surface tension down to the specific value, the critical micelle concentrations (cmc), which was found 5000 mg/L for C_6-CDs, 2,500 mg/L for C_8-CDs, and 100 mg/L for C_{12}-CDs with surface tension (γcmc) of 37.82 mN/m for C_6-CDs, 34.62 mN/m for C_8-CDs, and 27.45 mN/m for C_{12}-CDs. Hence, it may be concluded that an increase in the alkyl stem, lowers down the cmc and γcmc, thus enhancing the hydrophobic properties of NSs [34–37].

17.6.3 Micellization and Dispersion Stability of Nano-Surfactants

Asl *et al.* (2020) considered the pH, turbidity, and conductivity measurements to find out the critical micelle concentration of SiO2 nano-particles. Analysis of their dispersion ability was done by using visual observation and zeta potential within the surfactant solutions. Results of this study revealed that the maximum dispersion stability of SiO2 NPs in surfactants was attained at the concentration of 1000 ppm concentration whereas drastic change in the maximum dispersion stability was shown in the NSs solution of L-Arg and L-Cys with the concentration of 2000 and 4500 ppm, respectively, as CMCs [38].

In a study by Yang and his co-workers, the dynamic light scattering (DLS) technique was used to explore the behavior of surfactants to form different shapes of micelles like spheric [39], rod-shaped [40], and vesicles [41]. DLS was used to determine the CD aggregates at different conditions to recognize their micellization behavior. On the basis of an observation-based conclusion, it was stated that at 0.1 CMC, C_{12}-CDs exist predominantly as a single particle. Although, a few aggregates with an average size of ~200 nm were also depicted at 0.1CMC via peak intensity analysis but a number of aggregates was as low as negligible [42].

17.6.4 Colloidal Stability

Colloidal stability is one of the very important characteristics of NSs that make them potential candidates for numerous applications such as in surfactant EOR, lubricants, and nanofluids (NFs). Colloidal stability can be understood as the tendency of NPs that inhibit their aggregation and settling out at a substantial rate that is required for the NSs to retain their specific nano characteristics and their potential to get commercialized in various sectors [43]. Zeng *et al.* (2020) disclosed in a study that 2D flakes and 1D nanotubes along with the NSs of graphene QD showed high colloidal stability in water for months with high zeta potential values [44].

17.6.5 Size, Shape, and Type of Nanoparticle

The size of the NSs has a great influence on their dispersion stability which increases with a reduction in their size. NPs with a size less than 10 nm exhibited higher stability. Not only the size of nano surfactants but their shape also governs their overall functioning. For instance, NSs with rod-shaped or pleated NPs, e.g., carbon nanotubes or graphene, possess greater contact area, and hence, own a higher tendency to form agglomerates in comparison to the spherical NPs. An increase in the size followed by aggregation has an adverse impact on their dispersibility and increases their settling down. Moreover, the materials used in their synthesis also govern their stability and functioning. TiO_2-based NPs in composition with Heena-tragacanth as surfactant showed a greater reduction in the IFT of oil water system over graphene-based composition and claimed to have better applications in EOR [40].

17.7 Stratification of Nano-Surfactants

Stratification studies of NSs revealed that in the mixture of homogenous particles, they readily reached the interface and after solvent evaporation, stratified in an uninterrupted closely packed monolayer [45]. Experimental figures revealed that the rate of movements of NSs was three-fold higher than conventional Brownian motion, even across a noteworthy distance of about 200 μm from the interface. Though the comprehensive mechanism for the stratification of NS is not yet clear, but on the basis of the proposed hypothesis, it was stated in this work that the stratification of NSs has been inclined by a complex interaction between evaporation, change in the concentration, and the adsorption of the particles. NSs with a *Janus* geometry when mixed with homogenous particles, were found to have the ability of self-stratification to closely packed monolayer on the top of a drying film. This exclusive property qualifies them as an additive to enhance the water resistance of conventional coatings. Dissimilar to the homogeneous particle systems which stratify highly via diffusion approaches, the Janus particles display robust interfacial adsorption energy along with the characteristics to rapidly and fully accumulate at the air-water interface, though generating a self-stratified hydrophobic surface [46].

17.8 Applications of Nano-Surfactants

NSs have unique properties that make them suitable for a range of applications. Many researchers revealed their applicability in various sectors by performing experimental studies. One of the most promising applications is in EOR, where they can be used to increase the amount of oil that can be extracted from a reservoir. NSs can improve the wettability of the rock surface, reduce IFT, and increase the mobility of oil, making it easier to extract. Another potential application of NSs is in drug delivery systems. NSs can be used to improve the solubility and stability of drugs, as well as to control their release, which can lead to more effective and targeted treatments. NSs can also be used in cosmetics and personal care products, where they can improve the delivery of active ingredients, enhance the appearance of skin and hair, and increase the shelf life of products. NSs have also been studied for their potential use in environmental remediation. They can be used to enhance the biodegradation of pollutants, remove heavy metals from contaminated water, and improve the efficiency of wastewater treatment processes. In addition to these applications, NSs have potential uses in the food industry, cosmetic industry, and many more where they can improve the stability, non-toxicity, and shelf life of the emulsions or compositions. NSs are a promising area of research with a wide range of potential applications in various industries. Some of the potential applications validated by the scientific researchers are as follows:

17.8.1 Drug Release

The utilization of nanoencapsulation and targeted chemical delivery methods has brought about significant advancements in various domains, including medical treatment and diagnosis through pharmaceutical drug delivery. It has the potential to bring similar transformations to various operations within the upstream oil and gas sector. NSs are a promising tool in the field of drug delivery. These surfactants, which are typically composed of a hydrophobic tail and a hydrophilic head, can self-assemble into NPs with a high surface area and the ability to encapsulate drugs. This unique characteristic of NSs makes them ideal for delivering drugs with poor solubility, as they can increase the solubility of these drugs and enhance their bioavailability. NSs can also be functionalized with targeting ligands, allowing for specific delivery of drugs to particular cells or tissues. Furthermore, they can be modified to release drugs in response to specific stimuli such as pH, temperature, or enzymes, which can enhance drug efficacy while reducing toxicity. NSs have other advantages over traditional drug delivery methods. For example, they are biocompatible and biodegradable, which reduces the risk of adverse reactions and toxicity. They are also cost-effective and scalable, making them a viable option for large-scale drug production. Liu *et al.* (2020) [1] demonstrated the competence of NSs to get disperse and encapsulate homogeneous NPs and small molecules. The accumulated structures showed triggered release because of owing good response to the external magnetic field and solvent and pH changes. In this work, highly energetic plasmonic-magnetic NSs were established via were obtained through seeded-growth procedure via selective ligand exchange method, a novel acid activation approach of Au–Fe_3O4 dumbbell nanocrystals consisting of two lobes, one hydrophilic and another hydrophobic. Such a step drastically enhanced the surface energy of NS, which aids the strong adsorption at interfaces and reduction in the interfacial energy. These NSs were claimed to have high potential to be employed in catalysis, separation, and medicine [1]. The resulting magnetically labeled NF was designed to achieve two main objectives: to alter the wettability of rocks and reduce the IFT to enhance the mobilization of residual oil and to enable monitoring of injected fluids *in situ* through combination with EM surveys.

To produce NFs with different concentrations of 5-nm SPOINs, a two-step nanoencapsulation approach was used. In order to check the chemical and colloidal stability of the prepared nano-formulations, these compositions were evaluated by storing them for a year at the temperature of 90°C under close reservoir conditions, and results were found in favor of their remarkable stability. Transition electron microscopy (TEM) images also established their encapsulation. SPOINs-NFs confirmed more than three orders of magnitude reduction in IFT between crude oil and water, reducing it from ~ 25 mN/m to ~ 0.01 mN/m. The IFT and stability studies revealed a strong interaction among SPIONs and petroleum sulfonate surfactants. This encapsulation platform allows for the encapsulation of various NPs to create a library of multi-functional NFs that can support a range of upstream applications [47].

17.8.2 3D Printing

Zeng, *et al.* (2020) produced the colloidal NSs and studied their potential applications for 3D conformal printing of 2D van der Waals stuff. In this research work, the utilization of nanomaterials for printing techniques has been demonstrated as a flexible method for swift prototyping and remarkably big-scale fabrication of working gadgets. Surfactants have a critical role in numerous printing procedures because of their potential to lessen IFT among ink solvents and NPs, thus enhancing ink colloidal stability. In the current investigation, a colloidal graphene quantum dot (GQD)-based NS has claimed to steady a variety of 2D materials and offered better colloidal stability to a graphene ink via π–π stacking interaction, which enabled the printing of multiple high-resolution patterns on numerous substrates exploiting a single printing pass. As a result, NSs can remarkably improve the mechanical durability of the printed graphene films, as demonstrated by 100 taping, 100 scratching, and 1000 bending cycles in contrast to conventional molecular surfactants. Moreover, the printed composite film displays enhanced photoconductance when exposed to UV light with 400 nm, because of the excitation through the NS bandgap. Exploiting the 3D conformal aerosol jet printing procedure, a set of UV sensors with heterogeneous structures is straightaway printed on 2D flat and 3D spherical substrates, which exhibited the potential to create geometrically versatile devices based on NS inks [41].

17.8.3 Enhanced Oil Recovery

Traditional methods of using conventional surfactants for EOR have limitations, such as the adsorption of the surfactant on rock near the wellbore or its diffusion into small water-filled pores. To address these issues and improve the delivery of surfactants to the oil phase, a solution involves formulating the surfactant molecules into NPs as NSs.

Abdel-Fattah and colleagues (2017) developed petroleum sulfonate salt NP dispersions in seawater. These NS particles with a size of 10–60 nm along with co-surfactants consisting of petroleum sulfonate salts that are insoluble in seawater were found to be stabilized in seawater even at high temperatures. The colloidal approach and enormously small size of NS particles allowed them to wander and distribute petroleum sulfonates to residual and remaining oil deep in the reservoir without the requirement for big amounts. The NSs can recompense losses through adsorption onto rock surfaces and by diffusing into water-filled little pores. The researchers evaluated the performance of the NS formulations under reservoir conditions and found that they remained chemically and colloidal stable for over six months at 100°C. The NSs significantly reduced the seawater-crude oil IFT by two to three orders of magnitude at 100°C and formed oil-in-water emulsions without any mechanical mixing. The findings of this study suggested that NS particles are a promising tool for EOR applications and have the potential to significantly enhance oil recovery while being cost-effective and easy to prepare in the field [48].

Another research done on amphiphilic Janus silica nano-particles revealed them as a novel particle emulsifier. Janus silica nano-particles firmly adsorb at the surface of the droplets thus reducing oil-water IFT and significantly subsidizing the effective dispersion of several oils in artificial seawater (Figure 17.2) [49].

In 2018, Mashat and colleagues claimed the exploitation of NPs as a potential candidate to stabilize low-cost petroleum sulfonate surfactants in high salinity and temperature water, making them suitable for use in EOR operations in carbonate reservoirs. Their study involved the evaluation of three NS formulations through experimental investigations of their IFT with crude oil and phase behavior. To prepare the three NS formulations, a one-step nano-emulsification process was employed, which involved mixing high salinity water,

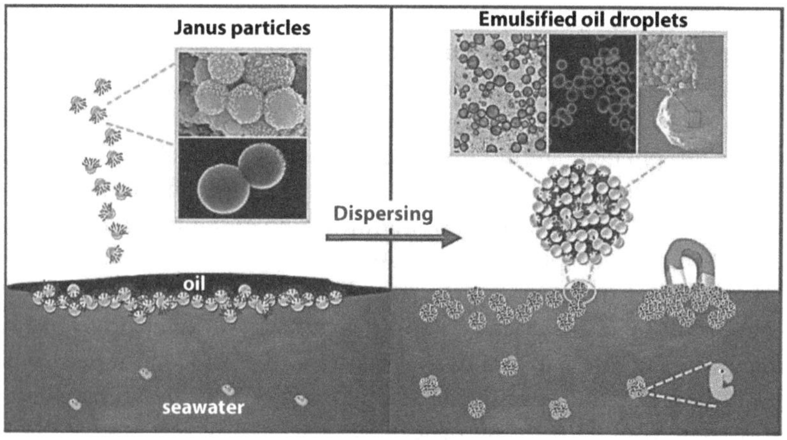

Figure 17.2 Janus silica nanoparticle in EOR (used with copyright) [49].

a 5 wt% petroleum sulfonate solution, and a low dose of three different 4 wt% co-surfactant solutions. The resultant compositions are with 0.2 wt% of total active constituents, and one of the formulations was claimed to be colloidally and chemically stable in water with high salinity (~ 56,000 ppm) at 100°C for over six months, although the rest two displayed the signs of unsteadiness after the duration of four months. The IFTs at the interphase of crude oil and NS solutions were quantified at 90°C via spinning drop interfacial tensiometer, and results revealed the cmc in the range of 10^{-2} to 10^{-3} mN/m, which was remarkably reduced, in comparison to those with high salinity water or solutions of corresponding co-surfactants at alike concentrations. Studies on phase behavior characterization displayed the improved formation of homogeneous oil-in-water emulsions at 100°C without mechanical mixing, demonstrating the capability of NS to mobilize oil in typical carbonate reservoir conditions. Due to their colloidal nature, NSs have the potential to migrate deeper into the reservoir compared to conventional micellar surfactants, as they are subject to size exclusion and chromatographic effects [50].

Chemically-driven tertiary mode oil mobilization has shown promise as an effective approach. Specifically, surfactants with or without polymers designed to suit the specific conditions of the reservoir were added to water while water flooding to isolate lingering oil from the reservoir rock's porous structure into production wells. To achieve successful mobilization while keeping costs manageable, it is important to use inexpensive, reservoir-compatible surfactants. Petroleum sulfonates are among the most affordable surfactants available and are suitable for oil mobilization, but their low solubility in harsh reservoir conditions, such as those with high salt content (greater than 50,000 ppm) and high temperature (over 100°C), makes them less useful in some cases. To address this, Alzobaidi *et al.* (2018) have developed a simple, one-step technique for encapsulating petroleum sulfonates in high-salinity injection water, creating nano-sized multi-emulsion droplets that remain stable in high temperatures over long periods. Additionally, the development of a continuous, economical synthetic procedure to create petroleum sulfonates straight through crude oil, further improving the sustainability and economics of the process was also reported in the same work. In-house-produced sulfonates were formulated into an NS solution using our encapsulation method and the results demonstrated that petroleum

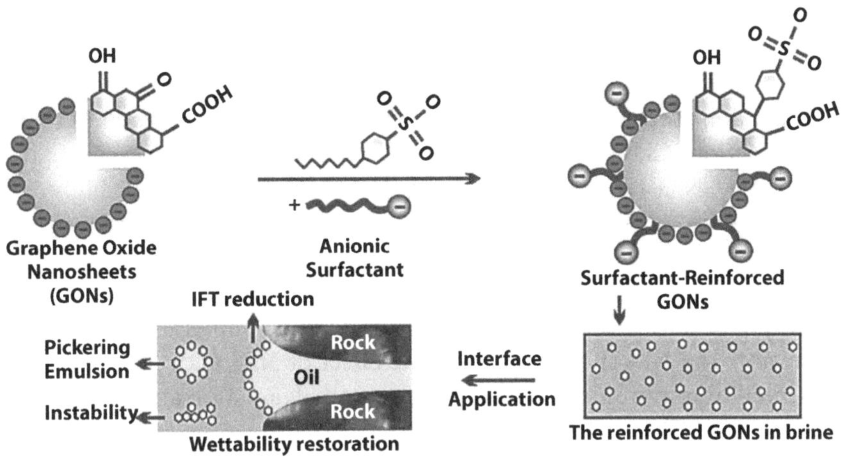

Figure 17.3 Nano-surfactants enhanced oil recovery (used with copyright permission) [51].

sulfonates were successfully synthesized and that the formulated NS solution persisted as stable in reservoir conditions, changed the wettability of oil-wet rock as reflected through contact angle measurements, and caused crude oil to be released from rock through spontaneous imbibition (Figure 17.3) [51, 52].

Studies on oil film removal efficiency and contact angle experiments demonstrated that C_{12}-CDs and C_{12}-SCDs exhibited superior wetting properties compared to the traditional surfactant SDBS. This is attributed to the abundance of hydroxyl, carboxyl, amine, and sulfonate groups in these surfactants. Specifically, C_{12}-SCDs were found to have better wetting properties than C_{12}-CDs because of having hydrophilic sulfonate moieties. In addition,

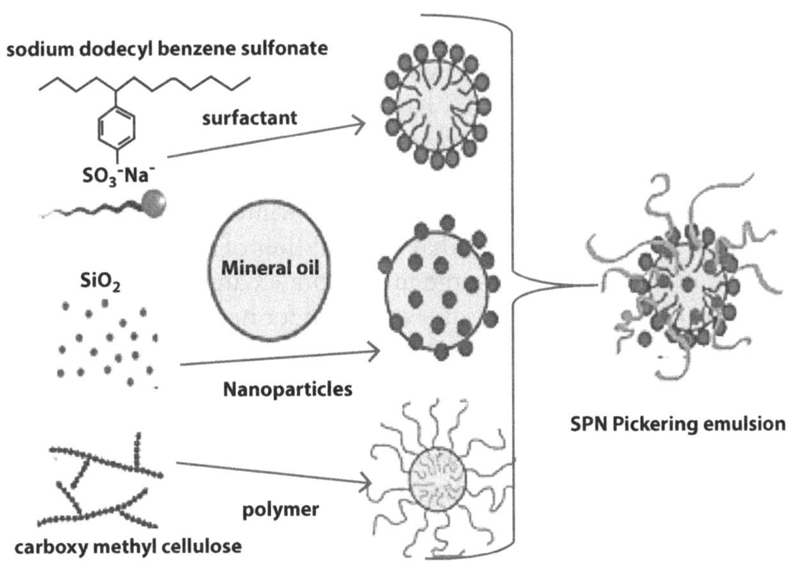

Figure 17.4 SPN pickering emulsion of NPs, surfactants, mineral oil, and polymer (used with copyright permission from reference) [55].

Pickering emulsions steadied through C_{12}-SCDs and displayed rapid response performance to pH and CO_2 changes. In contrast to their conventional counterparts SDBS and APG12, CDs were detected to have outstanding dispersion competencies for MWCNTs [53, 54]. Unlike the dispersion mechanism of cationic surfactant CPC, which depends on the strong electrostatic attraction, CDs rely on hydrophobic effects and π-π stacking to adsorb onto the surface of MWCNTs for effective dispersion. Notably, due to their high steric resistance and electrostatic repulsion, C_{12}-SCDs exhibited superior dispersion performance compared to C_{12}-CDs. Another investigation for the application of NSs in EOR involves the use of a stable oil-in-water surfactant–polymer–nanoparticle (SPN) Pickering emulsion through the exploitation of carboxy methyl cellulose, light mineral oil, and SiO_2 NPs in anionic surfactant. The NPs show the synergetic effect of NPS with polymer in surfactant has been observed thus inhibiting the droplets from amalgamation and reducing the IFT at the oil-water interface (Figure 17.4) [55–59].

17.8.4 In Agriculture

The study examined the toxicity effects of two distinct NSs synthesized via chemical methods. In these investigations, mixture of a negatively charged surfactant (N-(2-hydroxyethoxy)-O-(2-hydroxyethyl)-N-(2-(2-mono mercapto acetate)) hydroxyl ammonium-4-dodecyl benzene sulfonate) and a non-ionic surfactant (Di Phosphate monopoly ethylene glycol mercapto acetate, tri ethanol amine monophosphate polyethylene glycol oleiate) was assessed. Furthermore, evaluation has been done on the impact of these synthetic NSs on the physicochemical characteristics and efficacy of Nasrthrin 25% EC (Cypermethrin) and Tak 48% EC (Chlorpyrifos) pesticide compositions against 2nd instar larvae of cotton leafworm, Spodoptera littoralis (BOISD). The study presented the measured LC50s and LC90s values in a table, indicating that the formulations were as effective as traditional ones due to their capacity for controlled delivery of the active ingredients. The insecticidal activity of the tested pesticides was affected by changes in surface tension and pH values, which were also influenced by the surfactant type and the pesticide [54].

Two non-ionic organo-silicates and two surfactants containing silver NPs were separately added to commercial herbicides based on cycloxidim (C) or a combination of 2,4-dichlorophenoxyacetic acid and fluroxypyr (D) and the surface tension and contact angle of the liquid droplets from the resulting preparations were measured, and their effectiveness was tested by spraying winter wheat and winter rape crops with preparations C and D, respectively. Results showed that the addition of Ag-containing surfactant to the herbicides led to a significant upsurge in the contact angle and a small reduction in surface tension, in comparison to the values attained for herbicides in the absence of any adjuvant. The other adjuvants had minimal impact on the physicochemical parameters of the modified herbicides. The herbicidal effectiveness of the Ag-modified herbicide was found to be high when sprayed on wheat crops. On the other hand, modifications of herbicides with organosilicate surfactants had a much smaller effect on their herbicidal effectiveness for spraying wheat and rapeseed crops [60].

17.8.5 Others

A new method was used to create electro-responsive birefringence modulators, involving the synthesis of an amphiphilic fullerene NS (abbreviated FNS). The FNSs possess amphiphilic properties, allowing them to self-organize via phase separation in the LC mixture and form a 2D monolayer on the solid surface. It was confirmed via spectroscopic and morphological analyses that the FNSs acted as alignment layers in the LC cell. A multi-domain LC cell was designed and fabricated to test the electro-optical properties of the LC mixture. At the initial state, the LC cell exhibited vertically-aligned (VA) mode due to the automatically constructed vertical alignment of LC molecules. However, when voltage was applied, the LC molecules rearranged and changed to planar alignment (PA) mode. The system that can automatically induce the VA of LC molecules and switch the transmittance with low driving voltage has potential applications in the LC display field [61].

Asymmetric NS (C60BP) was synthesized using the Bingel reaction to create a macroscopic surface alignment layer for LC molecules. The C60BP consisted of mesogenic cyanobiphenyl groups attached to fullerene and possesses an amphiphilic nature due to the asymmetrical structural motif of LC-favoring and LC-repelling groups. When introduced into the anisotropic LC medium, the C60BP self-assembled into monolayered protrusions on the surface, resulting in the automatic construction of the alignment layer. This layer has potential applications in electrically controllable modulators [62–64].

17.9 Conclusions

NSs are gaining increasing attention in a variety of fields due to their unique properties and potential applications. NSs exhibit several advantages and enhanced performance compared to typical surfactants. They possess greater stability, both thermodynamically and against phase separation, due to their reduced size and increased interfacial area. This stability allows them to maintain their emulsified state for longer periods, even under challenging conditions such as high temperatures or high shear forces. NSs can solubilize hydrophobic compounds more effectively than conventional surfactants thus enabling improved dispersion and encapsulation of hydrophobic substances. NSs can enhance the bioavailability of poorly soluble drugs and nutrients. By forming nanoscale droplets, they increase the surface area available for interaction with biological tissues, thereby promoting better absorption and uptake of active compounds. This feature is particularly valuable in pharmaceutical applications where improving drug delivery and efficacy is crucial. NSs can achieve the desired emulsification or dispersion with lower concentrations compared to traditional surfactants. This benefit reduces the overall surfactant load, making them more cost-effective and potentially minimizing any adverse effects associated with high surfactant concentrations. NSs can contribute to environmental sustainability. Their improved performance and lower required concentrations can lead to reduced waste and pollution.

Despite their potential, there are also several challenges associated with the use of NSs. For instance, the long-term safety of these materials has not yet been fully established, and concerns remain regarding their potential toxicity and impact on the environment.

Additionally, the complex interactions between NSs and biological systems can be difficult to understand and predict, which may hinder their widespread adoption. New developments in materials science and nanotechnology are expected to lead to the creation of more advanced and functionalized NSs, which may overcome some of the current limitations.

References

1. LIU, F., *et al.*, Activation and assembly of plasmonic-magnetic nanosurfactants for encapsulation and triggered release. *Nano Lett.*, **20**, 12, 8773–8780, 2020.
2. BINKS, B.P., Colloidal particles at a range of fluid–fluid interfaces. *Langmuir*, **33**, 28, 6947–6963, 2017.
3. ZHENG, W., *et al.*, Micro-scale effects of nano-SiO2 modification with silane coupling agents on the cellulose/nano-SiO2 interface. *Nanotechnology*, **30**, 44, 445701, 2019.
4. BRADLEY, L.C., STEBE, K.J., LEE, D., Clickable janus particles. *J. Am. Chem. Soc.*, **138**, 36, 11437–11440, 2016.
5. JIANG, S. and GRANICK, S., Controlling the geometry (Janus balance) of amphiphilic colloidal particles. *Langmuir*, **24**, 6, 2438–2445, 2008.
6. TSYRENOVA, A., *et al.*, Unique Orientation of the Solid–Solid Interface at the Janus Particle Boundary Induced by Ionic Liquids. *J. Phys. Chem. Lett.*, **11**, 22, 9834–9841, 2020.
7. MILLER, K., *et al.*, Drying mediated orientation and assembly structure of amphiphilic Janus particles. *Soft Matter*, **14**, 33, 6793–6798, 2018.
8. TRIPATHY, D.B., *et al.*, *Surfactants from Renewable Raw Materials*, CRC Press, Abingdon, Oxon, OX14 4RN, 2021.
9. KESARWANI, S., *et al.*, Surfactants: Patent Landscape of the Most Versatile Class of Materials. *Surfactants Renewable Raw Mater.*, 227–256, 2021.
10. MCCURDY, R., High-rate hydraulic fracturing additives in non-Marcellus unconventional shales. *Hydraul. Fracturing Study: Chem. Anal. Methods*, 17, 1–114, 2011.
11. MURSALOVA, M.A. and SHAKHBAZOV, E.K., Investigation of Nano-Surfactor Compositions for Inhibition of Salt Deposits in Oil and Gas Production. *Sci. Works NIPI Neftegaz SOCAR*, **1**, 41–44, 2014.
12. LATTUADA, M. and HATTON, T.A., Synthesis, properties and applications of Janus nanoparticles. *Nano Today*, **6**, 3, 286–308, 2011.
13. CHAKRABORTY, S. and AGRAWAL, G., *Amphiphilic nanoparticle, composition comprising same and method of controlling oil spill using amphiphilic nanoparticle*, vol. 428, p. 383, 2016.
14. KIM, D.-Y., *et al.*, Asymmetric fullerene nanosurfactant: interface engineering for automatic molecular alignments. *Small*, **14**, 1, 1702439, 2018.
15. SAEEDI, L. Hossein, *et al.*, The Production and Evaluation of a Nano-biosurfactant. *Pet. Sci. Technol.*, **32**, 2, 125–132, 2014.
16. MOHAJERI, M., HEMMATI, M., SHEKARABI, A.S., An experimental study on using a nanosurfactant in an EOR process of heavy oil in a fractured micromodel. *J. Petroleum. Sci. Eng.*, **126**, 162–173, 2015.
17. ABDEL-FATTAH, A., II, *et al.*, NanoSurfactant for eor in carbonate reservoirs, in: *SPE Kingdom of Saudi Arabia Annual Technical Symposium and Exhibition*, OnePetro, Damman, Saudi Arabia, 2017.
18. CHERAGHIAN, G. and NEZHAD, S.S.K., Improvement of heavy oil recovery and role of nanoparticles of clay in the surfactant flooding process. *Pet. Sci. Technol.*, **34**, 15, 1397–1405, 2016.

19. HARBI, A.L., Ahmad, M., *et al.*, The study of nanosurfactant EOR in carbonates by advanced NMR technique, in: *Abu Dhabi International Petroleum Exhibition & Conference*, 2017.
20. NWIDEE, L.N., *et al.*, Nanoparticles influence on wetting behaviour of fractured limestone formation. *J. Pet. Sci. Eng.*, **149**, 782–788, 2017.
21. DAS, S.K., CHOI, S.US., PATEL, H.E., Heat transfer in nanofluids—a review. *Heat Transfer Eng.*, **27**, 10, 3–19, 2006.
22. NWIDEE, L.N., *et al.*, Nanofluids for enhanced oil recovery processes: wettability alteration using zirconium oxide, in: *Offshore Technology Conference Asia*, OTC, p. D011S002R003, 2016.
23. YU, W. and XIE, H., A review on nanofluids: preparation, stability mechanisms, and applications. *J. Nanomater.*, **2012**, 1–17, 2012.
24. SHARMA, T., KUMAR, G.S., SANGWAI, J.S., Comparative effectiveness of production performance of Pickering emulsion stabilized by nanoparticle–surfactant–polymerover surfactant–polymer (SP) flooding for enhanced oil recoveryfor Brownfield reservoir. *J. Pet. Sci. Eng.*, **129**, 221–232, 2015.
25. SHARMA, T., *et al.*, Thermal stability of oil-in-water Pickering emulsion in the presence of nanoparticle, surfactant, and polymer. *J. Ind. Eng. Chem.*, **22**, 324–334, 2015.
26. LIN, P., *et al.*, Dispersion and assembly of reduced graphene oxide in chiral nematic liquid crystals by charged two-dimensional nanosurfactants. *Chem. Eng. J.*, **334**, 1023–1033, 2018.
27. EMADI, S., *et al.*, Effect of nano silica particles on Interfacial Tension (IFT) and mobility control of natural surfactant (Cedr Extraction) solution in enhanced oil recovery process by nano-surfactant flooding. *J. Mol. Liq.*, **248**, 163–167, 2017.
28. Mashat, Afnan, Ayrat Gizzatov, and Amr Abdel-Fattah, Novel nano-surfactant formulation to overcome key drawbacks in conventional chemical EOR technologies. SPE Asia Pacific Oil and Gas Conference and Exhibition. *SPE*, 2018.
29. GU, D.H., CHOI, W., SON, J.S., Self-Assembly of Matchstick-Shaped Inorganic Nano-Surfactants with Controlled Surface Amphiphilicity. *JACS Au*, **2**, 10, 2307–2315, 2022.
30. YANG, T., *et al.*, Dumbbell-shaped bi-component mesoporous Janus solid nanoparticles for biphasic interface catalysis. *Angew. Chem. Int. Ed.*, **56**, 29, 8459–8463, 2017.
31. ZHAO, Y., *et al.*, Rational design and synthesis of multimorphology mesoporous carbon@ silica nanoparticles with tailored structure. *Carbon*, **183**, 912–928, 2021.
32. KHOWDIARY, M.M., *et al.*, Synthesis of novel nano-sulfonamide metal-based corrosion inhibitor surfactants. *Materials*, 15.3, 1146, 2022.
33. Park, M., Kim, J., Yoon, W.J., Lim, S.I., Rim, M., Wi, Y., ... & Jeong, K.U., Fullerene-Based Nanosurfactant for the Automatic Homeotropic Alignment of Liquid Crystals. *ACS Applied Nano Materials*, 6(9), 7942–7947, 2023.
34. DISCHER, D.E. and EISENBERG, A., Polymer vesicles. *Science*, 297.5583, 967–973, 2002.
35. CUI, P., *et al.*, Synthesis of a fluorescent cation surfactant derived from carbon quantum dots. *Mater. Lett.*, **235**, 161–163, 2019.
36. XI, F., *et al.*, Amphiphilic graphene quantum dots as a new class of surfactants. *Carbon*, **153**, 127–135, 2019.
37. LI, Y., *et al.*, A novel cationic surfactant synthesized from carbon quantum dots and the versatility. *Colloids Surf., A*, **626**, 127088, 2021.
38. HE, J., *et al.*, Aqueous synthesis of amphiphilic graphene quantum dots and their application as surfactants for preparing of fluorescent polymer microspheres. *Colloids Surf., A*, **563**, 77–83, 2019.
39. ASL, H.F., *et al.*, Effect of SiO2 nanoparticles on the performance of L-Arg and L-Cys surfactants for enhanced oil recovery in carbonate porous media. *J. Mol. Liq.*, **300**, 112290, 2020.
40. ZHAO, A., *et al.*, Amphiphilic PEG-based ether-anhydride terpolymers: Synthesis, characterization, and micellization. *J. Appl. Polym. Sci.*, **118**, 6, 3576–3585, 2010.

41. WU, Y., *et al.*, Rod-shaped micelles based on PHF-g-(PCL-PEG) with pH-triggered doxorubicin release and enhanced cellular uptake. *Biomacromolecules*, **20**, 3, 1167–1177, 2019.
42. TAN, H., *et al.*, Shape transformations of vesicles self-assembled from amphiphilic hyperbranched multiarm copolymers via simulation. *Langmuir*, **35**, 21, 6929–6938, 2018.
43. ZHAO, T., *et al.*, Spatial isolation of carbon and silica in a single Janus mesoporous nanoparticle with tunable amphiphilicity. *J. Am. Chem. Soc.*, **140**, 31, 10009–10015, 2018.
44. GUPTA, S.M., YADAV, P., SHARMA, S.K., Synthesis and Properties of Nano-Surfactants, in: *Surfactants from Renewable Raw Materials*, pp. 165–182, CRC Press, Abingdon, Oxon, OX14 4RN, 2021.
45. ZENG, M., *et al.*, Colloidal nanosurfactants for 3D conformal printing of 2D van der Waals materials. *Adv. Mater.*, **32**, 39, 2003081, 2020.
46. LI, Y., *et al.*, Self-stratification of amphiphilic Janus particles at coating surfaces. *Mater. Horiz.*, **7**, 8, 2047–2055, 2020.
47. LI, Y., *et al.*, Who wins the race near the interface? Stratification of colloids, nano-surfactants, and others. *J. Appl. Phys.*, **132**, 11, 110901, 2022.
48. ALJABRI, N., *et al.*, Magnetically Labelled Hybrid Nanosurfactant MLHNS for Upstream Oil and Gas Operations, in: *SPE Middle East Oil & Gas Show and Conference*, OnePetro, 2021.
49. Abuzaryan, Mohaddeseh, et al., Evaluation of SDS nano-surfactant for EOR application in a carbonate oil reservoir. *Arabian Journal of Geosciences*, 14, 21 , 2217, 2021.
50. HOU, Y., *et al.*, Amphiphilic Janus particles for efficient dispersion of oil contaminants in seawater. *J. Colloid Interface Sci.*, **556**, 54–64, 2019.
51. MASHAT, A., ABDEL-FATTAH, A., GIZZATOV, A., NanoSurfactant: A Novel Nanoparticle-Based EOR Approach, in: *SPE Europec featured at 80th EAGE Conference and Exhibition*, OnePetro, 2018.
52. KHORAMIAN, R., KHARRAT, R., GOLSHOKOOH, S., The development of novel nanofluid for enhanced oil recovery application. *Fuel*, **311**, 122558, 2022.
53. ALZOBAIDI, S., *et al.*, Continuous Flow Synthesis of Petroleum Sulfonates from Crude Oil and Formulation into Nano-Surfactants for Cost Efficient Oil Mobilization, in: *SPE Kingdom of Saudi Arabia Annual Technical Symposium and Exhibition*, OnePetro, 2018.
54. YANG, Y., ZHAO, M., LAI, L., Surface activity, micellization, and application of nano-surfactants—amphiphilic carbon dots. *Carbon*, **202**, 398–413, 2023.
55. Al-Azani, Khaled, et al., Oil recovery performance by surfactant flooding: a perspective on multiscale evaluation methods. *Energy & Fuels*, 36, 22, 13451–13478, 2022.
56. WANG, Q., *et al.*, Performance evaluation of temperature-resistant nano-surfactants, in: *E3S Web of Conferences*, p. 01009, EDP Sciences, Xi'an City, China, 2022.
57. SHI, N., MASHAT, A., ABDEL-FATTAH, A., Autonomous NanoSurfactant for Enhanced Oil Recovery Applications, in: *SPE Europec*, OnePetro, 2020.
58. WANG, J., *et al.*, Study on the Influence of Different Factors on Spontaneous Oil Recovery of Nanosurfactants in a Tight Reservoir. *ACS Omega*, **6**, 30, 19378–19385, 2021.
59. WEI, J., *et al.*, Stretchable and Reactive Membranes of Metal–Organic Framework Nanosurfactants on Liquid Droplets Enable Dynamic Control of Self-Propulsion, Cargo Pick-Up, and Drop-Off. *Adv. Intell. Syst.*, **1**, 7, 1900065, 2019.
60. KUMAR, N., GAUR, T., MANDAL, A., Characterization of SPN Pickering emulsions for application in enhanced oil recovery. *J. Ind. Eng. Chem.*, **54**, 304–315, 2017.
61. ELMASRY, N.S., HALA, E.M., EL-SHARKAWY, R.A., Improvement the efficacy and physical properties of some recommended pesticides on cotton leaf worm, Spodoptera littoralis (BOISD.) by using different nano surfactants, 45–50, 2019.

62. Yun, T. K., Lee, Y., Kim, M. J., Park, J., Kang, D., Kim, S., ... & Kim, K., Commensurate Assembly of C60 on Black Phosphorus for Mixed-Dimensional van der Waals Transistors. *Small*, 18, 10, 2105916, 2022.
63. Yoo, D.-M. and Jung, K.-W., Automatic Molecular Alignments with Amphiphilic Fullerene Nanosurfactant. *Korean Soc. Ind. Chem. Res. Pap. Abstr.*, **2022**, 376–376, 2022.
64. Youjin, C. and Kwangwoon, J., Asymmetric Fullerene Nanosurfactant for Automatic Molecular Alignments. *Korean Polymer Society Conference Abstracts*, vol. **44**, pp. 136–136, 2019.

18

Magnetization Dynamics of Ferromagnetic Nanostructures for Spintronics and Bio-Medical Applications

Monika Sharma[1,2], Ravi Kumar[1], Anjali Chauhan[1] and Bijoy K. Kuanr[1]*

[1]*Special Centre for Nanoscience, Jawaharlal Nehru University, New Delhi, Delhi, India*
[2]*Department of Physics, Deshbandhu College, University of Delhi, New Delhi, Delhi, India*

Abstract
Spintronics is an emerging field of research, which involves the transport of electron spin for future industrial applications. A complete understanding of high-frequency magnetization dynamics is essential to determine various intrinsic parameters of the nanostructures. In this chapter, we discuss a detailed study of the magnetization dynamics of nanostructures by Brillouin light scattering (BLS) and ferromagnetic resonance (FMR) technique. Various structures like ferromagnetic trilayers separated by nonmagnetic spacer layers, nanostrips, and one-dimensional nanocylinders were analyzed both by FMR and BLS measurements. We also observed that by varying the geometry of the nanostructures, one can improve the magnetization reversal characteristics. These nanostructures are ideal for zero-field operational frequency due to their shape anisotropy. The critical characteristics of the nanostructures such as saturation magnetization, intrinsic anisotropies, demagnetizing field, and exchange coupling energy were determined. These properties made magnetic nanostructures top contenders for biomedical applications like giant magnetoresistance biosensors, drug delivery, barcoder, and hyperthermia treatment. Finally, we will consider various future aspects for the development of high-frequency (microwave/mm-wave) spintronic devices.

Keywords: Spintronics, ferromagnetic resonance, Brillouin light scattering, GMR biosensors, hyperthermia treatment

18.1 Introduction

A great deal of recent study has been devoted to fifth-generation technology systems, which require high-performance devices for storing enormous information. Such novel devices can be achieved by replacing charge-based electronics with spintronics where the electron spin plays an important role [1–3]. The use of magnetic materials in spintronic devices provides unparalleled non-volatility, high-speed data storage, and resilience, surpassing existing memory technologies like resistive or phase-change memory [4–9]. The discovery

**Corresponding author*: bkkuanr@gmail.com

Divya Bajpai Tripathy, Anjali Gupta and Arvind Kumar Jain (eds.) Multifunctional Materials: Engineering and Biological Applications, (461–496) © 2025 Scrivener Publishing LLC

of giant magnetoresistance (GMR), which is extensively utilized in magnetic reading heads within hard disk drives, has garnered significant attention in research in the field of spintronics for the last few decades [10–18]. GMR is usually investigated in layered structures (trilayer) comprising a pair of ferromagnetic layers distinct by a nonmagnetic layer. The spin-dependent scattering causes the change in resistance at the ferromagnetic/normal metal interface [19–28]. Depending upon whether the magnetization is parallel/antiparallel in two neighboring ferromagnetic layers, the resistance will be low/high, respectively [11, 29–34]. Extensive study has been conducted during the past three decades to understand fundamental phenomena of spin dynamics and spin relaxation [35–44].

Recently, nanostructured magnetic materials especially one-dimensional nanowires (NWs) have attracted a lot of interest in several biological applications *viz.* hyperthermia therapy, drug delivery, cell separation and manipulation, barcoding, nano-swimmers, and GMR biosensors [45–51]. In hyperthermia, NWs provide more frictional reactive areas than nanoparticles leading to enhanced heating efficiency and decreased duration of cancer therapy [52]. One-dimensional nanostructures have large magnetic moments and shape anisotropy-enhancing forces compatible with cell culture as they do not interfere with the growth cycle of cells and may be modified using physiologically active substances. Furthermore, NWs have exhibited unique characteristics like improved dispersibility, reduced cytotoxicity, greater capacity to bond with other bioactive compounds, compatibility with biological systems, and increased thermal stability. These improved properties made NWs ideal candidates to couple bio-molecules to nanoscale devices which can be used for bio-sensing applications.

In 1986, the existence of antiferromagnetic (AFM) interlayer exchange couplings was revealed by Peter Grunberg and co-workers in Fe/Cr trilayers and multilayers [53]. These multilayer systems were found to be suitable for spintronic devices as it was possible to switch the magnetization orientation from anti-parallel to parallel in adjacent magnetic layers by applying a magnetic field. Also, such systems attracted intensive interest due to their fundamental properties which opens many possible applications for the detection of small magnetic fields. The magneto-resistance can be further enhanced when the metallic spacer layer is replaced by an insulating layer. This increase in magneto-resistance is due to spin-dependent tunneling effect which was realized in ferromagnetic tunnel junctions (MTJs) as tunneling magnetoresistance (TMR) [54–61]. In 1995, about 20% of TMR was first observed in disordered aluminum oxide tunnel barrier MTJs at room temperature [31]. Later immense research was performed, which led to TMR value reaching today as high as 400% in MTJs having MgO single crystal tunneling barrier [62]. Currently, a lot of researchers are still very active in TMR by changing the barrier layers. Certainly, these devices are promising candidates for many spintronic applications such as HDD read heads, magnetic random access memory (MRAM), and extremely sensitive field sensors [63–72].

High-speed spintronic devices increasingly elevated their operational frequency which results in the frequency properties of interest undergoing a change in the giga-Hertz range. This operational frequency is in proximity to the precessional frequency of magnetization in ferromagnetic materials. Usually, ferromagnetic materials often undergo ferromagnetic resonance (FMR) within this frequency range which causes the magnetic moment to precess in the presence of an external time-dependent magnetic field [73]. Many interesting phenomena such as microwave-induced magnetization reversal resulting from the nonlinear motion of magnetization, injection of spins in neighboring layers due to precession magnetization, etc. are observed near FMR [74–78]. Therefore, it is essential to understand how the magnetization reversal switching happens by the external magnetic field or applied electric (spin) currents. Magnetization dynamics can be an effective tool that provides qualitative as well as quantitative information about the magnetic nanostructures [79–81]. The experimental studies focused on such processes are often challenging, this chapter accentuates studying the connection between the physical structure and magnetic response of the system, leading to rich and varied dynamic phenomena. The finer details of magnetization dynamics can be probed by various techniques in frequency, time, and wave-vector domain, which includes conventional FMR, time-resolved magneto-optical Kerr effect (TR-MOKE) microscopy, and Brillouin light scattering (BLS), respectively [82–84]. In standard FMR, the sample is stimulated at a certain frequency while the external magnetic field is adjusted to study the changes in magnetization dynamics during resonance. To measure the magnetization dynamics in a wave-vector domain, one of the efficient techniques is the BLS technique. The progress of the space-resolved and time-resolved BLS approach has carried the recent measurements to advanced levels. Furthermore, the investigation of magnetization dynamics in the time domain is facilitated by TR-MOKE microscopy, which provides a high temporal resolution on the scale of sub-hundred femtoseconds and a spatial resolution in the sub-micron range.

The first section of this chapter is focused on the understanding of magnetization dynamics within ferromagnetic nanostructures. We will next discuss the experimental techniques namely, FMR and BLS which can be used to probe the magnetization dynamics in these nanostructures. The static and dynamic measurements can be done by these techniques to study spin wave phenomena. Within the BLS process, spin waves with a certain wave vector $\vec{k}$ can either be generated (Stokes process) or eliminated (anti-Stokes process). In FMR, only uniform precession modes are taken into account, where no momentum is transmitted. Only modes with a total momentum of zero, $\vec{k}=0$, are stimulated. The next section of this chapter considered various nanostructures namely ferromagnetic trilayers and multilayers separated by nonmagnetic spacer layers, nanostrips, and one-dimensional nanocylinders, which were analyzed both by FMR and BLS measurements. The critical factors of the nanostructures such as saturation magnetization, intrinsic anisotropies, demagnetizing field, and exchange coupling energy were determined. Finally, we will consider various future aspects for the development of high-frequency (microwave/mm-wave) spintronic devices.

18.2 Magnetization Dynamics in Ferromagnetic Nanostructures

For an atom possessing magnetic moment m and orbital angular momentum J, the motion of an electron in atomic orbit can be contemplated as a small current loop under the ascendancy of a magnetic field having a magnetic moment

$$m = IA\hat{n} \tag{18.1}$$

where $\hat{n}$ is the unit vector, which is normal to the area A encircled by the loop, and I is the current flowing in the loop. The magnetic moment and the total angular momentum can be related to each other and are given by [85, 86]

$$m = \gamma J \tag{18.2}$$

where γ is a constant usually referred as gyromagnetic ratio and is given by

$$\gamma = \frac{g\,|e|}{2m_e} \tag{18.3}$$

g is called Lande's splitting factor and its value depends upon many aspects. The value of g is approximately 2 for free spin and 1 for pure angular momenta. For combinations of **L** and **S**, it can take various other values indicating the protusion of m along the direction of J. The equation of motion for a single magnetic moment may be expressed as the rate of change of total angular momentum $\frac{dJ}{dt}$, which is directly proportional to the torque $m \times B_{eff}$ and has an expression as [85, 86]

$$\frac{dm}{dt} = -\gamma \mu_0 m \times H_{eff} \tag{18.4}$$

where $B_{eff} = \mu_0 H_{eff}$. As the magnetic flux in a magnetic sample may vary by an applied field due to the presence of demagnetizing fields, anisotropy fields, magnetostrictive fields, etc., H_{eff} represents the effective magnetic field. One may obtain the effective magnetic field by taking the negative gradient of the internal energy of the system with respect to the magnetization $(H_{eff} = -\nabla_M U)$. Magnetization M is the magnetic moment per unit volume since any magnetic system generally consists of a large number of atoms in which magnetic moments are coupled to each other by exchange interaction. Magnetization dynamics explains the time evolution of the magnetization vector M out of equilibrium. This leads to dynamical model as given by

$$\frac{dM}{dt} = -\mu_0 \gamma M \times H_{eff} \tag{18.5}$$

where $\gamma = g\left(\frac{e}{2m_e}\right)$ is the gyromagnetic ratio of the charge e and mass m_e of the electron. Constant g is Lande g-factor, also known as the spectroscopic splitting factor and its value for a free electron is $g = 2 \times 1.001159657$. Equation 18.5 defines the Larmor precession, with Larmor precession frequency $\omega_L = \gamma H$. This equation does not describe the relaxation of magnetization to H_{eff}. However, a real ferromagnetic system must include a damping term to elucidate the magnetization dynamics. Landau and Lifshitz (LL) amended this behavior by adding a phenomenological dissipation term [87]

$$\frac{dM}{dt} = -\mu_0 \gamma M \times H_{eff} - \frac{\mu_0 \lambda}{M_s^{\,2}} M \times (M \times H_{eff}) \tag{18.6}$$

where M_s is the saturation magnetization and constant λ is phenomenological damping constant, with $\lambda > 0$ in the damping term having the dimension of a frequency. Equation 18.6 explicates the magnetization precession around the effective magnetic field H_{eff}. It includes internal energy of the nanostructures (first term) and dissipation of energy or damping (second term). However, the most commonly employed equation to describe the motion of magnetization is Landau-Lifshitz-Gilbert (LLG) equation which includes the same precession terms of the LL equations [Equation 18.5] in addition to a different damping term given by Gilbert (1955) and hence the name [88]. The dissipation term represents a viscous damping effect that is influenced by the rate of change of magnetization over time. The modified LLG equation of motion [88, 89] has its form as

$$\frac{dM}{dt} = -\mu_0 \gamma M \times H_{eff} + \frac{\alpha}{M_s}\left(M \times \frac{dM}{dt}\right) \tag{18.7}$$

where α represents the dimensionless Gilbert damping parameter in the context of phenomenology. The presence of the damping term causes the magnetization to move in a helical path, as shown in Figure 18.1.

Equations 18.6 and 18.7 become equal for the limit $\alpha \ll 1$ with $\alpha = \frac{\lambda}{\gamma M_s}$. On the other hand, when $\alpha \gg 1$, LL equation predicts that the magnetic moments experience a fast depletion of energy to quickly reach their low energy state, but in the LLG equation, the dissipation of energy and the transition to the low energy state occur at a progressively slower pace. The time evolution of magnetization in the LLG equation results in $\frac{dM}{dt} \to 0$

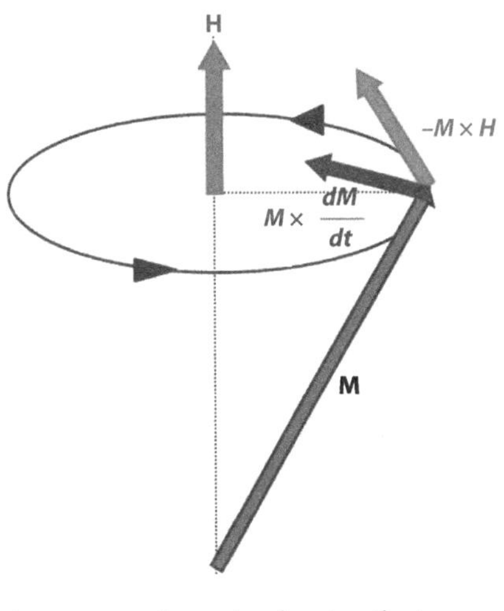

Figure 18.1 Schematic of magnetic moments dynamics showing the torques due to external magnetic field and damping [89].

for $\alpha \rightarrow \infty$ where in the LL equation it results in $\frac{dM}{dt} \rightarrow \infty$ for $\lambda \rightarrow \infty$. The LLG equation is more suitable to describe the precession of magnetization in a material as compared to the LL equation that has infinite change of magnetization. Damping constant α represents all the relaxation mechanisms and in magnetic materials its observed value is typically small, within the range of 0.01 and 0.1 [90]. It can be considered as a sum of intrinsic and extrinsic damping. Intrinsic damping is determined by inevitable phenomena arising from the interaction between electrons, phonons, and magnons in the lattice, through spin-orbit interaction and eddy currents in conducting electrons. On the other hand, extrinsic damping is caused by the scattering of two magnons due to structural defects. Conventional FMR provides an experimental means to access only intrinsic dampening.

18.2.1 Magnetic Damping

The theory of magnetization damping has mesmerized great interest over the past few years due to its technological pertinent in spintronics [91]. In ferromagnetic nanostructures, magnetization relaxation switching time is one of the most important parameters which can be determined by the measurements of resonance linewidth (ΔH). The FMR linewidth can be utilized to ascertain the impact of magnetic non-uniformity and to describe the involvement of the inherent loss mechanism. The relationship is determined to be directly proportional to the microwave frequency ω and inversely proportional to the saturation magnetization as [92]

$$\Delta H = \Delta H_0 + 1.16\alpha\left(\frac{\omega}{\gamma}\right) \tag{18.8}$$

where the first term represents contributions from inhomogeneous broadening, which is frequency independent, and the second term is due to frequency dependent extrinsic contribution generated from structural inhomogeneity and flaws presence in magnetic nanostructures. The linewidth quantifies the dispersion of several factors, such as uniaxial anisotropy and effective anisotropy field, inside each nanostructure. The linewidth increases with the Gilbert damping constant (α) [93, 94].

Intrinsic Damping Mechanism: The intrinsic Gilbert damping originates basically due to spin-orbit coupling since the mobile electrons first transfer energy to the spin system. It is also possible to explain the intrinsic damping due to the scattering of electrons by phonons and magnons. Spin-orbit interaction dominates in the phonon scattering process, while in the magnon scattering, angular momentum relaxes through exchange interaction as scattered electrons repopulate the magnetization direction. Afterward, the spin of the conduction electron relaxes to the lattice through spin-orbit interaction. The presence of eddy currents in a metallic ferromagnetic system acts as a damping mechanism by compensating for changes in magnetism.

Extrinsic Damping Mechanism: Two-magnon scattering is one of the most imperative extrinsic contributions to magnetic damping which arises due to surface defects and roughness and inhomogeneities in materials due to disorder at the atomic and molecular scales. The source for such inhomogeneities is quantized spin waves (magnons) which precess as a wave when excited with wave-number $\boldsymbol{k}$. Under uniform precession $\boldsymbol{k} = 0$, these spin waves are scattered to degenerate non-uniform states and dissipate energy into the lattice. Due to the generation and annihilation of one magnon, the total number of magnons is conserved. It also results in the increase of the FMR linewidth and, hence, damping.

18.2.2 Uniform Ferromagnetic Resonance Mode

As we move from bulk to nanoscale materials, the material properties change drastically such as surface-to-volume ratio, thermodynamic fluctuations, ballistic electron transport, and defects come into the picture [95]. Magnetic nanomaterials also exhibit distinct magnetic properties due to various anisotropies. Nowadays, magnetic nanostructures such as nanoparticles, thin films, and one-dimensional nanocylinders are the areas of interest for data storage applications [96], sensors [97], biomedical drugs [98], microwave devices [99], etc. The FMR technique involves the absorption of microwave energy when the magnetic moments of the system are excited by a spatially uniform microwave (rf) magnetic field. The resonance of the absorption will occur when the Larmor precession frequency coincides with the rf frequency causing an increase in energy of the system.

The linearized LLG equation can be used to solve the equation of motion of magnetization for uniform precessions in the lossless case ($\alpha = 0$). The nontrivial solutions of the linearized equation lead to the precession frequency as a function of the second derivative of the energy [89]:

$$\frac{\omega}{\gamma} = \frac{\gamma}{M_s sin\theta}\sqrt{\frac{\partial^2 E}{\partial\theta^2}\frac{\partial^2 E}{\partial\theta\varphi^2} - \left(\frac{\partial^2 E}{\partial\theta\partial\varphi}\right)^2} \tag{18.9}$$

where θ and φ represents the angles of magnetization (M) w.r.t. nanostructure axis in spherical coordinates. The transmission response of the magnetic nanostructures can be calculated using the transmission parameters. The microwave absorption power as a function of external magnetic field (H) and its derivative w.r.t. H are used in FMR measurements. The power density in the coplanar waveguide (CPW) is determined by evaluating the Poynting vector, which takes into account the high-frequency field components at all locations surrounding the center signal line. The power absorbed by the nanostructure matrix is determined by the provided equation:

$$P_{abs} = \frac{i\omega}{2} Re\left(h_X^* M_X + h_Z^* M_Z\right) \tag{18.10}$$

where h_X^* and h_Z^* are the components of the rf field along the appropriate coordinate's axes, and ω is the angular frequency. Here,

$$M_X = \chi_{eff}^{xx} h_X + \chi_{eff}^{xz} h_Z \tag{18.11}$$

and

$$M_Z = \chi_{eff}^{zx} h_Z + \chi_{eff}^{zz} h_X \tag{18.12}$$

χ_{eff}^{xx}, χ_{eff}^{xz}, χ_{eff}^{zx}, and χ_{eff}^{zz} are the constitutes of the susceptibility tensor of the effective medium, which are calculated from Equations 18.11 and 18.12. The power transmitted through the dielectric may be determined using the given equation:

$$\langle P_{abs} \rangle = \frac{c}{8\pi\sqrt{\varepsilon_d}} \left(\overrightarrow{h_d^*} . \overrightarrow{h_d} \right) \tag{18.13}$$

where ε_d is the relative permittivity of the dielectric medium. By applying the principle of power conservation, we can determine the transmission coefficients. For thin films, resonance frequency is governed by the demagnetizing field due to the dimensions of magnetic element and is given by

$$f_{res} = \left(\frac{\gamma}{2\pi} \right) \sqrt{H + (N_X - N_Z) 4\pi M_s + (N_Y - N_Z) 4\pi M_s} \tag{18.14}$$

where N_X, N_Y, and N_Z are the demagnetizing coefficients. The FMR frequency depends upon various factors such as shape anisotropy and magnetocrystalline anisotropy. The resonance frequency (f_r) for one-dimensional nanocylinders in an FMR measurement is written as follows:

$$\left(\frac{\omega}{\gamma} \right)^2 = \left[H_r cos(\theta - \theta_H) + H_{eff} \theta \right] \left[H_r cos(\theta - \theta_H) + H_{eff} cos 2\theta \right] \tag{18.15}$$

where H_{eff} is the effective anisotropy field that comes from a combination of effects including shape, magnetocrystalline and magnetoelastic anisotropy. H_r represents the external magnetic field whose orientation is defined by the angles θ_H and φ_H. The microwave field is considered to be perpendicular to the external field.

18.3 Experimental Techniques to Probe Magnetization Dynamics

18.3.1 Brillouin Light Scattering (BLS)

The mapping of spin waves in f and k space can be done by an optical technique known as BLS. In an inelastic scattering process, the spin waves quasi particles (magnons) interact with coherent photons. In this process, either the energy is loose by photons to create a magnon (Stokes process) or photons gain energy from an annihilated magnon (anti-Stokes process). This difference between the frequency of the Stokes and anti-Stokes process also known as the Brillouin shift gives the frequency shift of scattered photon. One can also determine the wave vector of the magnon by BLS in which the angle of the incident light is varied corresponding to sample surface. The transferred wave vector $\boldsymbol{k}$ can be written in terms of wavelength λ_L and incident angle as

$$\boldsymbol{k} = \frac{4\pi}{\lambda_L} sin\theta \tag{18.16}$$

Working principle: BLS uses multi-pass tandem Fabry–Pérot interferometers to obtain high resolution. To get the dispersion relations for transparent samples, one must measure the phonon frequencies at different scattering angles while varying wave vector k. In the case of materials that display multiple light scattering, such as dry colloidal crystals, the wave vector is not accurately defined. As a result, acoustic-like modes that rely on the wave vector (k) cannot be accessed. However, the BLS spectrum can record k-independent modes and was observed for sub-micron colloidal silica [100] and polymer crystals [101]. BLS has the advantage of measuring numerous thermally excited elastic resonances in a single measurement resulting in the identification of these k-independent frequencies at the resonance modes of individual nanostructure. Depending upon the geometrical and elastic parameters of the nanostructures, eigen-frequencies are uniquely defined. Thus, BLS is a powerful tool by which the elastic behavior of nanostructured materials can be investigated by choosing the suitable combination of k-independent and k-dependent scattering modes.

Figure 18.2 shows the BLS setup used for the measurements of the tri-layer samples indicating the path of the light inside the Fabry-Perot interferometer. A Fabry–Perot

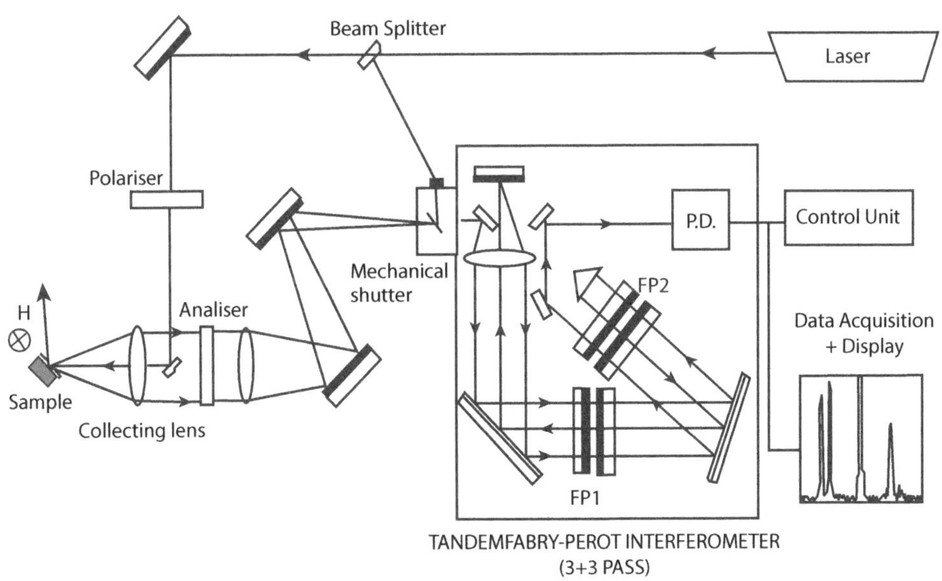

Figure 18.2 Schematic representation of the BLS setup used for the measurements including a (2×3) pass tandem Fabry-Perot interferometer.

interferometer with a pass tandem configuration and dimensions of 2×3 was utilized for BLS investigations at room temperature. The interferometer has an in-plane wave vector of $k_{\|}=1.65\times10^{5}$ cm^{-1}. The sample was subjected to an externally applied magnetic field, which was oriented both parallel and perpendicular to the investigated magnon wave vector. Surface acoustic and optic spin waves were detected on both the Stokes and anti-Stokes regions of the spectrum. The samples were positioned at the midpoint of a goniometer within a sample holder at a fixed temperature. A solid-state pumped frequency-doubled Nd:YAG laser was fixed on the goniometer and rotated for different scattering angles (0° and 160°) either in transmission or reflection geometry. The light passes through a Glan polarizer with vertical polarization before reaching the sample which ensures completely polarized incident light. After scattering through the sample, the light was gathered via an aperture and directed toward the entry pinhole of the tandem Fabry–Pérot interferometer. The transmitted light from the Fabry–Pérot interferometer was detected by a photo diode and processed with the help of analyzers. Finally, processing was done by computer software. To prevent mechanical disturbances, the entire arrangement was positioned on an optical bench equipped with active vibration dampening.

18.3.2 Conventional Ferromagnetic Resonance (FMR)

The conventional FMR is another important technique that can be used to study spin waves in ferromagnetic systems with wave vector $k=0$. In this technique, a magnetic system is excited by sinusoidal electromagnetic radiation at a fixed microwave frequency. The sample is kept in a resonant cavity which is coupled with microwave radiations at fixed frequency by a Gunn diode generating a standing microwave field. As discussed in Section 18.2.2, the resonance frequency depends upon the effective field which includes internal and external energies. Therefore, by sweeping the external magnetic field the resonance condition can be achieved usually measured in terms of the maximum power absorption of microwave radiations by the magnetic system. The external magnetic field is adjusted to allow for lock-in detection which enhances the signal-to-noise ratio (SNR). The measured FMR signal is proportional to the power absorbed or to the field derivative of the imaginary part of high-frequency susceptibility $\frac{\partial \chi''}{\partial H}$ averaged over the sample volume. Conventional FMR is a superlative technique for probing the spin waves in bulk samples demonstrating a single-domain magnetization configuration due to its high sensitivity. However, for materials showing complex domain configuration, this is not a relevant method. To avoid this issue, one can employ frequency sweeping instead of manipulating the external magnetic field. This can be achieved by utilizing a vector network analyzer (VNA).

18.3.2.1 Vector Network Analyzer Ferromagnetic Resonance (VNA-FMR)

As mentioned earlier, VNA-FMR is a more advanced technique than conventional FMR, as it exploits the dynamic response of magnetic structures by varying the frequency of the excited rf field at a constant external field (frequency-sweep) and by varying the field at a constant rf frequency (field-sweep). For magnetic systems having large damping constant and hence large linewidths, the VNA-FMR technique is more advantageous than the other mentioned techniques. The transmission and reflection coefficient of the FMR signal enables to measure both the amplitude and phase which can provide the real and imaginary susceptibility components [102, 103]. Commercially available VNAs provide high adaptability owing to their wide frequency range and sensitivity. Domain wall resonances may be measured in the frequency range of tens of MHz to FMR in ferromagnetic nanostructures, often ranging from 40 to 100 GHz. Kalarickal *et al.* [104] demonstrated that FMR linewidths can be extracted from the frequency linewidth measurements that can provide valuable information on the intrinsic and extrinsic material parameters.

The VNA-FMR approach enables the tracking of alterations in the micromagnetic structure caused by variations in the applied field. In VNA-FMR, it is also possible that the precession of the magnetization can be excited by sweeping the frequency of the oscillating rf magnetic field at a constant bias field. This offers the conservation of magnetization state which is advantageous for the inspection of complex magnetic nano-systems or which includes effects like exchange bias. This can be achieved by employing a high-bandwidth CPW as an alternative to a microwave cavity in which a fixed resonance frequency is used. The VNA sweeps the frequency of outgoing signal over a specified frequency range since it serves as both the source and detector of rf sinusoidal signal. At a particular bias field, maximum microwave power is absorbed when the excitation frequency matches the resonance frequency of the sample so that the magnetization precessional motion is maintained (Figure 18.3). Also at resonance, a destructive interference occurs as the phase of the signal is shifted by π. Due to the attenuation in the coaxial cables connected to VNA and CPW, the amplitude of the transmitted signals is decreased. The flip-chip test is conducted utilizing a CPW with a characteristic impedance of 50 Ω and an agilent vector network analyzer (VNA - model no. ZVB 20 Rohde & Schwarz). The VNA was linked to the CPW by coaxial cables featuring Teflon insulation and SMA connections. Noise is prohibited by using an optical bench with an air cushioning system.

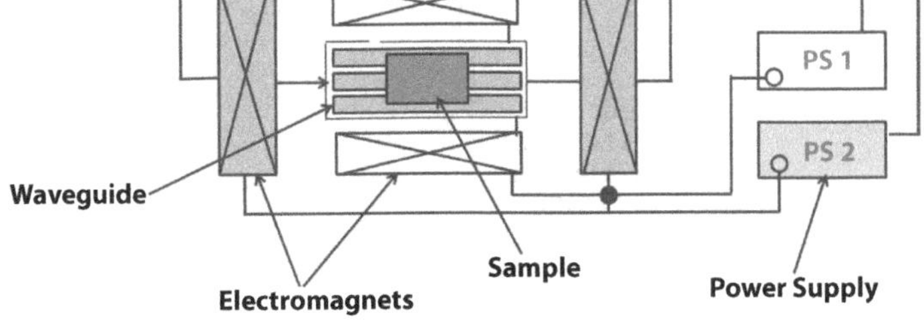

Figure 18.3 Schematic representation of the broadband VNA-FMR spectrometer. The vector network analyzer (VNA) quantifies the intricate scattering characteristics (S_{11} and S_{21}) of a specimen placed on a coplanar waveguide (CPW) that is positioned inside the center of electromagnets [89].

18.4 Dynamic Measurements of Magnetic Nanostructures

18.4.1 Fe/Al/Fe Trilayer Ultrathin Films

The trilayer ultrathin films of Fe/Al/Fe with varying interlayer thickness were epitaxially grown by a molecular beam epitaxiy (MBE) technique on (100)/Fe(1 nm)/Ag(150 nm) substrates with deposition parameters as pressure of $\leq 10^{-10}$ mbar and deposition rate of 0.1 Å/s at a temperature of 80 °C. The dynamic measurements of the samples were employed by the VNA-FMR technique in the frequency sweep mode. The acoustic and optic resonance frequencies were measured by the complex reflection coefficients S_{11} of the sample at various externally applied magnetic fields in the sample plane. A (2×3) pass tandem Fabry–Perot interferometer having in-plane wave vector $k_{\|} = 1.65\times10^{5}$ cm^{-1} was used to perform BLS experiments at room temperature. The sample was subjected to an externally applied magnetic field, which was oriented both parallel and perpendicular to the investigated magon wave vector. Surface acoustic and optic spin waves were detected on both the Stokes and anti-Stokes regions of the spectrum.

The complex reflection coefficient S_{11} as a function of frequency for a (50 Å)/Al(10 Å)/Fe(50 Å) trilayer sample at a fixed bias field of 0.1 kOe is shown in Figure 18.4. The resonance spectrum has two distinct peaks corresponding to the acoustic and optic resonances. The solid lines in Figure 18.4 show the Lorentzian fits to the spectrum to obtain the resonance frequencies and resonance linewidths (Δf) for acoustic (f_{ac}) and optic (f_{opt}) modes. The optic resonance was observed at a higher frequency as compared to the acoustic resonance which can be attributed to the antiparallel or canted alignment of magnetization corresponding to a non-saturated state in coupled layers. The linewidth of the optic mode exhibited an increase of one order of magnitude at around 540 MHz compared to the acoustic mode, which had a linewidth of roughly 65 MHz, across all magnetic fields. The acoustic mode exhibited excellent magnetic uniformity and a single-domain configuration, resulting in an exceptionally small linewidth of 65 MHz and a distinct absorption peak of

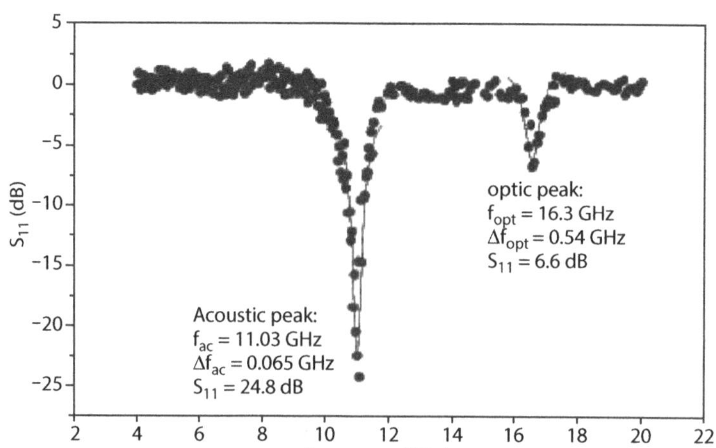

Figure 18.4 The reflection coefficient S_{11} as a function of frequency for a Fe(50Å)/Al(10Å)/Fe(50Å) trilayer measured at an external field of 0.1 kOe. Solid lines in the spectrum are the Lorentzian fits to the peaks [128].

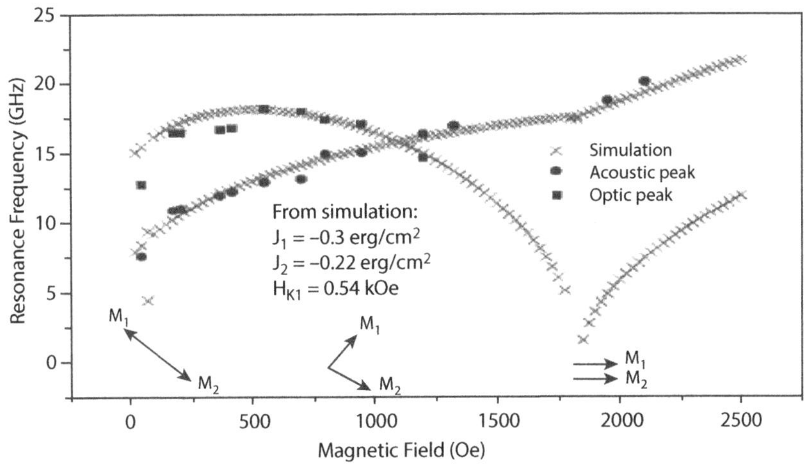

Figure 18.5 The VNA-FMR spectrum demonstrates the relationship between the applied external magnetic field along the hard axis and the optic (squares) and acoustic (dots) resonance frequencies for a trilayer consisting of a (50Å)/Al(10 Å)/Fe(50 Å). The crosses represent the theoretical simulated locations used to derive the coupling parameters [128].

225 dB. It was observed that the absorption amplitude for optic mode reduces sharply and becomes practically invisible for very high magnetic fields (>2 kOe). This is due to the fact that the magnetic Fe layers are of equal thickness (50 Å), and when the magnetization of the two layers reaches in the saturated state the out-of-phase amplitude for the optic mode dribbles considerably.

Figure 18.5 collates the experimental resonance frequencies of a (50Å)/Al(10Å)/Fe(50Å) trilayer for both acoustic (dots) and optic (squares) modes as a function of the externally applied magnetic field (H) along the hard axis of the sample. The crosses in Figure 18.5 are the theoretical fits from which the various coupling parameters are obtained. The variation of the acoustic and optic mode frequencies with external magnetic field H corresponds to different magnetization states present in the two Fe layers. It is observed that the magnetization is aligned in an antiparallel direction for two layers at very low fields ($0< H <0.15$ kOe). For the magnetic fields less than 150 Oe, an upward shift of resonance frequencies by 3.5 GHz occurs for both modes which corresponds to the transition to the spin-flop (SF) phase. In the SF phase for fields greater than 150 Oe, the angle decreases continuously from 180° to 0° between the two magnetizations. The optic frequency increases reaching a maximum at around 500 Oe and then it again decreases, whereas the acoustic frequency continuously increases starting from 150 Oe and near the saturation, it forms an eccentricity. The acoustic and optic frequencies increase linearly with field above the saturation field, i.e., beyond 1800 Oe and the difference in these frequencies indicates the coupling strength. The dispersion relation for the trilayer films is obtained by considering Rezende's approach of a thin-film approximation model leading to magnetic excitation frequencies as

$$\left(\frac{\omega}{\gamma}\right)^4 + a\omega^2 + b\omega + c = 0 \tag{18.17}$$

where parameters a, and c are constants which are derived from the equation of motion, $(\omega = 2\pi f)$ [13]. Only two modes of ω are significant for the FMR configuration from the four possible solutions of Equation 18.17, which results in the acoustic and optic modes. The crosses in Figure 18.5 represent the simulated curves from Equation 18.17. Since the optic mode is in a beeline susceptible to the coupling strength, the acoustic mode diverges away from it and the simulation without the J_2 term reproduces the acoustic mode frequencies. For the (50Å)/Al(10Å)/Fe(50Å) trilayer samples, the bilinear J_1 and biquadratic J_2 exchange constants derived from the fits are $J_1 = -0.3$ erg/cm^2 and $J_2 = -0.22$ erg/cm^2, respectively. The various magnetic parameters derived from the fits to Equation 18.17 are saturation magnetization as $4\pi M_s = 20.5$ kOe, fourfold cubic anisotropy field as $H_{K1} = 0.5$ kOe, and g-factor as $g = 2.09$.

Figure 18.6(a) shows the BLS spectrum for the Fe(50 Å)/Al(7 Å)/Fe(70 Å) trilayer structure revealing the (Stokes and anti-Stokes) acoustic and optic mode frequencies at various externally applied magnetic field. The AFM coupling can be easily observed from the asymmetry in Stokes and anti-Stokes plots. In the BLS spectrum, the mode frequencies as a function of magnetic field demonstrates three different magnetic states *viz.* AFM, spin flop, and saturation. A perfect antiparallel alignment occurs for the magnetic field below 1 kOe. The magnetization switches to a noncollinear or spin flop phase for fields higher than 1 kOe which may be resulted in a leap of both the acoustic and the optic mode frequency.

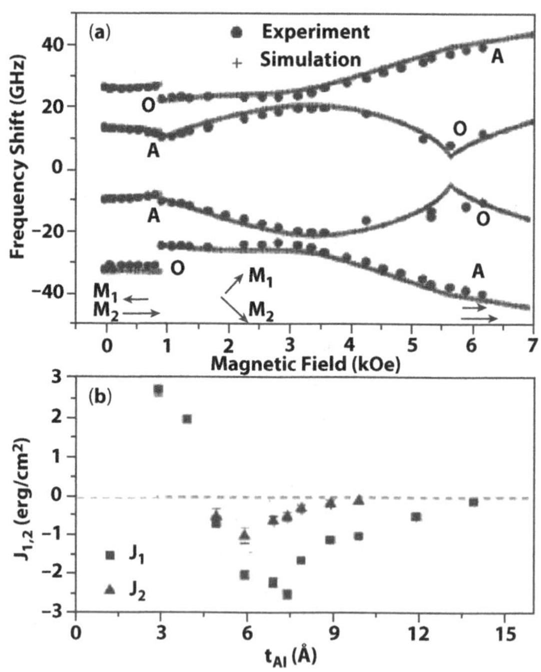

Figure 18.6 (a) The BLS spectrum of the frequency shift of acoustic and optic modes showing Stokes and anti-Stokes for a Fe(50Å)/Al(7Å)/Fe(50Å) trilayer structure. Solid circles are experimental data and crosses show the simulated fits of Equation 18.17. "A" represents the acoustic branches and "O" represents the optic branches. Magnetization orientation relative to $\boldsymbol{H}$ is shown in the bottom by arrows. (b) Coupling strengths bilinear J_1 and biquadratic J_2 as a function of Al spacer thickness for a Fe(50Å)/Al(0-15Å)/Fe(50Å) structure [128].

We derived the exchange coupling constants J_1 and J_2, from a Levenberg–Marquardt fitting using Equation 18.17 as -2.1 ± 0.06 and $-0.5 \pm 0.02 \, erg/cm^2$.

To study the nature and strength of coupling for different Al interlayer thicknesses, we used the focused laser beam embedded within the BLS setup which efficiently scans the whole Al cram (0–15 Å). Figure 18.6(b) shows the overall coupling strengths of J_1 and J_2, obtained from the field variation. We observe that for interlayer thicknesses less than 3 Å, strong ferromagnetic coupling of about 2.7 erg/cm^2, which may be originating from pinholes. It is also observed that the ferromagnetic coupling decreases with increasing Al interlayer thickness. The transition from ferromagnetic to AFM coupling occurs above 5 Å Al thickness having an AFM coupling of $J_1 = -0.02$ erg/cm^2. The AFM coupling becomes maximum at $t_{Al} = 7.5$Å, with J_1 and J_2 as -2.5 and -0.5 erg/cm^2, respectively. However, the biquadratic coupling J_2 occurs at the increasing slope of J_1 and has a maximum value of -1.0 erg/cm^2 which is at t_{Al} = 6 Å. The reason for this is to the strong fluctuation of ΔJ (variation of coupling). As the interlayer thickness approaches to 13 Å, the AFM coupling strength decreases to $J_1 = -0.15$ and $J_2 = -0.005$ erg/cm^2. There is no further change of the coupling strengths for the higher values of interlayer thickness which implies that the AFM coupling dominates for Al interlayer thickness until the end of the wedge structure. Neither J_1 nor J_2 showed any distinctive short-period oscillation.

We observed that the exchange coupling strength strongly depends upon the interlayer thickness. Slonczewski [105] tried to explain this by considering the "loose spin model," which assumes that the loose spins mediate interlayer coupling within the spacer or adjacent to Fe/Al interfaces. The presence of magnetic impurities at alloyed or roughened surfaces is responsible for the presence of unpaired spins in the Fe/Al/Fe system. These thin films have the potential to be utilized in the development of GMR biosensors. The GMR phenomenon is observable in a thin film structure composed of ferromagnetic layers that are separated by nonmagnetic layers. The GMR biosensors rely on the identification of biologically modified magnetic labels at the nano-scale, utilizing very sensitive microfabricated magnetic-field sensors. Djamal *et al.* (2011) reviewed GMR biosensors and their applications in clinical diagnosis [106]. Recently, Su *et al.* discussed about depositing magnetic resonance stacks on pliable substrates and moving the patterned magnetic resonance devices from stiff to flexible substrates [107].

18.4.2 Permalloy Nanostrips

In this section, we demonstrate at gigahertz frequencies the dynamic characteristics of Permalloy (Py) nanostrips. Using rf magnetron sputtering at a speed of 0.2 nm/s and an Ar pressure of 3 mTorr, a 100 nm thick Permalloy layer was deposited into a resin with an average composition of 81% Ni and 19% Fe. Electron beam nanolithography was employed for the fabrication of arrays of nanostrips. Arrays of nanosized strips were formed in each patch and had a width of 300 nm separated by a distance of 1 μm. A lift-off process was used to obtain the final structures, which involved the removal of resin followed by an ultrasonic bath of the overlaying film in acetone. We fabricated a series of six samples consisting of strip lengths of 300 nm, 600 nm, 900 nm, 1.5 μm, 3 μm, and 100 μm. The width to length which corresponds to the aspect ratio was given by 1:1, 1:2, 1:3, 1:5, 1:10, and 1:333.

Figure 18.7(a) shows the FMR spectra of Permalloy strips a continuous film obtained from the VNA (S_{21} versus frequency) at an externally applied magnetic field of 3 kOe. It is observed that the spectra for Permalloy nanostrips with aspect ratios of 1:1, 1:2, 1:3, 1:5, and 1:10 are shifted to higher frequencies by ~1, 2, 2.15, 2.2, and 2.5 GHz, respectively, when the field is applied along the long axis of the samples as compared to the Py film (f_r = 15 GHz at 3 kOe). Since the arrays have large spacing (1 μm) between the nanosized strips, one can consider the microwave response primarily due to the magnetic properties of the individual strips and the uniform wave vector $(k = 0)$. For longer strips, the rising behavior of resonance frequency is basically due to the increase in shape anisotropy.

The spins of the magnetic moment precess in uniform mode when the magnetic field is applied along the length of the strip in a microwave experiment. This results in the generation of a demagnetizing field in the "magnetic charges" which are present on the surface. The upshifting of the resonance frequency is predominantly due to these self-demagnetizing fields in Permalloy strips in comparison to a continuous film [108]. The absorption spectra of all the strips exhibit similar line forms. The 1:1 strip structure exhibits varying frequencies based on the direction of the magnetic field. From the AFM photos of the arrays, we noticed that the 1:1 strip structure is not perfectly square, but rather somewhat extended in one direction [109]. Furthermore, the resonance frequency shifts slightly due to the interactions among different nano-objects. Figure 18.7(b) reveals that when a magnetic field of 3 kOe is applied along the length and width, the strip resonance frequency changes to higher places. Specifically, the resonance frequency increases to 11.8 GHz along the length and 17.2 GHz along the width, with a ratio of 1:333. The resonant frequency shift is caused

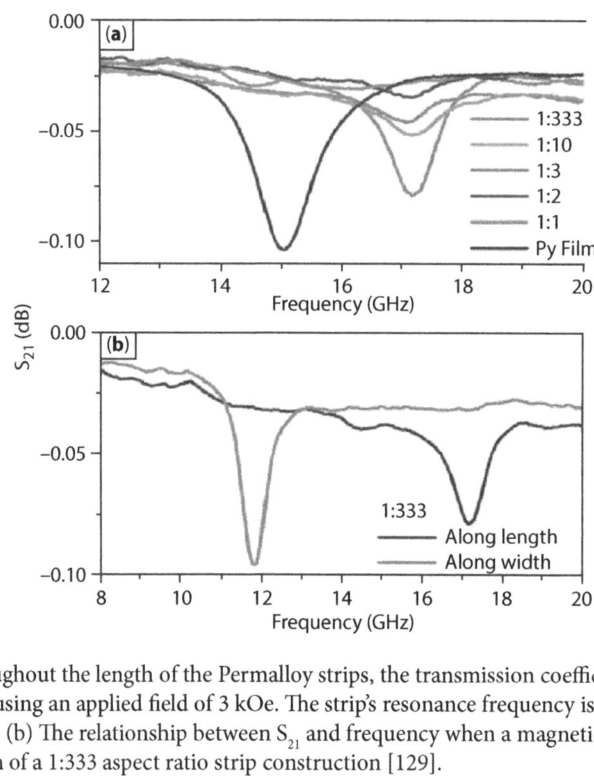

Figure 18.7 (a) Throughout the length of the Permalloy strips, the transmission coefficient was assessed by the network analyzer using an applied field of 3 kOe. The strip's resonance frequency is represented by the maximum absorption. (b) The relationship between S_{21} and frequency when a magnetic field is applied over the length and breadth of a 1:333 aspect ratio strip construction [129].

by the existence of uniaxial shape anisotropy, which leads to notable alterations in the frequency. The difference in resonance frequencies for the two orientations of the sample, i.e., either along the length or the width of the strip with the applied magnetic field gives the magnitude of the shape anisotropy.

The conglomeration of the resonance frequencies done by VNA-FMR for the Permalloy nanostrips and Permalloy film with the applied magnetic field along the length and width of the strips are shown in Figure 18.8(a). Predictably, we can tune the FMR frequencies with the externally applied magnetic field. It is observed that the resonance frequency increases with the external magnetic field for both the orientation of the samples, along the width and length of the strip. However, we noticed that for the 1:10 and 1:333 strips, when the field is applied along the width of the strip, there are two frequency-field regimes. The resonance frequency firstly decreases with the increase of the magnetic field and then again starts increasing linearly with the increase of the magnetic field. This particular point of magnetic field where the deviation of frequency-field data occurs corresponds that below this field, the sample is not in its saturation state. Moreover, this field gives the value of shape anisotropy. The magnetic characteristics, including the g-factor, saturation magnetization, and in-plane anisotropy, for Py, are determined by analyzing the frequency-field data of a continuous Py film and obtaining the best-fit readings. The saturation magnetization value is seen to be rather low, around 0.612 kG. The calculation of uniaxial shape anisotropy in strips of varying lengths is performed using the observed parameters in conjunction with the self-demagnetization fields.

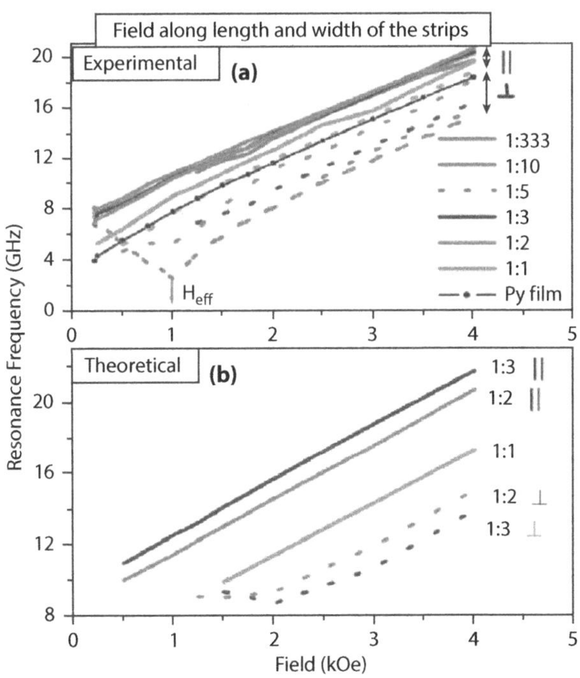

Figure 18.8 (a) Resonance frequency behavior for nanostrip structures of various aspect ratios and for a continuous Py film with the external magnetic field applied along the width and length of the strips. (b) The bottom portion shows the theoretical results obtained from micromagnetics calculations. The direction of the applied field w.r.t the length of the nanostrip is indicated by the parallel and perpendicular symbols [129].

These nanostrips can be considered as a segment of small cells having individual magnetization. We used this approach to analysis theoretically the dynamics of these structures by micromagnetic simulations for the uniform mode. The dynamical equation of motion of magnetization within an individual cell is obtained from the Landau-Lifshitz equation [88, 89]

$$\frac{\partial M}{\partial t}=-\frac{|\gamma|}{1+\alpha^2}\left[M\times H_{eff}-\frac{\alpha}{M_s}M\times\left(M\times\frac{dM}{dt}\right)\right] \tag{18.18}$$

where α is the gilbert damping constant. The internal energies which is the sum of effective exchange fields, the dipole fields, and zeeman field results in the effective field (H_{eff}) acting on a cell. The normal modes are found by using $M_s=0.68\,\text{kG}$, $A=1.3\times10^{(-6)}$, and $\alpha=.0001$. The thickness of individual cell size was 100 nm and 10 × 10 nm² in the plane of the sample. For larger structures, we considered a larger cell size for the calculations. Figure 18.8(b) demonstrates the theoretical fits to the experimental data of the resonance frequency. As discussed earlier, the nanostrip structure with 1:1 aspect ratio shows two different frequencies nearly 1 GHz apart with different orientations of the magnetic field and it is larger as compared to the experimental uncertainty of about of $\pm0.2\,\text{GHz}$. The theoretical calculations also revealed the soft mode like behavior in the sample undergoing a phase transition as discussed previously which has its origin at the edges of the sample when the field is applied perpendicular to the plane. This behavior is more pronounced in the longer structures than that for the shorter structures and seen both experimentally as well as theoretically.

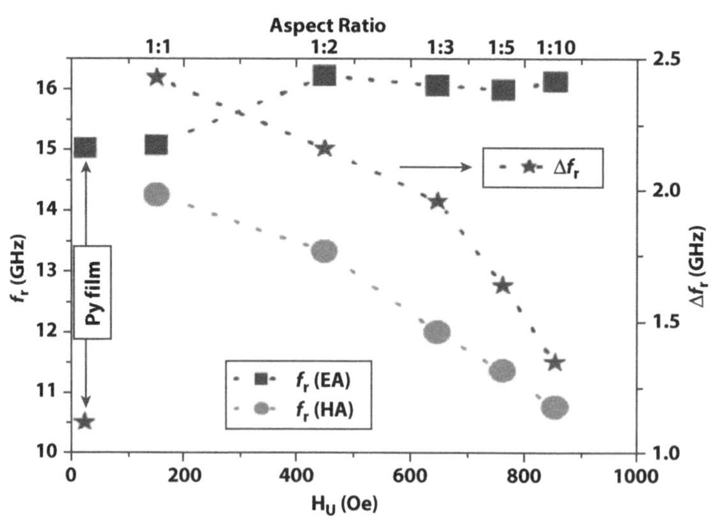

Figure 18.9 VNA-FMR response of the resonance frequency [right side] for easy axis (square) and hard axis (circles) and frequency linewidth (Δf_r) [left side] as a function of the shape anisotropy (H_u) at a magnetic field of 3 kOe. The aspect ratios of the strips (1:1, 1:2, 1:3, 1:5, and 1:10) are shown on the top of the X-axis [129].

Figure 18.9 shows that the strip exhibits resonance frequencies throughout its length (referred to as the easy axis) and breadth (referred to as the hard axis), as well as frequency linewidth, which are influenced by the shape anisotropy (H_u). The frequency linewidth (Δf_r) obtained from the absorption spectra of the strips using FMR reveals the signal originating from the entire sample. According to the data shown in Figure 18.9 (isolated star), it can be noted that as the length of the strip rises, the frequency linewidth of the strips decreases. The minimal thickness required for a continuous film to have a resonance frequency of $f_r = 1.12GHz$ with a magnetic field strength of 3 kOe. The decrease in linewidth can be attributable to the presence of shape anisotropy [89, 110].

18.4.3 One-Dimensional Magnetic Nanowires

This section is focused on the dynamic properties of arrays of 1-D nickel NWs with high aspect ratio. The arrays of NWs were fabricated by electrodeposition technique in porous anodic alumina template as reported elsewhere [111]. We used six different lengths of NWs: 5, 8.5, 16, 30, 40, and 60 μm. The dynamic characteristics of NWs were investigated using a flip-chip approach in conjunction with a CPW. The sample was placed on top of the transmission line, with the axes of the NWs aligned parallel to an external static magnetic field. The frequency was varied continuously from 0.05 to 20 GHz while maintaining a constant external magnetic field.

Figure 18.10(a to d) represents the transmission response measured for various samples of NW length (8.5, 16, 30, and 40 μm) with externally applied magnetic field parallel to the easy axes of NWs. The resonance frequency (f_{res}) and frequency linewidth (Δf_{res}) are derived from the Lorentzian fits to the experimental data as shown by the solid lines.

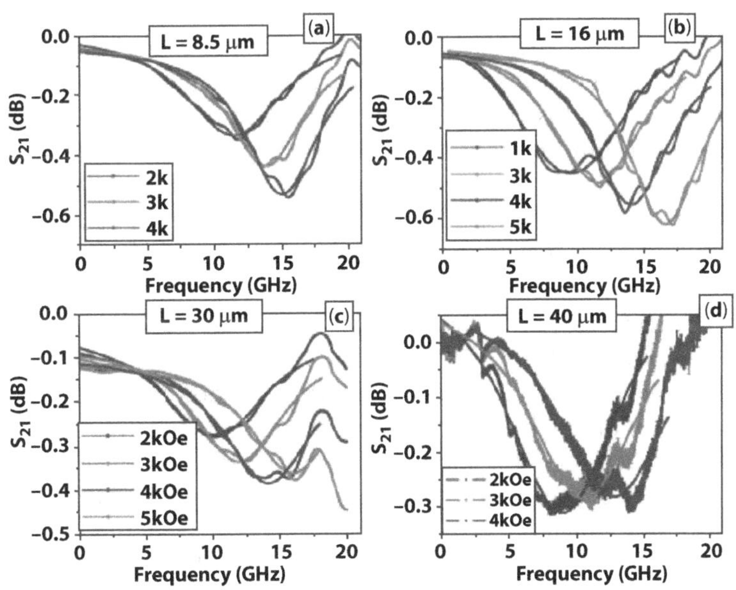

Figure 18.10 (a to d) Transmission response measured by the VNA-FMR as a function of frequency for different Ni NW samples of length 8.5, 16, 30, and 40 μm with fields applied parallel to the easy axis. Solid lines represent the Lorentzian fits to the experimental data [111].

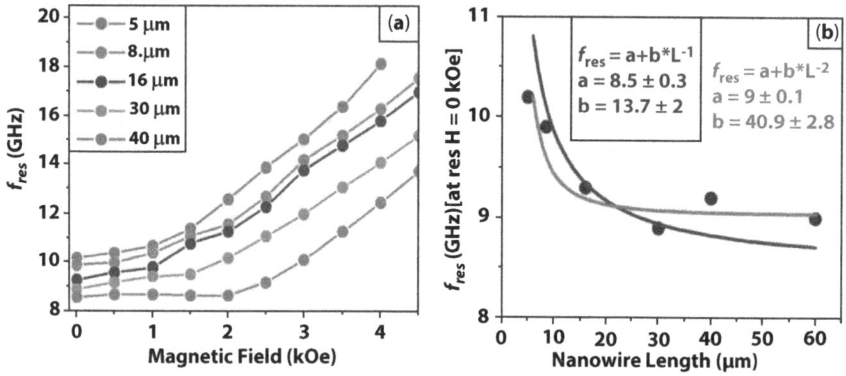

Figure 18.11 (a) Resonance frequency as a function of applied magnetic field for various length samples with field parallel to the easy axis of wires. (b) Zero field f_{res} versus nanowire lengths. The solid dot represents experimental data. The two solid lines are fit to L^{-1} (blue line) and L^{-2} (red line) function [111].

We observe from the results that absorption peak varies in position and intensity with the increase in the static magnetic field for the entire samples. For low applied fields specifically, there is a ~0.3 dB absorption dip at the FMR frequency which increases to nearly 1.0 dB for higher fields. The NW system mainly consists of magnetoelastic, magnetocrystalline, and shape anisotropy which contributes to total anisotropy field. Since the one-dimensional nanocylinders have eminently large aspect ratio and as a result of their significant anisotropy, which is formed by the wire's shape, these objects function as one-dimensional permanent nanomagnets when exposed to weak static magnetic fields. This trademark of NW system motivates the FMR to occur even without applying any external magnetic field [112]. In addition, these NWs may be adjusted by the application of a magnetic field, in addition to their use in zero field applications.

Figure 18.11(a) displays the $f_{res}(H_0)$ data for all the manufactured NW samples. The graph in Figure 18.11(b) displays the measured values of the resonance frequency $f_{res}(H=0)$ at different lengths of NWs. It has been noted that when the length of the NW increases from 5 to 60 μm, the resonance frequency $f_{res}(H=0)$ decreases from 10.2 to 8.8 GHz. The results indicate that decreasing the length (L) of the NWs from 40 to 5 μm increases the resonance frequency f_{res} by 1.25 times. Therefore, by customizing the length of the NWs, it is possible to finely tune the resonant frequency $f_{res}(H_0)$ within a narrow range of values.

The findings without an applied magnetic field [Figure 18.11(b)] indicate that the effective magnetic field H_{eff} reduced from 3.6 kOe for the Ni1 sample to 3.15 kOe for the Ni6 NW sample. The magnitude of H_{eff} exceeds the demagnetization field $H_d = 2\pi M_s = 3.05kOe$, for Ni in the case of an infinitely long cylinder. The high value of H_{eff} can be attributed to the Y-shaped NWs produced in the anodic alumina templates, which are in an unsaturated condition and not perfectly aligned with the applied H_0. One-dimensional NWs give rise to the shape anisotropy which further enhances the effective field. In Figure 18.11(b), solid lines show the L^{-1} and L^{-2} fits to the experimental data. The L^{-2} fit goes very closely to the experimental data. The f_{res} increases as a result of upward migration of H_{eff} since the applied magnetic field is fixed (as given in Equation 18.14). The experimental data observed for H_{eff} values as a function of the length of NWs are fitted to a power function. We can

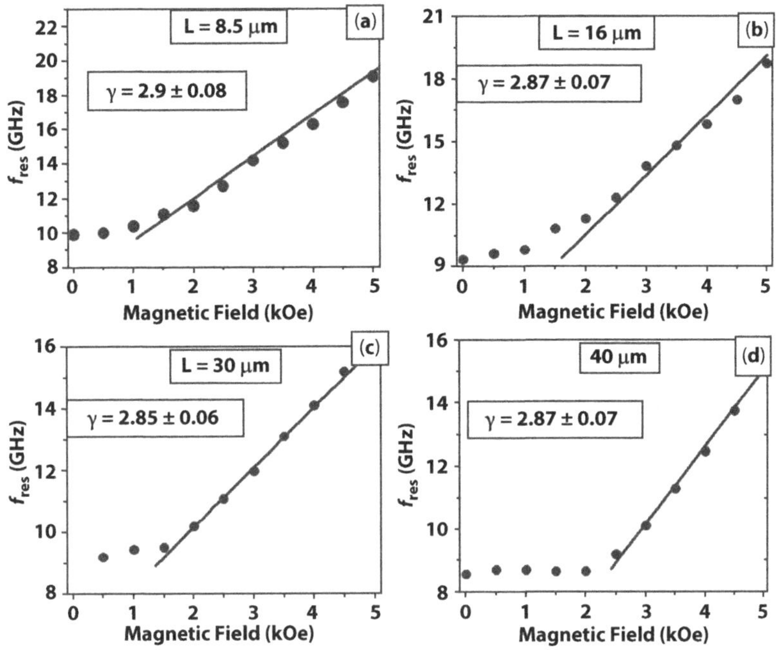

Figure 18.12 (a-d) The FMR resonance frequency for NWs samples of different length 8.5, 16, 30, and 40 μm as a function of magnetic field applied parallel to NWs axes (black symbols). Solid lines are theoretical fits to experimental data [111].

express the change of resonance frequency (δf_{res}) following Equation 18.15 as: $\delta f_{res} \propto H_{eff}$. The effective field varies as an L^{-2} function, therefore, δf_{res} also varies as L^{-2} function, given by $\delta f_{res} = a + b \times L^{-2}$ [see Figure 18.11(b)].

The resonance frequency f_{res} measured for Ni NW samples for the magnetic field applied from an external source parallel to the long axes of the NWs is shown in Figure 18.12(a to d). It is observed that at low fields the resonance frequency is relatively steady with the increase of H_0. After an applied field of ~ 1 kOe, the resonance frequency increases linearly with the magnetic field. We observed two regimes of frequency field data for parallel configuration of NWs axes which may be due to Y-shaped branches of NWs [69] as these are canted at an angle of 10° with the external magnetic field. A numerical calculation for parallel configuration of NWs shows a constant value of $f_{res}(H_0)$ with the magnetization vector $(\vec{M})$ canted at an angle of 10° with H_0 till 0.5 kOe field. Another factor may be the unsaturated state of magnetization due to random orientation of $\vec{M}$ with H_0 at very low fields resulting in constant data up to 1–1.5 kOe. Using Equation 18.15 we fitted the experimental data of $f_{res}(H_0)$ all NWs samples with fitting parameters as H_{eff} and g-factor. From the fitted results we observed that for an increase of NWs lengths from 5 to 60 μm, g-factor to increase from 2.0 to 2.1, respectively.

Figure 18.13(a) demonstrates the frequency linewidth behavior for different Ni NWs samples of various lengths with the magnetic field aligned parallel to the axis of the wire. We derived the frequency linewidths from the Lorentzian fits to the experimental data. In comparison to continuous Ni film, we noticed a wider frequency linewidth. Various factors

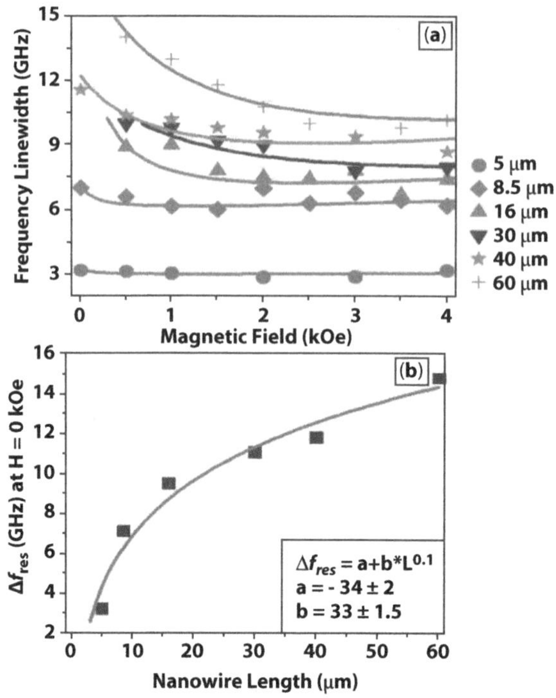

Figure 18.13 (a) Frequency linewidth for various length NW samples as a function of magnetic field. (b) Zero field Δf_{res} versus nanowire lengths. The solid line is a fit to power function [($\Delta f_{res} = \boldsymbol{a} + \boldsymbol{b} * \boldsymbol{L}^{0.1}$] [111].

contribute for the broadening of Δf_{res} such as distributions in NW length, Y-segment growth in AAO template and Y-segment diameters (d, d_1, and d_2 are branching diameters) as reported in our work [101]. We observed that with the increase of NWs length Δf_{res} increases. Frequency linewidth Δf_{res} is large at low field for all NW lengths samples. It can be due to:

(1) The sample is in unsaturated state at zero and very low fields. Therefore, despite the parallel alignment of the NWs' axis with the supplied static magnetic field (H), their magnetization is randomly oriented.
(2) For low to moderate fields, around 1 kOe, it may be inferred that the magnetization $\vec{M}$ aligns with the static magnetic field (H) direction in the elongated section of the NWs. For arms with a shorter length (d_1 and d_2), the magnetization is oriented at an angle relative to the static magnetic field. Consequently, at low frequency it corresponds to a secondary resonance that is extremely near the primary resonance frequency (f_{res}).
(3) One of other reason is that the interwire distance between these Y-segment NWs changes which causes the resonance frequency to be different in the branches.

The presence of several resonances in close proximity to the main resonance is the primary cause for the widening of the resonance linewidth in our samples. The resonance

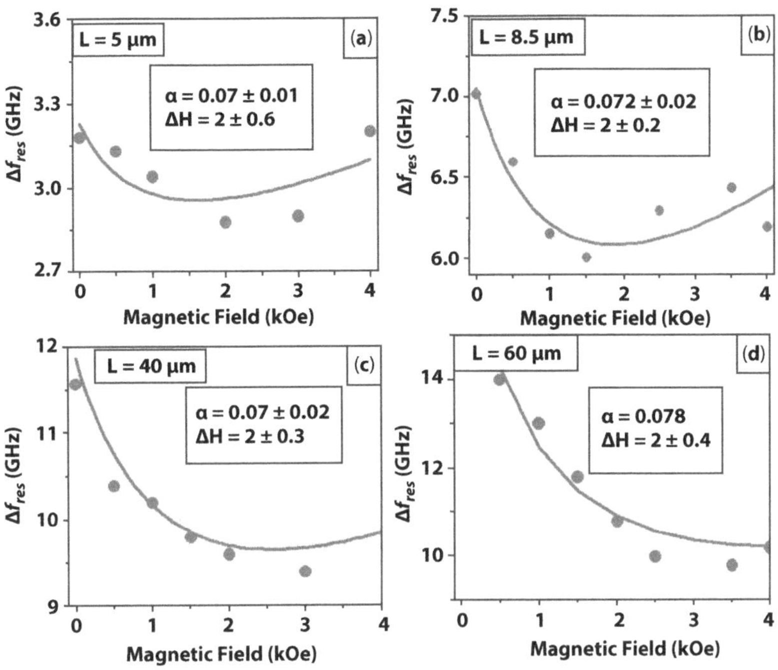

Figure 18.14 (a to d) Frequency linewidth for various NWs of length 5, 8.5, 40, and 60 µm for static magnetic field applied parallel to nanowire axes. The solid lines are theoretical fits from Equation 18.19 [111].

linewidth initially decreases with the increase of magnetic field. For higher length samples, it is observed that Δf_{res} value first falls as the field strength diminishes, but then gradually increases at high field strengths. Figure 18.13(b) shows Δf_{res} as a function of NWs length at 0 kOe magnetic field. The solid line corresponds to a fit to power function $[\Delta f_{res} = a + b * L^{0.1}]$.

The frequency linewidth with static magnetic field for various NWs sample is shown in Figure 18.14(a to d). Significantly larger linewidths have been observed for longer length NW samples. For all NW samples, at higher frequencies a distinctly narrower frequency linewidth is seen. For a wire-shaped magnetic element, the resonance frequency can be obtained accurately by the FMR condition as given in Equation 18.15. By differentiating Equation 18.15, one can obtain the relation between the frequency linewidth and field linewidth as [110]

$$\Delta f_{res} = \frac{\partial f_{res}}{\partial H} \Delta H_0 \tag{18.19}$$

Thus, the frequency linewidth depends basically on two factors: the slope $\frac{\partial f_{res}}{\partial H}$ and the precise form of ΔH_0. A larger $\Delta \omega$ would be expected if we consider an infinitely long Ni film with $N_y = 0$. In such a case, the slope will be large at $H_0 = 0$ and the resonance frequency being zero is the primary cause for the frequency linewidth to be significant at low fields and decrease at higher fields. This corresponds to a wire-shaped magnetic element, which agrees with the experimental results [111, 112]. The theoretical fits for the resonance

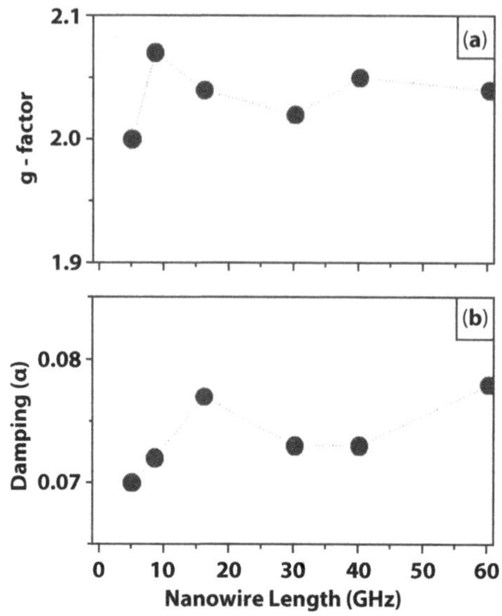

Figure 18.15 Gyromagnetic ratio (γ) and damping constant (α) as a function of length of nanowires. The lines are guide to the eye [111].

linewidth are obtained using Equation 18.19 in conjunction with Equation 18.15 and are shown by solid lines in Figure 18.14(a to d).

Various parameters like the Gilbert damping parameter α, zero-frequency offset ΔH_0 and the g-factor are derived from the theoretical fits. Figure 18.15 shows the variation of damping parameter α and g-factor as a function of NW length (L). It is observed that with the increase in length of NWs both parameters α and g-factor increases a little.

The extrinsic contribution to linewidth may be the reason for large value of α. The frequency linewidth can take a form

$$\Delta f_{res} = \Delta f_{int} + \Delta f_{ext} \tag{18.20}$$

where $= \Delta f_{int}$ and Δf_{ext} are intrinsic and extrinsic contribution to linewidth, respectively. Considering bulk value of Ni, Δf_{int} is observed to be 0.2 GHz. The various parameters such as g-factor = 2.2, Gilbert damping ($\alpha = 0.02$) and saturation magnetization $4\pi M_s = 6$ kOe are derived from the theoretical fittings to experimental data. The dominant contribution to linewidth may be attributed to the extrinsic factors that arises due to the distributions in NWs length and diameters (d, d_1, and d_2) in Y-shaped sections and variation of ΔH_{eff} and $\Delta\theta_H$. The variation in effective field ΔH_{eff} is due to the distribution of dipolar interaction field since both $4\pi M_s$ and shape anisotropy for the wire geometry is constant. Hence, the differences in the spacing between wires and the radius of the wires inside the alumina matrix might potentially influence the dipolar interactions. Shore *et al.* reported that with high saturation, magnetization SAR value increases [113]. Therefore, we observed that for low length NW sample Ni1, the SAR value is maximum and for Ni6 NW sample, SAR value decreases. The dipolar interaction between the NWs can also affect the magnetization reversal switching. Higher the dipolar interaction causes the NWs to become more closer and, hence, reducing the SAR value.

18.5 Biomedical Applications

Cancer being the leading cause of death becomes a curse on humanity, with 19.3 million new cancer cases occurred in 2020, as reported by GLOBOCAN [114]. For the treatment of cancer, many conventional techniques have been employed such as chemotherapy, radiation therapy, and surgery. Among these techniques, radiotherapy and chemotherapy methods faced severe side effects such as fatigue, alopecia, damage to healthy tissues, and multidrug resistance (MDR) [115–117]. The magnetic hyperthermia approach shows promise as a non-invasive strategy to improve cancer treatment. It involves injecting magnetic materials into the tumor tissue, which absorb radiofrequency power from an oscillating magnetic field [118, 119]. The heat produced due to Neel and Brownian relaxation loss processes destroys cancerous cells efficiently without harming the nearby healthy cells [120, 121]. For biomedical applications, iron oxide nanoparticles have been extensively studied for their high specific absorption rate (SAR) and low cytotoxicity in diverse applications such as magnetic resonance imaging (MRI) contrast enhancement, hyperthermia treatment, and, targeted drug administration [122–126]. However, bare iron oxide nanoparticles can accumulate to form bigger clumps preventing immobilization through the system. To overcome the above-mentioned issues, iron oxide nanoparticles have to be coated with some suitable surfactants which increase the colloidal stability. Graphene oxide (GO) is one of such surfactants that exhibit remarkable hydrophilic properties making it suitable for improving the colloidal stability and biocompatibility. We showed in the work that an easy and cost-effective electrochemical method for the fabrication of GO-Fe_3O_4 nanocomposites. We demonstrated that the SAR, heating efficiency, and cytotoxicity have been improved by immobilizing iron oxide on GO sheets. We showed *in-vivo* and *in vitro* measurements on A549 cells. The synthesis of nanocomposites has been reported elsewhere [127].

For biomedical applications, the hydro-dynamic diameter of particles, electronegativitym, and UV-visible absorption was determined using zeta potential, dynamic light

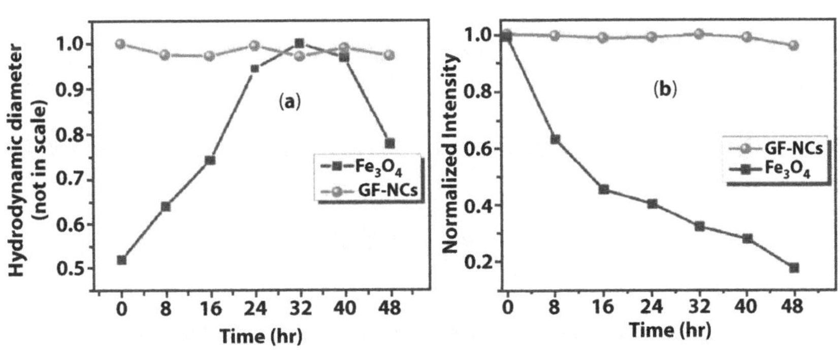

Figure 18.16 (a) Hydrodynamic diameter as a function of time for Fe_3O_4 nanoparticles and GO-Fe_3O_4 nanocomposites. (b) Normalized intensity as a function of time [127].

scattering (DLS), and UV-vis spectroscopy. Figure 18.16(a) shows the hydrodynamic diameter as a function of the time of Fe_3O_4 nanoparticles and composites. It is visible from the graph that the hydrodynamic diameter of composites remains constant for up to 48 hours, whereas for Fe_3O_4 nanoparticles, it increases more than double within 32 hours, hinting the agglomeration of iron oxide nanoparticles due to magneto-dipole interaction. We observed that with the increase of time, Fe_3O_4 diameter decreases due to sedimentation of accumulated particles under gravity. The normalized hydrodynamic diameter distribution peak intensity as a function of time for composite and iron oxide nanoparticles is shown in Figure 18.16(b). We see that the intensity of Fe_3O_4 dispersion decreases continuously due to rapid agglomeration whereas the intensity of nanocomposites does not change much even for 48 hours which depicts uniform distribution and colloidal stability.

The long-term stability of NPs is obtained by using zeta potential technique. The unstable suspension demonstrates zeta potential values in the range of −30 mV and +30 mV whereas for stable suspension zeta potential either higher than +30 mV or lower than −30 mV [130]. For Fe_3O_4 nanoparticles, the zeta potential value was observed to be +5 mV indicating unstable suspension. For GO-Fe_3O_4 nanocomposites, the zeta potential was obtained to be −40 mV indicating good stable dispersion. The Prussian blue staining method was to visualize the iron ions inside the cells, which revealed the internalization of GO-Fe_3O_4 nanocomposites. For this, the cells were imposed with nanocomposites with different incubation time (24 hours and 48 hours) and dose concentration. Figure 18.17 showed the intracellular uptake of nanocomposites for A549 cells seen with the help of light microscope at a magnification of 10X. It is observed as the incubation time is increased, more nanocomposites get internalized by the cells as seen in the formation of clusters of blue granules in the cytoplasm [Figure 18.17(b and e) and Figure 18.17(c and f)]. Also, more GO-Fe_3O_4 gets internalized by cells for higher concentration as shown in Figure 18.17(e and f).

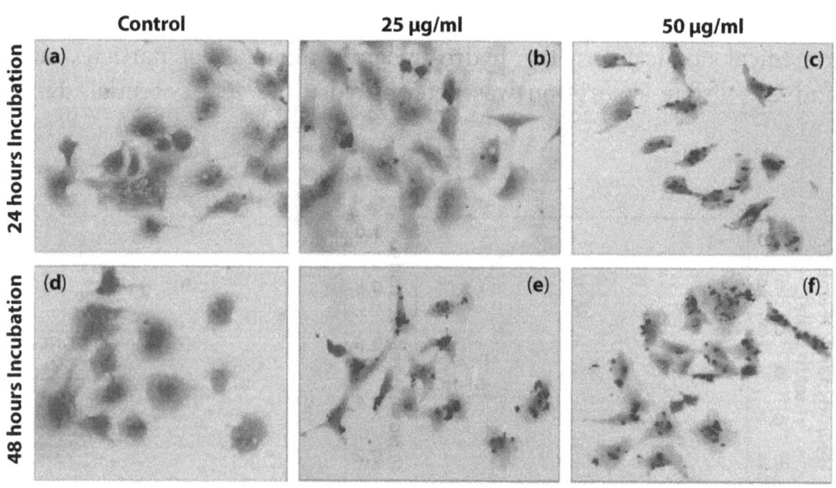

Figure 18.17 Microscope image of Prussian blue staining of A549 cells (a to d) unlabelled cells (b to e) GO-Fe_3O_4 nanocomposites incubated cells for 24 hours and 48 hours with 25 μg/ml concentrations, respectively. (c to f) GO-Fe_3O_4 nanocomposites incubated cells for 24 hours and 48 hours with 50 μg/ml concentrations, respectively [127].

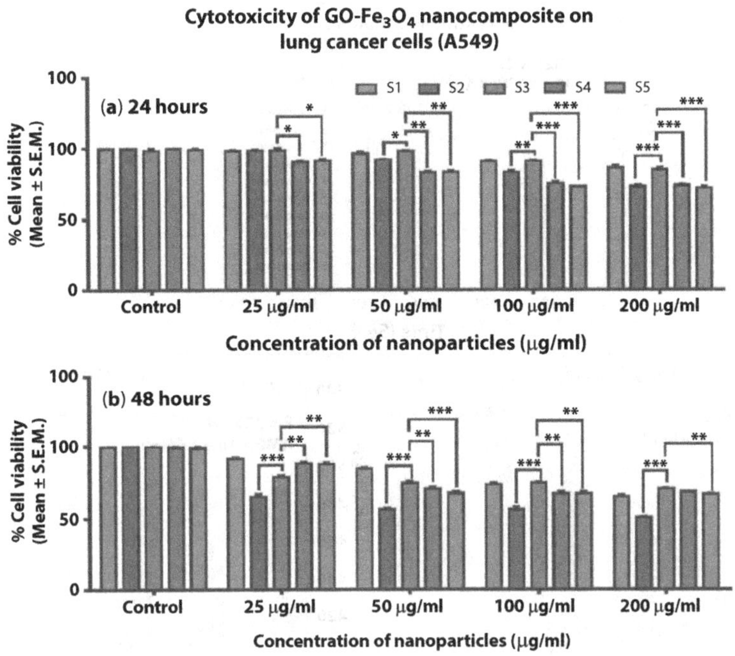

Figure 18.18 (a) Cell viability measurements of A549 cells imposed with nanocomposites of different concentrations (25, 50, 100, 150, and 200 μg/ml) for 24 hours. (b) A549 cells imposed with nanocomposites of different concentrations (25, 50, 100, 150, and 200 μg/ml) for 48 hours [127].

The cytotoxicity measurements were done using MTT assay using nanocomposite synthesized on A549 cells as shown in Figure 18.18. We observed low cytotoxicity for different concentrations (25–200 μg/ml) nanocomposites incubated for 24 hours and 48 hours. The cell viability of nanocomposites at different time period incubation at 200 μg/ml concentration was observed to be around 85% and 75% for most of the samples indicating good biocompatibility.

For the measurement of SAR values, the heating behavior of 200 μg/ml nanocomposites were subjected to alternating magnetic field (AMF) as shown in Figure 18.19. From the calorimetric measurements done by hyperthermia system which generates AMF we measured the time-dependent gradient increase in temperature for nanocomposite suspensions as shown in Figure 18.19(a). The observed SAR values calculated using linear slope method at fix concentration of 200 μg/ml for different samples to be 334, 351, 360, 380, and 407 W/gm [Figure 18.19(b)] and measured by Box-Lucas method to be 427, 449, 452, 518, and 543 W/gm [Figure 18.19(c)]. The Box-Lucas method gives higher values as it considers the temperature rise at higher time scale. The intrinsic loss power (IPL) which defines the heating transformation ability is shown in Figure 18.19 (b and c) for all nanocomposites. The IPL values were observed to be 3.56, 3.75, 3.85, 4.1, and 4.34 nHm^2Kg^{-1} from linear slope method and 4.7, 4.95, 4.96, 5.71, and 5.98 nHm^2Kg^{-1} from Box Lucas method. The retrieved IPL values are in agreement with FDA protocol. It is observed that as the crystallite size of nanocomposites increased, the IPL value also increased accordingly. Hence, we concluded that with proposed nanocomposites the heating efficiency is improved and up

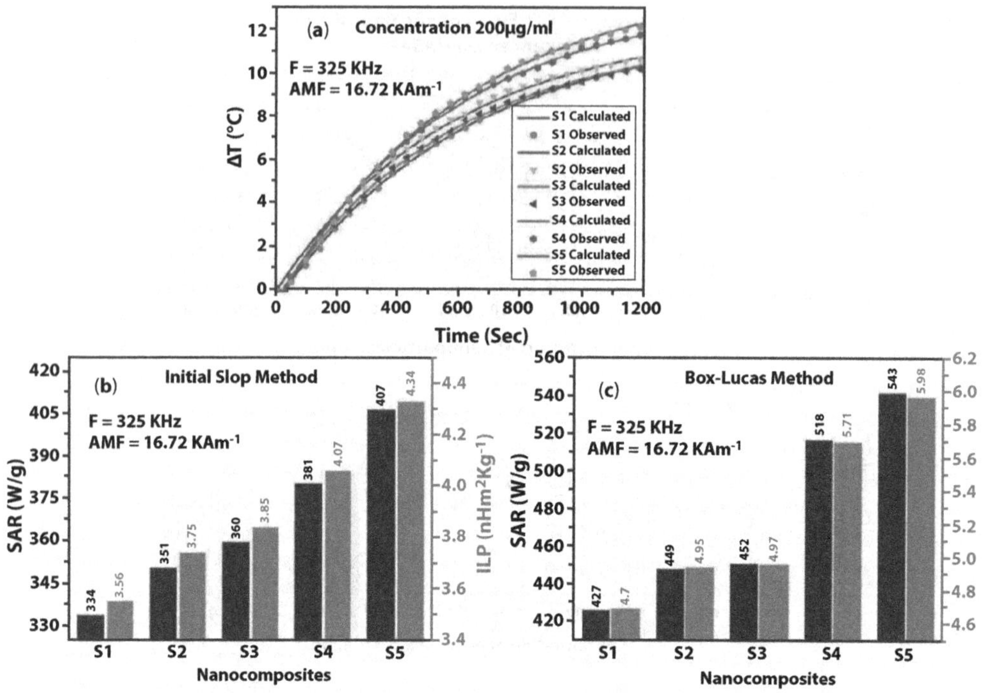

Figure 18.19 Temperature of GO-Fe_3O_4 nanocomposites of 200 μg/ml concentration under AC magnetic field (a) AMF dependent temperature kinetics (b) SAR and IPL values for GO-Fe_3O_4 nanocomposites in suspension using linear slope method (c) SAR and IPL values for GO-Fe_3O_4 nanocomposites in suspension using Box-Lucas method [127].

to 80% cancer cells death was observed in the *in vitro* experiments which are suitable for future cancer treatments.

18.6 Future Applications

Ferromagnetic nanostructures can be studied for exploring the field of magnonics where spin waves (magnons) are used to carry and process information of spin currents with no dissipative Joule effect. Periodically patterned magnetic devices can open fascinating outlooks for the operation of spin waves at nanoscale which can lead to the recognition of novel metamaterials. Various microwave devices such as circulators, gyrators, and resonators can be fabricated using these nanostructures. These devices can be useful in communication systems. The circulator ensures the isolation between different parts in transmitting and receiving modules. This circulation operation is achieved by the non-reciprocal behavior of ferromagnetic materials. Ferromagnetic nanostructures are thus promising candidates for miniaturization of devices in next-generation technology and advantageous for biomedical applications.

18.7 Conclusions

Recently, a lot of research has been focused on the magnetization dynamics of nanosystems due to its potential applications in spintronics, specifically aiming toward the design of GMR biosensors and cancer treatment through radio-frequency hyperthermia. These nanostructures can also be useful for signal processing communications, and magneto-electronic devices. The development of modern fabrication techniques has opened a broad area for new possibilities in future spintronics applications. Several methods, including as traditional FMR, TR-MOKE imaging, and Brillouin Light Scattering, can be used to investigate the ultrafast dynamics of magnetization. One of these methods is broadband FMR, which tracks how a change in the applied magnetic field affects the micromagnetic structure. A continuous external bias field can also be used to sweep the frequency of the oscillating magnetic field, which in turn induces the precession of the magnetization. Here, we have reviewed uniform modes of magnetization dynamics using the VNA-FMR technique in magnetic nanostructures such as trilayer thin films, nanosized stripes, and one-dimensional NWs. We discussed a theoretical background of magnetization dynamics in magnetic systems for better understanding. This involves the complete apprehension of LLG equations and the various damping mechanisms contributing to magnetic precession. We observed the magnetic excitation frequencies for Fe/Al/Fe trilayer films in the frequency domain with the help of VNA-FMR and BLS techniques. The acoustic and optic mode frequencies and their corresponding linewidths are determined by dispersion relation both at $k = 0$ (uniform precession mode) and $k \neq 0$ mode. We reported a very strong AFM exchange coupling Al spacers between ferromagnetic Fe layers. A detailed discussion of FMR in Permalloy nanostrips along the length and width of the strips is done. It is shown that for longer strips (1:10 and 1:333), when the field is applied along the width of the strip, a soft mode behavior occurs. The frequency linewidths also change with the change in aspect ratio. The magnetization dynamics of one-dimensional Ni NWs of various lengths were done by a flip-chip-based FMR technique. The resonance frequency has a linear relationship with the magnetic field when it is applied along the easy axis of NWs. The frequency linewidth is also larger for longer length samples. The gyromagnetic ratio, Gilbert damping, and magnetic anisotropies are derived from the aforementioned material properties. These observed phenomena in various nanostructures imply that magnetic nanosystems could be potential candidates for radio-frequency-assisted cancer treatment. In addition, these nanostructures can be helpful for microwave-assisted switching as well as magnetoelectronic devices.

References

1. Sarma, S.D., Spintronics: A new class of device based on electron spin, rather than on charge, may yield the next generation of microelectronics. *Am. Sci.*, 89, 6, 516–23, 2001.
2. Awschalom, D.D. and Flatté, M.E., Challenges for semiconductor spintronics. *Nat. Phys.*, 3, 3, 153, 2007.
3. Zhao, W. and Prenat, G., *Spintronics-based computing*, Springer, Berlin, Germany, 2015.

4. Wolf, S.A., Awschalom, D.D., Buhrman, R.A., Daughton, J.M., Von Molnar, S., Roukes, M.L., Chtchelkanova, A.Y., Treger, D.M., Spintronics: a spin-based electronics vision for the future. *Science*, 294, 546, 1488–95, 2015.
5. Liu, J.P., Fullerton, E., Gutfleisch, O. and Sellmyer, D.J. eds., Nanoscale magnetic materials and applications. Springer Science+ BusinessMedia, LLC, Dordrecht Heidelberg London New York, 2009.
6. Jiles, D.C., Recent advances and future directions in magnetic materials. *Acta Materialia*, 51, 19, 5907–39, 2003.
7. Brataas, A., Kent, A.D., Ohno, H., Current-induced torques in magnetic materials. *Nat. Mater.*, 11, 5, 372, 2012.
8. Žutić, I., Fabian, J., Sarma, S.D., Spintronics: Fundamentals and applications. *Rev. Mod. Phys.*, 76, 2, 323, 2004.
9. Chumak, A.V., Vasyuchka, V.I., Serga, A.A., Hillebrands, B., Magnon spintronics. *Nat. Phys.*, 11, 6, 453, 2015.
10. Baibich, M.N., Broto, J.M., Fert, A., Van Dau, F.N., Petroff, F., Etienne, P., Creuzet, G., Friederich, A., Chazelas, J., Giant magnetoresistance of (001) Fe/(001) Cr magnetic superlattices. *Phys. Rev. Lett.*, 61, 21, 2472, 1988.
11. Berkowitz, A.E., Mitchell, J.R., Carey, M.J., Young, A.P., Zhang, S., Spada, F.E., Parker, F.T., Hutten, A., Thomas, G., Giant magnetoresistance in heterogeneous Cu-Co alloys. *Phys. Rev. Lett.*, 68, 25, 3745, 1992.
12. Parkin, S.S., Li, Z.G., Smith, D.J., Giant magnetoresistance in antiferromagnetic Co/Cu multilayers. *Appl. Phys. Lett.*, 58, 23, 2710–2, 1991.
13. Van den Berg, H.A.M., Coehoorn, R., Gijs, M.A.M., Grünberg, P., Rasing, T. and Röll, K., Magnetic Multilayers and Giant Magnetoresistance: Fundamentals and Industrial Applications (Vol. 37). Springer Science & Business Media, New York, 2013.
14. Fullerton, E.E., Kelly, D.M., Guimpel, J., Schuller, I.K., Bruynseraede, Y., Roughness and giant magnetoresistance in Fe/Cr superlattices. *Phys. Rev. Lett.*, 68, 6, 859, 1992.
15. Ju, H.L., Kwon, C., Li, Q., Greene, R.L., Venkatesan, T., Giant magnetoresistance in La1–xSrxMnOz films near room temperature. *Appl. Phys. Lett.*, 65, 16, 2108–10, 1994.
16. Gijs, M.A., Lenczowski, S.K., Giesbers, J.B., Perpendicular giant magnetoresistance of microstructured Fe/Cr magnetic multilayers from 4.2 to 300 K. *Appl. Phys. Lett.*, 70, 21, 3343, 1993.
17. Tsymbal, E.Y. and Pettifor, D.G., Perspectives of giant magnetoresistance, in: *Solid state physics*, vol. 56, pp. 113–237, 2001.
18. White, R.L., Giant magnetoresistance: A primer. *IEEE Trans. Magn.*, 28, 5, 2482–7, 1992.
19. Kuanr, B.K., Maat, S., Chandrashekariaih, S., Veerakumar, V., Camley, R.E., Celinski, Z., Determination of exchange and rotational anisotropies in Ir Mn/ Fe (t)/ Ir Mn exchange coupled structures using dynamic and static techniques: Application to microwave devices. *J. Appl. Phys.*, 103, 7, 07C107, 2008 Apr 1.
20. Kuanr, A.V. and Kuanr, B.K., Influence of layering of NiO/NiFe on exchange bias, in: *2005 IEEE International Magnetics Conference (INTERMAG)*, IEEE, pp. 1631–1632, 2005 Apr 4.
21. Kuanr, B.K., Buchmeier, M., Buergler, D.E., Gruenberg, P., Camley, R., Celinski, Z., Dynamic and static measurements on epitaxial Fe/Si/Fe. *J. Vac. Sci. Technol. A: Vac. Surf. Films*, 21, 4, 1157–61, 2003 Jul 30.
22. Buchmeier, M., Kuanr, B.K., Gareev, R.R., Bürgler, D.E., Grünberg, P., Spin waves in magnetic double layers with strong antiferromagnetic interlayer exchange coupling: Theory and experiment. *Phys. Rev. B*, 67, 18, 184404, 2003 May 5.
23. Kuanr, B.K., Buchmeier, M., Gareev, R.R., Bürgler, D.E., Schreiber, R., Grünberg, P., Spin-wave modes and line broadening in strongly coupled epitaxial Fe/Al/Fe and Fe/Si/Fe trilayers observed by Brillouin light scattering. *J. Appl. Phys.*, 93, 6, 3427–34, 2003 Mar 15.

24. Grünberg, P., Bürgler, D.E., Gareev, R., Olligs, D., Buchmeier, M., Breidbach, M., Kuanr, B., Schreiber, R., Experiments on the relation between GMR and interface roughness and on the interlayer exchange coupling across semiconductors. *J. Phys. D: Appl. Phys.*, 35, 19, 2403, 2002 Sep 13.
25. Kuanr, B.K., Gokhale, S., Vedpathak, M., Kuanr, A.V., Nimtz, G., Temperature-dependent impedance anomaly, microwave GMR and exchange-coupling in thin Ni/Cu multilayered films. *J. Phys. D: Appl. Phys.*, 33, 1, 34, 2000 Jan 7.
26. Kuanr, B.K. and Gruenberg, P., Frequency Variation Of Microwave GMR And Exchange Coupling Of Epitaxial Fe/Au/Py trilayer ultra-this films, in: *1998 IEEE International Magnetics Conference (INTERMAG)*, IEEE, pp. 125–125, 1998 Jan 6.
27. Kuanr, B.K., Kuanr, A.V., Grünberg, P., Nimtz, G., Swept-frequency FMR on Fe/Cr/Fe trilayer ultrathin films; microwave giant magnetoresistance. *Phys. Lett. A*, 221, 3–4, 245–52, 1996 Sep 30.
28. Kuanr, B.K. and Kuanr, A.V., FMR studies on Fe/Cr/Fe trilayer ultrathin films; magneto-crystalline anisotropy and electromagnetic impedance. *J. Magn. Magn. Mater.*, 165, 1–3, 275–9, 1997 Jan 1.
29. Camley, R.E. and Stamps, R.L., Magnetic multilayers: spin configurations, excitations and giant magnetoresistance. *J. Phys.: Condens. Matter*, 5, 23, 3727, 1993.
30. Mosca, D.H., Petroff, F., Fert, A., Schroeder, P.A., Pratt Jr., W.P., Laloee, R., Oscillatory interlayer coupling and giant magnetoresistance in Co/Cu multilayers. *J. Magn. Magn. Mater.*, 94, 1–2, L1–5, 1991.
31. Miyazaki, T. and Tezuka, N., Giant magnetic tunneling effect in Fe/Al2O3/Fe junction. *J. Magn. Magn. Mater.*, 139, 3, L231–4, 1995.
32. Xiao, J.Q., Jiang, J.S., Chien, C.L., Giant magnetoresistance in nonmultilayer magnetic systems. *Phys. Rev. Lett.*, 68, 25, 3749, 1992.
33. Baumgart, P., Gurney, B.A., Wilhoit, D.R., Nguyen, T., Dieny, B., Speriosu, V.S., The role of spin-dependent impurity scattering in Fe/Cr giant magnetoresistance multilayers. *J. Appl. Phys.*, 69, 8, 4792–4, 1991.
34. Dieny, B., Speriosu, V.S., Parkin, S.S., Gurney, B.A., Wilhoit, D.R., Mauri, D., Giant magnetoresistive in soft ferromagnetic multilayers. *Phys. Rev. B*, 43, 1, 1297, 1991.
35. Beaurepaire, E., Merle, J.C., Daunois, A., Bigot, J.Y., Ultrafast spin dynamics in ferromagnetic nickel. *Phys. Rev. Lett.*, 76, 22, 4250, 1996.
36. Hillebrands, B. and Ounadjela, K., *Spin dynamics in confined magnetic structures I*, Springer Science & Business Media, 2003.
37. Braun, P.F., Marie, X., Lombez, L., Urbaszek, B., Amand, T., Renucci, P., Kalevich, V.K., Kavokin, K.V., Krebs, O., Voisin, P. and Masumoto, Y., Direct observation of the electron spin relaxation induced by nuclei in quantum dots. *Phys. Rev. Lett.*, 94, 11, 116601, 2005.
38. Ando, K., Takahashi, S., Harii, K., Sasage, K., Ieda, J., Maekawa, S., Saitoh, E., Electric manipulation of spin relaxation using the spin Hall effect. *Phys. Rev. Lett.*, 101, 3, 036601, 2008.
39. Kuanr, B.K., Kuanr, A.V., Camley, R.E., Celinski, Z., Magnetization relaxation in sputtered thin Fe films: An FMR study. *IEEE Trans. Magn.*, 42, 10, 2930–2, 2006.
40. Levitt, M.H., Spin dynamics: Basics of nuclear magnetic resonance. John Wiley & Sons, England, UK, Apr 21, 2008.
41. Hermenau, J., Ternes, M., Steinbrecher, M., Wiesendanger, R., Wiebe, J., Long Spin-Relaxation Times in a Transition-Metal Atom in Direct Contact to a Metal Substrate. *Nano letters*, 18, 3, 1978–83, 2018.
42. De Teresa, J.M., Ritter, C., Ibarra, M.R., Algarabel, P.A., Garcia-Munoz, J.L., Blasco, J., Garcia, J., Marquina, C., Charge localization, magnetic order, structural behavior, and spin dynamics of

(La–Tb)2/3Ca1/3MnO3 manganese perovskites probed by neutron diffraction and muon spin relaxation. *Phys. Rev. B*, 56, 6, 3317, 1997.
43. Leggett, A.J. and Takagi, S., Orientational dynamics of superfluid 3He: A "two-fluid" model. I. Spin dynamics with relaxation. *Ann. Phys.*, 106, 1, 79–133, 1977.
44. Schneider, D.J. and Freed, J.H., Spin relaxation and motional dynamics. *Lasers, Molecules and Methods*, vol. 73, pp. 387–528, 1989.
45. Nana, A.B., Marimuthu, T., Kondiah, P.P., Choonara, Y.E., Du Toit, L.C., Pillay, V., Multifunctional magnetic nanowires: Design, fabrication, and future prospects as cancer therapeutics. *Cancers*, 11, 12, 1956, 2019 Dec.
46. Anastasova, E.I., Prilepskii, A.Y., Fakhardo, A.F., Drozdov, A.S., Vinogradov, V.V., Magnetite nanocontainers: toward injectable highly magnetic materials for targeted drug delivery. *ACS Appl. Mater. Interfaces*, 10, 36, 30040–4, 2018 Aug 23.
47. Park, J.H., Yang, S.H., Lee, J., Ko, E.H., Hong, D., Choi, I.S., Nanocoating of single cells: From maintenance of cell viability to manipulation of cellular activities. *Adv. Mater.*, 26, 13, 2001–10, 2014 Apr.
48. Zamani Kouhpanji, M.R. and Stadler, B.J., Magnetic Nanowires for Nanobarcoding and Beyond. *Sensors*, 21, 13, 4573, 2021 Jan.
49. Ye, H., Wang, Y., Xu, D., Liu, X., Liu, S., Ma, X., Design and fabrication of micro/nano-motors for environmental and sensing applications. *Appl. Mater. Today*, 23, 101007, 2021 Jun 1.
50. Ambhorkar, P., Wang, Z., Ko, H., Lee, S., Koo, K.I., Kim, K., Cho, D.I., Nanowire-based biosensors: from growth to applications. *Micromachines*, 9, 12, 679, 2018 Dec.
51. Mukhtar, A., Wu, K., Cao, X., Magnetic nanowires in biomedical applications. *Nanotechnology*, 31, 43, 433001, 2020 Jul 29.
52. Lin, W.S., Lin, H.M., Chen, H.H., Hwu, Y.K. and Chiou, Y.J., Shape effects of iron nanowires on hyperthermia treatment. *J. Nanomater.*, 2013, 1, 237439, 2013.
53. Grünberg, P.E., Schreiber, R., Pang, Y., Brodsky, M.B., Sowers, H., Layered magnetic structures: Evidence for antiferromagnetic coupling of Fe layers across Cr interlayers. *Phys. Rev. Lett.*, 57, 19, 2442, 1986.
54. Xiong, Z.H., Wu, D., Vardeny, Z.V., Shi, J., Giant magnetoresistance in organic spin-valves. *Nature*, 427, 6977, 821, 2004.
55. Park, B.G., Wunderlich, J., Martí, X., Holý, V., Kurosaki, Y., Yamada, M., Yamamoto, H., Nishide, A., Hayakawa, J., Takahashi, H., Shick, A.B., A spin-valve-like magnetoresistance of an antiferromagnet-based tunnel junction. *Nat. Mater.*, 10, 5, 347, 2011.
56. Zhong, W., Liu, W., Au, C.T., Du, Y.W., Tunnelling magnetoresistance of double perovskite Sr2FeMoO6 enhanced by grain boundary adjustment. *Nanotechnology*, 17, 1, 250, 2005.
57. Borisov, K., Betto, D., Lau, Y.C., Fowley, C., Titova, A., Thiyagarajah, N., Atcheson, G., Lindner, J., Deac, A.M., Coey, J.M., Stamenov, P., Tunnelling magnetoresistance of the half-metallic compensated ferrimagnet Mn2RuxGa. *Appl. Phys. Lett.*, 108, 19, 192407, 2016.
58. Hanneken, C., Otte, F., Kubetzka, A., Dupé, B., Romming, N., Von Bergmann, K., Wiesendanger, R., Heinze, S., Electrical detection of magnetic skyrmions by tunnelling non-collinear magnetoresistance. *Nat. Nanotechnol.*, 10, 12, 1039, 2015.
59. Nag, A. and Ray, S., Magnetoresistance stories of double perovskites. *Pramana*, 84, 6, 967–75, 2015.
60. Ren, Y., Li, Z.Z., Xiao, M.W., Hu, A., Oscillation effect and sign-change behaviour of the bias-dependent tunnelling magnetoresistance in ferromagnetic junctions. *J. Phys.: Condens. Matter*, 17, 26, 4121, 2005.
61. Tsymbal, E.Y., Mryasov, O.N., LeClair, P.R., Spin-dependent tunnelling in magnetic tunnel junctions. *J. Phys.: Condens. Matter*, 15, 4, R109, 2003.

62. Deac, A.M., Fukushima, A., Kubota, H., Maehara, H., Suzuki, Y., Yuasa, S., Nagamine, Y., Tsunekawa, K., Djayaprawira, D.D., Watanabe, N., Bias-driven high-power microwave emission from MgO-based tunnel magnetoresistance devices. *Nat. Phys.*, 4, 10, 803, 2008.
63. Parkin, S.S., Kaiser, C., Panchula, A., Rice, P.M., Hughes, B., Samant, M., Yang, S.H., Giant tunnelling magnetoresistance at room temperature with MgO (100) tunnel barriers. *Nat. Mater.*, 3, 12, 862, 2004.
64. Bean, J.J., Saito, M., Fukami, S., Sato, H., Ikeda, S., Ohno, H., Ikuhara, Y., McKenna, K.P., Atomic structure and electronic properties of MgO grain boundaries in tunnelling magnetoresistive devices. *Sci. Rep.*, 7, 45594, 2017.
65. Graham, D.L., Ferreira, H.A., Freitas, P.P., Magnetoresistive-based biosensors and biochips. *Trends Biotechnol.*, 22, 9, 455–62, 2004.
66. Zhang, K., Li, M., Chen, Y.H., Wang, H.C., Zhao, T., inventors; Headway Technologies Inc, assignee. Novel way to reduce the ordering temperature for Co2MnSi-like Heusler alloys for CPP, TMR, MRAM, or other spintronics device applications, vol. 126, 2007, United States patent application US 11/472.
67. Tsang, D., Applied Spintronics Tech Inc, assignee. Magnetic tunneling junction cell array with shared reference layer for MRAM applications, 2005, United States patent US 6,963,500.
68. Zhang, K., Li, M., Zhou, Y., Oikawa, S., Yamada, K., Koui, K., Toshiba Corp, TDK Corp, assignee. Thin seeded Co/Ni multilayer film with perpendicular anisotropy for spintronic device applications, 2011, United States patent US 8,064,244.
69. Khvalkovskiy, A.V., Apalkov, D., Watts, S., Chepulskii, R., Beach, R.S., Ong, A., Tang, X., Driskill-Smith, A., Butler, W.H., Visscher, P.B., Lottis, D., Basic principles of STT-MRAM cell operation in memory arrays. *J. Phys. D: Appl. Phys.*, 46, 7, 074001, 2013.
70. Dieny, B., Sousa, R.C., Herault, J., Papusoi, C., Prenat, G., Ebels, U., Houssameddine, D., Rodmacq, B., Auffret, S., Buda-Prejbeanu, L.D., Cyrille, M.C., Spin-transfer effect and its use in spintronic components. *Int. J. Nanotechnol.*, 7, 4–8, 591–614, 2010.
71. Jabeur, K., Di Pendina, G., Bernard-Granger, F., Prenat, G., Spin orbit torque non-volatile flip-flop for high speed and low energy applications. *IEEE Electron Device Lett.*, 35, 3, 408–10, 2014.
72. Wang, K.L., Alzate, J.G., Amiri, P.K., Low-power non-volatile spintronic memory: STT-RAM and beyond. *J. Phys. D: Appl. Phys.*, 46, 7, 074003, 2013.
73. Kittel, C., On the theory of ferromagnetic resonance absorption. *Phys. Rev.*, 73, 2, 155, 1948.
74. Nozaki, Y., Ohta, M., Taharazako, S., Tateishi, K., Yoshimura, S., Matsuyama, K., Magnetic force microscopy study of microwave-assisted magnetization reversal in submicron-scale ferromagnetic particles. *Appl. Phys. Lett.*, 91, 8, 082510, 2007.
75. Yoshioka, T., Nozaki, T., Seki, T., Shiraishi, M., Shinjo, T., Suzuki, Y., Uehara, Y., Microwave-assisted magnetization reversal in a perpendicularly magnetized film. *Appl. Phys. Express*, 3, 1, 013002, 2009.
76. Nistor, C., Sun, K., Wang, Z., Wu, M., Mathieu, C., Hadley, M., Observation of microwave-assisted magnetization reversal in Fe65Co35 thin films through ferromagnetic resonance measurements. *Appl. Phys. Lett.*, 95, 1, 012504, 2009.
77. Wang, Z., Sun, K., Tong, W., Wu, M., Liu, M., Sun, N.X., Competition between pumping and damping in microwave-assisted magnetization reversal in magnetic films. *Phys. Rev. B*, 81, 6, 064402, 2010.
78. Taniguchi, T. and Kubota, H., Spin torque oscillator for microwave assisted magnetization reversal. *Japanese J. Appl. Phys.*, 57, 5, 053001, 2018.
79. Bigot, J.Y. and Vomir, M., Ultrafast magnetization dynamics of nanostructures. *Annalen der Physik*, 525, 1-2, 2–30, 2013.
80. Camley, R.E., Magnetization dynamics in thin films and multilayers. *J. Magn. Magn. Mater.*, 200, 1-3, 583–97, 1999.

81. Camara, I.S., Tacchi, S., Garnier, L.C., Eddrief, M., Fortuna, F., Carlotti, G., Marangolo, M., Magnetization dynamics of weak stripe domains in Fe-N thin films: a multi-technique complementary approach. *J. Phys.: Condens. Matter.*, 29, 46, 465803, 2017.
82. Cinchetti, M., Albaneda, M.S., Hoffmann, D., Roth, T., Wüstenberg, J.P., Krauß, M., Andreyev, O., Schneider, H.C., Bauer, M., Aeschlimann, M., Spin-flip processes and ultrafast magnetization dynamics in Co: Unifying the microscopic and macroscopic view of femtosecond magnetism. *Phys. Rev. Lett.*, 97, 17, 177201, 2006.
83. Adeyeye, A.O., Haldar, A., Shimon, G., Magnetization Dynamics of Reconfigurable 2D Magnonic Crystals, in: *Spin Wave Confinement*, pp. 99–137, 2017.
84. Barman, A. and Haldar, A., Time-domain study of magnetization dynamics in magnetic thin films and micro-and nanostructures. *Solid State Phys.*, 65, 1–108, Academic Press, 2014, Jan 1.
85. Tserkovnyak, Y., Brataas, A., Bauer, G.E., Halperin, B.I., Nonlocal magnetization dynamics in ferromagnetic heterostructures. *Rev. Mod. Phys.*, 77, 4, 1375, 2005.
86. Marko, D., *Magnetostatics and dynamics of ion irradiated NiFe/Ta multilayer films studied by vector network analyser ferromagnetic resonance*, Dresden University, 2010, Dissertation.
87. Landau, L.A. and Lifshitz, E., On the theory of the dispersion of magnetic permeability in ferromagnetic bodies. *Phys. Z. Sowjetunion*, 8, 153, 101–14, 1935.
88. Gilbert, T.L., A Lagrangian formulation of the gyromagnetic equation of the magnetization field. *Phys. Rev.*, 100, 1243, 1955.
89. Sharma, M., Pathak, S. and Sharm, M., FMR measurements of magnetic nanostructures. *Ferromagnetic Resonance-Theory and Applications*, Croatia, pp. 93–110, July 31, 2013.
90. Inaba, N., Uesaka, Y., Nakamura, A., Futamoto, M., Sugita, Y., Narishige, S., Damping constants of Co-Cr-Ta and Co-Cr-Pt thin films. *IEEE Trans. Magn.*, 33, 5, 2989–91, 1997.
91. Luo, C., Feng, Z., Fu, Y., Zhang, W., Wong, P.K., Kou, Z.X., Zhai, Y., Ding, H.F., Farle, M., Du, J., Zhai, H.R., Enhancement of magnetization damping coefficient of permalloy thin films with dilute Nd dopants. *Phys. Rev. B*, 89, 18, 184412, 2014.
92. Heinrich, B., Cochran, J.F., Hasegawa, R., FMR linebroadening in metals due to two-magnon scattering. *J. Appl. Phys.*, 57, 8, 3690–2, 1985.
93. Nibarger, J.P., Lopusnik, R., Silva, T.J., Damping as a function of pulsed field amplitude and bias field in thin film Permalloy. *Appl. Phys. Lett.*, 82, 13, 2112–4, 2003.
94. Skadsem, H.J., Tserkovnyak, Y., Brataas, A., Bauer, G.E., Magnetization damping in a local-density approximation. *Phys. Rev. B*, 75, 9, 094416, 2007.
95. Martın, J.I., Nogues, J., Liu, K., Vicent, J.L., Schuller, I.K., Ordered magnetic nanostructures: fabrication and properties. *J. Magn. Magn. Mater.*, 256, 1–3, 449–501, 2003.
96. Wang, S.X. and Taratorin, A.M., Magnetic Information Storage Technology: A Volume in the Electromagnetism Series. Elsevier, San Diego, California, USA, May 24, 1999.
97. Huang, X.J. and Choi, Y.K., Chemical sensors based on nanostructured materials. *Sens. Actuators, B*, 122, 2, 659–71, 2007.
98. Song, M.M., Song, W.J., Bi, H., Wang, J., Wu, W.L., Sun, J., Yu, M., Cytotoxicity and cellular uptake of iron nanowires. *Biomaterials*, 31, 7, 1509–17, 2010.
99. Huynen, I., Goglio, G., Vanhoenacker, D., Vander Vorst, A., A novel nanostructured microstrip device for tunable stopband filtering applications at microwaves. *IEEE Microwave Guided Wave Lett.*, 9, 10, 401–3, 1999.
100. Still, T., Retsch, M., Sainidou, R., Jonas, U., Fytas, G., High frequency acoustics in nanostructures by spontaneous Brillouin light scattering. *J. Acoust. Soc. Am.*, 123, 5, 3555, 2008.
101. Alonso-Redondo, E., Schmitt, M., Urbach, Z., Hui, C.M., Sainidou, R., Rembert, P., Matyjaszewski, K., Bockstaller, M.R., Fytas, G., A new class of tunable hypersonic phononic crystals based on polymer-tethered colloids. *Nat. Commun.*, 6, 8309, 2015.

102. Counil, G., Crozat, P., Devolder, T., Chappert, C., Zoll, S., Fournel, R., Lateral and longitudinal finite size effects in NA-FMR measurements. *IEEE Trans. Magn.*, 42, 10, 3321–2, 2006.
103. Bilzer, C., Devolder, T., Crozat, P., Chappert, C., Cardoso, S., Freitas, P.P., Vector network analyzer ferromagnetic resonance of thin films on coplanar waveguides: Comparison of different evaluation methods. *J. Appl. Phys.*, 101, 7, 074505, 2007.
104. Kalarickal, S.S., Krivosik, P., Wu, M., Patton, C.E., Schneider, M.L., Kabos, P., Silva, T.J., Nibarger, J.P., Ferromagnetic resonance linewidth in metallic thin films: Comparison of measurement methods. *J. Appl. Phys.*, 99, 9, 093909, 2006.
105. Slonczewski, : Origin of biquadratic exchange in magnetic multilayers. *J. Appl. Phys.*, 73, 10, 5957–62, 1993.
106. Djamal, M., Development of sensors based on giant magnetoresistance material. *Procedia Eng.*, 32, 60–8, 2012 Jan 1.
107. Su, D., Wu, K., Saha, R., Peng, C., Wang, J.P., Advances in magnetoresistive biosensors. *Micromachines*, 11, 1, 34, 2020 Jan.
108. Arias, R. and Mills, D.L., Theory of spin excitations and the microwave response of cylindrical ferromagnetic nanowires. *Phys. Rev. B*, 63, 13, 134439, 2001.
109. Malkinski, L.M., Yu, M., Vovk, A.Y., Scherer, D.J., Spinu, L., Zhou, W., Whittenburg, S., Davis, Z., Jung, J.S., Microwave absorption of patterned arrays of nanosized magnetic stripes with different aspect ratios. *J. Appl. Phys.*, 101, 9, 09J110, 2007.
110. Kuanr, B., Camley, R.E., Celinski, Z., Narrowing of the frequency-linewidth in structured magnetic strips: Experiment and theory. *Appl. Phys. Lett.*, 87, 1, 012502, 2005.
111. Sharma, M., Kuanr, B.K., Sharma, M., Basu, A., Relation between static and dynamic magnetization effects and resonance behavior in Ni nanowire arrays. *IEEE Trans. Magn.*, 50, 4, 1–0, 2014.
112. Darques, M., Spiegel, J., De la Torre Medina, J., Huynen, I., Piraux, L., Ferromagnetic nanowire-loaded membranes for microwave electronics. *J. Magn. Magn. Mater.*, 321, 14, 2055–65, 2009.
113. Shore, D., Ghemes, A., Dragos-Pinzaru, O., Gao, Z., Shao, Q., Sharma, A., Um, J., Tabakovic, I., Bischof, J.C., Stadler, B.J., Nanowarming using Au-tipped Co 35 Fe 65 ferromagnetic nanowires. *Nanoscale*, 11, 31, 14607–15, 2019.
114. Sung, H., Ferlay, J., Siegel, R.L., Laversanne, M., Soerjomataram, I., Jemal, A., Bray, F., Global cancer statistics 2020: GLOBOCAN estimates of incidence and mortality worldwide for 36 cancers in 185 countries. *CA: Cancer J. Clinicians*, 71, 3, 209–49, 2021 May.
115. Hildebrandt, B., Wust, P., Ahlers, O., Dieing, A., Sreenivasa, G., Kerner, T., Felix, R., Riess, H., The cellular and molecular basis of hyperthermia. *Crit. Rev. Oncology/Hematology*, 43, 1, 33–56, 2002 Jul 1.
116. Luderer, A.A., Borrelli, N.F., Panzarino, J.N., Mansfield, G.R., Hess, D.M., Brown, J.L., Barnett, E.H., Hahn, E.W., Glass-ceramic-mediated, magnetic-field-induced localized hyperthermia: response of a murine mammary carcinoma. *Radiat. Res.*, 94, 1, 190–8, 1983 Apr 1.
117. Wu, H., Song, L., Chen, L., Huang, Y., Wu, Y., Zang, F., An, Y., Lyu, H., Ma, M., Chen, J., Gu, N., Injectable thermosensitive magnetic nanoemulsion hydrogel for multimodal-imaging-guided accurate thermoablative cancer therapy. *Nanoscale*, 9, 42, 16175–82, 2017.
118. Shetake, N.G., Balla, M.M., Kumar, A., Pandey, B.N., Magnetic hyperthermia therapy: An emerging modality of cancer treatment in combination with radiotherapy. *J. Radiat. Can. Res.*, 7, 1, 13, 2016 Jan 1.
119. Xie, J., Yan, C., Yan, Y., Chen, L., Song, L., Zang, F., An, Y., Teng, G., Gu, N., Zhang, Y., Multimodal Mn–Zn ferrite nanocrystals for magnetically-induced cancer targeted hyperthermia: a comparison of passive and active targeting effects. *Nanoscale*, 8, 38, 16902–15, 2016.

120. Hervault, A., Dunn, A.E., Lim, M., Boyer, C., Mott, D., Maenosono, S., Thanh, N.T., Doxorubicin loaded dual pH-and thermo-responsive magnetic nanocarrier for combined magnetic hyperthermia and targeted controlled drug delivery applications. *Nanoscale*, 8, 24, 12152–61, 2016.
121. Barick, K.C., Singh, S., Jadhav, N.V., Bahadur, D., Pandey, B.N., Hassan, P.A., pH-responsive peptide mimic shell cross-linked magnetic nanocarriers for combination therapy. *Adv. Funct. Mater.*, 22, 23, 4975–84, 2012 Dec 5.
122. Venkatesha, N., Poojar, P., Qurishi, Y., Geethanath, S., Srivastava, C., Graphene oxide-Fe3O4 nanoparticle composite with high transverse proton relaxivity value for magnetic resonance imaging. *J. Appl. Phys.*, 117, 15, 154702, 2015 Apr 21.
123. Basel, M.T., Balivada, S., Wang, H., Shrestha, T.B., Seo, G.M., Pyle, M., Abayaweera, G., Dani, R., Koper, O.B., Tamura, M., Chikan, V., Cell-delivered magnetic nanoparticles caused hyperthermia-mediated increased survival in a murine pancreatic cancer model. *Int. J. Nanomed.*, 7, 297, 2012.
124. Balivada, S., Rachakatla, R.S., Wang, H., Samarakoon, T.N., Dani, R.K., Pyle, M., Kroh, F.O., Walker, B., Leaym, X., Koper, O.B., Tamura, M., A/C magnetic hyperthermia of melanoma mediated by iron (0)/iron oxide core/shell magnetic nanoparticles: a mouse study. *BMC cancer*, 10, 1, 1–9, 2010 Dec.
125. Chalkidou, A., Simeonidis, K., Angelakeris, M., Samaras, T., Martinez-Boubeta, C., Balcells, L., Papazisis, K., Dendrinou-Samara, C. and Kalogirou, O., *In vitro* application of Fe/MgO nanoparticles as magnetically mediated hyperthermia agents for cancer treatment. *J. Magn. Magn. Mater.*, 323, 6, 775–780, 2011.
126. Laurent, S., Dutz, S., Häfeli, U.O., Mahmoudi, M., Magnetic fluid hyperthermia: focus on superparamagnetic iron oxide nanoparticles. *Adv. Colloid Interface Sci.*, 166, 1–2, 8–23, 2011 Aug 10.
127. Kumar, R., Chauhan, A., Jha, S.K., Kuanr, B.K., Localized cancer treatment by radio-frequency hyperthermia using magnetic nanoparticles immobilized on graphene oxide: from novel synthesis to *in vitro* studies. *J. Mater. Chem. B*, 6, 33, 5385–99, 2018.
128. Kuanr, B.K., Buchmeier, M., Bürgler, D.E., Grünberg, P., Exchange coupling of molecular-beam-epitaxy-grown Fe/Al/Fe trilayers by dynamic techniques. *J. Appl. Phys.*, 91, 10, 7209–7211, 2002.
129. Kuanr, B.K., Lopusnik, R., Malkinski, L.M., Wenger, M., Yu, M., Scherer, D., Camley, R.E. and Celinski, Z., High-frequency characterization of Permalloy nanosized strips using network analyzer ferromagnetic resonance. *J. Appl. Phys.*, 103, 07C508(1–7), 2008.

Index

Printed in the USA/Agawam, MA
September 17, 2026

910661.175